PROBABILITY THEORY

THE LOGIC OF SCIENCE

This book goes beyond the conventional mathematics of probability theory, viewing the subject in a wider context. Now results are discussed, along with the application of probability theory to a wide variety of problems in physics, mathematics, economics, chemistry and biology. It contains many exercises and problems, and is suitable for use as a textbook on graduate level courses involving data analysis.

The material is aimed at readers who are already familiar with applied mathematics at an advanced undergraduate level or higher. The book is not restricted to one particular discipline but rather will be of interest to scientists working in any area where inference from incomplete information is necessary.

Edwin T Jaynes obtained his Ph.D in physics from Princeton University in 1950. He then went to Stanford University, where he stayed until 1960. He wrote extensively during this time and his papers reformulating statistical mechanics as a problem in inference were published while he was at Stanford.

In 1960 Jaynes moved to the University of Washington, St Louis where he became the Wayman Crow Professor of Physics. He retired in 1992, and died on April 30 1998.

PROBABILITY THEORY
THE LOGIC OF SCIENCE

E. T. Jaynes

edited by G. Larry Bretthorst

CAMBRIDGE
UNIVERSITY PRESS

CAMBRIDGE
UNIVERSITY PRESS

University Printing House, Cambridge CB2 8BS, United Kingdom

One Liberty Plaza, 20th Floor, New York, NY 10006, USA

477 Williamstown Road, Port Melbourne, VIC 3207, Australia

314-321, 3rd Floor, Plot 3, Splendor Forum, Jasola District Centre, New Delhi - 110025, India

79 Anson Road, #06-04/06, Singapore 079906

Cambridge University Press is part of the University of Cambridge.

It furthers the University's mission by disseminating knowledge in the pursuit of education, learning and research at the highest international levels of excellence.

www.cambridge.org
Information on this title: www.cambridge.org/ 9780521592710

First published 2003
Reprinted with corrections 2004
21st printing 2018

A catalogue record for this publication is available from the British Library

Library of Congress Cataloging in Publication data

Jaynes, E. T. (Edwin T.)
Probability theory: the logic of science / by E.T. Jaynes; edited by G. Larry Bretthorst.
p. cm.
Includes bibliographical references and index.
ISBN 0 521 59271 2
1. Probabilities. I. Bretthorst, G. Larry. II. Title.
QA273 .J36 2003
519.2–dc21   2002071486

ISBN 978-0-521-59271-0 Hardback

Dedicated to the memory of
Sir Harold Jeffreys, who
saw the truth and preserved it.

Contents

Editor's foreword

E. T. Jaynes died April 30, 1998. Before his death he asked me to finish and publish his book on probability theory. I struggled with this for some time, because there is no doubt in my mind that Jaynes wanted this book finished. Unfortunately, most of the later chapters, Jaynes' intended volume 2 on applications, were either missing or incomplete, and some of the early chapters also had missing pieces. I could have written these latter chapters and filled in the missing pieces, but if I did so, the work would no longer be Jaynes'; rather, it would be a Jaynes–Bretthorst hybrid with no way to tell which material came from which author. In the end, I decided the missing chapters would have to stay missing – the work would remain Jaynes'.

There were a number of missing pieces of varying length that Jaynes had marked by inserting the phrase 'MUCH MORE COMING'. I could have left these comments in the text, but they were ugly and they made the book look very incomplete. Jaynes intended this book to serve as both a reference and a text book. Consequently, there are question boxes (Exercises) scattered throughout most chapters. In the end, I decided to replace the 'MUCH MORE COMING' comments by introducing 'Editor's' Exercises. If you answer these questions, you will have filled in the missing material.

Jaynes wanted to include a series of computer programs that implemented some of the calculations in the book. I had originally intended to include these programs. But, as time went on, it became increasingly obvious that many of the programs were not available, and the ones that were were written in a particularly obscure form of BASIC (it was the programs that were obscure, not the BASIC). Consequently, I removed the references to these programs and, where necessary, inserted a few sentences to direct people to the necessary software tools to implement the calculations.

Numerous references were missing and had to be supplied. Usually the information available, a last name and date, was sufficient to find one or more probable references. When there were several good candidates, and I was unable to determine which Jaynes intended, I included multiple references and modified the citation. Sometimes the information was so vague that no good candidates were available. Fortunately, I was able to remove the citation with no detrimental effect. To enable readers to distinguish between cited works and other published sources, Jaynes' original annotated bibliography has been split into two sections: a Reference list and a Bibliography.

Finally, while I am the most obvious person who has worked on getting this book into publication, I am not the only person to do so. Some of Jaynes' closest friends have assisted me in completing this work. These include Tom Grandy, Ray Smith, Tom Loredo, Myron Tribus and John Skilling, and I would like to thank them for their assistance. I would also like to thank Joe Ackerman for allowing me to take the time necessary to get this work published.

G. Larry Bretthorst

Preface

The following material is addressed to readers who are already familiar with applied mathematics, at the advanced undergraduate level or preferably higher, and with some field, such as physics, chemistry, biology, geology, medicine, economics, sociology, engineering, operations research, etc., where inference is needed.[1] A previous acquaintance with probability and statistics is not necessary; indeed, a certain amount of innocence in this area may be desirable, because there will be less to unlearn.

We are concerned with probability theory and all of its conventional mathematics, but now viewed in a wider context than that of the standard textbooks. Every chapter after the first has 'new' (i.e. not previously published) results that we think will be found interesting and useful. Many of our applications lie outside the scope of conventional probability theory as currently taught. But we think that the results will speak for themselves, and that something like the theory expounded here will become the conventional probability theory of the future.

History

The present form of this work is the result of an evolutionary growth over many years. My interest in probability theory was stimulated first by reading the work of Harold Jeffreys (1939) and realizing that his viewpoint makes all the problems of theoretical physics appear in a very different light. But then, in quick succession, discovery of the work of R. T. Cox (1946), Shannon (1948) and Pólya (1954) opened up new worlds of thought, whose exploration has occupied my mind for some 40 years. In this much larger and permanent world of rational thinking in general, the current problems of theoretical physics appeared as only details of temporary interest.

The actual writing started as notes for a series of lectures given at Stanford University in 1956, expounding the then new and exciting work of George Pólya on 'Mathematics and Plausible Reasoning'. He dissected our intuitive 'common sense' into a set of elementary qualitative desiderata and showed that mathematicians had been using them all along to

---

[1] By 'inference' we mean simply: deductive reasoning whenever enough information is at hand to permit it; inductive or plausible reasoning when – as is almost invariably the case in real problems – the necessary information is not available. But if a problem can be solved by deductive reasoning, probability theory is not needed for it; thus our topic is the optimal processing of incomplete information.

guide the early stages of discovery, which necessarily precede the finding of a rigorous proof. The results were much like those of James Bernoulli's *Art of Conjecture* (1713), developed analytically by Laplace in the late 18th century; but Pólya thought the resemblance to be only qualitative.

However, Pólya demonstrated this qualitative agreement in such complete, exhaustive detail as to suggest that there must be more to it. Fortunately, the consistency theorems of R. T. Cox were enough to clinch matters; when one added Pólya's qualitative conditions to them the result was a proof that, if degrees of plausibility are represented by real numbers, then there is a uniquely determined set of quantitative rules for conducting inference. That is, any other rules whose results conflict with them will necessarily violate an elementary – and nearly inescapable – desideratum of rationality or consistency.

But the final result was just the standard rules of probability theory, given already by Daniel Bernoulli and Laplace; so why all the fuss? The important new feature was that these rules were now seen as uniquely valid principles of logic in general, making no reference to 'chance' or 'random variables'; so their range of application is vastly greater than had been supposed in the conventional probability theory that was developed in the early 20th century. As a result, the imaginary distinction between 'probability theory' and 'statistical inference' disappears, and the field achieves not only logical unity and simplicity, but far greater technical power and flexibility in applications.

In the writer's lectures, the emphasis was therefore on the quantitative formulation of Pólya's viewpoint, so it could be used for general problems of scientific inference, almost all of which arise out of incomplete information rather than 'randomness'. Some personal reminiscences about George Pólya and this start of the work are in Chapter 5.

Once the development of applications started, the work of Harold Jeffreys, who had seen so much of it intuitively and seemed to anticipate every problem I would encounter, became again the central focus of attention. My debt to him is only partially indicated by the dedication of this book to his memory. Further comments about his work and its influence on mine are scattered about in several chapters.

In the years 1957–1970 the lectures were repeated, with steadily increasing content, at many other universities and research laboratories.[2] In this growth it became clear gradually that the outstanding difficulties of conventional 'statistical inference' are easily understood and overcome. But the rules which now took their place were quite subtle conceptually, and it required some deep thinking to see how to apply them correctly. Past difficulties, which had led to rejection of Laplace's work, were seen finally as only misapplications, arising usually from failure to define the problem unambiguously or to appreciate the cogency of seemingly trivial side information, and easy to correct once this is recognized. The various relations between our 'extended logic' approach and the usual 'random variable' one appear in almost every chapter, in many different forms.

---

[2] Some of the material in the early chapters was issued in 1958 by the Socony-Mobil Oil Company as Number 4 in their series 'Colloquium Lectures in Pure and Applied Science'.

Eventually, the material grew to far more than could be presented in a short series of lectures, and the work evolved out of the pedagogical phase; with the clearing up of old difficulties accomplished, we found ourselves in possession of a powerful tool for dealing with new problems. Since about 1970 the accretion has continued at the same pace, but fed instead by the research activity of the writer and his colleagues. We hope that the final result has retained enough of its hybrid origins to be usable either as a textbook or as a reference work; indeed, several generations of students have carried away earlier versions of our notes, and in turn taught it to their students.

In view of the above, we repeat the sentence that Charles Darwin wrote in the Introduction to his *Origin of Species*: 'I hope that I may be excused for entering on these personal details, as I give them to show that I have not been hasty in coming to a decision.' But it might be thought that work done 30 years ago would be obsolete today. Fortunately, the work of Jeffreys, Pólya and Cox was of a fundamental, timeless character whose truth does not change and whose importance grows with time. Their perception about the nature of inference, which was merely curious 30 years ago, is very important in a half-dozen different areas of science today; and it will be crucially important in all areas 100 years hence.

Foundations

From many years of experience with its applications in hundreds of real problems, our views on the foundations of probability theory have evolved into something quite complex, which cannot be described in any such simplistic terms as 'pro-this' or 'anti-that'. For example, our system of probability could hardly be more different from that of Kolmogorov, in style, philosophy, and purpose. What we consider to be fully half of probability theory as it is needed in current applications – the principles for assigning probabilities by logical analysis of incomplete information – is not present at all in the Kolmogorov system.

Yet, when all is said and done, we find ourselves, to our own surprise, in agreement with Kolmogorov and in disagreement with his critics, on nearly all technical issues. As noted in Appendix A, each of his axioms turns out to be, for all practical purposes, derivable from the Pólya–Cox desiderata of rationality and consistency. In short, we regard our system of probability as not contradicting Kolmogorov's; but rather seeking a deeper logical foundation that permits its extension in the directions that are needed for modern applications. In this endeavor, many problems have been solved, and those still unsolved appear where we should naturally expect them: in breaking into new ground.

As another example, it appears at first glance to everyone that we are in very close agreement with the de Finetti system of probability. Indeed, the writer believed this for some time. Yet when all is said and done we find, to our own surprise, that little more than a loose philosophical agreement remains; on many technical issues we disagree strongly with de Finetti. It appears to us that his way of treating infinite sets has opened up a Pandora's box of useless and unnecessary paradoxes; nonconglomerability and finite additivity are examples discussed in Chapter 15.

Infinite-set paradoxing has become a morbid infection that is today spreading in a way that threatens the very life of probability theory, and it requires immediate surgical removal. In our system, after this surgery, such paradoxes are avoided automatically; they cannot arise from correct application of our basic rules, because those rules admit only finite sets and infinite sets that arise as well-defined and well-behaved limits of finite sets. The paradoxing was caused by (1) jumping directly into an infinite set without specifying any limiting process to define its properties; and then (2) asking questions whose answers depend on how the limit was approached.

For example, the question: 'What is the probability that an integer is even?' can have any answer we please in (0, 1), depending on what limiting process is used to define the 'set of all integers' (just as a conditionally convergent series can be made to converge to any number we please, depending on the order in which we arrange the terms).

In our view, an infinite set cannot be said to possess any 'existence' and mathematical properties at all – at least, in probability theory – until we have specified the limiting process that is to generate it from a finite set. In other words, we sail under the banner of Gauss, Kronecker, and Poincaré rather than Cantor, Hilbert, and Bourbaki. We hope that readers who are shocked by this will study the indictment of Bourbakism by the mathematician Morris Kline (1980), and then bear with us long enough to see the advantages of our approach. Examples appear in almost every chapter.

Comparisons

For many years, there has been controversy over 'frequentist' versus 'Bayesian' methods of inference, in which the writer has been an outspoken partisan on the Bayesian side. The record of this up to 1981 is given in an earlier book (Jaynes, 1983). In these old works there was a strong tendency, on both sides, to argue on the level of philosophy or ideology. We can now hold ourselves somewhat aloof from this, because, thanks to recent work, there is no longer any need to appeal to such arguments. We are now in possession of proven theorems and masses of worked-out numerical examples. As a result, the superiority of Bayesian methods is now a thoroughly demonstrated fact in a hundred different areas. One can argue with a philosophy; it is not so easy to argue with a computer printout, which says to us: 'Independently of all your philosophy, here are the facts of actual performance.' We point this out in some detail whenever there is a substantial difference in the final results. Thus we continue to argue vigorously for the Bayesian methods; but we ask the reader to note that our arguments now proceed by citing facts rather than proclaiming a philosophical or ideological position.

However, neither the Bayesian nor the frequentist approach is universally applicable, so in the present, more general, work we take a broader view of things. Our theme is simply: *probability theory as extended logic*. The 'new' perception amounts to the recognition that the mathematical rules of probability theory are not merely rules for calculating frequencies of 'random variables'; they are also the unique consistent rules for conducting inference (i.e. plausible reasoning) of any kind, and we shall apply them in full generality to that end.

It is true that all 'Bayesian' calculations are included automatically as particular cases of our rules; but so are all 'frequentist' calculations. Nevertheless, our basic rules are broader than either of these, and in many applications our calculations do not fit into either category.

To explain the situation as we see it presently: The traditional 'frequentist' methods which use only sampling distributions are usable and useful in many particularly simple, idealized problems; however, they represent the most proscribed special cases of probability theory, because they presuppose conditions (independent repetitions of a 'random experiment' but no relevant prior information) that are hardly ever met in real problems. This approach is quite inadequate for the current needs of science.

In addition, frequentist methods provide no technical means to eliminate nuisance parameters or to take prior information into account, no way even to use all the information in the data when sufficient or ancillary statistics do not exist. Lacking the necessary theoretical principles, they force one to 'choose a statistic' from intuition rather than from probability theory, and then to invent *ad hoc* devices (such as unbiased estimators, confidence intervals, tail-area significance tests) not contained in the rules of probability theory. Each of these is usable within the small domain for which it was invented but, as Cox's theorems guarantee, such arbitrary devices always generate inconsistencies or absurd results when applied to extreme cases; we shall see dozens of examples.

All of these defects are corrected by use of Bayesian methods, which are adequate for what we might call 'well-developed' problems of inference. As Harold Jeffreys demonstrated, they have a superb analytical apparatus, able to deal effortlessly with the technical problems on which frequentist methods fail. They determine the optimal estimators and algorithms automatically, while taking into account prior information and making proper allowance for nuisance parameters, and, being exact, they do not break down – but continue to yield reasonable results – in extreme cases. Therefore they enable us to solve problems of far greater complexity than can be discussed at all in frequentist terms. One of our main purposes is to show how all this capability was contained already in the simple product and sum rules of probability theory interpreted as extended logic, with no need for – indeed, no room for – any *ad hoc* devices.

Before Bayesian methods can be used, a problem must be developed beyond the 'exploratory phase' to the point where it has enough structure to determine all the needed apparatus (a model, sample space, hypothesis space, prior probabilities, sampling distribution). Almost all scientific problems pass through an initial exploratory phase in which we have need for inference, but the frequentist assumptions are invalid and the Bayesian apparatus is not yet available. Indeed, some of them never evolve out of the exploratory phase. Problems at this level call for more primitive means of assigning probabilities directly out of our incomplete information.

For this purpose, the Principle of maximum entropy has at present the clearest theoretical justification and is the most highly developed computationally, with an analytical apparatus as powerful and versatile as the Bayesian one. To apply it we must define a sample space, but do not need any model or sampling distribution. In effect, entropy maximization creates a model for us out of our data, which proves to be optimal by so many different criteria[3] that it is hard to imagine circumstances where one would not want to use it in a problem where we have a sample space but no model.

Bayesian and maximum entropy methods differ in another respect. Both procedures yield the optimal inferences from the information that went into them, but we may choose a model for Bayesian analysis; this amounts to expressing some prior knowledge – or some working hypothesis – about the phenomenon being observed. Usually, such hypotheses extend beyond what is directly observable in the data, and in that sense we might say that Bayesian methods are – or at least may be – speculative. If the extra hypotheses are true, then we expect that the Bayesian results will improve on maximum entropy; if they are false, the Bayesian inferences will likely be worse.

On the other hand, maximum entropy is a nonspeculative procedure, in the sense that it invokes no hypotheses beyond the sample space and the evidence that is in the available data. Thus it predicts only observable facts (functions of future or past observations) rather than values of parameters which may exist only in our imagination. It is just for that reason that maximum entropy is the appropriate (safest) tool when we have very little knowledge beyond the raw data; it protects us against drawing conclusions not warranted by the data. But when the information is extremely vague, it may be difficult to define any appropriate sample space, and one may wonder whether still more primitive principles than maximum entropy can be found. There is room for much new creative thought here.

For the present, there are many important and highly nontrivial applications where Maximum Entropy is the only tool we need. Part 2 of this work considers them in detail; usually, they require more technical knowledge of the subject-matter area than do the more general applications studied in Part 1. All of presently known statistical mechanics, for example, is included in this, as are the highly successful Maximum Entropy spectrum analysis and image reconstruction algorithms in current use. However, we think that in the future the latter two applications will evolve into the Bayesian phase, as we become more aware of the appropriate models and hypothesis spaces which enable us to incorporate more prior information.

We are conscious of having so many theoretical points to explain that we fail to present as many practical worked-out numerical examples as we should. Fortunately, three recent books largely make up this deficiency, and should be considered as adjuncts to the present work: *Bayesian Spectrum Analysis and Parameter Estimation* (Bretthorst, 1988), *Maximum Entropy in Action* (Buck and Macaulay, 1991), and *Data Analysis – A Bayesian Tutorial* (Sivia, 1996), are written from a viewpoint essentially identical to ours and present a wealth of real problems carried through to numerical solutions. Of course, these works do not contain nearly as much theoretical explanation as does the present one. Also, the Proceedings

---

[3] These concern efficient information handling; for example, (1) the model created is the simplest one that captures all the information in the constraints (Chapter 11); (2) it is the unique model for which the constraints would have been sufficient statistics (Chapter 8); (3) if viewed as constructing a sampling distribution for subsequent Bayesian inference from new data D, the only property of the measurement errors in D that are used in that subsequent inference are the ones about which that sampling distribution contained some definite prior information (Chapter 7). Thus the formalism automatically takes into account all the information we have, but avoids assuming information that we do not have. This contrasts sharply with orthodox methods, where one does not think in terms of information at all, and in general violates both of these desiderata.

volumes of the various annual MAXENT workshops since 1981 consider a great variety of useful applications.

Mental activity

As one would expect already from Pólya's examples, probability theory as extended logic reproduces many aspects of human mental activity, sometimes in surprising and even disturbing detail. In Chapter 5 we find our equations exhibiting the phenomenon of a person who tells the truth and is not believed, even though the disbelievers are reasoning consistently. The theory explains why and under what circumstances this will happen.

The equations also reproduce a more complicated phenomenon, divergence of opinions. One might expect that open discussion of public issues would tend to bring about a general consensus. On the contrary, we observe repeatedly that when some controversial issue has been discussed vigorously for a few years, society becomes polarized into two opposite extreme camps; it is almost impossible to find anyone who retains a moderate view. Probability theory as logic shows how two persons, given the same information, may have their opinions driven in opposite directions by it, and what must be done to avoid this.

In such respects, it is clear that probability theory is telling us something about the way our own minds operate when we form intuitive judgments, of which we may not have been consciously aware. Some may feel uncomfortable at these revelations; others may see in them useful tools for psychological, sociological, or legal research.

What is 'safe'?

We are not concerned here only with abstract issues of mathematics and logic. One of the main practical messages of this work is the great effect of prior information on the conclusions that one should draw from a given data set. Currently, much discussed issues, such as environmental hazards or the toxicity of a food additive, cannot be judged rationally if one looks only at the current data and ignores the prior information that scientists have about the phenomenon. This can lead one to overestimate or underestimate the danger.

A common error, when judging the effects of radioactivity or the toxicity of some substance, is to assume a linear response model without threshold (i.e. without a dose rate below which there is no ill effect). Presumably there is no threshold effect for cumulative poisons like heavy metal ions (mercury, lead), which are eliminated only very slowly, if at all. But for virtually every organic substance (such as saccharin or cyclamates), the existence of a finite metabolic rate means that there must exist a finite threshold dose rate, below which the substance is decomposed, eliminated, or chemically altered so rapidly that it causes no ill effects. If this were not true, the human race could never have survived to the present time, in view of all the things we have been eating.

Indeed, every mouthful of food you and I have ever taken contained many billions of kinds of complex molecules whose structure and physiological effects have never been determined – and many millions of which would be toxic or fatal in large doses. We cannot doubt that we are daily ingesting thousands of substances that are far more dangerous than saccharin – but in amounts that are safe, because they are far below the various thresholds of toxicity. At present, there are hardly any substances, except some common drugs, for which we actually know the threshold.

Therefore, the goal of inference in this field should be to estimate not only the slope of the response curve, but, *far more importantly*, to decide whether there is evidence for a threshold; and, if there is, to estimate its magnitude (the 'maximum safe dose'). For example, to tell us that a sugar substitute can produce a barely detectable incidence of cancer in doses 1000 times greater than would ever be encountered in practice, is hardly an argument against using the substitute; indeed, the fact that it is necessary to go to kilodoses in order to detect any ill effects at all, is rather conclusive evidence, not of the danger, but of the *safety*, of a tested substance. A similar overdose of sugar would be far more dangerous, leading not to barely detectable harmful effects, but to sure, immediate death by diabetic coma; yet nobody has proposed to ban the use of sugar in food.

Kilodose effects are irrelevant because we do not take kilodoses; in the case of a sugar substitute the important question is: What are the threshold doses for toxicity of a sugar substitute and for sugar, compared with the normal doses? If that of a sugar substitute is higher, then the rational conclusion would be that the substitute is actually safer than sugar, as a food ingredient. To analyze one's data in terms of a model which does not allow even the possibility of a threshold effect is to prejudge the issue in a way that can lead to false conclusions, however good the data. If we hope to detect any phenomenon, we must use a model that at least allows the *possibility* that it may exist.

We emphasize this in the Preface because false conclusions of just this kind are now not only causing major economic waste, but also creating unnecessary dangers to public health and safety. Society has only finite resources to deal with such problems, so any effort expended on imaginary dangers means that real dangers are going unattended. Even worse, the error is incorrectible by the currently most used data analysis procedures; a false premise built into a model which is never questioned cannot be removed by any amount of new data. Use of models which correctly represent the prior information that scientists have about the mechanism at work can prevent such folly in the future.

Such considerations are not the only reasons why prior information is essential in inference; the progress of science itself is at stake. To see this, note a corollary to the preceding paragraph: that new data that we insist on analyzing in terms of old ideas (that is, old models which are not questioned) *cannot lead us out of the old ideas*. However many data we record and analyze, we may just keep repeating the same old errors, missing the same crucially important things that the experiment was competent to find. That is what ignoring prior information can do to us; no amount of analyzing coin tossing data by a stochastic model could have led us to the discovery of Newtonian mechanics, which alone determines those data.

Old data, when seen in the light of new ideas, can give us an entirely new insight into a phenomenon; we have an impressive recent example of this in the Bayesian spectrum analysis of nuclear magnetic resonance data, which enables us to make accurate quantitative determinations of phenomena which were not accessible to observation at all with the previously used data analysis by Fourier transforms. When a data set is mutilated (or, to use the common euphemism, 'filtered') by processing according to false assumptions, important information in it may be destroyed irreversibly. As some have recognized, this is happening constantly from orthodox methods of detrending or seasonal adjustment in econometrics. However, old data sets, if preserved unmutilated by old assumptions, may have a new lease on life when our prior information advances.

Style of presentation

In Part 1, expounding principles and elementary applications, most chapters start with several pages of verbal discussion of the nature of the problem. Here we try to explain the constructive ways of looking at it, and the logical pitfalls responsible for past errors. Only then do we turn to the mathematics, solving a few of the problems of the genre to the point where the reader may carry it on by straightforward mathematical generalization. In Part 2, expounding more advanced applications, we can concentrate from the start on the mathematics.

The writer has learned from much experience that this primary emphasis on the logic of the problem, rather than the mathematics, is necessary in the early stages. For modern students, the mathematics is the easy part; once a problem has been reduced to a definite mathematical exercise, most students can solve it effortlessly and extend it endlessly, without further help from any book or teacher. It is in the conceptual matters (how to make the initial connection between the real-world problem and the abstract mathematics) that they are perplexed and unsure how to proceed.

Recent history demonstrates that anyone foolhardy enough to describe his own work as 'rigorous' is headed for a fall. Therefore, we shall claim only that we do not knowingly give erroneous arguments. We are conscious also of writing for a large and varied audience, for most of whom clarity of meaning is more important than 'rigor' in the narrow mathematical sense.

There are two more, even stronger, reasons for placing our primary emphasis on logic and clarity. Firstly, no argument is stronger than the premises that go into it, and, as Harold Jeffreys noted, those who lay the greatest stress on mathematical rigor are just the ones who, lacking a sure sense of the real world, tie their arguments to unrealistic premises and thus destroy their relevance. Jeffreys likened this to trying to strengthen a building by anchoring steel beams into plaster. An argument which makes it clear intuitively *why* a result is correct is actually more trustworthy, and more likely of a permanent place in science, than is one that makes a great overt show of mathematical rigor unaccompanied by understanding.

Secondly, we have to recognize that there are no really trustworthy standards of rigor in a mathematics that has embraced the theory of infinite sets. Morris Kline (1980, p. 351) came close to the Jeffreys simile: 'Should one design a bridge using theory involving infinite sets or the axiom of choice? Might not the bridge collapse?' The only real rigor we have today is in the operations of elementary arithmetic on finite sets of finite integers, and our own bridge will be safest from collapse if we keep this in mind.

Of course, it is essential that we follow this 'finite sets' policy whenever it matters for our results; but we do not propose to become fanatical about it. In particular, the arts of computation and approximation are on a different level than that of basic principle; and so once a result is derived from strict application of the rules, we allow ourselves to use any convenient analytical methods for evaluation or approximation (such as replacing a sum by an integral) without feeling obliged to show how to generate an uncountable set as the limit of a finite one.

We impose on ourselves a far stricter adherence to the mathematical rules of probability theory than was ever exhibited in the 'orthodox' statistical literature, in which authors repeatedly invoke the aforementioned intuitive *ad hoc* devices to do, arbitrarily and imperfectly, what the rules of probability theory would have done for them uniquely and optimally. It is just this strict adherence that enables us to avoid the artificial paradoxes and contradictions of orthodox statistics, as described in Chapters 15 and 17.

Equally important, this policy often simplifies the computations in two ways: (i) the problem of determining the sampling distribution of a 'statistic' is eliminated, and the evidence of the data is displayed fully in the likelihood function, which can be written down immediately; and (ii) one can eliminate nuisance parameters at the beginning of a calculation, thus reducing the dimensionality of a search algorithm. If there are several parameters in a problem, this can mean orders of magnitude reduction in computation over what would be needed with a least squares or maximum likelihood algorithm. The Bayesian computer programs of Bretthorst (1988) demonstrate these advantages impressively, leading in some cases to major improvements in the ability to extract information from data, over previously used methods. But this has barely scratched the surface of what can be done with sophisticated Bayesian models. We expect a great proliferation of this field in the near future.

A scientist who has learned how to use probability theory directly as extended logic has a great advantage in power and versatility over one who has learned only a collection of unrelated *ad hoc* devices. As the complexity of our problems increases, so does this relative advantage. Therefore we think that, in the future, workers in all the quantitative sciences will be obliged, as a matter of practical necessity, to use probability theory in the manner expounded here. This trend is already well under way in several fields, ranging from econometrics to astronomy to magnetic resonance spectroscopy; but, to make progress in a new area, it is necessary to develop a healthy disrespect for tradition and authority, which have retarded progress throughout the 20th century.

Finally, some readers should be warned not to look for hidden subtleties of meaning which are not present. We shall, of course, explain and use all the standard technical jargon of probability and statistics – because that is our topic. But, although our concern with the nature of logical inference leads us to discuss many of the same issues, our language differs greatly from the stilted jargon of logicians and philosophers. There are no linguistic tricks, and there is no 'meta-language' gobbledygook; only plain English. We think that this will convey our message clearly enough to anyone who seriously wants to understand it. In any event, we feel sure that no further clarity would be achieved by taking the first few steps down that infinite regress that starts with: 'What do you mean by "exists"?'

Acknowledgments

In addition to the inspiration received from the writings of Jeffreys, Cox, Pólya, and Shannon, I have profited by interaction with some 300 former students, who have diligently caught my errors and forced me to think more carefully about many issues. Also, over the years, my thinking has been influenced by discussions with many colleagues; to list a few (in the reverse alphabetical order preferred by some): Arnold Zellner, Eugene Wigner, George Uhlenbeck, John Tukey, William Sudderth, Stephen Stigler, Ray Smith, John Skilling, Jimmie Savage, Carlos Rodriguez, Lincoln Moses, Elliott Montroll, Paul Meier, Dennis Lindley, David Lane, Mark Kac, Harold Jeffreys, Bruce Hill, Mike Hardy, Stephen Gull, Tom Grandy, Jack Good, Seymour Geisser, Anthony Garrett, Fritz Fröhner, Willy Feller, Anthony Edwards, Morrie de Groot, Phil Dawid, Jerome Cornfield, John Parker Burg, David Blackwell, and George Barnard. While I have not agreed with all of the great variety of things they told me, it has all been taken into account in one way or another in the following pages. Even when we ended in disagreement on some issue, I believe that our frank private discussions have enabled me to avoid misrepresenting their positions, while clarifying my own thinking; I thank them for their patience.

E. T. Jaynes

July, 1996

Part 1

Principles and elementary applications

1

Plausible reasoning

The actual science of logic is conversant at present only with things either certain, impossible, or entirely doubtful, none of which (fortunately) we have to reason on. Therefore the true logic for this world is the calculus of Probabilities, which takes account of the magnitude of the probability which is, or ought to be, in a reasonable man's mind.

James Clerk Maxwell (1850)

Suppose some dark night a policeman walks down a street, apparently deserted. Suddenly he hears a burglar alarm, looks across the street, and sees a jewelry store with a broken window. Then a gentleman wearing a mask comes crawling out through the broken window, carrying a bag which turns out to be full of expensive jewelry. The policeman doesn't hesitate at all in deciding that this gentleman is dishonest. But by what reasoning process does he arrive at this conclusion? Let us first take a leisurely look at the general nature of such problems.

1.1 Deductive and plausible reasoning

A moment's thought makes it clear that our policeman's conclusion was not a logical deduction from the evidence; for there may have been a perfectly innocent explanation for everything. It might be, for example, that this gentleman was the owner of the jewelry store and he was coming home from a masquerade party, and didn't have the key with him. However, just as he walked by his store, a passing truck threw a stone through the window, and he was only protecting his own property.

Now, while the policeman's reasoning process was not logical deduction, we will grant that it had a certain degree of validity. The evidence did not make the gentleman's dishonesty *certain*, but it did make it extremely *plausible*. This is an example of a kind of reasoning in which we have all become more or less proficient, necessarily, long before studying mathematical theories. We are hardly able to get through one waking hour without facing some situation (e.g. will it rain or won't it?) where we do not have enough information to permit deductive reasoning; but still we must decide immediately what to do.

In spite of its familiarity, the formation of plausible conclusions is a very subtle process. Although history records discussions of it extending over 24 centuries, probably nobody has

3

ever produced an analysis of the process which anyone else finds completely satisfactory. In this work we will be able to report some useful and encouraging new progress, in which conflicting intuitive judgments are replaced by definite theorems, and *ad hoc* procedures are replaced by rules that are determined uniquely by some very elementary – and nearly inescapable – criteria of rationality.

All discussions of these questions start by giving examples of the contrast between deductive reasoning and plausible reasoning. As is generally credited to the *Organon* of Aristotle (fourth century BC)[1] deductive reasoning (*apodeixis*) can be analyzed ultimately into the repeated application of two strong syllogisms:

$$\text{if } A \text{ is true, then } B \text{ is true}$$

$$\frac{A \text{ is true}}{\text{therefore, } B \text{ is true,}} \tag{1.1}$$

and its inverse:

$$\text{if } A \text{ is true, then } B \text{ is true}$$

$$\frac{B \text{ is false}}{\text{therefore, } A \text{ is false.}} \tag{1.2}$$

This is the kind of reasoning we would like to use all the time; but, as noted, in almost all the situations confronting us we do not have the right kind of information to allow this kind of reasoning. We fall back on weaker syllogisms (*epagoge*):

$$\text{if } A \text{ is true, then } B \text{ is true}$$

$$\frac{B \text{ is true}}{\text{therefore, } A \text{ becomes more plausible.}} \tag{1.3}$$

The evidence does not prove that A is true, but verification of one of its consequences does give us more confidence in A. For example, let

$$A \equiv \text{it will start to rain by 10 AM at the latest;}$$
$$B \equiv \text{the sky will become cloudy before 10 AM.}$$

Observing clouds at 9:45 AM does not give us a logical certainty that the rain will follow; nevertheless our common sense, obeying the weak syllogism, may induce us to change our plans and behave *as if* we believed that it will, if those clouds are sufficiently dark.

This example shows also that the major premise, 'if A then B' expresses B only as a *logical* consequence of A; and not necessarily a causal physical consequence, which could be effective only at a later time. The rain at 10 AM is not the physical cause of the clouds at

---

[1] Today, several different views are held about the exact nature of Aristotle's contribution. Such issues are irrelevant to our present purpose, but the interested reader may find an extensive discussion of them in Lukasiewicz (1957).

9:45 AM. Nevertheless, the proper logical connection is not in the uncertain causal direction (clouds $\implies$ rain), but rather (rain $\implies$ clouds), which is certain, although noncausal.

We emphasize at the outset that we are concerned here with *logical* connections, because some discussions and applications of inference have fallen into serious error through failure to see the distinction between logical implication and physical causation. The distinction is analyzed in some depth by Simon and Rescher (1966), who note that all attempts to interpret implication as expressing physical causation founder on the lack of contraposition expressed by the second syllogism (1.2). That is, if we tried to interpret the major premise as 'A is the physical cause of B', then we would hardly be able to accept that 'not-B is the physical cause of not-A'. In Chapter 3 we shall see that attempts to interpret plausible inferences in terms of physical causation fare no better.

Another weak syllogism, still using the same major premise, is

If A is true, then B is true

_____

A is false   (1.4)

_____

therefore, B becomes less plausible.

In this case, the evidence does not prove that B is false; but one of the possible reasons for its being true has been eliminated, and so we feel less confident about B. The reasoning of a scientist, by which he accepts or rejects his theories, consists almost entirely of syllogisms of the second and third kind.

Now, the reasoning of our policeman was not even of the above types. It is best described by a still weaker syllogism:

If A is true, then B becomes more plausible

_____

B is true   (1.5)

_____

therefore, A becomes more plausible.

But in spite of the apparent weakness of this argument, when stated abstractly in terms of A and B, we recognize that the policeman's conclusion has a very strong convincing power. There is something which makes us believe that, in this particular case, his argument had almost the power of deductive reasoning.

These examples show that the brain, in doing plausible reasoning, not only decides whether something becomes more plausible or less plausible, but that it evaluates the *degree* of plausibility in some way. The plausibility for rain by 10 AM depends very much on the darkness of those clouds at 9:45. And the brain also makes use of old information as well as the specific new data of the problem; in deciding what to do we try to recall our past experience with clouds and rain, and what the weatherman predicted last night.

To illustrate that the policeman was also making use of the past experience of policemen in general, we have only to change that experience. Suppose that events like these happened several times every night to every policeman – and that in every case the gentleman turned out to be completely innocent. Very soon, policemen would learn to ignore such trivial things.

Thus, in our reasoning we depend very much on *prior information* to help us in evaluating the degree of plausibility in a new problem. This reasoning process goes on unconsciously, almost instantaneously, and we conceal how complicated it really is by calling it *common sense*.

The mathematician George Pólya (1945, 1954) wrote three books about plausible reasoning, pointing out a wealth of interesting examples and showing that there are definite rules by which we do plausible reasoning (although in his work they remain in qualitative form). The above weak syllogisms appear in his third volume. The reader is strongly urged to consult Pólya's exposition, which was the original source of many of the ideas underlying the present work. We show below how Pólya's principles may be made quantitative, with resulting useful applications.

Evidently, the deductive reasoning described above has the property that we can go through long chains of reasoning of the type (1.1) and (1.2) and the conclusions have just as much certainty as the premises. With the other kinds of reasoning, (1.3)–(1.5), the reliability of the conclusion changes as we go through several stages. But in their quantitative form we shall find that in many cases our conclusions can still approach the certainty of deductive reasoning (as the example of the policeman leads us to expect). Pólya showed that even a pure mathematician actually uses these weaker forms of reasoning most of the time. Of course, on publishing a new theorem, the mathematician will try very hard to invent an argument which uses only the first kind; but the reasoning process which led to the theorem in the first place almost always involves one of the weaker forms (based, for example, on following up conjectures suggested by analogies). The same idea is expressed in a remark of S. Banach (quoted by S. Ulam, 1957):

Good mathematicians see analogies between theorems; great mathematicians see analogies between analogies.

As a first orientation, then, let us note some very suggestive analogies to another field – which is itself based, in the last analysis, on plausible reasoning.

1.2 Analogies with physical theories

In physics, we learn quickly that the world is too complicated for us to analyze it all at once. We can make progress only if we dissect it into little pieces and study them separately. Sometimes, we can invent a mathematical model which reproduces several features of one of these pieces, and whenever this happens we feel that progress has been made. These models are called *physical theories*. As knowledge advances, we are able to invent better and better models, which reproduce more and more features of the real world, more and more accurately. Nobody knows whether there is some natural end to this process, or whether it will go on indefinitely.

In trying to understand common sense, we shall take a similar course. We won't try to understand it all at once, but we shall feel that progress has been made if we are able to construct idealized mathematical models which reproduce a few of its features. We expect that any model we are now able to construct will be replaced by more complete ones in the future, and we do not know whether there is any natural end to this process.

The analogy with physical theories is deeper than a mere analogy of method. Often, the things which are most familiar to us turn out to be the hardest to understand. Phenomena whose very existence is unknown to the vast majority of the human race (such as the difference in ultraviolet spectra of iron and nickel) can be explained in exhaustive mathematical detail – but all of modern science is practically helpless when faced with the complications of such a commonplace fact as growth of a blade of grass. Accordingly, we must not expect too much of our models; we must be prepared to find that some of the most familiar features of mental activity may be ones for which we have the greatest difficulty in constructing any adequate model.

There are many more analogies. In physics we are accustomed to finding that any advance in knowledge leads to consequences of great practical value, but of an unpredictable nature. Röntgen's discovery of X-rays led to important new possibilities of medical diagnosis; Maxwell's discovery of one more term in the equation for curl H led to practically instantaneous communication all over the earth.

Our mathematical models for common sense also exhibit this feature of practical usefulness. Any successful model, even though it may reproduce only a few features of common sense, will prove to be a powerful extension of common sense in some field of application. Within this field, it enables us to solve problems of inference which are so involved in complicated detail that we would never attempt to solve them without its help.

1.3 The thinking computer

Models have practical uses of a quite different type. Many people are fond of saying, 'They will never make a machine to replace the human mind – it does many things which no machine could ever do.' A beautiful answer to this was given by J. von Neumann in a talk on computers given in Princeton in 1948, which the writer was privileged to attend. In reply to the canonical question from the audience ('But of course, a mere machine can't really *think*, can it?'), he said:

You insist that there is something a machine cannot do. If you will tell me precisely what it is that a machine cannot do, then I can always make a machine which will do just that!

In principle, the only operations which a machine cannot perform for us are those which we cannot describe in detail, or which could not be completed in a finite number of steps. Of course, some will conjure up images of Gödel incompleteness, undecidability, Turing machines which never stop, etc. But to answer all such doubts we need only point to the existence of the human brain, which *does* it. Just as von Neumann indicated, the only real limitations on making 'machines which think' are our own limitations in not knowing exactly what 'thinking' consists of.

But in our study of common sense we shall be led to some very explicit ideas about the mechanism of thinking. Every time we can construct a mathematical model which reproduces a part of common sense by prescribing a definite set of operations, this shows us how to 'build a machine', (i.e. write a computer program) which operates on incomplete information and, by applying quantitative versions of the above weak syllogisms, does plausible reasoning instead of deductive reasoning.

Indeed, the development of such computer software for certain specialized problems of inference is one of the most active and useful current trends in this field. One kind of problem thus dealt with might be: given a mass of data, comprising 10 000 separate observations, determine in the light of these data and whatever prior information is at hand, the relative plausibilities of 100 different possible hypotheses about the causes at work.

Our unaided common sense might be adequate for deciding between two hypotheses whose consequences are very different; but, in dealing with 100 hypotheses which are not very different, we would be helpless without a computer *and* a well-developed mathematical theory that shows us how to program it. That is, what determines, in the policeman's syllogism (1.5), whether the plausibility for A increases by a large amount, raising it almost to certainty; or only a negligibly small amount, making the data B almost irrelevant? The object of the present work is to develop the mathematical theory which answers such questions, in the greatest depth and generality now possible.

While we expect a mathematical theory to be useful in programming computers, the idea of a thinking computer is also helpful psychologically in developing the mathematical theory. The question of the reasoning process used by actual human brains is charged with emotion and grotesque misunderstandings. It is hardly possible to say anything about this without becoming involved in debates over issues that are not only undecidable in our present state of knowledge, but are irrelevant to our purpose here.

Obviously, the operation of real human brains is so complicated that we can make no pretense of explaining its mysteries; and in any event we are not trying to explain, much less reproduce, all the aberrations and inconsistencies of human brains. That is an interesting and important subject; but it is not the subject we are studying here. Our topic is the *normative principles of logic*, and not the principles of psychology or neurophysiology.

To emphasize this, instead of asking, 'How can we build a mathematical model of human common sense?', let us ask, 'How could we build a machine which would carry out useful plausible reasoning, following clearly defined principles expressing an idealized common sense?'

1.4 Introducing the robot

In order to direct attention to constructive things and away from controversial irrelevancies, we shall invent an imaginary being. Its brain is to be designed *by us*, so that it reasons according to certain definite rules. These rules will be deduced from simple desiderata which, it appears to us, would be desirable in human brains; i.e. we think that a rational person, on discovering that they were violating one of these desiderata, would wish to revise their thinking.

In principle, we are free to adopt any rules we please; that is our way of *defining* which robot we shall study. Comparing its reasoning with yours, if you find no resemblance you are in turn free to reject our robot and design a different one more to your liking. But if you find a very strong resemblance, and decide that you want and trust this robot to help you in your own problems of inference, then that will be an accomplishment of the theory, not a premise.

Our robot is going to reason about propositions. As already indicated above, we shall denote various propositions by italicized capital letters, $\{A, B, C, \text{etc.}\}$, and for the time being we must require that any proposition used must have, to the robot, an unambiguous meaning and must be of the simple, definite logical type that must be either true or false. That is, until otherwise stated, we shall be concerned only with two-valued logic, or Aristotelian logic. We do not require that the truth or falsity of such an 'Aristotelian proposition' be ascertainable by any feasible investigation; indeed, our inability to do this is usually just the reason why we need the robot's help. For example, the writer personally considers both of the following propositions to be true:

$A \equiv$ Beethoven and Berlioz never met.

$B \equiv$ Beethoven's music has a better sustained quality than that of
Berlioz, although Berlioz at his best is the equal of anybody.

Proposition B is not a permissible one for our robot to think about at present, whereas proposition A is, although it is unlikely that its truth or falsity could be definitely established today.[2] After our theory is developed, it will be of interest to see whether the present restriction to Aristotelian propositions such as A can be relaxed, so that the robot might help us also with more vague propositions such as B (see Chapter 18 on the A_p-distribution).[3]

1.5 Boolean algebra

To state these ideas more formally, we introduce some notation of the usual symbolic logic, or Boolean algebra, so called because George Boole (1854) introduced a *notation* similar to the following. Of course, the principles of deductive logic itself were well understood centuries before Boole, and, as we shall see, all the results that follow from Boolean algebra were contained already as special cases in the rules of plausible inference given

[2] Their meeting is a chronological possibility, since their lives overlapped by 24 years; my reason for doubting it is the failure of Berlioz to mention any such meeting in his memoirs – on the other hand, neither does he come out and say definitely that they did *not* meet.

[3] The question of how one is to make a machine in some sense 'cognizant' of the conceptual meaning that a proposition like A has to humans, might seem very difficult, and much of the subject of artificial intelligence is devoted to inventing *ad hoc* devices to deal with this problem. However, we shall find in Chapter 4 that for us the problem is almost nonexistent; our rules for plausible reasoning automatically provide the means to do the mathematical equivalent of this.

by (1812). The symbol

$$AB, \tag{1.6}$$

called the *logical product* or the *conjunction*, denotes the proposition 'both A and B are true'. Obviously, the order in which we state them does not matter; AB and BA say the same thing. The expression

$$A + B, \tag{1.7}$$

called the *logical sum* or *disjunction*, stands for 'at least one of the propositions, A, B is true' and has the same meaning as $B + A$. These symbols are only a shorthand way of writing propositions, and do not stand for numerical values.

Given two propositions A, B, it may happen that one is true if and only if the other is true; we then say that they have the same *truth value*. This may be only a simple tautology (i.e. A and B are verbal statements which obviously say the same thing), or it may be that only after immense mathematical labor is it finally proved that A is the necessary and sufficient condition for B. From the standpoint of logic it does not matter; once it is established, by any means, that A and B have the same truth value, then they are logically equivalent propositions, in the sense that any evidence concerning the truth of one pertains equally well to the truth of the other, and they have the same implications for any further reasoning.

Evidently, then, it must be the most primitive axiom of plausible reasoning that two propositions with the same truth value are equally plausible. This might appear almost too trivial to mention, were it not for the fact that Boole himself (Boole, 1854, p. 286) fell into error on this point, by mistakenly identifying two propositions which were in fact different – and then failing to see any contradiction in their different plausibilities. Three years later, Boole (1857) gave a revised theory which supersedes that in his earlier book; for further comments on this incident, see Keynes (1921, pp. 167–168); Jaynes (1976, pp. 240–242).

In Boolean algebra, the equal sign is used to denote not equal numerical value, but equal truth value: $A = B$, and the 'equations' of Boolean algebra thus consist of assertions that the proposition on the left-hand side has the same truth value as the one on the right-hand side. The symbol '$\equiv$' means, as usual, 'equals by definition'.

In denoting complicated propositions we use parentheses in the same way as in ordinary algebra, i.e. to indicate the order in which propositions are to be combined (at times we shall use them also merely for clarity of expression although they are not strictly necessary). In their absence we observe the rules of algebraic hierarchy, familiar to those who use hand calculators: thus $AB + C$ denotes $(AB) + C$; and not $A(B + C)$.

The *denial* of a proposition is indicated by a bar:

$$\overline{A} \equiv A \text{ is false.} \tag{1.8}$$

The relation between A, $\overline{A}$ is a reciprocal one:

$$A = \overline{A} \text{ is false,} \tag{1.9}$$

and it does not matter which proposition we denote by the barred and which by the unbarred letter. Note that some care is needed in the unambiguous use of the bar. For example, according to the above conventions,

$$\overline{AB} = A B \text{ is false;} \tag{1.10}$$

$$\overline{A}\,\overline{B} = \text{both } A \text{ and } B \text{ are false.} \tag{1.11}$$

These are quite different propositions; in fact, $\overline{AB}$ is not the logical product $\overline{A}\,\overline{B}$, but the logical sum: $\overline{AB} = \overline{A} + \overline{B}$.

With these understandings, Boolean algebra is characterized by some rather trivial and obvious basic identities, which express the properties of:

Idempotence:
$$\begin{cases} AA = A \\ A + A = A \end{cases}$$

Commutativity:
$$\begin{cases} AB = BA \\ A + B = B + A \end{cases}$$

Associativity:
$$\begin{cases} A(BC) = (AB)C = ABC \\ A + (B + C) = (A + B) + C = A + B + C \end{cases} \tag{1.12}$$

Distributivity:
$$\begin{cases} A(B + C) = AB + AC \\ A + (BC) = (A + B)(A + C) \end{cases}$$

Duality:
$$\begin{cases} \text{If } C = AB, \text{ then } \overline{C} = \overline{A} + \overline{B} \\ \text{If } D = A + B, \text{ then } \overline{D} = \overline{A}\,\overline{B} \end{cases}$$

but by their application one can prove any number of further relations, some highly non-trivial. For example, we shall presently have use for the rather elementary theorem:

$$\text{if } \overline{B} = AD \text{ then } A\overline{B} = \overline{B} \text{ and } B\overline{A} = \overline{A}. \tag{1.13}$$

Implication

The proposition

$$A \Rightarrow B \tag{1.14}$$

to be read as '*A* implies *B*', does not assert that either *A* or *B* is true; it means only that $A\overline{B}$ is false, or, what is the same thing, $(\overline{A} + B)$ is true. This can be written also as the logical equation $A = AB$. That is, given (1.14), if *A* is true then *B* must be true; or, if *B* is false then *A* must be false. This is just what is stated in the strong syllogisms (1.1) and (1.2).

On the other hand, if A is false, (1.14) says nothing about B: and if B is true, (1.14) says nothing about A. But these are just the cases in which our weak syllogisms (1.3), (1.4) do say something. In one respect, then, the term 'weak syllogism' is misleading. The theory of plausible reasoning based on weak syllogisms is not a 'weakened' form of logic; it is an *extension* of logic with new content not present at all in conventional deductive logic. It will become clear in the next chapter (see (2.69) and (2.70)) that our rules include deductive logic as a special case.

A tricky point

Note carefully that in ordinary language one would take 'A implies B' to mean that B is logically deducible from A. But, in formal logic, 'A implies B' means only that the propositions A and AB have the same truth value. In general, whether B is logically deducible from A does not depend only on the propositions A and B; it depends on the totality of propositions $(A, A', A'', \ldots)$ that we accept as true and which are therefore available to use in the deduction. Devinatz (1968, p. 3) and Hamilton (1988, p. 5) give the truth table for the implication as a binary operation, illustrating that $A \Rightarrow B$ is false only if A is true and B is false; in all other cases $A \Rightarrow B$ is true!

This may seem startling at first glance; however, note that, indeed, if A and B are both true, then $A = AB$ and so $A \Rightarrow B$ is true; in formal logic every true statement implies every other true statement. On the other hand, if A is false, then AQ is also false for all Q, thus $A = AB$ and $A = A\overline{B}$ are both true, so $A \Rightarrow B$ and $A \Rightarrow \overline{B}$ are both true; a false proposition implies all propositions. If we tried to interpret this as logical deducibility (i.e. both B and $\overline{B}$ are deducible from A), it would follow that every false proposition is logically contradictory. Yet the proposition: 'Beethoven outlived Berlioz' is false but hardly logically contradictory (for Beethoven did outlive many people who were the same age as Berlioz).

Obviously, merely knowing that propositions A and B are both true does not provide enough information to decide whether either is logically deducible from the other, plus some unspecified 'toolbox' of other propositions. The question of logical deducibility of one proposition from a set of others arises in a crucial way in the Gödel theorem discussed at the end of Chapter 2. This great difference in the meaning of the word 'implies' in ordinary language and in formal logic is a tricky point that can lead to serious error if it is not properly understood; it appears to us that 'implication' is an unfortunate choice of word, and that this is not sufficiently emphasized in conventional expositions of logic.

1.6 Adequate sets of operations

We note some features of deductive logic which will be needed in the design of our robot. We have defined four operations, or 'connectives', by which, starting from two propositions A, B, other propositions may be defined: the logical product or conjunction AB, the logical sum or disjunction $A + B$, the implication $A \Rightarrow B$, and the negation $\overline{A}$. By combining these operations repeatedly in every possible way, one can generate any number of new propositions, such as

$$C \equiv (A + \overline{B})(\overline{A} + A\overline{B}) + \overline{A}B(A + B). \tag{1.15}$$

Many questions then occur to us: How large is the class of new propositions thus generated? Is it infinite, or is there a finite set that is closed under these operations? Can every proposition defined from A, B be thus represented, or does this require further connectives beyond the above four? Or are these four already overcomplete so that some might be dispensed with? What is the smallest set of operations that is adequate to generate all such 'logic functions' of A and B? If instead of two starting propositions A, B we have an arbitrary number $\{A_1, \ldots, A_n\}$, is this set of operations still adequate to generate all possible logic functions of $\{A_1, \ldots, A_n\}$?

All these questions are answered easily, with results useful for logic, probability theory, and computer design. Broadly speaking, we are asking whether, starting from our present vantage point, we can (1) increase the number of functions, (2) decrease the number of operations. The first query is simplified by noting that two propositions, although they may appear entirely different when written out in the manner (1.15), are not different propositions from the standpoint of logic if they have the same truth value. For example, it is left for the reader to verify that C in (1.15) is logically the same statement as the implication $C = (B \Rightarrow \overline{A})$.

Since we are, at this stage, restricting our attention to Aristotelian propositions, any logic function $C = f(A, B)$ such as (1.15) has only two possible 'values', true and false; and likewise the 'independent variables' A and B can take on only those two values.

At this point, a logician might object to our notation, saying that the symbol A has been defined as standing for some fixed proposition, whose truth cannot change; so if we wish to consider logic functions, then instead of writing $C = f(A, B)$ we should introduce new symbols and write $z = f(x, y)$, where x, y, z, are 'statement variables' for which various specific statements A, B, C may be substituted. But if A stands for some fixed but unspecified proposition, then it can still be either true or false. We achieve the same flexibility merely by the understanding that equations like (1.15) which define logic functions are to be true for all ways of defining A, B; i.e. instead of a statement variable we use a variable statement.

In relations of the form $C = f(A, B)$, we are concerned with logic functions defined on a discrete 'space' S consisting of only $2^2 = 4$ points; namely those at which A and B take on the 'values' {TT, TF, FT, FF}, respectively; and, at each point, the function $f(A, B)$ can take on independently either of two values {T, F}. There are, therefore, exactly $2^4 = 16$ different logic functions $f(A, B)$, and no more. An expression $B = f(A_1, \ldots, A_n)$ involving n propositions is a logic function on a space S of $M = 2^n$ points; and there are exactly 2^M such functions.

In the case $n = 1$, there are four logic functions $\{f_1(A), \ldots, f_4(A)\}$, which we can define by enumeration, listing all their possible values in a truth table:

A	T	F
$f_1(A)$	T	T
$f_2(A)$	T	F
$f_3(A)$	F	T
$f_4(A)$	F	F

But it is obvious by inspection that these are just

$$f_1(A) = A + \overline{A}$$
$$f_2(A) = A$$
$$f_3(A) = \overline{A} \qquad (1.16)$$
$$f_4(A) = A\,\overline{A},$$

so we prove by enumeration that the three operations: conjunction, disjunction, and negation are adequate to generate all logic functions of a single proposition.

For the case of general n, consider first the special functions, each of which is true at one and only one point of S. For $n = 2$ there are $2^n = 4$ such functions,

A, B	TT	TF	FT	FF
$f_1(A, B)$	T	F	F	F
$f_2(A, B)$	F	T	F	F
$f_3(A, B)$	F	F	T	F
$f_4(A, B)$	F	F	F	T

It is clear by inspection that these are just the four basic conjunctions,

$$f_1(A, B) = A\,B$$
$$f_2(A, B) = A\,\overline{B}$$
$$f_3(A, B) = \overline{A}\,B \qquad (1.17)$$
$$f_4(A, B) = \overline{A}\,\overline{B}.$$

Consider now any logic function which is true on certain specified points of S; for example, $f_5(A, B)$ and $f_6(A, B)$, defined by

A, B	TT	TF	FT	FF
$f_5(A, B)$	F	T	F	T
$f_6(A, B)$	T	F	T	T

We assert that each of these functions is the logical sum of the conjunctions (1.17) that are true on the same points (this is not trivial; the reader should verify it in detail). Thus,

$$\begin{aligned}
f_5(A, B) &= f_2(A, B) + f_4(A, B) \\
&= A\,\overline{B} + \overline{A}\,\overline{B} \\
&= (A + \overline{A})\,\overline{B} \\
&= \overline{B},
\end{aligned} \tag{1.18}$$

and, likewise,

$$\begin{aligned}
f_6(A, B) &= f_1(A, B) + f_3(A, B) + f_4(A, B) \\
&= AB + \overline{A}\,B + \overline{A}\,\overline{B} \\
&= B + \overline{A}\,\overline{B} \\
&= \overline{A} + B.
\end{aligned} \tag{1.19}$$

That is, $f_6(A, B)$ is the implication $f_6(A, B) = (A \Rightarrow B)$, with the truth table discussed above. Any logic function $f(A, B)$ that is true on at least one point of S can be constructed in this way as a logical sum of the basic conjunctions (1.17). There are $2^4 - 1 = 15$ such functions. For the remaining function, which is always false, it suffices to take the contradiction, $f_{16}(A, B) \equiv A\,\overline{A}$.

This method (called 'reduction to *disjunctive normal form*' in logic textbooks) will work for any n. For example, in the case $n = 5$ there are $2^5 = 32$ basic conjunctions,

$$\{ABCDE, \ ABCD\overline{E}, \ ABC\overline{D}E, \dots, \ \overline{A}\,\overline{B}\,\overline{C}\,\overline{D}\,\overline{E}\}, \tag{1.20}$$

and $2^{32} = 4\,294\,967\,296$ different logic functions $f_i(A, B, C, D, E)$; of which $4\,294\,967\,295$ can be written as logical sums of the basic conjunctions, leaving only the contradiction

$$f_{4294967296}(A, B, C, D, E) = A\,\overline{A}. \tag{1.21}$$

Thus one can verify by 'construction in thought' that the three operations

$$\{\text{conjunction, disjunction, negation}\}, \quad \text{i.e.} \quad \{\text{AND, OR, NOT}\}, \tag{1.22}$$

suffice to generate all possible logic functions; or, more concisely, they form an *adequate set*.

The duality property (1.12) shows that a smaller set will suffice; for disjunction of A, B is the same as denying that they are both false:

$$A + B = \overline{(\overline{A}\,\overline{B})}. \tag{1.23}$$

Therefore, the two operations (AND, NOT) already constitute an adequate set for deductive logic.[4] This fact will be essential in determining when we have an adequate set of rules for plausible reasoning; see Chapter 2.

---

[4] For you to ponder: Does it follow that these two commands are the only ones needed to write any computer program?

It is clear that we cannot now strike out either of these operations, leaving only the other; i.e. the operation 'AND' cannot be reduced to negations; and negation cannot be accomplished by any number of 'AND' operations. But this still leaves open the possibility that both conjunction and negation might be reducible to some third operation, not yet introduced, so that a single logic operation would constitute an adequate set.

It comes as a pleasant surprise to find that there is not only one but two such operations. The operation 'NAND' is defined as the negation of 'AND':

$$A \uparrow B \equiv \overline{AB} = \overline{A} + \overline{B} \qquad (1.24)$$

which we can read as 'A NAND B'. But then we have at once

$$\overline{A} = A \uparrow A$$
$$AB = (A \uparrow B) \uparrow (A \uparrow B) \qquad (1.25)$$
$$A + B = (A \uparrow A) \uparrow (B \uparrow B).$$

Therefore, every logic function can be constructed with NAND alone. Likewise, the operation NOR defined by

$$A \downarrow B \equiv \overline{A + B} = \overline{A} \, \overline{B} \qquad (1.26)$$

is also powerful enough to generate all logic functions:

$$\overline{A} = A \downarrow A$$
$$A + B = (A \downarrow B) \downarrow (A \downarrow B) \qquad (1.27)$$
$$AB = (A \downarrow A) \downarrow (B \downarrow B).$$

One can take advantage of this in designing computer and logic circuits. A 'logic gate' is a circuit having, besides a common ground, two input terminals and one output. The voltage relative to ground at any of these terminals can take on only two values; say $+3$ volts, or 'up', representing 'true'; and 0 volts or 'down', representing 'false'. A NAND gate is thus one whose output is up if and only if at least one of the inputs is down; or, what is the same thing, down if and only if both inputs are up; while for a NOR gate the output is up if and only if both inputs are down.

One of the standard components of logic circuits is the 'quad NAND gate', an integrated circuit containing four independent NAND gates on one semiconductor chip. Given a sufficient number of these and no other circuit components, it is possible to generate any required logic function by interconnecting them in various ways.

This short excursion into deductive logic is as far as we need go for our purposes. Further developments are given in many textbooks; for example, a modern treatment of Aristotelian logic is given by Copi (1994). For non-Aristotelian forms with special emphasis on Gödel incompleteness, computability, decidability, Turing machines, etc., see Hamilton (1988).

We turn now to our extension of logic, which is to follow from the conditions discussed next. We call them 'desiderata' rather than 'axioms' because they do not assert that anything is 'true' but only state what appear to be desirable goals. Whether these goals are attainable without contradictions, and whether they determine any unique extension of logic, are matters of mathematical analysis, given in Chapter 2.

1.7 The basic desiderata

To each proposition about which it reasons, our robot must assign some degree of plausibility, based on the evidence we have given it; and whenever it receives new evidence it must revise these assignments to take that new evidence into account. In order that these plausibility assignments can be stored and modified in the circuits of its brain, they must be associated with some definite physical quantity, such as voltage or pulse duration or a binary coded number, etc. – however our engineers want to design the details. For present purposes, this means that there will have to be some kind of association between degrees of plausibility and real numbers:

(I)   *Degrees of plausibility are represented by real numbers.*   (1.28)

Desideratum (I) is practically forced on us by the requirement that the robot's brain must operate by the carrying out of some definite physical process. However, it will appear (Appendix A) that it is also required theoretically; we do not see the possibility of any consistent theory without a property that is equivalent functionally to desideratum (I).

    We adopt a natural but nonessential convention: that a greater plausibility shall correspond to a greater number. It will also be convenient to assume a continuity property, which is hard to state precisely at this stage; to say it intuitively: an infinitesimally greater plausibility ought to correspond only to an infinitesimally greater number.

    The plausibility that the robot assigns to some proposition A will, in general, depend on whether we told it that some other proposition B is true. Following the notation of Keynes (1921) and Cox (1961), we indicate this by the symbol

$$A|B, \qquad (1.29)$$

which we may call 'the conditional plausibility that A is true, given that B is true' or just 'A given B'. It stands for some real number. Thus, for example,

$$A|BC \qquad (1.30)$$

(which we may read as 'A given BC') represents the plausibility that A is true, given that both B and C are true. Or,

$$A + B|CD \qquad (1.31)$$

represents the plausibility that at least one of the propositions A and B is true, given that both C and D are true; and so on. We have decided to represent a greater plausibility by a greater number, so

$$(A|B) > (C|B) \qquad (1.32)$$

says that, given B, A is more plausible than C. In this notation, while the symbol for plausibility is just of the form $A|B$ without parentheses, we often add parentheses for clarity of expression. Thus, (1.32) says the same thing as

$$A|B \; > \; C|B, \tag{1.33}$$

but its meaning is clearer to the eye.

In the interest of avoiding impossible problems, we are not going to ask our robot to undergo the agony of reasoning from impossible or mutually contradictory premises; there could be no 'correct' answer. Thus, we make no attempt to define $A|BC$ when B and C are mutually contradictory. Whenever such a symbol appears, it is understood that B and C are compatible propositions.

Also, we do not want this robot to think in a way that is directly opposed to the way you and I think. So we shall design it to reason in a way that is at least *qualitatively* like the way humans try to reason, as described by the above weak syllogisms and a number of other similar ones.

Thus, if it has old information C which gets updated to C' in such a way that the plausibility for A is increased:

$$(A|C') \; > \; (A|C); \tag{1.34}$$

but the plausibility for B given A is not changed:

$$(B|AC') \; = \; (B|AC). \tag{1.35}$$

This can, of course, produce only an increase, never a decrease, in the plausibility that both A and B are true:

$$(AB|C') \; \geq \; (AB|C); \tag{1.36}$$

and it must produce a decrease in the plausibility that A is false:

$$(\overline{A}|C') \; < \; (\overline{A}|C). \tag{1.37}$$

This qualitative requirement simply gives the 'sense of direction' in which the robot's reasoning is to go; it says nothing about *how much* the plausibilities change, except that our continuity assumption (which is also a condition for qualitative correspondence with common sense) now requires that if $A|C$ changes only infinitesimally, it can induce only an infinitesimal change in $AB|C$ and $\overline{A}|C$. The specific ways in which we use these qualitative requirements will be given in the next chapter, at the point where it is seen why we need them. For the present we summarize them simply as:

$$\text{(II)} \quad \textit{Qualitative correspondence with common sense.} \tag{1.38}$$

Finally, we want to give our robot another desirable property for which honest people strive without always attaining: that it always reasons *consistently*. By this we mean just the three common colloquial meanings of the word 'consistent':

(IIIa)   *If a conclusion can be reasoned out in more than one way, then every possible way must lead to the same result.*   (1.39a)

(IIIb)   *The robot always takes into account all of the evidence it has relevant to a question. It does not arbitrarily ignore some of the information, basing its conclusions only on what remains. In other words, the robot is completely nonideological.*   (1.39b)

(IIIc)   *The robot always represents equivalent states of knowledge by equivalent plausibility assignments. That is, if in two problems the robot's state of knowledge is the same (except perhaps for the labeling of the propositions), then it must assign the same plausibilities in both.*   (1.39c)

Desiderata (I), (II), and (IIIa) are the basic 'structural' requirements on the inner workings of our robot's brain, while (IIIb) and (IIIc) are 'interface' conditions which show how the robot's behavior should relate to the outer world.

At this point, most students are surprised to learn that our search for desiderata is at an end. The above conditions, it turns out, uniquely determine the rules by which our robot must reason; i.e. there is only one set of mathematical operations for manipulating plausibilities which has all these properties. These rules are deduced in Chapter 2.

(At the end of most chapters, we insert a section of informal Comments in which are collected various side remarks, background material, etc. The reader may skip them without losing the main thread of the argument.)

1.8 Comments

As politicians, advertisers, salesmen, and propagandists for various political, economic, moral, religious, psychic, environmental, dietary, and artistic doctrinaire positions know only too well, fallible human minds are easily tricked, by clever verbiage, into committing violations of the above desiderata. We shall try to ensure that they do not succeed with our robot.

We emphasize another contrast between the robot and a human brain. By Desideratum I, the robot's mental state about any proposition is to be represented by a real number. Now, it is clear that our attitude toward any given proposition may have more than one 'coordinate'. You and I form simultaneous judgments about a proposition not only as to whether it is plausible, but also whether it is desirable, whether it is important, whether it is useful, whether it is interesting, whether it is amusing, whether it is morally right, etc. If we assume that each of these judgments might be represented by a number, then a fully adequate description of a human state of mind would be represented by a vector in a space of a rather large number of dimensions.

Not all propositions require this. For example, the proposition 'The refractive index of water is less than 1.3' generates no emotions; consequently the state of mind which it produces has very few coordinates. On the other hand, the proposition, 'Your mother-in-law just wrecked your new car' generates a state of mind with many coordinates. Quite generally, the situations of everyday life are those involving many coordinates. It is just for this reason, we suggest, that the most familiar examples of mental activity are often the most difficult to reproduce by a model. Perhaps we have here the reason why science and mathematics are the most successful of human activities: they deal with propositions which produce the simplest of all mental states. Such states would be the ones least perturbed by a given amount of imperfection in the human mind.

Of course, for many purposes we would not want our robot to adopt any of these more 'human' features arising from the other coordinates. It is just the fact that computers do *not* get confused by emotional factors, do *not* get bored with a lengthy problem, do *not* pursue hidden motives opposed to ours, that makes them safer agents than men for carrying out certain tasks.

These remarks are interjected to point out that there is a large unexplored area of possible generalizations and extensions of the theory to be developed here; perhaps this may inspire others to try their hand at developing 'multidimensional theories' of mental activity, which would more and more resemble the behavior of actual human brains – not all of which is undesirable. Such a theory, if successful, might have an importance beyond our present ability to imagine.[5]

For the present, however, we shall have to be content with a much more modest undertaking. Is it possible to develop a consistent 'one-dimensional' model of plausible reasoning? Evidently, our problem will be simplest if we can manage to represent a degree of plausibility uniquely by a single real number, and ignore the other 'coordinates' just mentioned.

We stress that we are in no way asserting that degrees of plausibility in actual human minds have a unique numerical measure. Our job is not to postulate – or indeed to conjecture about – any such thing; it is to *investigate* whether it is possible, in our robot, to set up such a correspondence without contradictions.

But to some it may appear that we have already assumed more than is necessary, thereby putting gratuitous restrictions on the generality of our theory. Why must we represent degrees of plausibility by real numbers? Would not a 'comparative' theory based on a system of qualitative ordering relations such as $(A|C) > (B|C)$ suffice? This point is discussed further in Appendix A, where we describe other approaches to probability theory and note that some attempts have been made to develop comparative theories which it was thought would be logically simpler, or more general. But this turned out not to be the case; so, although it is quite possible to develop the foundations in other ways than ours, the final results will not be different.

---

[5] Indeed, some psychologists think that as few as five dimensions might suffice to characterize a human personality; that is, that we all differ only in having different mixes of five basic personality traits which may be genetically determined. But it seems to us that this must be grossly oversimplified; identifiable chemical factors continuously varying in both space and time (such as the distribution of glucose metabolism in the brain) affect mental activity but cannot be represented faithfully in a space of only five dimensions. Yet it may be that five numbers can capture enough of the truth to be useful for many purposes.

1.8.1 Common language vs. formal logic

We should note the distinction between the statements of formal logic and those of ordinary language. It might be thought that the latter is only a less precise form of expression; but on examination of details the relation appears different. It appears to us that ordinary language, carefully used, need not be less precise than formal logic; but ordinary language is more complicated in its rules and has consequently richer possibilities of expression than we allow ourselves in formal logic.

In particular, common language, being in constant use for other purposes than logic, has developed subtle nuances – means of implying something without actually stating it – that are lost on formal logic. Mr A, to affirm his objectivity, says, 'I believe what I see.' Mr B retorts: 'He doesn't see what he doesn't believe.' From the standpoint of formal logic, it appears that they have said the same thing; yet from the standpoint of common language, those statements had the intent and effect of conveying opposite meanings.

Here is a less trivial example, taken from a mathematics textbook. Let L be a straight line in a plane, and S an infinite set of points in that plane, each of which is projected onto L. Now consider the following statements:

> (I) The projection of the limit is the limit of the projections.
> (II) The limit of the projections is the projection of the limit.

These have the grammatical structures 'A is B' and 'B is A', and so they might appear logically equivalent. Yet in that textbook, (I) was held to be true, and (II) not true in general, on the grounds that the limit of the projections may exist when the limit of the set does not.

As we see from this, in common language – even in mathematics textbooks – we have learned to read subtle nuances of meaning into the exact phrasing, probably without realizing it until an example like this is pointed out. We interpret 'A is B' as asserting first of all, as a kind of major premise, that A exists; and the rest of the statement is understood to be conditional on that premise. Put differently, in common grammar the verb 'is' implies a distinction between subject and object, which the symbol '=' does not have in formal logic or in conventional mathematics. (However, in computer languages we encounter such statements as '$J = J + 1$', which everybody seems to understand, but in which the '=' sign has now acquired that implied distinction after all.)

Another amusing example is the old adage 'knowledge is power', which is a very cogent truth, both in human relations and in thermodynamics. An ad writer for a chemical trade journal[6] fouled this up into 'power is knowledge', an absurd – indeed, obscene – falsity.

These examples remind us that the verb 'is' has, like any other verb, a subject and a predicate; but it is seldom noted that this verb has two entirely different meanings. A person whose native language is English may require some effort to see the different meanings in the statements: 'The room is noisy' and 'There is noise in the room'. But in Turkish these meanings are rendered by different words, which makes the distinction so clear that a visitor

---

[6] *LC-CG Magazine*, March 1988, p. 211.

who uses the wrong word will not be understood. The latter statement is ontological, assert-ing the physical existence of something, while the former is epistemological, expressing only the speaker's personal perception.

Common language – or, at least, the English language – has an almost universal tendency to disguise epistemological statements by putting them into a grammatical form which sug-gests to the unwary an ontological statement. A major source of error in current probability theory arises from an unthinking failure to perceive this. To interpret the first kind of state-ment in the ontological sense is to assert that one's own private thoughts and sensations are realities existing externally in Nature. We call this the 'mind projection fallacy', and note the trouble it causes many times in what follows. But this trouble is hardly confined to prob-ability theory; as soon as it is pointed out, it becomes evident that much of the discourse of philosophers and Gestalt psychologists, and the attempts of physicists to explain quantum theory, are reduced to nonsense by the author falling repeatedly into the mind projection fallacy.

These examples illustrate the care that is needed when we try to translate the complex statements of common language into the simpler statements of formal logic. Of course, common language is often less precise than we should want in formal logic. But everybody expects this and is on the lookout for it, so it is less dangerous.

It is too much to expect that our robot will grasp all the subtle nuances of common language, which a human spends perhaps 20 years acquiring. In this respect, our robot will remain like a small child – it interprets all statements literally and blurts out the truth without thought of whom this may offend.

It is unclear to the writer how difficult – and even less clear how desirable – it would be to design a newer model robot with the ability to recognize these finer shades of meaning. Of course, the question of principle is disposed of at once by the existence of the human brain, which does this. But, in practice, von Neumann's principle applies; a robot designed by us cannot do it until someone develops a theory of 'nuance recognition', which reduces the process to a definitely prescribed set of operations. This we gladly leave to others.

In any event, our present model robot is quite literally real, because today it is almost universally true that any nontrivial probability evaluation is performed by a computer. The person who programmed that computer was necessarily, whether or not they thought of it that way, designing part of the brain of a robot according to some preconceived notion of how the robot should behave. But very few of the computer programs now in use satisfy all our desiderata; indeed, most are intuitive *ad hoc* procedures that were not chosen with any well-defined desiderata at all in mind.

Any such adhockery is presumably usable within some special area of application – that was the criterion for choosing it – but as the proofs of Chapter 2 will show, any adhockery which conflicts with the rules of probability theory must generate demonstrable inconsistencies when we try to apply it beyond some restricted area. Our aim is to avoid this by developing the general principles of inference once and for all, directly from the requirement of consistency, and in a form applicable to any problem of plausible inference that is formulated in a sufficiently unambiguous way.

1.8.2 Nitpicking

As is apparent from the above, in the present work we use the term 'Boolean algebra' in its long-established meaning as referring to two-valued logic in which symbols like '*A*' stand for propositions. A compulsive nitpicker has complained to us that some mathematicians have used the term in a slightly different meaning, in which '*A*' could refer to a class of propositions. But the two usages are not in conflict; we recognize the broader meaning, but just find no reason to avail ourselves of it.

The set of rules and symbols that we have called 'Boolean algebra' is sometimes called 'the propositional calculus'. The term seems to be used only for the purpose of adding that we need also another set of rules and symbols called 'the predicate calculus'. However, these new symbols prove to be only abbreviations for short and familiar phrases. The 'universal quantifier' is only an abbreviation for 'for all'; the 'existential quantifier' is an abbreviation for 'there is a'. If we merely write our statements in plain English, we are using automatically all of the predicate calculus that we need for our purposes, and doing it more intelligibly.

The validity of the second strong syllogism (in two-valued logic) is sometimes questioned. However, it appears that in current mathematics it is still considered valid reasoning to say that a supposed theorem is disproved by exhibiting a counterexample, that a set of statements is considered inconsistent if we can derive a contradiction from them, and that a proposition can be established by *reductio ad absurdum*, deriving a contradiction from its denial. This is enough for us; we are quite content to follow this long tradition. Our feeling of security in this stance comes from the conviction that, while logic may move forward in the future, it can hardly move backward. A new logic might lead to new results about which Aristotelian logic has nothing to say; indeed, that is just what we are trying to create here. But surely, if a new logic was found to conflict with Aristotelian logic in an area where Aristotelian logic is applicable, we would consider that a fatal objection to the new logic.

Therefore, to those who feel confined by two-valued deductive logic, we can say only: 'By all means, investigate other possibilities if you wish to; and please let us know about it as soon as you have found a new result that was not contained in two-valued logic or our extension of it, *and* is useful in scientific inference.' Actually, there are many different and mutually inconsistent multiple-valued logics already in the literature. But in Appendix A we adduce arguments which suggest that they can have no useful content that is not already in two-valued logic; that is, that an *n*-valued logic applied to one set of propositions is either equivalent to a two-valued logic applied to an enlarged set, or else it contains internal inconsistencies.

Our experience is consistent with this conjecture; in practice, multiple-valued logics seem to be used not to find new useful results, but rather in attempts to remove supposed difficulties with two-valued logic, particularly in quantum theory, fuzzy sets, and artificial intelligence. But on closer study, all such difficulties known to us have proved to be only examples of the mind projection fallacy, calling for direct revision of the concepts rather than a new logic.

2

The quantitative rules

Probability theory is nothing but common sense reduced to calculation.
Laplace, 1819

We have now formulated our problem, and it is a matter of straightforward mathematics to work out the consequences of our desiderata, which may be stated broadly as follows:

    (I) Representation of degrees of plausibility by real numbers;
    (II) Qualitative correspondence with common sense;
    (III) Consistency.

The present chapter is devoted entirely to deduction of the quantitative rules for inference which follow from these desiderata. The resulting rules have a long, complicated, and astonishing history, full of lessons for scientific methodology in general (see the Comments sections at the end of several chapters).

2.1 The product rule

We first seek a consistent rule relating the plausibility of the logical product AB to the plausibilities of A and B separately. In particular, let us find $AB|C$. Since the reasoning is somewhat subtle, we examine this from several different viewpoints.

As a first orientation, note that the process of deciding that AB is true can be broken down into elementary decisions about A and B separately. The robot can

    (1) decide that B is true; $(B|C)$
    (2) having accepted B as true, decide that A is true. $(A|BC)$

Or, equally well,

    (1′) decide that A is true; $(A|C)$
    (2′) having accepted A as true, decide that B is true. $(B|AC)$

In each case we indicate above the plausibility corresponding to that step.

Now let us describe the first procedure in words. In order for AB to be a true proposition, it is necessary that B is true. Thus the plausibility $B|C$ should be involved. In addition, if B

is true, it is further necessary that A should be true; so the plausibility $A|BC$ is also needed. But if B is false, then of course AB is false independently of whatever one knows about A, as expressed by $A|\overline{B}C$; if the robot reasons first about B, then the plausibility of A will be relevant only if B is true. Thus, if the robot has $B|C$ and $A|BC$ it will not need $A|C$. That would tell it nothing about AB that it did not have already.

Similarly, $A|B$ and $B|A$ are not needed; whatever plausibility A or B might have in the absence of information C could not be relevant to judgments of a case in which the robot knows that C is true. For example, if the robot learns that the earth is round, then in judging questions about cosmology today, it does not need to take into account the opinions it might have (i.e. the extra possibilities that it would need to take into account) if it did not know that the earth is round.

Of course, since the logical product is commutative, $AB = BA$, we could interchange A and B in the above statements; i.e. knowledge of $A|C$ and $B|AC$ would serve equally well to determine $AB|C = BA|C$. That the robot must obtain the same value for $AB|C$ from either procedure is one of our conditions of consistency, desideratum (IIIa).

We can state this in a more definite form. $(AB|C)$ will be some function of $B|C$ and $A|BC$:

$$(AB|C) = F[(B|C), (A|BC)]. \tag{2.1}$$

Now, if the reasoning we went through here is not completely obvious, let us examine some alternatives. We might suppose, for example, that

$$(AB|C) = F[(A|C), (B|C)] \tag{2.2}$$

might be a permissible form. But we can show easily that no relation of this form could satisfy our qualitative conditions of desideratum (II). Proposition A might be very plausible given C, and B might be very plausible given C; but AB could still be very plausible or very implausible.

For example, it is quite plausible that the next person you meet has blue eyes and also quite plausible that this person's hair is black; and it is reasonably plausible that both are true. On the other hand it is quite plausible that the left eye is blue, and quite plausible that the right eye is brown; but extremely implausible that both of those are true. We would have no way of taking such influences into account if we tried to use a formula of this kind. Our robot could not reason the way humans do, even qualitatively, with that kind of functional relation.

But other possibilities occur to us. The method of trying out all possibilities – a kind of 'proof by exhaustion' – can be organized as follows. Introduce the real numbers

$$u = (AB|C), \quad v = (A|C), \quad w = (B|AC), \quad x = (B|C), \quad y = (A|BC). \tag{2.3}$$

If u is to be expressed as a function of two or more of v, w, x, y, there are 11 possibilities. You can write out each of them, and subject each one to various extreme conditions, as in the brown and blue eyes (which was the abstract statement: A implies that B is false). Other extreme conditions are $A = B$, $A = C$, $C \Rightarrow \overline{A}$, etc. Carrying out this somewhat tedious analysis, Tribus (1969) finds that all but two of the possibilities can exhibit qualitative violations of common sense in some extreme case. The two which survive are $u = F(x, y)$ and $u = F(w, v)$, just the two functional forms already suggested by our previous reasoning.

We now apply the qualitative requirement discussed in Chapter 1. Given any change in the prior information $C \rightarrow C'$, such that B becomes more plausible but A does not change,

$$B|C' > B|C, \tag{2.4}$$

$$A|BC' = A|BC, \tag{2.5}$$

common sense demands that AB could only become more plausible, not less:

$$AB|C' \geq AB|C, \tag{2.6}$$

with equality if and only if $A|BC$ corresponds to impossibility. Likewise, given prior information C'' such that

$$B|C'' = B|C, \tag{2.7}$$

$$A|BC'' > A|BC, \tag{2.8}$$

we require that

$$AB|C'' \geq AB|C, \tag{2.9}$$

in which the equality can hold only if B is impossible, given C (for then AB might still be impossible given C'', although $A|BC$ is not defined). Furthermore, the function $F(x, y)$ must be continuous; for otherwise an arbitrarily small increase in one of the plausibilities on the right-hand side of (2.1) could result in a large increase in $AB|C$.

In summary, $F(x, y)$ must be a continuous monotonic increasing function of both x and y. If we assume it is differentiable (this is not necessary; see the discussion following (2.13)), then we have

$$F_1(x, y) \equiv \frac{\partial F}{\partial x} \geq 0 \tag{2.10a}$$

with equality if and only if y represents impossibility; and also

$$F_2(x, y) \equiv \frac{\partial F}{\partial y} \geq 0 \tag{2.10b}$$

with equality permitted only if x represents impossibility. Note for later purposes that, in this notation, F_i denotes differentiation with respect to the ith argument of F, whatever it may be.

Next we impose the desideratum (IIIa) of 'structural' consistency. Suppose we try to find the plausibility $(ABC|D)$ that three propositions would be true simultaneously. Because of the fact that Boolean algebra is associative: $ABC = (AB)C = A(BC)$, we can do this in two different ways. If the rule is to be consistent, we must get the same result for either order of carrying out the operations. We can say first that BC will be considered a single proposition, and then apply (2.1):

$$(ABC|D) = F[(BC|D), (A|BCD)], \tag{2.11}$$

and then in the plausibility $(BC|D)$ we can again apply (2.1) to give

$$(ABC|D) = F\{F[(C|D), (B|CD)], (A|BCD)\}. \tag{2.12a}$$

But we could equally well have said that AB shall be considered a single proposition at first. From this we can reason out in the other order to obtain a different expression:

$$(ABC|D) = F[(C|D), (AB|CD)] = F\{(C|D), F[(B|CD), (A|BCD)]\}. \tag{2.12b}$$

If this rule is to represent a consistent way of reasoning, the two expressions (2.12a) and (2.12b) must always be the same. A necessary condition that our robot will reason consistently in this case therefore takes the form of a functional equation,

$$F[F(x, y), z] = F[x, F(y, z)]. \tag{2.13}$$

This equation has a long history in mathematics, starting from the work of N. H. Abel (1826). Aczél (1966), in his monumental work on functional equations, calls it, very appropriately, 'The Associativity Equation', and lists a total of 98 references to works that discuss it or use it. Aczél derives the general solution (2.27), below, without assuming differentiability; unfortunately, the proof fills 11 pages (pp. 256–267) of his book (see also Aczél, 1987). We give here the shorter proof by R. T. Cox (1961), which assumes differentiability; see also the discussion in Appendix B.

It is evident that (2.13) has a trivial solution, $F(x, y) = \text{const.}$ But that violates our monotonicity requirement (2.10), and is in any event useless for our purposes. Unless (2.13) has a nontrivial solution, this approach will fail; so we seek the most general nontrivial solution. Using the abbreviations

$$u \equiv F(x, y), \qquad v \equiv F(y, z), \tag{2.14}$$

but still considering (x, y, z) the independent variables, the functional equation to be solved is

$$F(x, v) = F(u, z). \tag{2.15}$$

Differentiating with respect to x and y we obtain, in the notation of (2.10),

$$\begin{aligned} F_1(x, v) &= F_1(u, z)F_1(x, y) \\ F_2(x, v)F_1(y, z) &= F_1(u, z)F_2(x, y). \end{aligned} \tag{2.16}$$

Elimination of $F_1(u, z)$ from these equations yields

$$G(x, v)F_1(y, z) = G(x, y) \tag{2.17}$$

where we use the notation $G(x, y) \equiv F_2(x, y)/F_1(x, y)$. Evidently, the left-hand side of (2.17) must be independent of z. Now, (2.17) can be written equally well as

$$G(x, v)F_2(y, z) = G(x, y)G(y, z),$$

(2.18)

and denoting the left-hand sides of (2.17), (2.18) by U, V respectively, we verify that $\partial V/\partial y = \partial U/\partial z$. Thus, $G(x, y)G(y, z)$ must be independent of y. The most general function $G(x, y)$ with this property is

$$G(x, y) = r\frac{H(x)}{H(y)}$$

(2.19)

where r is a constant and the function $H(x)$ is arbitrary. In the present case, $G > 0$ by monotonicity of F, and so we require that $r > 0$, and $H(x)$ may not change sign in the region of interest. Using (2.19), (2.17) and (2.18) become

$$F_1(y, z) = \frac{H(v)}{H(y)}$$

(2.20)

$$F_2(y, z) = r\frac{H(v)}{H(z)}$$

(2.21)

and the relation $dv = dF(y, z) = F_1dy + F_2dz$ takes the form

$$\frac{dv}{H(v)} = \frac{dy}{H(y)} + r\frac{dz}{H(z)}$$

(2.22)

or, on integration,

$$w[F(y, z)] = w(v) = w(y)w^r(z),$$

(2.23)

where

$$w(x) \equiv \exp\left\{\int^x \frac{dx}{H(x)}\right\}.$$

(2.24)

The absence of a lower limit on the integral signifies an arbitrary multiplicative factor in w. But taking the function $w(\cdot)$ of (2.15) and applying (2.23), we obtain $w(x)w^r(v) = w(u)w^r(z)$; applying (2.23) again, our functional equation now reduces to

$$w(x)w^r(y)[w(z)]^{r^2} = w(x)w^r(y)w^r(z).$$

(2.25)

Thus we obtain a nontrivial solution only if $r = 1$, and our final result can be expressed in either of the two forms:

$$w[F(x, y)] = w(x)w(y)$$

(2.26)

or

$$F(x, y) = w^{-1}[w(x)w(y)].$$

(2.27)

Associativity and commutativity of the logical product thus require that the relation sought must take the functional form

$$w(AB|C) = w(A|BC)w(B|C) = w(B|AC)w(A|C), \qquad (2.28)$$

which we shall call henceforth the *product rule*. By its construction (2.24), $w(x)$ must be a positive continuous monotonic function, increasing or decreasing according to the sign of $H(x)$; at this stage it is otherwise arbitrary.

The result (2.28) has been derived as a necessary condition for consistency in the sense of desideratum (IIIa). Conversely, it is evident that (2.28) is also sufficient to ensure this consistency for any number of joint propositions. For example, there are an enormous number of different ways in which $(ABCDEFG|H)$ could be expanded by successive partitions in the manner of (2.12); but if (2.28) is satisfied, they will all yield the same result.

The requirements of qualitative correspondence with common sense impose further conditions on the function $w(x)$. For example, in the first given form of (2.28) suppose that A is certain, given C. Then in the 'logical environment' produced by knowledge of C, the propositions AB and B are the same, in the sense that one is true if and only if the other is true. By our most primitive axiom of all, discussed in Chapter 1, propositions with the same truth value must have equal plausibility:

$$AB|C = B|C, \qquad (2.29)$$

and also we will have

$$A|BC = A|C \qquad (2.30)$$

because if A is already certain given C (i.e. C implies A), then, given any other information B which does not contradict C, it is still certain. In this case, (2.28) reduces to

$$w(B|C) = w(A|C)w(B|C), \qquad (2.31)$$

and this must hold no matter how plausible or implausible B is to the robot. So our function $w(x)$ must have the property that

$$\text{certainty is represented by } w(A|C) = 1. \qquad (2.32)$$

Now suppose that A is impossible, given C. Then the proposition AB is also impossible given C:

$$AB|C = A|C, \qquad (2.33)$$

and if A is already impossible given C (i.e. C implies $\overline{A}$), then, given any further information B which does not contradict C, A would still be impossible:

$$A|BC = A|C. \qquad (2.34)$$

In this case, (2.28) reduces to

$$w(A|C) = w(A|C)w(B|C),\tag{2.35}$$

and again this equation must hold no matter what plausibility B might have. There are only two possible values of $w(A|C)$ that could satisfy this condition: it could be zero or $+\infty$ (the choice $-\infty$ is ruled out because then by continuity $w(B|C)$ would have to be capable of negative values; (2.35) would then be a contradiction).

In summary, qualitative correspondence with common sense requires that $w(x)$ be a positive continuous monotonic function. It may be either increasing or decreasing. If it is increasing, it must range from zero for impossibility up to one for certainty. If it is decreasing, it must range from ∞ for impossibility down to one for certainty. Thus far, our conditions say nothing at all about how it varies between these limits.

However, these two possibilities of representation are not different in content. Given any function $w_1(x)$ which is acceptable by the above criteria and represents impossibility by ∞, we can define a new function $w_2(x) \equiv 1/w_1(x)$, which will be equally acceptable and represents impossibility by zero. Therefore, there will be no loss of generality if we now adopt the choice $0 \le w(x) \le 1$ as a *convention*; that is, as far as content is concerned, all possibilities consistent with our desiderata are included in this form. (As the reader may check, we could just as well have chosen the opposite convention; and the entire development of the theory from this point on, including all its applications, would go through equally well, with equations of a less familiar form but exactly the same content.)

2.2 The sum rule

Since the propositions now being considered are of the Aristotelian logical type which must be either true or false, the logical product $A\overline{A}$ is always false, the logical sum $A + \overline{A}$ always true. The plausibility that A is false must depend in some way on the plausibility that it is true. If we define $u \equiv w(A|B)$, $v \equiv w(\overline{A}|B)$, there must exist some functional relation

$$v = S(u).\tag{2.36}$$

Evidently, qualitative correspondence with common sense requires that $S(u)$ be a continuous monotonic decreasing function in $0 \le u \le 1$, with extreme values $S(0) = 1$, $S(1) = 0$. But it cannot be just any function with these properties, for it must be consistent with the fact that the product rule can be written for either AB or $A\overline{B}$:

$$w(AB|C) = w(A|C)w(B|AC)\tag{2.37}$$

$$w(A\overline{B}|C) = w(A|C)w(\overline{B}|AC).\tag{2.38}$$

Thus, using (2.36) and (2.38), Eq. (2.37) becomes

$$w(AB|C) = w(A|C)S[w(\overline{B}|AC)] = w(A|C)S\left[\frac{w(A\overline{B}|C)}{w(A|C)}\right].\tag{2.39}$$

Again, we invoke commutativity: $w(AB|C)$ is symmetric in A, B, and so consistency requires that

$$w(A|C)S\left[\frac{w(A\overline{B}|C)}{w(A|C)}\right] = w(B|C)S\left[\frac{w(B\overline{A}|C)}{w(B|C)}\right]. \tag{2.40}$$

This must hold for all propositions A, B, C; in particular, (2.40) must hold when

$$\overline{B} = AD, \tag{2.41}$$

where D is any new proposition. But then we have the truth values noted before in (1.13):

$$A\overline{B} = \overline{B}, \qquad B\overline{A} = \overline{A}, \tag{2.42}$$

and in (2.40) we may write

$$w(A\overline{B}|C) = w(\overline{B}|C) = S[w(B|C)] \tag{2.43}$$
$$w(B\overline{A}|C) = w(\overline{A}|C) = S[w(A|C)].$$

Therefore, using the abbreviations

$$x \equiv w(A|C), \qquad y \equiv w(B|C), \tag{2.44}$$

(2.25) becomes a functional equation

$$xS\left[\frac{S(y)}{x}\right] = yS\left[\frac{S(x)}{y}\right], \qquad \begin{matrix} 0 \leq S(y) \leq x \\ 0 \leq x \leq 1 \end{matrix} \tag{2.45}$$

which expresses a scaling property that $S(x)$ must have in order to be consistent with the product rule. In the special case $y = 1$, this reduces to

$$S[S(x)] = x, \tag{2.46}$$

which states that $S(x)$ is a self-reciprocal function; $S(x) = S^{-1}(x)$. Thus, from (2.36) it follows also that $u = S(v)$. But this expresses only the evident fact that the relationship between A and $\overline{A}$ is a reciprocal one; it does not matter which proposition we denote by the simple letter, which by the barred letter. We noted this before in (1.8); if it had not been obvious before, we should be obliged to recognize it at this point.

The domain of validity given in (2.45) is found as follows. The proposition D is arbitrary, and so by various choices of D we can achieve all values of $w(D|AC)$ in

$$0 \leq w(D|AC) \leq 1. \tag{2.47}$$

But $S(y) = w(AD|C) = w(A|C)w(D|AC)$, and so (2.47) is just $(0 \leq S(y) \leq x)$, as stated in (2.45). This domain is symmetric in x, y; it can be written equally well with them interchanged. Geometrically, it consists of all points in the xy plane lying in the unit square $(0 \leq x, y \leq 1)$ and on or above the curve $y = S(x)$.

Indeed, the shape of that curve is determined already by what (2.45) says for points lying infinitesimally above it. For if we set $y = S(x) + \epsilon$, then as $\epsilon \to 0^+$ two terms in (2.45) tend to $S(1) = 0$, but at different rates. Therefore everything depends on the exact way in which $S(1 - \delta)$ tends to zero as $\delta \to 0$. To investigate this, we define a new variable $q(x, y)$ by

$$\frac{S(x)}{y} = 1 - \exp\{-q\}. \tag{2.48}$$

Then we may choose $\delta = \exp\{-q\}$, define the function $J(q)$ by

$$S(1 - \delta) = S(1 - \exp\{-q\} = \exp\{-J(q)\}, \tag{2.49}$$

and find the asymptotic form of $J(q)$ as $q \to \infty$.

Considering now x, q as the independent variables, we have from (2.48)

$$S(y) = S[S(x)] + \exp\{-q\}S(x)S'[S(x)] + O(\exp\{-2q\}). \tag{2.50}$$

Using (2.46) and its derivative $S'[S(x)]S'(x) = 1$, this reduces to

$$\frac{S(y)}{x} = 1 - \exp\{-(\alpha + q)\} + O(\exp\{-2q\}), \tag{2.51}$$

where

$$\alpha(x) \equiv \log\left[\frac{-xS'(x)}{S(x)}\right] > 0. \tag{2.52}$$

With these substitutions, our functional equation (2.45) becomes

$$J(q + \alpha) - J(q) = \log\left[\frac{x}{S(x)}\right] + \log(1 - \exp\{-q\}) + O(\exp\{-2q\}), \qquad \begin{array}{l} 0 < q < \infty \\ 0 < x \le 1 \end{array} \tag{2.53}$$

As $q \to \infty$ the last two terms go to zero exponentially fast, so $J(q)$ must be asymptotically linear,

$$J(q) \sim a + bq + O(\exp\{-q\}), \tag{2.54}$$

with positive slope

$$b = \alpha^{-1} \log\left[\frac{x}{S(x)}\right]. \tag{2.55}$$

In (2.54) there is no periodic term with period α, because (2.53) must hold for a continuum of different values of x, and therefore for a continuum of values of $\alpha(x)$. But, by definition, J is a function of q only, so the right-hand side of (2.55) must be independent of x. This gives, using (2.52),

$$\frac{x}{S(x)} = \left[\frac{-xS'(x)}{S(x)}\right]^b, \qquad 0 < b < \infty, \tag{2.56}$$

or, rearranging, $S(x)$ must satisfy the differential equation

$$S^{m-1}dS + x^{m-1}dx = 0, \tag{2.57}$$

where $m \equiv 1/b$ is some positive constant. The only solution of this satisfying $S(0) = 1$ is

$$S(x) = (1 - x^m)^{1/m}, \qquad \begin{array}{c} 0 \le x \le 1 \\ 0 < m < \infty \end{array} \qquad (2.58)$$

and, conversely, we verify at once that (2.58) is a solution of (2.45).

The result (2.58) was first derived by R. T. Cox (1946) by a different argument which assumed $S(x)$ twice differentiable. Again, Aczél (1966) derives the same result without assuming differentiability. (But to assume differentiability in the present application seems to us a very innocuous step, for if the functional equations had led us to nondifferentiable functions, we would have rejected this whole theory as a qualitative violation of common sense.) In any event, (2.58) is the most general function satisfying the functional equation (2.45) and the left boundary condition $S(0) = 1$; whereupon we are encouraged to find that it automatically satisfies the right boundary condition $S(1) = 0$.

Since our derivation of the functional equation (2.45) used the special choice (2.41) for B, we have shown thus far only that (2.58) is a necessary condition to satisfy the general consistency requirement (2.40). To check its sufficiency, substitute (2.58) into (2.40). We obtain

$$w^m(A|C) - w^m(A\overline{B}|C) = w^m(B|C) - w^m(B\overline{A}|C), \qquad (2.59)$$

a trivial identity by virtue of (2.28) and (2.38). Therefore, (2.58) is the necessary and sufficient condition on $S(x)$ for consistency in the sense (2.40).

Our results up to this point can be summarized as follows. Associativity of the logical product requires that some monotonic function $w(x)$ of the plausibility $x = A|B$ must obey the product rule (2.28). Our result (2.58) states that this same function must also obey a sum rule:

$$w^m(A|B) + w^m(\overline{A}|B) = 1 \qquad (2.60)$$

for some positive m. Of course, the product rule itself can be written equally well as

$$w^m(AB|C) = w^m(A|C)w^m(B|AC) = w^m(B|C)w^m(A|BC), \qquad (2.61)$$

but then we see that the value of m is actually irrelevant; for whatever value is chosen, we can define a new function

$$p(x) \equiv w^m(x), \qquad (2.62)$$

and our rules take the form

$$p(AB|C) = p(A|C)p(B|AC) = p(B|C)p(A|BC), \qquad (2.63)$$

$$p(A|B) + p(\overline{A}|B) = 1. \qquad (2.64)$$

In fact, this entails no loss of generality, for the only requirement we have imposed on the function $w(x)$ is that it is a continuous monotonic increasing function ranging from $w = 0$ for impossibility to $w = 1$ for certainty. But if $w(x)$ satisfies this, then so also does $w^m(x), 0 < m < \infty$. Therefore, to say that we could use different values of m does not give us any freedom that we did not have already in the arbitrariness of $w(x)$. All possibilities allowed by our desiderata are contained in (2.63) and (2.64), in which $p(x)$ is any continuous monotonic increasing function with the range $0 \leq p(x) \leq 1$.

Are further relations needed to yield a complete set of rules for plausible inference, adequate to determine the plausibility of any logic function $f(A_1, \ldots, A_n)$ from those of $\{A_1, \ldots, A_n\}$? We have, in the product rule (2.63) and sum rule (2.64), formulas for the plausibility of the conjunction AB and the negation $\overline{A}$. However, we have noted, in the discussion following (1.23), that conjunction and negation are an adequate set of operations, from which all logic functions can be constructed.

Therefore, one would conjecture that our search for basic rules should be finished; it ought to be possible, by repeated applications of the product rule and sum rule, to arrive at the plausibility of any proposition in the Boolean algebra generated by $\{A_1, \ldots, A_n\}$.

To verify this, we seek first a formula for the logical sum $A + B$. Applying the product rule and sum rule repeatedly, we have

$$
\begin{aligned}
p(A + B|C) &= 1 - p(\overline{AB}|C) = 1 - p(\overline{A}|C)p(\overline{B}|\overline{A}C) \\
&= 1 - p(\overline{A}|C)[1 - p(B|\overline{A}C)] = p(A|C) + p(\overline{A}B|C) \qquad (2.65) \\
&= p(A|C) + p(B|C)p(\overline{A}|BC) = p(A|C) + p(B|C)[1 - p(A|BC)]
\end{aligned}
$$

and finally

$$
p(A + B|C) = p(A|C) + p(B|C) - p(AB|C). \qquad (2.66)
$$

This generalized sum rule is one of the most useful in applications. Evidently, the primitive sum rule (2.64) is a special case of (2.66), with the choice $B = \overline{A}$.

Exercise 2.1. Is it possible to find a general formula for $p(C|A + B)$, analogous to (2.66), from the product and sum rules? If so, derive it; if not, explain why this cannot be done.

Exercise 2.2. Now suppose we have a set of propositions $\{A_1, \ldots, A_n\}$ which on information X are mutually exclusive: $p(A_i A_j|X) = p(A_i|X)\delta_{ij}$. Show that $p(C|(A_1 + A_2 + \cdots + A_n X)$ is a weighted average of the separate plausibilities $p(C|A_i X)$:

$$
p(C|(A_1 + \cdots + A_n X) = p(C|A_1 X + A_2 X + \cdots + A_n X) = \frac{\sum_i p(A_i|X)\, p(C|A_i X)}{\sum_i p(A_i|X)}.
$$
$$
\qquad (2.67)
$$

To extend the result (2.66), we noted following (1.17) that any logic function other than the trivial contradiction can be expressed in disjunctive normal form, as a logical sum of the basic conjunctions such as (1.17). Now the plausibility of any one of the basic conjunctions

$\{Q_i, \ 1 \le i \le 2^n\}$ is determined by repeated applications of the product rule; and then repeated application of (2.66) will yield the plausibility of any logical sum of the Q_i. In fact, these conjunctions are mutually exclusive, so we shall find (see (2.85) below) that this reduces to a simple sum $\sum_i p(Q_i|C)$ of at most $(2^n - 1)$ terms.

So, just as conjunction and negation are an adequate set of operations for deductive logic, the above product and sum rules are an adequate set for plausible inference, in the following sense. Whenever the background information is enough to determine the plausibilities of the basic conjunctions, our rules are adequate to determine the plausibility of every proposition in the Boolean algebra generated by $\{A_1, \ldots, A_n\}$. Thus, in the case $n = 4$ we need the plausibilities of $2^4 = 16$ basic conjunctions, whereupon our rules will determine the plausibility of each of the $2^{16} = 65\,536$ propositions in the Boolean algebra.

But this is almost always more than we need in a real application; if the background information is enough to determine the plausibility of a few of the basic conjunctions, this may be adequate for the small part of the Boolean algebra that is of concern to us.

2.3 Qualitative properties

Now let us check to see how the theory based on (2.63) and (2.64) is related to the theory of deductive logic and the various qualitative syllogisms from which we started in Chapter 1. In the first place it is obvious that in the limit as $p(A|B) \to 0$ or $p(A|B) \to 1$, the sum rule (2.64) expresses the primitive postulate of Aristotelian logic: if A is true, then $\overline{A}$ must be false, etc.

Indeed, all of that logic consists of the two strong syllogisms (1.1), (1.2) and all that follows from them; using now the implication sign (1.14) to state the major premise:

$$
\begin{array}{cc}
A \Rightarrow B & A \Rightarrow B \\
\dfrac{A \text{ is true}}{B \text{ is true}} & \dfrac{B \text{ is false}}{A \text{ is false}}
\end{array}
\tag{2.68}
$$

and the endless stream of their consequences. If we let C stand for their major premise:

$$
C \equiv A \Rightarrow B
\tag{2.69}
$$

then these syllogisms correspond to our product rule (2.63) in the forms

$$
p(B|AC) = \frac{p(AB|C)}{p(A|C)}, \qquad p(A|\overline{B}C) = \frac{p(A\overline{B}|C)}{p(\overline{B}|C)},
\tag{2.70}
$$

respectively. But from (2.68) we have $p(AB|C) = p(A|C)$ and $p(A\overline{B}|C) = 0$, and so (2.70) reduces to

$$
p(B|AC) = 1, \qquad p(A|\overline{B}C) = 0,
\tag{2.71}
$$

as stated in the syllogisms (2.68). Thus the relation is simply: *Aristotelian deductive logic is the limiting form of our rules for plausible reasoning, as the robot becomes more and more certain of its conclusions.*

But our rules have also what is not contained in deductive logic: a quantitative form of the weak syllogisms (1.3) and (1.4). To show that those original qualitative statements always follow from the present rules, note that the first weak syllogism

$$A \Rightarrow B$$

$$B \text{ is true} \tag{2.72}$$

$$\overline{}$$

therefore, A becomes more plausible

corresponds to the product rule (2.63) in the form

$$p(A|BC) = p(A|C)\frac{p(B|AC)}{p(B|C)}. \tag{2.73}$$

But from (2.68), $p(B|AC) = 1$, and since $p(B|C) \le 1$, (2.73) gives

$$p(A|BC) \ge p(A|C), \tag{2.74}$$

as stated in the syllogism. Likewise, the syllogism (1.4)

$$A \Rightarrow B$$

$$A \text{ is false} \tag{2.75}$$

$$\overline{}$$

therefore, B becomes less plausible

corresponds to the product rule in the form

$$p(B|\overline{A}C) = p(B|C)\frac{p(\overline{A}|BC)}{p(\overline{A}|C)}. \tag{2.76}$$

But from (2.74) it follows that $p(\overline{A}|BC) \le p(\overline{A}|C)$; and so (2.76) gives

$$p(B|\overline{A}C) \le p(B|C), \tag{2.77}$$

as stated in the syllogism.

Finally, the policeman's syllogism (1.5), which seemed very weak when stated abstractly, is also contained in our product rule, stated in the form (2.73). Letting C now stand for the background information (not noted explicitly in (1.5) because the need for it was not yet apparent), the major premise, 'If A is true, then B becomes more plausible', now takes the form

$$p(B|AC) > p(B|C), \tag{2.78}$$

and (2.73) gives at once

$$p(A|BC) > p(A|C), \tag{2.79}$$

as stated in the syllogism.

Now we have more than the mere qualitative statement (2.79). In Chapter 1 we wondered, without answering: What determines whether the evidence B elevates A almost to certainty, or has a negligible effect on its plausibility? The answer from (2.73) is that, since $p(B|AC)$

cannot be greater than unity, a large increase in the plausibility of A can occur only when $p(B|C)$ is very small. Observing the gentleman's behavior (B) makes his guilt (A) seem virtually certain, because that behavior is otherwise so very unlikely on the background information; no policeman has ever seen an innocent person behaving that way. On the other hand, if knowing that A is true can make only a negligible increase in the plausibility of B, then observing B can in turn make only a negligible increase in the plausibility of A.

We could give many more comparisons of this type; indeed, the complete qualitative correspondence of these rules with common sense has been noted and demonstrated by many writers, including Keynes (1921), Jeffreys (1939), Pólya (1945, 1954), R. T. Cox (1961), Tribus (1969), de Finetti (1974a,b), and Rosenkrantz (1977). The treatment of Pólya was described briefly in our Preface and Chapter 1, and we have just recounted that of Cox more fully. However, our aim now is to push ahead to quantitative applications; so we return to the basic development of the theory.

2.4 Numerical values

We have found so far the most general consistent rules by which our robot can manipulate plausibilities, granted that it must associate them with real numbers, so that its brain can operate by the carrying out of some definite physical process. While we are encouraged by the familiar formal appearance of these rules and their qualitative properties just noted, two evident circumstances show that our job of designing the robot's brain is not yet finished.

In the first place, while the rules (2.63), (2.64) place some limitations on how plausibilities of different propositions must be related to each other, it would appear that we have not yet found any *unique* rules, but rather an infinite number of possible rules by which our robot can do plausible reasoning. Corresponding to every different choice of a monotonic function $p(x)$, there seems to be a different set of rules, with different content.

Secondly, nothing given so far tells us what actual numerical values of plausibility should be assigned at the beginning of a problem, so that the robot can get started on its calculations. How is the robot to make its initial encoding of the background information into definite numerical values of plausibilities? For this we must invoke the 'interface' desiderata (IIIb), (IIIc) of (1.39), not yet used.

The following analysis answers both of these questions, in a way both interesting and unexpected. Let us ask for the plausibility $(A_1 + A_2 + A_3|B)$ that at least one of three propositions $\{A_1, A_2, A_3\}$ is true. We can find this by two applications of the extended sum rule (2.66), as follows. The first application gives

$$p(A_1 + A_2 + A_3|B) = p(A_1 + A_2|B) + p(A_3|B) - p(A_1 A_3 + A_2 A_3|B) \qquad (2.80)$$

where we first considered $(A_1 + A_2)$ as a single proposition, and used the logical relation

$$(A_1 + A_2)A_3 = A_1 A_3 + A_2 A_3. \qquad (2.81)$$

Applying (2.66) again, we obtain seven terms which can be grouped as follows:

$$p(A_1 + A_2 + A_3|B) = p(A_1|B) + p(A_2|B) + p(A_3|B)$$
$$- p(A_1 A_2|B) - p(A_2 A_3|B) - p(A_3 A_1|B) \qquad (2.82)$$
$$+ p(A_1 A_2 A_3|B).$$

Now suppose these propositions are mutually exclusive; i.e. the evidence B implies that no two of them can be true simultaneously:

$$p(A_i A_j|B) = p(A_i|B)\delta_{ij}. \qquad (2.83)$$

Then the last four terms of (2.82) vanish, and we have

$$p(A_1 + A_2 + A_3|B) = p(A_1|B) + P(A_2|B) + P(A_3|B). \qquad (2.84)$$

Adding more propositions A_4, A_5, etc., it is easy to show by induction that if we have n mutually exclusive propositions $\{A_1, \ldots, A_n\}$, (2.84) generalizes to

$$p(A_1 + \cdots + A_m|B) = \sum_{i=1}^{m} p(A_i|B), \qquad 1 \le m \le n, \qquad (2.85)$$

a rule which we will be using constantly from now on.

In conventional expositions, Eq. (2.85) is usually introduced first as the basic but, as far as one can see, arbitrary axiom of the theory. The present approach shows that this rule is deducible from simple qualitative conditions of consistency. The viewpoint which sees (2.85) as the primitive, fundamental relation is one which we are particularly anxious to avoid (see Comments section at the end of this chapter).

Now suppose that the propositions $\{A_1, \ldots, A_n\}$ are not only mutually exclusive but also exhaustive; i.e. the background information B stipulates that one and only one of them must be true. In that case, the sum (2.85) for $m = n$ must be unity:

$$\sum_{i=1}^{n} p(A_i|B) = 1. \qquad (2.86)$$

This alone is not enough to determine the individual numerical values $p(A_i|B)$. Depending on further details of the information B, many different choices might be appropriate, and in general finding the $p(A_i|B)$ by logical analysis of B can be a difficult problem. It is, in fact, an open-ended problem, since there is no end to the variety of complicated information that might be contained in B; and therefore no end to the complicated mathematical problems of translating that information into numerical values of $p(A_i|B)$. As we shall see, this is one of the most important current research problems; every new principle we can discover for translating information B into numerical values of $p(A_i|B)$ will open up a new class of useful applications of this theory.

There is, however, one case in which the answer is particularly simple, requiring only direct application of principles already given. But we are entering now into a very delicate area, a cause of confusion and controversy for over a century. In the early stages of this theory, as in elementary geometry, our intuition runs so far ahead of logical analysis that the point of the logical analysis is often missed. The trouble is that intuition leads us to the same final conclusions far more quickly, but without any correct appreciation of their range of validity. The result has been that the development of this theory has been retarded for some 150 years because various workers have insisted on debating these issues on the basis, not of demonstrative arguments, but of their conflicting intuitions.

At this point, therefore, we must ask the reader to suppress all intuitive feelings you may have, and allow yourself to be guided solely by the following logical analysis. The point we are about to make cannot be developed too carefully; and, unless it is clearly understood, we will be faced with tremendous conceptual difficulties from here on.

Consider two different problems. Problem I is the one just formulated: we have a given set of mutually exclusive and exhaustive propositions $\{A_1, \ldots, A_n\}$ and we seek to evaluate $p(A_i|B)_I$. Problem II differs in that the labels A_1, A_2 of the first two propositions have been interchanged. These labels are, of course, entirely arbitrary; it makes no difference which proposition we choose to call A_1 and which A_2. In Problem II, therefore, we also have a set of mutually exclusive and exhaustive propositions $\{A'_1, \ldots, A'_n\}$, given by

$$
\begin{aligned}
A'_1 &\equiv A_2, \\
A'_2 &\equiv A_1, \\
A'_k &\equiv A_k, \quad 3 \le k \le n,
\end{aligned}
\tag{2.87}
$$

and we seek to evaluate the quantities $p(A'_i|B)_{II}$, $i = 1, 2, \ldots, n$.

In interchanging the labels, we have generated a different but closely related problem. It is clear that, whatever state of knowledge the robot had about A_1 in Problem I, it must have the same state of knowledge about A'_2 in Problem II, for they are the same proposition, the given information B is the same in both problems, and it is contemplating the same totality of propositions $\{A_1, \ldots, A_n\}$ in both problems. Therefore we must have

$$
p(A_1|B)_I = p(A'_2|B)_{II},
\tag{2.88}
$$

and similarly

$$
p(A_2|B)_I = p(A'_1|B)_{II}.
\tag{2.89}
$$

We will call these the *transformation equations*. They describe only how the two problems are related to each other, and therefore they must hold whatever the information B might be; in particular, however plausible or implausible the propositions A_1, A_2 might seem to the robot in Problem I.

Now suppose that information B is indifferent between propositions A_1 and A_2; i.e. if it says something about one, it says the same thing about the other, and so it contains nothing that would give the robot any reason to prefer either one over the other. In this case, Problems I and II are not merely related, but entirely equivalent; i.e. the robot is in exactly the same state of knowledge about the set of propositions $\{A'_1, \ldots, A'_n\}$ in Problem II, *including their labeling*, as it is about the set $\{A_1, \ldots, A_n\}$ in Problem I.

Now we invoke our desideratum of consistency in the sense (IIIc) in (1.39). This stated that equivalent states of knowledge must be represented by equivalent plausibility assignments.

In equations, this statement is

$$p(A_i|B)_I = p(A_i'|B)_{II}, \qquad i = 1, 2, \ldots, n, \tag{2.90}$$

which we shall call the *symmetry equations*. But now, combining (2.88), (2.89), and (2.90), we obtain

$$p(A_1|B)_I = p(A_2|B)_I. \tag{2.91}$$

In other words, propositions A_1 and A_2 must be assigned equal plausibilities in Problem I (and, of course, also in Problem II).

At this point, depending on your personality and background in this subject, you will be either greatly impressed or greatly disappointed by the result (2.91). The argument we have just given is the first 'baby' version of the group invariance principle for assigning plausibilities; it will be extended greatly in Chapter 6, when we consider the general problem of assigning 'noninformative priors'.

More generally, let $\{A_1'', \ldots, A_n''\}$ be any permutation of $\{A_1, \ldots, A_n\}$ and let Problem III be that of determining the $p(A_i''|B)$. If the permutation is such that $A_k'' \equiv A_i$, there will be n transformation equations of the form

$$p(A_i|B)_I = p(A_k''|B)_{III} \tag{2.92}$$

which show how Problems I and III are related to each other; these relations will hold whatever the given information B.

But if information B is now indifferent between all the propositions A_i, then the robot is in exactly the same state of knowledge about the set of propositions $\{A_1'', \ldots, A_n''\}$ in Problem III as it was about the set $\{A_1, \ldots, A_n\}$ in Problem I; and again our desideratum of consistency demands that it assign equivalent plausibilities in equivalent states of knowledge, leading to the n symmetry conditions

$$p(A_k|B)_I = p(A_k''|B)_{III}, \qquad k = 1, 2, \ldots, n. \tag{2.93}$$

From (2.92) and (2.93) we obtain n equations of the form

$$p(A_i|B)_I = p(A_k|B)_I. \tag{2.94}$$

Now, these relations must hold whatever the particular permutation we used to define Problem III. There are $n!$ such permutations, and so there are actually $n!$ equivalent problems among which, for given i, the index k will range over all of the $(n-1)$ others in (2.94). Therefore, the only possibility is that all of the $p(A_i|B)_I$ be equal (indeed, this is required already by consideration of a single permutation if it is cyclic of order n). Since the $\{A_1, \ldots, A_n\}$ are exhaustive, (2.86) will hold, and the only possibility is therefore

$$p(A_i|B)_I = \frac{1}{n}, \qquad (1 \le i \le n), \tag{2.95}$$

and we have finally arrived at a set of definite numerical values! Following Keynes (1921), we shall call this result the *principle of indifference*.

Perhaps, in spite of our admonitions, the reader's intuition had already led to just this conclusion, without any need for the rather tortuous reasoning we have just been through. If so, then at least that intuition is consistent with our desiderata. But merely writing down (2.95) intuitively gives one no appreciation of the importance and uniqueness of this result. To see the uniqueness, note that if the robot were to assign any values different from (2.95), then by a mere permutation of labels we could exhibit a second problem in which the robot's state of knowledge is the same, but in which it is assigning different plausibilities.

To see the importance, note that (2.95) actually answers both of the questions posed at the beginning of this section. It shows – in one particular case which can be greatly generalized – how the information given the robot can lead to definite numerical values, so that a calculation can start. But it also shows something even more important because it is not at all obvious intuitively; the information given the robot determines the numerical values of the quantities $p(x) = p(A_i|B)$, and not the numerical values of the plausibilities $x = A_i|B$ from which we started. This, also, will be found to be true in general.

Recognizing this gives us a beautiful answer to the first question posed at the beginning of this section; after having found the product and sum rules, it still appeared that we had not found any unique rules of reasoning, because every different choice of a monotonic function $p(x)$ would lead to a different set of rules (i.e. a set with different content). But now we see that no matter what function $p(x)$ we choose, we shall be led to the same result (2.95), and the same numerical value of p. Furthermore, the robot's reasoning processes can be carried out entirely by manipulation of the quantities p, as the product and sum rules show; and the robot's final conclusions can be stated equally well in terms of the p's instead of the x's.

So, we now see that different choices of the function $p(x)$ correspond only to different ways we could design the robot's internal memory circuits. For each proposition A_i about which it is to reason, it will need a memory address in which it stores some number representing the degree of plausibility of A_i, on the basis of all the data it has been given. Of course, instead of storing the number p_i it could equally well store any strict monotonic function of p_i. But no matter what function it used internally, the externally observable behavior of the robot would be just the same.

As soon as we recognize this, it is clear that, instead of saying that $p(x)$ is an arbitrary monotonic function of x, it is much more to the point to turn this around and say that:

> *The plausibility $x \equiv A|B$ is an arbitrary monotonic function of p,*
> *defined in $(0 \leq p \leq 1)$.*

It is p that is rigidly fixed by the data, not x.

The question of uniqueness is therefore disposed of automatically by the result (2.95); in spite of first appearances, there is actually only one consistent set of rules by which our robot can do plausible reasoning, and, for all practical purposes, the plausibilities $x \equiv A|B$ from which we started have faded entirely out of the picture! We will just have no further use for them.

Having seen that our theory of plausible reasoning can be carried out entirely in terms of the quantities p, we finally introduce their technical names; from now on, we will call these quantities *probabilities*. The word 'probability' has been studiously avoided up to this point, because, whereas the word does have a colloquial meaning to the proverbial 'man on the street', it is for us a technical term, which ought to have a precise meaning. But until it had been demonstrated that these quantities are uniquely determined by the data of a problem, we had no grounds for supposing that the quantities p were possessed of any precise meaning.

We now see that the quantities p define a particular scale on which degrees of plausibility can be measured. Out of all possible monotonic functions which could, in principle, serve this purpose equally well, we choose this particular one, not because it is more 'correct', but because it is more convenient; i.e. it is the quantities p that obey the simplest rules of combination, the product and sum rules. Because of this, numerical values of p are directly determined by our information.

This situation is analogous to that in thermodynamics, where out of all possible empirical temperature scales t, which are monotonic functions of each other, we finally decide to use the Kelvin scale T; not because it is more 'correct' than others but because it is more convenient; i.e. the laws of thermodynamics take their simplest form [$dU = T dS - P dV$, $dG = -S dT + V dP$, etc.] in terms of this particular scale. Because of this, numerical values of temperatures on the kelvin scale are 'rigidly fixed' in the sense of being directly measurable in experiments, independently of the properties of any particular substance like water or mercury.

Another rule, equally appealing to our intuition, follows at once from (2.95). Consider the traditional 'Bernoulli urn' of probability theory; ours is known to contain ten balls of identical size and weight, labeled $\{1, 2, \ldots, 10\}$. Three balls (numbers 4, 6, 7) are black, the other seven are white. We are to shake the urn and draw one ball blindfolded. The background information B in (2.95) consists of the statements in the last two sentences. What is the probability that we draw a black one?

Define the propositions: $A_i \equiv$ 'the ith ball is drawn', $(1 \leq i \leq 10)$. Since the background information is indifferent to these ten possibilities, (2.95) applies, and the robot assigns

$$p(A_i|B) = \frac{1}{10}, \quad 1 \leq i \leq 10. \tag{2.96}$$

The statement that we draw a black ball is that we draw number 4, 6, or 7;

$$p(\text{black}|B) = p(A_4 + A_6 + A_7|B). \tag{2.97}$$

But these are mutually exclusive propositions (i.e. they assert mutually exclusive events), so (2.85) applies, and the robot's conclusion is

$$p(\text{black}|B) = \frac{3}{10}, \tag{2.98}$$

as intuition had told us already. More generally, if there are N such balls, and the proposition A is defined to be true on any specified subset of M of them, $(0 \leq M \leq N)$, false on the rest, we have

$$p(A|B) = \frac{M}{N}. \tag{2.99}$$

This was the original mathematical *definition* of probability, as given by James Bernoulli (1713) and used by most writers for the next 150 years. For example, Laplace's great *Théorie Analytique des Probabilités* (1812) opens with this sentence:

The Probability for an event is the ratio of the number of cases favorable to it, to the number of all cases possible when nothing leads us to expect that any one of these cases should occur more than any other, which renders them, for us, equally possible.

Exercise 2.3. As soon as we have the numerical values $a = P(A|C)$ and $b = P(B|C)$, the product and sum rules place some limits on the possible numerical values for their conjunction and disjunction. Supposing that $a \le b$, show that the probability for the conjunction cannot exceed that of the least probable proposition: $0 \le P(AB|C) \le a$, and the probability for the disjunction cannot be less than that of the most probable proposition: $b \le P(A + B|C) \le 1$. Then show that, if $a + b > 1$, there is a stronger inequality for the conjunction; and if $a + b < 1$ there is a stronger one for the disjunction. These necessary general inequalities are helpful in detecting errors in calculations.

2.5 Notation and finite-sets policy

Now we can introduce the notation to be used in the remainder of this work (discussed more fully in Appendix B). Henceforth, our formal probability symbols will use the capital P:

$$P(A|B), \tag{2.100}$$

which signifies that the arguments are *propositions*. Probabilities whose arguments are numerical values are generally denoted by other functional symbols, such as

$$f(r|np), \tag{2.101}$$

which denote ordinary mathematical functions. The reason for making this distinction is to avoid ambiguity in the meaning of our symbols, which has been a recent problem in this field. However, in agreement with the customary loose notation in the existing literature, we sometimes relax our standards enough to allow the probability symbols with small p: $p(x|y)$ or $p(A|B)$ or $p(x|B)$ to have arguments which can be either propositions or numerical values, in any mix. Thus the meaning of expressions with small p can be judged only from the surrounding context.

It is very important to note that our consistency theorems have been established only for probabilities assigned on *finite sets* of propositions. In principle, every problem must start with such finite-set probabilities; extension to infinite sets is permitted only when this is the result of a well-defined and well-behaved limiting process from a finite set. More generally, in any mathematical operations involving infinite sets, the safe procedure is the finite-sets policy:

> *Apply the ordinary processes of arithmetic and analysis only to expressions with a finite number of terms. Then, after the calculation is done, observe how the resulting finite expressions behave as the number of terms increases indefinitely.*

In laying down this rule of conduct, we are only following the policy that mathematicians from Archimedes to Gauss have considered clearly necessary for nonsense avoidance in all of mathematics. But, more recently, the popularity of infinite-set theory and measure theory have led some to disregard it and seek shortcuts which purport to use measure theory directly. Note, however, that this rule of conduct is consistent with the original Lebesgue definition of measure, and *when a well-behaved limit exists* it leads us automatically to correct 'measure theoretic' results. Indeed, this is how Lebesgue found his first results.

The danger is that the present measure theory notation presupposes the infinite limit already accomplished, but contains no symbol indicating which limiting process was used. Yet, as noted in our Preface, different limiting Processes – equally well-behaved – lead in general to different results. When there is no well-behaved limit, any attempt to go directly to the limit can result in nonsense, *the cause of which cannot be seen as long as one looks only at the limit, and not at the limiting process.*

This little 'sermon' is an introduction to Chapter 15 on infinite-set paradoxes, where we shall see some of the results that have been produced by those who ignored this rule of conduct, and tried to calculate probabilities directly on an infinite set without considering any limit from a finite set. The results are at best ambiguous, at worst nonsensical.

2.6 Comments

It has taken us two chapters of close reasoning to get back to the point (2.99) from which Laplace started some 180 years ago. We shall try to understand the intervening period, as a weird episode of history, throughout the rest of the present work. The story is so complicated that we can unfold it only gradually, over the next ten chapters. To make a start on this, let us consider some of the questions often raised about the use of probability theory as an extension of logic.

2.6.1 'Subjective' vs. 'objective'

These words are abused so much in probability theory that we try to clarify our use of them. In the theory we are developing, any probability assignment is necessarily 'subjective' in the sense that it describes only a state of knowledge, and not anything that could be measured in a physical experiment. Inevitably, someone will demand to know: '*Whose* state of knowledge?' The answer is always: 'That of the robot – or of anyone else who is given the same information and reasons according to the desiderata used in our derivations in this chapter.'

Anyone who has the same information, but comes to a different conclusion than our robot, is necessarily violating one of those desiderata. While nobody has the authority to forbid such violations, it appears to us that a rational person, should he discover that he was violating one of them, would wish to revise his thinking (in any event, he would surely have difficulty in persuading anyone else, who was aware of that violation, to accept his conclusions).

Now, it was just the function of our interface desiderata (IIIb), (IIIc) to make these probability assignments completely 'objective' in the sense that they are independent of the personality of the user. They are a means of describing (or, what is the same thing, of encoding) the *information* given in the statement of a problem, independently of whatever personal feelings (hopes, fears, value judgments, etc.) you or I might have about the propositions involved. It is 'objectivity' in this sense that is needed for a scientifically respectable theory of inference.

2.6.2 Gödel's theorem

To answer another inevitable question, we recapitulate just what has and what has not been proved in this chapter. The main constructive requirement which determined our product and sum rules was the desideratum (IIIa) of 'structural consistency'. Of course, this does not mean that our rules have been proved consistent; it means only that any other rules which represent degrees of plausibility by real numbers, but which differ in content from ours, will lead necessarily either to inconsistencies or violations of our other desiderata.

A famous theorem of Kurt Gödel (1931) states that no mathematical system can provide a proof of its own consistency. Does this prevent us from ever proving the consistency of probability theory as logic? We are not prepared to answer this fully, but perhaps we can clarify the situation a little.

Firstly, let us be sure that 'inconsistency' means the same thing to us and to a logician. What we had in mind was that if our rules were inconsistent, then it would be possible to derive contradictory results from valid application of them; for example, by applying the rules in two equally valid ways, one might be able to derive both $P(A|BC) = 1/3$ and $P(A|BC) = 2/3$. Cox's functional equations sought to guard against this. Now, when a logician says that a system of axioms $\{A_1, A_2, \ldots, A_n\}$ is inconsistent, he means that a contradiction can be deduced from them; i.e. some proposition Q and its denial $\overline{Q}$ are both deducible. Indeed, this is not really different from our meaning.

To understand the above Gödel result, the essential point is the principle of elementary logic that a contradiction $\overline{A} A$ implies all propositions, true and false. (Given any two propositions A and B, we have $A \Rightarrow (A + B)$, therefore $\overline{A} A \Rightarrow \overline{A}(A + B) = \overline{A} A + \overline{A} B \Rightarrow B$.) Then let $A = \{A_1, A_2, \ldots, A_n\}$ be the system of axioms underlying a mathematical theory and T any proposition, or theorem, deducible from them:[1]

$$A \Rightarrow T. \tag{2.102}$$

---

[1] In Chapter 1 we noted the tricky distinction between the weak property of formal implication and the strong one of logical deducibility; by 'implications of a proposition C' we really mean 'propositions logically deducible from C *and the totality of other background information*'. Conventional expositions of Aristotelian logic are, in our view, flawed by their failure to make explicit mention of background information, which is usually essential to our reasoning, whether inductive or deductive. But, in the present argument, we can understand A as including all the propositions that constitute that background information; then 'implication' and 'logical deducibility' are the same thing.

Now, whatever T may assert, the fact that T can be deduced from the axioms cannot prove that there is no contradiction in them, since, if there were a contradiction, T could certainly be deduced from them!

This is the essence of the Gödel theorem, as it pertains to our problems. As noted by Fisher (1956), it shows us the intuitive reason why Gödel's result is true. We do not suppose that any logician would accept Fisher's simple argument as a proof of the full Gödel theorem; yet for most of us it is more convincing than Gödel's long and complicated proof.[2]

Now suppose that the axioms contain an inconsistency. Then the opposite of T and therefore the contradiction $\overline{T}\, T$ can also be deduced from them:

$$A \Rightarrow \overline{T}. \tag{2.103}$$

So, if there is an inconsistency, its existence can be proved by exhibiting any proposition T and its opposite $\overline{T}$ that are both deducible from the axioms. However, in practice it may not be easy to find a T for which one sees how to prove both T and $\overline{T}$.

Evidently, we could prove the consistency of a set of axioms if we could find a feasible procedure which is guaranteed to locate an inconsistency if one exists; so Gödel's theorem seems to imply that no such procedure exists. Actually, it says only that no such procedure *derivable from the axioms of the system being tested* exists.

We shall find that probability theory comes close to this; it is a powerful analytical tool which can search out a set of propositions and detect a contradiction in them if one exists. The principle is that probabilities conditional on contradictory premises do not exist (the hypothesis space is reduced to the empty set). Therefore, put our robot to work; i.e. write a computer program to calculate probabilities $p(B|E)$ conditional on a set of propositions $E = (E_1\, E_2\, \ldots\, E_n)$. Even though no contradiction is apparent from inspection, if there is a contradiction hidden in E, the computer program will crash.

We discovered this 'empirically', and, after some thought, realized that it is not a reason for dismay, but rather a valuable diagnostic tool that warns us of unforeseen special cases in which our formulation of a problem can break down.

If the computer program does not crash, but prints out valid numbers, then we know that the conditioning propositions E_i are mutually consistent, and we have accomplished what one might have thought to be impossible in view of Gödel's theorem. But of course our use of probability theory appeals to principles not derivable from the propositions being tested, so there is no difficulty; it is important to understand what Gödel's theorem does and does not prove.

When Gödel's theorem first appeared, with its more general conclusion that a mathematical system may contain certain propositions that are undecidable within that system, it seems to have been a great psychological blow to logicians, who saw it at first as a devastating obstacle to what they were trying to achieve. Yet a moment's thought shows us

[2] The 1957 edition of Harold Jeffreys' *Scientific Inference* (see Jeffreys, 1931) has a short summary of Gödel's original reasoning which is far clearer and easier to read than any other 'explanation' we have seen. The full theorem refers to other matters of concern in 1931 but of no interest to us right now; the above discussion has abstracted the part of it that we need to understand for our present purposes.

that many quite simple questions are undecidable by deductive logic. There are situations in which one can prove that a certain property must exist in a finite set, even though it is impossible to exhibit any member of the set that has that property. For example, two persons are the sole witnesses to an event; they give opposite testimony about it and then both die. Then we know that one of them was lying, but it is impossible to determine which one.

In this example, the undecidability is not an inherent property of the proposition or the event; it signifies only the incompleteness of our own information. But this is equally true of abstract mathematical systems; when a proposition is undecidable in such a system, that means only that its axioms do not provide enough *information* to decide it. But new axioms, external to the original set, might supply the missing information and make the proposition decidable after all.

In the future, as science becomes more and more oriented to thinking in terms of information content, Gödel's result will be seen as more of a platitude than a paradox. Indeed, from our viewpoint 'undecidability' merely signifies that a problem is one that calls for *inference* rather than deduction. Probability theory as extended logic is designed specifically for such problems.

These considerations seem to open up the possibility that, by going into a wider field by invoking principles external to probability theory, one might be able to prove the consistency of our rules. At the moment, this appears to us to be an open question.

Needless to say, no inconsistency has ever been found from correct application of our rules, although some of our calculations will put them to a severe test. Apparent inconsistencies have always proved, on closer examination, to be misapplications of the rules. On the other hand, guided by Cox's theorems, which tell us where to look, we have never had the slightest difficulty in exhibiting the inconsistencies in the *ad hoc* rules which abound in the literature, which differ in content from ours and whose sole basis is the intuitive judgment of their inventors. Examples are found throughout this book, but particularly in Chapters 5, 15, and 17.

2.6.3 Venn diagrams

Doubtless, some readers will ask, 'After the rather long and seemingly unmotivated derivation of the extended sum rule (2.66), which in our new notation now takes the form

$$P(A + B|C) = P(A|C) + P(B|C) - P(AB|C), \qquad (2.104)$$

why did we not illustrate it by the Venn diagram? That makes its meaning so much clearer.' (Here we draw two circles labeled A and B, with intersection labeled AB, all within a circle C.)

The Venn diagram is indeed a useful device, illustrating – in one special case – why the negative term appears in (2.104). But it can also mislead, because it suggests to our intuition more than the actual content of (2.104). Looking at the Venn diagram, we are encouraged to ask, 'What do the points in the diagram mean?' If the diagram is intended to illustrate (2.104), then the probability for A is, presumably, represented by the area of circle A; for then the total area covered by circles A, B is the sum of their separate areas, minus the area of overlap, corresponding exactly to (2.104).

Now, the circle A can be broken down into nonoverlapping subregions in many different ways; what do these subregions mean? Since their areas are additive, if the Venn diagram is to remain applicable they must represent a refinement of A into the disjunction of some mutually exclusive subpropositions. We can – if we have no mathematical scruples about approaching infinite limits – imagine this subdivision carried down to the individual points in the diagram. Therefore these points must represent some ultimate 'elementary' propositions ω_i into which A can be resolved.[3] Of course, consistency then requires us to suppose that B and C can also be resolved into these same propositions ω_i.

We have already jumped to the conclusion that the propositions to which we assign probabilities correspond to sets of points in some space, that the logical disjunction $A + B$ stands for the union of the sets, the conjunction AB for their intersection, and that the probabilities are an additive measure over those sets. But the general theory we are developing has no such structure; all these things are properties only of the Venn diagram.

In developing our theory of inference we have taken special pains to avoid restrictive assumptions which would limit its scope; it is to apply, in principle, to any propositions with unambiguous meaning. In the special case where those propositions happen to be statements about sets, the Venn diagram is an appropriate illustration of (2.104). But most of the propositions about which we reason, for example,

$$A \equiv \text{it will rain today,} \tag{2.105}$$

$$B \equiv \text{the roof will leak,} \tag{2.106}$$

are simply declarative statements of fact, which may or may not be resolvable into a disjunction of more elementary propositions within the context of our problem.

Of course, one can always force such a resolution by introducing irrelevancies; for example, even though the above-defined B has nothing to do with penguins, we could still resolve it into the disjunction

$$B = BC_1 + BC_2 + BC_3 + \cdots + BC_N, \tag{2.107}$$

where $C_k \equiv$ the number of penguins in Antarctica is k. By choosing N sufficiently large, we will surely be making a valid statement of Boolean algebra; but this is idle, and it cannot help us to reason about a leaky roof.

Even if a meaningful resolution exists in our problem, it may not be of any use to us. For example, the proposition 'rain today' could be resolved into an enumeration of every conceivable trajectory of each individual raindrop; but we do not see how this could help a meteorologist trying to forecast rain. In real problems, there is a natural end to this resolving, beyond which it serves no purpose and degenerates into an empty formal exercise. We shall

---

[3] A physicist refuses to call them 'atomic' propositions, for obvious reasons.

give an explicit demonstration of this later (Chapter 8), in the scenario of 'Sam's broken thermometer': does the exact way in which it broke matter for the conclusions that Sam should draw from his corrupted data?

In some cases there is a resolution so relevant to the context of the problem that it becomes a useful calculational device; Eq. (2.98) was a trivial example. We shall be glad to take advantage of this whenever we can, but we cannot expect it in general.

Even when both A and B can be resolved in a way meaningful and useful in our problem, it would seldom be the case that they are resolvable into the *same* set of elementary propositions ω_i. And we always reserve the right to enlarge our context by introducing more propositions $D, E, F, \ldots$ into the discussion; and we could hardly ever expect that all of them would continue to be expressible as disjunctions of the *same* original set of elementary propositions ω_i. To assume this would be to place a quite unnecessary restriction on the generality of our theory.

Therefore, the conjunction AB should be regarded simply as the statement that both A and B are true; it is a mistake to try to read any more detailed meaning, such as an intersection of sets, into it in every problem. Then $p(AB|C)$ should also be regarded as an elementary quantity in its own right, not necessarily resolvable into a sum of still more elementary ones (although if it is so resolvable this may be a good way of calculating it). We have adhered to the original notation $A + B$, AB of Boole, instead of the more common $A \vee B$, $A \wedge B$, or $A \cup B$, $A \cap B$, which everyone associates with a set theory context, in order to head off this confusion as much as possible.

So, rather than saying that the Venn diagram justifies or explains (2.104), we prefer to say that (2.104) explains and justifies the Venn diagram, in one special case. But the Venn diagram has played a major role in the history of probability theory, as we note next.

2.6.4 The 'Kolmogorov axioms'

In 1933, A. N. Kolmogorov presented an approach to probability theory phrased in the language of set theory and measure theory (Kolmogorov, 1933). This language was just then becoming so fashionable that today many mathematical results are named, not for the discoverer, but for the one who first restated them in that language. For example, in the theory of continuous groups the term 'Hurwitz invariant integral' disappeared, to be replaced by 'Haar measure'. Because of this custom, some modern works – particularly by mathematicians – can give one the impression that probability theory started with Kolmogorov.

Kolmogorov formalized and axiomatized the picture suggested by the Venn diagram, which we have just described. At first glance, this system appears so totally different from ours that some discussion is needed to see the close relationship between them. In Appendix A we describe the Kolmogorov system and show that, for all practical purposes, the four axioms concerning his probability measure, first stated arbitrarily (for which Kolmogorov has been criticized), have all been derived in this chapter as necessary to meet our consistency requirements. As a result, we shall find ourselves defending Kolmogorov against his critics on many technical points. The reader who first learned probability theory on the Kolmogorov basis is urged to read Appendix A at this point.

Our system of probability, however, differs conceptually from that of Kolmogorov in that we do not interpret propositions in terms of sets, but we do interpret probability distributions as carriers of incomplete information. Partly as a result, our system has analytical resources not present at all in the Kolmogorov system. This enables us to formulate and solve many problems – particularly the so-called 'ill posed' problems and 'generalized inverse' problems – that would be considered outside the scope of probability theory according to the Kolmogorov system. These problems are just the ones of greatest interest in current applications.

3

Elementary sampling theory

At this point, the mathematical material we have available consists of the basic product and sum rules

$$P(AB|C) = P(A|BC)P(B|C) = P(B|AC)P(A|C) \qquad (3.1)$$

$$P(A|B) + P(\overline{A}|B) = 1 \qquad (3.2)$$

from which we derived the extended sum rule

$$P(A + B|C) = P(A|C) + P(B|C) - P(AB|C) \qquad (3.3)$$

and with the desideratum (IIIc) of consistency, the principle of indifference: if on background information B the hypotheses $(H_1, H_2, \ldots, H_N)$ are mutually exclusive and exhaustive, and B does not favor any one of them over any other, then

$$P(H_i|B) = \frac{1}{N}, \qquad 1 \le i \le N. \qquad (3.4)$$

From (3.3) and (3.4) we then derived the Bernoulli urn rule: if B specifies that A is true on some subset of M of the H_i, and false on the remaining $(N - M)$, then

$$P(A|B) = \frac{M}{N}. \qquad (3.5)$$

It is important to realize how much of probability theory can be derived from no more than this.

In fact, essentially all of conventional probability theory as currently taught, plus many important results that are often thought to lie beyond the domain of probability theory, can be derived from the above foundation. We devote the next several chapters to demonstrating this in some detail, and then in Chapter 11 we resume the basic development of our robot's brain, with a better understanding of what additional principles are needed for advanced applications.

The first applications of the theory given in this chapter are, to be sure, rather simple and naïve compared with the serious scientific inference that we hope to achieve later. Nevertheless, our reason for considering them in close detail is not mere pedagogical form. Failure to understand the logic of these simplest applications has been one of the major factors

51

retarding the progress of scientific inference – and therefore of science itself – for many decades. Therefore we urge the reader, even one who is already familiar with elementary sampling theory, to digest the contents of this chapter carefully before proceeding to more complicated problems.

3.1 Sampling without replacement

Let us make the Bernoulli urn scenario a little more specific by defining the following propositions.

$B \equiv$ An urn contains N balls, identical in every respect except that they carry numbers $(1, 2, \ldots, N)$ and M of them are colored red, with the remaining $(N - M)$ white, $0 \le M \le N$. We draw a ball from the urn blindfolded, observe and record its color, lay it aside, and repeat the process until n balls have been drawn, $0 \le n \le N$.

$R_i \equiv$ Red ball on the ith draw.

$W_i \equiv$ White ball on the ith draw.

Since, according to B, only red or white can be drawn, we have

$$P(R_i|B) + P(W_i|B) = 1, \qquad 1 \le i \le N, \tag{3.6}$$

which amounts to saying that, in the 'logical environment' created by knowledge of B, the propositions are related by negation:

$$\overline{R_i} = W_i, \qquad \overline{W_i} = R_i, \tag{3.7}$$

and, for the first draw, (3.5) becomes

$$P(R_1|B) = \frac{M}{N}, \tag{3.8}$$

$$P(W_1|B) = 1 - \frac{M}{N}. \tag{3.9}$$

Let us understand clearly what this means. The probability assignments (3.8) and (3.9) are not assertions of any physical property of the urn or its contents; they are a description of the *state of knowledge* of the robot prior to the drawing. Indeed, were the robot's state of knowledge different from B as just defined (for example, if it knew the actual positions of the red and white balls in the urn, or if it did not know the true values of N and M), then its probability assignments for R_1 and W_1 would be different; but the real properties of the urn would be just the same.

It is therefore illogical to speak of 'verifying' (3.8) by performing experiments with the urn; that would be like trying to verify a boy's love for his dog by performing experiments on the dog. At this stage, we are concerned with the logic of consistent reasoning from incomplete information; not with assertions of physical fact about what will be drawn from the urn (which are in any event impossible just because of the incompleteness of the information B).

Eventually, our robot will be able to make some very confident physical predictions which can approach, but (except in degenerate cases) not actually reach, the certainty of logical deduction; but the theory needs to be developed further before we are in a position to say what quantities can be well predicted, and what kind of information is needed for this. Put differently, relations between probabilities assigned by the robot in various states of knowledge, and observable facts in experiments, may not be assumed arbitrarily; we are justified in using only those relations that can be deduced from the rules of probability theory, as we now seek to do.

Changes in the robot's state of knowledge appear when we ask for probabilities referring to the second draw. For example, what is the robot's probability for red on the first two draws? From the product rule, this is

$$P(R_1 R_2 | B) = P(R_1 | B) P(R_2 | R_1 B). \tag{3.10}$$

In the last factor, the robot must take into account that one red ball has been removed at the first draw, so there remain $(N - 1)$ balls of which $(M - 1)$ are red. Therefore

$$P(R_1 R_2 | B) = \frac{M}{N} \frac{M - 1}{N - 1}. \tag{3.11}$$

Continuing in this way, the probability for red on the first r consecutive draws is

$$P(R_1 R_2 \cdots R_r | B) = \frac{M(M - 1) \cdots (M - r + 1)}{N(N - 1) \cdots (N - r + 1)}$$

$$= \frac{M!(N - r)!}{(M - r)!N!}, \quad r \leq M. \tag{3.12}$$

The restriction $r \leq M$ is not necessary if we understand that we define factorials by the gamma function relation $n! = \Gamma(n + 1)$, for then the factorial of a negative integer is infinite, and (3.12) is zero automatically when $r > M$.

The probability for white on the first w draws is similar but for the interchange of M and $(N - M)$:

$$P(W_1 W_2 \cdots W_w | B) = \frac{(N - M)!(N - w)!}{(N - M - w)!N!}. \tag{3.13}$$

Then, the probability for white on draws $(r + 1, r + 2, \ldots, r + w)$ given that we got red on the first r draws, is given by (3.13), taking into account that N and M have been reduced to $(N - r)$ and $(M - r)$, respectively:

$$P(W_{r+1} \cdots W_{r+w} | R_1 \cdots R_r B) = \frac{(N - M)!(N - r - w)!}{(N - M - w)!(N - r)!}, \tag{3.14}$$

and so, by the product rule, the probability for obtaining r red followed by $w = n - r$ white in n draws is, from (3.12) and (3.14),

$$P(R_1 \cdots R_r W_{r+1} \cdots W_n | B) = \frac{M!(N - M)!(N - n)!}{(M - r)!(N - M - w)!N!}, \tag{3.15}$$

a term $(N - r)!$ having cancelled out.

Although this result was derived for a particular order of drawing red and white balls, the probability for drawing exactly r red balls in any specified order in n draws is the same. To see this, write out the expression (3.15) more fully, in the manner

$$\frac{M!}{(M - r)!} = M(M - 1) \cdots (M - r + 1) \tag{3.16}$$

and similarly for the other ratios of factorials in (3.15). The right-hand side becomes

$$\frac{M(M - 1) \cdots (M - r + 1)(N - M)(N - M - 1) \cdots (N - M - w + 1)}{N(N - 1) \cdots (N - n + 1)}. \tag{3.17}$$

Now suppose that r red and $(n - r) = w$ white are drawn, in any other order. The probability for this is the product of n factors; every time red is drawn there is a factor (number of red balls in urn)/(total number of balls), and similarly for drawing a white one. The number of balls in the urn decreases by one at each draw; therefore for the kth draw a factor $(N - k + 1)$ appears in the denominator, whatever the colors of the previous draws.

Just before the kth red ball is drawn, whether this occurs at the kth draw or any later one, there are $(M - k + 1)$ red balls in the urn; thus, drawing the kth one places a factor $(M - k + 1)$ in the numerator. Just before the kth white ball is drawn, there are $(N - M - k + 1)$ white balls in the urn, and so drawing the kth white one places a factor $(N - M - k + 1)$ in the numerator, regardless of whether this occurs at the kth draw or any later one. Therefore, by the time all n balls have been drawn, of which r were red, we have accumulated exactly the same factors in numerator and denominator as in (3.17); different orders of drawing them only permute the order of the factors in the numerator. The probability for drawing exactly r balls in any specified order in n draws is therefore given by (3.15).

Note carefully that in this result the product rule was expanded in a particular way that showed us how to organize the calculation into a product of factors, each of which is a probability at one specified draw, *given the results of all the previous draws.* But the product rule could have been expanded in many other ways, which would give factors conditional on other information than the previous draws; the fact that all these calculations must lead to the same final result is a nontrivial consistency property, which the derivations of Chapter 2 sought to ensure.

Next, we ask: What is the robot's probability for drawing exactly r red balls in n draws, regardless of order? Different orders of appearance of red and white balls are mutually exclusive possibilities, so we must sum over all of them; but since each term is equal to (3.15), we merely multiply it by the binomial coefficient

$$\binom{n}{r} = \frac{n!}{r!(n - r)!}, \tag{3.18}$$

which represents the number of possible orders of drawing r red balls in n draws, or, as we shall call it, the *multiplicity* of the event r. For example, to get three red in three draws can happen in only

$$\binom{3}{3} = 1 \tag{3.19}$$

way, namely $R_1 R_2 R_3$; the event $r = 3$ has a multiplicity of 1. But to get two red in three draws can happen in

$$\binom{3}{2} = 3 \tag{3.20}$$

ways, namely $R_1 R_2 W_3$, $R_1 W_2 R_3$, $W_1 R_2 R_3$, so the event $r = 2$ has a multiplicity of 3.

Exercise 3.1. Why isn't the multiplicity factor (3.18) just $n!$? After all, we started this discussion by stipulating that the balls, in addition to having colors, also carry labels $(1, 2, \ldots, N)$, so that different permutations of the red balls among themselves, which give the $r!$ in the denominator of (3.18), are distinguishable arrangements.
Hint: In (3.15) we are not specifying which red balls and which white ones are to be drawn.

Taking the product of (3.15) and (3.18), the many factorials can be reorganized into three binomial coefficients. Defining $A \equiv$ 'Exactly r red balls in n draws, in any order' and the function

$$h(r|N, M, n) \equiv P(A|B), \tag{3.21}$$

we have

$$h(r|N, M, n) = \frac{\binom{M}{r}\binom{N - M}{n - r}}{\binom{N}{n}}, \tag{3.22}$$

which we shall usually abbreviate to $h(r)$. By the convention $x! = \Gamma(x + 1)$ it vanishes automatically when $r > M$, or $r > n$, or $(n - r) > (N - M)$, as it should.

We are here doing a little notational acrobatics for reasons explained in Appendix B. The point is that in our formal probability symbols $P(A|B)$ with the capital P, the arguments A, B always stand for propositions, which can be quite complicated verbal statements. If we wish to use ordinary numbers for arguments, then for consistency we should define new functional symbols such as $h(r|N, M, n)$. Attempts to try to use a notation like $P(r|NMn)$, thereby losing sight of the qualitative stipulations contained in A and B, have led to serious errors from misinterpretation of the equations (such as the marginalization paradox discussed later). However, as already indicated in Chapter 2, we follow the custom of most contemporary works by using probability symbols of the form $p(A|B)$, or $p(r|n)$ with small

p, in which we permit the arguments to be either propositions or algebraic variables; in this case, the meaning must be judged from the context.

The fundamental result (3.22) is called the *hypergeometric distribution* because it is related to the coefficients in the power series representation of the Gauss hypergeometric function

$$F(a, b, c; t) = \sum_{r=0}^{\infty} \frac{\Gamma(a+r)\Gamma(b+r)\Gamma(c)}{\Gamma(a)\Gamma(b)\Gamma(c+r)} \frac{t^r}{r!}. \tag{3.23}$$

If either a or b is a negative integer, the series terminates and this is a polynomial. It is easily verified that the *generating function*

$$G(t) \equiv \sum_{r=0}^{n} h(r|N, M, n)t^r \tag{3.24}$$

is equal to

$$G(t) = \frac{F(-M, -n, c; t)}{F(-M, -n, c; 1)}, \tag{3.25}$$

with $c = N - M - n + 1$. The evident relation $G(1) = 1$ is, from (3.24), just the statement that the hypergeometric distribution is correctly normalized. In consequence of (3.25), $G(t)$ satisfies the second-order hypergeometric differential equation and has many other properties useful in calculations.

Although the hypergeometric distribution $h(r)$ appears complicated, it has some surprisingly simple properties. The most probable value of r is found to within one unit by setting $h(r') = h(r' - 1)$ and solving for r'. We find

$$r' = \frac{(n+1)(M+1)}{N+2}. \tag{3.26}$$

If r' is an integer, then r' and $r' - 1$ are jointly the most probable values. If r' is not an integer, then there is a unique most probable value

$$\hat{r} = \text{INT}(r'), \tag{3.27}$$

that is, the next integer below r'. Thus, the most probable fraction $f = r/n$ of red balls in the sample drawn is nearly equal to the fraction $F = M/N$ originally in the urn, as one would expect intuitively. This is our first crude example of a physical prediction: a relation between a quantity F specified in our information and a quantity f measurable in a physical experiment derived from the theory.

The width of the distribution $h(r)$ gives an indication of the accuracy with which the robot can predict r. Many such questions are answered by calculating the *cumulative probability distribution*, which is the probability for finding R or fewer red balls. If R is an integer, this is

$$H(R) \equiv \sum_{r=0}^{R} h(r), \tag{3.28}$$

but for later formal reasons we define $H(x)$ to be a staircase function for all non-negative real x; thus $H(x) \equiv H(R)$, where $R = \text{INT}(x)$ is the greatest integer $\leq x$.

The *median* of a probability distribution such as $h(r)$ is defined to be a number m such that equal probabilities are assigned to the propositions $(r < m)$ and $(r > m)$. Strictly speaking, according to this definition a discrete distribution has in general no median. If there is an integer R for which $H(R-1) = 1 - H(R)$ and $H(R) > H(R-1)$, then R is the unique median. If there is an integer R for which $H(R) = 1/2$, then any r in $(R \leq r < R')$ is a median, where R' is the next higher jump point of $H(x)$; otherwise there is none.

But for most purposes we may take a more relaxed attitude and approximate the strict definition. If n is reasonably large, then it makes reasonably good sense to call that value of R for which $H(R)$ is closest to $1/2$, the 'median'. In the same relaxed spirit, the values of R for which $H(R)$ is closest to $1/4, 3/4$, may be called the 'lower quartile' and 'upper quartile', respectively, and if $n \gg 10$ we may call the value of R for which $H(R)$ is closest to $k/10$ the 'kth decile', and so on. As $n \to \infty$, these loose definitions come into conformity with the strict one.

Usually, the fine details of $H(R)$ are unimportant, and for our purposes it is sufficient to know the median and the quartiles. Then the (median) $\pm$ (interquartile distance) will provide a good enough idea of the robot's prediction and its probable accuracy. That is, on the information given to the robot, the true value of r is about as likely to lie in this interval as outside it. Likewise, the robot assigns a probability of $(5/6) - (1/6) = 2/3$ (in other words, odds of $2 : 1$) that r lies between the first and fifth hexile, odds of $8 : 2 = 4 : 1$ that it is bracketed by the first and ninth decile, and so on.

Although one can develop rather messy approximate formulas for these distributions which were much used in the past, it is easier today to calculate the exact distribution by computer. For example W. H. Press *et al.* (1986) list two routines that will calculate the generalized complex hypergeometric distribution for any values of a, b and c. Tables 3.1 and 3.2 give the hypergeometric distribution for $N = 100, M = 50, n = 10$, and $N = 100$, $M = 10, n = 50$, respectively. In the latter case, it is not possible to draw more than ten red balls, so the entries for $r > 10$ are all $h(r) = 0, H(r) = 1$, and are not tabulated. One is struck immediately by the fact that the entries for positive $h(r)$ are identical; the hypergeometric distribution has the symmetry property

$$h(r|N, M, n) = h(r|N, n, M) \qquad (3.29)$$

under interchange of M and n. Whether we draw ten balls from an urn containing 50 red ones, or 50 from an urn containing ten red ones, the probability for finding r red ones in the sample drawn is the same. This is readily verified by closer inspection of (3.22), and it is evident from the symmetry in a, b of the hypergeometric function (3.23).

Another symmetry evident from Tables 3.1 and 3.2 is the symmetry of the distribution about its peak: $h(r|100, 50, 10) = h(10 - r|100, 50, 10)$. However, this is not so in general; changing N to 99 results in a slightly unsymmetrical peak, as we see from Table 3.3. The symmetric peak in Table 3.1 arises as follows: if we interchange M and $(N - M)$ and at the same time interchange r and $(n - r)$ we have in effect only interchanged the words 'red'

Table 3.1. *Hypergeometric distribution;*
$N, M, n = 100, 10, 50.$

r	$h(r)$	$H(r)$
0	0.000593	0.000593
1	0.007237	0.007830
2	0.037993	0.045824
3	0.113096	0.158920
4	0.211413	0.370333
5	0.259334	0.629667
6	0.211413	0.841080
7	0.113096	0.954177
8	0.037993	0.992170
9	0.007237	0.999407
10	0.000593	1.000000

Table 3.2. *Hypergeometric distribution;*
$N, M, n = 100, 50, 10.$

r	$h(r)$	$H(r)$
0	0.000593	0.000593
1	0.007237	0.007830
2	0.037993	0.045824
3	0.113096	0.158920
4	0.211413	0.370333
5	0.259334	0.629667
6	0.211413	0.841080
7	0.113096	0.954177
8	0.037993	0.992170
9	0.007237	0.999407
10	0.000593	1.000000

and 'white', so the distribution is unchanged:

$$h(n - r|N, N - M, n) = h(r|N, M, n). \tag{3.30}$$

But when $M = N/2$, this reduces to the symmetry

$$h(n - r|N, M, n) = h(r|N, M, n) \tag{3.31}$$

observed in Table 3.1. By (3.29) the peak must be symmetric also when $n = N/2$.

Table 3.3. *Hypergeometric distribution;*
$N, M, n = 99, 50, 10$.

r	$h(r)$	$H(r)$
0	0.000527	0.000527
1	0.006594	0.007121
2	0.035460	0.042581
3	0.108070	0.150651
4	0.206715	0.357367
5	0.259334	0.616700
6	0.216111	0.832812
7	0.118123	0.950934
8	0.040526	0.991461
9	0.007880	0.999341
10	0.000659	1.000000

The hypergeometric distribution has two more symmetries not at all obvious intuitively or even visible in (3.22). Let us ask the robot for its probability $P(R_2|B)$ of red on the second draw. This is not the same calculation as (3.8), because the robot knows that, just prior to the second draw, there are only $(N-1)$ balls in the urn, not N. But it does not know what color of ball was removed on the first draw, so it does not know whether the number of red balls now in the urn is M or $(M-1)$. Then the basis for the Bernoulli urn result (3.5) is lost, and it might appear that the problem is indeterminate.

Yet it is quite determinate after all; the following is our first example of one of the useful techniques in probability calculations, which derives from the resolution of a proposition into disjunctions of simpler ones, as discussed in Chapters 1 and 2. The robot knows that either R_1 or W_1 is true; therefore using Boolean algebra we have

$$R_2 = (R_1 + W_1)R_2 = R_1 R_2 + W_1 R_2. \tag{3.32}$$

We apply the sum rule and the product rule to get

$$P(R_2|B) = P(R_1 R_2|B) + P(W_1 R_2|B)$$
$$= P(R_2|R_1 B)P(R_1|B) + P(R_2|W_1 B)P(W_1|B). \tag{3.33}$$

But

$$P(R_2|R_1 B) = \frac{M-1}{N-1}, \qquad P(R_2|W_1 B) = \frac{M}{N-1}, \tag{3.34}$$

and so

$$P(R_2|B) = \frac{M-1}{N-1}\frac{M}{N} + \frac{M}{N-1}\frac{N-M}{N} = \frac{M}{N}. \tag{3.35}$$

The complications cancel out, and we have the same probability for red on the first and second draws. Let us see whether this continues. For the third draw we have

$$R_3 = (R_1 + W_1)(R_2 + W_2)R_3 = R_1 R_2 R_3 + R_1 W_2 R_3 + W_1 R_2 R_3 + W_1 W_2 R_3, \quad (3.36)$$

and so

$$
\begin{aligned}
P(R_3|B) = {} & \frac{M}{N}\frac{M-1}{N-1}\frac{M-2}{N-2} + \frac{M}{N}\frac{N-M}{N-1}\frac{M-1}{N-2} \\
& + \frac{N-M}{N}\frac{M}{N-1}\frac{M-1}{N-2} + \frac{N-M}{N}\frac{N-M-1}{N-1}\frac{M}{N-2} \\
& = \frac{M}{N}.
\end{aligned}
\quad (3.37)
$$

Again all the complications cancel out. The robot's probability for red at any draw, *if it does not know the result of any other draw*, is always the same as the Bernoulli urn result (3.5). This is the first nonobvious symmetry. We shall not prove this in generality here, because it is contained as a special case of a still more general result; see Eq. (3.118) below.

The method of calculation illustrated by (3.32) and (3.36) is as follows: resolve the quantity whose probability is wanted into mutually exclusive subpropositions, then apply the sum rule and the product rule. If the subpropositions are well chosen (i.e. if they have some simple meaning in the context of the problem), their probabilities are often calculable. If they are not well chosen (as in the example of the penguins at the end of Chapter 2), then of course this procedure cannot help us.

3.2 Logic vs. propensity

The results of Section 3.1 present us with a new question. In finding the probability for red at the kth draw, knowledge of what color was found at some earlier draw is clearly relevant because an earlier draw affects the number M_k of red balls in the urn for the kth draw. Would knowledge of the color for a later draw be relevant? At first glance, it seems that it could not be, because the result of a later draw cannot influence the value of M_k. For example, a well-known exposition of statistical mechanics (Penrose, 1979) takes it as a fundamental axiom that probabilities referring to the present time can depend only on what happened earlier, not on what happens later. The author considers this to be a necessary physical condition of 'causality'.

Therefore we stress again, as we did in Chapter 1, that inference is concerned with *logical* connections, which may or may not correspond to causal physical influences. To show why knowledge of later events is relevant to the probabilities of earlier ones, consider an urn which is known (background information B) to contain only one red and one white ball: $N = 2$, $M = 1$. Given only this information, the probability for red on the first draw is $P(R_1|B) = 1/2$. But then if the robot learns that red occurs on the second draw, it becomes certain that it did not occur on the first:

$$P(R_1|R_2B) = 0. \tag{3.38}$$

More generally, the product rule gives us

$$P(R_jR_k|B) = P(R_j|R_kB)P(R_k|B) = P(R_k|R_jB)P(R_j|B). \tag{3.39}$$

But we have just seen that $P(R_j|B) = P(R_k|B) = M/N$ for all j, k, so

$$P(R_j|R_kB) = P(R_k|R_jB), \qquad \text{all } j, k. \tag{3.40}$$

Probability theory tells us that the results of later draws have precisely the same relevance as do the results of earlier ones! Even though performing the later draw does not physically affect the number M_k of red balls in the urn at the kth draw, *information* about the result of a later draw has the same effect on our *state of knowledge* about what could have been taken on the kth draw, as does information about an earlier one. This is our second nonobvious symmetry.

This result will be quite disconcerting to some schools of thought about the 'meaning of probability'. Although it is generally recognized that logical implication is not the same as physical causation, nevertheless there is a strong inclination to cling to the idea anyway, by trying to interpret a probability $P(A|B)$ as expressing some kind of partial causal influence of B on A. This is evident not only in the aforementioned work of Penrose, but more strikingly in the 'propensity' theory of probability expounded by the philosopher Karl Popper.[1]

It appears to us that such a relation as (3.40) would be quite inexplicable from a propensity viewpoint, although the simple example (3.38) makes its logical necessity obvious. In any event, the theory of logical inference that we are developing here differs fundamentally, in outlook and in results, from the theory of physical causation envisaged by Penrose and Popper. It is evident that logical inference can be applied in many problems where assumptions of physical causation would not make sense.

This does not mean that we are forbidden to introduce the notion of 'propensity' or physical causation; the point is rather that logical inference is applicable and useful whether or not a propensity exists. If such a notion (i.e. that some such propensity exists) is formulated as a well-defined hypothesis, then our form of probability theory can analyze its implications. We shall do this in Section 3.10 below. Also, we can test that hypothesis against alternatives

---

[1] In his presentation at the Ninth Colston Symposium, Popper (1957) describes his propensity interpretation as 'purely objective' but avoids the expression 'physical influence'. Instead, he would say that the probability for a particular face in tossing a die is not a physical property of the die (as Cramér (1946) insisted), but rather is an objective property of the whole experimental arrangement, the die plus the method of tossing. Of course, that the *result of the experiment* depends on the entire arrangement and procedure is only a truism. It was stressed repeatedly by Niels Bohr in connection with quantum theory, but presumably no scientist from Galileo on has ever doubted it. However, unless Popper really meant 'physical influence', his interpretation would seem to be supernatural rather than objective. In a later article (Popper, 1959) he defines the propensity interpretation more completely; now a propensity is held to be 'objective' and 'physically real' even when applied to the individual trial. In the following we see by mathematical demonstration some of the logical difficulties that result from a propensity interpretation. Popper complains that in quantum theory one oscillates between '... an *objective* purely statistical interpretation and a *subjective* interpretation in terms of our incomplete knowledge', and thinks that the latter is reprehensible and the propensity interpretation avoids any need for it. He could not possibly be more mistaken. In Chapter 9 we answer this in detail at the conceptual level; obviously, *incomplete knowledge is the only working material a scientist has*! In Chapter 10 we consider the detailed physics of coin tossing, and see just how the method of tossing affects the results by direct physical influence.

in the light of the evidence, just as we can test any well-defined hypothesis. Indeed, one of the most common and important applications of probability theory is to decide whether there is evidence for a causal influence: is a new medicine more effective, or a new engineering design more reliable? Does a new anticrime law reduce the incidence of crime? Our study of hypothesis testing starts in Chapter 4.

In all the sciences, logical inference is more generally applicable. We agree that physical influences can propagate only forward in time; but logical inferences propagate equally well in either direction. An archaeologist uncovers an artifact that changes his knowledge of events thousands of years ago; were it otherwise, archaeology, geology, and paleontology would be impossible. The reasoning of Sherlock Holmes is also directed to inferring, from presently existing evidence, what events must have transpired in the past. The sounds reaching your ears from a marching band 600 meters distant change your state of knowledge about what the band was playing two seconds earlier. Listening to a Toscanini recording of a Beethoven symphony changes your state of knowledge about the sounds Toscanini elicited from his orchestra many years ago.

As this suggests, and as we shall verify later, a fully adequate theory of nonequilibrium phenomena, such as sound propagation, also requires that backward logical inferences be recognized and used, although they do not express physical causes. The point is that the best inferences we can make about any phenomenon – whether in physics, biology, economics, or any other field – must take into account all the relevant information we have, regardless of whether that information refers to times earlier or later than the phenomenon itself; this ought to be considered a platitude, not a paradox. At the end of this chapter (Exercise 3.6), the reader will have an opportunity to demonstrate this directly, by calculating a backward inference that takes into account a forward causal influence.

More generally, consider a probability distribution $p(x_1 \cdots x_n | B)$, where x_i denotes the result of the ith trial, and could take on not just two values (red or white) but, say, the values $x_i = (1, 2, \ldots, k)$ labeling k different colors. If the probability is invariant under any permutation of the x_i, then it depends only on the sample numbers $(n_1 \cdots n_k)$ denoting how many times the result $x_i = 1$ occurs, how many times $x_i = 2$ occurs, etc. Such a distribution is called *exchangeable*; as we shall find later, exchangeable distributions have many interesting mathematical properties and important applications.

Returning to our urn problem, it is clear already from the fact that the hypergeometric distribution is exchangeable that every draw must have just the same relevance to every other draw, regardless of their time order and regardless of whether they are near or far apart in the sequence. But this is not limited to the hypergeometric distribution; it is true of any exchangeable distribution (i.e. whenever the probability for a sequence of events is independent of their order). So, with a little more thought, these symmetries, so inexplicable from the standpoint of physical causation, become obvious after all as propositions of logic.

Let us calculate this effect quantitatively. Supposing $j < k$, the proposition $R_j R_k$ (red at both draws j and k) is in Boolean algebra the same as

$$R_j R_k = (R_1 + W_1) \cdots (R_{j-1} + W_{j-1}) R_j (R_{j+1} + W_{j+1}) \cdots (R_{k-1} + W_{k-1}) R_k, \quad (3.41)$$

which we could expand in the manner of (3.36) into a logical sum of

$$2^{j-1} \times 2^{k-j-1} = 2^{k-2} \tag{3.42}$$

propositions, each specifying a full sequence, such as

$$W_1 R_2 W_3 \cdots R_j \cdots R_k \tag{3.43}$$

of k results. The probability $P(R_j R_k | B)$ is the sum of all their probabilities. But we know that, given B, the probability for any one sequence is independent of the order in which red and white appear. Therefore we can permute each sequence, moving R_j to the first position, and R_k to the second. That is, we can replace the sequence $(W_1 \cdots R_j \cdots)$ by $(R_1 \cdots W_j \cdots)$, etc. Recombining them, we have $(R_1 R_2)$ followed by every possible result for draws $(3, 4, \ldots, k)$. In other words, the probability for $R_j R_k$ is the same as that of

$$R_1 R_2 (R_3 + W_3) \cdots (R_k + W_k) = R_1 R_2, \tag{3.44}$$

and we have

$$P(R_j R_k | B) = P(R_1 R_2 | B) = \frac{M(M-1)}{N(N-1)}, \tag{3.45}$$

and likewise

$$P(W_j R_k | B) = P(W_1 R_2 | B) = \frac{(N-M)M}{N(N-1)}. \tag{3.46}$$

Therefore by the product rule

$$P(R_k | R_j B) = \frac{P(R_j R_k | B)}{P(R_j | B)} = \frac{M-1}{N-1} \tag{3.47}$$

and

$$P(R_k | W_j B) = \frac{P(W_j R_k | B)}{P(W_j | B)} = \frac{M}{N-1} \tag{3.48}$$

for all $j < k$. By (3.40), the results (3.47) and (3.48) are true for all $j \neq k$.

Since as noted this conclusion appears astonishing to many people, we shall belabor the point by explaining it still another time in different words. The robot knows that the urn originally contained M red balls and $(N - M)$ white ones. Then, learning that an earlier draw gave red, it knows that one less red ball is available for the later draws. The problem becomes the same as if we had started with an urn of $(N - 1)$ balls, of which $(M - 1)$ are red; (3.47) corresponds just to the solution (3.37) adapted to this different problem.

But why is knowing the result of a later draw equally cogent? Because if the robot knows that red will be drawn at any later time, then in effect one of the red balls in the urn must be 'set aside' to make this possible. The number of red balls which could have been taken in earlier draws is reduced by one, as a result of having this information. The above example (3.38) is an extreme special case of this, where the conclusion is particularly obvious.

3.3 Reasoning from less precise information

Now let us try to apply this understanding to a more complicated problem. Suppose the robot learns that red will be found at least once in later draws, but not at which draw or draws this will occur. That is, the new information is, as a proposition of Boolean algebra,

$$R_{\text{later}} \equiv R_{k+1} + R_{k+2} + \cdots + R_n. \tag{3.49}$$

This information reduces the number of red available for the kth draw by at least one, but it is not obvious whether R_{later} has exactly the same implications as does R_n. To investigate this we appeal again to the symmetry of the product rule:

$$P(R_k R_{\text{later}}|B) = P(R_k|R_{\text{later}}B)P(R_{\text{later}}|B) = P(R_{\text{later}}|R_k B)P(R_k|B), \tag{3.50}$$

which gives us

$$P(R_k|R_{\text{later}}B) = P(R_k|B)\frac{P(R_{\text{later}}|R_k B)}{P(R_{\text{later}}|B)}, \tag{3.51}$$

and all quantities on the right-hand side are easily calculated.

Seeing (3.49), one might be tempted to reason as follows:

$$P(R_{\text{later}}|B) = \sum_{j=k+1}^{n} P(R_j|B), \tag{3.52}$$

but this is not correct because, unless $M = 1$, the events R_j are not mutually exclusive, and, as we see from (2.82), many more terms would be needed. This method of calculation would be very tedious.

To organize the calculation better, note that the denial of R_{later} is the statement that white occurs at all the later draws:

$$\overline{R}_{\text{later}} = W_{k+1} W_{k+2} \cdots W_n. \tag{3.53}$$

So $P(\overline{R}_{\text{later}}|B)$ is the probability for white at all the later draws, regardless of what happens at the earlier ones (i.e. when the robot does not know what happens at the earlier ones). By exchangeability this is the same as the probability for white at the first $(n - k)$ draws, regardless of what happens at the later ones; from (3.13),

$$P(\overline{R}_{\text{later}}|B) = \frac{(N - M)!(N - n + k)!}{N!(N - M - n + k)!} = \binom{N - M}{n - k}\binom{N}{n - k}^{-1}. \tag{3.54}$$

Likewise, $P(\overline{R}_{\text{later}}|R_k B)$ is the same result for the case of $(N - 1)$ balls, $(M - 1)$ of which are red:

$$P(\overline{R}_{\text{later}}|R_k B) = \frac{(N - M)!}{(N - 1)!}\frac{(N - n + k - 1)!}{(N - M - n + k)!} = \binom{N - M}{n - k}\binom{N - 1}{n - k}^{-1}. \tag{3.55}$$

Now (3.51) becomes

$$P(R_k|R_{\text{later}}B) = \frac{M}{N-n+k} \times \frac{\binom{N-1}{n-k} - \binom{N-M}{n-k}}{\binom{N}{n-k} - \binom{N-M}{n-k}}. \tag{3.56}$$

As a check, note that if $n = k+1$, this reduces to $(M-1)/(N-1)$, as it should.

At the moment, however, our interest in (3.56) is not so much in the numerical values, but in understanding the logic of the result. So let us specialize it to the simplest case that is not entirely trivial. Suppose we draw $n = 3$ times from an urn containing $N = 4$ balls, $M = 2$ of which are white, and ask how knowledge that red occurs at least once on the second and third draws affects the probability for red at the first draw. This is given by (3.56) with $N = 4$, $M = 2$, $n = 3$, $k = 1$:

$$P(R_1|R_2 + R_3, B) = \frac{6-2}{12-2} = \frac{2}{5} = \left(\frac{1}{2}\right)\frac{1-1/3}{1-1/6}, \tag{3.57}$$

the last form corresponding to (3.51). Compare this to the previously calculated probabilities:

$$P(R_1|B) = \frac{1}{2}, \qquad P(R_1|R_2B) = P(R_2|R_1B) = \frac{1}{3}. \tag{3.58}$$

What seems surprising is that

$$P(R_1|R_{\text{later}}B) > P(R_1|R_2B). \tag{3.59}$$

Most people guess at first that the inequality should go the other way; i.e. knowing that red occurs at least once on the later draws ought to decrease the chances of red at the first draw more than does the information R_2. But in this case the numbers are so small that we can check the calculation (3.51) directly. To find $P(R_{\text{later}}|B)$ by the extended sum rule (2.82) now requires only one extra term:

$$P(R_{\text{later}}|B) = P(R_2|B) + P(R_3|B) - P(R_2R_3|B)$$
$$= \frac{1}{2} + \frac{1}{2} - \frac{1}{2} \times \frac{1}{3} = \frac{5}{6}. \tag{3.60}$$

We could equally well resolve R_{later} into mutually exclusive propositions and calculate

$$P(R_{\text{later}}|B) = P(R_2W_3|B) + P(W_2R_3|B) + P(R_2R_3|B)$$
$$= \frac{1}{2} \times \frac{2}{3} + \frac{1}{2} \times \frac{2}{3} + \frac{1}{2} \times \frac{1}{3} = \frac{5}{6}. \tag{3.61}$$

The denominator $(1 - 1/6)$ in (3.57) has now been calculated in three different ways, with the same result. If the three results were not the same, we would have found an inconsistency in our rules, of the kind we sought to prevent by Cox's functional equation arguments in Chapter 2. This is a good example of what 'consistency' means in practice, and it shows the trouble we would be in if our rules did not have it.

Likewise, we can check the numerator of (3.51) by an independent calculation:

$$P(R_{\text{later}}|R_1 B) = P(R_2|R_1 B) + P(R_3|R_1 B) - P(R_2 R_3|R_1 B)$$
$$= \frac{1}{3} + \frac{1}{3} - \frac{1}{3} \times 0 = \frac{2}{3}, \tag{3.62}$$

and the result (3.57) is confirmed. So we have no choice but to accept the inequality (3.59) and try to understand it intuitively. Let us reason as follows. The information R_2 reduces the number of red balls available for the first draw by one, and it reduces the number of balls in the urn available for the first draw by one, giving $P(R_1|R_2 B) = (M - 1)/(N - 1) = 1/3$. The information R_{later} reduces the 'effective number of red balls' available for the first draw by more than one, but it reduces the number of balls in the urn available for the first draw by two (because it assures the robot that there are two later draws in which two balls are removed). So let us try tentatively to interpret the result (3.57) as

$$P(R_1|R_{\text{later}} B) = \frac{(M)_{\text{eff}}}{N - 2}, \tag{3.63}$$

although we are not quite sure what this means. Given R_{later}, it is certain that at least one red ball is removed, and the probability that two are removed is, by the product rule:

$$P(R_2 R_3|R_{\text{later}} B) = \frac{P(R_2 R_3 R_{\text{later}}|B)}{P(R_{\text{later}}|B)} = \frac{P(R_2 R_3|B)}{P(R_{\text{later}}|B)}$$
$$= \frac{(1/2) \times (1/3)}{5/6} = \frac{1}{5} \tag{3.64}$$

because $R_2 R_3$ implies R_{later}; i.e. a relation of Boolean algebra is $(R_2 R_3 R_{\text{later}} = R_2 R_3)$. Intuitively, given R_{later} there is probability 1/5 that two red balls are removed, so the effective number removed is $1 + (1/5) = 6/5$. The 'effective' number remaining for draw one is 4/5. Indeed, (3.63) then becomes

$$P(R_1|R_{\text{later}} B) = \frac{4/5}{2} = \frac{2}{5}, \tag{3.65}$$

in agreement with our better motivated, but less intuitive, calculation (3.57).

3.4 Expectations

Another way of looking at this result appeals more strongly to our intuition and generalizes far beyond the present problem. We can hardly suppose that the reader is not already familiar with the idea of expectation, but this is the first time it has appeared in the present work, so we pause to define it. If a variable quantity X can take on the particular values $(x_1, \ldots, x_n)$ in n mutually exclusive and exhaustive situations, and the robot assigns corresponding probabilities $(p_1, p_2, \ldots, p_n)$ to them, then the quantity

$$\langle X \rangle = E(X) = \sum_{i=1}^{n} p_i x_i \tag{3.66}$$

is called the *expectation* (in the older literature, *mathematical expectation* or *expectation value*) of X. It is a weighted average of the possible values, weighted according to their probabilities. Statisticians and mathematicians generally use the notation $E(X)$; but physicists, having already pre-empted E to stand for energy and electric field, use the bracket notation $\langle X \rangle$. We shall use both notations here; they have the same meaning, but sometimes one is easier to read than the other.

Like most of the standard terms that arose out of the distant past, the term 'expectation' seems singularly inappropriate to us; for it is almost never a value that anyone 'expects' to find. Indeed, it is often known to be an impossible value. But we adhere to it because of centuries of precedent.

Given R_{later}, what is the expectation of the number of red balls in the urn for draw number one? There are three mutually exclusive possibilities compatible with R_{later}:

$$R_2 W_3, \ W_2 R_3, \ R_2 R_3 \tag{3.67}$$

for which M is $(1, 1, 0)$, respectively, and for which the probabilities are as in (3.64) and (3.65):

$$P(R_2 W_3 | R_{later} B) = \frac{P(R_2 W_3 | B)}{P(R_{later} | B)} = \frac{(1/2) \times (2/3)}{(5/6)} = \frac{2}{5}, \tag{3.68}$$

$$P(W_2 R_3 | R_{later} B) = \frac{2}{5}, \tag{3.69}$$

$$P(R_2 R_3 | R_{later} B) = \frac{1}{5}. \tag{3.70}$$

So

$$\langle M \rangle = 1 \times \frac{2}{5} + 1 \times \frac{2}{5} + 0 \times \frac{1}{5} = \frac{4}{5}. \tag{3.71}$$

Thus, what we called intuitively the 'effective' value of M in (3.63) is really the expectation of M.

We can now state (3.63) in a more cogent way: when the fraction $F = M/N$ of red balls is known, then the Bernoulli urn rule applies and $P(R_1 | B) = F$. When F is unknown, the probability for red is the expectation of F:

$$P(R_1 | B) = \langle F \rangle \equiv E(F). \tag{3.72}$$

If M and N are both unknown, the expectation is over the joint probability distribution for M and N.

That a probability is numerically equal to the expectation of a fraction will prove to be a general rule that holds as well in thousands of far more complicated situations, providing one of the most useful and common rules for physical prediction. We leave it as an exercise for the reader to show that the more general result (3.56) can also be calculated in the way suggested by (3.72).

3.5 Other forms and extensions

The hypergeometric distribution (3.22) can be written in various ways. The nine factorials can be organized into binomial coefficients also as follows:

$$h(r|N, M, n) = \frac{\binom{n}{r}\binom{N-n}{M-r}}{\binom{N}{M}}. \tag{3.73}$$

But the symmetry under exchange of M and n is still not evident; to see it we must write out (3.22) or (3.73) in full, displaying all the individual factorials.

We may also rewrite (3.22), as an aid to memory, in a more symmetric form: the probability for drawing exactly r red balls and w white ones in $n = r + w$ draws, from an urn containing R red and W white, is

$$h(r) = \frac{\binom{R}{r}\binom{W}{w}}{\binom{R+W}{r+w}}, \tag{3.74}$$

and in this form it is easily generalized. Suppose that, instead of only two colors, there are k different colors of balls in the urn, N_1 of color 1, N_2 of color 2, ..., N_k of color k. The probability for drawing r_1 balls of color 1, r_2 of color 2, ..., r_k of color k in $n = \sum r_i$ draws is, as the reader may verify, the generalized hypergeometric distribution:

$$h(r_1 \cdots r_k|N_1 \cdots N_k) = \frac{\binom{N_1}{r_1} \cdots \binom{N_k}{r_k}}{\binom{\sum N_i}{\sum r_i}}. \tag{3.75}$$

3.6 Probability as a mathematical tool

From the result (3.75) one may obtain a number of identities obeyed by the binomial coefficients. For example, we may decide not to distinguish between colors 1 and 2; i.e. a ball of either color is declared to have color 'a'. Then from (3.75) we must have, on the one hand,

$$h(r_a, r_3, \ldots, r_k|N_a, N_3, \ldots, N_k) = \frac{\binom{N_a}{r_a}\binom{N_3}{r_3} \cdots \binom{N_k}{r_k}}{\binom{\sum N_i}{\sum r_i}} \tag{3.76}$$

with

$$N_a = N_1 + N_2, \qquad r_a = r_1 + r_2. \tag{3.77}$$

But the event r_a can occur for any values of r_1, r_2 satisfying (3.77), and so we must have also, on the other hand,

$$h(r_a, r_3, \ldots, r_k | N_a, N_3, \ldots, N_k) = \sum_{r_1=0}^{r_a} h(r_1, r_a - r_1, r_3, \ldots, r_k | N_1, \ldots, N_k). \quad (3.78)$$

Then, comparing (3.76) and (3.78), we have the identity

$$\binom{N_a}{r_a} = \sum_{r_1=0}^{r_a} \binom{N_1}{r_1} \binom{N_2}{r_a - r_1}. \quad (3.79)$$

Continuing in this way, we can derive a multitude of more complicated identities obeyed by the binomial coefficients. For example,

$$\binom{N_1 + N_2 + N_3}{r_a} = \sum_{r_1=0}^{r_a} \sum_{r_2=0}^{r_1} \binom{N_1}{r_1} \binom{N_2}{r_2} \binom{N_3}{r_a - r_1 - r_2}. \quad (3.80)$$

In many cases, probabilistic reasoning is a powerful tool for deriving purely mathematical results; more examples of this are given by Feller (1950, Chap. 2 & 3) and in later chapters of the present work.

3.7 The binomial distribution

Although somewhat complicated mathematically, the hypergeometric distribution arises from a problem that is very clear and simple conceptually; there are only a finite number of possibilities and all the above results are exact for the problems as stated. As an introduction to a mathematically simpler, but conceptually far more difficult, problem, we examine a limiting form of the hypergeometric distribution.

The complication of the hypergeometric distribution arises because it is taking into account the changing contents of the urn; knowing the result of any draw changes the probability for red for any other draw. But if the number N of balls in the urn is very large compared with the number drawn ($N \gg n$), then this probability changes very little, and in the limit $N \to \infty$ we should have a simpler result, free of such dependencies. To verify this, we write the hypergeometric distribution (3.22) as

$$h(r|N, M, n) = \frac{\left[\frac{1}{N^r} \binom{M}{r}\right] \left[\frac{1}{N^{n-r}} \binom{N-M}{n-r}\right]}{\left[\frac{1}{N^n} \binom{N}{n}\right]}. \quad (3.81)$$

The first factor is

$$\frac{1}{N^r} \binom{M}{r} = \frac{1}{r!} \frac{M}{N} \left(\frac{M}{N} - \frac{1}{N}\right) \left(\frac{M}{N} - \frac{2}{N}\right) \cdots \left(\frac{M}{N} - \frac{r-1}{N}\right), \quad (3.82)$$

and in the limit $N \to \infty$, $M \to \infty$, $M/N \to f$, we have

$$\frac{1}{N^r} \binom{M}{r} \to \frac{f^r}{r!}. \tag{3.83}$$

Likewise,

$$\frac{1}{N^{n-r}} \binom{M-1}{n-r} \to \frac{(1-f)^{n-r}}{(n-r)!}, \tag{3.84}$$

$$\frac{1}{N^n} \binom{N}{n} \to \frac{1}{n!}. \tag{3.85}$$

In principle, we should, of course, take the limit of the product in (3.81), not the product of the limits. But in (3.81) we have defined the factors so that each has its own independent limit, so the result is the same; the hypergeometric distribution goes into

$$h(r|N, M, n) \to b(r|n, f) \equiv \binom{n}{r} f^r (1-f)^{n-r} \tag{3.86}$$

called the *binomial* distribution, because evaluation of the generating function (3.24) now reduces to

$$G(t) \equiv \sum_{r=0}^{n} b(r|n, f)t^r = (1 - f + ft)^n, \tag{3.87}$$

an example of Newton's binomial theorem.

Figure 3.1 compares three hypergeometric distributions with $N = 15, 30, 100$ and $M/N = 0.4$, $n = 10$ to the binomial distribution with $n = 10$, $f = 0.4$. All have their peak

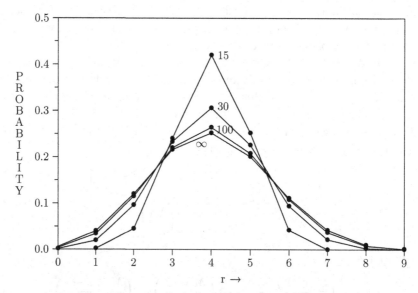

Fig. 3.1. The hypergeometric distribution for $N = 15, 30, 100, \infty$.

at $r = 4$, and all distributions have the same first moment $\langle r \rangle = E(r) = 4$, but the binomial distribution is broader.

The $N = 15$ hypergeometric distribution is zero for $r = 0$ and $r > 6$, since on drawing ten balls from an urn containing only six red and nine white, it is not possible to get fewer than one or more than six red balls. When $N > 100$ the hypergeometric distribution agrees so closely with the binomial that for most purposes it would not matter which one we used. Analytical properties of the binomial distribution are collected in Chapter 7. In Chapter 9 we find, in connection with significance tests, situations where the binomial distribution is exact for purely combinatorial reasons in a finite sample space, Eq. (9.46).

We can carry out a similar limiting process on the generalized hypergeometric distribution (3.75). It is left as an exercise to show that in the limit where all $N_i \rightarrow \infty$ in such a way that the fractions

$$f_i \equiv \frac{N_i}{\sum N_j} \tag{3.88}$$

tend to constants, (3.75) goes into the *multinomial distribution*

$$m(r_1 \cdots r_k | f_1 \cdots f_k) = \frac{r!}{r_1! \cdots r_k!} f_1^{r_1} \cdots f_k^{r_k}, \tag{3.89}$$

where $r \equiv \sum r_i$. And, as in (3.87), we can define a generating function of $(k-1)$ variables, from which we can prove that (3.89) is correctly normalized and derive many other useful results.

Exercise 3.2. Suppose an urn contains $N = \sum N_i$ balls, N_1 of color 1, N_2 of color 2, ..., N_k of color k. We draw m balls without replacement; what is the probability that we have at least one of each color? Supposing $k = 5$, all $N_i = 10$, how many do we need to draw in order to have at least a 90% probability for getting a full set?

Exercise 3.3. Suppose that in the previous exercise k is initially unknown, but we know that the urn contains exactly 50 balls. Drawing out 20 of them, we find three different colors; now what do we know about k? We know from deductive reasoning (i.e. with certainty) that $3 \leq k \leq 33$; but can you set narrower limits $k_1 \leq k \leq k_2$ within which it is highly likely to be?
Hint: This question goes beyond the sampling theory of this chapter because, like most real scientific problems, the answer depends to some degree on our common sense judgments; nevertheless, our rules of probability theory are quite capable of dealing with it, and persons with reasonable common sense cannot differ appreciably in their conclusions.

Exercise 3.4. The M urns are now numbered 1 to M, and M balls, also numbered 1 to M, are thrown into them, one in each urn. If the numbers of a ball and its urn are the same, we have a match. Show that the probability for at least one match is

$$h = \sum_{k=1}^{M} (-1)^{k+1}/k! \tag{3.90}$$

As $M \to \infty$, this converges to $1 - 1/e = 0.632$. The result is surprising to many, because, however large M is, there remains an appreciable probability for no match at all.

Exercise 3.5. N balls are tossed into M urns; there are evidently M^N ways this can be done. If the robot considers them all equally likely, what is the probability that each urn receives at least one ball?

3.8 Sampling with replacement

Up to now, we have considered only the case where we sample without replacement; and that is evidently appropriate for many real situations. For example, in a quality control application, what we have called simply 'drawing a ball' might consist of taking a manufactured item, such as an electric light bulb, from a carton of similar light bulbs and testing it to destruction. In a chemistry experiment, it might consist of weighing out a sample of an unknown protein, then dissolving it in hot sulfuric acid to measure its nitrogen content. In either case, there can be no thought of 'drawing that same ball' again.

But suppose now that, being less destructive, we sample balls from the urn and, after recording the 'color' (i.e. the relevant property) of each, we replace it in the urn before drawing the next ball. This case, of sampling with replacement, is enormously more complicated conceptually, but, with some assumptions usually made, ends up being simpler mathematically than sampling without replacement. Let us go back to the probability for drawing two red balls in succession. Denoting by B' the same background information as before, except for the added stipulation that the balls are to be replaced, we still have an equation like (3.9):

$$P(R_1 R_2 | B') = P(R_1 | B') P(R_2 | R_1 B') \tag{3.91}$$

and the first factor is still, evidently, (M/N); but what is the second one?

Answering this would be, in general, a very difficult problem, requiring much additional analysis if the background information B' includes some simple but highly relevant common sense information that we all have. What happens to that red ball that we put back in the urn? If we merely dropped it into the urn, and immediately drew another ball, then it was left lying on the top of the other balls (or in the top layer of balls), and so it is more likely to be drawn again than any other specified ball whose location in the urn is unknown. But this upsets the whole basis of our calculation, because the probability for drawing any particular (ith) ball is no longer given by the Bernoulli urn rule which led to (3.11).

3.8.1 Digression: a sermon on reality vs. models

The difficulty we face here is that many things which were irrelevant from symmetry, as long as the robot's state of knowledge was invariant under any permutation of the balls, suddenly become relevant, and, by one of our desiderata of rationality, the robot must take into account all the relevant information it has. But the probability for drawing any particular ball now depends on such details as the exact size and shape of the urn, the size of the balls, the exact way in which the first one was tossed back in, the elastic properties of balls and urn, the coefficients of friction between balls and between ball and urn, the exact way you reach in to draw the second ball, etc. In a symmetric situation, all of these details are irrelevant.

Even if all these relevant data were at hand, we do not think that a team of the world's best scientists and mathematicians, backed up by all the world's computing facilities, would be able to solve the problem; or would even know how to get started on it. Still, it would not be quite right to say that the problem is unsolvable *in principle*; only so complicated that it is not worth anybody's time to think about it. So what do we do?

In probability theory there is a very clever trick for handling a problem that becomes too difficult. We just solve it anyway by:

(1) making it still harder;
(2) redefining what we mean by 'solving' it, so that it becomes something we *can* do;
(3) inventing a dignified and technical-sounding word to describe this procedure, which has the psychological effect of concealing the real nature of what we have done, and making it appear respectable.

In the case of sampling with replacement, we apply this strategy as follows.

(1) Suppose that, after tossing the ball in, we shake up the urn. However complicated the problem was initially, it now becomes many orders of magnitude more complicated, because the solution now depends on every detail of the precise way we shake it, in addition to all the factors mentioned above.
(2) We now assert that the shaking has somehow made all these details irrelevant, so that the problem reverts back to the simple one where the Bernoulli urn rule applies.
(3) We invent the dignified-sounding word *randomization* to describe what we have done. This term is, evidently, a euphemism, whose real meaning is: *deliberately throwing away relevant information when it becomes too complicated for us to handle.*

We have described this procedure in laconic terms, because an antidote is needed for the impression created by some writers on probability theory, who attach a kind of mystical significance to it. For some, declaring a problem to be 'randomized' is an incantation with the same purpose and effect as those uttered by an exorcist to drive out evil spirits; i.e. it cleanses their subsequent calculations and renders them immune to criticism. We agnostics often envy the True Believer, who thus acquires so easily that sense of security which is forever denied to us.

However, in defense of this procedure, we have to admit that it often leads to a useful approximation to the correct solution; i.e. the complicated details, while undeniably relevant in principle, might nevertheless have little numerical effect on the answers to certain particularly simple questions, such as the probability for drawing r red balls in n trials when n is sufficiently small. But from the standpoint of principle, an element of vagueness necessarily enters at this point; for, while we may feel intuitively that this leads to a good approximation, we have no proof of this, much less a reliable estimate of the accuracy of the approximation, which presumably improves with more shaking.

The vagueness is evident particularly in the fact that different people have widely divergent views about how much shaking is required to justify step (2). Witness the minor furor surrounding a US Government-sponsored and nationally televized game of chance some years ago, when someone objected that the procedure for drawing numbers from a fish bowl to determine the order of call-up of young men for Military Service was 'unfair' because the bowl hadn't been shaken enough to make the drawing 'truly random', whatever that means. Yet if anyone had asked the objector: 'To *whom* is it unfair?' he could not have given any answer except, 'To those whose numbers are on top; I don't know who they are.' But after any amount of further shaking, this will still be true! So what does the shaking accomplish?

Shaking does not make the result 'random', because that term is basically meaningless as an attribute of the real world; it has no clear definition applicable in the real world. The belief that 'randomness' is some kind of real property existing in Nature is a form of the mind projection fallacy which says, in effect, 'I don't know the detailed causes – *therefore* – Nature does not know them.' What shaking accomplishes is very different. It does not affect *Nature's* workings in any way; it only ensures that no *human* is able to exert any wilful influence on the result. Therefore, nobody can be charged with 'fixing' the outcome.

At this point, you may accuse us of nitpicking, because you know that after all this sermonizing, we are just going to go ahead and use the randomized solution like everybody else does. Note, however, that our objection is not to the procedure itself, provided that we acknowledge honestly what we are doing; i.e. instead of solving the real problem, we are making a practical compromise and being, of necessity, content with an approximate solution. That is something we have to do in all areas of applied mathematics, and there is no reason to expect probability theory to be any different.

Our objection is to the belief that by randomization we somehow make our subsequent equations exact; so exact that we can then subject our solution to all kinds of extreme conditions and believe the results, when applied to the real world. The most serious and most common error resulting from this belief is in the derivation of limit theorems (i.e. when sampling with replacement, nothing prevents us from passing to the limit $n \rightarrow \infty$ and obtaining the usual 'laws of large numbers'). If we do not recognize the approximate nature of our starting equations, we delude ourselves into believing that we have proved things (such as the identity of probability and limiting frequency) that are just not true in real repetitive experiments.

The danger here is particularly great because mathematicians generally regard these limit theorems as the most important and sophisticated fruits of probability theory, and have a tendency to use language which implies that they are proving properties of the real world. Our point is that these theorems are valid properties *of the abstract mathematical model that was defined and analyzed*. The issue is: to what extent does that model resemble the real world? It is probably safe to say that no limit theorem is directly applicable in the real world, simply because no mathematical model captures every circumstance that is relevant in the real world. Anyone who believes that he is proving things about the real world, is a victim of the mind projection fallacy.

Let us return to the equations. What answer can we now give to the question posed after Eq. (3.91)? The probability $P(R_2|R_1B')$ of drawing a red ball on the second draw clearly depends not only on N and M, but also on the fact that a red one has already been drawn and replaced. But this latter dependence is so complicated that we can't, in real life, take it into account; so we shake the urn to 'randomize' the problem, and then declare R_1 to be irrelevant: $P(R_2|R_1B') = P(R_2|B') = M/N$. After drawing and replacing the second ball, we again shake the urn, declare it 'randomized,' and set $P(R_3|R_2R_1B') = P(R_3|B') = M/N$, etc. In this approximation, the probability for drawing a red ball at *any* trial is M/N.

This is not just a repetition of what we learned in (3.37); what is new here is that the result now holds *whatever information the robot may have about what happened in the other trials*. This leads us to write the probability for drawing exactly r red balls in n trials, regardless of order, as

$$\binom{n}{r} \left(\frac{M}{N}\right)^r \left(\frac{N-M}{N}\right)^{n-r}, \qquad (3.92)$$

which is just the binomial distribution (3.86). Randomized sampling with replacement from an urn with finite N has approximately the same effect as passage to the limit $N \rightarrow \infty$ without replacement.

Evidently, for small n, this approximation will be quite good; but for large n these small errors can accumulate (depending on exactly how we shake the urn, etc.) to the point where (3.92) is misleading. Let us demonstrate this by a simple, but realistic, extension of the problem.

3.9 Correction for correlations

Suppose that, from an intricate logical analysis, drawing and replacing a red ball increases the probability for a red one at the next draw by some small amount $\epsilon > 0$, while drawing and replacing a white one decreases the probability for a red one at the next draw by a (possibly equal) small quantity $\delta > 0$; and that the influence of earlier draws than the last one is negligible compared with ϵ or δ. You may call this effect a small 'propensity' if you like; at least it expresses a physical causation that operates only forward in time. Then, letting C stand for all the above background information, including the statements just made about correlations and the information that we draw n balls, we have

$$P(R_k|R_{k-1}C) = p + \epsilon, \quad P(R_k|W_{k-1}C) = p - \delta,$$

$$P(W_k|R_{k-1}C) = 1 - p - \epsilon, \quad P(W_k|W_{k-1}C) = 1 - p + \delta,$$

(3.93)

where $p \equiv M/N$. From this, the probability for drawing r red and $(n - r)$ white balls in any specified order is easily seen to be

$$p(p + \epsilon)^c(p - \delta)^{c'}(1 - p + \delta)^w(1 - p - \epsilon)^{w'}$$

(3.94)

if the first draw is red; whereas, if the first is white, the first factor in (3.94) should be $(1 - p)$. Here, c is the number of red draws preceded by red ones, c' the number of red preceded by white, w the number of white draws preceded by white, and w' the number of white preceded by red. Evidently,

$$c + c' = \begin{bmatrix} r - 1 \\ r \end{bmatrix}, \quad w + w' = \begin{bmatrix} n - r \\ n - r - 1 \end{bmatrix},$$

(3.95)

the upper and lower cases holding when the first draw is red or white, respectively.

When r and $(n - r)$ are small, the presence of ϵ and δ in (3.94) makes little difference, and the equation reduces for all practical purposes to

$$p^r(1 - p)^{n-r},$$

(3.96)

as in the binomial distribution (3.92). But, as these numbers increase, we can use relations of the form

$$\left(1 + \frac{\epsilon}{p}\right)^c \simeq \exp\left\{\frac{\epsilon c}{p}\right\},$$

(3.97)

and (3.94) goes into

$$p^r(1 - p)^{n-r}\exp\left\{\frac{\epsilon c - \delta c'}{p} + \frac{\delta w - \epsilon w'}{1 - p}\right\}.$$

(3.98)

The probability for drawing r red and $(n - r)$ white balls now depends on the order in which red and white appear, and, for a given ϵ, when the numbers c, c', w, w' become sufficiently large, the probability can become arbitrarily large (or small) compared with (3.92).

We see this effect most clearly if we suppose that $N = 2M$, $p = 1/2$, in which case we will surely have $\epsilon = \delta$. The exponential factor in (3.98) then reduces to

$$\exp\left\{2\epsilon[(c - c') + (w - w')]\right\}.$$

(3.99)

This shows that (i) as the number n of draws tends to infinity, the probability for results containing 'long runs' (i.e. long strings of red (or white) balls in succession), becomes arbitrarily large compared with the value given by the 'randomized' approximation; (ii) this effect becomes appreciable when the numbers (ϵc), etc., become of order unity. Thus, if $\epsilon = 10^{-2}$, the randomized approximation can be trusted reasonably well as long as $n <$ 100; beyond that, we might delude ourselves by using it. Indeed, it is notorious that in real repetitive experiments where conditions appear to be the same at each trial, such runs – although extremely improbable on the randomized approximation – are nevertheless observed to happen.

Now let us note how the correlations expressed by (3.93) affect some of our previous calculations. The probabilities for the first draw are of course the same as (3.8); we now use the notation

$$p = P(R_1|C) = \frac{M}{N}, \qquad q = 1 - p = P(W_1|C) = \frac{N - M}{N}. \qquad (3.100)$$

But for the second trial we have instead of (3.35)

$$\begin{aligned}
P(R_2|C) &= P(R_2 R_1|C) + P(R_2 W_1|C) \\
&= P(R_2|R_1 C) P(R_1|C) + P(R_2|W_1 C) P(W_1|C) \\
&= (p + \epsilon)p + (p - \delta)q \\
&= p + (p\epsilon - q\delta),
\end{aligned} \qquad (3.101)$$

and continuing for the third trial

$$\begin{aligned}
P(R_3|C) &= P(R_3|R_2 C)P(R_2|C) + P(R_3|W_2 C)P(W_2|C) \\
&= (p + \epsilon)(p + p\epsilon - q\delta) + (p - \delta)(q - p\epsilon + q\delta) \qquad (3.102) \\
&= p + (1 + \epsilon + \delta)(p\epsilon - q\delta).
\end{aligned}$$

We see that $P(R_k|C)$ is no longer independent of k; the correlated probability distribution is no longer exchangeable. But does $P(R_k|C)$ approach some limit as $k \to \infty$?

It would be almost impossible to guess the general $P(R_k|C)$ by induction, following the method in (3.101) and (3.102) a few steps further. For this calculation we need a more powerful method. If we write the probabilities for the kth trial as a vector,

$$V_k \equiv \begin{bmatrix} P(R_k|C) \\ P(W_k|C) \end{bmatrix}, \qquad (3.103)$$

then (3.93) can be expressed in matrix form:

$$V_k = M V_{k-1}, \qquad (3.104)$$

with

$$M = \begin{pmatrix} [p + \epsilon] & [p - \delta] \\ [q - \epsilon] & [q + \delta] \end{pmatrix}. \qquad (3.105)$$

This defines a *Markov chain* of probabilities, and M is called the *transition matrix*. Now the slow induction of (3.101) and (3.102) proceeds instantly to any distance we please:

$$V_k = M^{k-1} V_1. \qquad (3.106)$$

So, to have the general solution, we need only to find the eigenvectors and eigenvalues of M. The characteristic polynomial is

$$C(\lambda) \equiv \det(M_{ij} - \lambda\delta_{ij}) = \lambda^2 - \lambda(1 + \epsilon + \delta) + (\epsilon + \delta) \qquad (3.107)$$

so the roots of $C(\lambda) = 0$ are the eigenvalues

$$\begin{aligned} \lambda_1 &= 1 \\ \lambda_2 &= \epsilon + \delta. \end{aligned} \qquad (3.108)$$

Now, for any 2×2 matrix

$$M = \begin{pmatrix} a & b \\ c & d \end{pmatrix} \qquad (3.109)$$

with an eigenvalue λ, the corresponding (non-normalized) right eigenvector is

$$x = (b\lambda - a), \qquad (3.110)$$

for which we have at once $Mx = \lambda x$. Therefore, our eigenvectors are

$$x_1 = \begin{pmatrix} p - \delta \\ q - \epsilon \end{pmatrix}, \qquad x_2 = \begin{pmatrix} 1 \\ -1 \end{pmatrix}. \qquad (3.111)$$

These are not orthogonal, since M is not a symmetric matrix. Nevertheless, if we use (3.111) to define the transformation matrix

$$S = \begin{pmatrix} [p - \delta] & 1 \\ [q - \epsilon] & -1 \end{pmatrix}, \qquad (3.112)$$

we find its inverse to be

$$S^{-1} = \frac{1}{1 - \epsilon - \delta} \begin{pmatrix} 1 & 1 \\ [q - \epsilon] & -[p - \delta] \end{pmatrix}, \qquad (3.113)$$

and we can verify by direct matrix multiplication that

$$S^{-1}MS = \Lambda = \begin{pmatrix} \lambda_1 & 0 \\ 0 & \lambda_2 \end{pmatrix}, \qquad (3.114)$$

where Λ is the diagonalized matrix. Then we have for any r, positive, negative, or even complex:

$$M^r = S\Lambda^r S^{-1} \qquad (3.115)$$

or

$$M^r = \frac{1}{1 - \epsilon - \delta} \begin{pmatrix} p - \delta + [\epsilon + \delta]^r [q - \epsilon] & [p - \delta][1 - (\epsilon + \delta)^r] \\ [q - \epsilon][1 - (\epsilon + \delta)^r] & q - \epsilon + [\epsilon + \delta]^r [p - \delta] \end{pmatrix}, \qquad (3.116)$$

and since

$$V_1 = \begin{pmatrix} p \\ q \end{pmatrix} \qquad (3.117)$$

the general solution (3.106) sought is

$$P(R_k|C) = \frac{(p - \delta) - (\epsilon + \delta)^{k-1}(p\epsilon - q\delta)}{1 - \epsilon - \delta}.$$

(3.118)

We can check that this agrees with (3.100), (3.101) and (3.102). From examining (3.118) it is clear why it would have been almost impossible to guess the general formula by induction. When $\epsilon = \delta = 0$, this reduces to $P(R_k|C) = p$, supplying the proof promised after Eq. (3.37).

Although we started this discussion by supposing that ϵ and δ were small and positive, we have not actually used that assumption, and so, whatever their values, the solution (3.118) is exact for the abstract model that we have defined. This enables us to include two interesting extreme cases. If not small, ϵ and δ must be at least bounded, because all quantities in (3.93) must be probabilities (i.e. in [0, 1]). This requires that

$$-p \le \epsilon \le q, \qquad -q \le \delta \le p,$$

(3.119)

or

$$-1 \le \epsilon + \delta \le 1.$$

(3.120)

But from (3.119), $\epsilon + \delta = 1$ if and only if $\epsilon = q, \delta = p$, in which case the transition matrix reduces to the unit matrix

$$M = \begin{pmatrix} 1 & 0 \\ 0 & 1 \end{pmatrix}$$

(3.121)

and there are no 'transitions'. This is a degenerate case in which the positive correlations are so strong that whatever color happens to be drawn on the first trial is certain to be drawn also on all succeeding ones:

$$P(R_k|C) = p, \qquad \text{all } k.$$

(3.122)

Likewise, if $\epsilon + \delta = -1$, then the transition matrix must be

$$M = \begin{pmatrix} 0 & 1 \\ 1 & 0 \end{pmatrix}$$

(3.123)

and we have nothing but transitions; i.e. the negative correlations are so strong that the colors are certain to alternate after the first draw:

$$P(R_k|C) = \begin{cases} p, & k \text{ odd} \\ q, & k \text{ even} \end{cases}.$$

(3.124)

This case is unrealistic because intuition tells us rather strongly that ϵ and δ should be positive quantities; surely, whatever the logical analysis used to assign the numerical value of ϵ, leaving a red ball in the top layer must *increase*, not decrease, the probability of red on the next draw. But if ϵ and δ must not be negative, then the lower bound in (3.120) is really zero, which is achieved only when $\epsilon = \delta = 0$. Then M in (3.105) becomes singular, and we revert to the binomial distribution case already discussed.

In the intermediate and realistic cases where $0 < |\epsilon + \delta| < 1$, the last term of (3.118) attenuates exponentially with k, and in the limit

$$P(R_k|C) \to \frac{p - \delta}{1 - \epsilon - \delta}. \qquad (3.125)$$

But although these single-trial probabilities settle down to steady values as in an exchangeable distribution, the underlying correlations are still at work and the limiting distribution is not exchangeable. To see this, let us consider the conditional probabilities $P(R_k|R_jC)$. These are found by noting that the Markov chain relation (3.104) holds whatever the vector V_{k-1}; i.e. whether or not it is the vector generated from V_1 as in (3.106). Therefore, if we are given that red occurred on the jth trial, then

$$V_j = \begin{pmatrix} 1 \\ 0 \end{pmatrix}, \qquad (3.126)$$

and we have from (3.104)

$$V_k = M^{k-j} V_j, \qquad j \leq k, \qquad (3.127)$$

from which, using (3.115),

$$P(R_k|R_jC) = \frac{(p - \delta) + (\epsilon + \delta)^{k-j}(q - \epsilon)}{1 - \epsilon - \delta}, \qquad j < k, \qquad (3.128)$$

which approaches the same limit (3.125). The forward inferences are about what we might expect; the steady value (3.125) plus a term that decays exponentially with distance. But the backward inferences are different; note that the general product rule holds, as always:

$$P(R_kR_j|C) = P(R_k|R_jC)P(R_j|C) = P(R_j|R_kC)P(R_k|C). \qquad (3.129)$$

Therefore, since we have seen that $P(R_k|C) \neq P(R_j|C)$, it follows that

$$P(R_j|R_kC) \neq P(R_k|R_jC). \qquad (3.130)$$

The backward inference is still possible, but it is no longer the same formula as the forward inference as it would be in an exchangeable sequence.

As we shall see later, this example is the simplest possible 'baby' version of a very common and important physical problem: an irreversible process in the 'Markovian approximation'. Another common technical language would call it an *autoregressive model* of first order. It can be generalized greatly to the case of matrices of arbitrary dimension and many-step or continuous, rather than single-step, memory influences. But for reasons noted earlier (confusion of inference and causality in the literature of statistical mechanics), the backward inference part of the solution is almost always missed. Some try to do backward inference by extrapolating the forward solution backward in time, with quite bizarre and unphysical results. Therefore the reader is, in effect, conducting new research in doing the following exercise.

Exercise 3.6. Find the explicit formula $P(R_j|R_kC)$ for the backward inference corresponding to the result (3.128) by using (3.118) and (3.129). (a) Explain the reason for the difference between forward and backward inferences in simple intuitive terms. (b) In what way does the backward inference differ from the forward inference extrapolated backward? Which is more reasonable intuitively? (c) Do backward inferences also decay to steady values? If so, is a property somewhat like exchangeability restored for events sufficiently separated? For example, if we consider only every tenth draw or every hundredth draw, do we approach an exchangeable distribution on this subset?

3.10 Simplification

The above formulas (3.100)–(3.130) hold for any ϵ, δ satisfying the inequalities (3.119). But, on surveying them, we note that a remarkable simplification occurs if they satisfy

$$p\epsilon = q\delta. \tag{3.131}$$

For then we have

$$\frac{p-\delta}{1-\epsilon-\delta} = p, \qquad \frac{q-\epsilon}{1-\epsilon-\delta} = q, \qquad \epsilon+\delta = \frac{\epsilon}{q}, \tag{3.132}$$

and our main results (3.118) and (3.128) collapse to

$$P(R_k|C) = p, \quad \text{all } k, \tag{3.133}$$

$$P(R_k|R_jC) = P(R_j|R_kC) = p + q\left(\frac{\epsilon}{q}\right)^{|k-j|}, \qquad \text{all } k, j. \tag{3.134}$$

The distribution is still not exchangeable, since the conditional probabilities (3.134) still depend on the separation $|k - j|$ of the trials; but the symmetry of forward and backward inferences is restored, even though the causal influences ϵ, δ operate only forward. Indeed, we see from our derivation of (3.40) that this forward–backward symmetry is a necessary consequence of (3.133), whether or not the distribution is exchangeable.

What is the meaning of this magic condition (3.131)? It does not make the matrix M assume any particularly simple form, and it does not turn off the effect of the correlations. What it does is to make the solution (3.133) invariant; that is, the initial vector (3.117) is then equal but for normalization to the eigenvector x_1 in (3.111), so the initial vector remains unchanged by the matrix (3.105).

In general, of course, there is no reason why this simplifying condition should hold. Yet in the case of our urn, we can see a kind of rationale for it. Suppose that when the urn has initially N balls, they are in L layers. Then, after withdrawing one ball, there are about $n = (N - 1)/L$ of them in the top layer, of which we expect about np to be red, $nq = n(1 - p)$ white. Now we toss the drawn ball back in. If it was red, the probability of getting red at the next draw if we do not shake the urn is about

$$\frac{np+1}{n+1} = p + \frac{1-p}{n} + O\left(\frac{1}{n^2}\right), \tag{3.135}$$

and if it is white the probability for getting white at the next draw is about

$$\frac{n(1-p)+1}{n+1} = 1 - p + \frac{p}{n} + O\left(\frac{1}{n^2}\right). \tag{3.136}$$

Comparing with (3.93) we see that we could estimate ϵ and δ by

$$\epsilon \simeq q/n, \qquad\qquad \delta \simeq p/n \tag{3.137}$$

whereupon our magic condition (3.131) is satisfied. Of course, the argument just given is too crude to be called a derivation, but at least it indicates that there is nothing inherently unreasonable about (3.131). We leave it for the reader to speculate about what significance and use this curious fact might have, and whether it generalizes beyond the Markovian approximation.

We have now had a first glimpse of some of the principles and pitfalls of standard sampling theory. All the results we have found will generalize greatly, and will be useful parts of our 'toolbox' for the applications to follow.

3.11 Comments

In most real physical experiments we are not, literally, drawing from any 'urn'. Nevertheless, the idea has turned out to be a useful conceptual device, and in the 250 years since Bernoulli's *Ars Conjectandi* it has appeared to scientists that many physical measurements are very much like 'drawing from Nature's urn'. But to some the word 'urn' has gruesome connotations, and in much of the literature one finds such expressions as 'drawing from a population'.

In a few cases, such as recording counts from a radioactive source, survey sampling, and industrial quality control testing, one is quite literally drawing from a real, finite population, and the urn analogy is particularly apt. Then the probability distributions just found, and their limiting forms and generalizations noted in Chapter 7, will be appropriate and useful. In some cases, such as agricultural experiments or testing the effectiveness of a new medical procedure, our credulity can be strained to the point where we see a vague resemblance to the urn problem.

In other cases, such as flipping a coin, making repeated measurements of the temperature and wind velocity, the position of a planet, the weight of a baby, or the price of a commodity, the urn analogy seems so farfetched as to be dangerously misleading. Yet in much of the literature one still uses urn distributions to represent the data probabilities, and tries to justify that choice by visualizing the experiment as drawing from some 'hypothetical infinite population' which is entirely a figment of our imagination. Functionally, the main consequence of this is strict independence of successive draws, regardless of all other circumstances. Obviously, this is not sound reasoning, and a price must be paid eventually in erroneous conclusions.

This kind of conceptualizing often leads one to suppose that these distributions represent not just our prior state of knowledge about the data, but the *actual* long-run variability of the data in such experiments. Clearly, such a belief cannot be justified; anyone who claims to know in advance the long-run results in an experiment that has not been performed is drawing on a vivid imagination, not on any fund of actual knowledge of the phenomenon. Indeed, if that infinite population is only imagined, then it seems that we are free to imagine any population we please.

From a mere act of the imagination we cannot learn anything about the real world. To suppose that the resulting probability assignments have any real physical meaning is just another form of the mind projection fallacy. In practice, this diverts our attention to irrelevancies and away from the things that really matter (such as information about the real world that is not expressible in terms of any sampling distribution, or does not fit into the urn picture, but which is nevertheless highly cogent for the inferences we want to make). Usually, the price paid for this folly is missed opportunities; had we recognized that information, more accurate and/or more reliable inferences could have been made.

Urn-type conceptualizing is capable of dealing with only the most primitive kind of information, and really sophisticated applications require us to develop principles that go far beyond the idea of urns. But the situation is quite subtle, because, as we stressed before in connection with Gödel's theorem, an erroneous argument does not necessarily lead to a wrong conclusion. In fact, as we shall find in Chapter 9, highly sophisticated calculations sometimes lead us back to urn-type distributions, for purely mathematical reasons that have nothing to do conceptually with urns or populations. The hypergeometric and binomial distributions found in this chapter will continue to reappear, because they have a fundamental mathematical status quite independent of arguments that we used to find them here.[2]

On the other hand, we could imagine a different problem in which we would have full confidence in urn-type reasoning leading to the binomial distribution, although it probably never arises in the real world. If we had a large supply $\{U_1, U_2, \ldots, U_n\}$ of urns known to have identical contents, and those contents are known with certainty in advance – and then we used a fresh new urn for each draw – then we would assign $P(A) = M/N$ for every draw, strictly independently of what we know about any other draw. Such prior information would take precedence over any amount of data. If we did not know the contents (M, N) of the urns – but we knew they all had identical contents – this strict independence would be lost, because then every draw from one urn would tell us something about the contents of the other urns, although it does not physically influence them.

From this we see once again that logical dependence is in general very different from causal physical dependence. We belabor this point so much because it is not recognized at all in most expositions of probability theory, and this has led to errors, as is suggested

---

[2] In a similar way, exponential functions appear in all parts of analysis because of their fundamental mathematical properties, although their conceptual basis varies widely.

by Exercise 3.6. In Chapter 4 we shall see a more serious error of this kind (see the discussion following Eq. (4.29)). But even when one manages to avoid actual error, to restrict probability theory to problems of physical causation is to lose its most important applications. The extent of this restriction – and the magnitude of the missed opportunity – does not seem to be realized by those who are victims of this fallacy.

Indeed, most of the problems we have solved in this chapter are not considered to be within the scope of probability theory, and do not appear at all in those expositions which regard probability as a physical phenomenon. Such a view restricts one to a small subclass of the problems which can be dealt with usefully by probability theory as logic. For example, in the 'physical probability' theory it is not even considered legitimate to speak of the probability for an outcome at a specified trial; yet that is exactly the kind of thing about which it is necessary to reason in conducting scientific inference. The calculations of this chapter have illustrated this many times.

In summary: in each of the applications to follow, one must consider whether the experiment is really 'like' drawing from an urn; if it is not, then we must go back to first principles and apply the basic product and sum rules in the new context. This may or may not yield the urn distributions.

3.11.1 A look ahead

The probability distributions found in this chapter are called *sampling distributions*, or *direct probabilities*, which indicate that they are of the following form: Given some hypothesis H about the phenomenon being observed (in the case just studied, the contents (M, N) of the urn), what is the probability that we shall obtain some specified data D (in this case, some sequence of red and white balls)? Historically, the term 'direct probability' has long had the additional connotation of reasoning from a supposed physical cause to an observable effect. But we have seen that not all sampling distributions can be so interpreted. In the present work we shall not use this term, but use 'sampling distribution' in the general sense of *reasoning from some specified hypothesis to potentially observable data*, whether the link between hypothesis and data is logical or causal.

Sampling distributions make predictions, such as the hypergeometric distribution (3.22), about potential observations (for example, the possible values and relative probabilities of different values of r). If the correct hypothesis is indeed known, then we expect the predictions to agree closely with the observations. If our hypothesis is not correct, they may be very different; then the nature of the discrepancy gives us a clue toward finding a better hypothesis. This is, very broadly stated, the basis for scientific inference. Just how wide the disagreement between prediction and observation must be in order to justify our rejecting the present hypothesis and seeking a new one, is the subject of *significance tests*. It was the need for such tests in astronomy that led Laplace and Gauss to study probability theory in the 18th and 19th centuries.

Although sampling theory plays a dominant role in conventional pedagogy, in the real world such problems are an almost negligible minority. In virtually all real problems of scientific inference we are in just the opposite situation; the data D are known but the correct hypothesis H is not. Then the problem facing the scientist is of the inverse type: Given the data D, what is the probability that some specified hypothesis H is true? Exercise 3.3 above was a simple introduction to this kind of problem. Indeed, the scientist's motivation for collecting data is usually to enable him to learn something about the phenomenon in this way.

Therefore, in the present work our attention will be directed almost exclusively to the methods for solving the inverse problem. This does not mean that we do not calculate sampling distributions; we need to do this constantly and it may be a major part of our computational job. But it does mean that for us the finding of a sampling distribution is almost never an end in itself.

Although the basic rules of probability theory solve such inverse problems just as readily as sampling problems, they have appeared quite different conceptually to many writers. A new feature seems present, because it is obvious that the question: 'What do you know about the hypothesis H after seeing the data D?' cannot have any defensible answer unless we take into account: 'What did you know about H before seeing D?' But this matter of previous knowledge did not figure in any of our sampling theory calculations. When we asked: 'What do you know about the data given the contents (M, N) of the urn?' we did not seem to consider: 'What did you know about the data before you knew (M, N)?'

This apparent dissymmetry, it will turn out, is more apparent than real; it arises mostly from some habits of notation that we have slipped into, which obscure the basic unity of all inference. But we shall need to understand this very well before we can use probability theory effectively for hypothesis tests and their special cases, significance tests. In the next chapter we turn to this problem.

4

Elementary hypothesis testing

> I conceive the mind as a moving thing, and arguments as the motive forces
> driving it in one direction or the other.
>
> *John Craig (1699)*

John Craig was a Scottish mathematician, and one of the first scholars to recognize the merit in Isaac Newton's new invention of 'the calculus'. The above sentence, written some 300 years ago in one of the early attempts to create a mathematical model of reasoning, requires changing by only one word in order to describe our present attitude. We would like to think that our minds are swayed not by arguments, but by evidence. And if fallible humans do not always achieve this objectivity, our desiderata were chosen with the aim of achieving it in our robot. Therefore to see how our robot's mind is 'driven in one direction or the other' by new evidence, we examine some applications that, although simple mathematically, have proved to have practical importance in several different fields.

As is clear from the basic desiderata listed in Chapter 1, the fundamental principle underlying all probabilistic inference is:

> *To form a judgment about the likely truth or falsity of any proposition A,*
> *the correct procedure is to calculate the probability that A is true:*
>
> $$P(A|E_1 E_2 \cdots) \tag{4.1}$$
>
> *conditional on all the evidence at hand.*

In a sampling context (i.e. when *A* stands for some data set), this principle has seemed obvious to everybody from the start. We used it implicitly throughout Chapter 3 without feeling any need to state it explicitly. But when we turn to a more general context, the principle needs to be stressed because it has not been obvious to all workers (as we shall see repeatedly in later chapters).

The essence of 'honesty' or 'objectivity' demands that we take into account all the evidence we have, not just some arbitrarily chosen subset of it. Any such choice would amount either to ignoring evidence that we have, or presuming evidence that we do not have. This leads us to recognize at the outset that some information is always available to the robot.

4.1 Prior probabilities

Generally, when we give the robot its current problem, we will give it also some new information or 'data' D pertaining to the specific matter at hand. But almost always the robot will have other information which we denote, for the time being, by X. This includes, at the very least, all its past experience, from the time it left the factory to the time it received its current problem. That is always part of the information available, and our desiderata do not allow the robot to ignore it. If we humans threw away what we knew yesterday in reasoning about our problems today, we would be below the level of wild animals; we could never know more than we can learn in one day, and education and civilization would be impossible.

So to our robot there is no such thing as an 'absolute' probability; all probabilities are necessarily conditional on X at least. In solving a problem, its inferences should, according to the principle (4.1), take the form of calculating probabilities of the form $P(A|DX)$. Usually, part of X is irrelevant to the current problem, in which case its presence is unnecessary but harmless; if it is irrelevant, it will cancel out mathematically. Indeed, that is what we really mean by 'irrelevant'.

Any probability $P(A|X)$ that is conditional on X alone is called a *prior probability*. But we caution that the term 'prior' is another of those terms from the distant past that can be inappropriate and misleading today. In the first place, it does not necessarily mean 'earlier in time'. Indeed, the very concept of time is not in our general theory (although we may of course introduce it in a particular problem). The distinction is a purely logical one; any additional information beyond the immediate data D of the current problem is by definition 'prior information'.

For example, it has happened more than once that a scientist has gathered a mass of data, but before getting around to the data analysis he receives some surprising new information that completely changes his ideas of how the data should be analyzed. That surprising new information is, logically, 'prior information' because it is not part of the data. Indeed, the separation of the totality of the evidence into two components called 'data' and 'prior information' is an arbitrary choice made by us, only for our convenience in organizing a chain of inferences. Although all such organizations must lead to the same final results if they succeed at all, some may lead to much easier calculations than others. Therefore, we do need to consider the order in which different pieces of information shall be taken into account in our calculations.

Because of some strange things that have been thought about prior probabilities in the past, we point out also that it would be a big mistake to think of X as standing for some hidden major premise, or some universally valid proposition about Nature. Old misconceptions about the origin, nature, and proper functional use of prior probabilities are still common among those who continue to use the archaic term '*a-priori probabilities*'. The term '*a-priori*' was introduced by Immanuel Kant to denote a proposition whose truth can be known independently of experience; which is most emphatically what we do *not* mean here. X denotes simply whatever additional information the robot has beyond what we have chosen to call 'the data'. Those who are actively familiar with the use of prior probabilities in current real problems usually abbreviate further, and instead of saying 'the prior probability' or 'the prior probability distribution', they say simply, 'the *prior*'.

There is no single universal rule for assigning priors – the conversion of verbal prior information into numerical prior probabilities is an open-ended problem of logical analysis, to which we shall return many times. At present, four fairly general principles are known – group invariance, maximum entropy, marginalization, and coding theory – which have led to successful solutions of many different kinds of problems. Undoubtedly, more principles are waiting to be discovered, which will open up new areas of application.

In conventional sampling theory, the only scenario considered is essentially that of 'drawing from an urn', and the only probabilities that arise are those that presuppose the contents of the 'urn' or the 'population' already known, and seek to predict what 'data' we are likely to get as a result. Problems of this type can become arbitrarily complicated in the details, and there is a highly developed mathematical literature dealing with them. For example, the massive two-volume work of Feller (1950, 1966) and the weighty compendium of Kendall and Stuart (1977) are restricted entirely to the calculation of sampling distributions. These works contain hundreds of nontrivial solutions that are useful in all parts of probability theory, and every worker in the field should be familiar with what is available in them.

However, as noted in the preceding chapter, almost all real problems of scientific inference involve us in the opposite situation; we already know the data D, and want probability theory to help us decide on the likely contents of the 'urn'. Stated more generally, we want probability theory to indicate which of a given set of hypotheses $\{H_1, H_2, \ldots\}$ is most likely to be true in the light of the data and any other evidence at hand. For example, the hypotheses may be various suppositions about the physical mechanism that is generating the data. But fundamentally, as in Chapter 3, physical causation is not an essential ingredient of the problem; what is essential is only that there be some kind of *logical* connection between the hypotheses and the data.

To solve this problem does not require any new principles beyond the product rule (3.1) that we used to find conditional sampling distributions; we need only to make a different choice of the propositions. Let us now use the notation

$$X = \text{prior information,}$$
$$H = \text{some hypothesis to be tested,}$$
$$D = \text{the data,}$$

and write the product rule in the form

$$P(DH|X) = P(D|HX)P(H|X) = P(H|DX)P(D|X). \tag{4.2}$$

We recognize $P(D|HX)$ as the sampling distribution which we studied in Chapter 3, but now written in a more flexible notation. In Chapter 3 we did not need to take any particular note of the prior information X, because all probabilities were conditional on H, and so we could suppose implicitly that the general verbal prior information defining the problem was included in H. This is the habit of notation that we have slipped into, which has obscured the unified nature of all inference. Throughout all of sampling theory one can get away with this, and as a result the very term 'prior information' is absent from the literature of sampling theory.

Now, however, we are advancing to probabilities that are not conditional on H, but are still conditional on X, so we need separate notations for them. We see from (4.2) that to judge the likely truth of H in the light of the data, we need not only the sampling probability $P(D|HX)$ but also the prior probabilities for D and H:

$$P(H|DX) = P(H|X)\frac{P(D|HX)}{P(D|X)}. \qquad (4.3)$$

Although the derivation (4.2)–(4.3) is only the same mathematical result as (3.50)–(3.51), it has appeared to many workers to have a different logical status. From the start it has seemed clear how one determines numerical values of sampling probabilities, but not what determines the prior probabilities. In the present work we shall see that this was only an artifact of an unsymmetrical way of formulating problems, which left them ill-posed. One could see clearly how to assign sampling probabilities because the hypothesis H was stated very specifically; had the prior information X been specified equally well, it would have been equally clear how to assign prior probabilities.

When we look at these problems on a sufficiently fundamental level and realize how careful one must be to specify the prior information before we have a well-posed problem, it becomes evident that there is in fact no logical difference between (3.51) and (4.3); exactly the same principles are needed to assign either sampling probabilities or prior probabilities, and one man's sampling probability is another man's prior probability.

The left-hand side of (4.3), $P(H|DX)$, is generally called a '*posterior probability*', with the same *caveat* that this means only 'logically later in the particular chain of inference being made', and not necessarily 'later in time'. And again the distinction is conventional, not fundamental; one man's prior probability is another man's posterior probability. There is really only one kind of probability; our different names for them refer only to a particular way of organizing a calculation.

The last factor in (4.3) also needs a name, and it is called the *likelihood L(H)*. To explain current usage, we may consider a fixed hypothesis and its implications for different data sets; as we have noted before, the term $P(D|HX)$, in its dependence on D for fixed H, is called the 'sampling distribution'. But we may consider a fixed data set in the light of various different hypotheses $\{H, H', \ldots\}$; in its dependence on H for fixed D, $P(D|HX)$ is called the 'likelihood'.

A likelihood $L(H)$ is not itself a probability for H; it is a dimensionless numerical function which, when multiplied by a prior probability and a normalization factor, may become a probability. Because of this, constant factors are irrelevant, and may be struck out. Thus, the quantity $L(H_i) = y(D) P(D|H_i X)$ is equally deserving to be called the likelihood, where y is any positive number which may depend on D but is independent of the hypotheses $\{H_i\}$.

Equation (4.3) is then the fundamental principle underlying a wide class of scientific inferences in which we try to draw conclusions from data. Whether we are trying to learn the character of a chemical bond from nuclear magnetic resonance data, the effectiveness of a medicine from clinical data, the structure of the earth's interior from seismic data, the elasticity of a demand from economic data, or the structure of a distant galaxy from telescopic data, (4.3) indicates what probabilities we need to find in order to see what conclusions are justified by the totality of our evidence. If $P(H|DX)$ is very close to one (zero), then we may conclude that H is very likely to be true (false) and act accordingly. But if $P(H|DX)$ is not far from $1/2$, then the robot is warning us that the available evidence is not sufficient to justify any very confident conclusion, and we need to obtain more and better evidence.

4.2 Testing binary hypotheses with binary data

The simplest nontrivial problem of hypothesis testing is the one where we have only two hypotheses to test and only two possible data values. Surprisingly, this turns out to be a realistic and valuable model of many important inference and decision problems. Firstly, let us adapt (4.3) to this binary case. It gives us the probability that H is true, but we could have written it equally well for the probability that H is false:

$$P(\overline{H}|DX) = P(\overline{H}|X)\frac{P(D|\overline{H}X)}{P(D|X)}, \qquad (4.4)$$

and if we take the ratio of the two equations,

$$\frac{P(H|DX)}{P(\overline{H}|DX)} = \frac{P(H|X)}{P(\overline{H}|X)}\frac{P(D|H\,X)}{P(D|\overline{H}X)}, \qquad (4.5)$$

the term $P(D|X)$ will drop out. This may not look like any particular advantage, but the quantity that we have here, the ratio of the probability that H is true to the probability that it is false, has a technical name. We call it the '*odds*' on the proposition H. So if we write the 'odds on H, given D and X', as the symbol

$$O(H|D\,X) \equiv \frac{P(H|D\,X)}{P(\overline{H}|DX)}, \qquad (4.6)$$

then we can combine (4.3) and (4.4) into the following form:

$$O(H|D\,X) = O(H|X)\frac{P(D|HX)}{P(D|\overline{H}X)}. \qquad (4.7)$$

The posterior odds on H is (are?) equal to the prior odds multiplied by a dimensionless factor, which is also called a likelihood ratio. The odds are (is?) a strict monotonic function of the probability, so we could equally well calculate this quantity.[1]

---

[1] Our uncertain phrasing here indicates that 'odds' is a grammatically slippery word. We are inclined to agree with purists who say that it is, like 'mathematics' and 'physics', a singular noun in spite of appearances. Yet the urge to follow the vernacular and treat it as plural is sometimes irresistible, and so we shall be knowingly inconsistent and use it both ways, judging what seems euphonious in each case.

In many applications it is convenient to take the logarithm of the odds because of the fact that we can then add up terms. Now we could take logarithms to any base we please, and this cost the writer some trouble. Our analytical expressions always look neater in terms of natural (base e) logarithms. But back in the 1940s and 1950s when this theory was first developed, we used base 10 logarithms because they were easier to find numerically; the four-figure tables would fit on a single page. Finding a natural logarithm was a tedious process, requiring leafing through enormous old volumes of tables.

Today, thanks to hand calculators, all such tables are obsolete and anyone can find a ten-digit natural logarithm just as easily as a base 10 logarithm. Therefore, we started happily to rewrite this section in terms of the aesthetically prettier natural logarithms. But the result taught us that there is another, even stronger, reason for using base 10 logarithms. Our minds are thoroughly conditioned to the base 10 number system, and base 10 logarithms have an immediate, clear intuitive meaning to all of us. However, we just don't know what to make of a conclusion that is stated in terms of natural logarithms, until it is translated back into base 10 terms. Therefore, we re-rewrote this discussion, reluctantly, back into the old, ugly base 10 convention.

We define a new function, which we will call the *evidence* for H given D and X:

$$e(H|DX) \equiv 10 \log_{10} O(H|DX). \tag{4.8}$$

This is still a monotonic function of the probability. By using the base 10 and putting the factor 10 in front, we are now measuring evidence in *decibels* (hereafter abbreviated to db). The evidence for H, given D, is equal to the prior evidence plus the number of db provided by working out the log likelihood in the last term below:

$$e(H|DX) = e(H|X) + 10 \log_{10} \left[\frac{P(D|HX)}{P(D|\overline{H}X)} \right]. \tag{4.9}$$

Now suppose that this new information D actually consisted of several different propositions:

$$D = D_1 D_2 D_3 \cdots. \tag{4.10}$$

Then we could expand the likelihood ratio by successive applications of the product rule:

$$e(H|DX) = e(H|X) + 10 \log_{10} \left[\frac{P(D_1|H\,X)}{P(D_1|\overline{H}X)} \right] + 10 \log_{10} \left[\frac{P(D_2|D_1\,H\,X)}{P(D_2|D_1\,\overline{H}X)} \right] + \cdots. \tag{4.11}$$

But, in many cases, the probability for getting D_2 is not influenced by knowledge of D_1:

$$P(D_2|D_1HX) = P(D_2|HX). \tag{4.12}$$

One then says conventionally that D_1 and D_2 are *independent*. Of course, we should really say that the *probabilities which the robot assigns to them* are independent. It is a semantic confusion to attribute the property of 'independence' to propositions or events; for that implies, in common language, physical *causal* independence. We are concerned here with the very different quality of *logical* independence.

To emphasize this, note that neither kind of independence implies the other. Two events may be in fact causally dependent (i.e. one influences the other); but for a scientist who has not yet discovered this, the probabilities representing his state of knowledge – which determine the only inferences he is able to make – might be independent. On the other hand, two events may be causally independent in the sense that neither exerts any causal influence on the other (for example, the apple crop and the peach crop); yet we perceive a logical connection between them, so that new information about one changes our state of knowledge about the other. Then for us their probabilities are not independent.

Quite generally, as the robot's state of knowledge represented by H and X changes, probabilities conditional on them may change from independent to dependent or *vice versa*; yet the real properties of the events remain the same. Then one who attributed the property of dependence or independence to the events would be, in effect, claiming for the robot the power of psychokinesis. We must be vigilant against this confusion between reality and a state of knowledge about reality, which we have called the 'mind projection fallacy'.

The point we are making is not just pedantic nitpicking; we shall see presently (Eq. (4.29)) that it has very real, substantive consequences. In Chapter 3 we have discussed some of the conditions under which these probabilities might be independent, in connection with sampling from a very large known population and sampling with replacement. In the closing Comments section, we noted that whether urn probabilities do or do not factor can depend on whether we do or do not know that the contents of several urns are the same. In our present problem, as in Chapter 3, to interpret causal independence as logical independence, or to interpret logical dependence as causal dependence, has led some to nonsensical conclusions in fields ranging from psychology to quantum theory.

In case these several pieces of data are logically independent given $(H\,X)$ and also given $(\overline{H}\,X)$, (4.11) becomes

$$e(H|DX) = e(H|X) + 10 \sum_i \log_{10} \left[\frac{P(D_i|HX)}{P(D_i|\overline{H}X)} \right], \tag{4.13}$$

where the sum is over all the extra pieces of information that we obtain.

To get some feeling for numerical values here, let us construct Table 4.1. We have three different scales on which we can measure degrees of plausibility: evidence, odds, or probability; they are all monotonic functions of each other. Zero db of evidence corresponds to odds of 1 or to a probability of 1/2. Now, every physicist or electrical engineer knows that 3 db means a factor of 2 (nearly) and 10 db is a factor of 10 (exactly); and so if we go in steps of 3 db, or 10, we can construct this table very easily.

It is obvious from Table 4.1 why it is very cogent to give evidence in decibels. When probabilities approach one or zero, our intuition doesn't work very well. Does the difference between the probability of 0.999 and 0.9999 mean a great deal to you? It certainly doesn't to the writer. But after living with this for only a short while, the difference between evidence of plus 30 db and plus 40 db does have a clear meaning to us. It is now in a scale which our minds comprehend naturally. This is just another example of the Weber–Fechner law; intuitive human sensations tend to be logarithmic functions of the stimulus.

Table 4.1. *Evidence, odds, and probability.*

e	O	p
0	1:1	1/2
3	2:1	2/3
6	4:1	4/5
10	10:1	10/11
20	100:1	100/101
30	1000:1	0.999
40	10^4:1	0.9999
$-e$	$1/O$	$1-p$

Even the factor of 10 in (4.8) is appropriate. In the original acoustical applications, it was introduced so that a 1 db change in sound intensity would be, psychologically, about the smallest change perceptible to our ears. With a little familiarity and a little introspection, we think that the reader will agree that a 1 db change in evidence is about the smallest increment of plausibility that is perceptible to our intuition. Nobody claims that the Weber–Fechner law is a precise rule for all human sensations, but its general usefulness and appropriateness is clear; almost always it is not the absolute change, but more nearly the relative change, in some stimulus that we perceive. For an interesting account of the life and work of Gustav Theodor Fechner (1801–87), see Stigler (1986c).

Now let us apply (4.13) to a specific calculation, which we shall describe as a problem of industrial quality control (although it could be phrased equally well as a problem of cryptography, chemical analysis, interpretation of a physics experiment, judging two economic theories, etc.). Following the example of Good (1950), we assume numbers which are not very realistic in order to elucidate some points of principle. Let the prior information X consist of the following statements:

$X \equiv$ We have 11 automatic machines turning out widgets, which pour out of the machines into 11 boxes. This example corresponds to a very early stage in the development of widgets, because ten of the machines produce one in six defective. The 11th machine is even worse; it makes one in three defective. The output of each machine has been collected in an unlabeled box and stored in the warehouse.

We choose one of the boxes and test a few of the widgets, classifying them as 'good' or 'bad'. Our job is to decide whether we chose a box from the bad machine or not; that is, whether we are going to accept this batch or reject it.

Let us turn this job over to our robot and see how it performs. Firstly, it must find the prior evidence for the various propositions of interest. Let

$A \equiv$ we chose a bad batch (1/3 defective),
$B \equiv$ we chose a good batch (1/6 defective).

The qualitative part of our prior information X told us that there are only two possibilities; so in the 'logical environment' generated by X, these propositions are related by negation: given X, we can say that

$$\overline{A} = B, \qquad \overline{B} = A. \tag{4.14}$$

The only quantitative prior information is that there are 11 machines and we do not know which one made our batch, so, by the principle of indifference, $P(A|X) = 1/11$, and

$$e(A|X) = 10\log_{10}\frac{P(A|X)}{P(\overline{A}|X)} = 10\log_{10}\frac{(1/11)}{(10/11)} = -10\,\text{db}, \tag{4.15}$$

whereupon we have necessarily $e(B|X) = +10\,\text{db}$.

Evidently, in this problem the only properties of X that will be relevant for the calculation are just these numbers, ± 10 db. Any other kind of prior information which led to the same numbers would give us just the same mathematical problem from this point on. So, it is not necessary to say that we are talking only about a problem where there are 11 machines, and so on. There might be only one machine, and the prior information consists of our previous experience with it.

Our reason for stating the problem in terms of 11 machines was that we have, thus far, only one principle, indifference, by which we can convert raw information into numerical probability assignments. We interject this remark because of a famous statement by Feller (1950) about a single machine, which we consider in Chapter 17 after accumulating some more evidence pertaining to the issue he raised. To our robot, it makes no difference how many machines there are; the only thing that matters is the prior probability for a bad batch, however this information was arrived at.[2]

Now, from this box we take out a widget and test it to see whether it is defective. If we pull out a bad one, what will that do to the evidence for a bad batch? That will add to it

$$10\log_{10}\frac{P(\text{bad}|A\,X)}{P(\text{bad}|\overline{A}X)}\,\text{db} \tag{4.16}$$

where $P(\text{bad}|AX)$ represents the probability for getting a bad widget, given A, etc.; these are sampling probabilities, and we have already seen how to calculate them. Our procedure is very much 'like' drawing from an urn, and, as in Chapter 3, on one draw our datum D now consists only of a binary choice: (good/bad). The sampling distribution $P(D|HX)$

---

[2] Notice that in this observation we have the answer to a point raised in Chapter 1: How does one make the robot 'cognizant' of the semantic meanings of the various propositions that it is being called upon to deal with? The answer is that the robot does not need to be 'cognizant' of anything. If we give it, in addition to the model and the data, a list of the propositions to be considered, with their prior probabilities, this conveys all the 'meaning' needed to define the robot's mathematical problem for the applications now being considered. Later, we shall wish to design a more sophisticated robot which can also help us to assign prior probabilities by analysis of complicated but incomplete information, by the maximum entropy principle. But, even then, we can always define the robot's mathematical problem without going into semantics.

reduces to

$$P(\text{bad}|AX) = \frac{1}{3}, \qquad P(\text{good}|AX) = \frac{2}{3}, \qquad (4.17)$$

$$P(\text{bad}|BX) = \frac{1}{6}, \qquad P(\text{good}|BX) = \frac{5}{6}. \qquad (4.18)$$

Thus, if we find a bad widget on the first draw, this will increase the evidence for A by

$$10 \log_{10} \frac{(1/3)}{(1/6)} = 10 \log_{10} 2 = 3 \text{ db}. \qquad (4.19)$$

What happens now if we draw a second bad one? We are sampling without replacement, so as we noted in (3.11), the factor $(1/3)$ in (4.19) should be updated to

$$\frac{(N/3) - 1}{N - 1} = \frac{1}{3} - \frac{2}{3(N-1)}, \qquad (4.20)$$

where N is the number of widgets in the batch. But, to avoid this complication, we suppose that N is very much larger than any number that we contemplate testing; i.e. we are going to test such a negligible fraction of the batch that the proportion of bad and good ones in it is not changed appreciably by the drawing. Then the limiting form of the hypergeometric distribution (3.22) will apply, namely the binomial distribution (3.86). Thus we shall consider that, given A or B, the probability for drawing a bad widget is the same at every draw regardless of what has been drawn previously; so every bad one we draw will provide $+3$ db of evidence in favor of hypothesis A.

Now suppose we find a good widget. Using (4.14), we get evidence for A of

$$10 \log_{10} \frac{P(\text{good}|AX)}{P(\text{good}|BX)} = 10 \log_{10} \frac{(2/3)}{(5/6)} = -0.97 \text{ db}, \qquad (4.21)$$

but let's call it -1 db. Again, this will hold for any draw, if the number in the batch is sufficiently large. If we have inspected n widgets, of which we found n_b bad ones and n_g good ones, the evidence that we have the bad batch will be

$$e(A|DX) = e(A|X) + 3n_b - n_g. \qquad (4.22)$$

You see how easy this is to do once we have set up the logarithmic machinery. The robot's mind is 'driven in one direction or the other' in a very simple, direct way.

Perhaps this result gives us a deeper insight into why the Weber–Fechner law applies to intuitive plausible inference. Our 'evidence' function is related to the data that we have observed in about the most natural way imaginable; a given increment of evidence corresponds to a given increment of data. For example, if the first 12 widgets we test yield five bad ones, then

$$e(A|DX) = -10 + 3 \times 5 - 7 = -2 \text{ db}, \qquad (4.23)$$

or, the probability for a bad batch is raised by the data from $(1/11) = 0.09$ to $P(A|DX) \simeq 0.4$.

In order to get at least 20 db of evidence for proposition A, how many bad widgets would we have to find in a certain sequence of $n = n_b + n_g$ tests? This requires

$$3n_b - n_g = 4n_b - n = n(4f_b - 1) \geq 20, \tag{4.24}$$

so, if the fraction $f_b \equiv n_b/n$ of bad ones remains greater than 1/4, we shall accumulate eventually 20 db, or any other positive amount, of evidence for A. It appears that $f_b = 1/4$ is the threshold value at which the test can provide no evidence for either A or B over the other; but note that the $+3$ and -1 in (4.22) are only approximate. The exact threshold fraction of bad ones is, from (4.19) and (4.21),

$$f_t = \frac{\log\left(\frac{5}{4}\right)}{\log(2) + \log\left(\frac{5}{4}\right)} = 0.2435292, \tag{4.25}$$

in which the base of the logarithms does not matter. Sampling fractions greater (less) than this give evidence for A over B (B over A); but if the observed fraction is close to the threshold, it will require many tests to accumulate enough evidence.

Now all we have here is the probability or odds or evidence, whatever you wish to call it, of the proposition that we chose the bad batch. Eventually, we have to make a decision: we're going to accept it, or we're going to reject it. How are we going to do that? Well, we might decide beforehand: if the probability of proposition A reaches a certain level, then we'll decide that A is true. If it gets down to a certain value, then we'll decide that A is false.

There is nothing in probability theory *per se* which can tell us where to put these critical levels at which we make our decision. This has to be based on value judgments: what are the consequences of making wrong decisions, and what are the costs of making further tests? This takes us into the realm of decision theory, considered in Chapters 13 and 14. But for now it is clear that making one kind of error (accepting a bad batch) might be more serious than making the other kind of error (rejecting a good batch). That would have an obvious effect on where we place our critical levels.

So we could give the robot some instructions such as 'If the evidence for A is greater than $+0$ db, then reject this batch (it is more likely to be bad than good). If it goes as low as -13 db, then accept it (there is at least a 95% probability that it is good). Otherwise, continue testing.' We start doing the tests, and every time we find a bad widget the evidence for the bad batch goes up 3 db; every time we find a good one, it goes down 1 db. The tests terminate as soon as we enter either the accept or reject region for the first time.

The way described above is how our robot would do it if we told it to reject or accept on the basis that the *posterior probability* of proposition A reaches a certain level. This very useful and powerful procedure is called 'sequential inference' in the statistical literature, the term signifying that the number of tests is not determined in advance, but depends on the sequence of data values that we find; at each step in the sequence we make one of three choices: (a) stop with acceptance; (b) stop with rejection; (c) make another test. The term should not be confused with what has come to be called 'sequential analysis with nonoptional stopping', which is a serious misapplication of probability theory; see the discussions of optional stopping in Chapters 6 and 17.

4.3 Nonextensibility beyond the binary case

The binary hypothesis testing problem turned out to have such a beautifully simple solution that we might like to extend it to the case of more than two hypotheses. Unfortunately, the convenient independent additivity over data sets in (4.13) and the linearity in (4.22) do not generalize. By 'independent additivity' we mean that the increment of evidence from a given datum D_i depends only on D_i and H; not on what other data have been observed. As (4.11) shows, we always have additivity, but not independent additivity unless the probabilities are independent.

We state the reason for this nonextensibility in the form of an exercise for the reader; to prepare for it, suppose that we have n hypotheses $\{H_1, \ldots, H_n\}$ which on prior information X are mutually exclusive and exhaustive:

$$P(H_i H_j | X) = P(H_i | X)\delta_{ij}, \qquad \sum_{i=1}^{n} P(H_i | X) = 1. \qquad (4.26)$$

Also, we have acquired m data sets $\{D_1, \ldots, D_m\}$, and as a result the probabilities of the H_i become updated in odds form by (4.7), which now becomes

$$O(H_i | D_1, \ldots, D_m X) = O(H_i | X)\frac{P(D_1, \ldots, D_m | H_i X)}{P(D_1, \ldots, D_m | \overline{H}_i X)}. \qquad (4.27)$$

It is common that the numerator will factor because of the logical independence of the D_j, given H_i:

$$P(D_1, \ldots, D_m | H_i X) = \prod_j P(D_j | H_i X), \quad 1 \le i \le n. \qquad (4.28)$$

If the denominator should also factor,

$$P(D_1, \ldots, D_m | \overline{H}_i X) = \prod_j P(D_j | \overline{H}_i X), \quad 1 \le i \le n, \qquad (4.29)$$

then (4.27) would split into a product of the updates produced by each D_j separately, and the log-odds formula (4.9) would again take a form independently additive over the D_j as in (4.13).

Exercise 4.1. Show that there is no such nontrivial extension of the binary case. More specifically, prove that if (4.28) and (4.29) hold with $n > 2$, then at most one of the factors

$$\frac{P(D_1 | H_i X)}{P(D_1 | \overline{H}_i X)} \cdots \frac{P(D_m | H_i X)}{P(D_m | \overline{H}_i X)} \qquad (4.30)$$

is different from unity, therefore at most one of the data sets D_j can produce any updating of the probability for H_i.

This has been a controversial issue in the literature of artificial intelligence (Glymour, 1985; R. W. Johnson, 1985). Those who fail to distinguish between logical independence and causal independence would suppose that (4.29) is always valid, provided only that no D_i exerts a physical influence on any other D_j. But we have already noted the folly of such reasoning; this is an occasion when the semantic confusion can lead to serious numerical errors. When $n = 2$, (4.29) follows from (4.28). But when $n > 2$, (4.29) is such a strong condition that it would reduce the whole problem to a triviality not worth considering; we have left it (Exercise 4.1) for the reader to examine the equations to see why this is so. Because of Cox's theorems expounded in Chapter 2, the verdict of probability theory is that our conclusion about nonextensibility can be evaded only at the price of committing demonstrable inconsistencies in our reasoning.

To head off a possible misunderstanding of what is being said here, let us add the following. However many hypotheses we have in mind, it is of course always possible to pick out two of them and compare them only against each other. This reverts to the binary choice case already analyzed, and the independent additive property holds within that smaller problem (find the status of an hypothesis relative to a single alternative).

We could organize this by choosing A_1 as the standard 'null hypothesis' and comparing each of the others with it by solving $n - 1$ binary problems; whereupon the relative status of any two propositions is determined. For example, if A_5 and A_7 are favored over A_1 by 22.3 db and 31.9 db, respectively, then A_7 is favored over A_5 by $31.9 - 22.3 = 9.6$ db. If such binary comparisons provide all the information one wants, there is no need to consider multiple hypothesis testing at all.

But that would not solve our present problem; given the solutions of all these binary problems, it would still require a calculation as big as the one we are about to do to convert that information into the absolute status of any given hypothesis relative to the entire class of n hypotheses. Here we are going after the solution of the larger problem directly.

In any event, we need not base our stance merely on claims of authoritarian finality for an abstract theorem; more constructively, we now show that probability theory does lead us to a definite, useful procedure for multiple hypothesis testing, which gives us a much deeper insight and makes it clear why the independent additivity cannot, *and should not*, hold when $n > 2$. It would then ignore some very cogent information; that is the demonstrable inconsistency.

4.4 Multiple hypothesis testing

Suppose that something very remarkable happens in the sequential test just discussed: we tested 50 widgets and every one turned out to be bad. According to (4.22), that would give us 150 db of evidence for the proposition that we had the bad batch. $e(A|E)$ would end up at $+140$ db, which is a probability which differs from unity by one part in 10^{14}. Now, our common sense rejects this conclusion; some kind of innate skepticism rises in us. If you test 50 widgets and you find that all 50 are bad, you are not willing to believe that you have a batch in which only one in three are really bad. So what went wrong here? Why doesn't our robot work in this case?

We have to recognize that our robot is immature; it reasons like a four-year-old child does. The remarkable thing about small children is that you can tell them the most ridiculous things and they will accept it all with wide open eyes, open mouth, and it never occurs to them to question you. They will believe anything you tell them.

Adults learn to make mental allowance for the reliability of the source when told something hard to believe. One might think that, ideally, the information which our robot should have put into its memory was not that we had either 1/3 bad or 1/6 bad; the information it should have put in was that some unreliable human *said* that we had either 1/3 bad or 1/6 bad.

More generally, it might be useful in many problems if the robot could take into account the fact that the information it has been given may not be perfectly reliable to begin with. There is always a small chance that the prior information or data that we fed to the robot was wrong. In a real problem there are always hundreds of possibilities, and if you start out the robot with dogmatic initial statements which say that there are only two possibilities, then of course you must not expect its conclusions to make sense in every case.

To accomplish this skeptically mature behavior automatically in a robot is something that we can do, when we come to consider significance tests; but fortunately, after further reflection, we realize that for most problems the present immature robot is what we want after all, because we have better control over it.

We *do* want the robot to believe whatever we tell it; it would be dangerous to have a robot who suddenly became skeptical in a way not under our control when we tried to tell it some true but startling – and therefore highly important – new fact. But then the onus is on us to be aware of this situation, and when there is a good chance that skepticism will be needed, it is up to us to give the robot a hint about how to be skeptical for that particular problem.

In the present problem we can give the hint which makes the robot skeptical about *A* when it sees 'too many' bad widgets, by providing it with one more possible hypothesis, which notes that possibility and therefore, in effect, puts the robot on the lookout for it. As before, let proposition *A* mean that we have a box with 1/3 defective, and proposition *B* is the statement that we have a box with 1/6 bad. We add a third proposition, *C*, that something went entirely wrong with the machine that made our widgets, and it is turning out 99% defective.

Now we have to adjust our prior probabilities to take this new possibility into account. But we do not want this to be a major change in the nature of the problem; so let hypothesis *C* have a very low prior probability $P(C|X)$ of 10^{-6} (-60 db). We could write out *X* as a verbal statement which would imply this, but as in the previous footnote we can state what proposition *X* is, with no ambiguity at all for the robot's purposes, simply by giving it the probabilities conditional on *X*, of all the propositions that we're going to use in this problem. In that way we don't state everything about *X* that is important to us conceptually; but we state everything about *X* that is relevant to the robot's current mathematical problem.

So, suppose we start out with these initial probabilities:

$$P(A|X) = \frac{1}{11}(1 - 10^{-6}),$$

$$P(B|X) = \frac{10}{11}(1 - 10^{-6}), \qquad (4.31)$$

$$P(C|X) = 10^{-6},$$

where

$A \equiv$ we have a box with 1/3 defective,
$B \equiv$ we have a box with 1/6 defective,
$C \equiv$ we have a box with 99/100 defective.

The factors $(1 - 10^{-6})$ are practically negligible, and for all practical purposes we will start out with the initial values of evidence:

$$-10 \text{ db for } A,$$
$$+10 \text{ db for } B, \qquad (4.32)$$
$$-60 \text{ db for } C.$$

The data proposition D stands for the statement that 'm widgets were tested and every one was defective'. Now, from (4.9), the posterior evidence for proposition C is equal to the prior evidence plus ten times the logarithm of this probability ratio:

$$e(C|DX) = e(C|X) + 10\log_{10}\frac{P(D|CX)}{P(D|\overline{C}X)}. \qquad (4.33)$$

Our discussion of sampling with and without replacement in Chapter 3 shows that

$$P(D|CX) = \left(\frac{99}{100}\right)^m \qquad (4.34)$$

is the probability that the first m are all bad, given that 99% of the machine's output is bad, under our assumption that the total number in the box is large compared with the number m tested.

We also need the probability $P(D|\overline{C}X)$, which we can evaluate by two applications of the product rule (4.3):

$$P(D|\overline{C}X) = P(D|X)\frac{P(\overline{C}|DX)}{P(\overline{C}|X)}. \qquad (4.35)$$

In this problem, the prior information states dogmatically that there are only three possibilities, and so the statement $\overline{C} \equiv$ 'C is false' implies that either A or B must be true:

$$P(\overline{C}|DX) = P(A + B|DX) = P(A|DX) + P(B|DX), \qquad (4.36)$$

where we used the general sum rule (2.66), the negative term dropping out because A and B are mutually exclusive. Similarly,

$$P(\overline{C}|X) = P(A|X) + P(B|X). \tag{4.37}$$

Now, if we substitute (4.36) into (4.35), the product rule will be applicable again in the form

$$P(AD|X) = P(D|X)P(A|DX) = P(A|X)P(D|AX)$$
$$P(BD|X) = P(D|X)P(B|DX) = P(B|X)P(D|BX), \tag{4.38}$$

and so (4.35) becomes

$$P(D|\overline{C}X) = \frac{P(D|AX)P(A|X) + P(D|BX)P(B|X)}{P(A|X) + P(B|X)}, \tag{4.39}$$

in which all probabilities are known from the statement of the problem.

4.4.1 Digression on another derivation

Although we have the desired result (4.39), let us note that there is another way of deriving it, which is often easier than direct application of (4.3). The principle was introduced in our derivation of (3.33): resolve the proposition whose probability is desired (in this case D) into mutually exclusive propositions, and calculate the sum of their probabilities. We can carry out this resolution in many different ways by 'introducing into the conversation' any set of mutually exclusive and exhaustive propositions $\{P, Q, R, \ldots\}$ and using the rules of Boolean algebra:

$$D = D(P + Q + R + \cdots) = DP + DQ + DR + \cdots. \tag{4.40}$$

But the success of the method depends on our cleverness at choosing a particular set for which we can complete the calculation. This means that the propositions introduced must have a known kind of relevance to the question being asked; the example of penguins at the end of Chapter 2 will not be helpful if that question has nothing to do with penguins.

In the present case, for evaluation of $P(D|\overline{C}X)$, it appears that propositions A and B have this kind of relevance. Again, we note that proposition $\overline{C}$ implies $(A + B)$; and so

$$P(D|\overline{C}X) = P(D(A + B)|\overline{C}X) = P(DA + DB|\overline{C}X)$$
$$= P(DA|\overline{C}X) + P(DB|\overline{C}X). \tag{4.41}$$

These probabilities can be factored by the product rule:

$$P(D|\overline{C}X) = P(D|A\overline{C}X)P(A|\overline{C}X) + P(D|B\overline{C}X)P(B|\overline{C}X). \tag{4.42}$$

But we can abbreviate: $P(D|A\overline{C}X) \equiv P(D|AX)$ and $P(D|B\overline{C}X) \equiv P(D|BX)$, because, in the way we set up this problem, the statement that either A or B is true implies that C must be false. For this same reason, $P(\overline{C}|AX) = 1$, and so, by the product rule,

$$P(A|\overline{C}X) = \frac{P(A|X)}{P(\overline{C}|X)}, \tag{4.43}$$

and similarly for $P(B|\overline{C}X)$. Substituting these results into (4.42) and using (4.37), we again arrive at (4.39). This agreement provides another illustration – and a rather severe test – of the consistency of our rules for extended logic.

Returning to (4.39), we have the numerical value

$$P(D|\overline{C}X) = \left(\frac{1}{3}\right)^m \left(\frac{1}{11}\right) + \left(\frac{1}{6}\right)^m \frac{10}{11}, \tag{4.44}$$

and everything in (4.33) is now at hand. If we put all these things together, we find that the evidence for proposition C is:

$$e(C|DX) = -60 + 10\log_{10}\left[\frac{\left(\frac{99}{100}\right)^m}{\frac{1}{11}\left(\frac{1}{3}\right)^m + \frac{10}{11}\left(\frac{1}{6}\right)^m}\right]. \tag{4.45}$$

If $m > 5$, a good approximation is

$$e(C|DX) \simeq -49.6 + 4.73\,m, \qquad m > 5, \tag{4.46}$$

and if $m < 3$, a crude approximation is

$$e(C|DX) \simeq -60 + 7.73\,m, \qquad m < 3. \tag{4.47}$$

Proposition C starts out at -60 db, and the first few bad widgets we find will each give about 7.73 db of evidence in favor of C, so the graph of $e(C|DX)$ vs. m will start upward at a slope of 7.73. But then the slope drops, when $m > 5$, to 4.73. The evidence for C reaches 0 db when $m \simeq 49.6/4.73 = 10.5$. So, ten consecutive bad widgets would be enough to raise this initially very improbable hypothesis by 58 db, to the place where the robot is ready to consider it very seriously; and 11 consecutive bad ones would take it over the threshold, to where the robot considers it more likely to be true than false.

In the meantime, what is happening to our propositions A and B? As before, A starts off at -10 db, B starts off at $+10$ db, and the plausibility for A starts going up 3 db per defective widget. But after we've found too many bad ones, that skepticism would set in, and you and I would begin to doubt whether the evidence really supports proposition A after all; proposition C is becoming a much easier way to explain what is observed. Has the robot also learned to be skeptical?

After m widgets have been tested, and all proved to be bad, the evidence for propositions A and B, and the approximate forms, are as follows:

$$e(A|DX) = -10 + 10\log_{10}\left[\frac{\left(\frac{1}{3}\right)^m}{\left(\frac{1}{6}\right)^m + \frac{11}{10} \times 10^{-6}\left(\frac{99}{100}\right)^m}\right]$$

$$\simeq \begin{cases} -10 + 3m & \text{for } m < 7 \\ +49.6 - 4.73m & \text{for } m > 8 \end{cases} \tag{4.48}$$

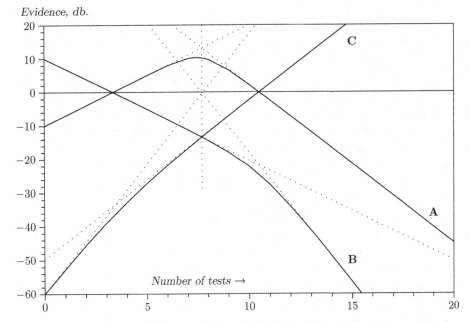

Fig. 4.1. A surprising multiple sequential test wherein a dead hypothesis (C) is resurrected.

$$e(B|DX) = +10 + 10\log_{10}\left[\frac{\left(\frac{1}{6}\right)^m}{\left(\frac{1}{3}\right)^m + 11 \times 10^{-6}\left(\frac{99}{100}\right)^m}\right]$$

(4.49)

$$\simeq \begin{cases} 10 - 3m & \text{for } m < 10 \\ 59.6 - 7.33m & \text{for } m > 11 \end{cases}.$$

The exact results are summarized in Figure 4.1. We can learn quite a lot about multiple hypothesis testing from studying this diagram. The initial straight line part of the A and B curves represents the solution as we found it before we introduced proposition C; the change in plausibility for propositions A and B starts off just the same as in the previous problem. The effect of proposition C does not appear until we have reached the place where C crosses B. At this point, suddenly the character of the A curve changes; instead of going on up, at $m = 7$ it has reached its highest value of 10 db. Then it turns around and comes back down; the robot has indeed learned how to become skeptical. But the B curve does *not* change at this point; it continues on linearly until it reaches the place where A and C have the same plausibility, and at this point it has a change in slope. From then on, it falls off more rapidly.

Most people find all this surprising and mysterious at first glance; but then a little meditation is enough to make us perceive what is happening and why. The change in plausibility for A due to one more test arises from the fact that we are now testing hypothesis A against two alternatives: B and C. But, initially, B is so much more plausible than C, that for all practical purposes we are simply testing A against B, and reproducing our previous solution (4.22). After enough evidence has accumulated to bring the plausibility for C up to the same level as B, then from that point on A is essentially being tested against C instead of B, which is a very different situation.

All of these changes in slope can be interpreted in this way. Once we see this principle, it is clear that the same thing is going to be true more generally. As long as we have a discrete set of hypotheses, a change in plausibility for any one of them will be approximately the result of a test of this hypothesis against a single alternative – the single alternative being that one of the remaining hypotheses which is most plausible at that time. As the relative plausibilities of the alternatives change, the slope of the A curve must also change; *this is the cogent information that would be lost* if we tried to retain the independent additive form (4.13) when $n > 2$.

Whenever the hypotheses are separated by about 10 db or more, then multiple hypothesis testing reduces approximately to testing each hypothesis against a single alternative. So, seeing this, you can construct curves of the sort shown in Fig. 4.1 very rapidly without even writing down the equations, because what would happen in the two-hypothesis case is easily seen once and for all. The diagram has a number of other interesting geometrical properties, suggested by drawing the six asymptotes and noting their vertical alignment (dotted lines), which we leave for the reader to explore.

All the information needed to construct fairly accurate charts resulting from any sequence of good and bad tests is contained in the 'plausibility flow diagrams' of Figure 4.2, which summarize the solutions of all those binary problems; every possible way to test one proposition against a single alternative. It indicates, for example, that finding a good widget raises the evidence for B by 1 db if B is being tested against A, and by 19.22 db if it is being tested against C. Similarly, finding a bad widget raises the evidence for A by 3 db if A is being tested against B, but lowers it by 4.73 db if it is being tested against C. Likewise, we see that finding a single good widget lowers the evidence for C by an amount that cannot be recovered by two bad ones; so there is a 'threshold of skepticism'. C will never attain an appreciable probability; i.e. the robot will never become skeptical about propositions A and B, as long as the observed fraction f of bad ones remains less than 2/3.

More precisely, we define a threshold fraction f_t thus: as the number of tests $m \to \infty$ with $f = m_b/m \to$ const., $e(C|DX)$ tends to $+\infty$ if $f > f_t$, and to $-\infty$ if $f < f_t$. The exact threshold turns out to be greater than 2/3: $f_t = 0.793951$ (Exercise 4.2). If the observed

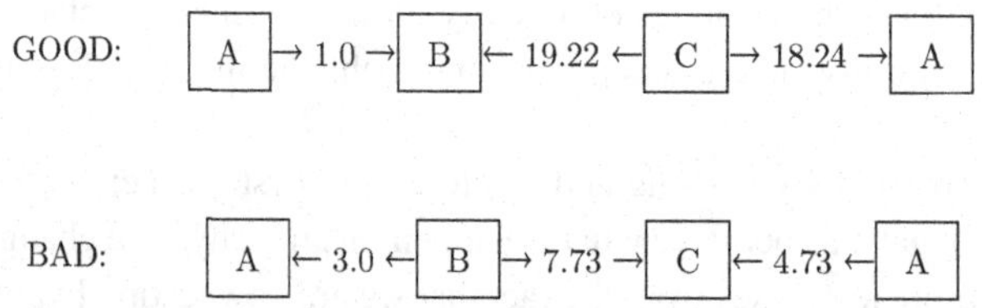

GOOD: $\boxed{A} \mapsto 1.0 \to \boxed{B} \leftarrow 19.22 \dashv \boxed{C} \mapsto 18.24 \to \boxed{A}$

BAD: $\boxed{A} \leftarrow 3.0 \dashv \boxed{B} \mapsto 7.73 \to \boxed{C} \leftarrow 4.73 \dashv \boxed{A}$

Fig. 4.2. Plausibility flow diagrams.

fraction of bad widgets remains above this value, the robot will be led eventually to prefer proposition C over A and B.

Exercise 4.2. Calculate the exact threshold of skepticism $f_t(x, y)$, supposing that proposition C has instead of 10^{-6} an arbitrary prior probability $P(C|X) = x$, and specifies instead of 99/100 an arbitrary fraction y of bad widgets. Then discuss how the dependence on x and y corresponds – or fails to correspond – to human common sense. *Hint:* In problems like this, always try first to get an analytic solution in closed form. If you are unable to do this, then you must write a short computer program which will display the correct numerical values in tables or graphs.

Exercise 4.3. Show how to make the robot skeptical about both unexpectedly high and unexpectedly low numbers of bad widgets in the observed sample. Give the full equations. Note particularly the following: if A is true, then we would expect, according to the binomial distribution (3.86), that the observed fraction of bad ones would tend to about 1/3 with many tests, while if B is true it should tend to 1/6. Suppose that it is found to tend to the threshold value (4.24), close to 1/4. On sufficiently large m, you and I would then become skeptical about A and B; but intuition tells us that this would require a much larger m than ten, which was enough to make us and the robot skeptical when we find them all bad. Do the equations agree with our intuition here, if a new hypothesis F is introduced which specifies $P(\text{bad}|FX) \simeq 1/4$?

In summary, the role of our new hypothesis C was only to be held in abeyance until needed, like a fire extinguisher. In a normal testing situation it is 'dead', playing no part in the inference because its probability is and remains far below that of the other hypotheses. But a dead hypothesis can be resurrected to life by very unexpected data. Exercises 4.2 and 4.3 ask the reader to explore the phenomenon of resurrection of dead hypotheses in more detail than we do in this chapter, but we return to the subject in Chapter 5.

Figure 4.1 shows an interesting thing. Suppose we had decided to stop the test and accept hypothesis A if the evidence for it reached +6 db. As we see, it would overshoot that value at the sixth trial. If we stopped the testing at that point, then we would never see the rest of this curve and see that it really goes down again. If we had continued the testing beyond this point, then we would have changed our minds again.

At first glance this seems disconcerting, but notice that it is inherent in all problems of hypothesis testing. If we stop the test at any finite number of trials, then we can never be absolutely sure that we have made the right decision. It is always possible that still more tests would have led us to change our decision. But note also that probability theory as logic has automatic built-in safety devices that can protect us against unpleasant surprises. Although it is always *possible* that our decision is wrong, this is extremely *improbable* if our critical level for decision requires $e(A|DX)$ to be large and positive. For example, if $e(A|DX) \geq 20$ db, then $P(A|DX) > 0.99$, and the total probability for all the alternatives is less than 0.01; then few of us would hesitate to decide confidently in favor of A.

In a real problem we may not have enough data to give such good evidence, and we might suppose that we could decide safely if the most likely hypothesis A is well separated from the alternatives, even though $e(A|DX)$ is itself not large. Indeed, if there are 1000 alternatives but the separation of A from the most likely alternative is more than 20 db, then the odds favor A by more than 100:1 over any one of the alternatives, and if we were obliged to make a definite choice of one hypothesis here and now, there could still be no hesitation in choosing A; it is clearly the best we can do with the information we have. Yet we cannot do it so confidently, for it is now very plausible that the decision is wrong, because the class of alternatives as a whole is about as probable as A. But probability theory warns us, by the numerical value of $e(A|DX)$, that this is the case; we need not be surprised by it.

In scientific inference our job is always to do the best we can with whatever information we have; there is no advance guarantee that our information will be sufficient to lead us to the truth. But many of the supposed difficulties arise from an inexperienced user's failure to recognize and use the safety devices that probability theory as logic always provides. Unfortunately, the current literature offers little help here because its viewpoint, concentrated mainly on sampling theory, directs attention to other things such as assumed sampling frequencies, as the following exercises illustrate.

Exercise 4.4.   Suppose that B is in fact true; estimate how many tests it will probably require in order to accumulate an additional 20 db of evidence (above the prior 10 db) in favor of B. Show that the sampling probability that we could ever obtain 20 db of evidence for A is negligibly small, even if we sample millions of times. In other words it is, for all practical purposes, impossible for a doctrinaire zealot to sample to a foregone false conclusion merely by continuing until he finally gets the evidence he wants.
Note: The calculations called for here are called 'random walk' problems; they are sampling theory exercises. Of course, the results are not wrong, only incomplete. Some essential aspects of inference in the real world are not recognized by sampling theory.

Exercise 4.5.   The estimate asked for in Exercise 4.4 is called the 'average sample number' (ASN), and the original rationale for the sequential procedure (Wald, 1947) was not our derivation from probability theory as logic, but Wald's conjecture (unproven at the time) that the sequential probability-ratio tests such as (4.19) and (4.21) minimize the ASN for a given reliability of conclusion. Discuss the validity of this conjecture; can one define the term 'reliability of conclusion' in such a way that the conjecture can be proved true?

Evidently, we could extend this example in many different directions. Introducing more 'discrete' hypotheses would be perfectly straightforward, as we have seen. More interesting would be the introduction of a continuous range of hypotheses, such as

$$H_f \equiv \text{the machine is putting out a fraction } f \text{ bad.}$$

Then, instead of a discrete prior probability distribution, our robot would have a continuous distribution in $0 \leq f \leq 1$, and it would calculate the posterior probabilities for various values of f on the basis of the observed samples, from which various decisions could be made. In fact, although we have not yet given a formal discussion of continuous probability distributions, the extension is so easy that we can give it as an introduction to this example.

4.5 Continuous probability distribution functions

Our rules for inference were derived in Chapter 2 only for the case of finite sets of discrete propositions $(A, B, \ldots)$. But this is all we ever need in practice. Suppose that f is any continuously variable real parameter of interest, then the propositions

$$\begin{aligned} F' &\equiv (f \leq q) \\ F'' &\equiv (f > q) \end{aligned} \tag{4.50}$$

are discrete, mutually exclusive, and exhaustive; so our rules will surely apply to them. Given some information Y, the probability for F' will in general depend on q, defining a function

$$G(q) \equiv P(F'|Y), \tag{4.51}$$

which is evidently monotonic increasing. Then what is the probability that f lies in any specified interval $(a < f \leq b)$? The answer is probably obvious intuitively, but it is worth noting that it is determined uniquely by the sum rule of probability theory, as follows. Define the propositions

$$A \equiv (f \leq a), \qquad B \equiv (f \leq b), \qquad W \equiv (a < f \leq b). \tag{4.52}$$

Then a relation of Boolean algebra is $B = A + W$, and since A and W are mutually exclusive, the sum rule reduces to

$$P(B|Y) = P(A|Y) + P(W|Y). \tag{4.53}$$

But $P(B|Y) = G(b)$, and $P(A|Y) = G(a)$, so we have the result

$$P(a < f \leq b|Y) = P(W|Y) = G(b) - G(a). \tag{4.54}$$

In the present case, $G(q)$ is continuous and differentiable, so we may write also

$$P(a < f \leq b|Y) = \int_a^b df \, g(f), \tag{4.55}$$

where $g(f) = G'(f) \geq 0$ is the derivative of G, generally called the *probability distribution function*, or the *probability density function* for f, given Y; either reading is consistent with the abbreviation pdf which we use henceforth, following the example of Zellner (1971). Its integral $G(f)$ may be called the *cumulative distribution function* for f.

Thus, limiting our basic theory to finite sets of propositions has not in any way hindered our ability to deal with continuous probability distributions; we have applied the basic product and sum rules only to discrete propositions in finite sets. As long as continuous distributions are defined as above (Eqs. (4.54), (4.55)) from a basis of finite sets of propositions, we are protected from inconsistencies by Cox's theorems. But if we become overconfident and try to operate directly on infinite sets without considering how they are to be generated from finite sets, this protection is lost and we stand at the mercy of all the paradoxes of infinite-set theory, as discussed in Chapter 15; we can then derive sense and nonsense with equal ease.

We must warn the reader about another semantic confusion which has caused error and controversy in probability theory for many decades. It would be quite wrong and misleading to call $g(f)$ the 'posterior distribution **of** f', because that verbiage would imply to the unwary that f itself is varying and is 'distributed' in some way. This would be another form of the mind projection fallacy, confusing reality with a state of knowledge about reality. In the problem we are discussing, f is simply an unknown constant parameter; what is 'distributed' is not the *parameter*, but the *probability*. Use of the terminology 'probability distribution **for** f' will be followed, in order to emphasize this constantly.

Of course, nothing in probability theory forbids us to consider the possibility that f might vary with time or with circumstance; indeed, probability theory enables us to analyze that case fully, as we shall see later. But then we should recognize that we are considering a *different* problem than the one just discussed; it involves different quantities with different states of knowledge about them, and requires a different calculation. Confusion of these two problems is perhaps the major occupational disease of those who fool themselves by using the above misleading terminology. The pragmatic consequence is that one is led to quite wrong conclusions about the accuracy and range of validity of the results.

Questions about what happens when $G(q)$ is discontinuous at a point q_0 are discussed further in Appendix B; for the present it suffices to note that, of course, approaching a discontinuous $G(q)$ as the limit of a sequence of continuous functions leads us to the correct results. As Gauss stressed long ago, any kind of singular mathematics acquires a meaning only as a limiting form of some kind of well-behaved mathematics, and it is ambiguous until we specify exactly what limiting process we propose to use. In this sense, singular mathematics has necessarily a kind of anthropomorphic character; the question is not what is it, but rather how shall we define it so that it is in some way useful to us?

In the present case, we approach the limit in such a way that the density function develops a sharper and sharper peak, going in the limit into a delta function $p_0 \, \delta(q - q_0)$ signifying a discrete hypothesis H_0, and enclosing a limiting area equal to the probability p_0 of that hypothesis; Eq. (4.65) below is an example.

But, in fact, if we become pragmatic we note that f is not really a continuously variable parameter. In its working lifetime, a machine will produce only a finite number of widgets; if it is so well built that it makes 10^8 of them, then the possible values of f are a finite set of integer multiples of 10^{-8}. Then our finite-set theory will apply, and consideration of a continuously variable f is only an approximation to the exact discrete theory. There is never any need to consider infinite sets or measure theory in the real, exact problem. Likewise, any data set that can actually be recorded and analyzed is digitized into multiples of some smallest element. Most cases of allegedly continuously variable quantities are like this when one takes note of the actual, real-world situation.

4.6 Testing an infinite number of hypotheses

In spite of the pragmatic argument just given, thinking of continuously variable parameters is often a natural and convenient approximation to a real problem (only we should not take it so seriously that we get bogged down in the irrelevancies for the real world that infinite sets and measure theory generate). So, suppose that we are now testing simultaneously an uncountably infinite number of hypotheses about the machine. As often happens in mathematics, this actually makes things simpler because analytical methods become available. However, the logarithmic form of the previous equations is now awkward, and so we will go back to the original probability form (4.3):

$$P(A|DX) = P(A|X)\frac{P(D|AX)}{P(D|X)}. \qquad (4.56)$$

Letting A now stand for the proposition 'The fraction of bad widgets is in the range $(f, f + df)$', there is a prior pdf

$$P(A|X) = g(f|X)df, \qquad (4.57)$$

which gives the probability that the fraction of bad widgets is in the range df; and let D stand for the results thus far of our experiment,

$D \equiv N$ widgets were tested and we found the results $GGBGBBG \cdots$, containing in all n bad ones and $(N - n)$ good ones.

Then the posterior pdf for f is given by

$$P(A|DX) = P(A|X)\frac{P(D|AX)}{P(D|X)} = g(f|DX)\,df, \qquad (4.58)$$

so the prior and posterior pdfs are related by

$$g(f|DX) = g(f|X)\frac{P(D|AX)}{P(D|X)}. \qquad (4.59)$$

The denominator is just a normalizing constant, which we could calculate directly; but usually it is easier to determine (if it is needed at all) from requiring that the posterior pdf satisfy the normalization condition

$$P(0 \leq f \leq 1 | DX) = \int_0^1 df\, g(f | DX) = 1, \tag{4.60}$$

which we should think of as an extremely good approximation to the exact formula, which has a sum over an enormous number of discrete values of f, instead of an integral.

The evidence of the data thus lies entirely in the f dependence of $P(D|AX)$. At this point, let us be very careful, in view of some errors that have trapped the unwary. In this probability, the conditioning statement A specifies an interval df, not a point value of f. Are we justified in taking an implied limit $df \to 0$ and replacing $P(D|AX)$ with $P(D|H_f X)$? Most writers have not hesitated to do this.

Mathematically, the correct procedure would be to evaluate $P(D|AX)$ exactly for positive df, and pass to the limit $df \to 0$ only afterward. But a tricky point is that if the problem contains another parameter θ in addition to f, then this procedure is ambiguous until we take the warning of Gauss very seriously, and specify exactly how the limit is to be approached (does df tend to zero at the same rate for all values of θ?). For example, if we set $df = \epsilon h(\theta)$ and pass to the limit $\epsilon \to 0$, our final conclusions may depend on which function $h(\theta)$ was used. Those who fail to notice this fall into the famous Borel–Kolmogorov paradox, in which a seemingly well-posed problem appears to have many different correct solutions. We shall discuss this in more detail later (Chapter 15), and show that the paradox is averted by strict adherence to our Chapter 2 rules.

In the present relatively simple problem, f is the only parameter present and $P(D|H_f X)$ is a continuous function of f; this is surely enough to guarantee that the limit is well-behaved and uneventful. But, just to be sure, let us take the trouble to demonstrate this by direct application of our Chapter 2 rules, keeping in mind that this continuum treatment is really an approximation to an exact discrete one. Then with $df > 0$, we can resolve A into a disjunction of a finite number of discrete propositions:

$$A = A_1 + A_2 + \cdots + A_n, \tag{4.61}$$

where $A_1 = H_f$ (f being one of the possible discrete values) and the A_i specify the discrete values of f in the interval $(f, f + df)$. They are mutually exclusive, so, as we noted in Chapter 2, Eq. (2.67), application of the product rule and the sum rule gives the general result

$$P(D|AX) = P(D|A_1 + A_2 + \cdots + A_n, X) = \frac{\sum_i P(A_i|X) P(D|A_i X)}{\sum_i P(A_i|X)}, \tag{4.62}$$

which is a weighted average of the separate probabilities $P(D|A_i X)$. This may be regarded also as a generalization of (4.39).

Then if all the $P(D|A_i X)$ were equal, (4.62) would become independent of their prior probabilities $P(A_i|X)$ and equal to $P(D|A_1 X) = P(D|H_f X)$; the fact that the conditioning statement on the left-hand side of (4.62) is a logical sum makes no difference, and $P(D|AX)$ would be rigorously equal to $P(D|H_f X)$. Even if the $P(D|A_i X)$ are not equal, as $df \to 0$, we have $n \to 1$ and eventually $A = A_1$, with the same result.

It may appear that we have gone to extraordinary lengths to argue for an almost trivially simple conclusion. But the story of the schoolboy who made a mistake in his sums and concluded that the rules of arithmetic are all wrong, is not fanciful. There is a long history of workers who did seemingly obvious things in probability theory without bothering to derive them by strict application of the basic rules, obtained nonsensical results – and concluded that probability theory as logic was at fault. The greatest, most respected mathematicians and logicians have fallen into this trap momentarily, and some philosophers spend their entire lives mired in it; we shall see some examples in the next chapter.

Such a simple operation as passing to the limit $df \rightarrow 0$ may produce results that seem to us obvious and trivial; or it may generate a Borel–Kolmogorov paradox. We have learned from much experience that this care is needed whenever we venture into a new area of applications; we must go back to the beginning and derive everything directly from first principles applied to finite sets. If we *obey* the Chapter 2 rules prescribed by Cox's theorems, we are rewarded by finding beautiful and useful results, free of contradictions.

Now, if we were given that f is the correct fraction of bad widgets, then the probability for getting a bad one at each trial would be f, and the probability for getting a good one would be $(1 - f)$. The probabilities at different trials are, by hypothesis (i.e. one of the many statements hidden there in X), logically independent given f, and so, as in our derivation of the binomial distribution (3.86),

$$P(D|H_f X) = f^n(1 - f)^{N-n} \tag{4.63}$$

(note that the experimental data D told us not only how many good and bad widgets were found, but also the order in which they appeared). Therefore, we have the posterior pdf

$$g(f|DX) = \frac{f^n(1 - f)^{N-n}g(f|X)}{\int_0^1 df\, f^n(1 - f)^{N-n}g(f|X)}. \tag{4.64}$$

You may be startled to realize that all of our previous discussion in this chapter is contained in this simple looking equation, as special cases. For example, the multiple hypothesis test starting with (4.43) and including the final results (4.45)–(4.49) is all contained in (4.64) corresponding to the particular choice of prior pdf:

$$g(f|X) = \frac{10}{11}(1 - 10^{-6})\delta\left(f - \frac{1}{6}\right) + \frac{1}{11}(1 - 10^{-6})\delta\left(f - \frac{1}{3}\right) + 10^{-6}\delta\left(f - \frac{99}{100}\right). \tag{4.65}$$

This is a case where the cumulative pdf, $G(f)$, is discontinuous. The three delta-functions correspond to the three discrete hypotheses B, A, C, respectively, of that example. They appear in the prior pdf (4.65) with coefficients which are the prior probabilities (4.31); and in the posterior pdf (4.64) with altered coefficients, which are just the posterior probabilities (4.45), (4.48) and (4.49).

Readers who have been taught to mistrust delta-functions as 'nonrigorous' are urged to read Appendix B at this point. The issue has nothing to do with mathematical rigor; it is simply one of notation appropriate to the problem. It would be difficult and awkward to express the information conveyed in (4.65) by a single equation in Lebesgue–Stieltjes type notation. Indeed, failure to use delta-functions where they are clearly called for has led mathematicians into elementary errors, as noted in Appendix B.

Suppose that at the start of this test our robot was fresh from the factory; it had no prior knowledge about the machines at all, except for our assurance that it is *possible* for a machine to make a good widget, and also *possible* for it to make a bad one. In this state of ignorance, what prior pdf $g(f|X)$ should it assign? If we have definite prior knowledge about f, this is the place to put it in; but we have not yet seen the principles needed to assign such priors. Even the problem of assigning priors to represent 'ignorance' will need much discussion later; but, for a simple result now, it may seem to the reader, as it did to Laplace 200 years ago, that in the present case the robot has no basis for assigning to any particular interval df a higher probability than to any other interval of the same size. Thus, the only honest way it can describe what it knows is to assign a uniform prior probability density, $g(f|X) = \text{const}$. This will receive a better theoretical justification later; to normalize it correctly as in (4.60) we must take

$$g(f|X) = 1, \quad 0 \le f \le 1. \tag{4.66}$$

The integral in (4.64) is then the well-known Eulerian integral of the first kind, today more commonly called the complete beta-function; and (4.64) reduces to

$$g(f|DX) = \frac{(N+1)!}{n!\,(N-n)!}\, f^n (1-f)^{N-n}. \tag{4.67}$$

4.6.1 Historical digression

It appears that this result was first found by an amateur mathematician, the Rev. Thomas Bayes (1763). For this reason, the kind of calculations we are doing are called 'Bayesian'. We shall follow this long-established custom, although it is misleading in several respects. The general result (4.3) is always called 'Bayes' theorem', although Bayes never wrote it; and it is really nothing but the product rule of probability theory which had been recognized by others, such as James Bernoulli and A. de Moivre (1718), long before the work of Bayes. Furthermore, it was not Bayes but Laplace (1774) who first saw the result in generality and showed how to use it in real problems of inference. Finally, the calculations we are doing – the direct application of probability theory as logic – are more general than mere application of Bayes' theorem; that is only one of several items in our toolbox.

The right-hand side of (4.67) has a single peak in $(0 \le f \le 1)$, located by differentiation at

$$f = \hat{f} \equiv \frac{n}{N}, \tag{4.68}$$

just the observed proportion, or relative frequency, of bad widgets. To find the sharpness of the peak, we write

$$L(f) \equiv \log g(f|DX) = n\log(f) + (N-n)\log(1-f) + \text{const.}, \qquad (4.69)$$

and expand $L(f)$ in a power series about $\hat{f}$. The first terms are

$$L(f) = L(\hat{f}) - \frac{(f-\hat{f})^2}{2\sigma^2} + \cdots, \qquad (4.70)$$

where

$$\sigma^2 \equiv \frac{\hat{f}(1-\hat{f})}{N}, \qquad (4.71)$$

and so, to this approximation, (4.67) is a *Gaussian*, or *normal*, distribution:

$$g(f|DX) \simeq K \exp\left\{ -\frac{(f-\hat{f})^2}{2\sigma^2} \right\} \qquad (4.72)$$

and K is a normalizing constant. Equations (4.71) and (4.72) constitute the de Moivre–Laplace theorem. It is actually an excellent approximation to (4.67) in the entire interval $(0 < f < 1)$ in the sense that the difference of the two sides tends to zero (although their ratio does not tend to unity), provided that $n \gg 1$ and $(N-n) \gg 1$. Properties of the Gaussian distribution are discussed in depth in Chapter 7.

Thus, after observing n bad widgets in N trials, the robot's state of knowledge about f can be described reasonably well by saying that it considers the most likely value of f to be just the observed fraction of bad widgets, and it considers the accuracy of this estimate to be such that the interval $\hat{f} \pm \sigma$ is reasonably likely to contain the true value. The parameter σ is called the *standard deviation* and σ^2 is the *variance* of the pdf (4.72). More precisely, from numerical analysis of (4.72), the robot assigns:

50% probability that the true value of f is contained in the interval $\hat{f} \pm 0.68\,\sigma$;
90% probability that it is contained in $\hat{f} \pm 1.65\,\sigma$;
99% probability that it is contained in $\hat{f} \pm 2.57\,\sigma$.

As the number N of tests increases, these intervals shrink, according to (4.71), proportional to $1/\sqrt{N}$, a common rule that arises repeatedly in probability theory.

In this way, we see that the robot starts in a state of 'complete ignorance' about f; but, as it accumulates information from the tests, it acquires more and more definite opinions about f, which correspond very nicely to common sense. Two cautions: (1) all this applies only to the case where, although the numerical value of f is initially unknown, it was one of the conditions defining the problem that f is known not to be changing with time, and (2) again we must warn against the error of calling σ the 'variance of f', which would imply that f is varying, and that σ is a real (i.e. measurable) physical property of f. That is one of the most common forms of the mind projection fallacy.

It is really necessary to belabor this point: σ is not a real property of f, but only a property of the *probability distribution* that the robot assigns to represent its state of knowledge about f. Two robots with different information would, naturally and properly, assign different pdfs for the same unknown quantity f, and the one which is better informed will probably – and deservedly – be able to estimate f more accurately; i.e., to use a smaller σ.

But, as noted, we may consider a different problem in which f is variable if we wish to do so. Then the mean-square variation s^2 of f over some class of cases will become a 'real' property, in principle measurable, and the question of its relation, if any, to the σ^2 of the robot's pdf for that problem can be investigated mathematically, as we shall do later in connection with time series. The relation will prove to be: if we know σ but have as yet no data *and no other prior information* about s, then the best prediction of s that we can make is essentially equal to σ; and if we do have the data but do not know σ *and have no other prior information* about σ, then the best estimate of σ that we can make is nearly equal to s. These relations are mathematically derivable consequences of probability theory as logic.

Indeed, it would be interesting, and more realistic for some quality-control situations, to introduce the possibility that f might vary with time, and the robot's job is to make the best possible inferences about whether a machine is drifting slowly out of adjustment, with the hope of correcting trouble before it became serious. Many other extensions of our problem occur to us: a simple classification of widgets as good and bad is not too realistic; there is likely a continuous gradation of quality, and by taking that into account we could refine these methods. There might be several important properties instead of just 'badness' and 'goodness' (for example, if our widgets are semiconductor diodes, forward resistance, noise temperature, rf impedance, low-level rectification efficiency, etc.), and we might also have to control the quality with respect to all of these. There might be a great many different machine characteristics, instead of just H_f, about which we need plausible inference.

It is clear that we could spend years and write volumes on all the further ramifications of this problem, and there is already a huge literature on it. Although there is no end to the complicated details that can be generated, there is in principle no difficulty in making whatever generalization we need. It requires no new principles beyond what we have given.

In the problem of detecting a drift in machine characteristics, we would want to compare our robot's procedure with the ones proposed long ago by Shewhart (1931). We would find that Shewhart's methods are intuitive approximations to what our robot would do; in some of the cases involving a normal distribution they are the same (but for the fact that Shewhart was not thinking sequentially; he considered the number of tests determined in advance). These are, incidentally, the only cases where Shewhart felt that his proposed methods were fully satisfactory.

This is really the same problem as that of detecting a signal in noise, which we shall study in more detail later on.

4.7 Simple and compound (or composite) hypotheses

The hypotheses (A, B, C, H_f) that we have considered thus far refer to a single parameter $f = M/N$, the unknown fraction of bad widgets in our box, and specify a sharply defined value for f (in H_f, it can be any prescribed number in $0 \leq f \leq 1$). Such hypotheses are called *simple*, because if we formalize this a bit more by defining an abstract 'parameter space' Ω consisting of all values of the parameter or parameters that we consider to be possible, such an hypothesis is represented by a single point in Ω.

Testing all the simple hypotheses in Ω, however, may be more than we need for our purposes. It may be that we care only whether our parameter lies in some subset $\Omega_1 \in \Omega$ or in the complementary set $\Omega_2 = \Omega - \Omega_1$, and the particular value of f in that subset is uninteresting (i.e. it would make no difference for what we plan to do next). Can we proceed directly to the question of interest, instead of requiring our robot to test every simple hypothesis in Ω_1?

The question is, to us, trivial; our starting point, Eq. (4.3), applies for all hypotheses H, simple or otherwise, so we have only to evaluate the terms in it for this case. But in (4.64) we have done almost all of that, and need only one more integration. Suppose that if $f > 0.1$ then we need to take some action (stop the machine and readjust it), but if $f \leq 0.1$ we should allow it to continue running. The space Ω then consists of all f in [0, 1], and we take Ω_1 as comprising all f in [0.1, 1], H as the hypothesis that f is in Ω_1. Since the actual value of f is not of interest, f is now called a *nuisance parameter*; and we want to get rid of it.

In view of the fact that the problem has no other parameter than f and different intervals df are mutually exclusive, the discrete sum rule $P(A_1 + \cdots + A_n | B) = \sum_i P(A_i | B)$ will surely generalize to an integral as the A_i become more and more numerous. Then the nuisance parameter f is removed by integrating it out of (4.64):

$$P(\Omega_1 | DX) = \frac{\int_{\Omega_1} df \, f^n (1 - f)^{N-n} g(f|X)}{\int_{\Omega} df \, f^n (1 - f)^{N-n} g(f|X)}. \tag{4.73}$$

In the case of a uniform prior pdf for f, we may use (4.64) and the result is the incomplete beta-function: the posterior probability that f is in any specified interval $(a < f < b)$ is

$$P(a < f < b | DX) = \frac{(N + 1)!}{n!(N - n)!} \int_a^b df \, f^n (1 - f)^{N-n}, \tag{4.74}$$

and in this form computer evaluation is easy.

More generally, when we have any composite hypothesis to test, probability theory tells us that the proper procedure is simply to apply the principle (4.1) by summing or integrating out, with respect to appropriate priors, whatever nuisance parameters it contains. The conclusions thus found take fully into account all of the evidence contained in the data and in the prior information about the parameters. Probability theory used as logic enables us to test, with a single principle, any number of hypotheses, simple or compound, in the light of the data and prior information. In later chapters we shall demonstrate these properties in many quantitatively worked out examples.

4.8 Comments

4.8.1 Etymology

Our opening quotation from John Craig (1699) is from a curious work on the probabilities of historical events, and how they change as the evidence changes. Craig's work was ridiculed mercilessly in the 19th century; and, indeed, his applications to religious issues do seem weird to us today. But Stigler (1986a) notes that Craig was writing at a time when the term 'probability' had not yet settled down to its present technical meaning, as referring to a (0–1) scale; and if we merely interpret Craig's 'probability for an hypothesis' as our log-odds measure (which we have seen to have in some respects a more primitive and intuitive meaning than probability), Craig's reasoning was actually quite good, and may be regarded as an anticipation of what we have done in this chapter.

Today, the logarithm-of-odds $\{u = \log[p/(1-p)]\}$ has proved to be such an important quantity that it deserves a shorter name; but we have had trouble finding one. Good (1950) was perhaps the first author to stress its importance in a published work, and he proposed the name *lods*, but the term has a leaden ring to our ears, as well as a nondescriptive quality, and it has never caught on.

Our same quantity (4.8) was used by Alan Turing and I. J. Good from 1941, in classified cryptographic work in England during World War II. Good (1980) later reminisced about this briefly, and noted that Turing coined the name 'deciban' for it. This has not caught on, presumably because nobody today can see any rationale for it.

The present writer, in his lectures of 1955–64 (for example, Jaynes, 1956), proposed the name *evidence*, which is intuitive and descriptive in the sense that, for given proportions, twice as many data provide twice as much evidence for an hypothesis. This was adopted by Tribus (1969), but it has not caught on either.

More recently, the term *logit* for $U \equiv \log[y/(a-y)]$, where $\{y_i\}$ are some items of data and a is chosen by some convention such as $a = 100$, has come into use. Likewise, graphs using U for one axis are called *logistic*. For example, in one commercial software graphics program, an axis on which values of U are plotted is called a 'logit axis' and regression on that graph is called '*logistic regression*'. There is at least a mathematical similarity to what we do here, but not any very obvious conceptual relation because U is not a measure of probability. In any event, the term 'logistic' had already an established usage dating back to Poincaré and Peano, as referring to the Russell–Whitehead attempt to reduce all mathematics to logic.[3]

---

[3] This terminology has a much longer historical basis. Alexander the Great sought to make all countries Greek in character, but he died before completing this goal, with the result that the countries he conquered had some Greek characteristics, but not

In the face of this confusion, we propose and use the following terminology. Note that we need two terms: the name of the quantity, and the name of the units in which it is measured. For the former we have retained the name *evidence*, which has at least the merit that it has been defined, and used consistently with the definition, in previously published works. One can then use various different units, with different names. In this chapter we have measured evidence in *decibels* because of its familiarity to scientists, the ease of finding numerical values, and the connection with the base ten number system which makes the results intuitively clear.

4.8.2 What have we accomplished?

The things which we have done in such a simple way in this chapter have been, in one sense, deceptive. We have had an introduction, in an atmosphere of apparent triviality, into almost every kind of problem that arises in the hypothesis testing business. But do not be deceived by the simplicity of our calculations into thinking that we have not reached the real nontrivial problems of the field. Those problems are only straightforward mathematical generalizations of what we have done here, and mathematically mature readers who have understood this chapter can now solve them for themselves, probably with less effort than it would require to find and understand the solutions available in the literature.

In fact, the methods of solution that we have indicated have far surpassed, in power to yield useful results, the methods available in the conventional non-Bayesian literature of hypothesis testing. To the best of our knowledge, no comprehension of the facts of multiple hypothesis testing, as illustrated in Figure 4.1, can be found in the orthodox literature (which explains why the principles of multiple hypothesis testing have been controversial in that literature). Likewise, our form of solution of the compound hypothesis problem (4.73) will not be found in the 'orthodox' literature of the subject.

It was our use of probability theory as logic that has enabled us to do so easily what was impossible for those who thought of probability as a physical phenomenon associated with 'randomness'. Quite the opposite; we have thought of probability distributions as *carriers of information*. At the same time, under the protection of Cox's theorems, we have avoided the inconsistencies and absurdities which are generated inevitably by those who try to deal with the problems of scientific inference by inventing *ad hoc* devices instead of applying the rules of probability theory. For a devastating criticism of these devices, see the book review by Pratt (1961).

It is not only in hypothesis testing, however, that the foundations of the theory matter for applications. As indicated in Chapter 1 and Appendix A, our formulation was chosen with the aim of giving the theory the widest possible range of useful applications. To drive home how much the scope of solvable problems depends on the chosen foundations, the reader may try Exercise 4.6.

all of them. So instead of calling them *Hellenic*, they were called *Hellenistic*. Thus, *logistic* implies something that has some properties of logic, but not all of them.

Exercise 4.6. In place of our product and sum rules, Ruelle (1991, p. 17) defines the 'mathematical presentation' of probability theory by three basic rules:

$$p(\overline{A}) = 1 - p(A);$$

if A and B are mutually exclusive, $p(A + B) = p(A) + p(B);$ (4.75)

if A and B are independent, $p(AB) = p(A)p(B).$

Survey the preceding two chapters, and determine how many of the applications that we solved in Chapters 3 and 4 could have been solved by application of these rules. *Hint:* If A and B are not independent, is $p(AB)$ determined by them? Is the notion of conditional probability defined? Ruelle makes no distinction between logical and causal independence; he defines 'independence' of A and B as meaning: 'the fact that one is realized has in the average no influence on the realization of the other'. It appears, then, that he would always accept (4.29) for all n.

This exercise makes it clear why conventional expositions do not consider scientific inference to be a part of probability theory. Indeed, orthodox statistical theory is helpless to deal with such problems because, thinking of probability as a physical phenomenon, it recognizes the existence only of sampling probabilities; thus it denies itself the technical tools needed to incorporate prior information, to eliminate nuisance parameters, or to recognize the information contained in a posterior probability. However, even most of the sampling theory results that we derived in Chapter 3 are beyond the scope of the mathematical and conceptual foundation given by Ruelle, as are virtually all of the parameter estimation results to be derived in Chapter 6.

We shall find later that our way of treating compound hypotheses illustrated here also generates automatically the conventional orthodox significance tests or superior ones; and at the same time gives a clear statement of what they are testing and their range of validity, previously lacking in the orthodox literature.

Now that we have seen the beginnings of this situation, before turning to more serious and mathematically more sophisticated problems, we shall relax and amuse ourselves in the next chapter by examining how probability theory as logic can clear up all kinds of weird errors, in the older literature, that arose from very simple misuse of probability theory, but whose consequences were relatively trivial. In Chapters 15 and 17 we consider some more complicated and serious errors that are causing major confusion in the current literature.

5

Queer uses for probability theory

I cannot conceal the fact here that in the specific application of these rules, I foresee many things happening which can cause one to be badly mistaken if he does not proceed cautiously.

James Bernoulli (1713, Part 4, Chapter III)

I. J. Good (1950) has shown how we can use probability theory backwards to measure our own strengths of belief about propositions. For example, how strongly do you believe in extrasensory perception?

5.1 Extrasensory perception

What probability would you assign to the hypothesis that Mr Smith has perfect extrasensory perception? More specifically, that he can guess right every time which number you have written down. To say zero is too dogmatic. According to our theory, this means that we are never going to allow the robot's mind to be changed by any amount of evidence, and we don't really want that. But where *is* our strength of belief in a proposition like this?

Our brains work pretty much the way this robot works, but we have an intuitive feeling for plausibility only when it's not too far from 0 db. We get fairly definite feelings that something is more than likely to be so or less than likely to be so. So the trick is to imagine an experiment. How much evidence would it take to bring your state of belief up to the place where you felt very perplexed and unsure about it? Not to the place where you believed it – that would overshoot the mark, and again we'd lose our resolving power. How much evidence would it take to bring you just up to the point where you were beginning to consider the possibility seriously?

So, we consider Mr Smith, who says he has extrasensory perception (ESP), and we will write down some numbers from one to ten on a piece of paper and ask him to guess which numbers we've written down. We'll take the usual precautions to make sure against other ways of finding out. If he guesses the first number correctly, of course we will all say 'you're a very lucky person, but I don't believe you have ESP'. And if he guesses two numbers correctly, we'll still say 'you're a very lucky person, but I still don't believe you have ESP'.

119

By the time he's guessed four numbers correctly – well, I still wouldn't believe it. So my state of belief is certainly lower than -40 db.

How many numbers would he have to guess correctly before you would really seriously consider the hypothesis that he has extrasensory perception? In my own case, I think somewhere around ten. My personal state of belief is, therefore, about -100 db. You could talk me into a ± 10 db change, and perhaps as much as ± 30 db, but not much more than that.

After further thought, we see that, although this result is correct, it is far from the whole story. In fact, if he guessed 1000 numbers correctly, I still would not believe that he has ESP, for an extension of the same reason that we noted in Chapter 4 when we first encountered the phenomenon of resurrection of dead hypotheses. An hypothesis A that starts out down at -100 db can hardly ever come to be believed, whatever the data, because there are almost sure to be alternative hypotheses $(B_1, B_2, \ldots)$ above it, perhaps down at -60 db. Then, when we obtain astonishing data that might have resurrected A, the alternatives will be resurrected instead. Let us illustrate this by two famous examples, involving telepathy and the discovery of Neptune. Also we note some interesting variants of this. Some are potentially useful, some are instructive case histories of probability theory gone wrong, in the way Bernoulli warned us about.

5.2 Mrs Stewart's telepathic powers

Before venturing into this weird area, the writer must issue a disclaimer. I was not there, and am not in a position to affirm that the experiment to be discussed actually took place; or, if it did, that the data were actually obtained in a valid way. Indeed, that is just the problem that you and I always face when someone tries to persuade us of the reality of ESP or some other marvellous thing – such things never happen to us or in our presence. All we are able to affirm is that the experiment and data have been reported in a real, verifiable reference (Soal and Bateman, 1954). This is the circumstance that we want to analyze now by probability theory. Lindley (1957) and Bernardo (1980) have also taken note of it from the standpoint of probability theory, and Boring (1955) discusses it from the standpoint of psychology.

In the reported experiment, from the experimental design the probability for guessing a card correctly should have been $p = 0.2$, independently in each trial. Let H_p be the 'null hypothesis' which states this, and supposes that only 'pure chance' is operating (whatever that means). According to the binomial distribution (3.86), H_p predicts that if a subject has no ESP, the number r of successful guesses in n trials should be about (mean $\pm$ standard deviation)

$$(r)_{\text{est}} = np \pm \sqrt{np(1-p)}. \tag{5.1}$$

For $n = 37\,100$ trials, this is 7420 ± 77.

But, according to the report, Mrs Gloria Stewart guessed correctly $r = 9410$ times in 37 100 trials, for a fractional success rate of $f = 0.2536$. These numbers constitute our data D. At first glance, they may not look very sensational; note, however, that her score was

$$\frac{9410 - 7420}{77} = 25.8 \tag{5.2}$$

standard deviations away from the chance expectation.

The probability for getting these data, on hypothesis H_p, is then the binomial

$$P(D|H_p) = \binom{n}{r} p^r (1 - p)^{n-r}. \tag{5.3}$$

But the numbers n, r are so large that we need the Stirling approximation to the binomial, derived in Chapter 9:

$$P(D|H_p) = A \exp\{nH(f, p)\}, \tag{5.4}$$

where

$$H(f, p) = f \log\left(\frac{p}{f}\right) + (1 - f) \log\left[\frac{1-p}{1-f}\right] = -0.008452 \tag{5.5}$$

is the entropy of the observed distribution $(f, 1 - f) = (0.2536, 0.7464)$ relative to the expected one, $(p, 1 - p) = (0.2000, 0.8000)$, and

$$A \equiv \sqrt{\left[\frac{n}{2\pi r(n - r)}\right]} = 0.00476. \tag{5.6}$$

Then we may take as the likelihood L_p of H_p, the sampling probability

$$L_p = P(D|H_p) = 0.00476 \exp\{-313.6\} = 3.15 \times 10^{-139}. \tag{5.7}$$

This looks fantastically small; however, before jumping to conclusions, the robot should ask: 'Are the data also fantastically improbable on the hypothesis that Mrs Stewart has telepathic powers?' If they are, then (5.7) may not be so significant after all.

Consider the Bernoulli class of alternative hypotheses H_q $(0 \leq q \leq 1)$, which suppose that the trials are independent, but that assign different probabilities of success q to Mrs Stewart $(q > 0.2$ if the hypothesis considers her to be telepathic). Out of this class, the hypothesis H_f that assigns $q = f = 0.2536$ yields the greatest $P(D|H_q)$ that can be attained in the Bernoulli class, and for this the entropy (5.5) is zero, yielding a maximum likelihood of

$$L_f = P(D|H_f) = A = 0.00476. \tag{5.8}$$

So, if the robot knew for a fact that Mrs Stewart is telepathic to the extent of $q = 0.2536$, then the probability that she could generate the observed data would not be particularly small. Therefore, the smallness of (5.7) is indeed highly significant; for then the likelihood ratio for the two hypotheses must be fantastically small. The relative likelihood depends only on the entropy factor:

$$\frac{L_p}{L_f} = \frac{P(D|H_p)}{P(D|H_f)} = \exp\{nH\} = \exp\{-313.6\} = 6.61 \times 10^{-137}, \tag{5.9}$$

and the robot would report: 'The data do indeed support H_f over H_p by an enormous factor.'

5.2.1 Digression on the normal approximation

Note, in passing, that in this calculation large errors could be made by unthinking use of the normal approximation to the binomial, also derived in Chapter 9 (or compare with (4.72)):

$$P(D|H_p, X) \simeq (\text{const.}) \times \exp\left\{\frac{-n(f-p)^2}{2p(1-p)}\right\}. \tag{5.10}$$

To use it here instead of the entropy approximation (5.4), amounts to replacing the entropy $H(f, p)$ by the first term of its power series expansion about the peak. Then we would have found instead a likelihood ratio $\exp\{-333.1\}$. Thus, the normal approximation would have made Mrs Stewart appear even more marvellous than the data indicate, by an additional odds ratio factor of

$$\exp\{333.1 - 313.6\} = \exp\{19.5\} = 2.94 \times 10^8. \tag{5.11}$$

This should warn us that, quite generally, normal approximations cannot be trusted far out in the tails of a distribution. In this case, we are 25.8 standard deviations out, and the normal approximation is in error by over eight orders of magnitude.

Unfortunately, this is just the approximation used by the chi-squared test discussed later, which can therefore lead us to wildly misleading conclusions when the 'null hypothesis' being tested fits the data very poorly. Those who use the chi-squared test to support their claims of marvels are usually helping themselves by factors such as (5.11). In practice, the entropy calculation (5.5) is just as easy and far more trustworthy (although the entropy and chi-squared test amount to the same thing within one or two standard deviations of the peak).

5.2.2 Back to Mrs Stewart

In any event, our present numbers are indeed fantastic; on the basis of such a result, ESP researchers would proclaim a virtual certainty that ESP is real. If we compare H_p and H_f by probability theory, the posterior probability that Mrs Stewart has ESP to the extent of $q = f = 0.2536$ is

$$P(H_f|DX) = P(H_f|X)\frac{P(D|H_f X)}{P(D|X)} = \frac{P_f L_f}{P_f L_f + P_p L_p}, \tag{5.12}$$

where P_p, P_f are the prior probabilities of H_p, H_f. But, because of (5.9), it hardly matters what these prior probabilities are; in the view of an ESP researcher who does not consider the prior probability $P_f = P(H_f|X)$ particularly small, $P(H_f|DX)$ is so close to unity that its decimal expression starts with over 100 nines.

He will then react with anger and dismay when, in spite of what he considers this overwhelming evidence, we persist in not believing in ESP. Why are we, as he sees it, so perversely illogical and unscientific?

The trouble is that the above calculations, (5.9) and (5.12), represent a very naïve application of probability theory, in that they consider only H_p and H_f, and no other hypotheses. If we really knew that H_p and H_f were the only possible ways the data (or, more precisely, the observable report of the experiment and data) could be generated, then the conclusions that follow from (5.9) and (5.12) would be perfectly all right. But, in the real world, our intuition is taking into account some additional possibilities that they ignore.

Probability theory gives us the results of consistent plausible reasoning from the information *that was actually used* in our calculation. It can lead us wildly astray, as Bernoulli noted in our opening quotation, if we fail to use all the information that our common sense tells us is relevant to the question we are asking. When we are dealing with some extremely implausible hypothesis, recognition of a seemingly trivial alternative possibility can make many orders of magnitude difference in the conclusions. Taking note of this, let us show how a more sophisticated application of probability theory explains and justifies our intuitive doubts.

Let H_p, H_f, and L_p, L_f, P_p, P_f be as above; but now we introduce some new hypotheses about how this report of the experiment and data might have come about, which will surely be entertained by the readers of the report even if they are discounted by its writers.

These new hypotheses $(H_1, H_2, \ldots, H_k)$ range all the way from innocent possibilities, such as unintentional error in the record keeping, through frivolous ones (perhaps Mrs Stewart was having fun with those foolish people, with the aid of a little mirror that they did not notice), to less innocent possibilities such as selection of the data (not reporting the days when Mrs Stewart was not at her best), to deliberate falsification of the whole experiment for wholly reprehensible motives. Let us call them all, simply, 'deception'. For our purposes, it does not matter whether it is we or the researchers who are being deceived, or whether the deception was accidental or deliberate. Let the deception hypotheses have likelihoods and prior probabilities $L_i, P_i, i = (1, 2, \ldots, k)$.

There are, perhaps, 100 different deception hypotheses that we could think of and are not too far-fetched to consider, although a single one would suffice to make our point.

In this new logical environment, what is the posterior probability for the hypothesis H_f that was supported so overwhelmingly before? Probability theory now tells us that

$$P(H_f|DX) = \frac{P_f L_f}{P_f L_f + P_p L_p + \sum P_i L_i}. \tag{5.13}$$

Introduction of the deception hypotheses has changed the calculation greatly; in order for $P(H_f|DX)$ to come anywhere near unity it is now necessary that

$$P_p L_p + \sum_i P_i L_i \ll P_f L_f. \tag{5.14}$$

Let us suppose that the deception hypotheses have likelihoods L_i of the same order as L_f in (5.8); i.e. a deception mechanism could produce the reported data about as easily as could a truly telepathic Mrs Stewart. From (5.7), $P_p L_p$ is completely negligible, so (5.14) is not greatly different from

$$\sum P_i \ll P_f. \tag{5.15}$$

But each of the deception hypotheses is, in my judgment, more likely than H_f, so there is not the remotest possibility that the inequality (5.15) could ever be satisfied.

Therefore, this kind of experiment can never convince me of the reality of Mrs Stewart's ESP; not because I assert $P_f = 0$ dogmatically at the start, but because the verifiable facts can be accounted for by many alternative hypotheses, every one of which I consider inherently more plausible than H_f, and none of which is ruled out by the information available to me.

Indeed, the very evidence which the ESP'ers throw at us to convince us, has the opposite effect on our state of belief; issuing reports of sensational data defeats its own purpose. For if the prior probability for deception is greater than that of ESP, then the more improbable the alleged data are on the null hypothesis of no deception and no ESP, the more strongly we are led to believe, not in ESP, but in deception. For this reason, the advocates of ESP (or any other marvel) will never succeed in persuading scientists that their phenomenon is real, until they learn how to eliminate the possibility of deception in the mind of the reader. As (5.15) shows, the reader's total prior probability for deception by all mechanisms must be pushed down below that of ESP.

It is interesting that Laplace perceived this phenomenon long ago. His *Essai Philosophique sur les Probabilités* (1814, 1819) has a long chapter on the 'Probabilities of testimonies', in which he calls attention to 'the immense weight of testimonies necessary to admit a suspension of natural laws'. He notes that those who make recitals of miracles,

decrease rather than augment the belief which they wish to inspire; for then those recitals render very probable the error or the falsehood of their authors. But that which diminishes the belief of educated men often increases that of the uneducated, always avid for the marvellous.

We observe the same phenomenon at work today, not only in the ESP enthusiast, but in the astrologer, reincarnationist, exorcist, fundamentalist preacher or cultist of any sort, who attracts a loyal following among the uneducated by claiming all kinds of miracles, but has zero success in converting educated people to his teachings. Educated people, taught to believe that a cause–effect relation requires a physical mechanism to bring it about, are scornful of arguments which invoke miracles; but the uneducated seem actually to prefer them.

Note that we can recognize the clear truth of this psychological phenomenon without taking any stand about the truth of the miracle; it is possible that the educated people are wrong. For example, in Laplace's youth educated persons did not believe in meteorites, but dismissed them as ignorant folklore because they are so rarely observed. For one familiar with the laws of mechanics the notion that 'stones fall from the sky' seemed preposterous, while those without any conception of mechanical law saw no difficulty in the idea. But the fall at Laigle in 1803, which left fragments studied by Biot and other French scientists, changed the opinions of the educated – including Laplace himself. In this case, the uneducated, avid for the marvellous, happened to be right: *c'est la vie*.

Indeed, in the course of writing this chapter, the writer found himself a victim of this phenomenon. In the 1987 Ph.D. thesis of G. L. Bretthorst, and more fully in Bretthorst (1988), we applied Bayesian analysis to estimation of frequencies of nonstationary sinusoidal signals, such as exponential decay in nuclear magnetic resonance (NMR) data, or chirp in oceanographic waves. We found – as was expected on theoretical grounds – an improved resolution over the previously used Fourier transform methods.

If we had claimed a 50% improvement, we would have been believed at once, and other researchers would have adopted this method eagerly. But, in fact, we found orders of magnitude improvement in resolution. It was, in retrospect, foolish of us to mention this at the outset, for in the minds of others the prior probability that we were irresponsible charlatans was greater than the prior probability that a new method could possibly be that good; and we were not at first believed.

Fortunately, we were able, by presenting many numerical analyses of data and distributing free computer programs so that doubters could check our claims for themselves on whatever data they chose, to eliminate the possibility of deception in the minds of our audience, and the method did find acceptance after all. The Bayesian analyses of free decay NMR signals now permits experimentalists to extract much more information from their data than was possible by taking Fourier transforms.

The reader should be warned, however, that our probability analysis (5.13) of Mrs Stewart's performance is still rather naïve in that it neglects correlations; having seen a persistent deviation from the chance expectation $p = 0.2$ in the first few hundred trials, common sense would lead us to form the hypothesis that some unknown systematic cause is at work, and we would come to expect the same deviation in the future. This would alter the numerical values given above, but not enough to change our general conclusions. More sophisticated probability models which are able to take such things into account are given in our discussions of advanced applications later; relevant topics are Dirichlet priors, exchangeable sequences, and autoregressive models.

Now let us return to that original device of I. J. Good, which started this train of thought. After all this analysis, why do we still hold that naïve first answer of -100 db for my prior probability for ESP, as recorded above, to be correct? Because Jack Good's imaginary device can be applied to whatever state of knowledge we choose to imagine; it need not be the real one. If I knew that true ESP and pure chance were the only possibilities, then the device would apply and my assignment of -100 db would hold. But, knowing that there are other possibilities in the real world does not change my state of belief about ESP; so the figure of -100 db still holds.

Therefore, in the present state of development of probability theory, the device of imaginary results is usable and useful in a very wide variety of situations, where we might not at first think it applicable. We shall find it helpful in many cases where our prior information seems at first too vague to lead to any definite prior probabilities; it stimulates our thinking and tells us how to assign them after all. Perhaps in the future we shall have more formal principles that make it unnecessary.

Exercise 5.1. By applying the device of imaginary results, find your own strength of belief in any three of the following propositions: (1) Julius Caesar is a real historical person (i.e. not a myth invented by later writers); (2) Achilles is a real historical person; (3) the Earth is more than a million years old; (4) dinosaurs did not die out; they are still living in remote places; (5) owls can see in total darkness; (6) the configuration of the planets influences our destiny; (7) automobile seat belts do more harm than good; (8) high interest rates combat inflation; (9) high interest rates cause inflation.

Hint: Try to imagine a situation in which the proposition H_0 being tested, and a single alternative H_1, would be the only possibilities, and you receive new 'data' D consistent with H_0: $P(D|H_0) \simeq 1$. The imaginary alternative and data are to be such that you can calculate the probability $P(D|H_1)$. Always use an H_0 that you are inclined not to believe; if the proposition as stated seems highly plausible to you, then for H_0 choose its denial.

Much more has been written about the Soal experiments in ESP. The deception hypothesis, already strongly indicated by our probability analysis, is supported by additional evidence (Hansel, 1980; Kurtz, 1985). Altogether, an appalling amount of effort has been expended on this incident, and it might appear that the only result was to provide a pedagogical example of the use of probability theory with very unlikely hypotheses. Can anything more useful be salvaged from it?

We think that this incident has some lasting value both for psychology and for probability theory, because it has made us aware of an important general phenomenon, which has nothing to do with ESP; a person may tell the truth and not be believed, even though the disbelievers are reasoning in a rational, consistent way. To the best of our knowledge it has not been noted before that probability theory as logic *automatically* reproduces and explains this phenomenon. This leads us to conjecture that it may generalize to other more complex and puzzling psychological phenomena.

5.3 Converging and diverging views

Suppose that two people, Mr A and Mr B have differing views (due to their differing prior information) about some issue, say the truth or falsity of some controversial proposition S. Now we give them both a number of new pieces of information or 'data', $D_1, D_2, \ldots, D_n$, some favorable to S, some unfavorable. As n increases, the totality of their information comes to be more nearly the same, therefore we might expect that their opinions about S will converge toward a common agreement. Indeed, some authors consider this so obvious that they see no need to demonstrate it explicitly, while Howson and Urbach (1989, p. 290) claim to have demonstrated it.

Nevertheless, let us see for ourselves whether probability theory can reproduce such phenomena. Denote the prior information by I_A, I_B, respectively, and let Mr A be initially a believer, Mr B a doubter:

$$P(S|I_A) \simeq 1, \qquad P(S|I_B) \simeq 0; \qquad (5.16)$$

after receiving data D, their posterior probabilities are changed to

$$P(S|DI_A) = P(S|I_A)\frac{P(D|SI_A)}{P(D|I_A)}$$
$$\qquad (5.17)$$
$$P(S|DI_B) = P(S|I_B)\frac{P(D|SI_B)}{P(D|I_B)}.$$

If D supports S, then since Mr A already considers S almost certainly true, we have $P(D|SI_A) \simeq P(D|I_A)$, and so

$$P(S|DI_A) \simeq P(S|I_A). \qquad (5.18)$$

Data D have no appreciable effect on Mr A's opinion. But now one would think that if Mr B reasons soundly, he must recognize that $P(D|SI_B) > P(D|I_B)$, and thus

$$P(S|DI_B) > P(S|I_B). \qquad (5.19)$$

Mr B's opinion should be changed in the direction of Mr A's. Likewise, if D had tended to refute S, one would expect that Mr B's opinions are little changed by it, whereas Mr A's will move in the direction of Mr B's. From this we might conjecture that, whatever the new information D, it should tend to bring different people into closer agreement with each other, in the sense that

$$|P(S|DI_A) - P(S|DI_B)| < |P(S|I_A) - P(S|I_B)|. \qquad (5.20)$$

Although this can be verified in special cases, it is not true in general.

Is there some other measure of 'closeness of agreement' such as $\log[P(S|DI_A)/P(S|DI_B)]$, for which this converging of opinions can be proved as a general theorem? Not even this is possible; the failure of probability theory to give this expected result tells us that convergence of views is not a general phenomenon. For robots and humans who reason according to the consistency desiderata of Chapter 1, something more subtle and sophisticated is at work.

Indeed, in practice we find that this convergence of opinions usually happens for small children; for adults it happens sometimes but not always. For example, new experimental evidence does cause scientists to come into closer agreement with each other about the explanation of a phenomenon.

Then it might be thought (and for some it is an article of faith in democracy) that open discussion of public issues would tend to bring about a general consensus on them. On the contrary, we observe repeatedly that when some controversial issue has been discussed vigorously for a few years, society becomes polarized into opposite extreme camps; it is almost impossible to find anyone who retains a moderate view. The Dreyfus affair in France, which tore the nation apart for 20 years, is one of the most thoroughly documented examples of this (Bredin, 1986). Today, such issues as nuclear power, abortion, criminal justice, etc., are following the same course. New information given simultaneously to different people may cause a convergence of views; but it may equally well cause a divergence.

This divergence phenomenon is observed also in relatively well-controlled psychological experiments. Some have concluded that people reason in a basically irrational way; prejudices seem to be strengthened by new information which ought to have the opposite effect. Kahneman and Tversky (1972) draw the opposite conclusion from such psychological tests, and consider them an argument against Bayesian methods.

But now, in view of the above ESP example, we wonder whether probability theory might also account for this divergence and indicate that people may be, after all, thinking in a reasonably rational, Bayesian way (i.e. in a way consistent with their prior information and prior beliefs). The key to the ESP example is that our new information was not

$$S \equiv \text{fully adequate precautions against error or deception were taken,} \atop \text{and Mrs Stewart did in fact deliver that phenomenal performance.} \qquad (5.21)$$

It was that some ESP researcher has *claimed* that S is true. But if our prior probability for S is lower than our prior probability that we are being deceived, hearing this claim has the opposite effect on our state of belief from what the claimant intended.

The same is true in science and politics; the new information a scientist gets is not that an experiment did in fact yield this result, with adequate protection against error. It is that some colleague has *claimed* that it did. The information we get from the TV evening news is not that a certain event actually happened in a certain way; it is that some news reporter has *claimed* that it did.[1]

Scientists can reach agreement quickly because we trust our experimental colleagues to have high standards of intellectual honesty and sharp perception to detect possible sources of error. And this belief is justified because, after all, hundreds of new experiments are reported every month, but only about once in a decade is an experiment reported that turns out later to have been wrong. So our prior probability for deception is very low; like trusting children, we believe what experimentalists tell us.

In politics, we have a very different situation. Not only do we doubt a politician's promises, few people believe that news reporters deal truthfully and objectively with economic, social, or political topics. We are convinced that virtually all news reporting is selective and distorted, designed not to report the facts, but to indoctrinate us in the reporter's socio-political views. And this belief is justified abundantly by the internal evidence in the reporter's own product – every choice of words and inflection of voice shifting the bias invariably in the same direction.

---

[1] Even seeing the event on our screens can no longer convince us, after recent revelations that all major US networks had faked some videotapes of alleged news events.

Not only in political speeches and news reporting, but wherever we seek for information on political matters, we run up against this same obstacle; we cannot trust anyone to tell us the truth, because we perceive that everyone who wants to talk about it is motivated either by self-interest or by ideology. In political matters, whatever the source of information, our prior probability for deception is always very high. However, it is not obvious whether this alone can prevent us from coming to agreement.

With this in mind, let us re-examine the equations of probability theory. To compare the reasoning of Mr A and Mr B, we could write Bayes' theorem (5.17) in the logarithmic form

$$\log\left[\frac{P(S|DI_A)}{P(S|DI_B)}\right] = \log\left[\frac{P(S|I_A)}{P(S|I_B)}\right] + \log\left[\frac{P(D|SI_A)\,P(D|I_B)}{P(D|I_A)\,P(D|SI_B)}\right], \tag{5.22}$$

which might be described by a simple hand-waving mnemonic like

$$\log \text{ posterior} = \log \text{ prior} + \log \text{ likelihood}. \tag{5.23}$$

Note, however, that (5.22) differs from our log-odds equations of Chapter 4, which might be described by the same mnemonic. There we compared different hypotheses, given the same prior information, and some factors $P(D|I)$ cancelled out. Here we are considering a fixed hypothesis S, in the light of different prior information, and they do not cancel, so the 'likelihood' term is different.

In the above, we supposed Mr A to be the believer, so log (prior) > 0. Then it is clear that on the log scale their views will converge as expected, the left-hand side of (5.22) tending to zero monotonically (i.e. Mr A will remain a stronger believer than Mr B) if

$$-\log(\text{prior}) < \log(\text{likelihood}) < 0, \tag{5.24}$$

and they will diverge monotonically if

$$\log(\text{likelihood}) > 0. \tag{5.25}$$

But they will converge with reversal (Mr B becomes a stronger believer than Mr A) if

$$-2\log(\text{prior}) < \log(\text{likelihood}) < -\log(\text{prior}), \tag{5.26}$$

and they will diverge with reversal if

$$\log(\text{likelihood}) < -2\log(\text{prior}). \tag{5.27}$$

Thus, probability theory appears to allow, in principle, that a single piece of new information D could have every conceivable effect on their relative states of belief.

But perhaps there are additional restrictions, not yet noted, which make some of these outcomes impossible; can we produce specific and realistic examples of all four types of behavior? Let us examine only the monotonic convergence and divergence by the following scenario, leaving it as an exercise for the reader to make a similar examination of the reversal phenomena.

The new information D is: 'Mr N has gone on TV with a sensational claim that a commonly used drug is unsafe', and three viewers, Mr A, Mr B, and Mr C, see this. Their prior probabilities $P(S|I)$ that the drug is safe are $(0.9, 0.1, 0.9)$, respectively; i.e. initially, Mr A and Mr C were believers in the safety of the drug, Mr B a disbeliever.

But they interpret the information D very differently, because they have different views about the reliability of Mr N. They all agree that, if the drug had really been proved unsafe, Mr N would be right there shouting it: that is, their probabilities $P(D|\overline{S}I)$ are $(1, 1, 1)$; but Mr A trusts his honesty while Mr C does not. Their probabilities $P(D|SI)$ that, if the drug is safe, Mr N would say that it is unsafe, are $(0.01, 0.3, 0.99)$, respectively.

Applying Bayes' theorem $P(S|DI) = P(S|I)\,P(D|SI)/P(D|I)$, and expanding the denominator by the product and sum rules, $P(D|I) = P(S|I)\,P(D|SI) + P(\overline{S}|I)\,P(D|\overline{S}I)$, we find their posterior probabilities that the drug is safe to be $(0.083, 0.032, 0.899)$, respectively. Put verbally, they have reasoned as follows:

A 'Mr N is a fine fellow, doing a notable public service. I had thought the drug to be safe from other evidence, but he would not knowingly misrepresent the facts; therefore hearing his report leads me to change my mind and think that the drug is unsafe after all. My belief in safety is lowered by 20.0 db, so I will not buy any more.'

B 'Mr N is an erratic fellow, inclined to accept adverse evidence too quickly. I was already convinced that the drug is unsafe; but even if it is safe he might be carried away into saying otherwise. So, hearing his claim does strengthen my opinion, but only by 5.3 db. I would never under any circumstances use the drug.'

C 'Mr N is an unscrupulous rascal, who does everything in his power to stir up trouble by sensational publicity. The drug is probably safe, but he would almost certainly claim it is unsafe whatever the facts. So hearing his claim has practically no effect (only 0.005 db) on my confidence that the drug is safe. I will continue to buy it and use it.'

The opinions of Mr A and Mr B converge in about the way we conjectured in (5.20) because both are willing to trust Mr N's veracity to some extent. But Mr A and Mr C diverge because their prior probabilities of deception are entirely different. So one cause of divergence is not merely that prior probabilities of deception are large, but that they are greatly different for different people.

This is not the only cause of divergence, however; to show this we introduce Mr X and Mr Y, who agree in their judgment of Mr N:

$$P(D|SI_X) = P(D|SI_Y) = a, \qquad P(D|\overline{S}I_X) = P(D|\overline{S}I_Y) = b. \qquad (5.28)$$

If $a < b$, then they consider him to be more likely to be telling the truth than lying. But they have different prior probabilities for the safety of the drug:

$$P(S|I_X) = x, \qquad P(S|I_Y) = y. \qquad (5.29)$$

Their posterior probabilities are then

$$P(S|DI_X) = \frac{ax}{ax + b(1-x)}, \qquad P(S|DI_Y) = \frac{ay}{ay + b(1-y)}, \qquad (5.30)$$

from which we see that not only are their opinions always changed in the same direction, on the evidence scale they are always changed by the same amount, $\log(a/b)$:

$$\log\left[\frac{P(S|DI_X)}{P(\overline{S}|DI_X)}\right] = \log\left[\frac{x}{1-x}\right] + \log\left[\frac{a}{b}\right]$$

$$\log\left[\frac{P(S|DI_Y)}{P(\overline{S}|DI_Y)}\right] = \log\left[\frac{y}{1-y}\right] + \log\left[\frac{a}{b}\right].$$

(5.31)

This means that, on the probability scale, they can either converge or diverge – see Exercise 5.2. These equations correspond closely to those in our sequential widget test in Chapter 4, but have now a different interpretation. If $a = b$, then they consider Mr N totally unreliable and their views are unchanged by his testimony. If $a > b$, they distrust Mr N so much that their opinions are driven in the opposite direction from what he intended. Indeed, if $b \to 0$, then $\log(a/b) \to \infty$; they consider it certain that he is lying, and so they are both driven to complete belief in the safety of the drug: $P(S|DI_X) = P(S|DI_Y) = 1$, independently of their prior probabilities.

Exercise 5.2. From these equations, find the exact conditions on (x, y, a, b) for divergence on the probability scale; that is,

$$|P(S|DI_X) - P(S|DI_Y)| > |P(S|I_X) - P(S|I_Y)|. \qquad (5.32)$$

Exercise 5.3. It is evident from (5.31) that Mr X and Mr Y can never experience a reversal of viewpoint; that is, if initially Mr X believes more strongly than Mr Y in the safety of the drug, this will remain true whatever the values of a, b. Therefore, a necessary condition for reversal must be that they have different opinions about Mr N; $a_x \neq a_y$ and/or $b_x \neq b_y$. But this does not prove that reversal is actually possible, so more analysis is needed. If reversal is possible, find a sufficient condition on $(x, y, a_x, a_y, b_x, b_y)$ for this to take place, and illustrate it by a verbal scenario like the above. If it is not possible, prove this and explain the intuitive reason why reversal cannot happen.

We see that divergence of opinions is readily explained by probability theory as logic, and that it is to be expected when persons have widely different prior information. But where was the error in the reasoning that led us to conjecture (5.20)? We committed a subtle form of the mind projection fallacy by supposing that the relation 'D supports S' is an absolute property of the propositions D and S. We need to recognize the relativity of it; whether D does or does not support S depends on our prior information. The same D that supports S for one person may refute it for another. As soon as we recognize this, then we no longer expect anything like (5.20) to hold in general. This error is very common; we shall see another example of it in Section 5.7.

Kahneman and Tversky (1972) claimed that we are not Bayesians, because in psychological tests people often commit violations of Bayesian principles. However, this claim is seen differently in view of what we have just noted. We suggest that people are reasoning according to a more sophisticated version of Bayesian inference than they had in mind.

This conclusion is strengthened by noting that similar things are found even in deductive logic. Wason and Johnson-Laird (1972) report psychological experiments in which subjects erred systematically in simple tests which amounted to applying a single syllogism. It seems that when asked to test the hypothesis '*A* implies *B*', they had a very strong tendency to consider it equivalent to '*B* implies *A*' instead of 'not-*B* implies not-*A*'. Even professional logicians could err in this way.[2]

Strangely enough, the nature of this error suggests a tendency toward Bayesianity, the opposite of the Kahneman–Tversky conclusion. For, if A supports B in the sense that for some X, $P(B|AX) > P(B|X)$, then Bayes' theorem states that B supports A in the same sense: $P(A|BX) > P(A|X)$. But it also states that $P(\overline{A}|\overline{B}X) > P(\overline{A}|X)$, corresponding to the syllogism. In the limit $P(B|AX) \to 1$, Bayes' theorem does not give $P(A|BX) \to 1$, but gives $P(\overline{A}|\overline{B}X) \to 1$, in agreement with the syllogism, as we noted in Chapter 2.

Errors made in staged psychological tests may indicate only that the subjects were pursuing different goals than the psychologists; they saw the tests as basically foolish, and did not think it worth making any mental effort before replying to the questions – or perhaps even thought that the psychologists would be more pleased to see them answer wrongly. Had they been faced with logically equivalent situations where their interests were strongly involved (for example, avoiding a serious accidental injury), they might have reasoned better. Indeed, there are stronger grounds – Darwinian natural selection – for expecting that we would reason in a basically Bayesian way.

5.4 Visual perception – evolution into Bayesianity?

Another class of psychological experiments fits nicely into this discussion. In the early 20th century, Adelbert Ames Jr was Professor of Physiological Optics at Dartmouth College. He devised ingenious experiments which fool one into 'seeing' something very different from the reality – one misjudges the size, shape, distance of objects. Some dismissed this as idle optical illusioning, but others who saw these demonstrations – notably including Alfred North Whitehead and Albert Einstein – saw their true importance as revealing surprising things about the mechanism of visual perception.[3] His work was carried on by Professor Hadley Cantril of Princeton University, who discussed these phenomena and produced movie demonstrations of them (Cantril, 1950).

---

[2] A possible complication of these tests – semantic confusion – readily suggests itself. We noted in Chapter 1 that the word 'implication' has a different meaning in formal logic than it has in ordinary language; '*A* implies *B*' does not have the usual colloquial meaning that *B* is logically deducible from *A*, as the subjects may have supposed.

[3] One of Ames' most impressive demonstrations has been recreated at the *Exploratorium* in San Francisco, the full-sized 'Ames room' into which visitors can look to see these phenomena at first hand.

The brain develops in infancy certain assumptions about the world based on all the sensory information it receives. For example, nearer objects appear larger, have greater parallax, and occlude distant objects in the same line of sight; a straight line appears straight from whatever direction it is viewed, etc. These assumptions are incorporated into the artist's rules of perspective and in three-dimensional computer graphics programs. We hold tenaciously onto them because they have been successful in correlating many different experiences. We will not relinquish successful hypotheses as long as they work; the only way to make one change these assumptions is to put one in a situation where they don't work. For example, in that Ames room where perceived size and distance correlate in the wrong way, a child walking across the room doubles in height.

The general conclusion from all these experiments is less surprising to our relativist generation than it was to the absolutist generation which made the discoveries. Seeing is not a direct apprehension of reality, as we often like to pretend. Quite the contrary: *seeing is inference from incomplete information*, no different in nature from the inference that we are studying here. The information that reaches us through our eyes is grossly inadequate to determine what is 'really there' before us. The failures of perception revealed by the experiments of Ames and Cantrell are not mechanical failures in the lens, retina, or optic nerve; they are the reactions of the subsequent inference process in the brain *when it receives new data that are inconsistent with its prior information*. These are just the situations where one is obliged to resurrect some alternative hypothesis; and that is what we 'see'. We expect that detailed analysis of these cases would show an excellent correspondence with Bayesian inference, in much the same way as in our ESP and diverging opinions examples.

Active study of visual perception has continued, and volumes of new knowledge have accumulated, but we still have almost no conception of how this is accomplished at the level of the neurons. Workers note the seeming absence of any organizing principle; we wonder whether the principles of Bayesian inference might serve as a start. We would expect Darwinian natural selection to produce such a result; after all, any reasoning format whose results conflict with Bayesian inference will place a creature at a decided survival disadvantage. Indeed, as we noted long ago (Jaynes, 1957b), in view of Cox's theorems, to deny that we reason in a Bayesian way is to assert that we reason in a deliberately inconsistent way; we find this very hard to believe. Presumably, a dozen other examples of human and animal perception would be found to obey a Bayesian reasoning format as its 'high level' organizing principle, for the same reason. With this in mind, let us examine a famous case history.

5.5 The discovery of Neptune

Another potential application for probability theory, which has been discussed vigorously by philosophers for over a century, concerns the reasoning process of a scientist, by which he accepts or rejects his theories in the light of the observed facts. We noted in Chapter 1 that this consists largely of the use of two forms of syllogism,

$$\text{one strong:} \left\{\begin{array}{c} \text{if } A, \text{ then } B \\ B \text{ false} \\ \hline A \text{ false} \end{array}\right\} \quad \text{and one weak:} \left\{\begin{array}{c} \text{if } A, \text{ then } B \\ B \text{ true} \\ \hline A \text{ more plausible} \end{array}\right\}. \quad (5.33)$$

In Chapter 2 we noted that these correspond to the use of Bayes' theorem in the forms

$$P(A|\overline{B}X) = P(A|X)\frac{P(\overline{B}|AX)}{P(\overline{B}|X)}, \qquad P(A|BX) = P(A|X)\frac{P(B|AX)}{P(B|X)}, \quad (5.34)$$

respectively, and that these forms do agree qualitatively with the syllogisms.

Interest here centers on the question of whether the second form of Bayes' theorem gives a satisfactory quantitative version of the weak syllogism, as scientists use it in practice. Let us consider a specific example given by Pólya (1954, Vol. II, pp. 130–132). This will give us a more useful example of the resurrection of alternative hypotheses.

The planet Uranus was discovered by Wm Herschel in 1781. Within a few decades (i.e. by the time Uranus had traversed about one-third of its orbit), it was clear that it was not following exactly the path prescribed for it by the Newtonian theory (laws of mechanics and gravitation). At this point, a naïve application of the strong syllogism might lead one to conclude that the Newtonian theory was demolished. However, its many other successes had established the Newtonian theory so firmly that in the minds of astronomers the probability for the hypothesis: 'Newton's theory is false' was already down at perhaps −50 db. Therefore, for the French astronomer Urbain Jean Joseph Leverrier (1811–1877) and the English scholar John Couch Adams (1819–1892) at St John's College, Cambridge, an alternative hypothesis down at perhaps −20 db was resurrected: there must be still another planet beyond Uranus, whose gravitational pull is causing the discrepancy.

Working unknown to each other and backwards, Leverrier and Adams computed the mass and orbit of a planet which could produce the observed deviation and predicted where the new planet would be found, with nearly the same results. The Berlin observatory received Leverrier's prediction on September 23, 1846, and, on the evening of the same day, the astronomer Johann Gottfried Galle (1812–1910) found the new planet (Neptune) within about one degree of the predicted position. For many more details, see Smart (1947) or Grosser (1979).

Instinctively, we feel that the plausibility for the Newtonian theory was increased by this little drama. The question is, how much? The attempt to apply probability theory to this problem will give us a good example of the complexity of actual situations faced by scientists, and also of the caution one needs in reading the rather confused literature on these problems.

Following Pólya's notation, let T stand for the Newtonian theory, N for the part of Leverrier's prediction that was verified. Then probability theory gives the posterior probability for T as

$$P(T|NX) = P(T|X)\frac{P(N|TX)}{P(N|X)}. \tag{5.35}$$

Suppose we try to evaluate $P(N|X)$. This is the prior probability for N, regardless of whether T is true or not. As usual, denote the denial of T by $\overline{T}$. Since $N = N(T + \overline{T}) = NT + N\overline{T}$, we have, by applying the sum and product rules,

$$\begin{aligned} P(N|X) &= P(NT + N\overline{T}|X) = P(NT|X) + P(N\overline{T}|X) \\ &= P(N|TX)P(T|X) + P(N|\overline{T}X)P(\overline{T}|X), \end{aligned} \tag{5.36}$$

and $P(N|\overline{T}X)$ has intruded itself into the problem. But in the problem as stated this quantity is not defined; the statement $\overline{T} \equiv$ 'Newton's theory is false' has no definite implications until we specify what alternative we have to put in place of Newton's theory.

For example, if there were only a single possible alternative according to which there could be no planets beyond Uranus, then $P(N|\overline{T}X) = 0$, and probability theory would again reduce to deductive reasoning, giving $P(T|NX) = 1$, independently of the prior probability $P(T|X)$.

On the other hand, if Einstein's theory were the only possible alternative, its predictions do not differ appreciably from those of Newton's theory for this phenomenon, and we would have $P(N|\overline{T}X) = P(N|TX)$, whereupon $P(T|NX) = P(T|X)$.

Thus, verification of the Leverrier–Adams prediction might elevate the Newtonian theory to certainty, or it might have no effect at all on its plausibility. It depends entirely on this: *against which specific alternatives are we testing Newton's theory?*

Now, to a scientist who is judging his theories, this conclusion is the most obvious exercise of common sense. We have seen the mathematics of this in some detail in Chapter 4, but all scientists see the same thing intuitively without any mathematics.

For example, if you ask a scientist, 'How well did the Zilch experiment support the Wilson theory?' you may get an answer like this: 'Well, if you had asked me last week I would have said that it supports the Wilson theory very handsomely; Zilch's experimental points lie much closer to Wilson's predictions than to Watson's. But, just yesterday, I learned that this fellow Woffson has a new theory based on more plausible assumptions, and his curve goes right through the experimental points. So now I'm afraid I have to say that the Zilch experiment pretty well demolishes the Wilson theory.'

5.5.1 Digression on alternative hypotheses

In view of this, working scientists will note with dismay that statisticians have developed *ad hoc* criteria for accepting or rejecting theories (chi-squared test, etc.) which make no reference to any alternatives. A practical difficulty of this was pointed out by Jeffreys (1939); there is not the slightest use in rejecting any hypothesis H_0 unless we can do it in favor of some definite alternative H_1 which better fits the facts.

Of course, we are concerned here with hypotheses which are not themselves statements of observable fact. If the hypothesis H_0 is merely that $x < y$, then a direct, error-free measurement of x and y which confirms this inequality constitutes positive proof of the correctness of the hypothesis, independently of any alternatives. We are considering hypotheses which might be called 'scientific theories' in that they are suppositions about what is not observable directly; only some of their consequences – logical or causal – can be observed by us.

For such hypotheses, Bayes' theorem tells us this: *Unless the observed facts are absolutely impossible on hypothesis H_0, it is meaningless to ask how much those facts tend 'in themselves' to confirm or refute H_0.* Not only the mathematics, but also our innate common sense (if we think about it for a moment) tell us that we have not asked any definite, well-posed question until we specify the possible alternatives to H_0. Then, as we saw in Chapter 4, probability theory can tell us how our hypothesis fares *relative to the alternatives that we have specified*; it does not have the creative imagination to invent new hypotheses for us.

Of course, as the observed facts approach impossibility on hypothesis H_0, we are led to worry more and more about H_0; but mere improbability, however great, cannot in itself be the reason for doubting H_0. We almost noted this after Eq. (5.7); now we are laying stress on it because it will be essential for our later general formulation of significance tests.

Early attempts to devise such tests foundered on the point we are making. Arbuthnot (1710) noted that in 82 years of demographic data more boys than girls were born in every year. On the 'null hypothesis' H_0 that the probability for a boy is $1/2$, he considered the probability for this result to be $2^{-82} = 10^{-24.7}$ (in our measure, -247 db), so small as to make H_0 seem to him virtually impossible, and saw in this evidence for 'Divine Providence'. He was, apparently, the first person to reject a statistical hypothesis on the grounds that it renders the data improbable. However, we can criticize his reasoning on several grounds.

Firstly, the alternative hypothesis $H_1 \equiv$ 'Divine Providence' does not seem usable in a probability calculation because it is not specific. That is, it does not make any definite predictions known to us, and so we cannot assign any probability for the data $P(D|H_1)$ conditional on H_1. (For this same reason, the mere logical denial $H_1 \equiv \overline{H_0}$ is unusable as an alternative.) In fact, it is far from clear why Divine Providence would wish to generate more boys than girls; indeed, if the number of boys and girls were exactly equal every year in a large population, that would seem to us much stronger evidence that some supernatural control mechanism must be at work.

Secondly, on the null hypothesis (independent and equal probability for a boy or girl at each birth) the probability $P(D|H_0)$ of finding the observed sequence would have been just as small whatever the data, so by Arbuthnot's reasoning the hypothesis would have been rejected whatever the data! Without having the probability $P(D|H_1)$ of the data on the alternative hypothesis *and* the prior probabilities of the hypotheses, there is just no well-posed problem and no rational basis for passing judgment.

Finally, having observed more boys than girls for ten consecutive years, rational inference might have led Arbuthnot to anticipate it for the 11th year. Thus his hypothesis H_0 was not only the numerical value $p = 1/2$; there was also an implicit assumption of logical independence for different years, of which he was probably unaware. On an hypothesis that allows for positive correlations, for example H_{ex}, which assigns an exchangeable sampling distribution, the probability $P(D|H_{ex})$ for the aggregated data could be very much greater than 2^{-82}. Thus, Arbuthnot took a small step in the right direction, but to get a usable significance test required a conceptual understanding of probability theory on a considerably higher level, as achieved by Laplace some 100 years later.

Another example occurred when Daniel Bernoulli won a French Academy prize of 1734 with an essay on the orbits of planets, in which he represented the orientation of each orbit by its polar point on the unit sphere and found them so close together as to make it very unlikely that the present distribution could result by chance. Although he too failed to state a specific alternative, we are inclined to accept his conclusion today because there seems to be a very clearly implied null hypothesis H_0 of 'chance' according to which the points should appear spread all over the sphere with no tendency to cluster together, and H_1 of 'attraction', which would make them tend to coincide; the evidence rather clearly supported H_1 over H_0.

Laplace (1812) did a similar analysis on comets, found their polar points much more scattered than those of the planets, and concluded that comets are not 'regular members' of the solar system like the planets. Here we finally had two fairly well-defined hypotheses being compared by a correct application of probability theory.[4]

Such tests need not be quantitative. Even when the application is only qualitative, probability theory is still useful to us in a normative sense; it is the means by which we can detect inconsistencies in our own qualitative reasoning. It tells us immediately what has not been intuitively obvious to all workers: that alternatives are needed before we have any rational criterion for testing hypotheses.

This means that if any significance test is to be acceptable to a scientist, we shall need to examine its rationale to see whether it has, like Daniel Bernoulli's test, some implied if unstated alternative hypotheses. Only when such hypotheses are identified are we in a position to say what the test accomplishes; i.e. what it is testing. But not to keep the reader in suspense: a statisticians' formal significance test can always be interpreted as a test of a specified hypothesis H_0 against a specified *class* of alternatives, and thus it is only a mathematical generalization of our treatment of multiple hypothesis tests in Chapter 4, Eqs. (4.31)–(4.49). However, the orthodox literature, which dealt with composite hypotheses by applying arbitrary *ad hockeries* instead of probability theory, never perceived this.

5.5.2 Back to Newton

Now we want to formulate a quantitative result about Newton's theory. In Pólya's discussion of the feat of Leverrier and Adams, once again no specific alternative to Newton's theory is stated; but from the numerical values used (Pólya, 1954, Vol. II, p. 131) we can infer that he had in mind a single possible alternative H_1 according to which it was known

---

[4] It is one of the tragedies of history that Cournot (1843), failing to comprehend Laplace's rationale, attacked it and reinstated the errors of Arbuthnot, thereby dealing scientific inference a setback from which it required a lifetime to recover.

that one more planet existed beyond Uranus, but all directions on the celestial sphere were considered equally likely. Then, since a cone of angle 1 degree fills in the sky a solid angle of about $\pi/(57.3)^2 = 10^{-3}$ steradian, $P(N|H_1X) \simeq 10^{-3}/4\pi = 1/13\,000$ is the probability that Neptune would have been within 1 degree of the predicted position.

Unfortunately, in the calculation no distinction was made between $P(N|X)$ and $P(N|\overline{T}X)$; that is, instead of the calculation (5.35) indicated by probability theory, the likelihood ratio actually calculated by Pólya was, in our notation,

$$\frac{P(N|TX)}{P(N|\overline{T}X)} = \frac{P(N|TX)}{P(N|H_1X)}. \tag{5.37}$$

Therefore, according to the analysis in Chapter 4, what Pólya obtained was not the ratio of posterior to prior probabilities, but the ratio of posterior to prior odds:

$$\frac{O(N|TX)}{O(N|X)} = \frac{P(N|TX)}{P(N|\overline{T}X)} = 13\,000. \tag{5.38}$$

The conclusions are much more satisfactory when we notice this. Whatever prior probability $P(T|X)$ we assign to Newton's theory, if H_1 is the only alternative considered, then verification of the prediction increased the *evidence* for Newton's theory by $10\log_{10}(13\,000) = 41$ db.

Actually, if there were a new planet it would be reasonable, in view of the aforementioned investigations of Daniel Bernoulli and Laplace, to adopt a different alternative hypothesis H_2, according to which its orbit would lie in the plane of the ecliptic, as Pólya again notes by implication rather than explicit statement. If, on hypothesis H_2, all values of longitude are considered equally likely, we might reduce this to about $10\log_{10}(180) = 23$ db. In view of the great uncertainty as to just what the alternative is (i.e. in view of the fact that the problem has not been defined unambiguously), any value between these extremes seems more or less reasonable.

There was a difficulty which bothered Pólya: if the *probability* of Newton's theory were increased by a factor of $13\,000$, then the prior probability was necessarily lower than $(1/13\,000)$; but this contradicts common sense, because Newton's theory was already very well established before Leverrier was born. Pólya interprets this in his book as revealing an inconsistency in Bayes' theorem, and the danger of trying to apply it numerically. Recognition that we are, in the above numbers, dealing with odds rather than probabilities, removes this objection and makes Bayes' theorem appear quite satisfactory in describing the inferences of a scientist.

This is a good example of the way in which objections to the Bayes–Laplace methods which you find in the literature disappear when you look at the problem more carefully. By an unfortunate slip in the calculation, Pólya was led to a misunderstanding of how Bayes' theorem operates. But I am glad to be able to close the discussion of this incident with a happier personal reminiscence.

In 1956, two years after the appearance of Pólya's work, I gave a series of lectures on these matters at Stanford University, and George Pólya attended them, sitting in the first row and paying the most strict attention to everything that was said. By then he understood this point very well – indeed, whenever a question was raised from the audience, Pólya would turn around and give the correct answer, before I could. It was very pleasant to have that kind of support, and I miss his presence today (George Pólya died, at the age of 97, in September 1985).

But the example also shows clearly that, in practice, the situation faced by the scientist is so complicated that there is little hope of applying Bayes' theorem to give quantitative results about the relative status of theories. Also there is no need to do this, because the real difficulty of the scientist is not in the reasoning process itself; his common sense is quite adequate for that. The real difficulty is in learning how to formulate new alternatives which better fit the facts. Usually, when one succeeds in doing this, the evidence for the new theory soon becomes so overwhelming that nobody needs probability theory to tell him what conclusions to draw.

Exercise 5.4. Our story has a curious sequel. In turn, it was noticed that Neptune was not following exactly its proper course, and so one naturally assumed that there is still another planet causing this. Percival Lowell, by a similar calculation, predicted its orbit, and Clyde Tombaugh proceeded to find the new planet (Pluto), although not so close to the predicted position. But now the story changes: modern data on the motion of Pluto's moon indicated that the mass of Pluto is too small to have caused the perturbation of Neptune which motivated Lowell's calculation. Thus, the discrepancies in the motions of Neptune and Pluto were unaccounted for. (We are indebted to Dr Brad Schaefer for this information.) Try to extend our probability analysis to take this new circumstance into account; at this point, where did Newton's theory stand? For more background information, see Hoyt (1980) or Whyte (1980). More recently, it appears that the mass of Pluto had been estimated wrongly and the discrepancies were after all not real; then it seems that the status of Newton's theory should revert to its former one. Discuss this sequence of pieces of information in terms of probability theory. Do we update by Bayes' theorem as each new fact comes in? Or do we just return to the beginning when we learn that a previous datum was false?

At present, we have no formal theory at all on the process of 'optimal hypothesis formulation', and we are dependent entirely on the creative imagination of individual persons such as Newton, Mendel, Einstein, Wegener, and Crick (1988). So, we would say that *in principle* the application of Bayes' theorem in the above way is perfectly legitimate; but *in practice* it is of very little use to a scientist.

However, we should not presume to give quick, glib answers to deep questions. The question of exactly how scientists do, in practice, pass judgment on their theories, remains complex and not well analyzed. Further comments on the validity of Newton's theory are offered in our closing Comments, Section 5.9.

5.6 Horse racing and weather forecasting

The preceding examples noted two different features common in problems of inference: (a) as in the ESP and psychological cases, the information we receive is often not a direct proposition like S in (5.21), but is an indirect claim that S is true, from some 'noisy' source that is itself not wholly reliable; (b) as in the example of Neptune, there is a long tradition of writers who have misapplied Bayes' theorem and concluded that Bayes' theorem is at fault. Both features are present simultaneously in a work of the Princeton philosopher Richard C. Jeffrey (1983), hereafter denoted by RCJ to avoid confusion with the Cambridge scholar Sir Harold Jeffreys.

RCJ considers the following problem. With only prior information I, we assign a probability $P(A|I)$ for A. Then we get new information B, and it changes as usual via Bayes' theorem to

$$P(A|BI) = P(A|I)P(B|AI)/P(B|I). \tag{5.39}$$

But then he decides that Bayes' theorem is not sufficiently general, because we often receive new information that is not certain; perhaps the probability for B is not unity but, say, q. To this we would reply: 'If you do not accept B as true, then why are you using it in Bayes' theorem this way?' But RCJ follows that long tradition and concludes, not that it is a misapplication of Bayes' theorem to use uncertain information as in (5.39), but that Bayes' theorem is itself faulty, and it needs to be generalized to take the uncertainty of new information into account.

His proposed generalization (denoting the denial of B by $\overline{B}$) is that the updated probability for A should be taken as a weighted average:

$$P(A)_J = qP(A|BI) + (1-q)P(A|\overline{B}I). \tag{5.40}$$

But this is an *ad hockery* that does not follow from the rules of probability theory unless we take q to be the *prior* probability $P(B|I)$, just the case that RCJ excludes (for then $P(A)_J = P(A|I)$, and there is no updating).

Since (5.40) conflicts with the rules of probability theory, we know that it necessarily violates one of the desiderata that we discussed in Chapters 1 and 2. The source of the trouble is easy to find, because those desiderata tell us where to look. The proposed 'generalization' (5.40) cannot hold generally because we could learn many different things, all of which indicate the same probability q for B; but which have different implications for A. Thus (5.40) violates desideratum (1.39b); it cannot take into account all of the new information, only the part of it that involves (i.e. is relevant to) B.

The analysis of Chapter 2 tells us that, if we are to salvage things and recover a well-posed problem with a defensible solution, *we must not depart in any way from Bayes' theorem*. Instead, we need to recognize the same thing that we stressed in the ESP example; if B is not known with certainty to be true, then B could not have been the new information; the actual information received must have been some proposition C such that $P(B|CI) = q$. But then, of course, we should be considering Bayes' theorem conditional on C, rather than B:

$$P(A|CI) = P(A|I)P(C|AI)/P(C|I). \tag{5.41}$$

If we apply it properly, Bayes' theorem automatically takes the uncertainty of new information into account. This result can be written, using the product and sum rules of probability theory, as

$$P(A|CI) = P(AB|CI) + P(A\overline{B}|CI) = P(A|BCI)P(B|CI) + P(A|\overline{B}CI)P(\overline{B}|CI),$$
(5.42)

and if we define $q \equiv P(B|CI)$ to be the updated probability for B, this can be written in the form

$$P(A|CI) = q P(A|BCI) + (1 - q)P(A|\overline{B}CI),$$
(5.43)

which resembles (5.40), but is not in general equal to it, unless we add the restriction that the probabilities $P(A|BCI)$ and $P(A|\overline{B}CI)$ are to be independent of C. Intuitively, this would mean that the logic flows thus:

$$(C \rightarrow B \rightarrow A)$$
(5.44)

rather than

$$(C \rightarrow A).$$
(5.45)

That is, C is relevant to A only through its intermediate relevance to B (C is relevant to B and B is relevant to A).

RCJ shows by example that this logic flow may be present in a real problem, but fails to note that his proposed solution (5.40) is then the same as the Bayesian result. Without that logic flow, (5.40) will be unacceptable in general because it does not take into account all of the new information. The information which is lost is indicated by the lack of an arrow going directly $(C \rightarrow A)$ in the logic flow diagram (5.45); information in C which is directly relevant to A, whether or not B is true.

If we think of the logic flow as something like the flow of light, we might visualize it thus. At night we receive sunlight only through its intermediate reflection from the moon; this corresponds to the RCJ solution. But in the daytime we receive light directly from the sun, whether or not the moon is there; this is what the RCJ solution has missed. (In fact, when we study the maximum entropy formalism in statistical mechanics and the phenomenon of 'generalized scattering', we shall find that this is more than a loose analogy; the process of conditional information flow is in almost exact mathematical correspondence with the Huygens principle of optics.)

Exercise 5.5. We might expect intuitively that when $q \rightarrow 1$ this difference would disappear; i.e. $P(A|BI) \rightarrow P(A|CI)$. Determine whether this is or is not generally true. If it is, indicate how small $1 - q$ must be in order to make the difference practically negligible. If it is not, illustrate by a verbal scenario the circumstances which can prevent this agreement.

We can illustrate this in a more down-to-earth way by one of RCJ's own scenarios:

> $A \equiv$ my horse will win the race tomorrow,
> $B \equiv$ the track will be muddy,
> $I \equiv$ whatever I know about my horse and jockey in particular, and about horses, jockeys, races, and life in general,

and the probability $P(A|I)$ is updated as a result of receiving a weather forecast. Then some proposition C such as:

> $C \equiv$ the TV weather forecaster showed us today's weather map, quoted some of the current meteorological data, and then by means unexplained assigned probability q' for rain tomorrow

is clearly present, but it is not recognized and stated by RCJ. Indeed, to do so would introduce much new detail, far beyond the gambit of propositions (A, B) of interest to horse racers.

If we recognize proposition C explicitly, then we must recall everything we know about the process of weather forecasting, what were the particular meteorological data leading to that forecast, how reliable weather forecasts are in the presence of such data, how the officially announced probability q' is related to what the forecaster really believes (i.e. what we think the forecaster perceives his own interest to be), etc.

If the above-defined C is the new information, then we must consider also, in the light of all our prior information, how C might affect the prospects for the race A through other circumstances than the muddiness B of the track; perhaps the jockey is blinded by bright sunlight, perhaps the rival horse runs poorly on cloudy days, whether or not the track is wet. These would be logical relations of the form $(C \rightarrow A)$ that (5.40) cannot take into account.

Therefore the full solution must be vastly more complicated than (5.40); but this is, of course, as it should be. Bayes' theorem, as always, is only telling us what common sense does; in general the updated probability for A must depend on far more than just the updated probability q for B.

5.6.1 Discussion

This example illustrates what we have noted before in Chapter 1; that familiar problems of everyday life may be more complicated than scientific problems, where we are often reasoning about carefully controlled situations. The most familiar problems may be so complicated – just because the result depends on so many unknown and uncontrolled factors – that a full Bayesian analysis, although correct in principle, is out of the question in practice. The cost of the computation is far more than we could hope to win on the horse.

Then we are necessarily in the realm of approximation techniques; but, since we cannot apply Bayes' theorem exactly, need we still consider it at all? Yes, because Bayes' theorem remains the normative principle telling us what we should aim for. Without it, we have nothing to guide our choices and no criterion for judging their success.

It also illustrates what we shall find repeatedly in later chapters: that generations of workers in this field have not comprehended the fact that Bayes' theorem is *a valid theorem*, required by elementary desiderata of rationality and consistency, and have made unbeliev-ably persistent attempts to replace it by all kinds of intuitive *ad hockeries*. Of course, we expect that any sincere intuitive effort will capture bits of the truth; yet all of these dozens of attempts have proved on analysis to be satisfactory only in those cases where they agree with Bayes' theorem after all.

We are at a loss, however, to understand what motivates these anti-Bayesian efforts, because we can see nothing unsatisfactory about Bayes' theorem, either in its theoretical foundations, its intuitive rationale, or its pragmatic results. The writer has devoted some 40 years to the analysis of thousands of separate problems by Bayes' theorem, and is still being impressed by the beautiful and important results it gives us, often in a few lines, and far beyond what those *ad hockeries* can produce. We have yet to find a case where it yields an unsatisfactory result (although the result is sometimes surprising at first glance, and it requires some meditation to educate our intuition and see that it is correct after all).

Needless to say, the cases where we are at first surprised are just the ones where Bayes' theorem is most valuable to us; because those are the cases where intuitive *ad hockeries* would never have found the result. Comparing Bayesian analysis with the *ad hoc* methods which saturate the literature, whenever there is any disagreement in the final conclusions, we have found it easy to exhibit the defect of the *ad hockery*, just as the analysis of Chapter 2 led us to expect, and as we saw in the above example.

In the past, many man-years of effort were wasted in futile attempts to square the circle; had Lindemann's theorem (that π is transcendental) been known and its implications recog-nized, all of this might have been averted. Likewise, had Cox's theorems been known, and their implications recognized, 100 years ago, many wasted careers might have been turned instead to constructive activity. This is our answer to those who have suggested that Cox's theorems are unimportant, because they only confirm what James Bernoulli and Laplace had conjectured long before.

Today, we have five decades of experience confirming what Cox's theorems tell us. It is clear that, not only is the quantitative use of the rules of probability theory as extended logic the only sound way to conduct inference; it is the *failure* to follow those rules strictly that has for many years been leading to unnecessary errors, paradoxes, and controversies.

5.7 Paradoxes of intuition

A famous example of this situation, known as Hempel's paradox, starts with the premise: 'A case of an hypothesis supports the hypothesis.' Then it observes: 'Now the hypothesis that all crows are black is logically equivalent to the statement that all non-black things are non-crows, and this is supported by the observation of a white shoe.' An incredible amount has been written about this seemingly innocent argument, which leads to an intolerable conclusion.

The error in the argument is apparent at once when one examines the equations of probability theory applied to it: the premise, which was not derived from any logical analysis,

is not generally true, and he prevents himself from discovering that fact by trying to judge
support of an hypothesis without considering any alternatives.

Good (1967), in a note entitled 'The white shoe is a red herring', demonstrated the error
in the premise by a simple counterexample. In World 1 there are one million birds, of which
100 are crows, all black. In World 2 there are two million birds, of which 200 000 are black
crows and 1 800 000 are white crows. We observe one bird, which proves to be a black crow.
Which world are we in?

Evidently, observation of a black crow gives evidence of

$$10 \log_{10} \left(\frac{200\,000/2\,000\,000}{100/1\,000\,000} \right) = 30 \text{ db}, \tag{5.46}$$

or an odds ratio of 1000:1, *against* the hypothesis that all crows are black; that is, for
World 2 against World 1. Whether an 'instance of an hypothesis' does or does not
support the hypothesis depends on the alternatives being considered and on the prior
information. We learned this in finding the error in the reasoning leading to (5.20). But,
incredibly, Hempel (1967) proceeded to reject Good's clear and compelling argument
on the grounds that it was unfair to introduce that background information about Worlds
1 and 2.

In the literature there are perhaps 100 'paradoxes' and controversies which are like this,
in that they arise from faulty intuition rather than faulty mathematics. Someone asserts
a general principle that seems to him intuitively right. Then, when probability analysis
reveals the error, instead of taking this opportunity to educate his intuition, he reacts by
rejecting the probability analysis. We shall see several more examples of this; in particular,
the marginalization paradox in Chapter 15.

As a colleague of the writer once remarked, 'Philosophers are free to do whatever they
please, because they don't have to do anything right.' But a responsible scientist does not
have that freedom; he will not assert the truth of a general principle, and urge others to adopt
it, merely on the strength of his own intuition. Some outstanding examples of this error,
which are not mere philosophers' toys like the RCJ tampering with Bayes' theorem and the
Hempel paradox, but have been actively harmful to Science and Society, are discussed in
Chapters 15 and 17.

5.8 Bayesian jurisprudence

It is interesting to apply probability theory in various situations in which we can't always reduce it to numbers very well, but still it shows automatically what kind of information would
be relevant to help us do plausible reasoning. Suppose someone in New York City has committed a murder, and you don't know at first who it is, but you know that there are 10 million
people in New York City. On the basis of no knowledge but this, $e(\text{guilty}|X) = -70$ db is
the plausibility that any particular person is the guilty one.

How much positive evidence for guilt is necessary before we decide that some man should
be put away? Perhaps +40 db, although your reaction may be that this is not safe enough,

and the number ought to be higher. If we raise this number we give increased protection to the innocent, but at the cost of making it more difficult to convict the guilty; and at some point the interests of society as a whole cannot be ignored.

For example, if 1000 guilty men are set free, we know from only too much experience that 200 or 300 of them will proceed immediately to inflict still more crimes upon society, and their escaping justice will encourage 100 more to take up crime. So it is clear that the damage to society as a whole caused by allowing 1000 guilty men to go free, is far greater than that caused by falsely convicting one innocent man.

If you have an emotional reaction against this statement, I ask you to think: if you were a judge, would you rather face one man whom you had convicted falsely; or 100 victims of crimes that you could have prevented? Setting the threshold at +40 db will mean, crudely, that on the average not more than one conviction in 10 000 will be in error; a judge who required juries to follow this rule would probably not make one false conviction in a working lifetime on the bench.

In any event, if we took +40 db starting out from −70 db, this means that in order to ensure a conviction you would have to produce about 110 db of evidence for the guilt of this particular person. Suppose now we learn that this person had a motive. What does that do to the plausibility for his guilt? Probability theory says

$$e(\text{guilty}|\text{motive}) = e(\text{guilty}|X) + 10\log_{10}\left[\frac{P(\text{motive}|\text{guilty})}{P(\text{motive}|\text{not guilty})}\right] \quad (5.47)$$

$$\simeq -70 - 10\log_{10} P(\text{motive}|\text{not guilty}),$$

since $P(\text{motive}|\text{guilty}) \simeq 1$, i.e. we consider it quite unlikely that the crime had no motive at all. Thus, the significance of learning that the person had a motive depends almost entirely on the probability $P(\text{motive}|\text{not guilty})$ that an innocent person would also have a motive.

This evidently agrees with our common sense, if we ponder it for a moment. If the deceased were kind and loved by all, hardly anyone would have a motive to do him in. Learning that, nevertheless, our suspect *did* have a motive, would then be very significant information. If the victim had been an unsavory character, who took great delight in all sorts of foul deeds, then a great many people would have a motive, and learning that our suspect was one of them is not so significant. The point of this is that we don't know what to make of the information that our suspect had a motive, unless we also know something about the character of the deceased. But how many members of juries would realize that, unless it was pointed out to them?

Suppose that a very enlightened judge, with powers not given to judges under present law, had perceived this fact and, when testimony about the motive was introduced, he directed his assistants to determine for the jury the *number* of people in New York City who had a motive. If this number is N_m then

$$P(\text{motive}|\text{not guilty}) = \frac{N_m - 1}{(\text{number of people in New York}) - 1} \simeq 10^{-7}(N_m - 1), \quad (5.48)$$

and (5.47) reduces, for all practical purposes, to

$$e(\text{guilty}|\text{motive}) \simeq -10 \log_{10}(N_m - 1). \tag{5.49}$$

You see that the population of New York has cancelled out of the equation; as soon as we know the number of people who had a motive, then it doesn't matter any more how large the city was. Note that (5.49) continues to say the right thing even when N_m is only 1 or 2.

You can go on this way for a long time, and we think you will find it both enlightening and entertaining to do so. For example, we now learn that the suspect was seen near the scene of the crime shortly before. From Bayes' theorem, the significance of this depends almost entirely on how many innocent persons were also in the vicinity. If you have ever been told not to trust Bayes' theorem, you should follow a few examples like this a good deal further, and see how infallibly it tells you what information would be relevant, what irrelevant, in plausible reasoning.[5] In recent years there has grown up a considerable literature on Bayesian jurisprudence; for a review with many references, see Vignaux and Robertson (1996).

Even in situations where we would be quite unable to say that numerical values should be used, Bayes' theorem still reproduces qualitatively just what your common sense (after perhaps some meditation) tells you. This is the fact that George Pólya demonstrated in such exhaustive detail that the present writer was convinced that the connection must be more than qualitative.

5.9 Comments

There has been much more discussion of the status of Newton's theory than we indicated above. For example, it has been suggested by Charles Misner that we cannot apply a theory with full confidence until we know its limits of validity – where it fails.

Thus, relativity theory, in showing us the limits of validity of Newtonian mechanics, also confirmed its accuracy within those limits; so it should increase our confidence in Newtonian theory when applied within its proper domain (velocities small compared with that of light). Likewise, the first law of thermodynamics, in showing us the limits of validity of the caloric theory, also confirmed the accuracy of the caloric theory within its proper domain (processes where heat flows but no work is done). At first glance this seems an attractive idea, and perhaps this is the way scientists really should think.

---

[5] Note that in these cases we are trying to decide, from scraps of incomplete information, on the truth of an Aristotelian proposition; whether the defendant did or did not commit some well-defined action. This is the situation – an issue of fact – for which probability theory as logic is designed. But there are other legal situations quite different; for example, in a medical malpractice suit it may be that all parties are agreed on the facts as to what the defendant actually did; the issue is whether he did or did not exercise reasonable judgment. Since there is no official, precise definition of 'reasonable judgment', the issue is not the truth of an Aristotelian proposition (however, if it were established that he wilfully violated one of our Chapter 1 desiderata of rationality, we think that most juries would convict him). It has been claimed that probability theory is basically inapplicable to such situations, and we are concerned with the partial truth of a non-Aristotelian proposition. We suggest, however, that in such cases we are not concerned with an issue of truth at all; rather, what is wanted is a value judgment. We shall return to this topic later (Chapters 13, 18).

Nevertheless, Misner's principle contrasts strikingly with the way scientists actually do think. We know of no case where anyone has avowed that his confidence in a theory was increased by its being, as we say, 'overthrown'. Furthermore, we apply the principle of conservation of momentum with full confidence, not because we know its limits of validity, but for just the opposite reason; we do not know of any such limits. Yet scientists believe that the principle of momentum conservation has real content; it is not a mere tautology.

Not knowing the answer to this riddle, we pursue it only one step further, with the observation that if we are trying to judge the validity of Newtonian mechanics, we cannot be sure that relativity theory showed us all its limitations. It is conceivable, for example, that it may fail not only in the limit of high velocities, but also in that of high accelerations. Indeed, there are theoretical reasons for expecting this; for Newton's $F = ma$ and Einstein's $E = mc^2$ can be combined into a perhaps more fundamental statement:

$$F = (E/c^2)a. \tag{5.50}$$

Why should the force required to accelerate a bundle of energy E depend on the velocity of light?

We see a plausible reason at once, if we adopt the – almost surely true – hypothesis that our allegedly 'elementary' particles cannot occupy mere mathematical points in space, but are extended structures of some kind. Then the velocity of light determines how rapidly different parts of the structure can 'communicate' with each other. The more quickly all parts can learn that a force is being applied, the more quickly they can all respond to it. We leave it as an exercise for the reader to show that one can actually derive Eq. (5.50) from this premise. (Hint: the force is proportional to the deformation that the particle must suffer before all parts of it start to move.)

But this embryonic theory makes further predictions immediately. We would expect that, when a force is applied suddenly, a short transient response time would be required for the acceleration to reach its Newtonian value. If so, then Newton's $F = ma$ is not an exact relation, only a final steady state condition, approached after the time required for light to cross the structure. It is conceivable that such a prediction could be tested experimentally.

Thus, the issue of our confidence in Newtonian theory is vastly more subtle and complex than merely citing its past predictive successes and its relationship to relativity theory; it depends also on our whole theoretical outlook.

It appears to us that actual scientific practice is guided by instincts that have not yet been fully recognized, much less analyzed and justified. We must take into account not only the logic of science, but also the sociology of science (perhaps also its soteriology). But this is so complicated that we are not even sure whether the extremely skeptical conservatism with which new ideas are invariably received is, in the long run, a beneficial stabilizing influence, or a harmful obstacle to progress.

5.9.1 What is queer?

In this chapter we have examined some applications of probability theory that seem 'queer' to us today, in the sense of being 'off the beaten track'. Any completely new application must presumably pass through such an exploratory phase of queerness. But in many cases, particularly the Bayesian jurisprudence and psychological tests with a more serious purpose than ESP, we think that queer applications of today may become respectable and useful applications of tomorrow. Further thought and experience will make us more aware of the proper formulation of a problem – better connected to reality – and then future generations will come to regard Bayesian analysis as indispensable for discussing it. Now we return to the many applications that are already advanced beyond the stage of queerness, into that of respectability and usefulness.

6

Elementary parameter estimation

A distinction without a difference has been introduced by certain writers
who distinguish 'Point estimation', meaning some process of arriving at
an estimate without regard to its precision, from 'Interval estimation' in
which the precision of the estimate is to some extent taken into account.

R. A. Fisher (1956)

Probability theory as logic agrees with Fisher in spirit; that is, it gives us automatically
both point and interval estimates from a single calculation. The distinction commonly
made between hypothesis testing and parameter estimation is considerably greater than
that which concerned Fisher; yet it too is, from our point of view, not a real difference.
When we have only a small number of discrete hypotheses $\{H_1, \ldots, H_n\}$ to consider,
we usually want to pick out a specific one of them as the most likely in that set, in the
light of the prior information and data. The cases $n = 2$ and $n = 3$ were examined in
some detail in Chapter 4, and larger n is in principle a straightforward and rather obvious
generalization.

   When the hypotheses become very numerous, however, a different approach seems called
for. A set of discrete hypotheses can always be classified by assigning one or more numerical
indices which identify them, as in H_t ($1 \leq t \leq n$), and if the hypotheses are very numerous
one can hardly avoid doing this. Then, deciding between the hypotheses H_t and estimating
the index t are practically the same thing, and it is a small step to regard the index, rather
than the hypotheses, as the quantity of interest; then we are doing parameter estimation. We
consider first the case where the index remains discrete.

6.1 Inversion of the urn distributions

In Chapter 3 we studied a variety of sampling distributions that arise in drawing from an
urn. There the number N of balls in the urn, and the number R of red balls and $N - R$
white ones, were considered known in the statement of the problem, and we were to make
'pre-data' inferences about what kind of mix of r red, $n - r$ white we were likely to get on
drawing n of them. Now we want to invert this problem, in the way envisaged by Bayes
and Laplace, to the 'post-data' problem: the data $D \equiv (n, r)$ are known, but the contents

(N, R) of the urn are not. From the data and our prior information about what is in the urn, what can we infer about its true contents? It is probably safe to say that every worker in probability theory is surprised by the results – almost trivial mathematically, yet deep and unexpected conceptually – that one finds in this inversion. In the following we note some of the surprises already well known in the literature, and add to them.

We found in Eq. (3.22) the sampling distribution for this problem; in our present notation this is the hypergeometric distribution

$$p(D|NRI) = h(r|NR, n) = \binom{N}{n}^{-1} \binom{R}{r} \binom{N-R}{n-r}, \tag{6.1}$$

where I now denotes the prior information, the general statement of the problem as given above.

6.2 Both N and R unknown

In general, neither N nor R is known initially, and the robot is to estimate both of them. If we succeed in drawing n balls from the urn, then of course we know deductively that $N \geq n$. It seems to us intuitively that the data could tell us nothing more about N; how could the number r of red balls drawn, or the order of drawing, be relevant to N? But this intuition is using a hidden assumption that we can hardly be aware of until we see the robot's answer to the question.

The joint posterior probability distribution for N and R is

$$p(NR|DI) = p(N|I)p(R|NI)\frac{p(D|NRI)}{p(D|I)}, \tag{6.2}$$

in which we have factored the joint prior probability by the product rule: $p(NR|I) = p(N|I)p(R|NI)$, and the normalizing denominator is a double sum,

$$p(D|I) = \sum_{N=0}^{\infty} \sum_{R=0}^{N} p(N|I)p(R|NI)p(D|NRI), \tag{6.3}$$

in which, of course, the factor $p(D|NRI)$ is zero when $N < n$, or $R < r$, or $N - R < n - r$. Then the marginal posterior probability for N alone is

$$p(N|DI) = \sum_{R=0}^{N} p(NR|DI) = p(N|I)\frac{\sum_R p(R|NI)p(D|NRI)}{p(D|I)}. \tag{6.4}$$

We could equally well apply Bayes' theorem directly:

$$p(N|DI) = p(N|I)\frac{p(D|NI)}{p(D|I)}, \tag{6.5}$$

and of course (6.4) and (6.5) must agree, by the product and sum rules.

These relations must hold whatever prior information I we may have about N, R that is to be expressed by $p(NR|I)$. In principle, this could be arbitrarily complicated, and conversion of verbally stated prior information into $p(N\,R|I)$ is an open-ended problem; you can always analyze your prior information more deeply. But usually our prior information is rather simple, and these problems are not difficult mathematically.

Intuition might lead us to expect further that, whatever prior $p(N|I)$ we had assigned, the data can only truncate the impossible values, leaving the relative probabilities of the possible values unchanged:

$$p(N|DI) = \begin{cases} Ap(N|I), & \text{if } N \geq n, \\ 0, & \text{if } 0 \leq N < n, \end{cases} \tag{6.6}$$

where A is a normalization constant. Indeed, the rules of probability theory tell us that this must be true if the data tell us only that $N \geq n$ and nothing else about N. For example, if

$$Z \equiv N \geq n, \tag{6.7}$$

then

$$p(Z|NI) = \begin{cases} 1 & \text{if } n \leq N \\ 0 & \text{if } n > N. \end{cases} \tag{6.8}$$

Bayes' theorem reads:

$$p(N|ZI) = p(N|I)\frac{p(Z|NI)}{p(Z|I)} = \begin{cases} Ap(N|I) & \text{if } N \geq n \\ 0 & \text{if } N < n. \end{cases} \tag{6.9}$$

If the data tell us only that Z is true, then we have (6.6) and the above normalization constant is $A = 1/p(Z|I)$. Bayes' theorem confirms that if we learn only that $N \geq n$, the relative probabilities of the possible values of N are not changed by this information; only the normalization must be readjusted to compensate for the values $N < n$ that now have zero probability. Laplace considered this result intuitively obvious, and took it as a basic principle of his theory.

However, the robot tells us in (6.5) that this will not be the case unless $p(D|NI)$ is independent of N for $N \geq n$. And, on second thought, we see that (6.6) need not be true if we have some kind of prior information linking N and R. For example, it is conceivable that one might know in advance that $R < 0.06N$. Then, necessarily, having observed the data $(n, r) = (10, 6)$, we would know not only that $N \geq 10$, but that $N > 100$. Any prior information that provides a logical link between N and R makes the datum r relevant to estimating N after all. But usually we lack any such prior information, and so estimation of N is uninteresting, reducing to the same result (6.6).

From (6.5), the general condition that the data can tell us nothing about N, except to truncate values less than n, is a nontrivial condition on the prior probability $p(R|NI)$:

$$p(D|NI) = \sum_{R=0}^{N} p(D|NRI)p(R|NI) = \begin{cases} f(n, r) & \text{if } N \geq n \\ 0 & \text{if } N < n, \end{cases} \tag{6.10}$$

where $f(n, r)$ may depend on the data, but is independent of N. Since we are using the standard hypergeometric urn sampling distribution (6.1), this is explicitly

$$\sum_{R=0}^{N} \binom{R}{r} \binom{N-R}{n-r} p(R|NI) = f(n, r) \binom{N}{n}, \qquad (N \geq n). \qquad (6.11)$$

This is that hidden assumption that our intuition could hardly have told us about. It is a kind of discrete integral equation[1] which the prior $p(R|NI)$ must satisfy as the necessary and sufficient condition for the data to be uninformative about N. The sum on the left-hand side is necessarily always zero when $N < n$, for the first binomial coefficient is zero when $R < r$, and the second is zero when $R \geq r$ and $N < n$. Therefore, the mathematical constraint on $p(R|NI)$ is only, rather sensibly, that $f(n, r)$ in (6.11) must be independent of N when $N \geq n$.

In fact, most 'reasonable' priors do satisfy this condition, and as a result estimation of N is relatively uninteresting. Then, factoring the joint posterior distribution (6.2) in the form

$$p(NR|DI) = p(N|DI)p(R|NDI), \qquad (6.12)$$

our main concern is with the factor $p(R|NDI)$, drawing inferences about R or about the ratio R/N with N supposed known. The posterior probability distribution for R is then, by Bayes' theorem,

$$p(R|DNI) = p(R|NI) \frac{p(D|NRI)}{p(D|NI)}. \qquad (6.13)$$

Different choices of the prior probability $p(R|NI)$ will yield many quite different results, and we now examine a few of them.

6.3 Uniform prior

Consider the state of prior knowledge denoted by I_0, in which we are, seemingly, as ignorant as we could be about R while knowing N: the uniform distribution

$$p(R|NI_0) = \begin{cases} \dfrac{1}{N+1} & \text{if } 0 \leq R \leq N \\ 0 & \text{if } R > N. \end{cases} \qquad (6.14)$$

Then a few terms cancel out, and (6.13) reduces to

$$p(R|DNI_0) = S^{-1} \binom{R}{r} \binom{N-R}{n-r}, \qquad (6.15)$$

---

[1] This peculiar name anticipates what we shall find later, in connection with marginalization theory; very general conditions of 'uninformativeness' are expressed by similar integral equations that the prior for one parameter must satisfy in order to make the data uninformative about another parameter.

where S is a normalization constant. For several purposes, we need the general summation formula

$$S \equiv \sum_{R=0}^{N} \binom{R}{r}\binom{N-R}{n-r} = \binom{N+1}{n+1},$$ (6.16)

whereupon the correctly normalized posterior distribution for R is

$$p(R|DNI_0) = \binom{N+1}{n+1}^{-1}\binom{R}{r}\binom{N-R}{n-r}.$$ (6.17)

This is not a hypergeometric distribution like (6.1) because the variable is now R instead of r.

The prior (6.14) yields, using (6.16),

$$\sum_{R=0}^{N} \frac{1}{N+1}\binom{R}{r}\binom{N-R}{n-r} = \frac{1}{N+1}\binom{N+1}{n+1} = \frac{1}{n+1}\binom{N}{n},$$ (6.18)

so the integral equation (6.11) is satisfied; with this prior the data can tell us nothing about N, beyond the fact that $N \geq n$.

Let us check (6.17) to see whether it satisfies some obvious common sense requirements. We see that it vanishes when $R < r$, or $R > N - n + r$, in agreement with what the data tell us by deductive reasoning. If we have sampled all the balls, $n = N$, then (6.17) reduces to Kronecker's delta, $\delta(R,r)$, again agreeing with deductive reasoning. This is another illustration of the fact that probability theory as extended logic automatically includes deductive logic as a special case.

If we obtain no data at all, $n = r = 0$, then (6.17) reduces, as it should, to the prior distribution: $p(R|DNI_0) = p(R|NI_0) = 1/(N+1)$. If we draw only one ball which proves to be red, $n = r = 1$, then (6.17) reduces to

$$p(R|DNI_0) = \frac{2R}{N(N+1)}.$$ (6.19)

The vanishing when $R = 0$ again agrees with deductive logic. From (6.1) the sampling probability $p(r = 1|n = 1, NR I_0) = R/N$ that our one ball would be red is our original Bernoulli urn result, proportional to R; and with a uniform prior the posterior probability for R must also be proportional to R. The numerical coefficient in (6.19) gives us an inadvertent derivation of the elementary sum rule,

$$\sum_{R=0}^{N} R = \frac{N(N+1)}{2}.$$ (6.20)

These results are only a few of thousands now known, indicating that probability theory as extended logic is an exact mathematical system. That is, results derived from correct application of our rules without approximation have the property of exact results in any other area of mathematics: you can subject them to arbitrary extreme conditions and they continue to make sense.[2]

What value of R does the robot estimate in general? The most probable value of R is found within one unit by setting $p(R') = p(R' - 1)$ and solving for R'. This yields

$$R' = (N + 1)\frac{r}{n}, \tag{6.21}$$

which is to be compared with (3.26) for the peak of the sampling distribution. If R' is not an integer, the most probable value is the next integer below R'. The robot anticipates that the fraction of red balls in the original urn should be about equal to the fraction in the observed sample, just as you and I would from intuition.

For a more refined calculation, let us find the mean value, or expectation, of R over this posterior distribution:

$$\langle R \rangle = E(R|DNI_0) = \sum_{R=0}^{N} R p(R|DNI_0). \tag{6.22}$$

To do the summation, note that

$$(R + 1)\binom{R}{r} = (r + 1)\binom{R + 1}{r + 1}, \tag{6.23}$$

and so, using (6.16) again,

$$\langle R \rangle + 1 = (r + 1)\binom{N + 1}{n + 1}^{-1}\binom{N + 2}{n + 2} = \frac{(N + 2)(r + 1)}{(n + 2)}. \tag{6.24}$$

When (n, r, N) are large, the expectation of R is very close to the most probable value (6.21), indicating either a sharply peaked posterior distribution or a symmetric one. This result becomes more significant when we ask: 'What is the expected fraction F of red balls left in the urn after this drawing?' This is

$$\langle F \rangle = \frac{\langle R \rangle - r}{N - n} = \frac{r + 1}{n + 2}. \tag{6.25}$$

6.4 Predictive distributions

Instead of using probability theory to estimate the unobserved contents of the urn, we may use it as well to predict future observations. We ask a different question: 'After having drawn a sample of r red balls in n draws, what is the probability that the next one drawn will be red?' Defining the propositions:

$$R_i \equiv \text{red on the } i\text{th draw}, \qquad 1 \le i \le N, \tag{6.26}$$

---

[2] By contrast, the intuitive *ad hockeries* of current 'orthodox' statistics generally give reasonable results within some 'safe' domain for which they were invented; but invariably they are found to yield nonsense in some extreme case. This, examined in Chapter 17, is what one expects of results which are only approximations to an exact theory; as one varies the conditions, the quality of the approximation varies.

this is

$$p(R_{n+1}|DNI_0) = \sum_{R=0}^{N} p(R_{n+1}R|DNI_0) = \sum_{R} p(R_{n+1}|RDNI_0)p(R|DNI_0), \quad (6.27)$$

or

$$p(R_{n+1}|DNI_0) = \sum_{R=0}^{N} \frac{R-r}{N-n} \binom{N+1}{n+1}^{-1} \binom{R}{r} \binom{N-R}{n-r}. \quad (6.28)$$

Using the summation formula (6.16) again, we find, after some algebra,

$$p(R_{n+1}|DNI_0) = \frac{r+1}{n+2}, \quad (6.29)$$

the same as (6.25). This agreement is another example of the rule noted before: a probability is not the same thing as a frequency; but, under quite general conditions, the *predictive probability* of an event at a single trial is numerically equal to the *expectation* of its frequency in some specified class of trials.

Equation (6.29) is a famous old result known as *Laplace's rule of succession*. It has played a major role in the history of Bayesian inference, and in the controversies over the nature of induction and inference. We shall find it reappearing many times; finally, in Chapter 18 we examine it carefully to see how it became controversial, but also how easily the controversies can be resolved today.

The result (6.29) has a greater generality than would appear from our derivation. Laplace first obtained it, not in consideration of drawing from an urn, but from considering a mixture of binomial distributions, as we shall do presently in (6.73). The above derivation in terms of urn sampling had been found as early as 1799 (see Zabell, 1989), but became well known only through its rediscovery in 1918 by C. D. Broad of Cambridge University, England, and its subsequent emphasis by Wrinch and Jeffreys (1919), W. E. Johnson (1924, 1932), and H. Jeffreys (1939). It was initially a great surprise to find that the urn result (6.29) is independent of N.

But this is only the point estimate; what accuracy does the robot claim for this estimate of R? The answer is contained in the same posterior distribution (6.17) that gave us (6.29); we may find its variance $\langle R^2 \rangle - \langle R \rangle^2$. Extending (6.23), note that

$$(R+1)(R+2)\binom{R}{r} = (r+1)(r+2)\binom{R+2}{r+2}. \quad (6.30)$$

The summation over R is again simple, yielding

$$\langle (R+1)(R+2) \rangle = (r+1)(r+2)\binom{N+1}{n+1}^{-1}\binom{N+3}{n+3}$$

$$= \frac{(r+1)(r+2)(N+2)(N+3)}{(n+2)(n+3)}. \quad (6.31)$$

Then, noting that $\text{var}(R) = \langle R^2 \rangle - \langle R \rangle^2 = \langle (R+1)^2 \rangle - \langle (R+1) \rangle^2$ and writing for brevity $p = \langle F \rangle = (r+1)/(n+2)$, from (6.24) and (6.31) we find

$$\text{var}(R) = \frac{p(1-p)}{n+3}(N+2)(N-n). \tag{6.32}$$

Therefore, our (mean) ± (standard deviation) combined point and interval estimate of R is

$$(R)_{\text{est}} = r + (N-n)p \pm \sqrt{\frac{p(1-p)}{n+3}(N+2)(N-n)}. \tag{6.33}$$

The factor $(N-n)$ inside the square root indicates that, as we would expect, the estimate becomes more accurate as we sample a larger fraction of the contents of the urn. Indeed, when $n = N$ the contents of the urn are known, and (6.33) reduces, as it should, to $(r \pm 0)$, in agreement with deductive reasoning.

Looking at (6.33), we note that $R - r$ is the number of red balls remaining in the urn, and $N - n$ is the total number of balls left in the urn; so an analytically simpler expression is found if we ask for the (mean) ± (standard deviation) estimate of the fraction of red balls remaining in the urn after the sample is drawn. This is found to be

$$(F)_{\text{est}} = \frac{(R-r)_{\text{est}}}{N-n} = p \pm \sqrt{\frac{p(1-p)}{n+3} \frac{N+2}{N-n}}, \qquad 0 \le n < N, \tag{6.34}$$

and this estimate becomes less accurate as we sample a larger portion of the balls. In the limit $N \to \infty$, this goes into

$$(F)_{\text{est}} = p \pm \sqrt{\frac{p(1-p)}{n+3}}, \tag{6.35}$$

which corresponds to the binomial distribution result. As an application of this, while preparing this chapter we heard a news report that a 'random poll' of 1600 voters was taken, indicating that 41% of the population favored a certain candidate in the next election, and claiming a ±3% margin of error for this result. Let us check the consistency of these numbers against our theory. To obtain $(F)_{\text{est}} = \langle F \rangle(1 \pm 0.03)$, we require, according to (6.35), a sample size n given by

$$n+3 = \frac{1-p}{p} \frac{1}{(0.03)^2} = \frac{1-0.41}{0.41} \times 1111 = 1598.9 \tag{6.36}$$

or $n \simeq 1596$. The close agreement suggests that the pollsters are using this theory (or at least giving implied lip service to it in their public announcements).

These results, found with a uniform prior for $p(R|NI_0)$ over $0 \le R \le N$, correspond very well with our intuitive common sense judgments. Other choices of the prior can affect the conclusions in ways which often surprise us at first glance; then, after some meditation, we see that they were correct after all. Let us put probability theory to a more severe test by considering some increasingly surprising examples.

6.5 Truncated uniform priors

Suppose our prior information had been different from the above I_0; our new prior information I_1 is that we know from the start that $0 < R < N$; there is at least one red and one white ball in the urn. Then the prior (6.14) must be replaced by

$$p(R|NI_1) = \begin{cases} \dfrac{1}{N-1} & \text{if } 1 \le R \le N-1 \\ 0 & \text{otherwise,} \end{cases} \tag{6.37}$$

and our summation formula (6.16) must be corrected by subtracting the two terms $R = 0, R = N$. Note that, if $R = 0$, then

$$\binom{R}{r} = \binom{R+1}{r+1} = \delta(r, 0), \tag{6.38}$$

and, if $R = N$, then

$$\binom{N-R}{n-r} = \delta(r, n), \tag{6.39}$$

so we have the summation formulas

$$S = \sum_{R=1}^{N-1} \binom{R}{r}\binom{N-R}{n-r} = \binom{N+1}{n+1} - \binom{N}{n}\delta(r, n) - \binom{N}{n}\delta(r, 0), \tag{6.40}$$

$$\sum_{R=1}^{N-1} \binom{R+1}{r+1}\binom{N-R}{n-r} = \binom{N+2}{n+2} - \binom{N+1}{n+1}\delta(r, n) - \binom{N}{n}\delta(r, 0). \tag{6.41}$$

What seems surprising at first is that, as long as the observed r is in $0 < r < n$, the new terms vanish, and so the previous posterior distribution (6.17) is unchanged:

$$p(R|DNI_1) = p(R|DNI_0), \qquad 0 < r < n. \tag{6.42}$$

Why does the new prior information make no difference? Indeed, it would certainly make a difference in any form of probability theory that uses only sampling distributions; for the sample space is changed by the new information.

Yet, on meditation, we see that the result (6.42) is correct, for in this case the data tell us by deductive reasoning that R cannot be zero or N; so, whether the prior information does or does not tell us the same thing cannot matter: our state of knowledge about R is the same and probability theory as logic so indicates. We discuss this further in Section 6.9.1.

Suppose that our data were $r = 0$; now the sum S in (6.15) is different:

$$S = \binom{N+1}{n+1} - \binom{N}{n}, \tag{6.43}$$

and in place of (6.17) the posterior probability distribution for R is found to be, after some calculation,

$$p(R|r = 0, NI_1) = \binom{N}{n+1}\binom{N-R}{n}, \qquad 1 \le R \le N - 1, \qquad (6.44)$$

and zero outside that range. But still, within that range, the relative probabilities of different values of R are not changed; we readily verify that the ratio

$$\frac{p(R|r = 0, NI_1)}{p(R|r = 0, NI_0)} = \frac{N+1}{N-n}, \qquad 1 \le R \le N - 1, \qquad (6.45)$$

is independent of R. What has happened here is that the datum $r = 0$ gives no evidence against the hypothesis that $R = 0$ and some evidence for it; so on prior information I_0 which allows this, $R = 0$ is the most probable value. But the prior information I_1 now makes a decisive difference; it excludes just that value, and thus forces all the posterior probability to be compressed into a smaller range, with an upward adjustment of the normalization coefficient. We learn from this example that different priors do not necessarily lead to different conclusions; and whether they do or do not can depend on which data set we happen to get – which is just as it should be.

Exercise 6.1. Find the posterior probability distribution $p(R|r = n, NI_1)$ by a derivation like the above. Then find the new (mean) $\pm$ (standard deviation) estimate of R from this distribution, and compare it with the above results from $p(R|r = n, NI_0)$. Explain the difference so that it seems obvious intuitively. Now show how well you understand this problem by describing in words, without doing the calculation, how the result would differ if we had prior information that ($3 \le R \le N$); i.e. the urn had initially at least three red balls, but there was no prior restriction on large values.

6.6 A concave prior

The rule of succession, based on the uniform prior $p(R|NI) \propto$ const. $(0 \le R \le N)$}, leads to a perhaps surprising numerical result, that the expected fraction (6.25) of red balls left in the urn is not the fraction r/n observed in the sample drawn, but slightly different, $(r + 1)/(n + 2)$. What is the reason for this small difference? The following argument is hardly a derivation, but only a line of free association. Note first that Laplace's rule of succession can be written in the form

$$\frac{r+1}{n+2} = \frac{n(r/n) + 2(1/2)}{n+2}, \qquad (6.46)$$

which exhibits the result as a weighted average of the observed fraction r/n and the prior expectation $1/2$, the data weighted by the number n of observations, the prior expectation by 2. It seems that the uniform prior carries a weight corresponding to two observations. Then could that prior be interpreted as a posterior distribution resulting from two observations

$(n, r) = (2, 1)$? If so, it seems that we must start from a still more uninformative prior than the uniform one. But is there any such thing as a still more uninformative prior?

Mathematically, this suggests that we try to apply Bayes' theorem backwards, to find whether there is any prior that would lead to a uniform posterior distribution. Denote this conjectured, still more primitive, state of 'pre-prior' information by I_{00}. Then Bayes' theorem would read:

$$p(R|DI_{00}) = p(R|I_{00})\frac{p(D|RI_{00})}{p(D|I_{00})} = \text{const.}, \qquad 0 \leq R \leq N, \qquad (6.47)$$

and the sampling distribution is still the hypergeometric distribution (6.1), because when R is specified it renders any vague information like I_{00} irrelevant: $p(D|RI_0) = p(D|RI_{00})$. With the assumed sample, $n = 2, r = 1$, the hypergeometric distribution reduces to

$$h(r = 1|N, R, n = 2) = \frac{R(N - R)}{N(N - 1)}, \qquad 0 \leq R \leq N, \qquad (6.48)$$

from which we see that there is no pre-prior that yields a constant posterior distribution over the whole range $(0 \leq R \leq N)$; it would be infinite for $R = 0$ and $R = N$. But we have just seen that the truncated prior, constant in $(1 \leq R \leq N - 1)$, yields the same results if it is known that the urn contains initially at least one red and one white ball. Since our presupposed data $(n, r) = (2, 1)$ guarantees this, we see that we have a solution after all. Consider the prior that emphasizes extreme values:

$$p(R|I_{00}) \equiv \frac{A}{R(N - R)}, \qquad 1 \leq R \leq N - 1, \qquad (6.49)$$

where A stands for a normalization constant, not necessarily the same in all the following equations. Given new data $D \equiv (n, r)$, if $1 \leq r \leq n - 1$ this yields, using (6.1), the posterior distribution

$$p(R|DNI_{00}) = \frac{A}{R(N - R)}\binom{R}{r}\binom{N - R}{n - r} = \frac{A}{r(n - r)}\binom{R - 1}{r - 1}\binom{N - R - 1}{n - r - 1}. \qquad (6.50)$$

From (6.16) we may deduce the summation formula

$$\sum_{r=1}^{N-1}\binom{R - 1}{r - 1}\binom{N - R - 1}{n - r - 1} = \binom{N - 1}{n - 1}, \qquad \begin{array}{l} 1 \leq R \leq N - 1, \\ 1 \leq r \leq n - 1, \end{array} \qquad (6.51)$$

so the correctly normalized posterior distribution is

$$p(R|DNI_{00}) = \binom{N - 1}{n - 1}^{-1}\binom{R - 1}{r - 1}\binom{N - R - 1}{n - r - 1}, \qquad \begin{array}{l} 1 \leq R \leq N - 1, \\ 1 \leq r \leq n - 1, \end{array} \qquad (6.52)$$

which is to be compared with (6.17). As a check, if $n = 2, r = 1$, this reduces to the desired prior (6.37):

$$p(R|DNI_{00}) = p(R|NI_1) = \frac{1}{N - 1}, \qquad 1 \leq R \leq N - 1. \qquad (6.53)$$

At this point, we can leave it as an exercise for the reader to complete the analysis for the concave prior with derivations analogous to (6.22)–(6.35).

Exercise 6.2. Using the general result (6.52), repeat the calculations analogous to (6.22)–(6.35) and prove three exact results: (a) the integral equation (6.11) is satisfied, so (6.6) still holds; (b) for general data compatible with the prior in the sense that $0 \leq n \leq N$, $1 \leq r \leq n - 1$ (so that the sample drawn includes at least one red and one white ball), the posterior mean estimated fractions R/N and $(R - r)/(N - n)$ are both equal to the observed fraction in the sample, $f = r/n$; the estimates now follow the data exactly so the concave prior (6.49) is given zero weight. Finally, (c) the (mean) $\pm$ (standard deviation) estimate is given by

$$\frac{(R)_{\text{est}}}{N} = f \pm \sqrt{\frac{f(1 - f)}{n + 1}} \left(1 - \frac{n}{N}\right), \tag{6.54}$$

also a simpler result than the analogous (6.33) found previously for the uniform prior.

Exercise 6.3. Now note that if $r = 0$ or $r = n$, the step (6.50) is not valid. Go back to the beginning and derive the posterior distribution for these cases. Show that if we draw one ball and find it not red, the estimated fraction of red in the urn now drops from $1/2$ to approximately $1/\log(N)$ (whereas with the uniform prior it drops to $(r + 1)/(n + 2) = 1/3$).

The exercises show that the concave prior gives many results simpler than those of the uniform one, but has also some near instability properties that become more pronounced with large N. Indeed, as $N \to \infty$ the concave prior approaches an improper (non-normalizable) one, which must give absurd answers to some questions, although it still gives reasonable answers to most questions (those in which the data are so informative that they remove the singularity associated with the prior).

6.7 The binomial monkey prior

Suppose prior information I_2 is that the urn was filled by a team of monkeys who tossed balls in at random, in such a way that each ball entering had independently the probability g of being red. Then our prior for R will be the binomial distribution (3.92): in our present notation,

$$p(R|N I_2) = \binom{N}{R} g^R (1 - g)^{N-R}, \qquad 0 \leq R \leq N, \tag{6.55}$$

and our prior estimate of the number of red balls in the urn will be the (mean) $\pm$ (standard deviation) over this distribution:

$$(R)_{\text{est}} = Ng \pm \sqrt{Ng(1 - g)}. \tag{6.56}$$

The sum (6.10) is readily evaluated for this prior, with the result that

$$p(D|NI) = \binom{n}{r} g^r (1-g)^{n-r}, \qquad N \geq n. \tag{6.57}$$

Since this is independent of N, this prior also satisfies our integral equation (6.11), so

$$p(NR|DI_2) = p(N|DI_2)p(R|NDI_2), \tag{6.58}$$

in which the first factor is the relatively uninteresting standard result (6.6). We are interested in the factor $p(R|NDI_2)$ in which N is considered known. We are interested also in the other factorization

$$p(NR|DI_2) = p(R|DI_2)p(N|RDI_2), \tag{6.59}$$

in which $p(R|DI)$ tells us what we know about R, regardless of N. (Try to guess intuitively how $p(R|DNI)$ and $p(R|DI)$ would differ for any I, before seeing the calculations.) Likewise, the difference between $p(N|RDI_2)$ and $p(N|DI_2)$ tells us how much we would learn about N if we were to learn the true R; and again our intuition can hardly anticipate the result of the calculation.

We have set up quite an agenda of calculations to do. Using (6.55) and (6.1), we find

$$p(R|DNI_2) = A \binom{N}{R} g^R (1-g)^{N-R} \binom{R}{r}\binom{N-R}{n-r}, \tag{6.60}$$

where A is another normalization constant. To evaluate it, note that we can rearrange the binomial coefficients:

$$\binom{N}{R}\binom{R}{r}\binom{N-R}{n-r} = \binom{N}{n}\binom{n}{r}\binom{N-n}{R-r}. \tag{6.61}$$

Therefore we find the normalization by

$$1 = \sum_R p(R|DNI_2) = A\binom{N}{n}\binom{n}{r}\sum_R \binom{N-n}{R-r} g^R (1-g)^{N-R}$$

$$= A\binom{N}{n}\binom{n}{r} g^r (1-g)^{n-r}, \qquad r \leq R \leq N-n+r, \tag{6.62}$$

and so our normalized posterior distribution for R is

$$p(R|DNI_2) = \binom{N-n}{R-r} g^{R-r} (1-g)^{N-R-n+r}, \tag{6.63}$$

from which we would make the (mean) $\pm$ (standard deviation) estimate

$$(R)_{\text{est}} = r + (N-n)g \pm \sqrt{g(1-g)(N-n)}. \tag{6.64}$$

But the resemblance to (6.33) suggests that we again look at it this way: we estimate the fraction of red balls left in the urn to be

$$\frac{(R-r)_{\text{est}}}{N-n} = g \pm \sqrt{\frac{g(1-g)}{N-n}}. \tag{6.65}$$

At first glance, (6.64) and (6.65) seem to be so much like (6.33) and (6.34) that it was hardly worth the effort to derive them. But on second glance we notice an astonishing fact: the parameter p in the former equations was determined entirely by the data; while g in the present ones is determined entirely by the prior information. In fact, (6.65) is exactly the prior estimate we would have made for the fraction of red balls in any subset of $N - n$ balls in the urn, *without any data at all*. It seems that the binomial prior has the magical property that it nullifies the data! More precisely, with that prior the data can tell us nothing at all about the unsampled balls.

Such a result will hardly commend itself to a survey sampler; the basis of his profession would be wiped out. Yet the result is correct, and there is no escape from the conclusion: if your prior information about the population is correctly described by the binomial prior, then sampling is futile (it tells you practically nothing about the population) unless you sample practically the whole population.

How can such a thing happen? Comparing the binomial prior with the uniform prior, one would suppose that the binomial prior, being moderately peaked, expresses more prior information about the proportion R/N of red balls; therefore by its use one should be able to improve his estimates of R. Indeed, we have found this effect, for the uncertainties in (6.64) and (6.65) are smaller than those in (6.33) and (6.34) by a factor of $\sqrt{(n+3)/(N+2)}$. What is intriguing is not the magnitude of the uncertainty, but the fact that (6.34) depends on the data, while (6.65) does not.

It is not surprising that the binomial prior is more informative about the unsampled balls than are the data of a small sample; but actually it is more informative about them than are *any amount* of data; even after sampling 99% of the population, we are no wiser about the remaining 1%.

So what is the invisible strange property of the binomial prior? It is in some sense so 'loose' that it destroys the logical link between different members of the population. But on meditation we see that this is just what was implied by our scenario of the urn being filled by monkeys tossing in balls in such a way that each ball had *independently* the probability g of being red. Given that filling mechanism, then knowing that any given ball is in fact red, gives one no information whatsoever about any other ball. That is, $P(R_1 R_2|I) = P(R_1|I)P(R_2|I)$. This logical independence in the prior is preserved in the posterior distribution.

Exercise 6.4. Investigate this apparent 'law of conservation of logical independence'. If the propositions: 'ith ball is red, $1 \leq i \leq N$' are logically independent in the prior information, what is the necessary and sufficient condition on the sampling distribution and the data that the factorization property is retained in the posterior distribution: $P(R_1 R_2|DI) = P(R_1|DI)P(R_2|DI)$?

This sets off another line of deep thought. In conventional probability theory, the binomial distribution is derived from the premise of causal independence of different tosses. In

Chapter 3 we found that consistency requires one to reinterpret this as logical independence. But now, can we reason in the opposite direction? *Does the appearance of a binomial distribution already imply logical independence of the separate events?* If so, then we could understand the weird result just derived, and anticipate many others like it. *We shall return to these questions in Part 2, after acquiring some more clues.*

6.8 Metamorphosis into continuous parameter estimation

As noted in the introduction to this chapter, if our hypotheses become so 'dense' that neighboring hypotheses (i.e. hypotheses with nearly the same values of the index t) are barely distinguishable in their observable consequences, then, whatever the data, their posterior probabilities cannot differ appreciably. So there cannot be one sharply defined hypothesis that is strongly favored over all others. Then it may be appropriate and natural to think of t as a continuously variable parameter θ, and to interpret the problem as that of making an estimate of the parameter θ, and a statement about the accuracy of the estimate.

A common and useful custom is to use Greek letters to denote continuously variable parameters, Latin letters for discrete indices or data values. We shall adhere to this except when it would conflict with a more deeply entrenched custom.[3]

The hypothesis testing problem has thus metamorphosed into a parameter estimation one. But it can equally well metamorphose back; for the hypothesis that a parameter θ lies in a certain interval $a < \theta < b$ is, of course, a compound hypothesis as defined in Chapter 4, so an interval estimation procedure (i.e. one where we specify the accuracy by giving the probability that the parameter lies in a given interval) is automatically a compound hypothesis testing procedure.

Indeed, we followed just this path in Chapter 4 and found ourselves, at Eq. (4.67), doing what is really parameter estimation. It seemed to us natural to pass from testing simple discrete hypotheses, to estimating continuous parameters, and finally to testing compound hypotheses at Eq. (4.74), because probability theory as logic does this automatically. As in our opening remarks, we do not see parameter estimation and hypothesis testing as fundamentally different activities – one aspect of the greater unity of probability theory as logic.

But this unity has not seemed at all natural to some others. Indeed, in orthodox statistics parameter estimation appears very different from hypothesis testing, both mathematically and conceptually, largely because it has no satisfactory way to deal with compound hypotheses or prior information. We shall see some specific consequences of this in Chapter 17. Of course, parameters need not be one-dimensional; but let us consider first some simple cases where they are.

6.9 Estimation with a binomial sampling distribution

We have already seen an example of a binomial estimation problem in Chapter 4, but we did not note its generality. There are hundreds of real situations in which each time a simple

---

[3] Thus for generations the charge on the electron and the velocity of light have been denoted by e, c, respectively. No scientist or engineer could bring himself to represent them by Greek letters, even when they are the parameters being estimated.

measurement or observation is made, there are only two possible results. The coin will show either heads or tails, the battery will or will not start the car, the baby will be a boy or a girl, the check will or will not arrive in the mail today, the student will pass or flunk the examination, etc. As we noted in Chapter 3, the first comprehensive sampling theory analysis of such an experiment was by James Bernoulli (1713) in terms of drawing balls from an urn, so such experiments are commonly called *Bernoulli trials*.

Traditionally, for any such binary experiment, we call one of the results, arbitrarily, a 'success' and the other a 'failure'. Generally, our data will be a record of the number of successes and the number of failures;[4] the order in which they occur may or may not be meaningful, and, if it is meaningful, it may or may not be known, and, if it is known, it may or may not be relevant to the question we are asking. Presumably, the conditions of the experiment will tell us whether the order is meaningful or known; and we expect probability theory to tell us whether it is relevant.

For example, if we toss ten coins simultaneously, then we have performed ten Bernoulli trials, but it is not meaningful to speak of their 'order'. If we toss one coin 100 times and record each result, then the order of the results is meaningful and known; but in trying to judge whether the coin is 'honest', common sense probably tells us that the order is not relevant. If we are observing patient recoveries from a disease and trying to judge whether resistance to the disease was improved by a new medicine introduced a month ago, this is much like drawing from an urn whose contents may have changed. Intuition then tells us that the order in which recoveries and nonrecoveries occur is not only highly relevant, it is the crucial information without which no inference about a change is possible.[5]

To set up the simple general binomial sampling problem, define

$$x_i \equiv \begin{cases} 1 & \text{if the } i\text{th trial yields success} \\ 0 & \text{otherwise.} \end{cases} \tag{6.66}$$

Then our data are $D \equiv \{x_1, \ldots, x_n\}$. The prior information I specifies that there is a parameter θ such that at each trial we have, independently of anything we know about other trials, the probability θ for a success, therefore the probability $(1 - \theta)$ for a failure. As discussed before, by 'independent' we mean logical independence. There may or may not be causal independence, depending on further details of I that do not matter at the moment. The sampling distribution is then (mathematically, this is our *definition* of the model to be studied):

$$p(D|\theta I) = \prod_{i=1}^{n} p(x_i|\theta I) = \theta^r (1 - \theta)^{n-r}, \tag{6.67}$$

---

[4] However, there are important problems involving censored data, to be considered later, in which only the successes can be recorded (or only the failures), and one does not know how many trials were performed. For example, a highway safety engineer knows from the public record how many lives were lost in spite of his efforts, but not how many were saved because of them.

[5] Of course, the final arbiter of relevance is not our intuition, but the equations of probability theory. But, as we shall see later, judging this can be a tricky business. Whether a given piece of information is or is not relevant depends not only on what question we are asking, but also on the totality of all of our other information.

in which r is the number of successes observed, $(n - r)$ the number of failures. Then, with any prior probability density function $p(\theta|I)$, we have immediately the posterior pdf

$$p(\theta|DI) = \frac{p(\theta|I)p(D|\theta I)}{\int d\theta \, p(\theta|I)p(D|\theta I)} = Ap(\theta|I)\theta^r(1 - \theta)^{n-r}, \tag{6.68}$$

where A is a normalizing constant. With a uniform prior for θ,

$$p(\theta|I) = 1, \qquad 0 \le \theta \le 1, \tag{6.69}$$

the normalization is determined by an Eulerian integral,

$$A^{-1} = \int_0^1 d\theta \, \theta^r(1 - \theta)^{n-r} = \frac{r!(n - r)!}{(n + 1)!} \tag{6.70}$$

and the normalized pdf is

$$p(\theta|DI) = \frac{(n + 1)!}{r!(n - r)!}\theta^r(1 - \theta)^{n-r}, \tag{6.71}$$

identical to Bayes' original result, noted in Chapter 4, Eq. (4.67). Its moments are

$$
\begin{aligned}
\langle \theta^m \rangle = E(\theta^m|DI) &= A \int_0^1 d\theta \, \theta^{r+m}(1 - \theta)^{n-r} = \frac{(n + 1)!}{(n + m + 1)!}\frac{(r + m)!}{r!} \\
&= \frac{(r + 1)(r + 2)\cdots(r + m)}{(n + 2)(n + 3)\cdots(n + m + 1)},
\end{aligned} \tag{6.72}
$$

leading to the predictive probability for success at the next trial of

$$p \equiv \langle \theta \rangle = \int_0^1 d\theta \, \theta p(\theta|DI) = \frac{r + 1}{n + 2}, \tag{6.73}$$

in which we see Laplace's rule of succession in its original derivation. Likewise, the (mean) $\pm$ (standard deviation) estimate of θ is

$$(\theta)_{\text{est}} = \langle \theta \rangle \pm \sqrt{\langle \theta^2 \rangle - \langle \theta \rangle^2} = p \pm \sqrt{\frac{p(1 - p)}{n + 3}}. \tag{6.74}$$

Indeed, the continuous results (6.73) and (6.74) must be derivable from the discrete ones (6.29) and (6.35) by passage to the limit $N \to \infty$; but since the latter equations are independent of N, the limit has no effect.

In this limit the concave pre-prior distribution (6.49) would go into an improper prior for θ:

$$\frac{A}{R(N - R)} \to \frac{d\theta}{\theta(1 - \theta)}, \tag{6.75}$$

for which some sums or integrals would diverge; but that is not the strictly correct method of calculation. For example, to calculate the posterior expectation of any function $f(R/N)$ in the limit of arbitrarily large N, we should take the limit of the ratio $\langle f(R/N) \rangle = \text{Num/Den}$,

where

$$\text{Num} \equiv \sum_{R=1}^{N-1} \frac{f(R/N)}{R(N-R)} p(D|NRI),$$

$$\text{Den} \equiv \sum_{R=1}^{N-1} \frac{1}{R(N-R)} p(D|NRI),$$

(6.76)

and, under very general conditions, this limit is well-behaved, leading to useful results. The limiting improper pre-prior (6.75) was advocated by Haldane (1932) and Jeffreys (1939), in the innocent days before the marginalization paradox, when nobody worried about such fine points. We were almost always lucky in that our integrals converged in the limit, so we used them directly, thus actually calculating the ratio of the limits rather than the limit of the ratio; but nevertheless getting the right answers. With this fine point now clarified, all this and its obvious generalizations seem perfectly straightforward; however, note the following point, important for a current controversy.

6.9.1 Digression on optional stopping

We did not include n in the conditioning statements in $p(D|\theta I)$ because, in the problem as defined, it is from the data D that we learn both n and r. But nothing prevents us from considering a different problem in which we decide in advance how many trials we shall make; then it is proper to add n to the prior information and write the sampling probability as $p(D|n\theta I)$. Or, we might decide in advance to continue the Bernoulli trials until we have achieved a certain number r of successes, or a certain log-odds $u = \log[r/(n-r)]$; then it would be proper to write the sampling probability as $p(D|r\theta I)$ or $p(D|u\theta I)$, and so on. Does this matter for our conclusions about θ?

In deductive logic (Boolean algebra) it is a triviality that $AA = A$; if you say: 'A is true' twice, this is logically no different from saying it once. This property is retained in probability theory as logic, since it was one of our basic desiderata that, in the context of a given problem, propositions with the same truth value are always assigned the same probability. In practice this means that there is no need to ensure that the different pieces of information given to the robot are independent; our formalism has automatically the property that redundant information is not counted twice.

Thus, in our present problem, the data, as defined, tell us n. Then, since $p(n|n\theta I) = 1$, the product rule may be written

$$p(nr\ \text{order}|n\theta I) = p(r\ \text{order}|n\theta I)p(n|n\theta I) = p(r\ \text{order}|n\theta I).$$

(6.77)

When something is known already from the prior information, then, whether the data do or do not tell us the same thing cannot matter; the likelihood function is the same. Likewise, write the product rule as

$$p(\theta n|DI) = p(\theta|nDI)p(n|DI) = p(n|\theta DI)p(\theta|DI),$$

(6.78)

or, since $p(n|\theta DI) = p(n|DI) = 1$,

$$p(\theta|nDI) = p(\theta|DI). \qquad (6.79)$$

In this argument we could replace n by any other quantity (such as r, or $(n-r)$, or $u \equiv \log r/(n-r)$) that was known from the data; if any part of the data happens to be included also in the prior information, then that part is redundant and it cannot affect our final conclusions.

Even so, some statisticians (for example, Armitage, 1960) who look only at sampling distributions, claim that the stopping rule *does* affect our inference. Apparently, they believe that if a statistic such as r is not known in advance, then parts of the sample space referring to false values of r remain relevant to our inferences, even after the true value of r becomes known from the data D, although they would not be relevant (they would not even be in the sample space) if the true value were known before seeing the data. Of course, that violates the principle $AA = A$ of elementary logic; it is astonishing that such a thing could be controversial in the 20th century.

It is evident that this same comment applies with equal force to any function $f(D)$ of the data, whether or not we are using it as an estimator. That is, whether f was or was not known in advance can have a major effect on our sample space and sampling distributions; but as redundant information it cannot have any effect on any rational inferences from the data. Furthermore, inference must depend on the data set that was observed, not on data sets that might have been observed but were not – because merely noting the possibility of unobserved data sets gives us no information that was not already in the prior information. Although this conclusion might have seemed obvious from the start, it is not recognized in orthodox statisticians because they do not think in terms of information. We shall see in Chapter 8 not only some irrational conclusions, but some absolutely spooky consequences (psychokinesis, black magic) this has had, and, in later applications, how much real damage this has caused.

What if a part of the data set was actually generated by the phenomenon being studied, but for whatever reason we failed to observe it? This is a major difficulty for orthodox statistics, because then the sampling distributions for our estimators are wrong, and the problem must be reconsidered from the start. But for us it is only a minor detail, easily taken into account. We show next that probability theory as logic tells us uniquely how to deal with true but unobserved data; they must be relevant in the sense that our conclusions must depend on whether they were or were not observed; so they have a mathematical status somewhat like that of a set of nuisance parameters.

6.10 Compound estimation problems

We now consider in some depth a class of problems more complicated in structure, where more than one process is occurring but not all the results are observable. We want to make inferences not only about parameters in the model, but about the unobserved data.

The mathematics to be developed next is applicable to a large number of quite different real problems. To form an idea of the scope of the theory, consider the following scenarios.

(A) In the general population, there is a frequency p at which any given person will contract a certain disease within the next year; and then another frequency θ that anyone with the disease will die of it within a year. From the observed variations $\{c_1, c_2, \ldots\}$ of deaths from the disease in successive years (which is a matter of public record), estimate how the incidence of the disease $\{n_1, n_2, \ldots\}$ is changing in the general population (which is not a matter of public record).

(B) Each week, a large number of mosquitos, N, is bred in a stagnant pond near this campus, and we set up a trap on the campus to catch some of them. Each mosquito lives less than a week, during which there is a frequency p of flying onto the campus and, once on the campus, a frequency θ of being caught in our trap. We count the numbers $\{c_1, c_2, \ldots\}$ caught each week. From these data and whatever prior information we have, what can we say about the numbers $\{n_1, n_2, \ldots\}$ on the campus each week, and what can we say about N?

(C) We have a radioactive source (say sodium 22 (^{22}Na), for example) which is emitting particles of some sort (say, positrons). There is a rate p, in particles per second, at which a radioactive nucleus sends particles through our counter; and each particle passing through produces counts at the rate θ. From measuring the number $\{c_1, c_2, \ldots\}$ of counts in different seconds, what can we say about the numbers $\{n_1, n_2, \ldots\}$ actually passing through the counter in each second, and what can we say about the strength of the source?

The common feature in these problems is that we have two 'binary games' played in succession, and we can observe only the outcome of the last one. From this, we are to make the best inferences we can about the original cause and the intermediate conditions. This could be described also as the problem of trying to recover, in one special case, censored data.

We want to show particularly how drastically these problems are changed by various changes in the prior information. For example, our estimates of the variation in incidence of a disease are greatly affected, not only by the data, but by our prior information about the process by which one contracts that disease.[6]

In our estimates we will want to (1) state the 'best' estimate possible on the data and prior information; and (2) make a statement about the accuracy of the estimate, giving again our versions of 'point estimation' and 'interval estimation' about which Fisher commented. We shall use the language of the radioactive source scenario, but it will be clear enough that the same arguments and the same calculations apply in many others.

6.11 A simple Bayesian estimate: quantitative prior information

Firstly, we discuss the parameter ϕ, which a scientist would call the 'efficiency' of the counter. By this we mean that, if ϕ is known, then each particle passing through the counter has *independently* the probability ϕ of making a count. Again we emphasize that this is not

---

[6] Of course, in this first venture into the following kind of analysis, we shall not take into account all the factors that operate in the real world, so some of our conclusions may be changed in a more sophisticated analysis. However, nobody would see how to do that unless he had first studied this simple introductory example.

mere causal independence (which surely always holds, as any physicist would assure us); we mean *logical* independence, i.e., if ϕ is known, then knowing anything about the number of counts produced by other particles would tell us nothing more about the probability for the next particle making a count.[7]

We have already stressed the distinction between logical and causal dependence many times; and now we have another case where failure to understand it could lead to serious errors. The point is that causal influences operate in the same way independently of your state of knowledge or mine; thus, if ϕ is not known, then everybody still believes that successive counts are *causally* independent. But they are no longer *logically* independent; for then knowing the number of counts produced by other particles tells us something about ϕ, and therefore modifies our probability that the next particle will produce a count. The situation is much like that of sampling with replacement, discussed in Chapter 3, where each ball drawn tells us something more about the contents of the urn.

From the independence, the probability that n particles will produce exactly c counts in any specified order is $\phi^c(1-\phi)^{n-c}$, and there are $\binom{n}{c}$ possible sequences producing c counts, so the probability of getting c counts regardless of order is the binomial distribution

$$b(c|n, \phi) = \binom{n}{c}\phi^c(1-\phi)^{n-c}, \qquad 0 \le c \le n. \tag{6.80}$$

From the standpoint of logical presentation in the real world, however, we have to carry out a kind of bootstrap operation with regard to the quantity ϕ; for how could it be known? Intuitively, you may have no difficulty in seeing the procedure you would use to determine ϕ from measurements with the counter. But, logically, we need to have the calculation about to be given before we can justify that procedure. So, for the time being, we'll just have to suppose that ϕ is a number given to us by our teacher in assigning us this problem, and have faith that, in the end, we shall understand how our teacher determined it.

Now let us introduce a quantity r which is the probability at which in any one second that any particular nucleus will emit a particle that passes through the counter. We assume the number of nuclei N to be so large and the half-life so long that we need not consider N as a variable for this problem. So there are N nuclei, each of which has independently the probability r of sending a particle through our counter in any one second. The quantity r is also, for present purposes, just a number given to us in the statement of the problem, because we have not yet seen, in terms of probability theory, the line of reasoning by which we could convert measurements into a numerical value of r (but again, you see intuitively, without any hesitation, that r is a way of describing the half-life of the source).

Suppose we were given N and r; what is the probability, on this evidence, that in any one second exactly n particles will pass through the counter? That is the same binomial

---

[7] In practice, there is a question of resolving time; if the particles come too close together we may not be able to see the counts as separate, because the counter experiences a 'dead time' after a count, during which it is unable to respond to another particle. We have disregarded those difficulties for this problem and imagined that we have infinitely good resolving time (or, what amounts to the same thing, that the counting rate is so low that there is negligible probability for missing a count). After we have developed the theory, the reader will be asked (Exercise 6.5) to generalize it to take these factors into account.

distribution problem, so the answer is

$$b(n|N,r) = \binom{N}{n} r^n (1-r)^{N-n}. \tag{6.81}$$

But in this case there's a good approximation to the binomial distribution, because the number N is enormously large and r is enormously small. In the limit $N \to \infty, r \to 0$ in such a way that $Nr \to s = \text{const.}$, what happens to (6.81)? To find this, write $r = s/N$, and pass to the limit $N \to \infty$. Then

$$\frac{N!}{(N-n)!} r^n = N(N-1)\cdots(N-n+1)\left(\frac{s}{N}\right)^n$$
$$= s^n \left(1 - \frac{1}{N}\right)\left(1 - \frac{2}{N}\right)\cdots\left(1 - \frac{n-1}{N}\right), \tag{6.82}$$

which goes into s^n in the limit. Likewise,

$$(1-r)^{N-n} = \left(1 - \frac{s}{N}\right)^{N-n} \to \exp\{-s\}, \tag{6.83}$$

and so the binomial distribution (6.81) goes over into the simpler *Poisson distribution*:

$$p(n|Nr) \to p(n|s) = \exp\{-s\}\frac{s^n}{n!}, \tag{6.84}$$

and it will be handy for us to take this limit. The number s is essentially what the experimenter calls his 'source strength', the expectation of the number of particles per second.

Now we have enough 'formalism' to start solving useful problems. Suppose we are not given the number of particles n in the counter, but only the source strength s. What is the probability, on this evidence, that we shall see exactly c counts in any one second? Using our method of resolving the proposition c into a set of mutually exclusive alternatives, then applying the sum rule and the product rule:

$$p(c|\phi s) = \sum_{n=0}^{\infty} p(cn|\phi s) = \sum_n p(c|n\phi s)p(n|\phi s) = \sum_n p(c|\phi n)p(n|s), \tag{6.85}$$

since $p(c|n\phi s) = p(c|\phi n)$; i.e. if we knew the actual number n of particles in the counter, it would not matter what s was. This is perhaps made clearer by the diagram in Figure 6.1, rather like the logic flow diagrams of Figure 4.3. In this case, we think of the diagram as indicating not only the logical connections, but also the causal ones; s is the physical cause which partially determines n; and then n in turn is the physical cause which partially determines c. To put it another way, s can influence c only through its intermediate influence on n. We saw the same logical situation in the Chapter 5 horse racing example.

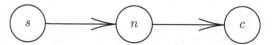

Fig. 6.1. The causal influences.

Since we have worked out both $p(c|\phi n)$ and $p(n|s)$, we need only substitute them into (6.85); after some algebra we have

$$p(c|\phi s) = \sum_{n=c}^{\infty} \left[\frac{n!}{c!(n-c)!} \phi^c (1-\phi)^{n-c} \right] \left[\frac{\exp\{-s\} s^n}{n!} \right] = \frac{\exp\{-s\phi\}(s\phi)^c}{c!}. \quad (6.86)$$

This is again a Poisson distribution with expectation

$$\langle c \rangle = \sum_{c=0}^{\infty} c p(c|\phi s) = s\phi. \quad (6.87)$$

Our result is hardly surprising. We have a Poisson distribution with a mean value which is the product of the source strength times the efficiency of the counter. Without going through the analysis, that is just the estimate of c that we would make intuitively, although it is unlikely that anyone could have guessed from intuition that the distribution still has the Poissonian form.

In practice, it is c that is known, and n that is unknown. If we knew the source strength s and also the number of counts c, what would be the probability, on that evidence, that there were exactly n particles passing through the counter during that second? This is a problem which arises all the time in physics laboratories, because we may be using the counter as a 'monitor', and have it set up so that the particles, after going through the counter, then initiate some other reaction which is the one we're really studying. It is important to get the best possible estimates of n, because that is one of the numbers we need in calculating the cross-section of this other reaction. Bayes' theorem gives

$$p(n|\phi cs) = p(n|s) \frac{p(c|n\phi s)}{p(c|\phi s)} = \frac{p(n|s)p(c|n\phi)}{p(c|\phi s)}, \quad (6.88)$$

and all these terms have been found above, so we just have to substitute (6.80) and (6.84)–(6.86) into (6.88). Some terms cancel, and we are left with:

$$p(n|\phi cs) = \frac{\exp\{-s(1-\phi)\}[s(1-\phi)]^{n-c}}{(n-c)!}. \quad (6.89)$$

It is interesting that we *still* have a Poisson distribution, now with parameter $s(1-\phi)$, but shifted upward by c; because, of course, n could not be less than c. The expectation over this distribution is

$$\langle n \rangle = \sum_{n} n p(n|\phi cs) = c + s(1-\phi). \quad (6.90)$$

So, now what is the best guess the robot can make as to the number of particles responsible for those c counts? Since this is the first time we have faced this issue in a serious way, let us take time for some discussion.

6.11.1 From posterior distribution function to estimate

Given the posterior pdf for some general parameter α, continuous or discrete, what 'best' estimate of α should the robot make, and what accuracy should it claim? There is no one 'right' answer; the problem is really one of decision theory which asks, 'What should we do?' This involves value judgments and therefore goes beyond the principles of inference, which ask only 'What do we know?' We shall return to this in Chapters 13 and 14, but for now we give a preliminary discussion adequate for the simple problems being considered.

Laplace (1774) already encountered this problem. The unknown true value of a parameter is α, and given some data D and prior information I we are to make an estimate $\alpha^*(D, I)$ which depends on them in some way. In the jargon of the trade, α^* is called an 'estimator', and nothing prevents us from considering any function of (D, I) whatsoever as a potential estimator. But which estimator is best? Our estimate will have an error $e = (\alpha^* - \alpha)$, and Laplace gave as a criterion that we should make that estimate which minimizes the expected magnitude $|e|$. He called this the 'most advantageous' method of estimation.

Laplace's criterion was generally rejected for 150 years in favor of the least squares method of Gauss and Legendre; we seek the estimate that minimizes the expected square of the error. In these early works it is not always clear whether this means expected over the sampling pdf for α^* or over the posterior pdf for α; the distinction was not always recognized, and the confusion was encouraged by the fact that in some cases considerations of symmetry lead us to the same final conclusion from either. Some of the bad consequences of using the former (sampling distribution) are noted in Chapter 13. It is clear today that the former ignores all prior information about α, while the latter takes it into account and is therefore what we want; taking expectations over the posterior pdf for α, the expected squared error of the estimate is

$$\langle(\alpha - \alpha^*)^2\rangle = \langle\alpha^2\rangle - 2\alpha^*\langle\alpha\rangle + \alpha^{*2}$$
$$= (\alpha^* - \langle\alpha\rangle)^2 + (\langle\alpha^2\rangle - \langle\alpha\rangle^2). \tag{6.91}$$

The choice

$$\alpha^* = \langle\alpha\rangle = \int d\alpha\, \alpha p(\alpha|DI), \tag{6.92}$$

that is, the posterior mean value, therefore always minimizes the expected square of the error, over the posterior pdf for α, and the minimum achievable value is the variance of the posterior pdf. The second term is the expected square of the deviation from the mean:

$$\text{var}(\alpha) \equiv \langle(\alpha - \langle\alpha\rangle)^2\rangle = (\langle\alpha^2\rangle - \langle\alpha\rangle^2), \tag{6.93}$$

often miscalled the *variance of* α; of course, it is really the variance of the *probability distribution* that the robot assigns for α. In any event, the robot can do nothing to minimize it. But the first term can be removed entirely by taking as the estimate just the mean value $\alpha^* = \langle\alpha\rangle$, which is the optimal estimator by the mean square error criterion.

Evidently, this result holds generally whatever the form of the posterior distribution $p(\alpha|DI)$; provided only that $\langle\alpha\rangle$ and $\langle\alpha^2\rangle$ exist, the mean square error criterion always leads to taking the mean value $\langle \alpha \rangle$ (i.e. the 'center of gravity' of the posterior distribution) as the 'best' guess. The posterior (mean $\pm$ standard deviation) then recommends itself to us as providing a more or less reasonable statement of what we know and how accurately we know it; and it is almost always the easiest to calculate. Furthermore, if the posterior pdf is sharp and symmetrical, this cannot be very different pragmatically from any other reasonable estimate. So, in practice, we use this more than any other. In the urn inversion problems, we simply adopted this procedure without comment.

But this may not be what we really want. We should be aware that there are valid arguments against the posterior mean, and cases where a different rule would better achieve what we want. The squared error criterion says that an error twice as great is considered four times as serious. Therefore, the mean value estimate, in effect, concentrates its attention most strongly on avoiding the very large (but also very improbable) errors, at the cost of possibly not doing as well as it might with the far more likely small errors.

Because of this, the posterior mean value estimate is quite sensitive to what happens far out in the tails of the pdf. If the tails are very unsymmetrical, our estimate could be pulled far away from the central region where practically all the probability lies and common sense tells us the parameter is most likely to be. In a similar way, a single very rich man in a poor village would pull the average wealth of the population far away from anything representative of the real wealth of the people. If we knew this was happening, then that average would be a quite irrational estimate of the wealth of any particular person met on the street.

This concentration on minimizing the large errors leads to another property that we might consider undesirable. Of course, by 'large errors' we mean errors that are large *on the scale of the parameter* α. If we redefined our parameter as some nonlinear function $\lambda = \lambda(\alpha)$ (for example, $\lambda = \alpha^3$, or $\lambda = \log(\alpha)$), an error that is large on the scale of α might seem small on the scale of λ, and vice versa. But then the posterior mean estimate

$$\lambda^* \equiv \langle \lambda \rangle = \int d\lambda \, \lambda p(\lambda|DI) = \int d\alpha \, \lambda(\alpha)p(\alpha|DI) \tag{6.94}$$

would not in general satisfy $\lambda^* = \lambda(\alpha^*)$. Minimizing the mean square error in α is not the same thing as minimizing the mean square error in $\lambda(\alpha)$.

Thus, the posterior mean value estimates lack a certain consistency under parameter changes. When we change the definition of a parameter, if we continue to use the mean value estimate, then we have changed the criterion of what we mean by a 'good' estimate.

Now let us examine Laplace's original criterion. If we choose an estimator $\alpha^+(D, I)$ by the criterion that it minimizes the expected absolute error

$$E \equiv \langle |\alpha^+ - \alpha| \rangle = \int_{-\infty}^{\alpha^+} d\alpha \, (\alpha^+ - \alpha)f(\alpha) + \int_{\alpha^+}^{\infty} d\alpha \, (\alpha - \alpha^+)f(\alpha), \tag{6.95}$$

we require

$$\frac{dE}{d\alpha^+} = \int_{-\infty}^{\alpha^+} d\alpha \, f(\alpha) - \int_{\alpha^+}^{\infty} d\alpha \, f(\alpha) = 0, \tag{6.96}$$

or $P(\alpha > \alpha^+ | DI) = 1/2$; Laplace's 'most advantageous' estimator is the *median* of the posterior pdf.

But what happens now on a change of parameters $\lambda = \lambda(\alpha)$? Suppose that λ is strictly a monotonic increasing function of α (so that α is in turn a single-valued function of λ and the transformation is reversible). Then it is clear from the above equation that the consistency is restored: $\lambda^+ = \lambda(\alpha^+)$.

More generally, all the percentiles have this invariance property: for example, if $\alpha 35$ is the 35 percentile value of α:

$$\int_{-\infty}^{\alpha 35} d\alpha\, f(\alpha) = 0.35, \tag{6.97}$$

then we have at once

$$\lambda_{35} = \lambda(\alpha 35). \tag{6.98}$$

Thus if we choose as our point estimate and accuracy claim the median and interquartile span over the posterior pdf, these statements will have an invariant meaning, independent of how we have defined our parameters. Note that this remains true even when $\langle \alpha \rangle$ and $\langle \alpha^2 \rangle$ diverge, so the mean square estimator does not exist.

Furthermore, it is clear from their derivation from variational arguments, that the median estimator considers an error twice as great to be only twice as serious, so it is less sensitive to what happens far out in the tails of the posterior pdf than is the mean value. In current technical jargon, one says that the median is more *robust* with respect to tail variations. Indeed, it is obvious that the median is entirely independent of all variations that do not move any probability from one side of the median to the other; and an analogous property holds for any percentile. One very rich man in a poor village has no effect on the median wealth of the population.

Robustness, in the general sense that the conclusions are insensitive to small changes in the sampling distribution or other conditions, is often held to be a desirable property of an inference procedure, and some authors criticize Bayesian methods because they suppose that they lack robustness. However, robustness in the usual sense of the word can always be achieved merely by *throwing away* cogent information! It is hard to believe that anyone could really want this if he were aware of it; but since those with only orthodox training do not think in terms of information content, they do not realize when they are wasting information. Evidently, the issue requires a much more careful discussion, to which we return later (Chapter 20) in connection with model comparison.[8]

In at least some problems, then, Laplace's 'most advantageous' estimates have indeed two significant advantages over the more conventional (mean $\pm$ standard deviation). But

---

[8] To anticipate our final conclusion: robustness with respect to sampling distributions is desirable only when we are not sure of the correctness of our model. But then a full Bayesian analysis will take into account all the models considered possible and their prior probabilities. The result automatically achieves the robustness previously sought in intuitive *ad hoc* devices; and some of those devices, such as the 'jackknife' and the 'redescending psi function' are derived from first principles, as first-order approximations to the Bayesian result. The Bayesian analysis of such problems gives us for the first time a clear statement of the circumstances in which robustness is desirable; and then, because Bayesian analysis never throws away information, it gives us more powerful algorithms for achieving robustness.

before the days of computers they were prohibitively difficult to calculate numerically, so the least squares philosophy prevailed as a matter of practical expedience.

Today, the computation problem is relatively trivial, and we can have whatever we want. It is easy to write computer programs which give us the option of displaying either the first and second moments or the quartiles (x_{25}, x_{50}, x_{75}), and only the force of long habit makes us continue to cling to the former.[9]

Still another principle for estimation is to take the peak $\hat{\alpha}$; or, as it is called, the 'mode' of the posterior pdf. If the prior pdf is a constant (or is at least constant in a neighborhood of this peak and not sufficiently greater elsewhere), the result is identical to the 'maximum likelihood' estimate (MLE) α' of orthodox statistics. It is usually attributed to R. A. Fisher, who coined that name in the 1920s, although Laplace and Gauss used the method routinely 100 years earlier without feeling any need to give it a special name other than 'most probable value'. As explained in Chapter 16, Fisher's ideology would not permit him to call it that. The merits and demerits of the MLE are discussed further in Chapter 13; for the present, we are not concerned with philosophical arguments, but wish only to compare the pragmatic results of MLE and other procedures.[10] This leads to some surprises, as we see next.

We will now return to our original problem of counting particles. At this point, a statistician of the 'orthodox' school of thought pays a visit to our laboratory. We describe the properties of the counter to him, and invite him to give us *his* best estimate as to the number of particles. He will, of course, use maximum likelihood because his textbooks have told him that (Cramér, 1946, p. 498): 'From a theoretical point of view, the most important general method of estimation so far known is the method of maximum likelihood.' His likelihood function is, in our notation, $p(c|\phi n)$ in its dependence on n. The value of n which maximizes it is found, within one unit, from setting

$$\frac{p(c|\phi n)}{p(c|\phi, n-1)} = \frac{n(1-\phi)}{n-c} = 1, \tag{6.99}$$

or

$$(n)_{\text{MLE}} = \frac{c}{\phi}. \tag{6.100}$$

You may find the difference between the two estimates (6.90) and (6.100) rather startling, particularly if we put in some numbers. Suppose our counter has an efficiency of 10%; in other words, $\phi = 0.1$, and the source strength is $s = 100$ particles per second, so that the expected counting rate according to (6.87) is $\langle c \rangle = s\phi = 10$ counts per second. But in this particular second, we got 15 counts. What should we conclude about the number of particles?

[9] In spite of all these considerations, the neat analytical results found in our posterior moments from urn and binomial models, contrasted with the messy appearance of calculations with percentiles, show that moments have some kind of theoretical significance that percentiles lack. This appears more clearly in Chapter 7.

[10] One evident pragmatic result is that the MLE fails altogether when the likelihood function has a flat top; then nothing in the data can give us a reason for preferring any point in that flat top over any other. But this is just the case we have in the 'generalized inverse' problems of current importance in applications; only prior information can resolve the ambiguity.

Probably the first answer one would give without thinking is that, if the counter has an efficiency of 10%, then in some sense each count must have been due to about ten particles; so, if there were 15 counts, there must have been about 150 particles. That is, as a matter of fact, exactly what the maximum likelihood estimate (6.100) would be in this case. But what does the robot tell us? Well, it says the best estimate by the mean square error criterion is only

$$\langle n \rangle = 15 + 100(1 - 0.1) = 15 + 90 = 105. \tag{6.101}$$

More generally, we could write (6.90) this way:

$$\langle n \rangle = s + (c - \langle c \rangle), \tag{6.102}$$

so if you see k more counts than you 'should have' in one second, according to the robot that is evidence for only k more particles, not $10k$.

This example turned out to be quite surprising to some experimental physicists engaged in work along these lines. Let's see if we can reconcile it with our common sense. If we have an average number of counts of 10 per second with this counter, then we would guess, by rules well known, that a fluctuation in counting rate of something like the square root of this, ± 3, would not be at all surprising, even if the number of incoming particles per second stayed strictly constant. On the other hand, if the average rate of flow of particles is $s = 100$ per second, the fluctuation in this rate which would not be surprising is $\pm\sqrt{100} = \pm 10$. But this corresponds to only ± 1 in the number of counts.

This shows that we cannot use a counter to measure fluctuations in the rate of arrival of particles, unless the counter has a very high efficiency. If the efficiency is high, then we know that practically every count corresponds to one particle, and we are reliably measuring those fluctuations. If the efficiency is low and we know that there is a definite, fixed source strength, then fluctuations in counting rate are much more likely to be due to things happening in the counter than to actual changes in the rate of arrival of particles.

The same mathematical result, in the disease scenario, means that if a disease is mild and unlikely to cause death, then variations in the observed number of deaths are not reliable indicators of variations in the incidence of the disease. If our prior information tells us that there is a constantly operating basic cause of the disease (such as a contaminated water supply), then a large change in the number of deaths from one year to the next is not evidence of a large change in the number of people having the disease. But if practically everyone who contracts the disease dies immediately, then of course the number of deaths tells us very reliably what the incidence of the disease was, whatever the means of contracting it.

What caused the difference between the Bayes and maximum likelihood solutions? The difference is due to the fact that we had *prior information* contained in this source strength s. The maximum likelihood estimate simply maximized the probability for getting c counts, given n particles, and that yields 150. In Bayes' solution, we will multiply this by a prior probability $p(n|s)$, which represents our knowledge of the antecedent situation, before maximizing, and we'll get an entirely different value for the estimate. As we saw in the inversion of urn distributions, *simple prior information can make a big change in the conclusions that we draw from a data set.*

Exercise 6.5. Generalize the above calculation to take the dead-time effect into account; that is, if we know that two or more particles incident on the counter within a short time interval Δt can produce at most only one count, how is our estimate of n changed? These effects are important in many practical situations, and there is a voluminous literature on the application of probability theory to them (see Bortkiewicz, 1898, 1913, and Takacs, 1958).

Now let's extend this problem a little. We are now going to use Bayes' theorem in four problems where there is no quantitative prior information, but only one qualitative fact, and again see the effect that prior information has on our conclusions.

6.12 Effects of qualitative prior information

The situation is depicted in Figure 6.2. Two robots, which we shall humanize by naming them Mr A and Mr B, have different prior information about the source of the particles. The source is hidden in another room which Mr A and Mr B are not allowed to enter. Mr A has no knowledge at all about the source of particles; for all he knows, it might be an accelerating machine which is being turned on and off in an arbitrary way, or the other room might be full of little men who run back and forth, holding first one radioactive source, then another, up to the exit window. Mr B has one additional qualitative fact: he knows that the source is a radioactive sample of long lifetime, in a fixed position. But he does not know

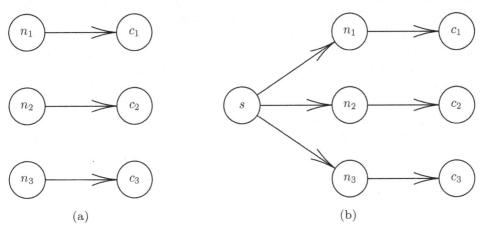

Fig. 6.2. (a) The structure of Mr A's problem; different intervals are logically independent. (b) Mr B's logical situation; knowledge of the existence of s makes n_2 relevant to n_1.

anything about its source strength (except, of course, that it is not infinite because, after all, the laboratory is not being vaporized by its presence; Mr A is also given assurance that he will not be vaporized during the experiment). They both know that the counter efficiency is 10%: $\phi = 0.1$. Again, we want them to estimate the number of particles passing through the counter from knowledge of the number of counts. We denote their prior information by I_A, I_B, respectively.

We commence the experiment. During the first second, $c_1 = 10$ counts are registered. What can Mr A and Mr B say about the number n_1 of particles? Bayes' theorem for Mr A reads

$$p(n_1|\phi c_1 I_A) = p(n_1|I_A)\frac{p(c_1|\phi n_1 I_A)}{p(c_1|\phi I_A)} = \frac{p(n_1|I_A)p(c_1|\phi n_1 I_A)}{p(c_1|\phi I_A)}. \tag{6.103}$$

The denominator is just a normalizing constant, and could also be written

$$p(c_1|\phi I_A) = \sum_{n_1} p(c_1|\phi n_1 I_A)p(n_1|I_A). \tag{6.104}$$

But now we seem to be stuck, for what is $p(n_1|I_A)$? The only information about n_1 contained in I_A is that n_1 is not large enough to vaporize the laboratory. How can we assign prior probabilities on this kind of evidence? This has been a point of controversy for a long time, for in any theory which regards probability as a real physical phenomenon, Mr A has no basis at all for determining the 'true' prior probabilities $p(n_1|I_A)$.

6.13 Choice of a prior

Now, of course, Mr A is programmed to recognize that there is no such thing as an 'objectively true' probability. As the notation $p(n_1|I_A)$ indicates, the purpose of assigning a prior is to describe his own state of knowledge I_A, and on this he is the final authority. So he does not need to argue the philosophy of it with anyone. We consider in Chapters 11 and 12 some of the general formal principles available to him for translating verbal prior information into prior probability assignments, but in the present discussion we wish only to demonstrate some pragmatic facts, by a prior that represents reasonably the information that n_1 is not infinite, and that for small n_1 there is no prior information that would justify any great variations in $p(n_1|I_A)$. For example, if as a function of n_1 the prior $p(n_1|I_A)$ exhibited features such as oscillations or sudden jumps, that would imply some very detailed prior information about n_1 that Mr A does not have.

Mr A's prior should, therefore, avoid all such structure; but this is hardly a formal principle, and so the result is not unique. But it is one of the points to be made from this example, noted by H. Jeffreys (1939), that it does not need to be unique because, in a sense, 'almost any' prior which is smooth in the region of high likelihood will lead to substantially the same final conclusions.[11]

---

[11] We have seen already that, in some circumstances, a prior can make a very large difference in the conclusions; but to do this it necessarily modulates the likelihood function in the region of its peak, not its tails.

So Mr A assigns a uniform prior probability out to some large but finite number N,

$$p(n_1|I_A) = \begin{cases} 1/N & \text{if } 0 \le n_1 < N \\ 0 & \text{if } N \le n_1, \end{cases} \qquad (6.105)$$

which seems to represent his state of knowledge tolerably well. The finite upper bound N is an admittedly *ad hoc* way of representing the fact that the laboratory is not being vaporized. How large could it be? If N were as large as 10^{60}, then not only the laboratory, but our entire galaxy, would be vaporized by the energy in the beam (indeed, the total number of atoms in our galaxy is of the order of 10^{60}). So Mr A surely knows that N is very much less than that. Of course, if his final conclusions depend strongly on N, then Mr A will need to analyze his exact prior information and think more carefully about the value of N and whether the abrupt drop in $p(n_1|I_A)$ at $n_1 = N$ should be smoothed out. Such careful thinking would not be wrong, but it turns out to be unnecessary, for it will soon be evident that details of $p(n_1|I_A)$ for large n_1 are irrelevant to Mr A's conclusions.

6.14 On with the calculation!

Nicely enough, the $1/N$ cancels out of (6.103) and (6.104), and we are left with

$$p(n_1|\phi c_1 I_A) = \begin{cases} Ap(c_1|\phi n_1) & \text{if } 0 \le n_1 < N \\ 0 & \text{if } N \le n_1, \end{cases} \qquad (6.106)$$

where A is a normalization factor:

$$A^{-1} = \sum_{n_1=0}^{N-1} p(c_1|\phi n_1). \qquad (6.107)$$

We have noted, in (6.100), that, as a function of n_1, $p(c_1|\phi n_1)$ attains its maximum at $n_1 = c_1/\phi$ (=100, in this problem). For $n_1 \phi \gg c_1$, $p(c_1|\phi n_1)$ falls off like $n_1^{c_1}(1-\phi)^{n_1} \simeq n_1^{c_1} \exp\{-n_1\phi\}$. Therefore, the sum (6.107) converges so rapidly that if N is as large as a few hundred, there is no appreciable difference between the exact normalization factor (6.107) and the sum to infinity.

In view of this, we may as well take advantage of a simplification; *after* applying Bayes' theorem, pass to the limit $N \to \infty$. But let us be clear about the rationale of this; we pass to the limit, not because we believe that N is infinite; we know that it is not. We pass to the limit rather because we know that this will simplify the calculation without affecting the final result; after this passage to the limit, all our calculations pertaining to this model can be performed exactly with the aid of the general summation formula

$$\sum_{m=0}^{\infty} \binom{m+a}{m} m^n x^m = \left(x\frac{\mathrm{d}}{\mathrm{d}x}\right)^n \frac{1}{(1-x)^{a+1}}, \qquad (|x| < 1). \qquad (6.108)$$

Thus, writing $m = n - c$, we replace (6.107) by

$$A^{-1} \simeq \sum_{n_1=0}^{\infty} p(c_1|\phi n_1) = \phi^c \sum_{m=0}^{\infty} \binom{m+c}{m}(1-\phi)^m = \phi^c \left\{ \frac{1}{[1-(1-\phi)]^{(c+1)}} \right\} = \frac{1}{\phi}.$$

(6.109)

Exercise 6.6. To better appreciate the quality of this approximation, denote the 'missing' terms in (6.107) by

$$S(N) \equiv \sum_{n_1=N}^{\infty} p(c_1|\phi n_1)$$

(6.110)

and show that the fractional discrepancy between (6.107) and (6.109) is about

$$\delta \equiv S(N)/S(0) \simeq \frac{\exp\{-N\phi\}(N\phi)^{c_1}}{c_1!}, \qquad \text{if } N\phi \gg 1.$$

(6.111)

From this, show that in the present case ($\phi = 0.1, c_1 = 10$), unless the prior information can justify an upper limit N less than about 270, the exact value of N – or indeed, all details of $p(n_1|I_A)$ for $n_1 > 270$ – can make less than one part in 10^4 difference in Mr A's conclusions. But it is hard to see how anyone could have any serious use for more than three figure accuracy in the final results; and so this discrepancy would have no effect at all on that final result. What happens for $n_1 \geq 340$ can affect the conclusions less than one part in 10^6, and for $n_1 \geq 400$ it is less than one part in 10^8.

This is typical of the way prior range matters in real problems, and it makes ferocious arguments over this seem rather silly. It is a valid question of principle, but its pragmatic consequences are not just negligibly small, but (usually) strictly nil because we are calculating only to a finite number of decimal places. Yet some writers have claimed that a fundamental qualitative change in the character of the problem occurs between $N = 10^{10}$ and $N = \infty$. The reader may be amused to estimate how much difference this makes in the final numerical results. To how many decimal places would we need to calculate before it made any difference at all?

Of course, if the prior information should start encroaching on the region $n_1 < 270$, it would then make a difference in the conclusions; but in that case the prior information was indeed cogent for the question being asked, and this is as it should be. Being thus reassured and using the approximation (6.109), we obtain the result

$$p(n_1|\phi c_1 I_A) = \phi p(c_1|\phi n_1) = \binom{n_1}{c_1} \phi^{c_1+1}(1-\phi)^{n_1-c_1}.$$

(6.112)

So, for Mr A, the most probable value of n_1 is the same as the maximum likelihood estimate

$$(\hat{n}_1)_A = \frac{c_1}{\phi} = 100,$$

(6.113)

while the posterior mean value estimate is calculated as follows:

$$\langle n_1 \rangle_A - c_1 = \sum_{n_1=c_1}^{\infty} (n_1 - c_1) p(n_1 | \phi c_1 I_A)$$

$$= \phi^{c_1+1}(1-\phi)(c_1+1) \sum_{n_1} \binom{n_1}{n_1 - c_1 - 1}(1-\phi)^{n_1-c_1-1}. \tag{6.114}$$

From (6.108) the sum is equal to

$$\sum_{m=0}^{\infty} \binom{m+c_1+1}{m}(1-\phi)^m = \frac{1}{\phi^{2+c_1}}, \tag{6.115}$$

and we get

$$\langle n_1 \rangle_A = c_1 + (c_1+1)\frac{1-\phi}{\phi} = \frac{c_1 + 1 - \phi}{\phi} = 109. \tag{6.116}$$

Now, how about the other robot, Mr B? Does his extra knowledge help him here? He knows that there is some definite fixed source strength s. And, because the laboratory is not being vaporized, he knows that there is some upper limit S_0. Suppose that he assigns a uniform prior probability density for $0 \le s < S_0$. Then he will obtain

$$p(n_1|\theta I_B) = \int_0^{\infty} ds\, p(n_1|s)p(s|\theta I_B) = \frac{1}{S_0}\int_0^{S_0} p(n_1|s)ds = \frac{1}{S_0}\int_0^{S_0} ds\, \frac{s^{n_1}\exp\{-s\}}{n_1!}. \tag{6.117}$$

Now, if n_1 is appreciably less than S_0, the upper limit of integration can for all practical purposes be taken as infinity, and the integral is just unity. So, we have

$$p(n_1|\theta I_B) = p(s|\theta I_B) = \frac{1}{S_0} = \text{const.}, \qquad n_1 < S_0. \tag{6.118}$$

In putting this into Bayes' theorem with $c_1 = 10$, the significant range of values of n_1 will be of the order of 100, and unless the prior information indicates a value of S_0 lower than about 300, we will have the same situation as before; Mr B's extra knowledge didn't help him at all, and he comes out with the same posterior distribution and the same estimates:

$$p(n_1|c_1 I_B) = p(n_1|\phi c_1 I_A) = \phi p(c_1|\phi n_1). \tag{6.119}$$

6.15 The Jeffreys prior

Harold Jeffreys (1939, Chap. 3) proposed a different way of handling this problem. He suggests that the proper way to express 'complete ignorance' of a continuous variable known to be positive is to assign uniform prior probability to its logarithm; i.e. the prior probability density is

$$p(s|I_J) \propto \frac{1}{s}, \qquad (0 \le s < \infty). \tag{6.120}$$

Of course, we cannot normalize this, but that does not stop us from using it. In many cases, including the present one, it can be used directly because all the integrals involved converge. In almost all cases we can approach this prior as the limit of a sequence of proper (normalizable) priors, with mathematically well-behaved results. If even that does not yield a proper posterior distribution, then the robot is warning us that the data are too uninformative about either very large s or very small s to justify any definite conclusions, and we need to obtain more evidence before any useful inferences are possible.

Jeffreys justified (6.120) on the grounds of invariance under certain changes of parameters; i.e. instead of using the parameter s, what prevents us from using $t \equiv s^2$, or $u \equiv s^3$? Evidently, to assign a uniform prior probability density to s is not at all the same thing as assigning a uniform prior probability to t; but if we use the Jeffreys prior, we are saying the same thing whether we use s or any power s^m as the parameter.

There is the germ of an important principle here, but it was only recently that the situation has been fairly well understood. When we take up the theory of transformation groups in Chapter 12, we will see that the real justification of Jeffreys' rule cannot lie merely in the fact that the parameter is positive, but that our desideratum of consistency, in the sense that equivalent states of knowledge should be represented by equivalent probability assignments, uniquely determines the Jeffreys rule in the case when s is a 'scale parameter'. Then, marginalization theory will reinforce this by deriving it Uniquely – without appealing to any principles beyond the basic product and sum rules of probability theory – as the only prior for a scale parameter that is completely uninformative about other parameters that may be in the model.

These arguments and others equally cogent all lead to the same conclusion: the Jeffreys prior is the only correct way to express complete ignorance of a scale parameter. The question then reduces to whether s can properly be regarded as a scale parameter in this problem. However, this line of thought has taken us beyond the present topic; in the spirit of our current problem, we shall just put (6.120) to the test and see what results it gives. The calculations are all very easy, and we find these results:

$$p(n_1|I_J) = \frac{1}{n_1}, \qquad p(c_1|I_J) = \frac{1}{c_1}, \qquad p(n_1|\phi c_1 I_J) = \frac{c_1}{n_1} p(c_1|\phi n_1). \qquad (6.121)$$

This leads to the most probable and mean value estimates:

$$(\hat{n}_1)_J = \frac{c_1 - 1 + \phi}{\phi} = 91, \qquad \langle n_1 \rangle_J = \frac{c}{\phi} = 100. \qquad (6.122)$$

The amusing thing emerges that Jeffreys' prior probability rule just lowers the most probable and posterior mean value estimates by nine each, bringing the mean value right back to the maximum likelihood estimate!

This comparison is valuable in showing us how little difference there is numerically between the consequences of different prior probability assignments which are not sharply peaked, and helps to put arguments about them into proper perspective. We made a rather drastic change in the prior probabilities, in a problem where there was really very little information contained in the meager data, and it still made less than 10% difference in the result. This is, as we shall see, small compared with the probable error in the estimate which was inevitable in any event. In a more realistic problem where we have more data, the difference would be even smaller.

A useful rule of thumb, illustrated by the comparison of (6.113), (6.116) and (6.122), is that changing the prior probability $p(\alpha|I)$ for a parameter by one power of α has in general about the same effect on our final conclusions as does having one more data point. This is because the likelihood function generally has a relative width $1/\sqrt{n}$, and one more power of α merely adds an extra small slope in the neighborhood of the maximum, thus shifting the maximum slightly. Generally, if we have effectively n independent observations, then the fractional error in an estimate that was inevitable in any event is $1/\sqrt{n}$, approximately,[12] while the fractional change in the estimate due to one more power of α in the prior is about $1/n$.

In the present case, with ten counts, thus ten independent observations, changing from a uniform to Jeffreys prior made just under 10% difference. If we had 100 counts, the error which is inevitable in any event would be about 10%, while the difference from the two priors would be less than 1%.

So, from a pragmatic standpoint, arguments about which prior probabilities correctly express a state of 'complete ignorance', like those over prior ranges, usually amount to quibbling over pretty small peanuts.[13] From the standpoint of principle, however, they are important and need to be thought about a great deal, as we shall do in Chapter 12 after becoming familiar with the numerical situation. While the Jeffreys prior is the theoretically correct one, it is in practice a small refinement that makes a difference only in the very small sample case. In the past, these issues were argued back and forth endlessly on a foggy philosophical level, without taking any note of the pragmatic facts of actual performance; that is what we are trying to correct here.

6.16 The point of it all

Now we are ready for the interesting part of this problem. For during the next second, we see $c_2 = 16$ counts. What can Mr A and Mr B now say about the numbers n_1, n_2 of particles responsible for c_1, c_2? Well, Mr A has no reason to expect any relationship between what happened in the two time intervals, and so to him the increase in counting rate is evidence only of an increase in the number of incident particles. His calculation for the second time interval is the same as before, and he will give us the most probable value,

$$(\hat{n}_2)_A = \frac{c_2}{\phi} = 160, \tag{6.123}$$

---

[12] However, as we shall see later, there are two special cases where the $1/\sqrt{n}$ rule fails: if we are trying to estimate the location of a discontinuity in an otherwise continuous probability distribution, and if different data values are strongly correlated.

[13] This is most definitely *not* true if the prior probabilities are to describe a definite piece of prior knowledge, as the next example shows.

and his mean value estimate,

$$\langle n_2 \rangle_A = \frac{c_2 + 1 - \phi}{\phi} = 169. \tag{6.124}$$

Knowledge of c_2 doesn't help him to make any improved estimate of n_1, which stays the same as before.

Mr B is in an entirely different position than Mr A; his extra qualitative information suddenly becomes very important. For knowledge of c_2 enables him to improve his previous estimate of n_1. Bayes' theorem now gives

$$p(n_1|\phi c_2 c_1 I_B) = p(n_1|\phi c_1 I_B)\frac{p(c_2|\phi n_1 c_1 I_B)}{p(c_2|\phi c_1 I_B)} = p(n_1|\phi c_1 I_B)\frac{p(c_2|\phi n_1 I_B)}{p(c_2|\phi c_1 I_B)}. \tag{6.125}$$

Again, the denominator is a normalizing constant, which we can find by summing the numerator over n_1. We see that the significant thing is $p(c_2|\phi n_1 I_B)$. Using our method of resolving c_2 into mutually exclusive alternatives, this is

$$\begin{aligned} p(c_2|\phi n_1 I_B) &= \int_0^\infty ds\, p(c_2 s|\phi n_1 I_B) \\ &= \int_0^\infty ds\, p(c_2|s\phi n_1 I_B)p(s|\phi n_1 I_B) \\ &= \int_0^\infty ds\, p(c_2|\phi s I_B)p(s|\phi n_1 I_B). \end{aligned} \tag{6.126}$$

We have already found $p(c_2|\phi s I_B)$ in (6.86), and we need only

$$p(s|\phi n_1 I_B) = p(s|\phi I_B)\frac{p(n_1|\phi s I_B)}{p(n_1|\phi I_B)} = p(n_1|\phi s), \quad \text{if } n_1 \ll S_0, \tag{6.127}$$

where we have used (6.118). We have found $p(n_1|s)$ in (6.84), so we have

$$p(c_2|\phi n_1 I_B) = \int_0^\infty ds \left[\frac{\exp\{-s\phi\}(s\phi)^{c_2}}{c_2!}\right]\left[\frac{\exp\{-s\}s^{n_1}}{n_1!}\right] = \binom{n_1 + c_2}{c_2}\frac{\phi^{c_2}}{(1+\phi)^{n_1+c_2+1}}. \tag{6.128}$$

Substituting (6.119) and (6.128) into (6.125) and carrying out an easy summation to get the denominator, the result is (*not* a binomial distribution):

$$p(n_1|\phi c_2 c_1 I_B) = \binom{n_1 + c_2}{c_1 + c_2}\left(\frac{2\phi}{1+\phi}\right)^{c_1+c_2+1}\left(\frac{1-\phi}{1+\phi}\right)^{n_1-c_1}. \tag{6.129}$$

Note that we could have derived this equally well by direct application of the resolution method:

$$\begin{aligned} p(n_1|\phi c_2 c_1 I_B) &= \int_0^\infty ds\, p(n_1 s|\phi c_2 c_1 I_B) \\ &= \int_0^\infty ds\, p(n_1|\phi s c_2 c_1 I_B)p(s|\phi c_2 c_1 I_B) \\ &= \int_0^\infty ds\, p(n_1|\phi s c_1 I_B)p(s|\phi c_2 c_1 I_B). \end{aligned} \tag{6.130}$$

We have already found $p(n_1|\phi sc_1 I_B)$ in (6.89), and it is easily shown that $p(s|\phi c_2 c_1 I_B) \propto$ $p(c_2|\phi s I_B)p(c_1|\phi s I_B)$, which is therefore given by the Poisson distribution (6.86). This, of course, leads to the same rather complicated result (6.129); thus providing another – and rather severe – test of the consistency of our rules.

To find Mr B's new most probable value of n_1, we set

$$\frac{p(n_1|\phi c_2 c_1 I_B)}{p(n_1 - 1|\phi c_2 c_1 I_B)} = \frac{n_1 + c_2}{n_1 - c_1}\frac{1 - \phi}{1 + \phi} = 1, \tag{6.131}$$

or

$$(\hat{n}_1)_B = \frac{c_1}{\phi} + (c_2 - c_1)\frac{1 - \phi}{2\phi} = \frac{c_1 + c_2}{2\phi} + \frac{c_1 - c_2}{2} = 127. \tag{6.132}$$

His new posterior mean value is also readily calculated, and is equal to

$$\langle n_1\rangle_B = \frac{c_1 + 1 - \phi}{\phi} + (c_2 - c_1 - 1)\frac{1 - \phi}{2\phi} = \frac{c_1 + c_2 + 1 - \phi}{2\phi} + \frac{c_1 - c_2}{2} = 131.5. \tag{6.133}$$

Both estimates are considerably raised, and the difference between most probable and mean value is only half what it was before, suggesting a narrower posterior distribution, as we shall confirm presently. If we want Mr B's estimates for n_2, then from symmetry we just interchange the subscripts 1 and 2 in the above equations. This gives for his most probable and mean value estimates, respectively,

$$(\hat{n}_2)_B = 133, \qquad \langle n_2\rangle_B = 137.5. \tag{6.134}$$

Now, can we understand what is happening here? Intuitively, the reason why Mr B's extra qualitative prior information makes a difference is that knowledge of both c_1 and c_2 enables him to make a better estimate of the source strength s, which in turn is relevant for estimating n_1. The situation is indicated more clearly by the diagrams in Figure 6.2. By hypothesis, to Mr A each sequence of events $n_i \rightarrow c_i$ is logically independent of the others, so knowledge of one doesn't help him in reasoning about any other. In each case he must reason from c_i directly to n_i, and no other route is available. But to Mr B, there are two routes available: he can reason directly from c_1 to n_1 as Mr A does, as described by $p(n_1|\phi c_1 I_A) = p(n_1|\phi c_1 I_B)$; but, because of his knowledge that there is a fixed source strength s 'presiding over' both n_1 and n_2, he can also reason along the route $c_2 \rightarrow n_2 \rightarrow s \rightarrow n_1$. If this were the *only* route available to him (i.e. if he didn't know c_1), he would obtain the distribution

$$p(n_1|\phi c_2 I_B) = \int_0^\infty ds\, p(n_1|s)p(s|c_2 I_B) = \frac{\phi^{c_2+1}}{c_2!(1 + \phi)^{c_2+1}}\frac{(n_1 + c_2)!}{n_1!(1 + \phi)^{n_1}}, \tag{6.135}$$

and, comparing the above relations, we see that Mr B's final distribution (6.129) is, except for normalization, just the product of the ones found by reasoning along his two routes:

$$p(n_1|\phi c_1 c_2 I_B) = (\text{const.}) \times p(n_1|\phi c_1 I_B)p(n_1|\phi c_2 I_B) \tag{6.136}$$

in consequence of the fact that $p(c_1 c_2|\phi n_1) = p(c_1|\phi n_1)p(c_2|\phi n_1)$. The information (6.135) about n_1 obtained by reasoning along the new route $c_2 \rightarrow n_2 \rightarrow s \rightarrow n_1$ thus introduces a 'correction factor' in the distribution obtained from the direct route $c_1 \rightarrow n_1$, enabling Mr B to improve his estimates.

This suggests that, if Mr B could obtain the number of counts in a great many different seconds, $(c_3, c_4, \ldots, c_m)$, he would be able to do better and better; and perhaps in the limit $m \rightarrow \infty$ his estimate of n_1 might be as good as the one we found when the source strength was known exactly. We will check this surmise presently by working out the degree of reliability of these estimates, and by generalizing these distributions to arbitrary m, from which we can obtain the asymptotic forms.

6.17 Interval estimation

There is still an essential feature missing in the comparison of Mr A and Mr B in our particle counter problem. We would like to have some measure of the degree of reliability which they attach to their estimates, especially in view of the fact that their estimates are so different. Clearly, the best way of doing this would be to draw the entire probability distributions

$$p(n_1|\phi c_2 c_1 I_A) \quad \text{and} \quad p(n_1|\phi c_2 c_1 I_B) \tag{6.137}$$

and from this make statements of the form, '90% of the posterior probability is concentrated in the interval $\alpha < n_1 < \beta$'. But, for present purposes, we will be content to give the standard deviations (i.e., square root of the variance as defined in Eq. (6.93)) of the various distributions we have found. An inequality due to Tchebycheff then asserts that, if σ is the standard deviation of any probability distribution over n_1, then the amount P of probability concentrated between the limits $\langle n_1 \rangle \pm t\sigma$ satisfies[14]

$$P \geq 1 - \frac{1}{t^2}. \tag{6.139}$$

This tells us nothing when $t \leq 1$, but it tells us more and more as t increases beyond unity. For example, in any probability distribution with finite $\langle n \rangle$ and $\langle n^2 \rangle$, at least 3/4 of the probability is contained in the interval $\langle n \rangle \pm 2\sigma$, and at least 8/9 is in $\langle n \rangle \pm 3\sigma$.

6.18 Calculation of variance

The variances σ^2 of all the distributions we have found above are readily calculated. In fact, calculation of any moment of these distributions is easily performed by the general formula (6.108). For Mr A and Mr B, and the Jeffreys prior probability distribution, we

---

[14] Proof: Let $p(x)$ be a probability density over $(-\infty < x < \infty)$, a any real number, and $y \equiv x - \langle x \rangle$. Then

$$a^2(1 - P) = a^2 p(|y| > a) = a^2 \int_{|y|>a} dx \, p(x) \leq \int_{|y|>a} dx \, y^2 p(x) \leq \int_{-\infty}^{\infty} dx \, y^2 p(x) = \sigma^2. \tag{6.138}$$

Writing $a = t\sigma$, this is $t^2(1 - P) \leq 1$, the same as Eq. (6.139). This proof includes the discrete cases, since then $p(x)$ is a sum of delta-functions. A large collection of useful Tchebycheff-type inequalities is given by I. R. Savage (1961).

Table 6.1. *The effect of prior information on estimates of* n_1 *and* n_2.

		Problem 1	Problem 2	
		n_1	n_1	n_2
A	most probable	100	100	160
	mean $\pm$ s.d.	109 ± 31	109 ± 31	169 ± 39
B	most probable	100	127	133
	mean $\pm$ s.d.	109 ± 31	131.5 ± 25.9	137.5 ± 25.9
Jeffreys	most probable	91	121.5	127.5
	mean $\pm$ s.d.	100 ± 30	127 ± 25.4	133 ± 25.4

find the variances

$$\text{var}(n_1 | \phi c_1 I_A) = \frac{(c_1 + 1)(1 - \phi)}{\phi^2}, \tag{6.140}$$

$$\text{var}(n_1 | \phi c_2 c_1 I_B) = \frac{(c_1 + c_2 + 1)(1 - \phi^2)}{4\phi^2}, \tag{6.141}$$

$$\text{var}(n_1 | \phi c_1 I_J) = \frac{c_1(1 - \phi)}{\phi^2}, \tag{6.142}$$

and the variances for n_2 are found from symmetry.

This has been a rather long discussion, so let's summarize all our results so far in Table 6.1. We give, for problem 1 ($c_1 = 10$) and problem 2 ($c_1 = 10$, $c_2 = 16$), the most probable values of the number of particles found by Mr A and Mr B, and also the (mean value) $\pm$ (standard deviation) estimates. From Table 6.1 we see that Mr B's extra information not only has led him to change his estimates considerably from those of Mr A, but it has enabled him to make an appreciable decrease in his probable error. *Even purely qualitative prior information which has nothing to do with frequencies can greatly alter the conclusions we draw from a given data set.* Now, in virtually every real problem of scientific inference, we do have qualitative prior information of more or less the kind supposed here. Therefore, any method of inference which fails to take prior information into account is capable of misleading us, in a potentially dangerous way. The fact that it yields a reasonable result in one problem is no guarantee that it will do so in the next.

It is also of interest to ask how good Mr B's estimate of n_1 would be if he knew only c_2; and therefore had to use the distribution (6.135) representing reasoning along the route $c_2 \rightarrow n_2 \rightarrow s \rightarrow n_1$ of Figure 6.2. From (6.135) we find the most probable, and the (mean) $\pm$ (standard deviation) estimates

$$\hat{n}_1 = \frac{c_2}{\phi} = 160, \tag{6.143}$$

$$\text{mean} \pm \text{s.d.} = \frac{c_2 + 1}{\phi} \pm \frac{\sqrt{(c_2 + 1)(\phi + 1)}}{\phi} = 170 \pm 43.3. \tag{6.144}$$

In this case, he would obtain a slightly poorer estimate (i.e. a larger probable error) than Mr A even if the counts $c_1 = c_2$ were the same, because the variance (6.140) for the direct route contains a factor $(1 - \phi)$, which gets replaced by $(1 + \phi)$ if we have to reason over the indirect route. Thus, if the counter has low efficiency, the two routes give nearly equal reliability for equal counting rates; but if it has high efficiency, $\phi \simeq 1$, then the direct route $c_1 \to n_1$ is far more reliable. Our common sense will tell us that this is just as it should be.

6.19 Generalization and asymptotic forms

We conjectured above that Mr B might be helped a good deal more in his estimate of n_1 by acquiring still more data $\{c_3, c_4, \ldots, c_m\}$. Let's investigate that further. The standard deviation of the distribution (6.89) in which the source strength was known exactly is only $\sqrt{s(1 - \phi)} = 10.8$ for $s = 130$; and from Table 6.1 Mr B's standard deviation for his estimate of n_1 is now about 2.5 times this value. What would happen if we gave him more and more data from other time intervals, such that his estimate of s approached 130? To answer this, note that, if $1 \leq k \leq m$, we have:

$$\begin{aligned} p(n_k | \phi c_1 \cdots c_m I_B) &= \int_0^\infty ds \, p(n_k s | \phi c_1 \cdots c_m I_B) \\ &= \int_0^\infty ds \, p(n_k | \phi s c_k I_B) p(s | \phi c_1 \cdots c_m I_B), \end{aligned} \tag{6.145}$$

in which we have put $p(n_k | \phi s c_1 \cdots c_m I_B) = p(n_k | \phi s c_k I_B)$ because, from Figure 6.2, if s is known, then all the c_i with $i \neq k$ are irrelevant for inferences about n_k. The second factor in the integrand of (6.145) can be evaluated by Bayes' theorem:

$$\begin{aligned} p(s | \phi c_1 \cdots c_m I_B) &= p(s | \phi I_B) \frac{p(c_1 \cdots c_m | \phi s I_B)}{p(c_1 \cdots c_m | \phi I_B)} \\ &= (\text{const.}) \times p(s | \phi I_B) p(c_1 | \phi s I_B) \cdots p(c_m | \phi s I_B). \end{aligned} \tag{6.146}$$

Using (6.86) and normalizing, this reduces to

$$p(s | \phi c_1 \cdots c_m I_B) = \frac{(m\phi)^{c+1}}{c!} s^c \exp\{-ms\phi\}, \tag{6.147}$$

where $c \equiv c_1 + \cdots + c_m$ is the total number of counts in the m seconds. The mode, mean and variance of the distribution (6.147) are, respectively,

$$\hat{s} = \frac{c}{m\phi}, \qquad \langle s \rangle = \frac{c + 1}{m\phi}, \qquad \text{var}(s) = \langle s^2 \rangle - \langle s \rangle^2 = \frac{c + 1}{m^2 \phi^2} = \frac{\langle s \rangle}{m\phi}. \tag{6.148}$$

So it turns out, as we might have expected, that as $m \to \infty$, the distribution $p(s | c_1 \cdots c_m)$ becomes sharper and sharper, the most probable and mean value estimates of s get closer and closer together, and it appears that in the limit we would have just a delta-function:

$$p(s|\phi c_1 \cdots c_m I_B) \to \delta(s - s'), \tag{6.149}$$

where

$$s' \equiv \lim_{m \to \infty} \frac{c_1 + c_2 + \cdots + c_m}{m\phi}. \tag{6.150}$$

But the limiting form (6.149) was found a bit abruptly, as was James Bernoulli's first limit theorem. We might like to see in more detail how the limit is approached, in analogy to the de Moivre–Laplace limit theorem for the binomial (5.10), or the limit (4.72) of the beta distribution.

For example, expanding the logarithm of (6.147) about its peak $\hat{s} = c/m\phi$, and retaining only through the quadratic terms, we find for the asymptotic formula a Gaussian distribution:

$$p(s|\phi c_1 \cdots c_m I_B) \to A \exp\left\{-\frac{c(s - \hat{s})^2}{2\hat{s}^2}\right\}, \tag{6.151}$$

which is actually valid for all s, in the sense that the difference between the left-hand side and right-hand side is small for all s (although their ratio is not close to unity for all s). This leads to the estimate, as $c \to \infty$,

$$(s)_{\text{est}} = \hat{s}\left(1 \pm \frac{1}{\sqrt{c}}\right). \tag{6.152}$$

Quite generally, posterior distributions go into a Gaussian form as the data increase, because any function with a single rounded maximum, raised to a higher and higher power, goes into a Gaussian function. In the next chapter we shall explore the basis of Gaussian distributions in some depth.

So, in the limit, Mr B does indeed approach exact knowledge of the source strength. Returning to (6.145), both factors in the integrand are now known from (6.89) and (6.147), and so

$$p(n_k|\phi c_1 \cdots c_m I_B) = \int_0^\infty ds \, \frac{\exp\{-s(1 - \phi)\}[s(1 - \phi)]^{n_k - c_k}}{(n_k - c_k)!} \frac{(m\phi)^{c+1}}{c!} s^c \exp\{-ms\phi\}, \tag{6.153}$$

or

$$p(n_k|c_1 \cdots c_m I_B) = \frac{(n_k - c_k + c)!}{(n_k - c_k)!c!} \frac{(m\phi)^{c+1}(1 - \phi)^{n_k - c_k}}{(1 + m\phi - \phi)^{n_k - c_k + c + 1}}, \tag{6.154}$$

which is the promised generalization of (6.135). In the limit $m \to \infty$, $c \to \infty$, $(c/m\phi) \to s' = \text{const.}$, this goes into the Poisson distribution

$$p(n_k|c_1 \cdots c_m I_B) \to \frac{\exp\{-s'(1 - \phi)\}}{(n_k - c_k)!} [s'(1 - \phi)]^{n_k - c_k}, \tag{6.155}$$

which is identical to (6.89). We therefore confirm that, given enough additional data, Mr B's standard deviation can be reduced from 26 to 10.8, compared with Mr A's value of 31. For finite m, the mean value estimate of n_k from (6.154) is

$$\langle n_k \rangle = c_k + \langle s \rangle (1 - \phi), \tag{6.156}$$

where $\langle s \rangle = (c+1)/m\phi$ is the mean value estimate of s from (6.148). Equation (6.156) is to be compared with (6.90). Likewise, the most probable value of n_k, according to (6.154), is

$$\hat{n}_k = c_k + \hat{s}(1 - \phi), \tag{6.157}$$

where $\hat{s}$ is given by (6.148).

Note that Mr B's revised estimates in problem 2 still lie within the range of reasonable error assigned by Mr A. It would be rather disconcerting if this were not the case, as it would then appear that probability theory is giving Mr A an over-optimistic picture of the reliability of his estimates. There is, however, no theorem which guarantees this; for example, if the counting rate had jumped to $c_2 = 80$, then Mr B's revised estimate of n_1 would be far outside Mr A's limits of reasonable error. But, in this case, Mr B's common sense would lead him to doubt the reliability of his prior information I_B; we would have another example like that in Chapter 4, of a problem where one of those 'something else' alternative hypotheses down at -100 db, which we don't even bother to formulate until they are needed, is resurrected by very unexpected new evidence.

Exercise 6.7. The above results were found using the language of the particle counter scenario. Summarize the final conclusions in the language of the disease incidence scenario, as one or two paragraphs of advice for a medical researcher who is trying to judge whether public health measures are reducing the incidence of a disease in the general population, but has data only on the number of deaths from it. This should, of course, include something about judging under what conditions our model corresponds well to the real world; and what to do if it does not.

Now we turn to a different kind of problem to see some new features that can appear when we use a sampling distribution that is continuous except at isolated points of discontinuity.

6.20 Rectangular sampling distribution

The following 'taxicab problem' has been part of the orally transmitted folklore of this field for several decades, but orthodoxy has no way of dealing with it, and we have never seen it mentioned in the orthodox literature. You are traveling on a night train; on awakening from sleep, you notice that the train is stopped at some unknown town, and all you can see is a taxicab with the number 27 on it. What is then your guess as to the number N of taxicabs in the town, which would in turn give a clue as to the size of the town? Almost everybody answers intuitively that there seems to be something about the choice $N_{\text{est}} = 2 \times 27 = 54$ that recommends itself; but few can offer a convincing rationale for this. The obvious 'model' that forms in our minds is that there will be N taxicabs, numbered, respectively,

$(1, \ldots, N)$, and, given N, the one we see is equally likely to be any of them. Given that model, we would then know deductively that $N \geq 27$; but, from that point on, our reasoning depends on our statistical indoctrination.

Here we study a continuous version of the same problem, in which more than one taxi may be in view, leaving it as an exercise for the reader to write down the parallel solution to the above taxicab problem, and then state the exact relationship between the continuous and discrete problems. We consider a rectangular sampling distribution in $[0, \alpha]$, where the width α of the distribution is the parameter to be estimated, and finally suggest further exercises which will extend what we learn from it.

We have a data set $D \equiv \{x_1, \ldots, x_n\}$ of n observations thought of as 'drawn from' this distribution, urn-wise; that is, each datum x_i is assigned independently the pdf

$$p(x_i | \alpha I) = \begin{cases} \alpha^{-1} & \text{if } 0 \leq x_i \leq \alpha < \infty \\ 0 & \text{otherwise.} \end{cases} \tag{6.158}$$

Then our entire sampling distribution is

$$p(D | \alpha I) = \prod_i p(x_i | \alpha I) = \alpha^{-n}, \qquad 0 \leq \{x_1, \ldots, x_n\} \leq \alpha, \tag{6.159}$$

where for brevity we suppose, in the rest of this section, that when the inequalities following an equation are not all satisfied, the left-hand side is zero. It might seem at first glance that this situation is too trivial to be worth analyzing; yet, if one does not see in advance exactly how every detail of the solution will work itself out, there is always something to be learned from studying it. In probability theory, the most trivial looking problems reveal deep and unexpected things.

The posterior pdf for α is, by Bayes' theorem,

$$p(\alpha | DI) = p(\alpha | I) \frac{p(D | \alpha I)}{p(D | I)}, \tag{6.160}$$

where $p(\alpha | I)$ is our prior. Now it is evident that any Bayesian problem with a proper (normalizable) prior and a bounded likelihood function must lead to a proper, well-behaved posterior distribution, whatever the data – as long as the data do not themselves contradict any of our other information. If any datum was found to be negative, $x_i < 0$, the model (6.159) would be known deductively to be wrong (put better, the data contradict the prior information I that led us to choose that model). Then the robot crashes, both (6.159) and (6.160) vanishing identically. But any data set for which the inequalities in (6.159) are satisfied is a possible one *according to the model*. Must it then yield a reasonable posterior pdf?

Not necessarily! The data could be compatible with the model, but still incompatible with the other prior information. Consider a proper rectangular prior

$$p(\alpha | I) = (\alpha_1 - \alpha_{00})^{-1}, \qquad \alpha_{00} \leq \alpha \leq \alpha_1, \tag{6.161}$$

where α_{00}, α_1 are fixed numbers satisfying $0 \leq \alpha_{00} \leq \alpha_1 < \infty$, given to us in the statement of the problem. If any datum were found to exceed the upper prior bound: $x_i > \alpha_1$, then the data and the prior information would again be logically contradictory.

But this is just what we anticipated already in Chapters 1 and 2; we are trying to reason from two pieces of information D, I, each of which may be actually a logical conjunction of many different propositions. If there is a contradiction hidden anywhere in the totality of this, there can be no solution (in a set theory context, the set of possibilities that we have prescribed is the empty set) and the robot crashes, in one way or another. So in the following we suppose that the data are consistent with all the prior information – including the prior information that led us to choose this model.[15] Then the above rules should yield the correct and exact answer to the question we have posed.

The denominator of (6.160) is

$$p(D|I) = \int_R d\alpha \, (\alpha_1 - \alpha_{00})^{-1} \alpha^{-n}, \qquad (6.162)$$

where the region R of integration must satisfy two conditions:

$$R \equiv \left\{ \begin{array}{c} \alpha_{00} \leq \alpha \leq \alpha_1 \\ x_{max} \leq \alpha \leq \alpha_1 \end{array} \right\} \qquad (6.163)$$

and $x_{max} \equiv \max\{x_1, \ldots, x_n\}$ is the greatest datum observed. If $x_{max} \leq \alpha_{00}$, then in (6.163) we need only the former condition; the numerical values of the data x_i are entirely irrelevant (although the number n of observations remains relevant). If $\alpha_{00} \leq x_{max}$, then we need only the latter inequality; the prior lower bound α_{00} has been superseded by the data, and is irrelevant to the problem from this point on.

Substituting (6.159), (6.161) and (6.162) into (6.160), the factor $(\alpha_1 - \alpha_{00})$ cancels out, and if $n > 1$ our general solution reduces to

$$p(\alpha|DI) = \frac{(n-1)\alpha^{-n}}{\alpha_0^{1-n} - \alpha_1^{1-n}}, \qquad \alpha_0 \leq \alpha \leq \alpha_1, \qquad n > 1, \qquad (6.164)$$

where $\alpha_0 \equiv \max(\alpha_{00}, x_{max})$.

6.21 Small samples

Small values of n often present special situations that might be overlooked in a general derivation. In orthodox statistics, as we shall see in Chapter 17, they can lead to weird pathological results (like an estimator for a parameter which lies outside the parameter space, and so is known deductively to be impossible). In any other area of mathematics, when a contradiction like this appears, one concludes at once that an error has been made. But curiously, in the literature of orthodox statistics, such pathologies are never interpreted as revealing an error in the orthodox reasoning. Instead they are simply passed over; one proclaims his concern only with large n. But small n proves to be very interesting for us, just

---

[15] Of course, in the real world we seldom have prior information that would justify such sharp bounds on x and α, and so such sharp contradictions would not arise; but that signifies only that we are studying an ideal limiting case. There is nothing strange about this; in elementary geometry, our attention is directed first to such things as perfect triangles and circles, although no such things exist in the real world. There, also, we are really studying ideal limiting cases of reality; but what we learn from that study enables us to deal successfully with thousands of real situations that arise in such diverse fields as architecture, engineering, astronomy, geodesy, stereochemistry, and the artist's rules of perspective. It is the same here.

because of the fact that Bayesian analysis has no pathological, exceptional cases. As long as we avoid outright logical contradictions in the statement of a problem and use proper priors, the solutions do not break down but continue to make good sense.

It is very instructive to see how Bayesian analysis always manages to accomplish this, which also makes us aware of a subtle point in practical calculation. Thus, in the present case, if $n = 1$, then (6.164) appears indeterminate, reducing to $(0/0)$. But if we repeat the derivation from the start for the case $n = 1$, the properly normalized posterior pdf for α is found to be, instead of (6.164),

$$p(\alpha|DI) = \frac{\alpha^{-1}}{\log(\alpha_1/\alpha_0)}, \qquad \alpha_0 \leq \alpha \leq \alpha_1, \qquad n = 1. \qquad (6.165)$$

The case $n = 0$ can hardly be of any use; nevertheless, Bayes' theorem still gives the obviously right answer. For then $D =$ 'no data at all', and $p(D|\alpha I) = p(D|I) = 1$; that is, if we take no data, we shall have no data, whatever the value of α. Then the posterior distribution (6.160) reduces, as common sense demands, to the prior distribution

$$p(\alpha|DI) = p(\alpha|I), \qquad \alpha_0 \leq \alpha \leq \alpha_1, \qquad n = 0. \qquad (6.166)$$

6.22 Mathematical trickery

Now we see a subtle point: the last two results are contained already in (6.164) without any need to go back and repeat the derivation from the start. We need to understand the distinction between the real world problem and the abstract mathematics. Although *in the real problem n* is by definition a non-negative integer, the *mathematical expression* (6.164) is well-defined and meaningful when n is any complex number. Furthermore, as long as $\alpha_1 < \infty$, it is an entire function of n (that is, bounded and analytic everywhere except the point at infinity). Now in a purely mathematical derivation we are free to make use of whatever analytical properties our functions have, whether or not they would make sense in the real problem. Therefore, since (6.164) can have no singularity at any finite point, we may evaluate it at $n = 1$ by taking the limit as $n \to 1$. But

$$\begin{aligned}\frac{n-1}{\alpha_0^{1-n} - \alpha_1^{1-n}} &= \frac{n-1}{\exp\{-(n-1)\log(\alpha_0)\} - \exp\{-(n-1)\log(\alpha_1)\}} \\ &= \frac{n-1}{[1-(n-1)\log(\alpha_0)+\cdots] - [1-(n-1)\log(\alpha_1)+\cdots]} \\ &\to \frac{1}{\log(\alpha_1/\alpha_0)},\end{aligned} \qquad (6.167)$$

leading to (6.165). Likewise, putting $n = 0$ into (6.164), it reduces to (6.166), because now we have necessarily $\alpha_0 = \alpha_{00}$. Even in extreme, degenerate cases, Bayesian analysis continues to yield the correct results.[16] And it is evident that all moments and percentiles of

---

[16] Under the influence of early orthodox teaching, the writer became fully convinced of this only after many years of experimentation with hundreds of such cases, and his total failure to produce any pathology as long as the Chapter 2 rules were followed strictly.

the posterior distribution are also entire functions of n, so they may be calculated once and for all for all n, taking limiting values whenever the general expression reduces to $(0/0)$ or (∞/∞); this will always yield the same result that we obtain by going back to the beginning and repeating the calculation for that particular n value.[17]

If $\alpha_1 < \infty$, the posterior distribution is confined to a finite interval, and so it has necessarily moments of all orders. In fact,

$$\langle \alpha^m \rangle = \frac{n-1}{\alpha_0^{1-n} - \alpha_1^{1-n}} \int_{\alpha_0}^{\alpha_1} d\alpha\, \alpha^{m-n} = \frac{n-1}{n-m-1} \frac{\alpha_0^{1+m-n} - \alpha_1^{1+m-n}}{\alpha_0^{1-n} - \alpha_1^{1-n}}, \quad (6.168)$$

and when $n \to 1$ or $m \to n-1$, we are to take the limit of this expression in the manner of (6.167), yielding the more explicit forms:

$$\langle \alpha^m \rangle = \begin{cases} \dfrac{\alpha_1^m - \alpha_0^m}{m \log(\alpha_1/\alpha_0)} & \text{if } n = 1 \\[3mm] \dfrac{(n-1)\log(\alpha_1/\alpha_0)}{\alpha_0^{1-n} - a_1^{1-n}} & \text{if } m = n-1. \end{cases} \quad (6.169)$$

In the above results, the posterior distribution is confined to a finite region ($a_0 \le \alpha \le \alpha_1$) and there can be no singular result.

Exercise 6.8. Complete this example, finding explicit values for the estimates of α and their accuracy. Discuss how these results correspond or fail to correspond to common sense judgments.

Finally, we leave it as an exercise for the reader to consider what happens as $\alpha_1 \to \infty$ and we pass to an infinite domain.

Exercise 6.9. When $\alpha_1 \to \infty$, some moments must cease to exist, so some inferences must cease to be possible; others remain possible. Examine the above equations to find under what conditions a posterior (mean $\pm$ standard deviation) or (median $\pm$ interquartile span) remains possible, considering in particular the case of small n. State how the results correspond to common sense.

---

[17] Recognizing this, we see that, whenever a mathematical expression is an analytic function of some parameter, we can exploit that fact as a tool for calculation with it, whatever meaning it might have in the original problem. For example, the *numbers* 2 and π often appear, and it is often in an expression $Q(2)$ or $Q(\pi)$ which is an analytic function of the *symbol* '2' or 'π'. Then, if it is helpful, we are free to replace '2' or 'π' by 'x' and evaluate quantities involving Q by such operations as differentiating with respect to x, or complex integration in the x plane, etc., setting $x = 2$ or $x = \pi$ at the end; and this is perfectly rigorous. Once we have distilled the real problem into one of abstract mathematics, our symbols mean whatever we say they mean; the writer learned this trick from Professor W. W. Hansen of Stanford University, who would throw a class into an uproar when he evaluated an integral, correctly, by differentiating another integral with respect to π.

6.23 Comments

The calculations which we have done here with ease – in particular, (6.129) and (6.152) – cannot be done with any version of probability theory which does not permit the use of the prior and posterior probabilities needed, and consequently does not allow us to integrate out a nuisance parameter with respect to a prior. It appears to us that Mr *B*'s results are beyond the reach of orthodox methods. Yet at every stage, probability theory as logic has followed the procedures that are determined uniquely by the basic product and sum rules of probability theory; and it has yielded well-behaved, reasonable, and useful results. In some cases, the prior information was absolutely essential, even though it was only qualitative. Later we shall see even more striking examples of this.

It should not be supposed that this recognition of the need to use prior information is a new discovery. It was emphasized very strongly by Bertrand (1889); he gave several examples, of which we quote the last (he wrote in very short paragraphs):

The inhabitants of St. Malo [a small French town on the English channel] are convinced; for a century, in their village, the number of deaths at the time of high tide has been greater than at low tide. We admit the fact.

On the coast of the English channel there have been more shipwrecks when the wind was from the northwest than for any other direction. The number of instances being supposed the same and equally reliably reported, still one will not draw the same conclusions.

While we would be led to accept as a certainty the influence of the wind on shipwrecks, common sense demands more evidence before considering it even plausible that the tide influences the last hour of the Malouins.

The problems, again, are identical; the impossibility of accepting the same conclusions shows the necessity of taking into account the prior probability for the cause.

Clearly, Bertrand cannot be counted among those who advocate R. A. Fisher's maxim: 'Let the data speak for themselves!' which has so dominated statistics in this century. The data *cannot* speak for themselves; and they never have, in any real problem of inference.

For example, Fisher advocated the method of maximum likelihood for estimating a parameter; in a sense, this is the value that is indicated most strongly by the data alone. But that takes note of only one of the factors that probability theory (and common sense) requires. For, if we do not supplement the maximum likelihood method with some prior information about which hypotheses we shall consider reasonable, then it will always lead us inexorably to favor the 'sure thing' hypothesis H_S, according to which every tiny detail of the data was inevitable; nothing else could possibly have happened. For the data always have a much higher probability (namely $p(D|H_S) = 1$), on H_S than on any other hypothesis; H_S is always the maximum likelihood solution over the class of all hypotheses. Only our extremely low prior probability for H_S can justify our rejecting it.[18]

---

[18] Small children, when asked to account for some observed fact such as the exact shape of a puddle of spilled milk, have a strong tendency to invent 'sure thing' hypotheses; they have not yet acquired the worldly experience that makes educated adults consider them too unlikely to be considered seriously. But a scientist, who knows that the shape is determined by the laws of hydrodynamics and has vast computing power available, is no more able than the child to predict that shape, because he lacks the requisite prior information about the exact initial conditions.

Orthodox practice deals with this in part by the device of specifying a model, which is, of course, a means of incorporating some prior information about the phenomenon being observed. But this is incomplete, defining only the parameter space within which we shall seek that maximum; without a prior probability over that parameter space, we have no way of incorporating further prior information about the likely values of the parameter, which we almost always have and which is often highly cogent for any rational inference. For example, although a parameter space may extend formally to infinity, in virtually every real problem we know in advance that the parameter is enormously unlikely to be outside some finite domain. This information may or may not be crucial, depending on what data set we happen to get.

As the writer can testify from his student days, steadfast followers of Fisher often interpret 'Let the data speak for themselves' as implying that it is reprehensible – a violation of 'scientific objectivity' – to allow one's self to be influenced at all by prior information. It required a few years of experience to perceive, with Bertrand, what a disastrous error this is in real problems. Fisher was able to manage without mentioning prior information only because, in the problems he chose to work on, he had no very important prior information anyway, and plenty of data. Had he worked on problems with cogent prior information and sparse data, we think that his ideology would have changed rather quickly.

Scientists in all fields see this readily enough – as long as they rely on their own common sense instead of orthodox teaching. For example, Stephen J. Gould (1989) describes the bewildering variety of soft-bodied animals that lived in early Cambrian times, preserved perfectly in the famous Burgess shale of the Canadian Rockies. Two paleontologists examined the same fossil, named *Aysheaia*, and arrived at opposite conclusions regarding its proper taxonomic classification. One who followed Fisher's maxim would be obliged to question the competence of one of them; but Gould does not make this error. He concludes (p. 172), 'We have a reasonably well-controlled psychological experiment here. The data had not changed, so the reversal of opinion can only record a revised presupposition about the most likely status of Burgess organisms.'

Prior information is essential also for a different reason, if we are trying to make inferences concerning which mechanism is at work. Fisher would, presumably, insist as strongly as any other scientist that a cause–effect relationship requires a physical mechanism to bring it about. But as in St Malo, the data alone are silent on this; they do not speak for themselves.[19] Only prior information can tell us whether some hypothesis provides a possible mechanism for the observed facts, consistent with the known laws of physics. If it does not, then the fact that it accounts well for the data may give it a high likelihood, but cannot give it any credence. A fantasy that invokes the labors of hordes of little invisible elves and pixies running about to generate the data would have just as high a likelihood; but it would still have no credence for a scientist.

---

[19] Statisticians, even those who profess themselves disciples of Fisher, have been obliged to develop adages about this, such as 'correlation does not imply causation'. or 'a good fit is no substitute for a reason'. to discourage the kind of thinking that comes automatically to small children, and to adults with untrained minds.

It is not only orthodox statisticians who have denigrated prior information in the 20th century. The fantasy writer H. P. Lovecraft once defined 'common sense' as 'merely a stupid absence of imagination and mental flexibility'. Indeed, it is just the accumulation of unchanging prior information about the world that gives the mature person the mental stability that rejects arbitrary fantasies (although we may enjoy diversionary reading of them).

Today, the question whether our present information does or does not provide credible evidence for the existence of a causal effect is a major policy issue, arousing bitter political, commercial, medical, and environmental contention, resounding in courtrooms and legislative halls.[20] Yet cogent prior information – without which the issue cannot possibly be judged – plays little role in the testimony of 'expert witnesses' with orthodox statistical training, because their standard procedures have no place to use it. We note that Bertrand's clear and correct insight into this appeared the year before Fisher was born; the progress of scientific inference has not always been forward.

Thus, this chapter begins and ends with a glance back at Fisher, about whom the reader may find more in Chapter 16.

---

[20] For some frightening examples, see Gardner (1957, 1981). Deliberate suppression of inconvenient prior information is also the main tool of the scientific charlatans.

7

The central, Gaussian or normal distribution

> My own impression ... is that the mathematical results have outrun their interpretation and that some simple explanation of the force and meaning of the celebrated integral ... will one day be found ... which will at once render useless all the works hitherto written.
>
> *Augustus de Morgan (1838)*

Here, de Morgan was expressing his bewilderment at the 'curiously ubiquitous' success of methods of inference based on the Gaussian, or normal, 'error law' (sampling distribution), even in cases where the law is not at all plausible as a statement of the actual frequencies of the errors. But the explanation was not forthcoming as quickly as he expected.

In the middle 1950s the writer heard an after-dinner speech by Professor Willy Feller, in which he roundly denounced the practice of using Gaussian *probability* distributions for errors, on the grounds that the *frequency* distributions of real errors are almost never Gaussian. Yet in spite of Feller's disapproval, we continued to use them, and their ubiquitous success in parameter estimation continued. So, 145 years after de Morgan's remark, the situation was still unchanged, and the same surprise was expressed by George Barnard (1983): *'Why have we for so long managed with normality assumptions?'*

Today we believe that we can, at last, explain (1) the inevitably ubiquitous use, and (2) the ubiquitous success, of the Gaussian error law. Once seen, the explanation is indeed trivially obvious; yet, to the best of our knowledge, it is not recognized in any of the previous literature of the field, because of the universal tendency to think of probability distributions in terms of frequencies. We cannot understand what is happening until we learn to think of probability distributions in terms of their demonstrable *information content* instead of their imagined (and, as we shall see, irrelevant) frequency connections.

A simple explanation of these properties – stripped of past irrelevancies – has been achieved only very recently, and this development changed our plans for the present work. We decided that it is so important that it should be inserted at this somewhat early point in the narrative, even though we must then appeal to some results that are established only later. In the present chapter, then, we survey the historical basis of Gaussian distributions and present a quick preliminary understanding of their functional role in inference. This understanding will then guide us directly – without the usual false starts and blind alleys – to the computational procedures which yield the great majority of the useful applications of probability theory.

7.1 The gravitating phenomenon

We have noted an interesting phenomenon several times in previous chapters; in probability theory, there seems to be a central, universal distribution

$$\varphi(x) \equiv \frac{1}{\sqrt{2\pi}} \exp\left\{-\frac{x^2}{2}\right\} \tag{7.1}$$

toward which all others gravitate under a very wide variety of different operations – and which, once attained, remains stable under an even wider variety of operations. The famous 'central limit theorem' concerns one special case of this. In Chapter 4, we noted that a binomial or beta sampling distribution goes asymptotically into a Gaussian when the number of trials becomes large. In Chapter 6 we noted a virtually universal property, that posterior distributions for parameters go into Gaussian when the number of data values increases.

In physics, these gravitating and stability properties have made this distribution the universal basis of kinetic theory and statistical mechanics; in biology, it is the natural tool for discussing population dynamics in ecology and evolution. We cannot doubt that it will become equally fundamental in economics, where it already enjoys ubiquitous use, but somewhat apologetically, as if there were some doubt about its justification. We hope to assist this development by showing that its range of validity for such applications is far wider than is usually supposed.

Figure 7.1 illustrates this distribution. Its general shape is presumably already well known to the reader, although the numerical values attached to it may not be. The cumulative

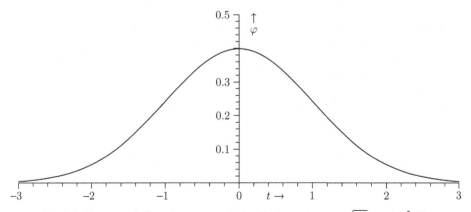

Fig. 7.1. The central, Gaussian or normal distribution: $\varphi(t) = 1/\sqrt{2\pi}\ \exp(-t^2/2)$.

Gaussian, defined as

$$\Phi(x) \equiv \int_{-\infty}^{x} dt \, \varphi(t),$$

$$= \int_{-\infty}^{0} dt \, \varphi(t), + \int_{0}^{x} dt \, \varphi(t), \qquad (7.2)$$

$$= \frac{1}{2} \left[1 + \text{erf}(x) \right],$$

will be used later in this chapter for solving some problems. Numerical values for this function are easily calculated using the error function, $\text{erf}(x)$.

This distribution is called the Gaussian, or normal, distribution, for historical reasons discussed below. Both names are inappropriate and misleading today; all the correct connotations would be conveyed if we called it, simply, the *central distribution* of probability theory.[1] We consider first three derivations of it that were important historically and conceptually, because they made us aware of three important properties of the Gaussian distribution.

7.2 The Herschel–Maxwell derivation

One of the most interesting derivations, from the standpoint of economy of assumptions, was given by the astronomer John Herschel (1850). He considered the two-dimensional probability distribution for errors in measuring the position of a star. Let x be the error in the longitudinal (east–west) direction and y the error in the declination (north–south) direction, and ask for the joint probability distribution $\rho(x, y)$. Herschel made two postulates (P1, P2) that seemed required intuitively by conditions of geometrical homogeneity.

(P1) Knowledge of x tells us nothing about y

That is, probabilities of errors in orthogonal directions should be independent; so the undetermined distribution should have the functional form

$$\rho(x, y) dx dy = f(x) dx \times f(y) dy. \qquad (7.3)$$

We can write the distribution equally well in polar coordinates r, θ defined by $x = r \cos \theta$, $y = r \sin \theta$:

$$\rho(x, y) dx dy = g(r, \theta) r \, dr \, d\theta. \qquad (7.4)$$

---

[1] It is general usage outside probability theory to denote any function of the general form $\exp\{-ax^2\}$ as a *Gaussian function*, and we shall follow this.

(P2) This probability should be independent of the angle: $g(r, \theta) = g(r)$

Then (7.3) and (7.4) yield the functional equation

$$f(x)f(y) = g\left(\sqrt{x^2 + y^2}\right), \tag{7.5}$$

and, setting $y = 0$, this reduces to $g(x) = f(x)f(0)$, so (7.5) becomes the functional equation

$$\log\left[\frac{f(x)}{f(0)}\right] + \log\left[\frac{f(y)}{f(0)}\right] = \log\left[\frac{f(\sqrt{x^2 + y^2})}{f(0)}\right]. \tag{7.6}$$

But the general solution of this is obvious; a function of x plus a function of y is a function only of $x^2 + y^2$. The only possibility is that $\log[f(x)/f(0)] = ax^2$. We have a normalizable probability only if a is negative, and then normalization determines $f(0)$; so the general solution can only have the form

$$f(x) = \sqrt{\frac{\alpha}{\pi}} \exp\left\{-\alpha x^2\right\}, \qquad \alpha > 0, \tag{7.7}$$

with one undetermined parameter. The only two-dimensional probability density satisfying Herschel's invariance conditions is a circular symmetric Gaussian:

$$\rho(x, y) = \frac{\alpha}{\pi} \exp\left\{-\alpha(x^2 + y^2)\right\}. \tag{7.8}$$

Ten years later, James Clerk Maxwell (1860) gave a three-dimensional version of this same argument to find the probability distribution $\rho(v_x, v_y, v_z) \propto \exp\{-\alpha(v_x^2 + v_y^2 + v_z^2)\}$ for velocities of molecules in a gas, which has become well known to physicists as the 'Maxwellian velocity distribution law' fundamental to kinetic theory and statistical mechanics.

The Herschel–Maxwell argument is particularly beautiful because two qualitative conditions, incompatible in general, become compatible for just one quantitative distribution, which they therefore uniquely determine. Einstein (1905a,b) used the same kind of argument to deduce the Lorentz transformation law from his two qualitative postulates of relativity theory.[2]

The Herschel–Maxwell derivation is economical also in that it does not actually make any use of probability theory; only geometrical invariance properties which could be applied equally well in other contexts. Gaussian functions are unique objects in their own right, for purely mathematical reasons. But now we give a famous derivation that makes explicit use of probabilistic intuition.

---

[2] These are: (1) the laws of physics take the same form for all moving observers; and (2) the velocity of light has the same constant numerical value for all such observers. These are also contradictory in general, but become compatible for one particular quantitative law of transformation of space and time to a moving coordinate system.

7.3 The Gauss derivation

We estimate a location parameter θ from $(n+1)$ observations $(x_0, \ldots, x_n)$ by maximum likelihood. If the sampling distribution factors: $p(x_0, \ldots, x_n|\theta) = f(x_0|\theta) \cdots f(x_n|\theta)$, the likelihood equation is

$$\sum_{i=0}^{n} \frac{\partial}{\partial \theta} \log f(x_i|\theta) = 0, \tag{7.9}$$

or, writing

$$\log f(x|\theta) = g(\theta - x) = g(u), \tag{7.10}$$

the maximum likelihood estimate $\hat{\theta}$ will satisfy

$$\sum_{i} g'(\hat{\theta} - x_i) = 0. \tag{7.11}$$

Now, intuition may suggest to us that the estimate ought to be also the arithmetic mean of the observations:

$$\hat{\theta} = \bar{x} = \frac{1}{n+1} \sum_{i=0}^{n} x_i, \tag{7.12}$$

but (7.11) and (7.12) are in general incompatible ((7.12) is not a root of (7.11)). Nevertheless, consider a possible sample, in which only one observation x_0 is nonzero: if in (7.12) we put

$$x_0 = (n+1)u, \qquad x_1 = x_2 = \cdots = x_n = 0, \qquad (-\infty < u < \infty), \tag{7.13}$$

then $\hat{\theta} = u$, $\hat{\theta} - x_0 = -nu$, whereupon eqn. (7.11) becomes $g'(-nu) + ng'(u) = 0$, $n = 1, 2, 3, \ldots$. The case $n = 1$ tells us that $g'(u)$ must be an antisymmetric function: $g'(-u) = -g'(u)$, so this reduces to

$$g'(nu) = ng'(u), \qquad (-\infty < u < \infty), \quad n = 1, 2, 3, \ldots. \tag{7.14}$$

Evidently, the only possibility is a linear function:

$$g'(u) = au, \qquad g(u) = \frac{1}{2}au^2 + b. \tag{7.15}$$

Converting back by (7.10), a normalizable distribution again requires that a be negative, and normalization then determines the constant b. The sampling distribution must have the form

$$f(x|\theta) = \sqrt{\frac{\alpha}{2\pi}} \exp\left\{-\frac{1}{2}\alpha(x - \theta)^2\right\} \qquad (0 < \alpha < \infty). \tag{7.16}$$

Since (7.16) was derived assuming the special sample (7.13), we have shown thus far only that (7.16) is a necessary condition for the equality of maximum likelihood estimate and sample mean. Conversely, if (7.16) is satisfied, then the likelihood equation (7.9) always has the unique solution (7.12); and so (7.16) is the necessary and sufficient condition for this agreement. The only freedom is the unspecified scale parameter α.

7.4 Historical importance of Gauss's result

This derivation was given by Gauss (1809), as little more than a passing remark in a work concerned with astronomy. It might have gone unnoticed but for the fact that Laplace saw its merit and the following year published a large work calling attention to it and demonstrating the many useful properties of (7.16) as a sampling distribution. Ever since, it has been called the 'Gaussian distribution'.

Why was the Gauss derivation so sensational in effect? Because it put an end to a long – and, it seems to us today, scandalous – psychological hang up suffered by some of the greatest mathematicians of the time. The distribution (7.16) had been found in a more or less accidental way already by de Moivre (1733), who did not appreciate its significance and made no use of it. Throughout the 18th century, it would have been of great value to astronomers faced constantly with the problem of making the best estimates from discrepant observations; yet the greatest minds failed to see it. Worse, even the qualitative fact underlying data analysis – cancellation of errors by averaging of data – was not perceived by so great a mathematician as Leonhard Euler.

Euler (1749), trying to resolve the 'Great Inequality of Jupiter and Saturn', found himself with what was at the time a monstrous problem (described briefly in our closing Comments, Section 7.27). To determine how the longitudes of Jupiter and Saturn had varied over long times, he made use of 75 observations from a 164 year period (1582–1745), and eight orbital parameters to estimate from them.

Today, a desk-top microcomputer could solve this problem by an algorithm to be given in Chapter 19, and print out the best estimates of the eight parameters and their accuracies, in about one minute (the main computational job is the inversion of an (8×8) matrix). Euler failed to solve it, but not because of the magnitude of this computation; he failed even to comprehend the principle needed to solve it. Instead of seeing that by combining many observations their errors tend to cancel, he thought that this would only 'multiply the errors' and make things worse. In other words, Euler concentrated his attention entirely on the worst possible thing that could happen, as if it were certain to happen – which makes him perhaps the first really devout believer in Murphy's Law.[3]

Yet, practical people, with experience in actual data taking, had long perceived that this worst possible thing does *not* happen. On the contrary, averaging our observations has the great advantage that the errors tend to cancel each other.[4] Hipparchus, in the second century BC, estimated the precession of the equinoxes by averaging measurements on several stars. In the late 16th century, taking the average of several observations was the routine procedure of Tycho Brahe. Long before it had any formal theoretical justification from mathematicians, intuition had told observational astronomers that this averaging of data was the right thing to do.

Some 30 years after Euler's effort, another competent mathematician, Daniel Bernoulli (1777), still could not comprehend the procedure. Bernoulli supposes that an archer is

---

[3] 'If anything *can* go wrong, it *will* go wrong.'

[4] If positive and negative errors are equally likely, then the probability that ten errors all have the same sign is $(0.5)^9 \simeq 0.002$.

shooting at a vertical line drawn on a target, and asks how many shots land in various vertical bands on either side of it:

> Now is it not self-evident that the hits must be assumed to be thicker and more numerous on any given band the nearer this is to the mark? If all the places on the vertical plane, whatever their distance from the mark, were equally liable to be hit, the most skillful shot would have no advantage over a blind man. That, however, is the tacit assertion of those who use the common rule (the arithmetic mean) in estimating the value of various discrepant observations, when they treat them all indiscriminately. In this way, therefore, the degree of probability of any given deviation could be determined to some extent *a posteriori*, since there is no doubt that, for a large number of shots, the probability is proportional to the number of shots which hit a band situated at a given distance from the mark.

We see that Daniel Bernoulli (1777), like his uncle James Bernoulli (1713), saw clearly the distinction between probability and frequency. In this respect, his understanding exceeded that of John Venn 100 years later, and Jerzy Neyman 200 years later. Yet he fails completely to understand the basis for taking the arithmetic mean of the observations as an estimate of the true 'mark'. He takes it for granted (although a short calculation, which he was easily capable of doing, would have taught him otherwise) that, if the observations are given equal weight in calculating the average, then one must be assigning equal probability to all errors, however great. Presumably, others made intuitive guesses like this, unchecked by calculation, making this part of the folklore of the time. Then one can appreciate how astonishing it was when Gauss, 32 years later, proved that the condition

$$\text{(maximum likelihood estimate)} = \text{(arithmetic mean)} \qquad (7.17)$$

uniquely determines the Gaussian error law, not the uniform one.

In the meantime, Laplace (1783) had investigated this law as a limiting form of the binomial distribution, derived its main properties, and suggested that it was so important that it ought to be tabulated; yet, lacking the above property demonstrated by Gauss, he still failed to see that it was the natural error law (the Herschel derivation was still 77 years in the future). Laplace persisted in trying to use the form $f(x) \propto \exp\{-a|x|\}$, which caused no end of analytical difficulties. But he did understand the qualitative principle that combination of observations improves the accuracy of estimates, and this was enough to enable him to solve, in 1787, the problem of Jupiter and Saturn, on which the greatest minds had been struggling since before he was born.

Twenty-two years later, when Laplace saw the Gauss derivation, he understood it all in a flash – doubtless mentally kicked himself for not seeing it before – and hastened (Laplace, 1810, 1812) to give the central limit theorem and the full solution to the general problem of reduction of observations, which is still how we analyze it today. Not until the time of Einstein did such a simple mathematical argument again have such a great effect on scientific practice.

7.5 The Landon derivation

A derivation of the Gaussian distribution that gives us a very lively picture of the process by which a Gaussian frequency distribution is built up in Nature was given in 1941 by Vernon D. Landon, an electrical engineer studying properties of noise in communication circuits. We give a generalization of his argument, in our current terminology and notation.

The argument was suggested by the empirical observation that the variability of the electrical noise voltage $v(t)$ observed in a circuit at time t seems always to have the same general properties, even though it occurs at many different levels (say, mean square values) corresponding to different temperatures, amplifications, impedance levels, and even different kinds of sources – natural, astrophysical, or man-made by many different devices such as vacuum tubes, neon signs, capacitors, resistors made of many different materials, etc. Previously, engineers had tried to characterize the noise generated by different sources in terms of some 'statistic' such as the ratio of peak to RMS (root mean square) value, which it was thought might identify its origin. Landon recognized that these attempts had failed, and that the samples of electrical noise produced by widely different sources '... cannot be distinguished one from the other by any known test'.[5]

Landon reasoned that if this frequency distribution of noise voltage is so universal, then it must be better determined theoretically than empirically. To account for this universality but for magnitude, he visualized not a single distribution for the voltage at any given time, but a hierarchy of distributions $p(v|\sigma)$ characterized by a single scale parameter σ^2, which we shall take to be the expected square of the noise voltage. The stability seems to imply that, if the noise level σ^2 is increased by adding a small increment of voltage, the probability distribution still has the same functional form, but is only moved up the hierarchy to the new value of σ. He discovered that for only one functional form of $p(v|\sigma)$ will this be true.

Suppose the noise voltage v is assigned the probability distribution $p(v|\sigma)$. Then it is incremented by a small extra contribution ϵ, becoming $v' = v + \epsilon$, where ϵ is small compared with σ, and has a probability distribution $q(\epsilon)d\epsilon$, independent of $p(v|\sigma)$. Given a specific ϵ, the probability for the new noise voltage to have the value v' would be just the previous probability that v should have the value $(v' - \epsilon)$. Thus, by the product and sum rules of probability theory, the new probability distribution is the convolution

$$f(v') = \int d\epsilon \, p(v' - \epsilon|\sigma)q(\epsilon). \tag{7.18}$$

Expanding this in powers of the small quantity ϵ and dropping the prime, we have

$$f(v) = p(v|\sigma) - \frac{\partial p(v|\sigma)}{\partial v}\int d\epsilon \, \epsilon q(\epsilon) + \frac{1}{2}\frac{\partial^2 p(v|\sigma)}{\partial v^2}\int d\epsilon \, \epsilon^2 q(\epsilon) + \cdots, \tag{7.19}$$

---

[5] This universal, stable type of noise was called 'grass' because that is what it looks like on an oscilloscope. To the ear, it sounds like a smooth hissing without any discernible pitch; today this is familiar to everyone because it is what we hear when a television receiver is tuned to an unused channel. Then the automatic gain control turns the gain up to the maximum, and both the hissing sound and the flickering 'snow' on the screen are the greatly amplified noise generated by random thermal motion of electrons in the antenna according to the Nyquist law noted below.

or, now writing for brevity $p \equiv p(v|\sigma)$,

$$f(v) = p - \langle \epsilon \rangle \frac{\partial p}{\partial v} + \frac{1}{2} \langle \epsilon^2 \rangle \frac{\partial^2 p}{\partial v^2} + \cdots. \tag{7.20}$$

This shows the general form of the expansion; but now we assume that the increment is as likely to be positive as negative:[6] $\langle \epsilon \rangle = 0$. At the same time, the expectation of v^2 is increased to $\sigma^2 + \langle \epsilon^2 \rangle$, so Landon's invariance property requires that $f(v)$ should be equal also to

$$f(v) = p + \langle \epsilon^2 \rangle \frac{\partial p}{\partial \sigma^2}. \tag{7.21}$$

Comparing (7.20) and (7.21), we have the condition for this invariance:

$$\frac{\partial p}{\partial \sigma^2} = \frac{1}{2} \frac{\partial^2 p}{\partial v^2}. \tag{7.22}$$

But this is a well-known differential equation (the 'diffusion equation'), whose solution with the obvious initial condition $p(v|\sigma = 0) = \delta(v)$ is

$$p(v|\sigma) = \frac{1}{\sqrt{2\pi\sigma^2}} \exp\left\{ -\frac{v^2}{2\sigma^2} \right\}, \tag{7.23}$$

the standard Gaussian distribution. By minor changes in the wording, the above mathematical argument can be interpreted either as *calculating* a probability distribution, or as *estimating* a frequency distribution; in 1941 nobody except Harold Jeffreys and John Maynard Keynes took note of such distinctions. As we shall see, this is, in spirit, an incremental version of the central limit theorem; instead of adding up all the small contributions at once, it takes them into account one at a time, requiring that at each step the new probability distribution has the same functional form (to second order in ϵ).

This is just the process by which noise is produced in Nature – by addition of many small increments, one at a time (for example, collisions of individual electrons with atoms, each collision radiating another tiny pulse of electromagnetic waves, whose sum is the observed noise). Once a Gaussian form is attained, it is preserved; this process can be stopped at any point, and the resulting final distribution still has the Gaussian form. What is at first surprising is that this stable form is independent of the distributions $q(\epsilon)$ of the small increments; that is why the noise from different sources could not be distinguished by any test known in 1941.[7]

Today we can go further and recognize that the reason for this independence was that only the second moment $\langle \epsilon^2 \rangle$ of the increments mattered for the updated point distribution (that

---

[6] If the small increments all had a systematic component in the same direction, one would build up a large 'DC' noise voltage, which is manifestly not the present situation. But the resulting solution might have other applications; see Exercise 7.1.

[7] Landon's original derivation concerned only a special case of this, in which $q(\epsilon) = [\pi\sqrt{a^2 - \epsilon^2}]^{-1}$, $|\epsilon| < a$, corresponding to an added sinusoid of amplitude a and unknown phase. But the important thing was his *idea* of the derivation, which anyone can generalize once it is grasped. In essence he had discovered independently, in the expansion (7.20), what is now called the Fokker–Planck equation of statistical mechanics, a powerful method which we shall use later to show how a nonequilibrium probability distribution relaxes into an equilibrium one. It is now known to have a deep meaning, in terms of continually remaximized entropy.

is, the probability distribution for the voltage *at a given time* that we were seeking). Even the magnitude of the second moment did not matter for the functional form; it determined only how far up the σ^2 hierarchy we moved. But if we ask a more detailed question, involving time-dependent correlation functions, then noise samples from different sources are no longer indistinguishable. The second-order correlations of the form $\langle \epsilon(t)\epsilon(t')\rangle$ are related to the power spectrum of the noise through the Wiener–Khinchin theorem, which was just in the process of being discovered in 1941; they give information about the duration in time of the small increments. But if we go to fourth-order correlations $\langle \epsilon(t_1)\epsilon(t_2)\epsilon(t_3)\epsilon(t_4)\rangle$ we obtain still more detailed information, different for different sources, even though they all have the same Gaussian point distribution and the same power spectrum.[8]

> **Exercise 7.1.** The above derivation established the result to order $\langle \epsilon^2\rangle$. Now suppose that we add n such small increments, bringing the variance up to $\sigma^2 + n\langle \epsilon^2\rangle$. Show that in the limit $n \to \infty$, $\langle \epsilon^2\rangle \to 0$, $n\langle \epsilon^2\rangle \to$ const., the Gaussian distribution (7.23) becomes exact (the higher terms in the expansion (7.19) become vanishingly small compared with the terms in $\langle \epsilon^2\rangle$).

> **Exercise 7.2.** Repeat the above derivation without assuming that $\langle \epsilon\rangle = 0$ in (7.20). The resulting differential equation is a Fokker–Planck equation. Show that there is now a superimposed steady drift, the solutions having the form $\exp\{-(v - a\sigma^2)^2/2\sigma^2\}$. Suggest a possible useful application of this result.
> *Hint:* σ^2 and v may be given other interpretations, such as time and distance.

7.6 Why the ubiquitous use of Gaussian distributions?

We started this chapter by noting the surprise of de Morgan and Barnard at the great and ubiquitous success that is achieved in inference – particularly, in parameter estimation – through the use of Gaussian sampling distributions, and the reluctance of Feller to believe that such success was possible. It is surprising that to understand this mystery requires almost no mathematics – only a conceptual reorientation toward the idea of probability theory as logic.

Let us think in terms of the *information* that is conveyed by our equations. Whether or not the long-run frequency distribution of errors is in fact Gaussian is almost never

---

[8] Recognition of this invalidates many naïve arguments by physicists who try to prove that 'Maxwell demons' are impossible by assuming that thermal radiation has a universal character, making it impossible to distinguish the source of the radiation. But only the second-order correlations are universal; a demon who perceives fourth-order correlations in thermal radiation is far from blind about the details of his surroundings. Indeed, the famous Hanbury Brown–Twiss interferometer (1956), invokes just such a fourth-order demon, in space instead of time and observing $\langle \epsilon^2(x_1)\epsilon^2(x_2)\rangle$ to measure the angular diameters of stars. Conventional arguments against Maxwell demons are logically flawed and prove nothing.

known empirically; what the scientist knows about them (from past experience or from theory) is almost always simply their general magnitude. For example, today most accurate experiments in physics take data electronically, and a physicist usually knows the mean square error of those measurements because it is related to the temperature by the well-known Nyquist thermal fluctuation law.[9] But he seldom knows any other property of the noise. If he assigns the first two moments of a noise probability distribution to agree with such information, but has no further information and therefore imposes no further constraints, then a Gaussian distribution fit to those moments will, according to the principle of maximum entropy as discussed in Chapter 11, represent most honestly his state of knowledge about the noise.

But we must stress a point of logic concerning this. It represents most honestly the physicist's state of knowledge *about the particular samples of noise for which he had data.* This never includes the noise in the measurement which he is about to make! If we suppose that knowledge about some past samples of noise applies also to the specific sample of noise that we are about to encounter, then we are making an inductive inference that might or might not be justified; and honesty requires that we recognize this. Then past noise samples are relevant for predicting future noise only through those aspects that we believe should be reproducible in the future.

In practice, common sense usually tells us that any observed fine details of past noise are irrelevant for predicting fine details of future noise, but that coarser features, such as past mean square values, may be expected reasonably to persist, and thus be relevant for predicting future mean square values. Then our probability assignment for future noise should make use only of those coarse features of past noise which we believe to have this persistence. That is, it should have maximum entropy subject to the constraints of the coarse features that we retain because we expect them to be reproducible. Probability theory becomes a much more powerful reasoning tool when guided by a little common sense judgment of this kind about the real world, as expressed in our choice of a model and assignment of prior probabilities.

Thus we shall find in studying maximum entropy below that, when we use a Gaussian sampling distribution for the noise, we are in effect telling the robot: 'The only thing I know about the noise is its first two moments, so please take that into account in assigning your probability distribution, but be careful *not* to assume anything else about the noise.' We shall see presently how well the robot obeys this instruction.[10]

---

[9] A circuit element of resistance $R(\omega)$ ohms at angular frequency ω develops across its terminals in a small frequency band $\Delta\omega = 2\pi\,\Delta f$ a fluctuating mean square open-circuit voltage $V^2 = 4kTR\Delta f$, where f is the frequency in hertz (cycles per second), $k \equiv 1.38 \times 10^{-23}$ joule/degree is Boltzmann's constant, and T is the Kelvin temperature. Thus it can deliver to another circuit element the maximum noise power $P = V^2/4R = kT\Delta f$. At room temperature, $T = 300\,\mathrm{K}$, this is about 4×10^{-15} watt/megahertz bandwidth. Any signal of lower intensity than this will be lost in the thermal noise and cannot be recovered, ordinarily, by any amount of amplification. But prior information about the kind of signal to be expected will still enable a Bayesian computer program to extract weaker signals, as the work of Bretthorst (1988) demonstrates. We study this in Part 2.

[10] If we have further pieces of information about the noise, such as a fourth moment or an upper bound, the robot can take these into account also by assigning generalized Gaussian – that is, general maximum entropy – noise probability distributions. Examples of the use of fourth-moment constraints in economics and physical chemistry are given by Gray and Gubbins (1984) and Zellner (1988).

This does not mean that the full frequency distribution of the past noise is to be ignored if it happens to be known. Probability theory as logic does not conflict with conventional orthodox theory if we actually have the information (that is, perfect knowledge of limiting frequencies, and no other information) that orthodox theory presupposes; but it continues to operate using whatever information we have. In the vast majority of real problems we lack this frequency information but have other information (such as mean square value, digitizing interval, power spectrum of the noise); and a correct probability analysis readily takes this into account, by using the technical apparatus that orthodoxy lacks.

Exercise 7.3. Suppose that the long-run frequency distribution of the noise has been found empirically to be the function $f(e)$ (never mind how one could actually obtain that information), and that we have no other information about the noise. Show, by reasoning like that leading to (4.55) and using Laplace's Rule of Succession (6.73), that, in the limit of a very large amount of frequency data, our *probability* distribution for the noise becomes numerically equal to the observed frequency distribution: $p(e|I) \to f(e)$. This is what Daniel Bernoulli conjectured in Section 7.4. But state very carefully the exact conditions for this to be true.

In other fields, such as analysis of economic data, knowledge of the noise may be more crude, consisting of its approximate general magnitude and nothing else. But for reasons noted below (the central limit theorem), we still have good reasons to expect a Gaussian functional form; so a Gaussian distribution fit to that magnitude is still a good representation of one's state of knowledge. If even that knowledge is lacking, we still have good reason to expect the Gaussian functional form, so a sampling distribution with σ an undetermined nuisance parameter to be estimated from the data is an appropriate and useful starting point. Indeed, as Bretthorst (1988) demonstrates, this is often the safest procedure, even in a physics experiment, because the noise may not be the theoretically well understood Nyquist noise. No source has ever been found which generates noise below the Nyquist value – and from the second law of thermodynamics we do not expect to find such a source, because the Nyquist law is only the low-frequency limit of the Planck black-body radiation law – but a defective apparatus may generate noise far above the Nyquist value. One can still conduct the experiment with such an apparatus, taking into account the greater noise magnitude; but, of course, a wise experimenter who knows that this is happening will try to improve his apparatus before proceeding.

We shall find, in the central limit theorem, still another strong justification for using Gaussian error distributions. But if the Gaussian law is nearly always a good representation of our state of knowledge about the errors *in our specific data set*, it follows that inferences made from it are nearly always the best ones that could have been made from the information that we actually have.

Now, as we note presently, the data give us a great deal of information about the noise, not usually recognized. But Bayes' theorem automatically takes into account whatever can be inferred about the noise from the data; to the best of our knowledge, this has not been recognized in the previous literature. Therefore Bayesian inferences using a Gaussian sampling distribution could be improved upon only by one who had additional information about the actual errors in his specific data set, *beyond its first two moments and beyond what is known from the data*.

For this reason, whether our inferences are successful or not, unless such extra information is at hand, there is no justification for adopting a different error law; and, indeed, no principle to tell us which different one to adopt. This explains the ubiquitous use. Since the time of Gauss and Laplace, the great majority of all inference procedures with continuous probability distributions have been conducted – necessarily and properly – with Gaussian sampling distributions. Those who disapproved of this, whatever the grounds for their objection, have been unable to offer any alternative that was not subject to a worse objection; so, already in the time of de Morgan, some 25 years after the work of Laplace, use of the Gaussian rule had become ubiquitous by default, and this continues today.

Recognition of this considerably simplifies our expositions of Bayesian inference; 95% of our analysis can be conducted with a Gaussian sampling distribution, and only in special circumstances (unusual prior information such as that the errors are pure digitizing errors or that there is an upper bound to the possible error magnitude) is there any reason for adopting a different one. But even in those special circumstances, the Gaussian analysis usually leads to final conclusions so near to the exact ones that the difference is hardly worth the extra effort.

It is now clear that the most ubiquitous reason for using the Gaussian sampling distribution is not that the error frequencies are known to be – or assumed to be – Gaussian, but rather because those frequencies are *unknown*. One sees what a totally different outlook this is than that of Feller and Barnard; 'normality' was not an *assumption* of physical fact at all. It was a *valid description* of our state of knowledge. In most cases, had we done anything different, we would be making an unjustified, gratuitous assumption (violating one of our Chapter 1 desiderata of rationality). But this still does not explain why the procedure is so successful.

7.7 Why the ubiquitous success?

By 'ubiquitous success' we mean that, for nearly two centuries, the Gaussian sampling distribution has continued to be, in almost all problems, much easier to use and to yield better results (more accurate parameter estimates) than any alternative sampling distribution that anyone has been able to suggest. To explain this requires that analysis that de Morgan predicted would one day be found. But why did it require so long to find that analysis?

As a start toward answering this, note that we are going to use some function of the data as our estimate; then, whether our present inference – here and now – is or is not successful, depends entirely on what that function is, and on the actual errors that are present *in the one specific data set that we are analyzing.* Therefore to explain its success requires that we examine that specific data set. The frequency distribution of errors in other data sets that we might have got but did not – and which we are therefore not analyzing – is irrelevant, unless (a) it is actually known, not merely imagined; (b) it tells us something about the errors in our specific data set that we would not know otherwise.

We have never seen a real problem in which these conditions were met; those who emphasized frequencies most strongly merely *assumed* them without pointing to any actual measurement. They persisted in trying to justify the Gaussian distribution in terms of assumed frequencies in imaginary data sets that have never been observed; thus they continued to dwell on fantasies instead of the information that was actually relevant to the inference; and so we understand why they were unable to find any explanation of the success of that distribution.

Thus, Feller, thinking exclusively in terms of sampling distributions for estimators, thought that, unless our sampling distribution for the e_i correctly represented the actual frequencies of errors, our estimates would be in some way unsatisfactory; in exactly what way seems never to have been stated by Feller or anyone else. Now there is a closely related truth here: *If our estimator is a given, fixed function of the data, then the actual variability of the estimate in the long-run over all possible data sets, is indeed determined by the actual long-run frequency distribution of the errors, if such a thing exists.*

But does it follow that our assigned sampling distribution must be equal to that frequency distribution in order to get satisfactory estimates? To the best of our knowledge, orthodoxy has never attempted to give any such demonstration, or even recognized the need for it. But this makes us aware of another, equally serious, difficulty.

7.8 What estimator should we use?

In estimating a parameter μ from data D, the orthodoxian would almost surely use the maximum likelihood estimator; that is, the value of μ for which $p(D|\mu)$ is a maximum. If the prior information is unimportant (that is, if the prior probability density $p(\mu|I)$ is essentially constant over the region of high likelihood), the Bayesian might do this also. But is there any proof that the maximum likelihood estimator yields the most accurate estimates? Might not the estimates of μ be made still better in the long-run (i.e., more closely concentrated about the true value μ_0) by a different choice of estimator? This question also remains open; there are two big gaps in the logic here.

More fundamental than the logical gaps is the conceptual disorientation; the scenario envisaged by Feller is not the real problem facing a scientist. As John Maynard Keynes (1921) emphasized long ago, his job is not to fantasize about an imaginary 'long-run' which will never be realized, but to estimate the parameters in the one real case before him, from the one real data set that he actually has.[11]

---

[11] Curiously, in that same after-dinner speech, Feller also railed against those who fail to distinguish between the long-run and the individual case, yet it appears to us that it was Feller who failed to make that distinction properly. He would judge the merit of

To raise these issues is not mere nitpicking; let us show that in general there actually is a better estimator, by the long-run sampling theory criterion, than the maximum likelihood estimator. As we have just seen, Gauss proved that the condition

$$\text{(maximum likelihood estimator)} = \text{(arithmetic mean of the observations)} \qquad (7.24)$$

uniquely determines the Gaussian sampling distribution. Therefore, if our sampling distribution is not Gaussian, these two estimators are different. Then, which is better?

Almost all sampling distributions used are of the 'independent, identically distributed' (iid) form:

$$p(x_1 \cdots x_n | \mu I) = \prod_{i=1}^{n} f(x_i - \mu). \qquad (7.25)$$

Bayesian analysis has the theoretical principles needed to determine the optimal estimate for each data set whatever the sampling distribution; it will lead us to make the posterior mean estimate as the one that minimizes the expected square of the error, the posterior median as the one that minimizes the absolute error, etc. If the sampling distribution is not Gaussian, the estimator proves typically to be a linear combination of the observations $(\mu)_{\text{est}} = \sum w_i y_i$, but *with variable weighting coefficients w_i depending on the data configuration* $(y_i - y_j)$, $1 \le i, j \le n$. Thus the estimate is, in general, a nonlinear function of the observations.[12]

In contrast, consider a typical real problem from the orthodox viewpoint which has no prior probabilities or loss functions. We are trying to estimate a location parameter μ, and our data D consist of n observations: $D = \{y_1, \ldots, y_n\}$. But they have errors that vary in a way that is uncontrolled by the experimenter and unpredictable from his state of knowledge.[13] In the following we denote the unknown true value by μ_0, and use μ as a general running variable. Then our model is

---

an individual case inference by its imagined long-run properties. But it is not only possible, but common as soon as we depart from Gaussian sampling distributions, that an estimator which is proved to be as good as can be obtained, as judged by its long-run success over all data sets, may nevertheless be very poor for our particular data set and should not be used for it. Then the sampling distribution for any particular estimator (i.e. any particular function $f(y_1 \cdots y_n)$ of the data) becomes irrelevant because *with different data sets we shall use different estimators*. Thus, to suppose that a procedure that works satisfactorily with Gaussian distributions should be used also with others, can lead one to be badly mistaken in more than one way. This introduces us to the phenomena of sufficiency and ancillarity, pointed out by R. A. Fisher in the 1930s and discussed in Chapter 8. But it is now known that Bayes' theorem automatically detects these situations and does the right thing here, choosing for each data set the optimal estimator *for that data set*. In other words, the correct solution to the difficulties pointed out by Fisher is just to return to the original Bayesian analysis of Laplace and Jeffreys, that Fisher thought to be wrong.

[12] The reader may find it instructive to verify this in detail for the simple looking Cauchy sampling distribution

$$p(y_i | \mu I) = \frac{1}{\pi} \left[\frac{1}{1 + (y_i - \mu)^2} \right] \qquad (7.26)$$

for which the nonlinear functions are surprisingly complicated.

[13] This does not mean that they are 'not determined by anything' as is so often implied by those suffering from the mind projection fallacy; it means only that they are not determined by any circumstances that the experimenter is controlling or observing. Whether the determining factors could or could not be observed in principle is irrelevant to the present problem, which is to reason as best we can *in the state of knowledge that we have specified*.

$$y_i = \mu_0 + e_i, \qquad (1 \le i \le n), \tag{7.27}$$

where e_i is the actual error in the ith measurement. Now, if we assign an independent Gaussian sampling distribution for the errors $e_i = y_i - \mu_0$:

$$p(D|\mu_0 \sigma I) = \left(\frac{1}{2\pi\sigma^2}\right)^{n/2} \exp\left\{-\frac{\sum(y_i - \mu_0)^2}{2\sigma^2}\right\}, \tag{7.28}$$

we have

$$\sum_{i=1}^{n}(y_i - \mu_0)^2 = n[(\mu_0 - \overline{y})^2 + s^2], \tag{7.29}$$

where

$$\overline{y} \equiv \frac{1}{n}\sum y_i = \mu_0 + \overline{e}, \qquad s^2 \equiv \overline{y^2} - \overline{y}^2 = \overline{e^2} - \overline{e}^2 \tag{7.30}$$

are the only properties of the data that appear in the likelihood function. Thus the consequence of assigning the Gaussian error distribution is that *only the first two moments* of the data are going to be used for inferences about μ_0 (and about σ, if it is unknown). They are called the *sufficient statistics*. From (7.30) it follows that only the first two moments of the noise values $\{e_1, \ldots, e_n\}$,

$$\overline{e} = \frac{1}{n}\sum_i e_i, \qquad \overline{e^2} = \frac{1}{n}\sum_i e_i^2, \tag{7.31}$$

can matter for the error in our estimate. We have, in a sense, the simplest possible connection between the errors in our data and the error in our estimate.

If we estimate μ by the arithmetic mean of the observations, the actual error we shall make in the estimate is the average of the individual errors in our specific data set:[14]

$$\Delta \equiv \overline{y} - \mu_0 = \overline{e}. \tag{7.32}$$

Note that $\overline{e}$ is not an average over any probability distribution; it is the average of the *actual* errors, and this result holds however the actual errors e_i are distributed. For example, whether a histogram of the e_i closely resembles the assigned Gaussian (7.28) or whether all of the error happens to be in e_1 does not matter in the least; (7.32) remains correct.

7.9 Error cancellation

An important reason for the success of the Gaussian sampling distribution lies in its relation to the aforementioned error cancellation phenomenon. Suppose we estimate μ by some linear combination of the data values:

$$(\mu)_{\text{est}} = \sum_{i=1}^{n} w_i y_i, \tag{7.33}$$

---

[14] Of course, probability theory tells us that this is the best estimate we can make if, as supposed, the only information we have about μ comes from this one data set. If we have other information (previous data sets, other prior information) we should take it into account; but then we are considering a different problem.

where the weighting coefficients w_i are real numbers satisfying $\sum w_i = 1$, $w_i \geq 0$, $1 \leq i \leq n$. Then with the model (7.27), the square of the error we shall make in our estimate is

$$\Delta^2 = [(\mu)_{\text{est}} - \mu_0]^2 = \left(\sum_i w_i e_i\right)^2 = \sum_{i,j=1}^{n} w_i w_j e_i e_j, \tag{7.34}$$

and the expectation of this over whatever sampling distribution we have assigned is

$$\langle \Delta^2 \rangle = \sum_{i,j} w_i w_j \langle e_i e_j \rangle. \tag{7.35}$$

But if we have assigned identical and independent probabilities to each e_i separately, as is almost always supposed, then $\langle e_i e_j \rangle = \sigma^2 \delta_{ij}$, and so

$$\langle \Delta^2 \rangle = \sigma^2 \sum_i w_i^2. \tag{7.36}$$

Now set $w_i = n^{-1} + q_i$, where the $\{q_i\}$ are real numbers constrained only by $\sum w_i = 1$, or $\sum q_i = 0$. The expected square of the error is then

$$\langle \Delta^2 \rangle = \sigma^2 \sum_i \left(\frac{1}{n^2} + \frac{2q_i}{n} + q_i^2\right) = \sigma^2 \left(\frac{1}{n} + \sum_i q_i^2\right), \tag{7.37}$$

from which it is evident that $\langle \Delta^2 \rangle$ reaches its absolute minimum

$$\langle \Delta^2 \rangle_{\text{min}} = \frac{\sigma^2}{n} \tag{7.38}$$

if and only if all $q_i = 0$. We have the result that uniform weighting, $w_i = 1/n$, leading to the arithmetic mean of the observations as our estimate, achieves a smaller expected square of the error than any other; in other words, it affords the maximum possible opportunity for that error cancellation to take place. Note that the result is independent of what sampling distribution $p(e_i|I)$ we use for the individual errors. But highly cogent prior information about μ (that is, the prior density $p(\mu|I)$ varies greatly within the high likelihood region) would lead us to modify this somewhat.

If we have no important prior information, use of the Gaussian sampling distribution automatically leads us to estimate μ by the arithmetic mean of the observations; and Gauss proved that the Gaussian distribution is the only one which does this. Therefore, among all sampling distributions which estimate μ by the arithmetic mean of the observations, the Gaussian distribution is uniquely determined as the one that gives maximum error cancellation.

This finally makes it very clear why the Gaussian sampling distribution has enjoyed that ubiquitous success over the years compared with others, fulfilling de Morgan's prediction:

When we assign an independent Gaussian sampling distribution to additive noise, what we achieve is not that the error frequencies are correctly represented, but that those frequencies are made irrelevant to the inference, in two respects. (1) All other aspects of the noise beyond $\bar{e}$ and $\overline{e^2}$ contribute nothing to the numerical value or the accuracy of our estimates. (2) Our estimate is more accurate than that from any other sampling distribution that estimates a location parameter by a linear combination of the observations, because it has the maximum possible error cancellation.

Exercise 7.4. More generally, one could contemplate a sampling distribution $p(e_1, \ldots, e_n | I)$ which assigns different marginal distributions $p(e_i | I)$ to the different e_i, and allows arbitrary correlations between different e_i. Then the covariance matrix $C_{ij} \equiv \langle e_i e_j \rangle$ is a general $n \times n$ positive definite matrix. In this case, prove that the minimum $\langle \Delta^2 \rangle$ is achieved by the weighting coefficients

$$w_i = \sum_j K_{ij} / \sum_{ij} K_{ij}, \qquad (7.39)$$

where $K = C^{-1}$ is the inverse covariance matrix; and that the minimum achievable $\langle \Delta^2 \rangle$ is then

$$\langle \Delta^2 \rangle_{\min} = \left(\sum_{ij} K_{ij} \right)^{-1}. \qquad (7.40)$$

In the case $C_{ij} = \sigma^2 \delta_{ij}$, this reduces to the previous result (7.38).

In view of the discovery of de Groot and Goel (1980) that 'Only normal distributions have linear posterior expectations', it may be that we are discussing an empty case. We need the solution to another mathematical problem: 'What is the most general sampling distribution that estimates a location parameter by a linear function of the observations?' The work of de Groot and Goel suggests, but in our view does not prove, that the answer is again a Gaussian distribution. Note that we are considering two different problems here, (7.38) is the 'risk', or expected, square of the error over the sampling distribution; while de Groot and Goel were considering expectations over the posterior distribution.

7.10 The near irrelevance of sampling frequency distributions

Another way of looking at this is helpful. As we have seen before, in a repetitive situation the probability of any event is usually the same as its expected frequency (using, of course, the same basic probability distribution for both). Then, given a sampling distribution $f(y|\theta)$, it tells us that $\int_R dy \, f(y|\theta)$ is the expected frequency, *before the data are known* of the event $y \in R$.

But if, as always supposed in elementary parameter estimation, the parameters are held fixed throughout the taking of a data set, then the variability of the data *is also, necessarily,* the variability of the actual errors in that data set. If we have defined our model to have the form $y_i = f(x_i) + e_i$, in which the noise is additive, then the exact distribution of the errors is known from the data to within a uniform translation: $e_i - e_j = y_i - y_j$. We know from the data y that the exact error in the ith observation has the form $e_i = y_i - e_0$, where

e_0 is an unknown constant. Whether the frequency distribution of the errors does or does not have the Gaussian functional form is *known from the data*. Then what use remains for the sampling distribution, which in orthodox theory yields only the prior expectations of the error frequencies? Whatever form of frequency distribution we might have expected before seeing the data, is rendered irrelevant by the information in the data! What remains significant for inference is the likelihood function – how the probability of the observed data set varies with the parameters θ.

Although all these results are mathematically trivial, we stress their nontrivial consequences by repeating them in different words. A Gaussian distribution has a far deeper connection with the arithmetic mean than that shown by Gauss. If we assign the independent Gaussian error distribution, then the error in our estimate is always the arithmetic mean of the true errors in our data set; and whether the frequency distribution of those errors is or is not Gaussian is totally irrelevant. Any error vector $\{e_1, \ldots, e_n\}$ with the same first moment $\bar{e}$ will lead us to the same estimate of μ; and any error vector with the same first two moments will lead us to the same estimates of both μ and σ and the same accuracy claims, *whatever the frequency distributions of the individual errors*. This is a large part of the answer to de Morgan, Feller, and Barnard.

This makes it clear that what matters to us functionally – that is, what determines the actual error of our estimate – is not whether the Gaussian error law correctly describes the limiting frequency distribution of the errors; but rather whether that error law correctly describes our *prior information* about the actual errors in our data set. If it does, then the above calculations are the best we can do with the information we have; and there is nothing more to be said.

The only case where we should – or indeed, could – do anything different is when we have additional prior information about the errors beyond their first two moments. For example, if we know that they are simple digitizing errors with digitizing interval δ, then we know that there is a rigid upper bound to the magnitude of any error: $|e_i| \leq \delta/2$. Then if $\delta < \sigma$, use of the appropriate truncated sampling distribution instead of the Gaussian (7.28) will almost surely lead to more accurate estimates of μ. This kind of prior information can be very helpful (although it complicates the analytical solution, this is no deterrent to a computer), and we consider a problem of this type in Section 7.17.

Closer to the present issue, in what sense and under what conditions does the Gaussian error law 'correctly describe' our information about the errors?

7.11 The remarkable efficiency of information transfer

Again, we anticipate a few results from later chapters in order to obtain a quick, preliminary view of what is happening, which will improve our judgment in setting up real problems. The noise probability distribution $p(e|\alpha\beta)$ which has maximum entropy $H = -\int de\, p(e) \log p(e)$ subject to the constraints of prescribed expectations

$$\langle e \rangle = \alpha, \qquad \langle e^2 \rangle = \alpha^2 + \beta^2, \tag{7.41}$$

in which the brackets ⟨ ⟩ now denote averages over the probability distribution $p(e|\alpha\beta)$, is the Gaussian

$$p(e|\alpha\beta) = \frac{1}{\sqrt{2\pi\beta^2}} \exp\left\{-\frac{(e-\alpha)^2}{2\beta^2}\right\}. \qquad (7.42)$$

So a state of prior information which leads us to prescribe the expected first and second moments of the noise – and nothing else – uniquely determines the Gaussian distribution. Then it is eminently satisfactory that this leads to inferences that depend on the noise only through the first and second moments of the actual errors. When we assign error probabilities by the principle of maximum entropy, *the only properties of the errors that are used in our Bayesian inference are the properties about which we specified some prior information.* This is a very important second part of that answer.

In this example, we have stumbled for the first time onto a fundamental feature of probability theory as logic: if we assign probabilities to represent our information, then circumstances about which we have no information, are not used in our subsequent inferences. But it is not only true of this example; we shall find when we study maximum entropy that it is a general theorem that any sampling distribution assigned by maximum entropy leads to Bayesian inferences that depend only on the information that we incorporated as constraints in the entropy maximization.[15]

Put differently, our rules for extended logic automatically use all the information that we have, and avoid assuming information that we do not have. Indeed, our Chapter 1 desiderata require this. In spite of its extremely simple formal structure in the product and sum rules, probability theory as logic has a remarkable sophistication in applications. It perceives instantly what generations of statisticians and probabilists failed to see; for a probability calculation to have a useful and reliable function in the real world, it is by no means required that the probabilities have any relation to frequencies.[16]

Once this is pointed out, it seems obvious that circumstances about which we have no information *cannot* be of any use to us in inference. Rules for inference which fail to recognize this and try to introduce such quantities as error frequencies into the calculation as *ad hoc* assumptions, even when we have no information about them, are claiming, in effect, to get something for nothing (in fact, they are injecting arbitrary – and therefore almost certainly false – information). Such devices may be usable in some small class of problems; but they are guaranteed to yield wrong and/or misleading conclusions if applied outside that class.

On the other hand, probability theory as logic is always safe and conservative, in the following sense: it always spreads the probability out over the full range of conditions

---

[15] Technically (Chapter 8), the class of sampling distributions which have sufficient statistics is precisely the class generated by the maximum entropy principle; and the resulting sufficient statistics are precisely the constraints which determined that maximum entropy distribution.

[16] This is not to say that probabilities are *forbidden* to have any relation to frequencies; the point is rather that whether they do or do not depends on the problem, and probability theory as logic works equally well in either case. We shall see, in the work of Galton below, an example where a clear frequency connection is present, and analysis of the general conditions for this will appear in Chapter 9.

allowed by the information used; our basic desiderata require this. Thus it always yields the conclusions that are justified by the information *which was put into it*. The robot can return vague estimates if we give it vague or incomplete information; but then *it warns us of that fact by returning posterior distributions so wide that they still include the true value of the parameter*. It cannot actually mislead us – in the sense of assigning a high probability to a false conclusion – unless we have given it false information.

For example, if we assign a sampling distribution which supposes the errors to be far smaller than the actual errors, then we have put false information into the problem, and the consequence will be, not necessarily bad estimates of parameters, but false claims about the accuracy of those estimates and – often more serious – the robot can hallucinate, artifacts of the noise being misinterpreted as real effects. As de Morgan (1872, p. 113) put it, this is the error of 'attributing to the motion of the moon in her orbit all the tremors which she gets from a shaky telescope'.

Conversely, if we use a sampling distribution which supposes the errors to be much larger than the actual errors, the result is not necessarily bad estimates, but overly conservative accuracy claims for them and – often more serious – blunt perception, failing to recognize effects that are real, by dismissing them as part of the noise. This would be the opposite error of attributing to a shaky telescope the real and highly important deviation of the moon from her expected orbit. If we use a sampling distribution that reflects the true average errors and the true mean square errors, we have the maximum protection against both of these extremes of misperception, steering the safest possible middle course between them. These properties are demonstrated in detail later.

7.12 Other sampling distributions

Once we understand the reasons for the success of Gaussian inference, we can also see very rare special circumstances where a different sampling distribution would better express our state of knowledge. For example, if we know that the errors are being generated by the unavoidable and uncontrollable rotation of some small object, in such a way that when it is at angle θ, the error is $e = \alpha \cos \theta$ but the actual angle is unknown, a little analysis shows that the prior probability assignment $p(e|I) = (\pi \sqrt{\alpha^2 - e^2})^{-1}$, $e^2 < \alpha^2$, correctly describes our state of knowledge about the error. Therefore it should be used instead of the Gaussian distribution; since it has a sharp upper bound, it may yield appreciably better estimates than would the Gaussian – even if α is unknown and must therefore be estimated from the data (or perhaps it is the parameter of interest to be estimated).

Or, if the error is known to have the form $e = \alpha \tan \theta$ but θ is unknown, we find that the prior probability is the Cauchy distribution $p(e|I) = \pi^{-1}\alpha/(\alpha^2 + e^2)$. Although this case is rare, we shall find it an instructive exercise to analyze inference with a Cauchy sampling distribution, because qualitatively different things can happen. Orthodoxy regards this as 'a pathological, exceptional case' as one referee put it, but it causes no difficulty in Bayesian analysis, which enables us to understand it.

7.13 Nuisance parameters as safety devices

As an example of this principle, if we do not have actual knowledge about the magnitude σ of our errors, then it could be dangerous folly to assume some arbitrary value; the wisest and safest procedure is to adopt a model which honestly acknowledges our ignorance by allowing for various possible values of σ; we should assign a prior $p(\sigma|I)$ which indicates the range of values that σ might reasonably have, consistent with our prior information. Then in the Bayesian analysis we shall find first the joint posterior pdf for both parameters:

$$p(\mu\sigma|DI) = p(\mu\sigma|I)\frac{p(D|\mu\sigma I)}{p(D|I)}. \tag{7.43}$$

But now notice how the product rule rearranges this:

$$p(\mu\sigma|DI) = p(\sigma|I)p(\mu|\sigma I)\frac{p(D|\sigma I)p(\mu|\sigma DI)}{p(D|I)p(\mu|\sigma I)} = p(\mu|\sigma DI)p(\sigma|DI). \tag{7.44}$$

So, if we now integrate out σ as a nuisance parameter, we obtain the marginal posterior pdf for μ alone in the form:

$$p(\mu|DI) = \int d\sigma \, p(\mu|\sigma DI)p(\sigma|DI), \tag{7.45}$$

a weighted average of the pdfs $p(\mu|\sigma DI)$ for all possible values of σ, weighted according to the marginal posterior pdf $p(\sigma|DI)$ for σ, which represents everything we know about σ.

Thus when we integrate out a nuisance parameter, we are not throwing away any information relevant to the parameters we keep; on the contrary, probability theory automatically estimates the nuisance parameter for us from all the available evidence, and takes that information fully into account in the marginal posterior pdf for the interesting parameters (but it does this in such a slick, efficient way that one may not realize that this is happening, and think that he is losing something). In the limit where the data are able to determine the true value $\sigma = \sigma_0$ very accurately, $p(\sigma|DI) \rightarrow \delta(\sigma - \sigma_0)$ and $p(\mu|DI) \rightarrow p(\mu|\sigma_0 DI)$; the theory yields, as it should, the same conclusions that we would have if the true value were known from the start.

This is just one example illustrating that, as noted above, whatever question we ask, probability theory as logic automatically takes into account all the possibilities *allowed by our model* and our information. Then, of course, the onus is on us to choose a model wisely so that the robot is given the freedom to estimate for itself, from the totality of its information, any parameter that we do not know. If we fail to recognize the existence of a parameter which is uninteresting but nevertheless affects our data – and so leave it out of the model – then the robot is crippled and cannot return the optimal inferences to us. The marginalization paradox, discussed in Chapter 15, and the data pooling paradox of Chapter 8, exhibit some of the things that can happen then; the robot's conclusions are still the best ones that could have been made *from the information we gave it*, but they are not the ones that simple common sense would make, using extra information that we failed to give it.

In practice, we find that recognition of a relevant, but unknown and uninteresting, parameter by including it in the model and then integrating it out again as a nuisance parameter, can greatly improve our ability to extract the information we want from our data – often by orders of magnitude. By this means we are forewarning the robot about a possible disturbing complication, putting it on the lookout for it; and the rules of probability theory then lead the robot to make the optimal allowance for it.

This point is extremely important in some current problems of estimating environmental hazards or the safety of new machines, drugs or food additives, where inattention to all of the relevant prior information that scientists have about the phenomenon – and therefore failure to include that information in the model and prior probabilities – can cause the danger to be grossly overestimated or underestimated. For example, from knowledge of the engineering design of a machine, one knows a great deal about its possible failure modes and their consequences, that could not be obtained from any feasible amount of reliability testing by 'random experiments'. Likewise, from knowledge of the chemical nature of a food additive, one knows a great deal about its physiological effects that could not be obtained from any feasible amount of mere toxicity tests.

Of course, this is not to say that reliability tests and toxicity tests should not be carried out; the point is rather that random experiments are very inefficient ways of obtaining information (we learn, so to speak, only like the square root of the number of trials), and rational conclusions cannot be drawn from them unless the equally cogent – often far more cogent – prior information is also taken into account. We saw some examples of this phenomenon in Chapter 6, (6.123)–(6.144). The real function of the random experiment is to guard against completely unexpected bad effects, about which our prior information gave us no warning.

7.14 More general properties

Although the Gauss derivation was of the greatest historical importance, it does not satisfy us today because it depends on intuition; *why* must the 'best' estimate of a location parameter be a linear function of the observations? Evidently, in view of the Gauss derivation, if our assigned sampling distribution is not Gaussian, the best estimate of the location parameter will *not* be the sample mean. It could have a wide variety of other functional forms; then, under what circumstances, is Laplace's prescription the one to use?

We have just seen the cogent pragmatic advantages of using a Gaussian sampling distribution. Today, anticipating a little from later chapters, we would say that its unique theoretical position derives not from the Gauss argument, but rather from four mathematical stability properties, which have fundamentally nothing to do with probability theory or inference, and a fifth, which has everything to do with them, but was not discovered until the mid-20th century:

(A) Any smooth function with a single rounded maximum, if raised to higher and higher powers, goes into a Gaussian function. We saw this in Chapter 6.
(B) The product of two Gaussian functions is another Gaussian function.

(C) The convolution of two Gaussian functions is another Gaussian function.

(D) The Fourier transform of a Gaussian function is another Gaussian function.

(E) A Gaussian probability distribution has higher entropy than any other with the same variance; therefore any operation on a probability distribution which discards information, but conserves variance, leads us inexorably closer to a Gaussian. The central limit theorem, derived below, is the best known example of this, in which the operation being performed is convolution.

Properties (A) and (E) explain why a Gaussian form is approached more and more closely by various operations; properties (B), (C) and (D) explain why that form, once attained, is preserved.

7.15 Convolution of Gaussians

The convolution property (C) is shown as follows. Expanding now the notation[17] of (7.1)

$$\varphi(x - \mu|\sigma) \equiv \frac{1}{\sigma}\varphi\left(\frac{x-\mu}{\sigma}\right) = \sqrt{\frac{1}{2\pi\sigma^2}}\exp\left\{-\frac{(x-\mu)^2}{2\sigma^2}\right\} = \sqrt{\frac{w}{2\pi}}\exp\left\{-\frac{w}{2}(x-\mu)^2\right\}$$

(7.46)

in which we introduce the 'weight' $w \equiv 1/\sigma^2$ for convenience, the product of two such functions is

$$\varphi(x - \mu_1|\sigma_1)\varphi(y - x - \mu_2|\sigma_2) = \frac{1}{2\pi\sigma_1\sigma_2}\exp\left\{-\frac{1}{2}\left[\left(\frac{x-\mu_1}{\sigma_1}\right)^2 + \left(\frac{y-x-\mu_2}{\sigma_2}\right)^2\right]\right\};$$

(7.47)

but we bring out the dependence on x by rearranging the quadratic form:

$$\left(\frac{x-\mu_1}{\sigma_1}\right)^2 + \left(\frac{y-x-\mu_2}{\sigma_2}\right)^2 = (w_1 + w_2)(x - \hat{x})^2 + \frac{w_1w_2}{w_1+w_2}(y - \mu_1 - \mu_2)^2,$$

(7.48)

where $\hat{x} \equiv (w_1\mu_1 + w_2y - w_2\mu_2)/(w_1 + w_2)$. The product is still a Gaussian with respect to x; on integrating out x we have the convolution law:

$$\int_{-\infty}^{\infty} dx\, \varphi(x - \mu_1|\sigma_1)\varphi(y - x - \mu_2|\sigma_2) = \varphi(y - \mu|\sigma),$$

(7.49)

where $\mu \equiv \mu_1 + \mu_2$, $\sigma^2 \equiv \sigma_1^2 + \sigma_2^2$. Two Gaussians convolve to make another Gaussian, the means μ and variances σ^2 being additive. Presently we shall see some important applications that require only the single convolution formula (7.49). Now we turn to the famous theorem, which results from repeated convolutions.

---

[17] This notation is not quite inconsistent, since $\varphi(\ )$ and $\varphi(\ |\ )$ are different functional symbols.

7.16 The central limit theorem

The question whether non-Gaussian distributions also have parameters additive under convolution leads us to the notion of *cumulants* discussed in Appendix C. The reader who has not yet studied this should do so now.

Editor's Exercise 7.5. Jaynes never actually derived the central limit theorem in this section; rather he is deriving the only known exception to the central limit theorem. In Appendix C he comes close to deriving the central limit theorem. Defining

$$\phi(\alpha) = \int_{-\infty}^{\infty} f(x) \exp\{i\alpha x\}, \tag{7.50}$$

and a repeated convolution gives

$$h_n(y) = f * f * f * \cdots * f = \frac{1}{2\pi} \int_{-\infty}^{\infty} dy \phi(y)^n \exp\{-i\alpha y\}, \tag{7.51}$$

$$[\phi(\alpha)]^n = \exp\left\{ n\left(C_0 + \alpha C_1 - \frac{\alpha^2 C_2}{2} + \cdots \right) \right\}, \tag{7.52}$$

where the cumulants, C_n, are defined in Appendix C. If cumulants higher than C_2 are ignored, one obtains

$$h_n(y) \approx \frac{1}{2\pi} \int_{-\infty}^{\infty} d\alpha \exp\left\{ in\alpha \langle x \rangle - \frac{n\sigma^2 \alpha^2}{2} - i\alpha y \right\},$$

$$= \frac{1}{2\pi} \int_{-\infty}^{\infty} d\alpha \exp\left\{ -\frac{n\sigma^2 \alpha^2}{2} \right\} \exp\{ -i\alpha(n\langle x \rangle - y) \}, \tag{7.53}$$

$$= \frac{1}{\sqrt{2\pi n\sigma^2}} \exp\left\{ -\frac{(y - n\langle x \rangle x)^2}{2n\sigma^2} \right\},$$

and this completes the derivation of the central limit theory. What are the conditions under which this is a good approximation? Is this derivation valid when one is computing the ratios of probabilities?

If the functions $f_i(x)$ to which we apply that theory are probability distributions, then they are necessarily non-negative and normalized: $f_i(x) \geq 0$, $\int dx\, f_i(x) = 1$. Then the zeroth moments are all $Z_i = 1$, and the Fourier transforms

$$\mathcal{F}_i(\alpha) \equiv \int_{-\infty}^{\infty} dx\, f_i(x) \exp\{i\alpha x\} \tag{7.54}$$

are absolutely convergent for real α. Note that all this remains true if the f_i are discontinuous, or contain delta-functions; therefore the following derivation will apply equally well to the continuous or discrete case or any mixture of them.[18]

Consider two variables to which are assigned probability distributions conditional on some information I:

$$f_1(x_1) = p(x_1|I), \qquad f_2(x_2) = p(x_2|I). \tag{7.55}$$

We want the probability distribution $f(y)$ for the sum $y = x_1 + x_2$. Evidently, the cumulative probability density for y is

$$P(y' \le y|I) = \int_{-\infty}^{\infty} dx_1 \, f_1(x_1) \int_{-\infty}^{y-x_1} dx_2 \, f_2(x_2), \tag{7.56}$$

where we integrated over the region R defined by $(x_1 + x_2 \le y)$. Then the probability density for y is

$$f(y) = \left[\frac{d}{dy} P(y' \le y|I) \right]_{y=y'} = \int dx_1 \, f_1(x_1) f_2(y - x_1), \tag{7.57}$$

just the convolution, denoted by $f(y) = f_1 * f_2$ in Appendix C. Then the probability density for the variable $z = y + x_3$ is

$$g(z) = \int dy \, f(y) f_3(z - y) = f_1 * f_2 * f_3 \tag{7.58}$$

and so on by induction: the probability density for the sum $y = x_1 + \cdots + x_n$ of n variables is the multiple convolution $h_n(y) = f_1 * \cdots * f_n$.

In Appendix C we found that convolution in the x space corresponds to simple multiplication in the Fourier transform space: introducing the *characteristic function* for $f_k(x)$

$$\varphi_k(\alpha) \equiv \langle \exp\{i\alpha x\} \rangle = \int_{-\infty}^{\infty} dx \, f_k(x) \exp\{i\alpha x\} \tag{7.59}$$

and the inverse Fourier transform

$$f_k(x) = \frac{1}{2\pi} \int_{-\infty}^{\infty} d\alpha \, \varphi_k(\alpha) \exp\{-i\alpha x\}, \tag{7.60}$$

we find that the probability density for the sum of n variables x_i is

$$h_n(q) = \frac{1}{2\pi} \int d\alpha \, \varphi_1(\alpha) \cdots \varphi_n(\alpha) \exp\{-i\alpha q\}, \tag{7.61}$$

or, if the probability distributions $f_i(x)$ are all the same,

$$h_n(q) = \frac{1}{2\pi} \int d\alpha \, [\varphi(\alpha)]^n \exp\{-i\alpha q\}. \tag{7.62}$$

---

[18] At this point, the reader who has been taught to distrust or disbelieve in delta-functions must unlearn that by reading Appendix B on the concept of a 'function'. This is explained also by Lighthill (1957) and Dyson (1958). Without the free use of delta-functions and other generalized functions, real applications of Fourier analysis are in an almost helpless, crippled condition compared with what can be done by using them.

The probability density for the arithmetic mean $\bar{x} = q/n$ is evidently, from (7.62),

$$p(\bar{x}) = nh_n(n\bar{x}) = \frac{n}{2\pi} \int d\alpha \, [\varphi(\alpha) \exp\{-i\alpha\bar{x}\}]^n. \tag{7.63}$$

It is easy to prove that there is only one probability distribution with this property. If the probability distribution $p(x|I)$ for a single observation x has the characteristic function

$$\varphi(\alpha) = \int dx \, p(x|I) \exp\{i\alpha x\}, \tag{7.64}$$

then the one for the average of n observations, $\bar{x} = n^{-1} \sum x_i$, has a characteristic function of the form $\varphi^n(n^{-1}\alpha)$. The necessary and sufficient condition that x and $\bar{x}$ have the same probability distribution is therefore that $\varphi(\alpha)$ satisfy the functional equation $\varphi^n(n^{-1}\alpha) = \varphi(\alpha)$. Now, substituting $\alpha' = n^{-1}\alpha$, and recognizing that one dummy argument is as good as another, one obtains

$$n \log \varphi(\alpha) = \log \varphi(n\alpha), \qquad -\infty < \alpha < \infty, \qquad n = 1, 2, 3, \ldots. \tag{7.65}$$

Evidently, this requires a linear relation on the positive real line:

$$\log \varphi(\alpha) = C\alpha, \qquad 0 \le \alpha < \infty, \tag{7.66}$$

where C is some complex number. Writing $C = -k + i\theta$, the most general solution satisfying the reality condition $\varphi(-\alpha) = \varphi^*(\alpha)$ is

$$\varphi(\alpha) = \exp\{i\alpha\theta - k|\alpha|\}, \qquad -\infty < \theta < \infty, \qquad 0 < k < \infty, \tag{7.67}$$

which yields

$$p(x|I) = \frac{1}{2\pi} \int_{-\infty}^{\infty} d\alpha \, \exp\{-k|\alpha|\} \exp\{i\alpha(\theta - x)\} = \frac{1}{\pi} \left[\frac{k}{k^2 + (x - \theta)^2} \right], \tag{7.68}$$

the Cauchy distribution with median θ, quartiles $\theta \pm k$. Now we turn to some important applications of the above mathematical results.

7.17 Accuracy of computations

As a useful application of the central limit theorem, consider a computer programmer deciding on the accuracy to be used in a program. This is always a matter of compromise between misleading, inaccurate results on the one hand, and wasting computation facilities with more accuracy than needed on the other.

Of course, it is better to err on the side of a little more accuracy than really needed. Nevertheless, it is foolish (and very common) to tie up a large facility with a huge computation to double precision (16 decimal places) or even higher, when the user has no use for anything like that accuracy in the final result. The computation might have been done in less time but with the same result on a desktop microcomputer, had it been programmed for an accuracy that is reasonable for the problem.

Programmers can speed up and simplify their creations by heeding what the central limit theorem tells us. In probability calculations we seldom have any serious need for more than three-figure accuracy in our final results, so we shall be well on the safe side if we strive to get four-figure accuracy reliably in our computations.

As a simple example, suppose we are computing the sum

$$S \equiv \sum_{n=1}^{N} a_n \tag{7.69}$$

of N terms a_n, each one positive and of order unity. To achieve a given accuracy in the sum, what accuracy do we need in the individual terms?

Our computation program or lookup table necessarily gives each a_n digitized to some smallest increment ϵ, so this will be actually the true value plus some error e_n. If we have a_n to six decimal digits, then $\epsilon = 10^{-6}$; if we have it to 16 binary digits, then $\epsilon = 2^{-16} = 1/65\,536$. The error in any one entry is in the range $(-\epsilon/2 < e_n \le \epsilon/2)$, and in adding N such terms the maximum possible error is $N\epsilon/2$. Then it might be thought that the programmer should ensure that this is acceptably small.

But if N is large, this maximum error is enormously unlikely; this is just the point that Euler failed to see. The individual errors are almost certain to be positive and negative roughly equally often, giving a high degree of mutual cancellation, so that the net error should tend to grow only as $\sqrt{N}$.

The central limit theorem tells us what is essentially a simple combinatorial fact, that out of all conceivable error vectors $\{e_1, \ldots, e_N\}$ that could be generated, the overwhelming majority have about the same degree of cancellation, which is the reason for the $\sqrt{N}$ rule. If we consider each individual error equally likely to be anywhere in $(-\epsilon/2, \epsilon/2)$, this corresponds to a rectangular probability distribution on that interval, leading to an expected square error per datum of

$$\frac{1}{\epsilon} \int_{-\epsilon/2}^{\epsilon/2} dx\, x^2 = \frac{\epsilon^2}{12}. \tag{7.70}$$

Then by the central limit theorem the probability distribution for the sum S will tend to a Gaussian with a variance $N\epsilon^2/12$, while S is approximately N. If N is large so that the central limit theorem is accurate, then the probability that the magnitude of the net error will exceed $\epsilon\sqrt{N}$, which is $\sqrt{12} = 3.46$ standard deviations, is about

$$2[1 - \Phi(3.46)] \simeq 0.0006, \tag{7.71}$$

where $\Phi(x)$ is the cumulative normal distribution. One will almost never observe an error that great. Since $\Phi(2.58) = 0.995$, there is about a 1% chance that the net error magnitude will exceed $0.74\epsilon\sqrt{N} = 2.58$ standard deviations.

Therefore if we strive, not for certainty, but for 99% or greater probability, that our sum S is correct to four figures, this indicates the value of ϵ that can be tolerated in our algorithm or lookup table. We require $0.74\epsilon\sqrt{N} \leq 10^{-4}N$, or

$$\epsilon \leq 1.35 \times 10^{-4}\sqrt{N}. \tag{7.72}$$

The perhaps surprising result is that if we are adding $N = 100$ roughly equal terms, to achieve a virtual certainty of four-figure accuracy in the sum we require only three-figure accuracy in the individual terms! Under favorable conditions, the mutual cancellation phenomenon can be effective far beyond Euler's dreams. Thus we can get by with a considerably shorter computation for the individual terms, or a smaller lookup table, than might be supposed.

This simple calculation can be greatly generalized, as indicated by Exercise 7.5. But we note an important proviso to be investigated in Exercise 7.6; this holds only when the individual errors e_n are logically independent. Given ϵ in advance, if knowing e_1 then tells us anything about any other e_n, then there are correlations in our probability assignment to errors, the central limit theorem no longer applies, and a different analysis is required. Fortunately, this is almost never a serious limitation in practice because the individual a_n are determined by some continuously variable algorithm and differ among themselves by amounts large compared with ϵ, making it impossible to determine any e_i given any other e_j.

Exercise 7.6. Suppose that we are to evaluate a Fourier series $S(\theta) = \sum a_n \sin n\theta$. Now the individual terms vary in magnitude and are themselves both positive and negative. In order to achieve four-figure accuracy in $S(\theta)$ with high probability, what accuracy do we now require in the individual values of a_n and $\sin n\theta$?

Exercise 7.7. Show that if there is a positive correlation in the probabilities assigned to the e_i, then the error in the sum may be much greater than indicated by the central limit theorem. Try to make a more sophisticated probability analysis taking correlations into account, which would be helpful to a computer programmer who has some kind of information about mutual properties of errors leading to such correlations, but is still striving for the greatest efficiency for a given accuracy.

The literature of orthodox statistics contains some quite different recommendations than ours concerning accuracy of numerical calculations. For example, the textbook of McClave and Benson (1988, p. 99) considers calculation of a sample standard deviation s of $n = 50$ observations $\{x_1, \ldots, x_n\}$ from that of $s^2 = \overline{x^2} - \overline{x}^2$. McClave and Benson state that: 'You should retain twice as many decimal places in s^2 as you want in s. For example, if you want to calculate s to the nearest hundredth, you should calculate s^2 to the nearest ten-thousandth.' When we studied calculus (admittedly many years ago) it was generally thought that small increments are related by $\delta(s^2) = 2s\delta s$, or $\delta s/s = (1/2)\delta(s^2)/s^2$. So, if s^2 is calculated to four significant figures, this determines s not to two significant figures, but to somewhat better than four. But, in any event, McClave and Benson's practice of inserting a gratuitous extra factor $n/(n-1)$ in the symbol which they denote by 's^2' makes a joke of any pretense of four-figure accuracy in either when $n = 100$.

7.18 Galton's discovery

The single convolution formula (7.49) led to one of the most important applications of probability theory in biology. Although from our present standpoint (7.49) is only a straightforward integration formula, which we may write for present purposes in the form

$$\int_{-\infty}^{\infty} \mathrm{d}x\, \varphi(x|\sigma_1)\varphi(y - ax|\sigma_2) = \varphi(y|\sigma), \tag{7.73}$$

where we have made the scale changes $x \to ax$, $\sigma_1 \to a\sigma_1$, and so now

$$\sigma = \sqrt{a^2\sigma_1^2 + \sigma_2^2}, \tag{7.74}$$

it became in the hands of Francis Galton (1886) a major revelation about the mechanism of biological variation and stability.[19] We use the conventional language of that time, which did not distinguish between the notions of probability and frequency, using the words interchangeably. But this is not a serious matter because his data were, in fact, frequencies, and, as we shall see in Chapter 9, strict application of probability theory as logic would then lead to *probability* distributions that are substantially equal to the *frequency* distributions (exactly equal in the limit where we have an arbitrarily large amount of frequency data and no other relevant prior information). Consider, for example, the frequency distribution of heights h of adult males in the population of England. Galton found that this could be represented fairly well by a Gaussian

$$\varphi(h - \mu|\sigma)\mathrm{d}h = \varphi\left(\frac{h - \mu}{\sigma}\right)\frac{\mathrm{d}h}{\sigma} \tag{7.75}$$

with $\mu = 68.1$ inches, $\sigma = 2.6$ inches. Then he investigated whether children of tall parents tend to be tall, etc. To keep the number of variables equal to two, in spite of the fact that each person has two parents, he determined that the average height of men was about 1.08 times that of women, and defined a person's 'midparent' as an imaginary being of height

$$h_{\mathrm{mid}} \equiv \frac{1}{2}(h_{\mathrm{father}} + 1.08 h_{\mathrm{mother}}). \tag{7.76}$$

---

[19] A photograph of Galton, with more details of his work and a short biographical sketch, may be found in Stigler (1986c). His autobiography (Galton, 1908) has additional details.

He collected data on 928 adults born of 205 midparents and found, as expected, that children of tall parents do indeed tend to be tall, etc., but that children of tall parents still show a spread in heights, although less than the spread ($\pm\sigma$) of the entire population.

If the children of each selected group of parents still spread in height, why does the spread in height of the entire population not increase continually from one generation to the next? Because of the phenomenon of 'reversion'; the children of tall parents tend to be taller than the average person, but less tall than their parents. Likewise, children of short parents are generally shorter than the average person, but taller than their parents. If the population as a whole is to be stable, this 'systematic' tendency to revert back to the mean of the entire population must exactly balance the 'random' tendency to spreading. Behind the smooth facade of a constant overall distribution of heights, an intricate little time-dependent game of selection, drift, and spreading is taking place constantly.

In fact, Galton (with some help from mathematicians) could predict the necessary rate of reversion theoretically, and verify it from his data. If $x \equiv (h - \mu)$ is the deviation from the mean height of the midparents, let the population as a whole have a height distribution $\varphi(x|\sigma_1)$, while the sub-population of midparents of height $(x + \mu)$ tend to produce children of height $(y + \mu)$ with a frequency distribution $\varphi[(y - ax)|\sigma_2]$. Then the height distribution of the next generation will be given by (7.73). If the population as a whole is to be stable, it is necessary that $\sigma = \sigma_1$, or the reversion rate must be

$$ a = \pm\sqrt{1 - \frac{\sigma_2^2}{\sigma_1^2}}, \tag{7.77} $$

which shows that a need not be positive; if tall parents tended to 'compensate' by producing unusually short children, this would bring about an alternation from one generation to the next, but there would still be equilibrium for the population as a whole.

We see that equilibrium is not possible if $|a| > 1$; the population would explode. Although (7.73) is true for all a, equilibrium would then require $\sigma_2^2 < 0$. The boundary of stability is reached at $\sigma_2 = 0$, $|a| = 1$; then each sub-population breeds true, and whatever initial distribution of heights happened to exist would be maintained thereafter. An economist might call the condition $a = 1$ a 'unit root' situation; there is no reversion and no spreading.[20]

Of course, this analysis is in several obvious respects an oversimplified model of what happens in actual human societies. But that involves only touching up of details; Galton's analysis was, historically, of the greatest importance in giving us a general understanding of the kind of processes at work. For this, its freedom from nonessential details was a major merit.

---

[20] It is a currently popular theory among some economists that many economic processes, such as the stock market, are very close to the unit root behavior, so that the effects of momentary external perturbations like wars and droughts tend to persist instead of being corrected. There is no doubt that phenomena like this exist, at least in some cases; in the 1930s John Maynard Keynes noted what he called 'the stickiness of prices and wages'. For a discussion of this from a Bayesian viewpoint, see Sims (1988).

Exercise 7.8. Galton's device of the midparent was only to reduce the computational burden, which would otherwise have been prohibitive in the 1880s, by reducing the problem to a two-variable one (midparent and children). But today computing power is so plentiful and cheap that one can easily analyze the real four-variable problem, in which the heights of father, mother, son, and daughter are all taken into account. Reformulate Galton's problem to take advantage of this; what hypotheses about spreading and reversion might be considered and tested today? As a class project, one might collect new data (perhaps on faster-breeding creatures like fruit-flies) and write the computer program to analyze them and estimate the new spreading and reversion coefficients. Would you expect a similar program to apply to plants? Some have objected that this problem is too biological for a physics class, and too mathematical for a biology class; we suggest that, in a course dedicated to scientific inference in general, the class should include both physicists and biologists, working together.

Twenty years later this same phenomenon of selection, drift, and spreading underlying equilibrium was perceived independently by Einstein (1905a,b) in physics. The steady thermal Boltzmann distribution for molecules at temperature T to have energy E is $\exp\{-E/kT\}$. Being exponential in energies $E = u + (mv^2/2)$, where $u(x)$ is potential energy, this is Gaussian in particle velocities v. This generates a time-dependent drift in position; a particle which is at position x at time $t = 0$ has at time t the conditional probability to be at y of

$$p(y|xt) \propto \exp\left\{-\frac{(y-x)^2}{4Dt}\right\} \tag{7.78}$$

from random drift alone, but this is countered by a steady drift effect of external forces $F = -\nabla u$, corresponding to Galton's reversion rate.

Although the details are quite different, Galton's equation (7.77) is the logical equivalent of Einstein's relation $D = \lambda kT$ connecting diffusion coefficient D, representing random spreading of particles, with the temperature T and the mobility λ (velocity per unit force) representing the systematic reversion rate counteracting the diffusion. Both express the condition for equilibrium as a balance between a 'random spreading' tendency, and a systematic counter-drift that holds it in check.

7.19 Population dynamics and Darwinian evolution

Galton's type of analysis can explain much more than biological equilibrium. Suppose the reversion rate does not satisfy (7.77). Then the height distribution in the population will not be static, but will change slowly. Or, if short people tend to have fewer children than do tall people, then the average height of the population will drift slowly upward.[21] Do we have here the mechanism for Darwinian evolution? The question could hardly go unasked, since Francis Galton was a cousin of Charles Darwin.

A new feature of probability theory has appeared here that is not evident in the works of Laplace and Gauss. Being astronomers, their interests were in learning facts of astronomy, and telescopes were only a tool toward that end. The vagaries of telescopes themselves were for them only 'errors of observation' whose effects were to be eliminated as much as possible; and so the sampling distribution was called by them an 'error law'.

But a telescope maker might see it differently. For him, the errors it produces are the objects of interest to study, and a star is only a convenient fixed object on which to focus his instrument for the purpose of determining those errors. Thus a given data set might serve two entirely different purposes; one man's 'noise' is another man's 'signal'.

But then, in any science, the 'noise' might prove to be not merely something to get rid of, but the essential phenomenon of interest. It seems curious (at least, to a physicist) that this was first seen clearly not in physics, but in biology. In the late 19th century many biologists saw it as the major task confronting them to confirm Darwin's theory by exhibiting the detailed mechanism by which evolution takes place. For this purpose, the journal *Biometrika* was founded by Karl Pearson and Walter Frank Raphael Weldon, in 1901. It started (Volume 1, page 1) with an editorial setting forth the journal's program, in which Weldon wrote:

The starting point of Darwin's theory of evolution is precisely the existence of those differences between individual members of a race or species which morphologists for the most part rightly neglect. The first condition necessary, in order that a process of Natural Selection may begin among a race, or species, is the existence of differences among its members; and the first step in an enquiry into the possible effect of a selective process upon any character of a race must be an estimate of the frequency with which individuals, exhibiting any degree of abnormality with respect to that character, occur.

Weldon had here reached a very important level of understanding. Morphologists, thinking rather like astronomers, considered individual variations as only 'noise' whose effects must be eliminated by averaging, in order to get at the significant 'real' properties of the species as a whole. Weldon, learning well from the example of Galton, saw it in just the opposite light; those individual variations *are the engine that drives the process of evolutionary change*, which will be reflected eventually in changes in the morphologists' averages. Indeed, without individual variations, the mechanism of natural selection has nothing to operate on. So, to demonstrate the mechanism of evolution at its source, and not merely the final result, it is the frequency distribution of individual variations that must be studied.

---

[21] It is well known that, in developed nations, the average height of the population has, in fact, drifted upward by a substantial amount in the past 200 years. This is commonly attributed to better nutrition in childhood; but it is worth noting that if tall people tended to have more or longer-lived children than did short people for sociological reasons, the same average drift in height would be observed, having nothing to do with nutrition. This would be true Darwinian evolution, powered by individual variations. It appears to us that more research is needed to decide on the real cause of this upward drift.

Of course, at that time scientists had no conception of the physical mechanism of mutations induced by radioactivity (much less by errors in DNA replication), and they expected that evolution would be found to take place gradually, via nearly continuous changes.[22] Nevertheless, the program of studying the individual variations would be the correct one to find the fundamental mechanism of evolution, whatever form it took. The scenario is somewhat like the following.

7.20 Evolution of humming-birds and flowers

Consider a population of humming-birds in which the 'noise' consists of a distribution of different beak lengths. The survival of birds is largely a matter of finding enough food; a bird that finds itself with the mutation of an unusually long beak will be able to extract nectar from deeper flowers. If such flowers are available it will be able to nourish itself and its babies better than others because it has a food supply not available to other birds; so the long-beak mutation will survive and become a greater portion of the bird population, in more or less the way Darwin imagined.

But this influence works in two directions; a bird is inadvertently fertilizing flowers by carrying a few grains of pollen from one to the next. A flower that happens to have the mutation of being unusually deep will find itself sought out preferentially by long-beaked birds because they need not compete with other birds for it. Therefore its pollen will be carried systematically to other flowers of the same species and mutation where it is effective, instead of being wasted on the wrong species. As the number of long-beaked birds increases, deep flowers thus have an increasing survival advantage, ensuring that their mutation is present in an increasing proportion of the flower population; this in turn gives a still greater advantage to long-beaked birds, and so on. We have a positive feedback situation.

Over millions of years, this back-and-forth reinforcement of mutations goes through hundreds of cycles, resulting eventually in a symbiosis so specialized – a particular species of bird and a particular species of flower that seem designed specifically for each other – that it appears to be a miraculous proof of a guiding purpose in Nature, at least to those who do not think as deeply as did Darwin and Galton.[23] Yet short-beaked birds do not die out, because birds patronizing deep flowers leave the shallow flowers for them. By itself, the process would tend to an equilibrium distribution of populations of short- and long-beaked birds, coupled to distributions of shallow and deep flowers. But if they breed

---

[22] The necessity for evolution to be particulate (by discrete steps) was perceived later by several people, including Fisher (1930b). Evolutionary theory taking this into account, and discarding the Lamarckian notion of inheritance of acquired characteristics, is often called *neo-Darwinism*. However, the discrete steps are usually small, so Darwin's notion of 'gradualism' remains quite good pragmatically.

[23] The unquestioned belief in such a purpose pervades even producers of biological research products who might be expected to know better. In 1993 there appeared in biological trade journals a full-page ad with a large color photograph of a feeding humming-bird and the text: '*Specific purpose.* The sharply curved bill of the white-tipped sickle-billed humming-bird is specifically adapted to probe the delicate tubular flowers of heliconia plants for the nectar on which the creature survives.' Then this is twisted somehow into a plug for a particular brand of DNA polymerase – said to be produced for an equally specific purpose. This seems to us a dangerous line of argument; since the bird bills do not, in fact, have a specific purpose, what becomes of the alleged purpose of the polymerase?

independently, over long periods other mutations will take place independently in the two types, and eventually they would be considered as belonging to two different species.

As noted, the role of 'noise' as the mechanism driving a slow change in a system was perceived independently by Einstein (of course, he knew about Darwin's theory, but we think it highly unlikely that he would have known about the work of Galton or Weldon in Switzerland in 1905). 'Random' thermal fluctuations caused by motion of individual atoms are not merely 'noise' to be averaged out in our predictions of mass behavior; they are *the engine that drives irreversible processes in physics*, and eventually brings about thermal equilibrium. Today this is expressed very specifically in the many 'fluctuation-dissipation theorems' of statistical mechanics, which we derive in generality from the maximum entropy principle in Chapter 11. They generalize the results of Galton and Einstein. The aforementioned Nyquist fluctuation law was, historically, the first such theorem to be discovered in physics.

The visions of Weldon and Einstein represented such a major advance in thinking that today, some 100 years later, many have not yet comprehended them or appreciated their significance in either biology or physics. We still have biologists[24] who try to account for evolution by a quite unnecessary appeal to the second law of thermodynamics, and physicists[25] who try to account for the second law by appealing to quite unnecessary modifications in the equations of motion. The operative mechanism of evolution is surely Darwin's original principle of natural selection, and any effects of the second law can only hinder it.[26]

Natural selection is a process entirely different from the second law of thermodynamics. The purposeful intervention of man can suspend or reverse natural selection – as we observe in wars, medical practice, and dog breeding – but it can hardly affect the second law. Furthermore, as Stephen J. Gould has emphasized, the second law always follows the same course, but evolution in Nature does not. Whether a given mutation makes a creature better adapted or less adapted to its environment depends on the environment. A mutation that causes a creature to lose body heat more rapidly would be beneficial in Brazil but fatal in Finland; and so the same actual sequence of mutations can result in entirely different

---

[24] For example, see Weber, Depew and Smith (1988). Here the trouble is that the second law of thermodynamics goes in the wrong direction; if the second law were the driving principle, evolution would proceed inexorably back to the primordial soup, which has a much higher entropy than would any collection of living creatures that might be made from the same atoms. This is easily seen as follows. What is the difference between a gram of living matter and a gram of primordial soup made of the same atoms? Evidently, it is that the living matter is far from thermal equilibrium, and it is obeying thousands of additional constraints on the possible reactions and spatial distribution of atoms (from cell walls, osmotic pressures, etc.) that the primordial soup is not obeying. But removing a constraint always has the effect of making a larger phase space available, thus increasing the entropy. The primordial soup represents the thermal equilibrium, resulting from removal of all the biological constraints; indeed, our present chemical thermodynamics is based on (derivable from) the Gibbs principle that thermal equilibrium is the macrostate of maximum entropy subject to only the physical constraints (energy, volume, mole numbers).

[25] Several writers have thought that Liouville's theorem (conservation of phase volume in classical mechanics or unitarity of time development in quantum theory) is in conflict with the second law. On the contrary, in Jaynes (1963b, 1965) we demonstrate that, far from being in conflict, the second law is an immediate elementary *consequence* of Liouville's theorem, and in Jaynes (1989) we give a simple application of this to biology: calculation of the maximum theoretical efficiency of a muscle.

[26] This is not to say that natural selection is the *only* process at work; random drift is still an operative cause of evolution with or without subsequent selection. Presumably, this is the reason for the fantastic color patterns of such birds as parrots, which surely have no survival value; the black bird is even more successful at surviving. For an extensive discussion of the evidence and later research efforts by many experts, see the massive three-volume work *Evolution After Darwin* (Tax, 1960) produced to mark the centenary of the publication of Darwin's *Origin of Species*, or the more informal work of Dawkins (1987).

creatures in different environments – each appearing to be adapting purposefully to its surroundings.

7.21 Application to economics

The remarkable – almost exact – analogy between the processes that bring about equilibrium in physics and in biology surely has other important implications, particularly for theories of equilibrium and stability in economics, not yet exploited. It seems likely, for example, that the 'turbulence' of individual variations in economic behavior is the engine that drives macroeconomic change in the direction of the equilibrium envisaged by Adam Smith. The existence of this turbulence was recognized by John Maynard Keynes (1936), who called it 'animal spirits' which cause people to behave erratically; but he did not see in this the actual cause that prevents stagnation and keeps the economy on the move.

In the next level of understanding we see that Adam Smith's equilibrium is never actually attained in the real world because of what a physicist would call 'external perturbations', and what an economist would call 'exogenous variables' which vary on the same time scale. That is, wars, droughts, taxes, tariffs, bank reserve requirements, discount rates and other disturbances come and go on about the same time scale as would the approach to equilibrium in a perfectly 'calm' society.

The effect of small disturbances may be far greater than one might expect merely from the 'unit root' hypothesis noted above. If small individual decisions (like whether to buy a new car or open a savings account instead) take place independently, their effects on the macroeconomy should average out according to the $\sqrt{N}$ rule, to show only small ripples with no discernible periodicity. But seemingly slight influences (like a month of bad weather or a 1% change in the interest rate) might persuade many to do this a little sooner or later than they would otherwise. That is, a very slight influence may be able to pull many seemingly independent agents into phase with each other so they generate large organized waves instead of small ripples.

Such a phase-locked wave, once started, can itself become a major influence on other individual decisions (of buyers, retailers, and manufacturers), and if these secondary influences are in the proper phase with the original ones, we could have a positive feedback situation; the wave may grow and perpetuate itself by mutual reinforcement, as did the humming-birds and flowers. Thus, one can see why a macroeconomy may be inherently unstable for reasons that have nothing to do with capitalism or socialism. Classical equilibrium theory may fail not just because there is no 'restoring force' to bring the system back to equilibrium; relatively small fortuitous events may set up a big wave that goes instead into an oscillating limit cycle – perhaps we are seeing this in business cycles. To stop the oscillations and move back toward the equilibrium predicted by classical theory, the macroeconomy would be dependent on the erratic behavior of individual people, spreading the phases out again. Contrarians may be necessary for a stable economy!

As we see it, these are the basic reasons why economic data are very difficult to interpret; even if relevant and believable data were easy to gather, the rules of the game and the conditions of play are changing constantly. But we think that important progress can still be made by exploiting what is now known about entropy and probability theory as tools of logic. In particular, the conditions for instability should be predictable from this kind of analysis, just as they are in physics, meteorology, and engineering. A very wise government might be able to make and enforce regulations that prevent phase locking – just as it now prevents wild swings in the stock market by suspending trading. We are not about to run out of important things to do in theoretical economics.

7.22 The great inequality of Jupiter and Saturn

An outstanding problem for 18th century science was noted by Edmund Halley in 1676. Observation showed that the mean motion of Jupiter (30.35 deg/yr) was slowly accelerating, that of Saturn (12.22 deg/yr) decelerating. But this was not just a curiosity for astronomers; it meant that Jupiter was drifting closer to the Sun, Saturn farther away. If this trend were to continue indefinitely, then eventually Jupiter would fall into the Sun, carrying with it the Earth and all the other inner planets. This seemed to prophesy the end of the world – and in a manner strikingly like the prophesies of the Bible.

Understandably, this situation was of more than ordinary interest, and to more people than astronomers. Its resolution called forth some of the greatest mathematical efforts of 18th century savants, either to confirm the coming end; or preferably to show how the Newtonian laws would eventually put a stop to the drift of Jupiter and save us.

Euler, Lagrange, and Lambert made heroic attacks on the problem without solving it. We noted above how Euler was stopped by a mass of overdetermined equations; 75 simultaneous but inconsistent equations for eight unknown orbital parameters. If the equations were all consistent, he could choose any eight of them and solve (this would still involve inversion of an 8×8 matrix), and the result would be the same whatever eight he chose. But the observations all had unknown errors of measurement, and so there were

$$\binom{75}{8} \simeq 1.69 \times 10^{10} \tag{7.79}$$

possible choices; i.e. over 16 billion different sets of estimates for the parameters, with apparently nothing to choose between them.[27] At this point, Euler managed to extract reasonably good estimates of two of the unknowns (already an advance over previous knowledge), and simply gave up on the others. For this work (Euler, 1749), he won the French Academy of Sciences prize.

The problem was finally solved in 1787 by one who was born that same year. Laplace (1749–1827) 'saved the world' by using probability theory to estimate the parameters accurately enough to show that the drift of Jupiter was not secular after all; the observations

---

[27] Our algorithm for this in Chapter 19, Eqs. (19.24) and (19.37), actually calculates a weighted average over all these billions of estimates; but in a manner so efficient that one is unaware that all this is happening. What probability theory determines for us – and what Euler and Daniel Bernoulli never comprehended – is the optimal weighting coefficients in this average, leading to the greatest possible reliability for the estimate and the accuracy claims.

at hand had covered only a fraction of a cycle of an oscillation with a period of about 880 years. This is caused by an 'accidental' near resonance in their orbital periods:

$$2 \times \text{(period of Saturn)} \simeq 5 \times \text{(period of Jupiter)}. \qquad (7.80)$$

Indeed, from the above mean motion data we have

$$2 \times \frac{360}{12.22} = 58.92 \text{ yr}, \qquad 5 \times \frac{360}{30.35} = 59.32 \text{ yr}. \qquad (7.81)$$

In the time of Halley, their difference was only about 0.66% and decreasing.

So, long before it became a danger to us, Jupiter indeed reversed its drift – just as Laplace had predicted – and it is returning to its old orbit. Presumably, Jupiter and Saturn have repeated this seesaw game several million times since the solar system was formed. The first half-cycle of this oscillation to be observed by man will be completed in about the year 2012.

7.23 Resolution of distributions into Gaussians

The tendency of probability distributions to gravitate to the Gaussian form suggests that we might view the appearance of a Gaussian, or 'normal', frequency distribution as loose evidence (but far from proof) that some kind of equilibrium has been reached. This view is also consistent with (but by no means required by) the results of Galton and Einstein. In the first attempts to apply probability theory in the biological and social sciences (for example, Quetelet, 1835, 1869), serious errors were made through supposing firstly that the appearance of a normal distribution in data indicates that one is sampling from a homogeneous population, and secondly that any departure from normality indicates an inhomogeneity in need of explanation. By resolving a non-normal distribution into Gaussians, Quetelet thought that one would be discovering the different sub-species, or varieties, that were present in the population. If this were true reliably, we would indeed have a powerful tool for research in many different fields. But later study showed that the situation is not that simple.

We have just seen how one aspect of it was corrected finally by Galton (1886), in showing that a normal frequency distribution by no means proves homogeneity; from (7.73), a Gaussian of width σ can arise inhomogeneously – and in many different ways – from the overlapping of narrower Gaussian distributions of various widths σ_1, σ_2. But those subpopulations are in general merely mathematical artifacts like the sine waves in a Fourier transform; they have no individual significance for the phenomenon unless one can show that a particular set of subpopulations has a real existence and plays a real part in the mechanism underlying stability and change. Galton was able to show this from his data by measuring those widths.

The second assumption, that non-normal distributions can be resolved into Gaussian subdistributions, turns out to be not actually wrong (except in a nitpicking mathematical sense); but without extra prior information it is ambiguous in what it tells us about the phenomenon.

We have here an interesting problem, with many useful applications: is a non-Gaussian distribution explainable as a mixture of Gaussian ones? Put mathematically, if an observed data histogram is well described by a distribution $g(y)$, can we find a mixing function $f(x) \geq 0$ such that $g(y)$ is seen as a mixture of Gaussians:

$$\int dx\, \varphi(y - x|\sigma)f(x) = g(y), \qquad -\infty \leq y \leq \infty. \tag{7.82}$$

Neither Quetelet nor Galton was able to solve this problem, and today we understand why. Mathematically, does this integral equation have solutions, or unique solutions? It appears from (7.73) that we cannot expect unique solutions in general, for, in the case of Gaussian $g(y)$, many different mixtures (many different choices of a, σ_1, σ_2) will all lead to the same $g(y)$. But perhaps if we specify the width σ of the Gaussian kernel in (7.82) there is a unique solution for $f(x)$.

Solution of such integral equations is rather subtle mathematically. We give two arguments: the first depends on the properties of Hermite polynomials and yields a class of exact solutions; the second appeals to Fourier transforms and yields an understanding of the more general situation.

7.24 Hermite polynomial solutions

The rescaled Hermite polynomials $R_n(x)$ may be defined by the displacement of a Gaussian distribution $\varphi(x)$, which gives the generating function

$$\frac{\varphi(x - a)}{\varphi(x)} = \exp\{xa - a^2/2\} = \sum_{n=0}^{\infty} R_n(x)\frac{a^n}{n!}, \tag{7.83}$$

or, solving for R_n, we have the Rodriguez form

$$R_n(x) = \frac{d^n}{da^n}\left[\exp\{xa - a^2/2\}\right]_{a=0} = (-1)^n \exp\{x^2/2\}\frac{d^n}{dx^n}\exp\{-x^2/2\}. \tag{7.84}$$

The first few of these polynomials are: $R_0 = 1$, $R_1 = x$, $R_2 = x^2 - 1$, $R_3 = x^3 - 3x$, $R_4 = x^4 - 6x^2 + 3$. The conventional Hermite polynomials $H_n(x)$ differ only in scaling: $H_n(x) = 2^{n/2}R_n(x\sqrt{2})$.

Multiplying (7.83) by $\varphi(x)\exp\{xb - b^2/2\}$ and integrating out x, we have the orthogonality relation

$$\int_{-\infty}^{\infty} dx\, R_m(x)R_n(x)\varphi(x) = n!\delta_{mn}, \tag{7.85}$$

and in consequence these polynomials have the remarkable property that convolution with a Gaussian function reduces simply to

$$\int_{-\infty}^{\infty} dx\, \varphi(y - x)R_n(x) = y^n. \tag{7.86}$$

Therefore, if $g(y)$ is represented by a power series,

$$g(y) = \sum_n a_n y^n, \tag{7.87}$$

we have immediately a formal solution of (7.82):

$$f(x) = \sum_n a_n \sigma^n R_n \left(\frac{x}{\sigma}\right). \tag{7.88}$$

Since the coefficient of x^n in $R_n(x)$ is unity, the expansions (7.87) and (7.88) converge equally well. So, if $g(y)$ is any polynomial or entire function (i.e. one representable by a power series (7.87) with infinite radius of convergence), the integral equation has the unique solution (7.88).

We can see the solution (7.88) a little more explicitly if we invoke the expansion of R_n, deducible from (7.83) by expanding $\exp\{xa - a^2/2\}$ in a power series in x:

$$R_n \left(\frac{x}{\sigma}\right) = \sum_{m=0}^{M} (-1)^m \frac{n!}{2^m m! \, (n - 2m)!} \left(\frac{x}{\sigma}\right)^{n-2m}, \tag{7.89}$$

where $M = (n-1)/2$ if n is odd, $M = n/2$ if n is even. Then, noting that

$$\frac{n!}{(n - 2m)!} \left(\frac{x}{\sigma}\right)^{n-2m} = \sigma^{2m-n} \frac{\mathrm{d}^{2m}}{\mathrm{d}x^{2m}} x^n, \tag{7.90}$$

we have the formal expansion

$$f(x) = \sum_{m=0}^{\infty} \frac{(-1)^m \sigma^{2m}}{2^m m!} \frac{\mathrm{d}^{2m}}{\mathrm{d}x^{2m}} g(x) = g(x) - \frac{\sigma^2}{2} \frac{\mathrm{d}^2 g(x)}{\mathrm{d}x^2} + \frac{\sigma^4}{8} \frac{\mathrm{d}^4 g(x)}{\mathrm{d}x^4} - \cdots . \tag{7.91}$$

An analytic function is differentiable any number of times, and if $g(x)$ is an entire function this will converge to the unique solution. If $g(x)$ is a very smooth function, it converges very rapidly, so the first two or three terms of (7.91) are already a good approximation to the solution. This gives us some insight into the workings of the integral equation; as $\sigma \to 0$, the solution (7.91) relaxes into $f(x) \to g(x)$, as it should. The first two terms of (7.91) are what would be called, in image reconstruction, 'edge detection'; for small σ the solution goes into this. The larger σ, the more the higher-order derivatives matter; that is, the more fine details of the structure of $g(y)$ contribute to the solution. Intuitively, the broader the Gaussian kernel, the more difficult it is to represent fine structure of $g(y)$ in terms of that kernel.

Evidently, we could continue this line of thought with much more analytical work, and it might seem that the problem is all but solved; but now the subtlety starts. Solutions like (7.88) and (7.91), although formally correct in a mathematical sense, ignore some facts of the real world; is $f(x)$ non-negative when $g(y)$ is? Is the solution stable, a small change in $g(y)$ inducing only a small change in $f(x)$? What if $g(x)$ is not an entire function but is piecewise continuous; for example, rectangular?

7.25 Fourier transform relations

For some insight into these questions, let us look at the integral equation from the Fourier transform viewpoint. Taking the transform of (7.82) according to

$$\mathcal{F}(k) \equiv \int_{-\infty}^{\infty} dx \, f(x) \exp\{ikx\}, \tag{7.92}$$

(7.82) reduces to

$$\exp\left\{-\frac{k^2\sigma^2}{2}\right\} \mathcal{F}(k) = \mathcal{G}(k), \tag{7.93}$$

which illustrates that the Fourier transform of a Gaussian function is another Gaussian function, and shows us at once the difficulty of finding more general solutions than (7.88). If $g(y)$ is piecewise continuous, then, as $k \to \infty$, from the Riemann–Lebesgue lemma $\mathcal{G}(k)$ will fall off only as $1/k$. Then $\mathcal{F}(k)$ must blow up violently, like $\exp\{+k^2\sigma^2/2\}/k$, and one shudders to think what the function $f(x)$ must look like (infinitely violent oscillations of infinitely high frequency?) If $g(y)$ is continuous, but has discontinuous first derivatives like a triangular distribution, then $\mathcal{G}(k)$ falls off as k^{-2}, and we are in a situation about as bad. Evidently, if $g(y)$ has a discontinuity in any derivative, there is no solution $f(x)$ that would be acceptable in the physical problem. This is evident also from (7.91); the formal solution would degenerate into infinitely high derivatives of a delta-function.

In order that we can interpret $g(y)$ as a mixture of possible Gaussians, $f(x)$ must be non-negative. But we must allow the possibility that the $f(x)$ sought is a sum of delta-functions; indeed, to resolve $g(y)$ into a discrete mixture of Gaussians $g(y) = \sum a_j \varphi(x - x_j)$ was the real goal of Quetelet and Galton. If this could be achieved uniquely, their interpretation might be valid. Then $\mathcal{F}(k)$ does not fall off at all as $k \to \pm\infty$, so $\mathcal{G}(k)$ must fall off as $\exp\{-k^2\sigma^2/2\}$. In short, in order to be resolvable into Gaussians of width σ with positive mixture function $f(x)$, the function $g(y)$ must itself be at least as smooth as a Gaussian of width σ. This is a formal difficulty.

There is a more serious practical difficulty. If $g(y)$ is a function determined only empirically, we do not have it in the form of an analytic function; we have only a finite number of approximate values g_i at discrete points y_i. We can find many analytic functions which appear to be good approximations to the empirical one. But because of the instability evident in (7.88) and (7.91) they will lead to greatly different final results $f(x)$. Without a stability property and a criterion for choosing that smooth function, we really have no definite solution in the sense of inversion of an integral equation.[28]

In other words, finding the appropriate mixture $f(x)$ to account for an empirically determined distribution $g(y)$ is not a conventional mathematical problem of inversion; *it is itself a problem of inference, requiring the apparatus of probability theory.* In this way, a problem in probability theory can generate a hierarchy of subproblems, each involving probability theory again but on a different level.

---

[28] For other discussions of the problem, see Andrews and Mallows (1974) and Titterington, Smith and Makov (1985).

7.26 There is hope after all

Following up the idea in Section 7.2.5, the original goal of Quetelet has now been very nearly realized by analysis of the integral equation as a problem of Bayesian inference instead of mathematical inversion; and useful examples of analysis of real data by this have now been found. Sivia and Carlile (1992) report the successful resolution of noisy data into as many as nine different Gaussian components, representing molecular excitation lines, by a Bayesian computer program.[29]

It is hardly surprising that Quetelet and Galton could not solve this problem in the 19th century; but it is very surprising that today many scientists, engineers, and mathematicians still fail to see the distinction between inversion and inference, and struggle with problems like this that have no deductive solutions, only inferential ones. The problem is, however, very common in current applications; it is known as a 'generalized inverse' problem, and today we can give unique and useful inferential solutions to such problems by specifying the (essential, but hitherto unmentioned) prior information to be used, converting an ill-posed problem into a straightforward Bayesian exercise.

This suggests another interesting mathematical problem; for a given entire function $g(y)$, over what range of σ is the solution (7.88) non-negative? There are some evident clues: when $\sigma \to 0$, we have $\varphi(x - y|\sigma) \to \delta(x - y)$ and so, as noted above, $f(x) \to g(x)$; so, for σ sufficiently small, $f(x)$ will be non-negative if $g(y)$ is. But when $\sigma \to \infty$ the Gaussians in (7.82) become very broad and smooth; so, if $f(x)$ is non-negative, the integral in (7.82) must be at least as broad. Thus, when $g(y)$ has detailed structure on a scale smaller than σ, there can be no solution with non-negative $f(x)$; and it is not obvious whether there can be any solution at all.

Exercise 7.9. From the above arguments one would conjecture that there will be some upper bound σ_{max} such that the solution $f(x)$ is non-negative when and only when $0 \le \sigma < \sigma_{max}$. It will be some functional $\sigma_{max}[g(y)]$ of $g(y)$. Prove or disprove this conjecture; if it is true, give a verbal argument by which we could have seen this without calculation; if it is false, give a specific counter-example showing why.

Hint. It appears that (7.91) might be useful in this endeavor.

---

[29] We noted in Chapter 1 that most of the computer programs used in this field are only intuitive *ad hoc* devices that make no use of the principles of probability theory; therefore in general they are usable in some restricted domain, but they fail to extract all the relevant information from the data and are subject to both the errors of hallucination and blunt perception. One commercial program for resolution into Gaussians or other functions simply reverts to empirical curve fitting. It is advertised (*Scientific Computing*, July 1993, p. 15) with a provocative message, which depicts two scientists with the same data curve showing two peaks; by hand drawing one could resolve it very crudely into two Gaussians. The ad proclaims: 'Dr Smith found two peaks. . . . Using [our program] Dr Jones found *three* peaks . . .'. Guess who got the grant? We are encouraged to think that we can extract money from the Government by first allowing the software company to extract $500 from us for this program, whose output would indeed be tolerable for noiseless data. But it would surely degenerate quickly into dangerous, unstable nonsense as the noise level increases. The problem is not, basically, one of inversion or curve fitting; it is a problem of *inference*. A Bayesian inference program like those of Bretthorst (1988) will continue to return the best resolution possible from the data and the model, without instability, whatever the noise level. If the noise level becomes so high as to make the data useless, the Bayesian estimates just relax back into the prior estimates, as they should.

This suggests that the original goal of Quetelet and Galton was ambiguous; any sufficiently smooth non-Gaussian distribution may be generated by many different superpositions of different Gaussians of different widths. Therefore a given set of subpopulations, even if found mathematically, would have little biological significance unless there were additional prior information pointing to Gaussians of that particular width σ as having a 'real' existence and playing some active role in the phenomena. Of course, this *caveat* applies equally to the aforementioned Bayesian solution; but Sivia and Carlile did have that prior information.

7.27 Comments

7.27.1 Terminology again

As we are obliged to point out so often, this field seems to be cursed more than any other with bad and misleading terminology which seems impossible to eradicate. The electrical engineers have solved this problem very effectively; every few years, an official committee issues a revised standard terminology, which is then enforced by editors of their journals (witness the meek acceptance of the change from 'megacycles' to 'megahertz' which was accomplished almost overnight a few years ago).

In probability theory there is no central authority with the power to bring about dozens of needed reforms, and it would be self-defeating for any one author to try to do this by himself; he would only turn away readers. But we can offer tentative suggestions in the hope that others may see merit in them.

The literature gives conflicting evidence about the origin of the term 'normal distribution'. Karl Pearson (1920) claimed to have introduced it 'many years ago', in order to avoid an old dispute over priority between Gauss and Legendre; but he gives no reference. Hilary Seal (1967) attributes it instead to Galton; but again fails to give a reference, so it would require a new historical study to decide this. However, the term had long been associated with the general topic: given a linear model $y = X\beta + e$, where the vector y and the matrix X are known, the vector of parameters β and the noise vector e unknown, Gauss (1823) called the system of equations $X'X\hat{\beta} = X'y$, which give the least squares parameter estimates $\hat{\beta}$, the 'normal equations', and the ellipsoid of constant probability density was called the 'normal surface'. It appears that somehow the name was transferred from the equations to the sampling distribution that leads to those equations.

Presumably, Gauss meant 'normal' in its mathematical sense of 'perpendicular', expressing the geometric meaning of those equations. The minimum distance from a point (the estimate) to a plane (the constraint) is the length of the perpendicular. But, as Pearson himself observes, the term 'normal distribution' is a bad one because the common colloquial meaning of 'normal' is *standard* or *sane*, implying a value judgment. This leads many to think – consciously or subconsciously – that all other distributions are in some way abnormal.

Actually, it is quite the other way; it is the so-called 'normal' distribution that is abnormal in the sense that it has many unique properties not possessed by any other. Almost all of our experience in inference has been with this abnormal distribution, and much of the folklore that we must counter here was acquired as a result. For decades, workers in statistical inference have been misled, by that abnormal experience, into thinking that methods such as confidence intervals, that happen to work satisfactorily with this distribution, should work as well with others.

The alternative name 'Gaussian distribution' is equally bad for a different reason, although there is no mystery about its origin. Stigler (1980) sees it as a general law of eponymy that *no discovery is named for its original discoverer*. Our terminology is in excellent compliance with this law, since the fundamental nature of this distribution and its main properties were noted by Laplace when Gauss was six years old; and the distribution itself had been found by de Moivre before Laplace was born. But, as we noted, the distribution became popularized by the work of Gauss (1809), who gave a derivation of it that was simpler than previous ones and seemed very compelling intuitively at the time. This is the derivation that we gave above, Eq. (7.16), and which resulted in his name becoming attached to it.

The term 'central distribution' would avoid both of these objections while conveying a correct impression; it is the final 'stable' or 'equilibrium' distribution toward which all others gravitate under a wide variety of operations (large number limit, convolution, stochastic transformation, etc.), and which, once attained, is maintained through an even greater variety of transformations, some of which are still unknown to statisticians because they have not yet come up in their problems.

For example, in the 1870s Ludwig Boltzmann gave a compelling, although heuristic, argument indicating that collisions in a gas tend to bring about a 'Maxwellian', or Gaussian, frequency distribution for velocities. Then Kennard (1938, Chap. 3) showed that this distribution, once attained, is maintained automatically, without any help from collisions, as the molecules move about, constantly changing their velocities, in any conservative force field (that is, forces $f(x)$ derivable from a potential $\phi(x)$ by gradients: $f(x) = -\nabla\phi(x)$). Thus, this distribution has stability properties considerably beyond anything yet utilized by statisticians, or yet demonstrated in the present work.

While venturing to use the term 'central distribution' in a cautious, tentative way, we continue to use also the bad but traditional terms, preferring 'Gaussian' for two reasons. Ancient questions of priority are no longer of interest; far more important today, 'Gaussian' does not imply any value judgment. Use of emotionally loaded terms appears to us a major cause of the confusion in this field, causing workers to adhere to principles with noble-sounding names like 'unbiased' or 'admissible' or 'uniformly most powerful', in spite of the nonsensical results they can yield in practice. But also, we are writing for an audience that includes both statisticians and scientists. Everybody understands what 'Gaussian distribution' means; but only statisticians are familiar with the term 'normal distribution'.

The fundamental Boltzmann distribution of statistical mechanics, exponential in energies, is of course Gaussian or Maxwellian in particle velocities. The general central tendency of probability distributions toward this final form is now seen as a consequence of their maximum entropy properties (Chapter 11). If a probability distribution is subjected to some transformation that discards information but leaves certain quantities invariant, then, under very general conditions, if the transformation is repeated, the distribution tends to the one with maximum entropy, subject to the constraints of those conserved quantities.

This brings us to the term 'central limit theorem', which we have derived as a special case of the phenomenon just noted – the behavior of probability distributions under repeated convolutions, which conserve first and second moments. This name was introduced by George Pólya (1920), with the intention that the adjective 'central' was to modify the noun 'theorem'; i.e. it is the limit theorem which is *central to probability theory*. Almost universally, students today think that 'central' modifies 'limit', so that it is instead a theorem about a '*central limit*', whatever that means.[30]

In view of the equilibrium phenomenon, it appears that Pólya's choice of words was after all fortunate in a way that he did not foresee. Our suggested terminology takes advantage of this; looked at in this way, the terms 'central distribution' and 'central limit theorem' both convey the right connotations to one hearing them for the first time. One can read 'central limit' as meaning a limit toward a central distribution, and will be invoking just the right intuitive picture.

---

[30] The confusion does not occur in the original German, where Pólya's words were: *Über den zentralen Grenzwertsatz der Wahrscheinlichkeitsrechnung*, an interesting example where the German habit of inventing compound words removes an ambiguity in the literal English rendering.

8

Sufficiency, ancillarity, and all that

In the preceding five chapters we have examined the use of probability theory in problems that, although technically elementary, illustrated a fairly good sample of typical current applications. Now we are in a position to look back over these examples and note some interesting features that they have brought to light. It is useful to understand these features, for tactical reasons. Many times in the past when one tried to conduct inference by applying intuitive *ad hoc* devices instead of probability theory, they would not work acceptably unless some special circumstances were present, and others absent. Thus they were of major theoretical importance in orthodox statistics.

None of the material of the present chapter, however, is really needed in our applications; for us, these are incidental details that take care of themselves as long as we obey the rules. That is, if we merely apply the rules derived in Chapter 2, strictly and consistently in every problem, they lead us to do the right thing and arrive at the optimal inferences for that problem automatically, without our having to take any special note of these things. For us, they have rather a 'general cultural value' in helping us to understand better the inner workings of probability theory. One can see much more clearly why it is necessary to obey the Chapter 2 rules, and the predictable consequences of failure to do so.

8.1 Sufficiency

In our examples of parameter estimation, probability theory sometimes does not seem to use all the data that we offer it. In Chapter 6, when we estimated the parameter θ of a binomial distribution from data on n trials, the posterior pdf for θ depended on the data only through the number n of trials and the number r of successes; all information about the order in which success and failure occurred was ignored.

With a rectangular sampling distribution in $\alpha \le x \le \beta$, the joint posterior pdf for α, β used only the extreme data values $(x_{\min}, x_{\max})$ and ignored the intermediate data.

Likewise, in Chapter 7, with a Gaussian sampling distribution and a data set $D \equiv \{x_1, \ldots, x_n\}$, the posterior pdf for the parameters μ, σ depended on the data only through n and their first two moments $(\bar{x}, \overline{x^2})$. The $(n-2)$ other properties of the data convey a great deal of additional information of some kind; yet our use of probability theory ignored them.

243

Is probability theory failing to do all it could here? No, the proofs of Chapter 2 have precluded that possibility; the rules being used are the only ones that can yield unique answers while agreeing with the qualitative desiderata of rationality and consistency. It seems, then, that the unused parts of the data must be *irrelevant* to the question we are asking.[1] But can probability theory itself confirm this conjecture for us in a more direct way?

This introduces us to a quite subtle theoretical point about inference. Special cases of the phenomenon were noted by Laplace (1812, 1824 edn, Supp. V). It was generalized and given its present name 100 years later by Fisher (1922), and its significance for Bayesian inference was noted by H. Jeffreys (1939). Additional understanding of its role in inference was achieved only recently, in the resolution of the 'marginalization paradox' discussed in Chapter 15.

If certain aspects of the data are not used when they are known, then presumably it would not matter (we should come to the same final conclusion) if they were unknown. Thus, if the posterior pdf for a parameter θ is found to depend on the data $D = \{x_1, \ldots, x_n\}$ only through a function $r(x_1, \ldots, x_n)$ (call it 'property R'), then it seems plausible that given r alone we should be able to draw the same inferences about θ. This would confirm that the unused parts of the data were indeed irrelevant in the sense just conjectured.

With a sampling density function $p(x_1 \ldots x_n | \theta)$ and prior $p(\theta | I) = f(\theta)$, the posterior pdf using all the data is

$$p(\theta|DI) = h(\theta|D) = \frac{f(\theta)p(x_1 \ldots x_n|\theta)}{\int d\theta' \, f(\theta')p(x_1 \ldots x_n|\theta')}. \tag{8.1}$$

Note that we are not assuming independent or exchangeable sampling here; the sampling pdf need not factor in the form $p(x_1 \ldots x_n|\theta) = \Pi_i p(x_i|\theta)$ and the marginal probabilities $p(x_i|\theta) = k_i(x_i, \theta)$ and $p(x_j|\theta) = k_j(x_j, \theta)$ need not be the same function. Now carry out a change of variables $(x_1, \ldots, x_n) \rightarrow (y_1, \ldots, y_n)$ in the sample space S_x, such that $y_1 = r(x_1, \ldots, x_n)$, and choose $(y_2, \ldots, y_n)$ so that the Jacobian

$$J = \frac{\partial(y_1, \ldots, y_n)}{\partial(x_1, \ldots, x_n)} \tag{8.2}$$

is bounded and nonvanishing everywhere on S_x. Then the change of variables is a 1:1 mapping of S_x onto S_y, and the sampling density

$$g(y_1, \ldots, y_n|\theta) = J^{-1} p(x_1 \ldots x_n|\theta) \tag{8.3}$$

may be used just as well as $p(x_1 \ldots x_n|\theta)$ in the posterior pdf:

$$h(\theta|D) = \frac{f(\theta)g(y_1, \ldots, y_n|\theta)}{\int d\theta' \, f(\theta')g(y_1, \ldots, y_n|\theta')} \tag{8.4}$$

since the Jacobian, being independent of θ, cancels out.

Then property R is the statement that for all $\theta \in S_\theta$, (8.4) is independent of $(y_2, \ldots, y_n)$. Writing this condition out as derivatives set to zero, we find that it defines a set of $n - 1$

---

[1] Of course, when we say that some information is 'irrelevant' we mean only that we don't need it *for our present purpose*; it might be crucially important for some other purpose that we shall have tomorrow.

simultaneous integral equations (actually, only orthogonality conditions) that the prior $f(\theta)$ must satisfy:

$$\int_{S_\theta} d\theta' \, K_i(\theta, \theta') f(\theta') = 0 \qquad \left\{ \begin{array}{l} \theta \in S_\theta \\ 2 \le i \le n \end{array} \right\}, \tag{8.5}$$

where the ith kernel is

$$K_i(\theta, \theta') \equiv g(y|\theta) \frac{\partial g(y|\theta')}{\partial y_i} - g(y|\theta') \frac{\partial g(y|\theta)}{\partial y_i}, \tag{8.6}$$

and we used the abbreviation $y \equiv (y_1, \ldots, y_n)$, etc. It is antisymmetric: $K_i(\theta, \theta') = -K_i(\theta', \theta)$.

8.2 Fisher sufficiency

If (8.5) holds only for some particular prior $f(\theta)$, then $K_i(\theta, \theta')$ need not vanish; in its dependence on θ' it needs only to be orthogonal to that particular function. But if (8.5) is to hold for all $f(\theta)$, as Fisher (1922) required by implication – by failing to mention $f(\theta)$ – then $K_i(\theta, \theta')$ must be orthogonal to a complete set of functions $f(\theta')$; thus zero almost everywhere for $(2 \le i \le n)$. Noting that the kernel may be written in the form

$$K_i(\theta, \theta') = g(y|\theta) \, g(y|\theta') \frac{\partial}{\partial y_i} \log \left[\frac{g(y|\theta')}{g(y|\theta)} \right], \tag{8.7}$$

this condition may be stated as: given any (θ, θ'), then for all possible samples (that is, all values of $\{y_1, \ldots, y_n; \theta; \theta'\}$ for which $g(y|\theta) \, g(y|\theta') \ne 0$), the ratio $[g(y|\theta')/g(y|\theta)]$ must be independent of the components $(y_2, \ldots, y_n)$. Thus to achieve property R independently of the prior, $g(y|\theta)$ must have the functional form

$$g(y_1, \ldots, y_n|\theta) = q(y_1|\theta) m(y_2, \ldots, y_n). \tag{8.8}$$

Integrating $(y_2, \ldots, y_n)$ out of (8.8), we see that the function denoted by $q(y_1|\theta)$ is, to within a normalization constant, the marginal sampling pdf for y_1.

Transforming back to the original variables, Fisher sufficiency requires that the sampling pdf has the form

$$p(x_1 \ldots x_n|\theta) = p(r|\theta) b(x_1, \ldots, x_n), \tag{8.9}$$

where $p(r|\theta)$ is the marginal sampling density for $r(x_1, \ldots, x_n)$.

Equation (8.9) was given by Fisher (1922). If a sampling distribution factors in the manner (8.8), (8.9), then the sampling pdf for $(y_2, \ldots, y_n)$ is independent of θ. This being the case, he felt intuitively that the values of $(y_2, \ldots, y_n)$ can convey no information about θ; full information should be conveyed by the single quantity r, which he then termed a *sufficient statistic*. But Fisher's reasoning was only a conjecture referring to a sampling theory context. We do not see how it could be proved in that limited context, which did not use the concepts of prior and posterior probabilities.

Probability theory as logic can demonstrate this property directly without any need for conjecture. Indeed, using (8.9) in (8.1), the function $b(x)$ cancels out, and we find immediately the relation

$$h(\theta|D) \propto f(\theta)p(r|\theta). \tag{8.10}$$

Thus, if (8.10) holds, then $r(x_1, \ldots, x_n)$ is a sufficient statistic in the sense of Fisher, and in Bayesian inference with the assumed model (8.1), knowledge of the single quantity r does indeed tell us everything about θ that is contained in the full data set $(x_1, \ldots, x_n)$; and this will be true for all priors $f(\theta)$.

The idea generalizes at once to more variables. Thus, if the sampling distribution factors in the form $g(y_1, \ldots, y_n|\theta) = h(y_1, y_2|\theta)m(y_3, \ldots, y_n)$, we would say that $y_1(x_1, \ldots, x_n)$ and $y_2(x_1, \ldots, x_n)$ are jointly sufficient statistics for θ and, in this, θ could be multidimensional. If there are two parameters θ_1, θ_2 such that there is a coordinate system $\{y_i\}$ in which

$$g(y_1, \ldots, y_n|\theta_1\theta_2) = h(y_1|\theta_1)k(y_2|\theta_2)m(y_3, \ldots, y_n), \tag{8.11}$$

then $y_1(x_1, \ldots, x_n)$ is a sufficient statistic for θ_1, and y_2 is a sufficient statistic for θ_2; and so on.

8.2.1 Examples

Our discussion of the Gaussian distribution in Chapter 7 has already demonstrated that it has sufficient statistics [Eqs. (7.25)–(7.30)]. If the data $D = \{y_1, \ldots, y_n\}$ consist of n independent observations y_i, then the sampling distribution with mean and variance μ, σ^2 could be written as

$$p(D|\mu\sigma I) = \left(\frac{1}{2\pi\sigma^2}\right)^{n/2} \exp\left\{-\frac{n}{2\sigma^2}[(\mu - \overline{y})^2 + s^2]\right\}, \tag{8.12}$$

where $\overline{y}, s^2$ are the observed sample mean and variance, Eq. (7.29). Since these are the only properties of the data that appear in the sampling distribution (8.12) – and therefore are the only properties of the data that occur in the joint posterior distribution $p(\mu\sigma|DI)$ – they are jointly sufficient statistics for estimation of μ, σ. The test for sufficiency via Bayes' theorem is often easier to carry out than is the test for factorization (8.11), although of course they amount to the same thing.

Let us examine sufficiency for the separate parameters. If σ is known, then we would find the posterior distribution for μ alone:

$$p(\mu|\sigma DI) = A\frac{p(\mu|I)p(D|\mu\sigma I)}{\int d\mu\, p(\mu|I)p(D|\mu I)}, \tag{8.13}$$

$$p(x_1 \ldots x_n|\mu\sigma I) = A \exp\left\{-\frac{1}{\sigma^2}\sum_{i=1}^{n}(x_i - \mu)^2\right\},$$

$$= A \exp\left\{-\frac{ns^2}{2\sigma^2}\right\} \times \exp\left\{-\frac{n}{2\sigma^2}(\overline{x} - \mu)^2\right\}, \tag{8.1}$$

where

$$\bar{x} \equiv \frac{1}{n} \sum_{i=1}^{n} x_i, \qquad \overline{x^2} \equiv \frac{1}{n} \sum_i x_i^2, \qquad s^2 \equiv \overline{x^2} - \bar{x}^2, \qquad (8.15)$$

$$p(\mu|\sigma DI) \propto p(u|I) \exp\left\{-\frac{n}{2\sigma^2}(\bar{x} - \mu)\right\} \qquad (8.16)$$

are the sample mean, mean square, and variance, respectively. Since now the factor $\exp\{-ns^2/2\bar{s}^2\}$ appears in both numerator and denominator, it cancels out.

Likewise, if μ is known, then the posterior pdf for σ alone is found to be

$$p(\sigma|\mu DI) \propto p(\sigma|I)\sigma^{-n} \exp\left\{-\frac{n}{2\sigma^2}(\overline{x^2} - 2\mu\bar{x} + \mu^2)\right\}. \qquad (8.17)$$

Fisher sufficiency was of major importance in orthodox (non-Bayesian) statistics, because it had so few criteria for choosing an estimator. It had, moreover, a fundamental status lacking in other criteria because, for the first time, the notion of *information* appeared in orthodox thinking. If a sufficient statistic for θ exists, it is hard to justify using any other for inference about θ. From a Bayesian standpoint one would be, deliberately, throwing away some of the information in the data that is relevant to the problem.[2]

8.2.2 The Blackwell–Rao theorem

Arguments in terms of information content had almost no currency in orthodox theory, but a theorem given by D. Blackwell and C. R. Rao in the 1940s did establish a kind of theoretical justification for the use of sufficient statistics in orthodox terms. Let $r(x_1, \ldots, x_n)$ be a Fisher sufficient statistic for θ, and let $\beta(x_1, \ldots, x_n)$ be any proposed estimator for θ. By (8.9) the joint pdf for the data conditional on r:

$$p(x_1 \ldots x_n|r\theta) = b(x)p(r|x\theta) = b(x)\delta(r - r(x)) \qquad (8.18)$$

is independent of θ. Then the conditional expectation

$$\beta_0(r) \equiv \langle \beta|r\theta \rangle = E(\beta|r\theta) \qquad (8.19)$$

is also independent of θ, so β_0 is a function only of the x_i, and so is itself a conceivable estimator for θ, which depends on the observations only through the sufficient statistic: $\beta_0 = E(\beta|r)$. The theorem is then that the 'quadratic risk'

$$R(\theta, \beta) \equiv E[(\beta - \theta)^2|\theta] = \int dx_1 \cdots dx_n \, [\beta(x_1, \ldots, x_n) - \theta]^2 \qquad (8.20)$$

satisfies the inequality

$$R(\theta, \beta_0) \le R(\theta, \beta), \qquad (8.21)$$

---

[2] This rather vague statement becomes a definite theorem when we learn that, if we measure information in terms of entropy, then zero information loss in going from the full data set D to a statistic r is equivalent to sufficiency of r. The beginnings of this appeared long ago, in the Pitman–Koopman theorem (Koopman, 1936; Pitman, 1936); we give a modern version in Chapter 11.

for all θ. If $R(\theta, \beta)$ is bounded, there is equality if and only if $\beta_0 = \beta$; that is, if β itself depends on the data only through the sufficient statistic r.

In other words, given any estimator β for θ, if a sufficient statistic r exists, then we can find another estimator β_0 that achieves a lower or equal risk and depends only on r. Thus the best estimator we can find by the criterion of quadratic risk can always be chosen so that it depends on the data only through r. A proof is given by de Groot (1975, 1986 edn, p. 373); the orthodox notion of risk is discussed further in Chapters 13 and 14. But if a sufficient statistic does not exist, orthodox estimation theory is in real trouble because it wastes information; no single estimator can take note of all the relevant information in the data.

The Blackwell–Rao argument is not compelling to a Bayesian, because the criterion of risk is a purely sampling theory notion that ignores prior information. But Bayesians have a far better justification for using sufficient statistics; it is straightforward mathematics, evident from (8.9) and (8.10) that, if a sufficient statistic exists, Bayes' theorem will lead us to it *automatically*, without our having to take any particular note of the idea. Indeed, far more is true: from the proofs of Chapter 2, Bayes' theorem will lead us to the optimal inferences,[3] whether or not a sufficient statistic exists. So, in Bayesian inference, sufficiency is a valid concept; but it is not a fundamental theoretical consideration, only a pleasant convenience affecting the amount of computation but not the quality of the inference.

We have seen that sufficient statistics exist for the binomial, rectangular, and Gaussian sampling distributions. But consider the Cauchy distribution

$$p(x_1 \ldots x_n | \theta I) = \prod_{i=1}^{n} \frac{1}{\pi} \frac{1}{1 + (x_i - \theta)^2}. \tag{8.22}$$

This does not factor in the manner (8.9), and so there is no sufficient statistic. With a Cauchy sampling distribution, it appears that no part of the data is irrelevant; every scrap of it is used in Bayesian inference, and it makes a difference in our inferences about θ (that is, in details of the posterior pdf for θ). Then there can be no satisfactory orthodox estimator for θ; a single function conveys only one piece of information concerning the data, and misses $(n - 1)$ others, all of which are relevant and used by Bayesian methods.

8.3 Generalized sufficiency

What Fisher could not have realized, because of his failure to use priors, is that the proviso *for all priors* is essential here. Fisher sufficiency, Eq. (8.9), is the strong condition necessary to achieve property R independently of the prior. But what was realized only recently is that property R may hold under weaker conditions that depend on which prior we assign. Thus, the notion of sufficiency, which originated in the Bayesian considerations of Laplace, actually has a wider meaning in Bayesian inference than in sampling theory.

---

[3] That is, optimal in the aforementioned sense that no other procedure can yield unique results while agreeing with our desiderata of rationality.

To see this, note that, since the integral equations (8.5) are linear, we may think in terms of linear vector spaces. Let the class of all priors span a function space (Hilbert space) H of functions on the parameter space S_θ. If property R holds only for some subclass of priors $f(\theta) \in H'$ that span a subspace $H' \subset H$, then in (8.5) it is required only that the projection of $K_i(\theta, \theta')$ onto that subspace vanishes. Then $K_i(\theta, \theta')$ may be an arbitrary function on the complementary function space $(H - H')$ of functions orthogonal to H'.

This new understanding is that, for some priors, it is possible to have 'effective sufficient statistics', even though a sufficient statistic in the sense of Fisher does not exist. Given any specified function $r(x_1, \ldots, x_n)$ and sampling density $p(x_1 \ldots x_n|\theta)$, this determines a kernel $K_i(\theta, \theta')$ which we may construct by (8.6). If this kernel is incomplete (i.e. as (θ, θ', i) vary over their range, the kernel, thought of as a set of functions of θ' parameterized by (θ, i), does not span the entire function space H), then the set of simultaneous integral equations (8.5) has nonvanishing solutions $f(\theta)$. If there are non-negative solutions, they will determine a subclass of priors $f(\theta)$ for which r would play the role of a sufficient statistic.

Then the possibility seems open that, for different priors, different functions $r(x_1, \ldots, x_n)$ of the data may take on the role of sufficient statistics. This means that *use of a particular prior may make certain particular aspects of the data irrelevant. Then a different prior may make different aspects of the data irrelevant.* One who is not prepared for this may think that a contradiction or paradox has been found.

This phenomenon is mysterious only for those who think of probability in terms of frequencies; as soon as we think of probability distributions as *carriers of information* the reason for it suddenly seems trivial and obvious. It really amounts to no more than the principle of Boolean algebra $AA = A$; redundant information is not counted twice. A piece of information in the prior makes a difference in our conclusions only when it tells us something that the data do not tell us. Conversely, a piece of information in the data makes a difference in our conclusions only when it tells us something that the prior information does not. Any information that is conveyed by both is redundant, and can be removed from either one without affecting our conclusions. Thus in Bayesian inference a prior can make some aspect of the data irrelevant simply by conveying some information that is also in the data.

But is this new freedom expressing trivialities, or potentially useful new capabilities for Bayesian inference, which Fisher and Jeffreys never suspected? To show that we are not just speculating about an empty case, note that we have already seen an extreme example of this phenomenon, in the strange properties that use of the binomial monkey prior had in urn sampling (Chapter 6); it made all of the data irrelevant, although with other priors all of the data were relevant.

8.4 Sufficiency plus nuisance parameters

In Section 8.2, the parameter θ might have been multidimensional, and the same general arguments would go through in the same way. The question becomes much deeper if we now suppose that there are two parameters θ, η in the problem, but we are not interested in η, so for us the question of sufficiency concerns only the marginal posterior pdf for θ. Factoring the prior $p(\theta\eta|I) = f(\theta)\,g(\eta|\theta)$, we may write the desired posterior pdf as

$$h(\theta|D) = \frac{\int d\eta\, p(\theta\eta)f(x_1,\ldots,x_n|\theta\eta)}{\int\int d\theta d\eta\, p(\theta\eta)f(x_1,\ldots,x_n|\theta\eta)} = \frac{f(\theta)F(x_1,\ldots,x_n|\theta)}{\int d\theta\, f(\theta)F(x_1,\ldots,x_n|\theta)}, \qquad (8.23)$$

where

$$F(x_1,\ldots,x_n|\theta) \equiv \int d\eta\, p(\eta|\theta I)f(x_1,\ldots,x_n|\theta,\eta). \qquad (8.24)$$

Since this has the same mathematical form as (8.1), the steps (8.5)–(8.9) may be repeated and the same result must follow; given any specified $p(\eta|\theta I)$ for which the integral (8.24) converges, if we then find that the marginal distribution for θ has property R for all priors $f(\theta)$, then $F(x_1,\ldots,x_n|\theta)$ must factorize in the form

$$F(x_1,\ldots,x_n|\theta) = F^*(r|\theta)B(x_1,\ldots,x_n). \qquad (8.25)$$

But the situation is entirely different because $F(x_1,\ldots,x_n|\theta)$ no longer has the meaning of a sampling density, being a different function for different priors $p(\eta|\theta I)$. Now $\{F, F^*, B\}$ are all functionals of $p(\eta|\theta I)$.[4] Thus the presence of nuisance parameters changes the details, but the general phenomenon of sufficiency is retained.

8.5 The likelihood principle

In applying Bayes' theorem, the posterior pdf for a parameter θ is always a product of a prior $p(\theta|I)$ and a likelihood function $L(\theta) \propto p(D|\theta I)$; the only place where the data appear is in the latter. Therefore it is manifest that

> *Within the context of the specified model*, the likelihood function $L(\theta)$ from data D contains all the information about θ that is contained in D.

For us, this is an immediate and mathematically trivial consequence of the product rule of probability theory, and is no more to be questioned than the multiplication table. Put differently, two data sets D, D' that lead to the same likelihood function to within normalization: $L(\theta) = aL'(\theta)$, where 'a' is a constant independent of θ, have just the same import for any inferences about θ, whether it be point estimation, interval estimation, or hypothesis testing. But for those who think of a probability distribution as a physical phenomenon arising from 'randomness' rather than a carrier of incomplete information, the above quoted statement – since it involves only the sampling distribution – has a meaning independent of the product rule and Bayes' theorem. They call it the 'likelihood principle', and its status as a valid principle of inference has been the subject of long controversy, still continuing today.

An elementary argument for the principle, given by George Barnard (1947), is that irrelevant data ought to cancel out of our inferences. He stated it thus: Suppose that in

---

[4] In orthodox statistics, $F^*(r|\theta)$ would be interpreted as the sampling density to be expected in a compound experiment in which θ is held fixed but η is varied at random from one trial to the next, according to the distribution $p(\eta|\theta I)$.

addition to obtaining the data D we flip a coin and record the result $Z = H$ or T. Then the sampling probability for all our data becomes, as Barnard would have written it,

$$p(DZ|\theta) = p(D|\theta)p(Z).$$ (8.26)

Then he reasoned that, obviously, the result of a coin flip can tell us nothing more about the parameter θ beyond what the data D have to say; and so inference about θ based on DZ ought to be exactly the same as inference based on D alone. From this he drew the conclusion that constant factors in the likelihood must be irrelevant to inferences; that is, inferences about θ may depend only on the ratios of likelihoods for different values:

$$\frac{L_1}{L_2} = \frac{p(DZ|\theta_1 I)}{p(DZ|\theta_2 I)} = \frac{p(D|\theta_1 I)}{p(D|\theta_2 I)},$$ (8.27)

which are the same whether Z is or is not included. This is commonly held to be the first statement of the likelihood principle by an orthodox statistician. It is just what we considered obvious already back in Chapter 4, when we noted that a likelihood is not a probability because its normalization is arbitrary. But not all orthodoxians found Barnard's argument convincing.

Alan Birnbaum (1962) gave the first attempted proof of the likelihood principle to be generally accepted by orthodox statisticians. From the enthusiastic discussion following his paper, we see that many regarded this as a major historical event in statistics. He again appeals to coin tossing, but in a different way, through the principle of Fisher sufficiency plus a 'conditionality principle' which appeared to him more primitive:

Conditionality principle

Suppose we can estimate θ from either of two experiments, E_1 and E_2. If we flip a coin to decide which to do, then the information we get about θ should depend only on the experiment that was actually performed. That is, recognition of an experiment that might have been performed, but was not, cannot tell us anything about θ.

But Birnbaum's argument was not accepted by all orthodox statisticians, and Birnbaum himself seems to have had later doubts. One can criticize the conditionality principle by asking: 'How did you choose the experiments E_1, E_2?' Presumably, they were chosen with some knowledge of their properties. For example, we may know that one kind of experiment may be very good for small θ, a different one for large θ. Suppose that both E_1 and E_2 are most accurate for small θ and that there is a third experiment E_3 which is accurate for large θ. We assume that we chose E_1 and E_2, and the coin flip chose E_1. Then the fact that the coin flip did not choose E_2 need not make recognition of E_2 irrelevant to the inference; the very fact that we included it in our enumeration of experiments worth considering implies some prior knowledge favoring small θ.

In any event, Kempthorne and Folks (1971) and Fraser (1980) continued to attack the likelihood principle and deny its validity. From his failure to attack it when he was attacking almost every other principle of inference, we may infer that R. A. Fisher probably accepted the likelihood principle, although his own procedures did not respect it. But he continued to denounce the use of Bayes' theorem on other ideological grounds. For further discussion, see A. W. F. Edwards (1974), or Berger and Wolpert (1988). The issue becomes even more complex and confusing in connection with the notion of ancillarity, discussed below.

Orthodoxy is obliged to violate the likelihood principle for three different reasons: (1) its central dogma that 'The merit of an estimator is determined by its long-run sampling properties', which makes no reference to the likelihood function; (2) its secondary dogma that the accuracy of an estimate is determined by the width of the sampling distribution for the estimator, which again takes note of the likelihood principle; and (3) procedures in which 'randomization' is held to generate the probability distribution *used in the inference*! These are still being taught, and defended vigorously, by people who do not seem to comprehend that their conclusions are then determined, not by the relevant evidence in the data, but by irrelevant artifacts of the randomization. In Chapter 17 we shall examine the so-called 'randomization tests' of orthodoxy and see how Bayesian analysis deals with the same problems.

Indeed, even coin flip arguments cannot be accepted unconditionally if they are to be taken literally; particularly by a physicist who is aware of all the complicated things that happen in real coin flips, as described in Chapter 10. If there is any logical connection between θ and the coin, so that knowing θ would tell us anything about the coin flip, then knowing the result of the coin flip must tell us something about θ. For example, if we are measuring a gravitational field by the period of a pendulum, but the coin is tossed in that same gravitational field, there is a clear logical connection. Both Barnard's argument and Birnbaum's conditionality principle contain an implicit hidden assumption that this is not the case. Presumably, they would reply that, without saying so explicitly, they really meant 'coin flip' in a more abstract sense of some binary experiment totally detached from θ and the means of measuring it. But then, the onus was on them to define exactly what that binary experiment was, and they never did this.

In our view, this line of thought takes us off into an infinite regress of irrelevancies; in our system, the likelihood principle is already proved as an immediate consequence of the product rule of probability theory, independently of all considerations of coin flips or any other auxiliary experiment. But for those who ignore Cox's theorems, *ad hoc* devices continue to take precedence over the rules of probability theory, and there is a faction in orthodoxy that still militantly denies the validity of the likelihood principle.

It is important to note that the likelihood principle, like the likelihood function, refers only to the context of *a specified model which is not being questioned*; seen in a wider context, it may or may not contain all the information in the data that we need to make the best estimate of θ, or to decide whether to take more data or stop the experiment now. Is there additional external evidence that the apparatus is deteriorating? Or, is there reason to suspect that our model may not be correct? Perhaps a new parameter λ is needed. But to claim that the need for additional information like this is a refutation of the likelihood principle, is only to display a misunderstanding of what the likelihood principle is; it is a 'local' principle, not a 'global' one.

8.6 Ancillarity

Consider estimation of a location parameter θ from a sampling distribution $p(x|\theta I) = f(x - \theta|I)$.[5] Fisher (1934) perceived a strange difficulty with orthodox procedures. Choosing some function of the data $\theta^*(x_1, \ldots, x_n)$ as our estimator, two different data sets might yield the same estimate for θ, yet have very different configurations (such as range, fourth central moments, etc.), and must leave us in a very different state of knowledge concerning θ. In particular, it seemed that a very broad range and a sharply clustered one might lead us to the same actual estimate, but they ought to yield very different conclusions as to the accuracy of that estimate. Yet if we hold that the accuracy of an estimate is determined by the width of the sampling distribution for the estimator, one is obliged to conclude that all estimates from a given estimator have the same accuracy, regardless of the configuration of the sample.

Fisher's proposed remedy was not to question the orthodox reasoning which caused this anomaly, but rather to invent still another *ad hockery* to patch it up: use sampling distributions conditional on some 'ancillary' statistic $z(x_1, \ldots, x_n)$ that gives some information about the data configuration that is not contained in the estimator. In general, a single statistic cannot describe the data configuration fully; this could require as many as $(n - 1)$ ancillary statistics. But Fisher could not always supply them; often they do not exist, because he also demanded that the sampling distribution $p(z|\theta I) = p(z|I)$ for an ancillary statistic must be independent of θ. We do not know Fisher's private reason for imposing this independence, but from a Bayesian viewpoint we can see easily what it accomplishes.

The conditional sampling distribution for the data that Fisher would use is then $p(D|z\theta I)$. In orthodox statistics, this changed sampling distribution can in general lead to different conclusions about θ. But we process this by Bayes' theorem:

$$p(D|z\theta I) = \frac{p(zD|\theta I)}{p(z|\theta I)} = p(D|\theta I)\frac{p(z|D\theta I)}{p(z|\theta I)}. \tag{8.28}$$

Now if $z = z(D)$ is a function only of the data, then $p(z|D\theta I)$ is just a delta-function $\delta[z - z(D)]$; so, if $p(z|\theta I)$ is independent of θ, the conditioned sampling distribution $p(D|z\theta I)$ has the same θ dependence (that is, it yields the same likelihood function) as does the unconditional sampling distribution $p(D|\theta I)$. Put differently, from a Bayesian standpoint what Fisher's procedure accomplishes is nothing at all; the likelihood $L(\theta)$ is unchanged, so any method of inference – whether for point estimation, interval estimation, or hypothesis testing – that respects the likelihood principle will lead to just the same inferences about θ, whether or not we condition on an ancillary statistic. Indeed, in Bayesian analysis, if z is a function only of the data, then the value of z is known from the data, so it is redundant information; whether it is or is not included also in the prior information cannot matter. This is, again, just the principle $AA = A$ of elementary logic that we are obliged to stress so often because orthodoxy does not seem to comprehend its implications.

---

[5] For example, if the mean of a set of samples is used as the estimator, then, given a set of samples, the observed variation of the mean is called the sampling distribution of the mean.

The fact that Fisher obtained different estimates, depending on whether he did or did not condition on ancillary statistics, indicates only that his unconditioned procedure violated the likelihood principle. On the other hand, if we condition on a quantity Z that is not just a function of the data, then Z conveys additional information that is not in the data; and we must expect that in general this *will* alter our inferences about θ.

Orthodoxy, when asked for the accuracy of the estimate, departs from the likelihood principle a second time by appealing not to any property of the likelihood function from our data set, but rather to the width of the sampling distribution for the estimator – a property of that imaginary collection of data sets that one thought might have been observed but were not. For us, adhering to the likelihood principle, it is the width of the likelihood function, from the one data set that we actually have, that tells us the accuracy of the estimate from that data set; imaginary data sets that were not seen are irrelevant to the question we are asking.[6] Thus, for a Bayesian the question of ancillarity never comes up at all; we proceed directly from the statement of the problem to the solution that obeys the likelihood principle.

8.7 Generalized ancillary information

Now let us take a broader view of the notion of ancillary information, as referring not to Fisher ancillarity (in which the ancillary statistic z is part of the data), but to any additional quantity Z that we do not consider part of the prior information or the data. As before, we define

$$
\begin{aligned}
\theta &= \text{parameters (interesting or uninteresting)} \\
E &= e_1, \ldots, e_n, \quad \text{noise} \\
D &= d_1, \ldots, d_n, \quad \text{data} \\
d_i &= f(t_i\theta) + e_i, \quad \text{model.}
\end{aligned}
\tag{8.29}
$$

But now we add

$$
Z = z_1, \ldots, z_m \qquad \text{ancillary data.}
\tag{8.30}
$$

We want to estimate θ from the posterior pdf, $p(\theta|DZI)$, and direct application of Bayes' theorem gives

$$
p(\theta|DZI) = p(\theta|I)\frac{p(DZ|\theta I)}{p(DZ|I)},
\tag{8.31}
$$

in which Z appears as part of the data. But now we suppose that Z has, by itself, no direct relevance to θ:

$$
p(\theta|ZI) = p(\theta|I).
\tag{8.32}
$$

This is the essence of what Fisher meant by the term 'ancillary', although his ideology did not permit him to state it this way (since he admitted only sampling distributions, he was

---

[6] The width of the sampling distribution for the estimator is the answer to a very different question: How would the estimates vary over the class of all different data sets that we think might have been seen?

obliged to define all properties in terms of sampling distributions). He would say instead that ancillary data have a sampling distribution independent of θ:

$$p(Z|\theta I) = p(Z|I), \tag{8.33}$$

which he would interpret as: θ exerts no causal physical influence on Z. But from the product rule,

$$p(\theta Z|I) = p(\theta|ZI)p(Z|I) = p(Z|\theta I)p(\theta|I), \tag{8.34}$$

we see that from the standpoint of probability theory as logic, (8.32) and (8.33) are equivalent; either implies the other. Expanding the likelihood ratio by the product rule and using (8.33),

$$\frac{p(DZ|\theta I)}{p(DZ|I)} = \frac{p(D|\theta ZI)}{p(D|ZI)}. \tag{8.35}$$

Then, in view of (8.32), we can rewrite (8.31) equally well as

$$p(\theta|DZI) = p(\theta|ZI)\frac{p(D|\theta ZI)}{p(D|ZI)}, \tag{8.36}$$

and now the generalized ancillary information appears to be part of the prior information.

A peculiar property of generalized ancillary information is that the relationship between θ and Z is a reciprocal one; had we been interested in estimating Z but knew θ, then θ would appear as a 'generalized ancillary statistic'. To see this most clearly, note that the definitions (8.32) and (8.33) of an ancillary statistic are equivalent to the factorization:

$$p(\theta Z|I) = p(\theta|I)p(Z|I). \tag{8.37}$$

Now recall how we handled this before, when our likelihood was only

$$L_0(\theta) \propto p(D|\theta I). \tag{8.38}$$

Because of the model equation (8.29), if θ is known, then the probability of getting any datum d_i is just the probability that the noise would have made up the difference:

$$e_i = d_i - f(t_i, \theta). \tag{8.39}$$

So if the prior pdf for the noise is a function

$$p(E|\theta I) = u(e_1, \ldots, e_n, \theta) = u(\{e_i\}, \theta) \tag{8.40}$$

we have

$$p(D|\theta I) = u(\{d_i - f(t_i, \theta)\}, \theta), \tag{8.41}$$

the same function of $\{d_i - f(t_i, \theta)\}$. In the special case of a white Gaussian noise pdf independent of θ, this led to Eq. (7.28).

Our new likelihood function (8.35) can be dealt with in the same way, only in place of (8.41) we shall have a different noise pdf, conditional on Z. Thus the effect of ancillary data is simply to update the original noise pdf:

$$p(E|\theta I) \rightarrow p(E|\theta Z I), \tag{8.42}$$

and in general ancillary data that have any relevance to the noise will affect our estimates of all parameters through this changed estimate of the noise.

In (8.40)–(8.42) we have included θ in the conditioning statement to the right of the vertical stroke to indicate the most general case. But in all the cases examined in the orthodox literature, knowledge of θ would not be relevant to estimating the noise, so what they actually did was the replacement

$$p(E|I) \rightarrow p(E|Z I) \tag{8.43}$$

instead of (8.42).

Also, in the cases we have analyzed, this updating is naturally regarded as arising from a joint sampling distribution, which is a function

$$p(DZ|I) = w(e_1, \ldots, e_n, z_1, \ldots, z_m). \tag{8.44}$$

The previous noise pdf (8.40) is then a marginal distribution of (8.44):

$$p(D|I) = u(e_1 \cdots e_n) = \int dz_1 \cdots dz_m \, w(e_1, \ldots, e_n, z_1, \ldots, z_m), \tag{8.45}$$

the prior pdf for the ancillary data is another marginal distribution:

$$p(Z|I) = \int de_1 \cdots de_n \, w(e_1, \ldots, e_n, z_1, \ldots, z_m), \tag{8.46}$$

and the conditional distribution is

$$p(D|ZI) = \frac{p(DZ|I)}{p(Z|I)} = \frac{w(e_i, z_j)}{v(z_j)}. \tag{8.47}$$

Fisher's original application, and the ironic lesson it had for the relation of Bayesian and sampling theory methods, is explained in the Comments at the end of this chapter, Section 8.12.

8.8 Asymptotic likelihood: Fisher information

Given a data set $D \equiv \{x_1, \ldots, x_n\}$, the log likelihood is

$$\frac{1}{n} \log L(\theta) = \frac{1}{n} \sum_{i=1}^{n} \log p(x_i|\theta). \tag{8.48}$$

What happens to this function as we accumulate more and more data? The usual assumption is that, as $n \rightarrow \infty$, the sampling distribution $p(x|\theta)$ is actually equal to the limiting relative frequencies of the various data values x_i. We know of no case where one could actually know this to be true in the real world; so the following heuristic argument is all that is justified. If this assumption were true, then we would have asymptotically, as $n \to \infty$,

$$\frac{1}{n} \log L(\theta) \to \int dx \, p(x|\theta_0) \log p(x|\theta), \qquad (8.49)$$

where θ_0 is the 'true' value, presumed unknown. Denoting the entropy of the 'true' density by

$$H_0 = - \int dx \, p(x|\theta_0) \log p(x|\theta_0), \qquad (8.50)$$

we have for the asymptotic likelihood function

$$\frac{1}{n} \log L(\theta) + H_0 = \int dx \, p(x|\theta_0) \log \left[\frac{p(x|\theta)}{p(x|\theta_0)} \right] \leq 0, \qquad (8.51)$$

where, letting $q \equiv p(x|\theta_0)/p(x|\theta)$, we used the fact that, for positive real q, we have $\log(q) \leq q - 1$, with equality if and only if $q = 1$. Thus we have equality in (8.51) if and only if $p(x|\theta) = p(x|\theta_0)$ for all x for which $p(x|\theta_0) > 0$. But if two different values θ, θ_0 of the parameter lead to identical sampling distributions, then they are confounded: the data cannot distinguish between them. If the parameter is always 'identified', in the sense that different values of θ always lead to different sampling distributions for the data, then we have equality in (8.51) if and only if $\theta = \theta_0$, so the asymptotic likelihood function $L(\theta)$ reaches its maximum at the unique point $\theta = \theta_0$.

Supposing the parameter multidimensional: $\theta \equiv \{\theta_1, \ldots, \theta_m\}$ and expanding about this maximum, we have

$$\log p(x|\theta) = \log p(x|\theta_0) - \frac{1}{2} \sum_{i,j=1}^{m} \frac{\partial^2 \log p(x|\theta)}{\partial \theta_i \, \partial \theta_j} \delta\theta_i \delta\theta_j \qquad (8.52)$$

or

$$\frac{1}{n} \log \left[\frac{L(\theta)}{L(\theta_0)} \right] = -\frac{1}{2} \sum_{ij} I_{ij} \delta\theta_i \delta\theta_j, \qquad (8.53)$$

where

$$I_{ij} \equiv \int d^n x \, p(x|\theta_0) \frac{\partial^2 \log p(x|\theta)}{\partial \theta_i \, \partial \theta_j} \qquad (8.54)$$

is called the *Fisher information matrix*. It is a useful measure of the 'resolving power' of the experiment; that is, considering two close values θ, θ', how big must the separation $|\theta - \theta'|$ be in order that the experiment can distinguish between them?

8.9 Combining evidence from different sources

We all know that there are good and bad experiments. The latter accumulate in vain. Whether there are a hundred or a thousand, one single piece of work by a real master – by a Pasteur, for example – will be sufficient to sweep them into oblivion.

Henri Poincaré (1904, p. 141)

We all feel intuitively that the totality of evidence from a number of experiments ought to enable better inferences about a parameter than does the evidence of any one experiment. But intuition is not powerful enough to tell us when this is valid. One might think naïvely that if we have 25 experiments, each yielding conclusions with an accuracy of $\pm 10\%$, then by averaging them we get an accuracy of $\pm 10/\sqrt{25} = \pm 2\%$. This seems to be supposed by a method currently in use in psychology and sociology, called meta-analysis (Hedges and Olkin, 1985). Probability theory as logic shows clearly how and under what circumstances it is safe to combine this evidence.

The classical example showing the error of uncritical reasoning here is the old fable about the height of the Emperor of China. Supposing that each person in China surely knows the height of the Emperor to an accuracy of at least ± 1 meter; if there are $N = 1\,000\,000\,000$ inhabitants, then it seems that we could determine his height to an accuracy at least as good as

$$\frac{1}{\sqrt{1\,000\,000\,000}}\, \mathrm{m} = 3 \times 10^{-5}\,\mathrm{m} = 0.03\,\mathrm{mm}, \tag{8.55}$$

merely by asking each person's opinion and averaging the results.

The absurdity of the conclusion tells us rather forcefully that the $\sqrt{N}$ rule is not always valid, even when the separate data values are causally independent; it is essential that they be *logically* independent. In this case, we know that the vast majority of the inhabitants of China have never seen the Emperor; yet they have been discussing the Emperor among themselves, and some kind of mental image of him has evolved as folklore. Then, knowledge of the answer given by one does tell us something about the answer likely to be given by another, so they are not logically independent. Indeed, folklore has almost surely generated a systematic error, which survives the averaging; thus the above estimate would tell us something about the folklore, but almost nothing about the Emperor.

We could put it roughly as follows:

$$\text{error in estimate} = S \pm \frac{R}{\sqrt{N}}, \tag{8.56}$$

where S is the common systematic error in each datum, R is the RMS 'random' error in the individual data values. Uninformed opinions, even though they may agree well among themselves, are nearly worthless as evidence. Therefore sound scientific inference demands that, when this is a possibility, we use a form of probability theory (i.e., a probabilistic model) which is sophisticated enough to detect this situation and make allowances for it.

As a start on this, (8.56) gives us a crude but useful rule of thumb; it shows that, unless we *know* that the systematic error is less than about one-third of the random error, we cannot be sure that the average of one million data values is any more accurate or reliable than the average of ten. As Henri Poincaré put it: 'The physicist is persuaded that one good measurement is worth many bad ones.' Indeed, this has been well recognized by experimental physicists for generations; but warnings about it are conspicuously missing from textbooks written by statisticians, and so it is not sufficiently recognized in the 'soft' sciences whose practitioners are educated from those textbooks.

Let us investigate this more carefully using probability theory as logic. Firstly we recall the chain consistency property of Bayes' theorem. Suppose we seek to judge the truth of some hypothesis H, and we have two experiments which yield data sets A, B, respectively. With prior information I, from the first we would conclude

$$p(H|AI) = p(H|I)\frac{p(A|HI)}{p(A|I)}. \tag{8.57}$$

Then this serves as the prior probability when we obtain the new data B:

$$p(H|ABI) = p(H|AI)\frac{p(B|AHI)}{p(B|AI)} = p(H|I)\frac{p(A|HI)p(B|AHI)}{p(A|I)p(B|AI)}. \tag{8.58}$$

But

$$p(A|HI)p(B|AHI) = p(AB|HI) \tag{8.59}$$
$$p(A|I)p(B|AI) = p(AB|I),$$

so (8.58) reduces to

$$p(H|ABI) = p(H|I)\frac{p(AB|HI)}{p(AB|I)}, \tag{8.60}$$

which is just what we would have found had we used the total evidence $C = AB$ in a single application of Bayes' theorem. This is the chain consistency property. We see from this that it is valid to combine the *evidence* from several experiments if:

(1) the prior information I is the same in all;
(2) the prior for each experiment includes also the results of the earlier ones.

To study one condition a time, let us leave it as an exercise for the reader to examine the effect of violating (1), and suppose for now that we obey (1) but not (2), but we have from the second experiment alone the conclusion

$$p(H|BI) = p(H|I)\frac{p(B|HI)}{p(B|I)}. \tag{8.61}$$

Is it possible to combine the conclusions (8.57) and (8.61) of the two experiments into a single more reliable conclusion? It is evident from (8.58) that this cannot be done in general; it is not possible to obtain $p(H|ABI)$ as a function of the form

$$p(H|ABI) = f\,[p(H|AI), p(H|BI)], \tag{8.62}$$

because this requires information not contained in either of the arguments of that function. But if it is true that $p(B|AHI) = p(B|HI)$, then from the product rule written in the form

$$p(AB|I) = p(A|BHI)p(B|HI) = p(B|AHI)p(A|HI), \tag{8.63}$$

Table 8.1. *Experiment A.*

	Failures	Successes	Success (%)
Old	16 519	4343	20.8 ± 0.28
New	742	122	14.1 ± 1.10

Table 8.2. *Experiment B.*

	Failures	Successes	Success (%)
Old	3876	14 488	78.9 ± 0.30
New	1233	3907	76.0 ± 0.60

it follows that $p(A|BHI) = p(A|HI)$, and this will work. For this, the data sets A, B must be logically independent in the sense that, given H and I, *knowing either data set would tell us nothing about the other*.

If we do have this logical independence, then it is valid to combine the results of the experiments in the above naïve way, and we will in general improve our inferences by so doing. Meta-analysis, applied without regard to these necessary conditions, can be utterly misleading.

At this point, we are beginning to see the kind of dangerous nonsense that can be produced by those who fail to distinguish between causal independence and logical independence. But the situation is still more subtle and dangerous; suppose one tried to circumvent this by pooling all the data before analyzing them; that is, using (8.60). Let us see what could happen to us.

8.10 Pooling the data

The following data are real, but the circumstances were more complicated than supposed in the following scenario. Patients were given either of two treatments, the old one and a new one, and the number of successes (recoveries) and failures (deaths) were recorded. In experiment A the data were as given in Table 8.1. In which the entries in the last column are of the form $100 \times [p \pm \sqrt{p(1-p)/n}]$, indicating the standard deviation to be expected from binomial sampling. Experiment B, conducted two years later, yielded the data given in Table 8.2. In each experiment, the old treatment appeared slightly but significantly better (that is, the differences in p were greater than the standard deviations). The results were very discouraging to the researchers.

But then one of them had a brilliant idea: let us pool the data, simply adding up in the manner $4343 + 14\,488 = 18\,831$, etc. Then we have the contingency table, Table 8.3. Now the new treatment appears much better with overwhelmingly high significance (the difference is over 20 times the sum of the standard deviations)! They eagerly publish this

Table 8.3. *Pooled data.*

	Failures	Successes	Success (%)
Old	20 395	18 831	48.0 ± 0.25
New	1975	4029	67.1 ± 0.61

gratifying conclusion, presenting only the pooled data; and become (for a short time) famous as great discoverers.

How is such an anomaly possible with such innocent looking data? How can two data sets, each supporting the same conclusion, support the opposite conclusion when pooled? Let the reader, before proceeding, ponder these tables and form your own opinion of what is happening.

The point is that an extra parameter is clearly present. Both treatments yielded much better results two years later. This unexpected fact is, evidently, far more important than the relatively small differences in the treatments. Nothing in the data *per se* tells us the reason for this (better control over procedures, selection of promising patients for testing, etc.) and only prior information about further circumstances of the tests can suggest a reason.

Pooling the data under these conditions introduces a very misleading bias; the new treatment appears better simply because, in the second experiment, six times as many patients were given the new treatment, while fewer were given the old one. The correct conclusion from these data is that the old treatment remains noticeably better than the new one; but another factor is present that is vastly more important than the treatment.

We conclude from this example that pooling the data to estimate a parameter θ is not permissible if the separate experiments involve other parameters $(\alpha, \beta, \ldots)$ which can be different in different experiments. In (8.61)–(8.63) we supposed (by failing to mention them) that no such parameters were present, but real experiments almost always have nuisance parameters which are eliminated separately in drawing conclusions.

In summary, the meta-analysis *procedure* is not necessarily wrong; but when applied without regard to these necessary qualifications it can lead to disaster. But we do not see how anybody could have found all these qualifications by intuition alone. Without the Bayesian analysis there is almost no chance that one could apply meta-analysis safely; the safe procedure is not to mention meta-analysis at all as if it were a new principle, but simply to apply probability theory with *strict adherence to our Chapter 2 rules*. Whenever meta-analysis is appropriate, the full Bayesian procedure automatically reduces to meta-analysis.

8.10.1 Fine-grained propositions

One objection that has been raised to probability theory as logic notes a supposed technical difficulty in setting up problems. In fact, many seem to be perplexed by it, so let us examine the problem and its resolution.

The Venn diagram mentality, noted at the end of Chapter 2, supposes that every probability must be expressed as an additive measure on some set; or, equivalently, that every proposition to which we assign a probability must be resolved into a disjunction of elementary 'atomic' propositions. Carrying this supposition over into the Bayesian field has led some to reject Bayesian methods on the grounds that, in order to assign a meaningful prior probability to some proposition such as $W \equiv$ the dog walks, we would be obliged to resolve it into a disjunction $W = W_1 + W_2 + \cdots$ of every conceivable subproposition about how the dog does this, such as

$W_1 \equiv$ first it moves the right forepaw, then the left hindleg, then ...

$W_2 \equiv$ first it moves the right forepaw, then the right hindleg, then...

. . .

This can be done in any number of different ways, and there is no principle that tells us which resolution is 'right'. Having defined these subpropositions somehow, there is no evident element of symmetry that could tell us which ones should be assigned equal prior probabilities. Even the professed Bayesian L. J. Savage (1954, 1961, 1962) raised this objection, and thought that it made it impossible to assign priors by the principle of indifference. Curiously, those who reasoned this way seem never to have been concerned about how the orthodox probabilist is to define *his* 'universal set' of atomic propositions, which performs for him the same function as would that infinitely fine-grained resolution of the dog's movements.

8.11 Sam's broken thermometer

If Sam, in analyzing his data to test his pet theory, wants to entertain the possibility that his thermometer is broken, does he need to enumerate every conceivable way in which it could be broken? The answer is not intuitively obvious at first glance, so let

$A \equiv$ Sam's pet theory,

$H_o \equiv$ the thermometer is working properly,

$H_i \equiv$ the thermometer is broken in the ith way, $1 \leq i \leq n$,

where, perhaps, $n = 1000$. Then, although

$$p(A|DH_0I) = p(A|H_0I)\frac{p(D|AH_0I)}{p(D|H_0I)} \tag{8.64}$$

is the Bayesian calculation Sam would like to do, it seems that honesty compels him to note 1000 other possibilities $\{H_1, \ldots, H_n\}$, and so he must do the calculation

$$p(A|DI) = \sum_{i=0}^{n} p(AH_i|DI) = p(A|H_0DI)p(H_0|I) + \sum_{i=1}^{n} p(A|H_iDI)p(H_i|DI).$$
$$\tag{8.65}$$

Now expand the last term by Bayes' theorem:

$$p(A|H_i DI) = p(A|H_i I)\frac{p(D|AH_i I)}{p(D|H_i I)} \tag{8.66}$$

$$p(H_i|DI) = p(H_i|I)\frac{p(D|H_i I)}{p(D|I)}. \tag{8.67}$$

Presumably, knowing the condition of his thermometer does not in itself tell Sam anything about the status of his pet theory, so

$$p(A|H_i I) = p(A|I), \qquad 0 \le i \le n. \tag{8.68}$$

But if he knew the thermometer was broken, then the data would tell him nothing about his pet theory (all this is supposed to be contained in the prior information I):

$$p(A|H_i DI) = p(A|H_i I) = p(A|I), \qquad 1 \le i \le n. \tag{8.69}$$

Then from (8.66), (8.68) and (8.69) we have

$$p(D|AH_i I) = p(D|H_i I), \qquad 1 \le i \le n. \tag{8.70}$$

That is, if he knows the thermometer is broken, and as a result the data can tell him nothing about his pet theory, then his probability of getting those data cannot depend on whether his pet theory is true. Then (8.65) reduces to

$$p(A|DI) = \frac{p(A|I)}{p(D|I)}\left[p(D|AH_0 I)p(H_0 I) + \sum_{i=1}^{n} p(D|H_i I)p(H_i|I) \right]. \tag{8.71}$$

From this, we see that if the different ways of being broken do not in themselves tell him different things about the data,

$$p(D|H_i I) = p(D|H_1 I), \qquad 1 \le i \le n, \tag{8.72}$$

then enumeration of the n different ways of being broken is unnecessary; the calculation reduces to finding the likelihood

$$L \equiv p(D|AH_0 I)p(H_0|I) + p(D|H_1 I)[1 - p(H_0|I)] \tag{8.73}$$

and only the total probability of being broken,

$$p(\overline{H}_0|I) = \sum_{i=1}^{n} p(H_i|I) = 1 - p(H_0|I), \tag{8.74}$$

is relevant. Sam does not need to enumerate 1000 possibilities. But if $p(D|H_i I)$ can depend on i, then the sum in (8.71) should be over those H_i that lead to different $p(D|H_i I)$. That is, information contained in the variations of $p(D|H_i I)$ would be relevant to his inference, and so they should be taken into account in a full calculation.

Contemplating this argument, common sense now tells us that this conclusion should have been 'obvious' from the start. Quite generally, enumeration of a large number of 'fine-grained' propositions and assigning prior probabilities to all of them is necessary only if the breakdown into those fine details contains information relevant to the question being asked. If they do not, then only the disjunction of all of the propositions is relevant to our problem, and we need only assign a prior probability directly to it.

In practice, this means that in a real problem there will be some natural end to the process of introducing finer and finer subpropositions; not because it is wrong to introduce them, but because it is unnecessary and it contributes nothing to the solution of the problem. The difficulty feared by Savage does not arise in real problems; and this is one of the many reasons why our policy of assigning probabilities on finite sets succeeds in the real world.

8.12 Comments

There are still a number of interesting special circumstances, less important technically but calling for short discussions.

Trying to conduct inference by inventing intuitive *ad hoc* devices instead of applying probability theory has become such a deeply ingrained habit among those with conventional training that, even after seeing the Cox theorems and the applications of probability theory as logic, many fail to appreciate what has been shown, and persist in trying to improve the results – without acquiring any more information – by adding further *ad hoc* devices to the rules of probability theory. We offer here three observations intended to discourage such efforts, by noting what *information* is and is not contained in our equations.

8.12.1 The fallacy of sample re-use

Richard Cox's theorems show that, given certain data and prior information D, I, any procedure which leads to a different conclusion than that of Bayes' theorem, will necessarily violate some very elementary desideratum of consistency and rationality. This implies that a *single* application of Bayes' theorem with given D, I will extract all the information that is in D, I, relevant to the question being asked. Furthermore, we have already stressed that, if we apply probability theory correctly, there is no need to check whether the different pieces of information used are logically independent; any redundant information will cancel out and will not be used twice.[7]

The feeling persists that, somehow, using the same data again in some other procedure might extract still more information from D that Bayes' theorem has missed the first time, and thus improve our ultimate inferences from D. Since there is no end to the conceivable arbitrary devices that might be invented, we see no way to prove once and for all that no such attempt will succeed, other than pointing to Cox's theorems. But for any particular device we can always find a direct proof that it will not work; that is, the device cannot

---

[7] Indeed, this is a property of any algorithm, in or out of probability theory, which can be derived from a constrained variational principle, because adding a new constraint cannot change the solution if the old solution already satisfied that constraint.

change our conclusions unless it also violates one of our Chapter 2 desiderata of rationality. We consider one commonly encountered example.

Having applied Bayes' theorem with given D, I to find the posterior probability

$$p(\theta|DI) = p(\theta|I)\frac{p(D|\theta I)}{p(D|I)} \tag{8.75}$$

for some parameter θ, suppose we decide to introduce some additional evidence E. Then another application of Bayes' theorem updates that conclusion to

$$p(\theta|EDI) = p(\theta|DI)\frac{p(E|\theta DI)}{p(E|DI)}, \tag{8.76}$$

so the necessary and sufficient condition that the new information will change our conclusions is that, on some region of the parameter space of positive measure, the likelihood ratio in (8.76) differs from unity:

$$p(E|\theta DI) \neq p(E|DI). \tag{8.77}$$

But if the evidence E was something already implied by the data and prior information, then

$$p(E|\theta DI) = p(E|DI) = 1, \tag{8.78}$$

and Bayes' theorem confirms that re-using redundant information cannot change the results. This is really only the principle of elementary logic: $AA = A$.

There is a famous case in which it appeared at first glance that one actually did get important improvement in this way; this leads us to recognize that the meaning of 'logical independence' is subtle and crucial. Suppose we take $E = D$; we simply use the same data set twice. But we act as if the second D were logically independent of the first D; that is, although they are the same data, let us call them D^* the second time we use them. Then we simply ignore the fact that D and D^* are actually one and the same data set, and instead of (8.76)–(8.78) we take, in violation of the rules of probability theory,

$$p(D^*|DI) = p(D^*|I) \quad \text{and} \quad p(D^*|\theta DI) = p(D^*|\theta I). \tag{8.79}$$

Then the likelihood ratio in (8.76) is the same as in the first application of Bayes' theorem, (8.75). We have squared the likelihood function, thus achieving a sharper posterior distribution with apparently more accurate estimate of θ!

It is evident that a fraud is being perpetrated here; by the same argument we could re-use the same data any number of times, thus raising the likelihood function to an arbitrarily high power, and seemingly getting arbitrarily accurate estimates of θ – all from the same original data set D which might consist of only one or two observations.

If we actually had two different data sets D, D^* which were *logically independent*, in the sense that knowing one would tell us nothing about the other – but which happened to be numerically identical – then indeed (8.79) would be valid, and the correct likelihood function from the two data sets *would* be the square of the likelihood from one of them.

Therefore the fraudulent procedure is, in effect, claiming to have twice as many observations as we really have. One can find this procedure actually used and advocated in the literature, in the guise of a 'data dependent prior' (Akaike, 1980). This is also close to the topic of 'meta-analysis' discussed earlier, where ludicrous errors can result from failure to perceive the logical dependence of different data sets which are causally independent.

The most egregious example of attempted sample re-use is in the aforementioned 'randomization tests', in which every one of the $n!$ permutations of the data is thought to contain new evidence relevant to the problem! We examine this astonishing view and its consequences in Chapter 17.

8.12.2 A folk theorem

In ordinary algebra, suppose that we have a number of unknowns $\{x_1, \ldots, x_n\}$ in some domain X to be determined, and are given the values of m functions of them:

$$
\begin{aligned}
y_1 &= f_1(x_1, \ldots, x_n) \\
y_2 &= f_2(x_1, \ldots, x_n) \\
&\cdots \\
y_m &= f_m(x_1, \ldots, x_n).
\end{aligned}
\tag{8.80}
$$

If $m = n$ and the Jacobian $\partial(y_1, \ldots, y_n)/\partial(x_1, \ldots, x_n)$ is not zero, then we can in principle solve for the x_i uniquely. But if $m < n$ the system is underdetermined; one cannot find all the x_i because the information is insufficient.

It appears that this well-known theorem of algebra has metamorphosed into a popular folk theorem of probability theory. Many authors state, as if it were an evident truth, that from m observations one cannot estimate more than m parameters. Authors with the widest divergence of viewpoints in other matters seem to be agreed on this. Therefore we almost hesitate to point out the obvious; that nothing in probability theory places any such limitation on us. In probability theory, as our data tend to zero, the effect is not that fewer and fewer parameters can be estimated; given a single observation, nothing prevents us from estimating a million different parameters. What happens as our data tend to zero is that those estimates just relax back to the prior estimates, as common sense tells us they must.

There may still be a grain of truth in this, however, if we consider a slightly different scenario; instead of varying the amount of data for a fixed number of parameters, suppose we vary the number of parameters for a fixed amount of data. Then does the accuracy of our estimate of one parameter depend on how many other parameters we are estimating? We note verbally what one finds, leaving it as an exercise for the reader to write down the detailed equations. The answer depends on how the sampling distributions change as we add new parameters; are the posterior pdfs for the parameters independent? If so, then our estimate of one parameter cannot depend on how many others are present.

But if in adding new parameters they all get correlated in the posterior pdf, then the estimate of one parameter θ might be greatly degraded by the presence of others (uncertainty in the values of the other parameters could then 'leak over' and contribute to the uncertainty in θ). In that case, it may be that some function of the parameters can be estimated more accurately than can any one of them. For example, if two parameters have a high negative correlation in the posterior pdf, then their sum can be estimated much more accurately than can their difference.[8] All these subtleties are lost on orthodox statistics, which does not recognize even the concept of correlations in a posterior pdf.

8.12.3 *Effect of prior information*

As we noted above, it is obvious, from the general principle of non-use of redundant information $AA = A$, that our data make a difference only when they tell us something that our prior information does not. It should be (but apparently is not) equally obvious that prior information makes a difference only when it tells us something that the data do not. Therefore, whether our prior information is or is not important can depend on which data set we get. For example, suppose we are estimating a general parameter θ, and we know in advance that $\theta < 6$. If the data lead to a negligible likelihood in the region $\theta > 6$, then that prior information has no effect on our conclusions. Only if the data alone would have indicated appreciable likelihood in $\theta > 6$ does the prior information matter.

But consider the opposite extreme: if the data placed practically all the likelihood in the region $\theta > 6$, then the prior information would have overwhelming importance and the robot would be led to an estimate very nearly $\theta^* = 6$, determined almost entirely by the prior information. But in that case the evidence of the data strongly contradicts the prior information, and we would become skeptical about the correctness of the prior information, the model, or the data. This is another case where astonishing new information may cause resurrection of alternative hypotheses that we always have lurking somewhere in our minds.

The robot, by design, has no creative imagination and always believes literally what we tell it; and so, if we fail to tell it about any alternative hypotheses, it will continue to give us the best estimates based on unquestioning acceptance of the hypothesis space that we gave it – right up to the point where the data and the prior information become logically contradictory – at which point, as noted at the end of Chapter 2, the robot crashes.

In principle, a single data point could determine accurate values of a million parameters. For example, if a function $f(x_1, x_2, \ldots)$ of one million variables takes on the value $\sqrt{2}$ only at a single point, and we learn that $f = \sqrt{2}$ exactly, then we have determined one million variables exactly. Or, if a single parameter is determined to an accuracy of 12 decimal digits, a simple mapping can convert this into estimates of six parameters to two digits each. But this gets us into the subject of 'algorithmic complexity', which is not our present topic.

8.12.4 *Clever tricks and gamesmanship*

Two very different attitudes toward the technical workings of mathematics are found in the literature. In 1761, Leonhard Euler complained about isolated results which 'are not based

---

[8] We shall see this in Chapter 18, in the theory of seasonal adjustment in economics. The phenomenon is demonstrated and discussed in detail in Jaynes (1985e); conventional non-Bayesian seasonal adjustment loses important information here.

on a systematic method' and therefore whose 'inner grounds seem to be hidden'. Yet in the 20th century, writers as diverse in viewpoint as Feller and de Finetti are agreed in considering computation of a result by direct application of the systematic rules of probability theory as dull and unimaginative, and revel in the finding of some isolated clever trick by which one can see the answer to a problem without any calculation.

For example, Peter and Paul toss a coin alternately starting with Peter, and the one who first tosses 'heads' wins. What are the probabilities p, p' for Peter or Paul to win? The direct, systematic computation would sum $(1/2)^n$ over the odd and even integers:

$$p = \sum_{n=0}^{\infty} \frac{1}{2^{2n+1}} = \frac{2}{3}, \qquad p' = \sum_{n=1}^{\infty} \frac{1}{2^{2n}} = \frac{1}{3}. \qquad (8.81)$$

The clever trick notes instead that Paul will find himself in Peter's shoes if Peter fails to win on the first toss: *ergo*, $p' = p/2$, so $p = 2/3$, $p' = 1/3$.

Feller's perception was so keen that in virtually every problem he was able to see a clever trick; and then gave only the clever trick. So his readers get the impression that:

(1) probability theory has no systematic methods; it is a collection of isolated, unrelated clever tricks, each of which works on one problem but not on the next one;
(2) Feller was possessed of superhuman cleverness;
(3) only a person with such cleverness can hope to find new useful results in probability theory.

Indeed, clever tricks do have an aesthetic quality that we all appreciate at once. But we doubt whether Feller, or anyone else, was able to see those tricks on first looking at the problem.

We solve a problem for the first time by that (perhaps dull to some) direct calculation applying our systematic rules. *After* seeing the solution, we may contemplate it and see a clever trick that would have led us to the answer much more quickly. Then, of course, we have the opportunity for gamesmanship by showing others only the clever trick, scorning to mention the base means by which we first found the answer. But while this may give a boost to our ego, it does not help anyone else.

Therefore we shall continue expounding the systematic calculation methods, because they are the only ones which are guaranteed to find the solution. Also, we try to emphasize *general* mathematical techniques which will work not only on our present problem, but on hundreds of others. We do this even if the current problem is so simple that it does not require those general techniques. Thus we develop the very powerful algorithms involving group invariance, partition functions, entropy, and Bayes' theorem, that do not appear at all in Feller's work. For us, as for Euler, these are the solid meat of the subject, which make it unnecessary to discover a different new clever trick for each new problem.

We learned this policy from the example of George Pólya. For a century, mathematicians had been, seemingly, doing their best to conceal the fact that they were finding their theorems first by the base methods of plausible conjecture, and only afterward finding the 'clever trick' of an effortless, rigorous proof. Pólya (1954) gave away the secret in his *Mathematics and Plausible Reasoning*, which was a major stimulus for the present work.

Clever tricks are always pleasant diversions, and useful in a temporary way, when we want only to convince someone as quickly as possible. Also, they can be valuable in understanding a result; having found a solution by tedious calculation, if we can then see a simple way of looking at it that would have led to the same result in a few lines, this is almost sure to give us a greater confidence in the correctness of the result, and an intuitive understanding of how to generalize it. We point this out many times in the present work. But the road to success in probability theory goes first through mastery of the general, systematic methods of permanent value. For a teacher, therefore, maturity is largely a matter of overcoming the urge to gamesmanship.

9

Repetitive experiments: probability and frequency

> The essence of the present theory is that no probability, direct, prior, or posterior, is simply a frequency.
>
> *H. Jeffreys (1939)*

We have developed probability theory as a generalized logic of plausible inference which should apply, in principle, to any situation where we do not have enough information to permit deductive reasoning. We have seen it applied successfully in simple prototype examples of nearly all the current problems of inference, including sampling theory, hypothesis testing, and parameter estimation.

Most of probability theory, however, as treated in the past 100 years, has confined attention to a special case of this, in which one tries to predict the results of, or draw inferences from, some experiment that can be repeated indefinitely under what appear to be identical conditions; but which nevertheless persists in giving different results on different trials. Indeed, virtually all application-oriented expositions *define* probability as meaning 'limiting frequency in independent repetitions of a random experiment' rather than as an element of logic. The mathematically oriented often define it more abstractly, merely as an additive measure, without any specific connection to the real world. However, when they turn to applications, they too tend to think of probability in terms of frequency. It is important that we understand the exact relationship between these conventional treatments and the theory being developed here.

Some of these relationships have been seen already; in the preceding five chapters we have shown that probability theory as logic can be applied consistently in many problems of inference that do not fit into the frequentist preconceptions, and so would be considered beyond the scope of probability theory. Evidently, the problems that can be solved by frequentist probability theory form a subclass of those that are amenable to probability theory as logic, but it is not yet clear just what that subclass is. In the present chapter we seek to clarify this, with some surprising results, including a better understanding of the role of induction in science.

There are also many problems where the attempt to use frequentist probability theory in inference leads to nonsense or disaster. We postpone examination of this pathology to later chapters, particularly Chapter 17.

9.1 Physical experiments

Our first example of such a repetitive experiment appeared in Chapter 3, where we considered sampling with replacement from an urn, and noted that even there great complications arise. But we managed to muddle our way through them by the conceptual device of 'randomization' which, although ill-defined, had enough intuitive force to overcome the fundamental lack of logical justification.

Now we want to consider general repetitive experiments where there need not be any resemblance to drawing from an urn, and for which those complications may be far greater and more diverse than they were for the urn. But at least we know that any such experiment is subject to physical law. If it consists of tossing a coin or die, it will surely conform to the laws of Newtonian mechanics, well known for 300 years. If it consists of giving a new medicine to a variety of patients, the principles of biochemistry and physiology, only partially understood at present, surely determine the possible effects that can be observed. An experiment in high-energy elementary particle physics is subject to physical laws about which we are about equally ignorant; but even here well-established general principles (conservation of charge, angular momentum, etc.) restrict the possibilities.

Clearly, competent inferences about any such experiment must take into account whatever is currently known concerning the physical laws that apply to the situation. Generally, this knowledge will determine the 'model' that we prescribe in the statement of the problem. If one fails to take account of the real physical situation and the known physical laws that apply, then the most impeccably rigorous mathematics from that point on will not guard against producing nonsense or worse. The literature gives much testimony to this.

In any repeatable experiment or measurement, some relevant factors are the same at each trial (whether or not the experimenter is consciously trying to hold them constant – or is even consciously aware of them), and some vary in a way not under the control of the experimenter. Those factors that are the same (whether from the experimenter's good control of conditions or from his failure to influence them at all) are called *systematic*. Those factors which vary in an uncontrolled way are often called *random*, a term which we shall usually avoid, because in current English usage it carries some very wrong connotations.[1] We should call them, rather, *irreproducible by the experimental technique used*. They might become reproducible by an improved technique; indeed, the progress of all areas of experimental science involves the continual development of more powerful techniques that exert finer control over conditions, making more effects reproducible. Once a phenomenon becomes reproducible, as has happened in molecular biology, it emerges from the cloud of speculation and fantasy to become a respectable part of 'hard' science.

In this chapter we examine in detail how our robot reasons about a repetitive experiment. Our aim is to find the logical relations between the information it has and the kind of

---

[1] To many, the term 'random' signifies on the one hand lack of physical determination of the individual results, *but, at the same time*, operation of a physically real 'propensity' rigidly fixing long-run frequencies. Naturally, such a self-contradictory view of things gives rise to endless conceptual difficulties and confusion throughout the literature of every field that uses probability theory. We note some typical examples in Chapter 10, where we confront this idea of 'randomness' with the laws of physics.

predictions it is able to make. Let our experiment consist of n trials, with m possible results at each trial; if it consists of tossing a coin, then $m = 2$; for a die, $m = 6$. If we are administering a vaccine to a sequence of patients, then m is the number of distinguishable reactions to the treatment, n is the number of patients, etc.

At this point, one would say, conventionally, something like: 'Each trial is capable of giving any one of m possible results, so in n trials there are $N = m^n$ different conceivable outcomes.' However, the exact meaning of this is not clear: is it a statement or an assumption of physical fact, or only a description of the robot's information? The content and range of validity of what we are doing depends on the answer.

The number m may be regarded, always, as a description of the state of knowledge in which we conduct a probability analysis; but this may or may not correspond to the number of real possibilities actually existing in Nature. On examining a cubical die, we feel rather confident in taking $m = 6$; but in general we cannot know in advance how many different results are possible. Some of the most important problems of inference are of the 'Charles Darwin' type.

Exercise 9.1.   When Charles Darwin first landed on the Galapagos Islands in September 1835, he had no idea how many different species of plants he would find there. Having examined $n = 122$ specimens, and finding that they can be classified into $m = 19$ different species, what is the probability that there are still more species, as yet unobserved? At what point does one decide to stop collecting specimens because it is unlikely that anything more will be learned? This problem is much like that of the sequential test of Chapter 4, although we are now asking a different question. It requires judgment about the real world in setting up the mathematical model (that is, in the prior information used in choosing the appropriate hypothesis space), but persons with reasonably good judgment will be led to substantially the same conclusions.

In general, then, far from being a known physical fact, the number m should be understood to be simply the number of results per trial *that we shall take into account in the present calculation*. Then it is perhaps being stated most defensibly if we say that when we specify m we are defining *a tentative working hypothesis*, whose consequences we want to learn. In any event, we are concerned with two different sample spaces; the space S for a single trial, consisting of m points, and the extension space

$$S^n = S \otimes S \otimes \cdots \otimes S, \tag{9.1}$$

the direct product of n copies of S, which is the sample space for the experiment as a whole. For clarity, we use the word 'result' for a single trial referring to space S, while 'outcome' refers to the experiment as a whole, defined on space S^n. Thus, one outcome consists of the enumeration of n results (including their order if the experiment is conducted in such a way that an order is defined). Then we may say that the number of results *being considered in the present calculation* is m, while the number of outcomes being considered is $N = m^n$.

Denote the result of the ith trial by r_i ($1 \leq r_i \leq m$, $1 \leq i \leq n$). Then any outcome of the experiment can be indicated by specifying the numbers $\{r_1, \ldots, r_n\}$, which constitute a conceivable data set D. Since the different outcomes are mutually exclusive and exhaustive, if our robot is given any information I about the experiment, the most general probability assignment it can make is a function of the r_i:

$$P(D|I) = p(r_1 \ldots r_n) \tag{9.2}$$

satisfying the sums over all possible data sets

$$\sum_{r_1=1}^{m} \sum_{r_2=1}^{m} \cdots \sum_{r_n=1}^{m} p(r_1 \ldots r_n) = 1. \tag{9.3}$$

As a convenience, since the r_i are non-negative integers, we may regard them as digits (*modulo m*) in a number R expressed in the base m number system; $0 \leq R \leq N - 1$. Our robot, however poorly informed it may be about the real world, is an accomplished manipulator of numbers, so we may instruct it to communicate with us in the base m number system instead of the decimal (base ten) system that you and I were trained to use because of an anatomical peculiarity of humans.

For example, suppose that our experiment consists of tossing a die four times; there are $m = 6$ possible results at each trial, and $N = 6^4 = 1296$ possible outcomes for the experiment, which can be indexed (1 to 1296). Then to indicate the outcome that is designated as number 836 in the decimal system, the robot notes that

$$836 = (3 \times 6^3) + (5 \times 6^2) + (1 \times 6^1) + (2 \times 6^0) \tag{9.4}$$

and so, in the base six system the robot displays this as outcome number 3512.

Unknown to the robot, this has a deeper meaning to you and me; for us, this represents the outcome in which the first toss gave three spots up, the second gave five spots, the third gave one spot, and the fourth toss gave two spots (since in the base six system the individual digits r_i have meaning only *modulo 6*, the display $5024 \equiv 5624$ represents an outcome in which the second toss yielded six spots up).

More generally, for an experiment with m possible results at each trial, repeated n times, we communicate with the robot in the base m number system, whereupon each number displayed will have exactly n digits, and for us the ith digit will represent, *modulo m*, the result of the ith trial. By this device we trick our robot into taking instructions and giving its conclusions in a format which has for us an entirely different meaning. We can now ask the robot for its predictions on any question we care to ask about the digits in the display number, and this will never betray to the robot that it is really making predictions about a repetitive physical experiment (for the robot, by construction as discussed in Chapter 4, always accepts what we tell it as the literal truth).

With the conceptual problem defined as carefully as we know how to do, we may turn finally to the actual calculations. We noted in the discussion following Eq. (2.86) that, depending on details of the information I, many different probability assignments (9.2) might be appropriate; consider first the obvious simplest case of all.

9.2 The poorly informed robot

Suppose we tell the robot only that there are N possibilities, and give no other information. That is, the robot is not only ignorant about the relevant physical laws; it is not even told that the full experiment consists of n repetitions of a simpler one. For it, the situation is as if there were only a single trial, with N possible results, the 'mechanism' being completely unknown.

At this point, you might object that we have withheld from the robot some very important information that must be of crucial importance for rational inferences about the experiment; and so we have. Nevertheless, it is important that we understand the surprising consequences of neglecting that information.

What meaningful predictions about the experiment could the robot possibly make, when it is in such a primitive state of ignorance that it does not even know that there is any repetitive experiment involved? Actually, the poorly informed robot is far from helpless; although it is hopelessly naïve in some respects, nevertheless it is already able to make a surprisingly large number of correct predictions for purely combinatorial reasons (this should give us some respect for the cogency of multiplicity factors, which can mask a lot of ignorance).

Let us see first just what those poorly informed predictions are; then we can give the robot additional pertinent pieces of information and see how its predictions are revised as it comes to know more and more about the real physical experiment. In this way we can follow the robot's education step by step, until it reaches a level of sophistication comparable to (in some cases, exceeding) that displayed by real scientists and statisticians discussing real experiments.

Denote this initial state of ignorance (the robot knows only the number N of possible outcomes and nothing else) by I_0. The principle of indifference (2.95) then applies; the robot's 'sample space' or 'hypothesis space' consists of $N = m^n$ discrete points, and to each it assigns probability N^{-1}. Any proposition A that is defined to be true on a subset $S' \subset S^n$ and false on the complementary subset $S^n - S'$ will, by the rule (2.99), then be assigned the probability

$$P(A|I_0) = \frac{M(n, A)}{N},$$ (9.5)

where $M(n, A)$ is the multiplicity of A (number of points of S^n on which A is true). This trivial looking result summarizes everything the robot can say on the prior information I_0, and it illustrates again that, whenever they are relevant to the problem, connections between probability and frequency appear automatically, as mathematical consequences of our rules.

Consider n tosses of a die, $m = 6$; the probability (9.2) of any completely specified outcome is

$$p(r_1 \ldots r_n | I_0) = \frac{1}{6^n}, \qquad 1 \le r_i \le 6, \quad 1 \le i \le n. \tag{9.6}$$

Then what is the probability that the first toss gives three spots, regardless of what happens later? We ask the robot for the probability that the first digit $r_1 = 3$. Then the 6^{n-1} propositions

$$A(r_2, \ldots, r_n) \equiv r_1 = 3 \text{ and the remaining digits are } r_2, \ldots, r_n \tag{9.7}$$

are mutually exclusive, and so (2.85) applies:

$$
\begin{aligned}
P(r_1 = 3 | I_0) &= \sum_{r_2=1}^{6} \cdots \sum_{r_n=1}^{6} p(3 \, r_2 \ldots r_n | I_0) \\
&= 6^{n-1} p(r_1 \ldots r_n | I_0) \\
&= 1/6.
\end{aligned}
\tag{9.8}
$$

(Note that '$r_1 = 3$' is a proposition, so by our notational rules in Appendix B we are allowed to put it in a formal probability symbol with capital P.) But by symmetry, if we had asked for the probability that any specified (ith) toss gives any specified (kth) result, the calculation would have been the same:

$$P(r_i = k | I_0) = 1/6, \qquad 1 \le k \le 6, \quad 1 \le i \le n. \tag{9.9}$$

Now, what is the probability that the first toss gives k spots, and the second gives j spots? The robot's calculation is just like the above; the results of the remaining tosses comprise 6^{n-2} mutually exclusive possibilities, and so

$$
\begin{aligned}
P(r_1 = k, r_2 = j | I_0) &= \sum_{r_3=1}^{6} \cdots \sum_{r_n=1}^{6} p(k, j, r_3 \ldots r_n | I_0) \\
&= 6^{n-2} p(r_1 \ldots r_n | I_0) = 1/6^2 \\
&= 1/36,
\end{aligned}
\tag{9.10}
$$

and by symmetry the answer would have been the same for any two different tosses. Similarly, the robot will tell us that the probability for any specified outcome at any three different tosses is

$$p(r_i r_j r_k | I_0) = 1/6^3 = 1/216, \tag{9.11}$$

and so on!

Let us now try to educate the robot. Suppose we give it the additional information that, to you and me, means that the first toss gave three spots. But we tell this to the robot in the form: out of the originally possible N outcomes, the correct one belongs to the subclass for which the first digit is $r_1 = 3$. With this additional information, what probability will it now assign to the proposition $r_2 = j$? This conditional probability is determined by the product rule (2.63):

$$p(r_2|r_1 I_0) = \frac{p(r_1 r_2|I_0)}{p(r_1|I_0)}, \tag{9.12}$$

or, using (9.9) and (9.10),

$$p(r_2|r_1 I_0) = \frac{1/36}{1/6} = 1/6 = p(r_2|I_0). \tag{9.13}$$

The robot's prediction is unchanged. If we tell it the result of the first two tosses and ask for its predictions about the third, we have from (9.11) the same result:

$$p(r_3|r_1 r_2 I_0) = \frac{p(r_3 r_1 r_2|I_0)}{p(r_1 r_2|I_0)} = \frac{1/216}{1/36} = 1/6 = p(r_3|I_0). \tag{9.14}$$

We can continue in this way, and will find that if we tell the robot the results of any number of tosses, this will have no effect at all on its predictions for the remaining ones. It appears that the robot is in such a profound state of ignorance I_0 that it cannot be educated. However, if it does not respond to one kind of instruction, perhaps it will respond to another. But first we need to understand the cause of the difficulty.

9.3 Induction

In what way does the robot's behavior surprise us? Its reasoning here is different from the way you and I would reason, in that the robot does not seem to learn from the past. If we were told that the first dozen digits were all 3, you and I would take the hint and start placing our bets on 3 for the next digit. But the poorly informed robot does not take the hint, no matter how many times it is given.

More generally, if you or I could perceive any regular pattern in the previous results, we would more or less expect it to continue; this is the reasoning process called *induction*. The robot does not yet see how to reason inductively. However, the robot must do all things quantitatively, and you and I would have to admit that we are not certain whether the regularity will continue. It only seems somewhat likely, but our intuition does not tell us how likely. So our intuition, as in Chapters 1 and 2, gives us only a qualitative 'sense of direction' in which we feel the robot's quantitative reasoning ought to go.

Note that what we are calling induction is a very different process from what is called, confusingly, 'mathematical induction'. The latter is a rigorous deductive process, and we are not concerned with it here.

The problem of 'justifying induction' has been a difficult one for the conventional formulations of probability theory, and the nemesis of some philosophers beginning with David Hume (1739, 1777) in the 18th century. For example, the philosopher Karl Popper (1974) has gone so far as to flatly deny the possibility of induction. He asked the rhetorical question: 'Are we rationally justified in reasoning from repeated instances of which we have experience to instances of which we have no experience?' This is, quite literally, the poorly informed robot speaking to us, and wanting us to answer '**No!**' But we want to show that a better informed robot will answer: 'Yes, if we have prior information providing a log-ical connection between the different trials' and give specific circumstances that enable induction to be made.

The difficulty has seemed particularly acute in the theory of survey sampling, which corresponds closely to our equations above. Having questioned 1000 people and found that 672 of them favor proposition A in the next election, by what right do the pollsters jump to the conclusion that about $67 \pm 3\%$ of the millions not surveyed also favor proposition A? For the poorly informed robot (and, apparently, for Popper too), learning the opinions of any number of persons tells it nothing about the opinions of anyone else.

The same logical problem appears in many other scenarios. In physics, suppose we measured the energies of 1000 atoms, and found that 672 of them were in excited states, the rest in the ground state. Do we have any right to conclude that about 67% of the 10^{23} other atoms not measured are also in excited states? Or, 1000 cancer patients were given a new treatment and 672 of them recovered; then in what sense is one justified in predicting that this treatment will also lead to recovery in about 67% of future patients? On prior information I_0, there is no justification at all for such inferences.

As these examples show, the problem of logical justification of induction (i.e., of clarifying the exact meaning of the statements, and the exact sense in which they can be supported by logical analysis) is important as well as difficult.

9.4 Are there general inductive rules?

What is shown by (9.13) and (9.14) is that, on the information I_0, the results of different tosses are, logically, completely independent propositions; giving the robot any information whatsoever about the results of specified tosses tells it nothing relevant to any other toss. The reason for this was stressed above: the robot does not yet know that the successive digits $\{r_1, r_2, \ldots\}$ represent successive repetitions of the *same* experiment. It can be educated out of this state only by giving it some kind of information that has relevance to all tosses; for example, if we tell it something, however slight, about some property – physical or logical – that is common to all trials.

Perhaps, then, we might learn by introspection: What is that extra 'hidden' information, common to all trials, that you and I are using, unconsciously, when we do inductive reason-ing? Then we might try giving this hidden information to the robot (i.e., incorporate it into our equations).

A very little introspection is enough to make us aware that there is no one piece of hidden information; there are many different kinds. Indeed, the inductive reasoning that we all do varies widely, even for identical data, as our prior knowledge about the experiment varies. Sometimes we 'take the hint' immediately, and sometimes we are as slow to do it as the poorly informed robot.

For example, suppose the data are that the first three tosses of a coin have all yielded 'heads': $D = H_1 H_2 H_3$. What is our intuitive probability $P(H_4|DI)$ for heads on the fourth toss? This depends very much on what that prior information I is. On prior information

I_0 the answer is always $p(H_4|DI_0) = 1/2$, whatever the data. Two other possibilities are:

> $I_1 \equiv$ We have been allowed to examine the coin carefully and observe the tossing. We know that the coin has a head and a tail and is perfectly symmetrical, with its center of gravity in the right place, and we saw nothing peculiar in the way it was tossed.
>
> $I_2 \equiv$ We were not allowed to examine the coin, and we are very dubious about the 'honesty' of either the coin or the tosser.

On information I_1, our intuition will probably tell us that the prior evidence of the symmetry of the coin far outweighs the evidence of three tosses; so we shall ignore the data and again assign $P(H_4|DI_1) = 1/2$.

But on information I_2 we would consider the data to have some cogency: we would feel that the fact of three heads and no tails constitutes some evidence (although certainly not proof) that some systematic influence is at work favoring heads, and so we would assign $P(H_4|DI_2) > 1/2$. Then we would be doing real inductive reasoning.

Now we seem to be facing a paradox. For I_1 represents a great deal more information than does I_2; yet it is $P(H_4|DI_1)$ that agrees with the poorly informed robot! In fact, it is easy to see that all our inferences based on I_1 agree with those of the poorly informed robot, as long as the prior evidence of symmetry outweighs the evidence of the data.

This is only an example of something that we have surely noted many times in other contexts. The fact that one person has far greater knowledge than another does not mean that they necessarily disagree; an idiot might guess the same truth that a scholar labored for years to discover. All the same, it does call for some deep thought to understand why knowledge of perfect symmetry could leave us making the same inferences as does the poorly informed robot.

As a start on this, note that we would not be able to assign any definite numerical value to $P(H_4|DI_2)$ until that vague information I_2 is specified much more clearly. For example, consider the extreme case:

> $I_3 \equiv$ We know that the coin is a trick one, that has either two heads or two tails; but we do not know which.

Then we would, of course, assign $P(H_4|DI_3) = 1$; in this state of prior knowledge, the evidence of a single toss is already conclusive. It is not possible to take the hint any more strongly than this.

As a second clue, note that our robot did seem, at first glance, to be doing inductive reasoning of a kind back in Chapter 3; for example in (3.14), where we examined the hypergeometric distribution. But on second glance it was doing 'reverse induction'; the more red balls that had been drawn, the lower its probability for red in the future. And this reverse induction disappeared when we went on to the limit of the binomial distribution.

But you and I could also be persuaded to do reverse induction in coin tossing. Consider the prior information:

$I_4 \equiv$ The coin has a concealed inner mechanism that constrains it to give exactly 50 heads and 50 tails in the next 100 tosses.

On this prior information, we would say that tossing the coin is, for the next 100 times, equivalent to drawing from an urn that contains initially 50 red balls and 50 white ones. We could then use the product rule as in (9.12) but with the hypergeometric distribution $h(r|N, M, n)$ of (3.22):

$$P(H_4|DI_4) = \frac{h(4|100, 50, 4)}{h(3|100, 50, 3)} = \frac{0.05873}{0.12121} = 0.4845 < \frac{1}{2}. \qquad (9.15)$$

But in this case it is easier to reason it out directly: $P(H_4|DI_4) = (M - 3)/(N - 3) = 47/97 = 0.4845$.

The great variety of different conclusions that we have found from the same data makes it clear that there can be no such thing as a single universal inductive rule and, in view of the unlimited variety of different kinds of conceivable prior information, makes it seem dubious that there could exist even a classification of all inductive rules by some system of parameters.

Nevertheless, such a classification was attempted by the philosopher R. Carnap (1891–1970), who found (Carnap, 1952) a continuum of rules identified by a single parameter λ $(0 < \lambda < \infty)$. But ironically, Carnap's rules turned out to be identical with those given, on the basis of entirely different reasoning, by Laplace in the 18th century (the 'rule of succession' and its generalizations) that had been rejected as metaphysical nonsense by statisticians and philosophers.[2]

Laplace was not considering the general problem of induction, but was only finding the consequences of a certain type of prior information, so the fact that he did not obtain every conceivable inductive rule never arose and would have been of no concern to him. In the meantime, superior analyses of Laplace's problem had been given by W. E. Johnson (1932), de Finetti (1937) and Harold Jeffreys (1939), of which Carnap seemed unaware.

Carnap was seeking the general inductive rule (i.e., the rule by which, given the record of past results, one can make the best possible prediction of future ones). But he suffered from one of the standard occupational diseases of philosophers; his exposition wanders off into abstract symbolic logic without ever considering a specific real example. So he never rises to the level of seeing that different inductive rules correspond to *different prior information*. It seems to us obvious, from arguments like the above, that this is the primary fact controlling induction, without which the problem cannot even be stated, much less solved; there is no 'general inductive rule'. Yet neither the term 'prior information' nor the concept ever appears in Carnap's exposition.

---

[2] Carnap (1952, p. 35), like Venn (1866), claims that Laplace's rule is inconsistent (in spite of the fact that it is identical with his own rule); we examine these claims in Chapter 18 and find, in agreement with Fisher (1956), that they have misapplied Laplace's rule by ignoring the necessary conditions required for its derivation.

This should give a good idea of the level of confusion that exists in this field, and the reason for it; conventional frequentist probability theory simply ignores prior information and – just for that reason – it is helpless to account for induction. Fortunately, probability theory as logic is able to deal with the full problem.

9.5 Multiplicity factors

In spite of the formal simplicity of (9.5), the actual numerical evaluation of $P(A|I_0)$ for a complicated proposition A may involve immense combinatorial calculations. For example, suppose we toss a die twelve times. The number of conceivable outcomes is

$$6^{12} = 2.18 \times 10^9,$$ \hfill (9.16)

which is about equal to the number of minutes since the Great Pyramid was built. The geologists and astrophysicists tell us that the age of the universe is of the order of 10^{10} years, or 3×10^{17} seconds. Thus, in 30 tosses of a die, the number of possible outcomes ($6^{30} = 2.21 \times 10^{23}$) is about equal to the number of microseconds in the age of the universe. Yet we shall be particularly interested in evaluating quantities like (9.5) pertaining to a famous experiment involving 20 000 tosses of a die!

It is true that we are concerned with finite sets; but they can be rather large and we need to learn how to calculate on them. An exact calculation will generally involve intricate number-theoretic details (such as whether n is a prime number, whether it is odd or even, etc.), and may require many different analytical expressions for different n. While we could make some further progress by elementary methods, any real facility in these calculations requires some more sophisticated mathematical techniques. We digress to collect some of the basic mathematical facts needed for them. These were given, for the most part, by Laplace, J. Willard Gibbs, and Claude Shannon. In view of the large numbers, there turn out to be extremely good approximations which are easy to calculate.

A large class of problems may be fit into the following scheme, for which we can indicate the exact calculation that should, in principle, be done. Let $\{g_1, g_2, \ldots, g_m\}$ be any set of m finite real numbers. For concreteness, one may think of g_j as the 'value' or the 'gain' of observing the jth result in any trial (perhaps the number of pennies we win whenever that result occurs), but the following considerations are independent of whatever meaning we attach to the $\{g_j\}$, with the proviso that they are additive; i.e., sums such as $g_1 + g_2$ are to be, like sums of pennies, meaningful to us. We could, equally well, make it more abstract by saying simply that we are concerned with predicting linear functions of the n_j. The total amount of G generated by the experiment is then

$$G = \sum_{i=1}^{n} g(r_i) = \sum_{j=1}^{m} n_j g_j,$$ \hfill (9.17)

where the sample number n_j is the number of times the jth result occurred. If we ask the robot for the probability for obtaining this amount, it will answer, from (9.5),

$$p(G|n, I_0) = f(G|n, I_0) = \frac{M(n, G)}{N}, \tag{9.18}$$

where $N = m^n$ and $M(n, G)$ is the multiplicity of the event G; i.e., the number of different outcomes which yield the value G (we now indicate in it also the number of trials n – to the robot, the number of digits needed to define an outcome – because we want to allow this to vary). Many probabilities are determined by this multiplicity factor, in its dependence on n and G.

9.6 Partition function algorithms

Expanding $M(n, G)$ according to the result of the nth trial gives the recursion relation

$$M(n, G) = \sum_{j=1}^{m} M(n - 1, G - g_j). \tag{9.19}$$

For small n, a computer could apply this n times for direct evaluation of $M(n, G)$, but this would be impractical for very large n. Equation (9.19) is a linear difference equation with constant coefficients in both n and G, so it must have elementary solutions of exponential form:

$$\exp\{\alpha n + \lambda G\}. \tag{9.20}$$

On substitution into (9.19), we find that this is a solution of the difference equation if α and λ are related by

$$\exp\{\alpha\} = Z(\lambda) \equiv \sum_{j=1}^{m} \exp\{-\lambda g_j\}. \tag{9.21}$$

The function $Z(\lambda)$ is called the *partition function*, and it will have a fundamental importance throughout all of probability theory. An arbitrary superposition of such elementary solutions:

$$H(n, G) = \int d\lambda \, Z^n(\lambda) \exp\{\lambda G\} h(\lambda) \tag{9.22}$$

is, from linearity, a formal solution of (9.19). However, the true $M(n, G)$ also satisfies the initial condition $M(0, G) = \delta(G, 0)$, and is defined only for certain discrete values of $G = \sum n_j g_j$, the values that are possible results of n trials. Further elaboration of (9.22) leads to analytical methods of calculation that will be used in the advanced applications in the later chapters; but for the present let us note the remarkable things that can be done just with algebraic methods.

Equation (9.22) has the form of an inverse Laplace transform. To find the discrete Laplace transform of $M(n, G)$ multiply $M(n, G)$ by $\exp\{-\lambda G\}$ and sum over all possible values of G. This sum contains a contribution from every possible outcome of the experiment, and so it can be expressed equally well as a sum over all possible sample numbers:

$$\sum_G \exp\{-\lambda G\} M(n, G) = \sum_{n_j \in U} W(n_1, \ldots, n_m) \exp\left\{-\lambda \sum n_j g_j\right\}, \qquad (9.23)$$

where the multinomial coefficient

$$W(n_1, \ldots, n_m) \equiv \frac{n!}{n_1! \cdots n_m!} \qquad (9.24)$$

is the number of outcomes which lead to the sample numbers $\{n_j\}$. If $x_j^{n_j} = \exp\{-n_j g_j\}$ then $\exp\{-\sum_j^m n_j g_j\} = x_1^{n_1} x_2^{n_2} \ldots x_m^{n_m}$. The multinomial expansion is defined by

$$(x_1 + \cdots + x_m)^n = \sum_{n_j \in U} W(n_1, \ldots, n_m) x_1^{n_1} \ldots x_m^{n_m}. \qquad (9.25)$$

In (9.23) we sum over the 'universal set' U, defined by

$$\left\{U : n_j \geq 0, \quad \sum_{j=1}^m n_j = n\right\}, \qquad (9.26)$$

which consists of all possible sample numbers in n trials. But, comparing (9.23) with (9.25), this is just

$$\sum_G \exp\{-\lambda G\} M(n, G) = Z^n(\lambda). \qquad (9.27)$$

Equation (9.27) says that the number of ways $M(n, G)$ in which a particular value G can be realized is just the coefficient of $\exp\{-\lambda G\}$ in $Z^n(\lambda)$; in other words, $Z(\lambda)$ raised to the nth power displays the exact way in which all the possible outcomes in n trials are partitioned among the possible values of G, which indicates why the name 'partition function' is appropriate.

9.6.1 Solution by inspection

In some simple problems, this observation gives us the solution by mere inspection of $Z^n(\lambda)$. For example, if we make the choice

$$g_j \equiv \delta(j, 1), \qquad (9.28)$$

then the total G is just the first sample number:

$$G = \sum_j n_j g_j = n_1. \qquad (9.29)$$

The partition function (9.21) is then

$$Z(\lambda) = \exp\{-\lambda\} + m - 1 \qquad (9.30)$$

and, from Newton's binomial expansion,

$$Z^n(\lambda) = \sum_{s=0}^{n} \binom{n}{s} \exp\{-\lambda s\}(m-1)^{n-s}. \tag{9.31}$$

$M(n, G) = M(n, n_1)$ is then the coefficient of $\exp\{-\lambda n_1\}$ in this expression:

$$M(n, G) = M(n, n_1) = \binom{n}{n_1}(m-1)^{n-n_1}. \tag{9.32}$$

In this simple case, the counting could have been done also as: $M(n, n_1) =$ (number of ways of choosing n_1 trials out of n) $\times$ (number of ways of allocating the remaining $m-1$ trial results to the remaining $n - n_1$ trials). However, the partition function method works just as well in more complicated problems; and even in this example the partition function method, once understood, is easier to use.

In the choice (9.28) we separated off the trial result $j = 1$ for special attention. More generally, suppose we separate the m trial results comprising the sample space S arbitrarily into a subset S' containing s of them, and the complementary subset $\overline{S'}$ consisting of the $(m-s)$ remaining ones, where $1 < s < m$. Call any result in the subset S' a 'success', any in $\overline{S'}$ a 'failure'. Then we replace (9.28) by

$$g_j = \begin{cases} 1 & j \in S' \\ 0 & \text{otherwise,} \end{cases} \tag{9.33}$$

and (9.29)–(9.32) are generalized as follows. G is now the total number of successes, called traditionally r:

$$G = \sum_{j=1}^{m} n_j g_j \equiv r, \tag{9.34}$$

and the partition function now becomes

$$Z(\lambda) = s \exp\{-\lambda\} + m - s, \tag{9.35}$$

from which

$$Z^n(\lambda) = \sum_{r=0}^{n} \binom{n}{r} s^r \exp\{-\lambda r\}(m-s)^{n-r}, \tag{9.36}$$

and so the coefficient of $\exp\{-\lambda r\}$ is

$$M(n, G) = M(n, r) = \binom{n}{r} s^r (m-s)^{n-r}. \tag{9.37}$$

From (9.18), the poorly informed robot's probability for r successes is therefore

$$P(G = r | I_0) = \binom{n}{r} p^r (1-p)^{n-r}, \qquad 0 \le r \le n, \tag{9.38}$$

where $p = s/m$. But this is just the binomial distribution $b(r|n, p)$, whose derivation cost us so much conceptual agonizing in Chapter 3. There we found the binomial distribution (3.86) as the limiting form in drawing from an infinitely large urn, and again as a randomized approximate form (3.92) in drawing with replacement from a finite urn; but in neither case was it exact for a finite urn. Now we have found a case where the binomial distribution arises for a different reason, and it is exact for a finite sample space.

This quantitative exactness is a consequence of our making the problem more abstract; there is now, in the prior information I_0, no mention of complicated physical properties such as those of urns, balls, and hands reaching in. But more important, and surprising, is simply the qualitative fact that the binomial distribution, ostensibly arising out of repeated sampling, has appeared in the inferences of a robot so poorly informed that it does not even have the concept of repetitions! In other words, the binomial distribution has an exact *combinatorial* basis, completely independent of the notion of 'repetitive sampling'.

This gives us a clue toward understanding how the poorly informed robot functions. In conventional probability theory, starting with James Bernoulli (1713), the binomial distribution has always been derived from the postulate that the probability for any result is to be the same at each trial, *strictly independent of what happens at any other trial*. But, as we have noted already, that is exactly what the poorly informed robot would say – not out of its knowledge of the physical conditions of the experiment, but out of its complete *ignorance* of what is happening.

Now we could go through many other derivations and we would find that this agreement persists: the poorly informed robot will find not only the binomial but also its generalization, the multinomial distribution, as combinatorial theorems.

Exercise 9.2. Derive the multinomial distribution found in Chapter 3, Eq. (3.77), as a generalization or extension of our derivation of (9.38).

Then all the usual probability distributions of sampling theory (Poisson, gamma, Gaussian, chi-squared, etc.) will follow as limiting forms of these. All the results that conventional probability theory has been obtaining from the frequency definition, and the assumption of strict independence of different trials, are just what the poorly informed robot would find in the same problems. In other words, *frequentist probability theory is, functionally, just the reasoning of the poorly informed robot*.

Then, since the poorly informed robot is unable to do inductive reasoning, we begin to understand why conventional probability theory has trouble with it. Until we learn how to introduce some kind of logical connection between the results of different trials, the results of any trials cannot tell us anything about any other trial, and it will be impossible to 'take the hint'.

Frequentist probability theory seems to be stuck with independent trials because it lays great stress on limit theorems, and examination of them shows that their derivation depends entirely on the strict independence of different trials. The slightest positive correlation between the results of different trials will render those theorems qualitatively wrong. Indeed, without that strict independence, not only limit theorems, but virtually all of the sampling distributions for estimators, on which orthodox statistics depends, would be incorrect.

Here the poorly informed robot would seem to have the tactical advantage; for all those limit theorems and sampling distributions for estimators are valid exactly on information I_0. There is another important difference; in conventional probability theory that 'independence' is held to mean causal physical independence; but how is one to judge this as a property of the real world? We have seen no discussion of this in the orthodox literature. To the robot it means logical independence, a stronger condition, but one that makes its calculations cleaner and simpler.

Solution by inspection of $Z^n(\lambda)$ has the merit that it yields exact results. However, only relatively simple problems can be solved in this way. We now note a much more powerful algebraic method.

9.7 Entropy algorithms

We return to the problem of calculating multiplicities as in (9.18)–(9.37), but in a little more general formulation. Consider a proposition $A(n_1, \ldots, n_m)$ which is a function of the sample numbers n_j; it is defined to be true when $(n_1, \ldots, n_m)$ are in some subset $R \in U$, where U is the universal set (9.26), and false when they are in the complementary set $\overline{R} = U - R$. If A is linear in the n_j, then it is the same as our G in (9.17). The multiplicity of A (number of outcomes for which it is true) is

$$M(n, A) = \sum_{n_j \in R} W(n_1, \ldots, n_m), \tag{9.39}$$

where the multinomial coefficient W was defined in (9.24).

How many terms $T(n, m)$ are in the sum (9.39)? This is a well-known combinatorial problem for which the reader will easily find the solution[3]

$$T(n, m) = \binom{n + m - 1}{n} = \frac{(n + m - 1)!}{n!(m - 1)!}, \tag{9.40}$$

and we note that, as $n \to \infty$,

$$T(n, m) \sim \frac{n^{m-1}}{(m - 1)!}. \tag{9.41}$$

The number of terms grows as a finite $[(m - 1)\text{th}]$ power of n (as can be seen intuitively by thinking of the n_j as Cartesian coordinates in an m-dimensional space and noting the geometrical meaning of the conditions (9.26) defining U). Denote the greatest term in the

---

[3] Physicists will recognize $T(n, m)$ as the 'Bose–Einstein multiplicity factor' of statistical mechanics (the number of linearly independent quantum states which can be generated by putting n Bose–Einstein particles into m single-particle states). Finding $T(n, m)$ is the same combinatorial problem.

region R by

$$W_{\max} \equiv \mathrm{Max}_R W(n_1, \ldots, n_m). \tag{9.42}$$

Then the sum (9.39) cannot be less than $W_{\max}$, and the number of terms in (9.39) cannot be greater than $T(n, m)$, so

$$W_{\max} \leq M(n, A) \leq W_{\max} T(n, m) \tag{9.43}$$

or

$$\frac{1}{n} \log(W_{\max}) \leq \frac{1}{n} \log M(n, A) \leq \frac{1}{n} \log(W_{\max}) + \frac{1}{n} \log T(n, m). \tag{9.44}$$

But as $n \to \infty$, from (9.41), we have

$$\frac{1}{n} \log T(n, m) \to 0, \tag{9.45}$$

and so

$$\frac{1}{n} \log M(n, A) \to \frac{1}{n} \log(W_{\max}). \tag{9.46}$$

The multinomial coefficient W grows so rapidly with n that in the limit the single maximum term in the sum (9.39) dominates it. The logarithm of T grows less rapidly than n, so in the limit it makes no difference in (9.44).

Then how does $\log(W/n)$ behave in the limit? The limit we want is the one in which the sample frequencies $f_j = n_j/n$ tend to constants; in other words, the limit as $n \to \infty$ of

$$\frac{1}{n} \log \left[\frac{n!}{(n f_1)! \cdots (n f_m)!} \right] \tag{9.47}$$

as the f_j are held constant. But, from the Stirling asymptotic approximation,

$$\log(n!) \sim n \log(n) - n + \log \sqrt{2\pi n} + O\left(\frac{1}{n}\right), \tag{9.48}$$

we find that, in the limit, $\log(W/n)$ tends to a finite constant value independent of n:

$$\frac{1}{n} \log(W) \to H \equiv - \sum_{j=1}^{m} f_j \log(f_j), \tag{9.49}$$

which is just what we call the *entropy* of the frequency distribution $\{f_1, \ldots, f_m\}$. We have the result that, for very large n, if the sample frequencies f_j tend to constants, the multiplicity of A goes into a surprisingly simple expression:

$$M(n, A) \sim \exp\{nH\}, \tag{9.50}$$

in the sense that the ratio of the two sides in (9.50) tends to unity (although their difference does not tend to zero; but they are both growing so rapidly that this makes no percentage difference in the limit). From (9.46) it is understood that in (9.50) the frequencies $f_j = n_j/n$ to be used in H are the ones which maximize H over the region R for which A is defined.

We now see what was not evident before; that this multiplicity is to be found by determining the *frequency* distribution $\{f_1, \ldots, f_m\}$ which has maximum entropy subject to whatever constraints define R.[4]

It requires some thought and analysis to appreciate what we have in (9.50). Note first that we now have the means to complete the calculations which require explicit values for the multiplicities $M(n, G)$. Before proceeding to calculate the entropy, let us note briefly how this will go. If A is linear in the n_j, then the multiplicity (9.50) is also equal asymptotically to

$$M(n, G) = \exp\{nH\}, \tag{9.51}$$

and so the probability for realizing the total G is, from (9.18),

$$p(G|n, I_0) = m^{-n} \exp\{nH\} = \exp\{-n(H_0 - H)\}, \tag{9.52}$$

where $H_0 = \log(m)$ is the absolute maximum of the entropy, derived below in (9.74). Often, the quantity most directly relevant is not the entropy, but the difference between the entropy and its maximum possible value; this is a direct measure of how strong are the constraints R. For many purposes it would have been better if entropy had been defined as that difference; but the historical precedence would be very hard to change now. In any event, (9.52) has some deep intuitive meaning that we develop in later chapters.

Let us note the effect of acquiring new information; now we learn that a specified trial yielded the amount g_j. This new information changes the multiplicity of A because now the remaining $(n - 1)$ trials must have yielded the total amount $(G - g_j)$, and the number of ways this could happen is $M(n - 1, G - g_j)$. Also, the frequencies are slightly changed, because one trial that yielded g_j is now absent from the counting. Instead of $f_k = n_k/n$ in (9.18), we now have the frequencies $\{f'_1, \ldots, f'_m\}$, where

$$f'_k = \frac{n_k - \delta_{jk}}{n - 1}, \qquad 1 \le k \le m, \tag{9.53}$$

or writing $f'_k = f_k + \delta f_k$, the change is

$$\delta f_k = \frac{f_k - \delta_{jk}}{n - 1}, \tag{9.54}$$

which is exact; as a check, note that $\sum f'_k = 1$ and $\sum \delta f_k = 0$, as it should be.

This small change in frequencies induces a small change in the entropy; writing the new value as $H' = H + \delta H$, we find

$$\delta H = \sum_k \frac{\partial H}{\partial f_k} \delta f_k + O\left(\frac{1}{n^2}\right) = \left[\frac{H + \log(f_j)}{n - 1}\right] + O\left(\frac{1}{n^2}\right), \tag{9.55}$$

---

[4] We now also see that, not only is the notion of entropy inherent in probability theory independently of the work of Shannon, the maximum entropy principle is also, at least in this case, derivable directly from the rules of probability theory without additional assumptions.

and so

$$H' = \frac{nH + \log(f_j)}{n-1} + O\left(\frac{1}{n^2}\right). \tag{9.56}$$

Then the new multiplicity is, asymptotically,

$$M(n-1, G-g_j) = \exp\{(n-1)H'\} = f_j \exp\{nH\}\left[1 + O\left(\frac{1}{n}\right)\right]. \tag{9.57}$$

The asymptotic forms of multiplicity are astonishingly simple compared with the exact expressions. This means that, contrary to first appearances when we noted the enormous size of the extension set S^n, the large n limit is by far the *easiest* thing to calculate when we have the right mathematical machinery. Indeed, the set S^n has disappeared from our considerations; the remaining problem is to calculate the f_k that maximize the entropy (9.49) over the domain R. But this is a problem that is solved on the sample space S of a single trial!

The probability for getting the total gain G is changed from (9.52) to

$$p(G|r_i = j, nI_0) = \frac{M(n-1, G-g_j)}{m^{n-1}}, \tag{9.58}$$

and, given only I_0, the prior probability for the event $r_i = j$ is, from (9.5),

$$p(r_i = j|nI_0) = \frac{1}{m}. \tag{9.59}$$

This gives us everything we need to apply Bayes' theorem conditional on G:

$$p(r_i = j|GnI_0) = p(r_i = j|nI_0)\frac{p(G|r_i = j, nI_0)}{p(G|nI_0)}, \tag{9.60}$$

or

$$p(r_i = j|GnI_0) = \frac{1}{m}\frac{[M(n-1, G-g_j)/m^{n-1}]}{[M(n, G)/m^n]} = \frac{M(n-1, G-g_j)}{M(n, G)} = f_j. \tag{9.61}$$

Knowledge of G therefore changes the robot's probability for the jth result from the uniform prior probability $1/m$ to the observed frequency f_j of that result. Intuition might have expected this connection between probability and frequency to appear eventually, but it may seem surprising that this requires only that the total G be known. Note, however, that specifying G determines the maximum entropy frequency distribution $\{f_1, \ldots, f_m\}$, so there is no paradox here.

Exercise 9.3. Extend this result to derive the joint probability

$$p(r_i = j, r_s = t|GnI_0) = M(n-2, G-g_j-g_t)/M(n, G) \tag{9.62}$$

as a ratio of multiplicities and give the resulting probability. Are the trials still independent, or does knowledge of G induce correlations between different trials?

These results show the gratifyingly simple and reasonable things that the poorly informed robot can do. In conventional frequentist probability theory, these connections are only postulated arbitrarily; the poorly informed robot derives them as consequences of the rules of probability theory.

Now we return to the problem of carrying out the entropy maximization to obtain explicit expressions for the entropies H and frequencies f_j.

9.8 Another way of looking at it

The following observation gives us a better intuitive understanding of the partition function method. Unfortunately, it is only a number-theoretic trick, useless in practice. From (9.28) and (9.29) we see that the multiplicity of ways in which the total G can be realized can be written as

$$M(n, G) = \sum_{\{n_j\}} W(n_1, \ldots, n_m), \qquad (9.63)$$

where we are to sum over all sets of non-negative integers $\{n_j\}$ satisfying

$$\sum n_j = n, \qquad \sum n_j g_j = G. \qquad (9.64)$$

Let $\{n_j\}$ and $\{n'_j\}$ be two such different sets which yield the same total: $\sum n_j g_j = \sum n'_j g_j = G$. Then it follows that

$$\sum_{j=1}^{m} k_j g_j = 0, \qquad (9.65)$$

where by hypothesis the integers $k_j \equiv n_j - n'_j$ cannot all be zero.

Two numbers f, g are said to be *incommensurable* if their ratio is not a rational number; i.e., if (f/g) cannot be written as (r/s), where r and s are integers (but, of course, any ratio may be thus approximated arbitrarily close by choosing r, s large enough). Likewise, we shall call the numbers $(g_1, \ldots, g_m)$ *jointly incommensurable* if no one of them can be written as a linear combination of the others with rational coefficients. But if this is so, then (9.65) implies that all $k_j = 0$:

$$n_j = n'_j, \qquad 1 \leq j \leq m, \qquad (9.66)$$

so if the $\{g_1, \ldots, g_m\}$ are jointly incommensurable, then *in principle* the solution is immediate; for then a given value of $G = \sum n_j g_j$ can be realized by only one set of sample numbers n_j; i.e., if G is specified exactly, this determines the exact values of all the $\{n_j\}$. Then we have only one term in (9.63):

$$M(n, G) = W(n_1, \ldots, n_m) \qquad (9.67)$$

and

$$M(n - 1, G - g_j) = W(n'_1, \ldots, n'_m), \qquad (9.68)$$

where, necessarily, $n'_i = n_i - \delta_{ij}$. Then the exact result (9.61) reduces to

$$p(r_k = j | Gn I_0) = \frac{W(n'_1, \ldots, n'_m)}{W(n_1, \ldots, n_m)} = \frac{(n-1)!}{n!} \frac{n_j!}{(n_j - 1)!} = \frac{n_j}{n}. \tag{9.69}$$

In this case the result could have been found in a different way: whenever by any means the robot knows the sample number n_j (i.e., the number of digits $\{r_1, \ldots, r_n\}$ equal to j) but does not know at which trials the jth result occurred (i.e., which digits are equal to j), it can apply James Bernoulli's rule (9.18) directly:

$$P(r_k = j | n_j I_0) = \frac{n_j}{\text{(total number of digits)}}. \tag{9.70}$$

Again, the *probability* for any proposition A is equal to the *frequency* with which it is true in the relevant set of equally possible hypotheses. So again, our robot, even if poorly informed, is nevertheless producing the standard results that current conventional treatments all assure us are correct. Conventional writers appear to regard this as a kind of law of physics; but we need not invoke any 'law' to account for the fact that a measured frequency often approximates an assigned probability (to a relative accuracy something like $1/\sqrt{n}$, where n is the number of trials). If the information used to assign that probability includes all of the systematic effects at work in the real experiment, then the great majority of all things that *could* happen in the experiment correspond to frequencies remaining in such a shrinking interval; this is simply a combinatorial theorem, which in essence was given already by de Moivre and Laplace in the 18th century, in their asymptotic formula. In virtually all of current probability theory this strong connection between probability and frequency is taken for granted for all probabilities, but without any explanation of the mechanism that produces it; for us, this connection is only a special case.

9.9 Entropy maximization

The above derivation (9.50) of $M(n, A)$ is valid for a proposition A that is defined by some arbitrary function of the sample numbers n_j. In general, one might need many different algorithms for this maximization. But in the case $A = G$, where we are concerned with a linear function $G = \sum n_j g_j$, the domain R is defined by specifying just the *average* of G over the n trials:

$$\overline{G} = \frac{G}{n} = \sum_{j=1}^{m} f_j g_j, \tag{9.71}$$

which is also an average over the frequency distribution. Then the maximization problem has a solution that was given once and for all by J. Willard Gibbs (1902) in his work on statistical mechanics.

It required another lifetime for Gibbs' algorithm to be generally appreciated; for 75 years it was rejected and attacked by some because, for those who thought of probability as a real physical phenomenon, it appeared arbitrary. Only through the work of Claude Shannon (1948) was it possible to understand what the Gibbs' algorithm was accomplishing. This was pointed out first in Jaynes (1957a) in suggesting a new interpretation of statistical mechanics (as an example of logical inference rather than as a physical theory), and this led rather quickly to a generalization of Gibbs' equilibrium theory to nonequilibrium statistical mechanics.

In Chapter 11 we set down the complete mathematical apparatus generated by the maximum entropy principle; for the present it will be sufficient to give the solution for the case at hand. An inequality given by Gibbs leads to an elegant solution to our maximization problem.

Let $\{f_1, \ldots, f_m\}$ be any possible frequency distribution on m points, satisfying ($f_j \geq 0$, $\sum_j f_j = 1$), and let $\{u_1, \ldots, u_m\}$ be any other frequency distribution satisfying the same conditions. Then using the fact that on the positive real line $\log(x) \leq (x - 1)$ with equality if and only if $x = 1$, we have

$$\sum_{j=1}^{m} f_j \log\left(\frac{u_j}{f_j}\right) \leq 0, \tag{9.72}$$

with equality if and only if $f_j = u_j$ for all j. In this we recognize the entropy expression (9.49), so the Gibbs inequality becomes

$$H(f_1, \ldots, f_m) \leq - \sum_{j=1}^{m} f_j \log(u_j), \tag{9.73}$$

from which various conclusions can be drawn. Making the choice $u_j = 1/m$ for all j, it becomes

$$H \leq \log(m), \tag{9.74}$$

so the maximum possible value of H is $\log(m)$, attained if and only if f_j is the uniform distribution $f_j = 1/m$ for all j. Now make the choice

$$u_j = \frac{\exp\{-\lambda g_j\}}{Z(\lambda)}, \tag{9.75}$$

where the normalizing factor $Z(\lambda)$ is just the partition function (9.21). Choose the constant λ so that some specified average $\overline{G} = \sum u_j g_j$ is attained; we shall see presently how to do this. The Gibbs inequality becomes

$$H \leq \sum f_j g_j + \log Z(\lambda). \tag{9.76}$$

Now let f_j vary over the class of all frequency distributions that yield the wanted average (9.71). The right-hand side of (9.76) remains constant, and H attains its maximum value on R:

$$H_{\max} = \overline{G} + \log(Z), \tag{9.77}$$

if and only if $f_j = u_j$. It remains only to choose λ so that the average value $\overline{G}$ is realized. But it is evident from (9.75) that

$$\overline{G} = \frac{\partial}{\partial \lambda} \log(Z) \qquad (9.78)$$

so this is to be solved for λ. It is easy to see that this has only one real root (on the real axis, the right-hand side of (9.78) is a continuous, strictly decreasing monotonic function of λ), so the solution is unique.

We have just derived the 'Gibbs canonical ensemble' formalism, which in quantum statistics is able to determine all equilibrium thermodynamic properties of a closed system (that is, no particles enter it or leave it); but now its generality far beyond that application is evident.

9.10 Probability and frequency

In our terminology, a *probability* is something that we assign, in order to represent a state of knowledge, or that we calculate from previously assigned probabilities according to the rules of probability theory. A *frequency* is a factual property of the real world that we measure or estimate. The phrase 'estimating a probability' is just as much a logical incongruity as 'assigning a frequency' or 'drawing a square circle'.

The fundamental, inescapable distinction between probability and frequency lies in this relativity principle: probabilities change when we change our state of knowledge; frequencies do not. It follows that the probability $p(E)$ that we assign to an event E can be equal to its frequency $f(E)$ only for certain particular states of knowledge. Intuitively, one would expect this to be the case when the only information we have about E consists of its observed frequency; and the mathematical rules of probability theory confirm this in the following way.

We note the two most familiar connections between probability and frequency. Under the assumption of exchangeability and certain other prior information (Jaynes, 1968), the rule for translating an observed frequency in a binary experiment into an assigned probability is Laplace's rule of succession. We have encountered this already in Chapter 6 in connection with urn sampling, and we analyze it in detail in Chapter 18. Under the assumption of independence, the rule for translating an assigned probability into an estimated frequency is James Bernoulli's weak law of large numbers (or, to get an error estimate, the de Moivre–Laplace limit theorem).

However, many other connections exist. They are contained, for example, in the principle of maximum entropy (Chapter 11), the principle of transformation groups (Chapter 12), and in the theory of fluctuations in exchangeable sequences (Jaynes, 1978).

If anyone wished to research this matter, we think he could find a dozen logically distinct connections between probability and frequency that have appeared in various applications. But these connections always appear automatically, whenever they are relevant to the problem, as mathematical consequences of probability theory as logic; there is never any need to define a probability as a frequency. Indeed, Bayesian theory may justifiably claim to use the notion of frequency more effectively than does the 'frequency' theory. For the latter admits only one kind of connection between probability and frequency, and has trouble in cases where a different connection is appropriate.

R. A. Fisher, J. Neyman, R. von Mises, W. Feller, and L. J. Savage denied vehemently that probability theory is an extension of logic, and accused Laplace and Jeffreys of committing metaphysical nonsense for thinking that it is. It seems to us that, if Mr A wishes to study properties of frequencies in random experiments and publish the results for all to see and teach them to the next generation, he has every right to do so, and we wish him every success. But in turn Mr B has an equal right to study problems of logical inference that have no necessary connection with frequencies or random experiments, and to publish his conclusions and teach them. The world has ample room for both.

Then why should there be such unending conflict, unresolved after over a century of bitter debate? Why cannot both coexist in peace? What we have never been able to comprehend is this: If Mr A wants to talk about frequencies, then why can't he just use the *word* 'frequency'? Why does he insist on appropriating the word 'probability' and using it in a sense that flies in the face of both historical precedent and the common colloquial meaning of that word? By this practice he guarantees that his meaning will be misunderstood by almost every reader who does not belong to his inner circle clique. It seems to us that he would find it easy – and very much in his own self-interest – to avoid these constant misunderstandings, simply by saying what he means. (H. Cramér (1946) did this fairly often, although not with 100% reliability, so his work is today easier to read and comprehend.)

Of course, von Mises, Feller, Fisher, and Neyman would not be in full agreement among themselves on anything. Nevertheless, whenever any of them uses the word 'probability', if we merely substitute the word 'frequency' we shall go a long way toward clearing up the confusion by producing a statement that means more nearly what they had in mind.

We think it is obvious that the vast majority of the real problems of science fall into Mr B's category, and therefore, in the future, science will be obliged to turn more and more toward his viewpoint and results. Furthermore, Mr B's use of the word 'probability' as expressing human information enjoys not only the historical precedent going back to James Bernoulli (1713), but it is also closer to the modern colloquial meaning of the word.

9.11 Significance tests

The rather subtle interplay between the notions of probability and frequency appears again in the topic of significance tests, or 'tests of goodness of fit'. In Chapter 5 we discussed such problems as assessing the validity of Newtonian celestial mechanics, and noted that orthodox significance tests purport to accept and reject hypotheses without considering any alternatives. Then we demonstrated why we cannot say how the observed facts affect the status of some hypothesis H until we state the specific alternative(s) against which H is to be tested. Common sense tells all scientists that a given piece of observational evidence E might demolish Newton's theory, or elevate it to certainty, or anything in between. It depends entirely on this: against which alternative(s) is it being tested? Bayes' theorem sends us the same message; for example, suppose we wish to consider only two hypotheses, H and H'. Then on any data D and prior information I, we must always have $P(H|DI) + P(H'|DI) = 1$, and in terms of our logarithmic measure of plausibility in decibels as discussed in Chapter 4, Bayes' theorem becomes

$$e(H|DI) = e(H|I) + 10\log_{10}\left[\frac{P(D|H)}{P(D|H')}\right], \qquad (9.79)$$

which we might describe in words by saying that 'Data D supports hypothesis H relative to H' by $10\log_{10}[P(D|H)/P(D|H')]$ decibels'. The phrase '*relative to H'*' is essential here, because relative to some other alternative H'' the change in evidence, $[e(H|DI) - e(H|I)]$, might be entirely different; it does not make sense to ask how much the observed facts tend 'in themselves' to support or refute H (except, of course, when data D are impossible on hypothesis H, so deductive reasoning can take over).

Now as long as we talk only in these generalities, our common sense readily assents to this need for alternatives. But if we consider specific problems, we may have some doubts. For example, in the particle counter problem of Chapter 6 we had a case (known source strength and counter efficiency s, ϕ) where the probability for getting c counts in any one second is a Poisson distribution with mean value $\lambda = s\phi$:

$$p(c|s\phi) = \exp\{-\lambda\}\frac{\lambda^c}{c!}, \qquad 0 \le c \le \infty. \qquad (9.80)$$

Although it wasn't necessary for the problem we were considering then, we can still ask: What can we infer from this about the relative *frequencies* with which we would see c counts if we repeat the measurement in many different seconds, with the resulting data set $D \equiv \{c_1, c_2, \ldots, c_n\}$? If the assigned probability for any particular event (say the event $c = 12$) is independently equal to

$$p = \exp\{-\lambda\}\frac{\lambda^{12}}{12!} \qquad (9.81)$$

at each trial, then the probability that the event will occur exactly r times in n trials is the binomial distribution (9.38):

$$b(r|n, p) = \binom{n}{r}p^r(1 - p)^{n-r}. \qquad (9.82)$$

There are several ways of calculating the moments of this distribution; one, easy to remember, is that the first moment is

$$\langle r \rangle = E(r)$$

$$= \sum_{r=0}^{n} r b(r|n, p)$$

$$= \left[p \frac{d}{dp} \sum_{r} \binom{n}{r} p^r q^{(n-r)} \right]_{q=1-p} \tag{9.83}$$

$$= \left(p \frac{d}{dp} \right) \times (p+q)^n$$

$$= np;$$

likewise,

$$\langle r^2 \rangle = \left(p \frac{d}{dp} \right)^2 (p+q)^n = np + n(n-1)p^2,$$

$$\langle r^3 \rangle = \left(p \frac{d}{dp} \right)^3 (p+q)^n = np + 2n(n-1)p^2 + n(n-1)(n-2)p^3, \tag{9.84}$$

and so on! For each higher moment we merely apply the operator (pd/dp) one more time, setting $p + q = 1$ at the end.

Our (mean) $\pm$ (standard deviation) estimate, over the sampling distribution, of r is then

$$(r)_{\text{est}} = \langle r \rangle \pm \sqrt{\langle r^2 \rangle - \langle r \rangle^2}$$

$$= np \pm \sqrt{np(1-p)}, \tag{9.85}$$

and our estimate of the frequency $f = r/n$ with which the event $c = 12$ will occur in n trials, is

$$(f)_{\text{est}} = p \pm \sqrt{\frac{p(1-p)}{n}}. \tag{9.86}$$

These relations and their generalizations give the most commonly encountered connection between probability and frequency; it is the original connection given by James Bernoulli (1713).

In the 'long run', therefore, we expect that the actual frequencies of various counts will be distributed in a manner approximating the Poisson distribution (9.80) to within the tolerances indicated by (9.86). Now we can perform the experiment, and the experimental frequencies either will or will not resemble the predicted values. If, by the time we have observed a few thousand counts, the observed frequencies are wildly different from a Poisson distribution (i.e. far outside the limits (9.86)), our intuition will tell us that the arguments which led to the Poisson prediction must be wrong; either the functional form of (9.80), or the independence at different trials, must not represent the real conditions in which the experiment was done.

Yet we have not said anything about any alternatives! Is our intuitive common sense wrong here, or is there some way we can reconcile it with probability theory? The question is not about probability theory but about psychology; it concerns what our intuition is doing here.

9.11.1 Implied alternatives

Let's look again at (9.79). No matter what H' is, we must have $p(D|H') \leq 1$, and therefore a statement which is independent of any alternative hypotheses is

$$e(H|DI) \geq e(H|I) + 10\log_{10} p(D|H) = e(H|I) - \psi_{\infty}, \qquad (9.87)$$

where

$$\psi_{\infty} \equiv -10\log_{10} p(D|H) \geq 0. \qquad (9.88)$$

Thus, *there is no possible alternative which data D could support, relative to H, by more than ψ_{∞} decibels.*

This suggests the solution to our paradox: in judging the amount of agreement between theory and observation, the proper question to ask is not, 'How well do data D support hypothesis H?' without mentioning any alternatives. A much better question is, 'Are there any alternatives H' which data D would support relative to H, and how much support is possible?' Probability theory can give no meaningful answer to the first question because it is not well-posed; but it can give a very definite (quantitative and unambiguous) answer to the second.

We might be tempted to conclude that the proper criterion of 'goodness of fit' is simply ψ_{∞}; or what amounts to the same thing, just the probability $p(D|H)$. This is not so, however, as the following argument shows. As we noted at the end of Chapter 6, after we have obtained D, it is always possible to invent a strange, 'sure thing' hypothesis H_S according to which every detail of D was inevitable: $p(D|H_S) = 1$, and H_S will always be supported relative to H by exactly ψ_{∞} decibels. Let us see what this implies. Suppose we toss a die $n = 10\,000$ times and record the detailed results. Then, on the hypothesis $H \equiv$ 'the die is honest', each of the 6^n possible outcomes has probability 6^{-n}, or

$$\psi_{\infty} = 10\log_{10}(6^n) = 77\,815 \text{ db}. \qquad (9.89)$$

No matter what we observe in the 10 000 tosses, there is always an hypothesis H_S that will be supported relative to H by this enormous amount. If, after observing 10 000 tosses, we still believe the die is honest, it can be only because we considered the prior probability for H_S to be even lower than $-77\,815$ db. Otherwise, we are reasoning inconsistently.

This is, if startling, all quite correct. The prior probability for H_S was indeed much lower than 6^{-n}, simply because there were 6^n different 'sure thing' hypotheses which were all on the same footing before we observed the data D. But it is obvious that in practice we don't want to bother with H_S; even though it is supported by the data more than any other, its prior probability is so low that we know in advance that we are not going to accept it anyway.

In practice, we are not interested in comparing H to all conceivable alternatives, but only to all those in some restricted class Ω, consisting of hypotheses which we consider in some sense 'reasonable'. Let us note one example (by far the most common and useful one) of a test relative to such a restricted class of alternatives.

Consider again the above experiment which has m possible results $\{A_1, \ldots, A_m\}$ at each trial. Define the quantities

$$x_i \equiv k, \qquad \text{if } A_k \text{ is true at the } i\text{th trial;} \qquad (9.90)$$

thus, each x_i can take on independently the values $1, 2, \ldots, m$. Now we wish to take into account only the hypotheses belonging to the 'Bernoulli class' B_m in which there are m possible results at each trial and the probabilities of the A_k on successive repetitions of the experiment are considered independent and stationary; thus, when H is in B_m, the probability conditional on H of any specific sequence $\{x_1, \ldots, x_n\}$ of observations has the form

$$p(x_1 \ldots x_n | H) = p_1^{n_1} \ldots p_m^{n_m}, \qquad (9.91)$$

where n_k is the above sample number. To every hypothesis in B_m there corresponds a set of numbers $\{p_1 \ldots p_m\}$ such that $p_k \geq 0$, $\sum_k p_k = 1$, and for our present purposes these numbers completely characterize the hypothesis. Conversely, every such set of numbers defines an hypothesis belonging to the Bernoulli class B_m.

Now we note an important lemma, given by J. Willard Gibbs (1902). Letting $x = n_k/np_k$, and using the fact that on the positive real line $\log(x) \geq (1 - x^{-1})$ with equality if and only if $x = 1$, we find at once that

$$\sum_{k=1}^{m} n_k \log \left(\frac{n_k}{np_k} \right) \geq 0, \qquad (9.92)$$

with equality if and only if $p_k = n_k/n$ for all k. This inequality is the same as

$$\log p(x_1 \ldots x_n | H) \leq n \sum_{k=1}^{m} f_k \log(f_k), \qquad (9.93)$$

where $f_k = n_k/n$ is the observed frequency of result A_k. The right-hand side of (9.88) depends only on the observed sample D, so if we consider various hypotheses $\{H_1, H_2, \ldots\}$ in B_m, the quantity (9.88) gives us a measure of how well the different hypotheses fit the data; the nearer to equality, the better the fit.

For convenience in numerical work, we express (9.88) in decibel units as in Chapter 4:

$$\psi_B \equiv 10 \sum_{k=1}^{m} n_k \log_{10} \left(\frac{n_k}{np_k} \right). \qquad (9.94)$$

To see the meaning of ψ_B, suppose we apply Bayes' theorem in the form of (9.79). Only two hypotheses, $H = \{p_1, \ldots, p_m\}$ and $H' = \{p'_1, \ldots, p'_m\}$ are being considered. Let the values of ψ_B according to H and H' be ψ_B, ψ'_B, respectively. Then Bayes' theorem reads

$$e(H|x_1 \ldots x_n) = e(H|I) + 10 \log_{10} \left[\frac{p(x_1 \ldots x_n|H)}{p(x_1 \ldots x_n|H')} \right]$$

$$= e(H|I) + \psi'_B - \psi_B. \qquad (9.95)$$

Now we can always find an hypothesis H' in B_m for which $p'_k = n_k/n$, and so $\psi'_B = 0$; therefore ψ_B has the following meaning:

> Given an hypothesis H and the observed data $D \equiv \{x_1, \ldots, x_n\}$, compute ψ_B from (9.94). Then, given any ψ in the range $0 \leq \psi \leq \psi_B$, it is possible to find an alternative hypothesis H' in B_m such that the data support H' relative to H by ψ decibels. There is no H' in B_m which is supported relative to H by more than ψ_B decibels.

Thus, although ψ_B makes no reference to any specific alternative, it is nevertheless exactly the appropriate measure of 'goodness of fit' *relative to the class B_m of Bernoulli alternatives*. It searches out B_m and locates the best alternative in that class.

Now we can understand the seeming paradox with which our discussion of significance tests started; the ψ-test is just the quantitative version of what our intuition has been, unconsciously, doing. We have already noted in Chapter 5, Section 5.4, that natural selection in exactly the sense of Darwin would tend to evolve creatures that reason in a Bayesian way because of its survival value.

We can also interpret ψ_B in this manner: we may regard the observed results $\{x_1, \ldots, x_n\}$ as a 'message' consisting of n symbols chosen from an alphabet of m letters. On each repetition of the experiment, Nature transmits to us one more letter of the message. How much information is transmitted by this message under the Bernoulli probability assignment? Note that

$$\psi_B = 10n \sum_{k=1}^{m} f_k \log_{10}(f_k/p_k) \qquad (9.96)$$

with $f_k = n_k/n$. Thus, $(-\psi_B/n)$ is the entropy per symbol $H(f; p)$ of the observed message distribution $\{f_1, \ldots, f_m\}$ relative to the 'expected distribution' $\{p_1, \ldots, p_m\}$. This shows that the notion of entropy was always inherent in probability theory; independently of Shannon's theorems, entropy or some monotonic function of entropy appears automatically in the equations of anyone who is willing to use Bayes' theorem for hypothesis testing.

Historically, a different criterion was introduced by Karl Pearson early in the 20th century. We expect that, if hypothesis H is true, then n_k will be close to np_k, in the sense that the difference $|n_k - np_k|$ will grow with n only as $\sqrt{n}$. Call this 'condition A'. Then using the expansion $\log(x) = (x-1) - (x-1)^2/2 + \cdots$, we find that

$$\sum_{k=1}^{m} n_k \log \left[\frac{n_k}{np_k} \right] = \frac{1}{2} \sum_{k} \frac{(n_k - np_k)^2}{np_k} + O\left(\frac{1}{\sqrt{n}} \right), \qquad (9.97)$$

the quantity designated as $O(1/\sqrt{n})$ tending to zero as indicated *provided that* the observed sample does in fact satisfy condition A. The quantity

$$\chi^2 \equiv \sum_{k=1}^{m} \frac{(n_k - np_k)^2}{np_k} = n \sum_k \frac{(f_k - p_k)^2}{p_k} \tag{9.98}$$

is thus very nearly proportional to ψ_B if the sample frequencies are close to the expected values:

$$\psi_B = [10 \log_{10}(e)] \times \frac{1}{2}\chi^2 + O\left(\frac{1}{\sqrt{n}}\right) = 4.343\chi^2 + O\left(\frac{1}{\sqrt{n}}\right). \tag{9.99}$$

Pearson suggested that χ^2 be used as a criterion of 'goodness of fit', and this has led to the 'chi-squared test', one of the most used techniques of orthodox statistics. Before describing the test, we examine its theoretical basis and suitability as a criterion. Evidently, $\chi^2 \geq 0$, with equality if and only if the observed frequencies agree exactly with those expected if the hypothesis is true. So, larger values of χ^2 correspond to greater deviation between prediction and observation, and too large a value of χ^2 should lead us to doubt the truth of the hypothesis. But these qualitative properties are possessed also by ψ_B and by any number of other quantities we could define. We have seen how probability theory determines directly the theoretical basis, and precise quantitative meaning, of ψ_B; so we ask whether there exists any connected theoretical argument pointing to χ^2 as the optimal measure of goodness of fit, by some well-defined criterion.

The results of a search for this connected argument are disappointing. Scanning a number of orthodox textbooks, we find that χ^2 is usually introduced as a straight *deus ex machina*; but Cramér (1946) does attempt to prepare the way for the idea, in these words:

It will then be in conformity with the general principle of least squares to adopt as measure of deviation an expression of the form $\sum c_i (n_i/n - p_i)^2$ where the coefficients c_i may be chosen more or less arbitrarily. It was shown by K. Pearson that if we take $c_i = n/p_i$, we shall obtain a deviation measure with particularly simple properties.

In other words, χ^2 is adopted, not because it is demonstrated to have good performance by any criterion, but only because it has simple properties!

We have seen that in some cases χ^2 is nearly a multiple of ψ_B, and then they must, of course, lead to essentially the same conclusions. But let us try to understand the quantitative difference in these criteria by a technique introduced in Jaynes (1976), which we borrowed from Galileo. Galileo's telescope was able to reveal the moons of Jupiter because it could *magnify* what was too small to be perceived by the unaided eye, up to the point where it could be seen by everybody. Likewise, we often find a quantitative difference in the Bayesian and orthodox results, so small that our common sense is unable to pass judgment on which result is preferable. But when this happens, we can find some extreme case where the difference is magnified to the point where common sense *can* tell us which method is giving sensible results, and which is not.

As an example of this magnification technique, we compare ψ_B and χ^2 to see which is the more reasonable *criterion* of goodness of fit.

9.12 Comparison of psi and chi-squared

A coin toss can give *three* different results: (1) heads, (2) tails, (3) it may stand on edge if it is sufficiently thick. Suppose that Mr A's knowledge of the thick English pound coin is such that he assigns probabilities $p_1 = p_2 = 0.499$, $p_3 = 0.002$ to these cases. We are in communication with Mr B on the planet Mars, who has never seen a coin and doesn't have the slightest idea what a coin is. So, when told that there are three possible results at each trial, and nothing more, he can only assign equal probabilities, $p'_1 = p'_2 = p'_3 = 1/3$.

Now we want to test Mr A's hypothesis against Mr B's by doing a 'random' experiment. We toss the coin 29 times and observe the results ($n_1 = n_2 = 14$, $n_3 = 1$). Then if we use the ψ criterion, we would have for the two hypotheses

$$\psi_A = 10\left[28\log_{10}\left(\frac{14}{29 \times 0.499}\right) + \log_{10}\left(\frac{1}{29 \times 0.002}\right)\right]$$

$$= 8.34 \text{ db},$$

$$\psi_B = 10\left[28\log_{10}\left(\frac{14 \times 3}{29}\right) + \log_{10}\left(\frac{3}{29}\right)\right]$$

$$= 35.19 \text{ db}.$$

(9.100)

From this experiment, Mr B learns two things: (a) that there is another hypothesis about the coin that is 35.2 db better than his (this corresponds to odds of over 3 300:1), and so, unless he can justify an extremely low prior probability for that alternative, he cannot reasonably adhere to his first hypothesis; and (b) that Mr A's hypothesis is better than his by some 26.8 db, and in fact is within about 8 db of the best hypothesis in the Bernoulli class B_3. Here the ψ test tells us pretty much what our common sense does.

Suppose that the man on Mars knew only about orthodox statistical principles as usually taught; and therefore believed that χ^2 was the proper criterion of goodness of fit. He would find that

$$\chi_A^2 = 2\frac{(14 - 29 \times 0.499)^2}{29 \times 0.499} + \frac{(1 - 29 \times 0.002)^2}{29 \times 0.002}$$

$$= 15.33,$$

$$\chi_B^2 = 2\frac{(14 - 29 \times 0.333)^2}{29 \times 0.333} + \frac{(1 - 29 \times 0.333)^2}{29 \times 0.333}$$

$$= 11.66,$$

(9.101)

and he would report back delightedly: 'My hypothesis, by the accepted statistical test, is shown to be slightly preferable to yours!'

Many persons trained to use χ^2 will find this comparison startling, and will try immediately to find the error in our numerical work above. We have here still another fulfilment of what Cox's theorems predict. The ψ criterion is exactly derivable from the rules of probability theory; therefore any criterion which is only an approximation to it must contain either an inconsistency or a qualitative violation of common sense, which can be exhibited by producing special cases.

We can learn an important lesson about the practical use of χ^2 by looking more closely at what is happening here. On hypothesis A, the 'expected' number of heads or tails in 29 tosses was $np_1 = 14.471$. The actual observed number must be an integer, and we supposed that in each case it was the closest possible integer, namely 14. Yet this small discrepancy between expected and observed sample numbers, in a sense the smallest it could possibly be, nevertheless had an enormous effect on χ^2. The spook lies in the fact that χ_A^2 turned out so much larger than seems reasonable; there is nothing surprising about the other numerical values. Evidently, it is the last term in χ_A^2, which refers to the fact the coin stood on edge once in 29 tosses, that is causing the trouble. On hypothesis A, the probability that this would happen exactly r times in n tosses is our binomial distribution (9.57), and with $n = 29$, $p = 0.002$, we find that the probability for seeing the coin on edge one or more times in 29 trials is $1 - b(0|n, p) = 1 - 0.998^{29} = 1/17.73$; i.e. the fact that we saw it even once is a bit unexpected, and constitutes some evidence against A. But this amount of evidence is certainly not overwhelming; if our travel guide tells us that London has fog, on the average, one day in 18, we are hardly astonished to see fog on the day we arrive. Yet this contributes an amount 15.30, almost all of the value of $\chi_A^2 = 15.33$.

It is the $(1/p_i)$ weighting factor in the summand of χ^2 that causes this anomaly. Because of it, the χ^2 criterion essentially concentrates its attention on the extremely unlikely possibilities if the hypothesis contains them; and the slightest discrepancy between expected and observed sample number for the unlikely events grotesquely over-penalizes the hypothesis. The ψ-test also contains this effect, but in a much milder form, the $1/p_i$ term appearing only in the logarithm.

To see this effect more clearly, suppose now that the experiment had yielded instead the results $n_1 = 14$, $n_2 = 15$, $n_3 = 0$. Evidently, by either the χ^2 or ψ criterion, this ought to make hypothesis A look better, B worse, than in the first example. Repeating the calculations, we now find

$$\psi_A = 0.30 \text{ db} \qquad \chi_A^2 = 0.0925$$
$$\psi_B = 51.2 \text{ db} \qquad \chi_B^2 = 14.55. \qquad (9.102)$$

You see that by far the greatest relative change was in χ_A^2; both criteria now agree that hypothesis A is far superior to B, as far as this experiment indicates.

This shows what can happen through uncritical use of χ^2. Professor Q believes in extrasensory perception, and undertakes to prove it to us poor benighted, intransigent doubters. So he plays card games. As in Chapter 5, on the 'null hypothesis' that only chance is operating, it is extremely unlikely that the subject will guess many cards correctly. But Professor Q is determined to avoid the tactical errors of his predecessors, and is alert to the phenomenon of deception hypotheses discussed in Chapter 5; so he averts that possibility by making videotape recordings of every detail of the experiments. The first few hundred times he plays, the results are disappointing; but these are readily explained away on the grounds that the subject is not in a 'receptive' mood. Of course, the tapes recording these experiments are erased.

One day, providence smiles on Professor Q; the subject comes through handsomely and he has the incontrovertible record of it. Immediately he calls in the statisticians, the mathematicians, the notary publics, and the newspaper reporters. An extremely improbable event has at last occurred; and χ^2 is enormous. Now he can publish the results and assert: 'The validity of the data is certified by reputable, disinterested persons, the statistical analysis has been under the supervision of recognized statisticians, the calculations have been checked by competent mathematicians. By the accepted statistical test, the null hypothesis has been decisively rejected.' And everything he has said is absolutely true!

Moral

For testing hypotheses involving moderately large probabilities, which agree moderately well with observation, it will not make much difference whether we use ψ or χ^2. But for testing hypotheses involving extremely unlikely events, we had better use ψ; or life might become too exciting for us.

9.13 The chi-squared test

Now we examine briefly the chi-squared test as done in practice. We have the so-called 'null hypothesis' H to be tested, and no alternative is stated. The null hypothesis predicts certain relative frequencies $\{f_1, \ldots, f_m\}$ and corresponding sample numbers $n_k = n f_k$, where n is the number of trials. We observe the actual sample numbers $\{n_1, \ldots, n_m\}$. But if the n_k are very small, we group categories together, so that each n_k is at least, say, five. For example, in a case with $m = 6$, if the observed sample numbers were $\{6, 11, 14, 7, 3, 2\}$ we would group the last two categories together, making it a problem with $m = 5$ distinguishable results per trial, with sample numbers $\{6, 11, 14, 7, 5\}$, and null hypothesis H which assigns probabilities $\{p_1, p_2, p_3, p_4, p_5 + p_6\}$. We then calculate the observed value of χ^2:

$$\chi^2_{\text{obs}} = \sum_{k=1}^{m} \frac{(n_k - n p_k)^2}{n p_k} \tag{9.103}$$

as our measure of deviation of observation from prediction. Evidently, it is very unlikely that we would find $\chi^2_{\text{obs}} = 0$ even if the null hypothesis is true. So, goes the orthodox reasoning, we should calculate the probability that χ^2 would have various values, and reject H if the probability $P(\chi^2_{\text{obs}})$ of finding a deviation as great as *or greater than* χ^2_{obs} is sufficiently small; this is the 'tail area' criterion, and one usually takes 5% (that is, $P(\chi^2_{\text{obs}}) = 0.05$) as the threshold of rejection.

Now the n_k are integers, so χ^2 is capable of taking on only a discrete set of numerical values, at most $(n + m - 1)!/n!(m - 1)!$ different values if the p_k are all different and incommensurable. Therefore, the exact χ^2 distribution is necessarily discrete and defined at only a finite number of points. However, for sufficiently large n, the number and density of points becomes so large that we may approximate the true χ^2 distribution by a continuous one. The 'simple property' referred to by Cramér is then the fact, at first glance surprising, that, in the limit of large n, we obtain a *universal* distribution law: the sampling probability that χ^2 will lie in the interval $d(\chi^2)$ is

$$g(\chi^2)d(\chi^2) = \frac{\chi^{f-2}}{2^{f/2}(f/2-1)!} \exp\left\{-\frac{1}{2}\chi^2\right\} d(\chi^2), \qquad (9.104)$$

where f is called the 'number of degrees of freedom' of the distribution. If the null hypothesis H is completely specified (i.e., if it contains no variable parameters), then $f = m - 1$, where m is the number of categories used in the sum of (9.98). But if H contains unspecified parameters which must be estimated from the data, we take $f = m - 1 - r$, where r is the number of parameters estimated.[5]

We readily calculate the expectation and variance over this distribution: $\langle\chi^2\rangle = f$, $\text{var}(\chi^2) = 2f$, so the (mean) $\pm$ (standard deviation) estimate of the χ^2 that we expect to see is just

$$\left(\chi^2\right)_{\text{est}} = f \pm \sqrt{2f}. \qquad (9.105)$$

The reason usually given for grouping categories for which the sample numbers are small, is that the approximation (9.104) would otherwise be bad. But grouping inevitably throws away some of the relevant information in the data, and there is never any reason to do this when using the exact ψ.

The probability that we would see a deviation as great as or greater than χ^2_{obs} is then

$$P(\chi^2_{\text{obs}}) = \int_{\chi^2_{\text{obs}}}^{\infty} d(\chi^2)\, g(\chi^2) = \int_{q_{\text{obs}}}^{\infty} dq\, \frac{q^k}{k!} \exp\{-q\}, \qquad (9.106)$$

where $q \equiv (1/2)\chi^2$, $k \equiv (f-2)/2$. If $P(\chi^2_{\text{obs}}) < 0.05$, we reject the null hypothesis at the 5% 'significance level'. Tables of χ^2 for which $P = 0.01, 0.05, 0.10, 0.50$, for various numbers of degrees of freedom, are given in most orthodox textbooks and collections of statistical tables (for example, Crow, Davis, and Maxfield, 1960).

Note the traditional procedure here; we chose some basically arbitrary significance level, then reported only whether the null hypothesis was or was not rejected at that level. Evidently this doesn't tell us very much about the real import of the data; if you tell me that the hypothesis was rejected at the 5% level, then I can't tell from that whether it would have been rejected at the 1%, or 2%, level. If you tell me that it was not rejected at the 5% level, then I don't know whether it would have been rejected at the 10%, or 20%, level. The orthodox statistician would tell us far more about what the data really indicate if he would report instead *the significance level $P(\chi^2)$ at which the null hypothesis is just barely rejected*; for then we know what the verdict would be at all levels. This is the practice of reporting

---

[5] The need for this correction was perceived by the young R. A. Fisher but not comprehended by Karl Pearson; and this set off the first of their fierce controversies, described in Chapter 16.

so-called 'P-values', a major improvement over the original custom. Unfortunately, the orthodox χ^2 and other tables are still so constructed that you cannot use them to report the conclusions in this more informative way, because they give numerical values only at such widely separated values of the significance level that interpolation is not possible.

How does one find numerical P-values without using the chi-squared tables? Writing $q = q_0 + t$, (9.106) becomes

$$
\begin{aligned}
P &= \int_0^\infty dt \, \frac{(q_0 + t)^k}{k!} \exp\{-(q_0 + t)\} \\
&= \frac{1}{k!} \sum_{k=0}^m \binom{m}{k} \int_0^\infty dt \, q_0^k t^{m-k} \exp\{-(q_0 + t)\} \\
&= \sum_{k=0}^m \exp\{-q_0\} \frac{q_0^k}{k!}.
\end{aligned}
\tag{9.107}
$$

But this is just the cumulative Poisson distribution and easily computed.

If you use the ψ-test instead, however, you don't need any tables. The evidential meaning of the sample is then described simply by the *numerical value* of ψ, and not by a further arbitrary construct such as tail areas. Of course, the numerical value of ψ doesn't in itself tell you whether to reject the hypothesis (although we could, with just as much justification as in the chi-squared test, prescribe some definite 'level' at which to reject). From the Bayesian point of view, there is simply no use in rejecting any hypothesis unless we can replace it with a definite alternative known to be better; and, obviously, whether this is justified must depend not only on ψ, but also on the prior probability for the alternative and on the consequences of making wrong decisions. Common sense tells us that this is, necessarily, a problem not just of inference, but of decision theory.

In spite of the vast difference in viewpoints, there is not necessarily much difference in the actual conclusions reached. For example, as the number of degrees of freedom f increases, the orthodox statistician will accept a higher value of χ^2 (roughly proportional to f, as (9.105) indicates) on the grounds that such a high value is quite likely to occur if the hypothesis is true; but the Bayesian, who will reject it only in favor of a definite alternative, must also accept a proportionally higher value of ψ, because the number of reasonable alternatives is increasing exponentially with f, and the prior probability for any one of them is correspondingly decreasing. So, in either case we end up rejecting the hypothesis if ψ or χ^2 exceeds some critical limit, with an enormous difference in the philosophy of how we choose that limit, but not necessarily a big difference in its actual location.

For many more details about chi-squared, see Lancaster (1969); and for some curious views that Bayesian methods fail to give proper significance tests, see Box and Tiao (1973).

9.14 Generalization

Although the point is not made in the orthodox literature, which does not mention alternatives at all, we see from the preceding section that χ^2 is not a measure of goodness of fit relative to all conceivable alternatives, but only relative to those in the same Bernoulli class. Until this is recognized, one really does not know what the χ^2-test is testing.

The procedure by which we constructed the ψ-test generalizes at once to the rule for constructing the exact test which compares the null hypothesis to any well-defined class C of alternatives. Just write Bayes' theorem describing the effect of data D on the relative plausibility of two hypotheses H_1, H_2 in that class, in the form

$$e(H_1|DI) - e(H_2|DI) = \psi_2 - \psi_1, \tag{9.108}$$

where ψ_i depends only on the data and H_i is non-negative over C, and vanishes for some H_i in C. Then we can always find an H_2 in C for which $\psi_2 = 0$, and so we have constructed the appropriate ψ_1 which measures goodness of fit relative to the class of alternatives C, and has the same meaning as that defined after (9.95). ψ_1 is the maximum amount by which any hypothesis in C can be supported relative to H_1 by the data D.

Thus, if we want a Bayesian test that is exact but operates in a similar way to orthodox significance tests, it can be produced quite easily. But we shall see in Chapter 17 that a different viewpoint has advantages; the format of orthodox significance tests can be replaced, as was done already by Laplace, by a parameter estimation procedure, which yields even more useful information.

Anscombe (1963) held it to be a weakness of the Bayesian method that we had to introduce a specific class of alternatives. We have answered that sufficiently here and in Chapters 4 and 5. We would hold it to be a great merit of the Bayesian approach that it forces us to recognize these essential features of inference, which have not been apparent to all orthodox statisticians. Our discussion of significance tests is a good example of what, we suggest, is the general situation; if an orthodox method is usable in some problem, then the Bayesian approach to inference supplies the missing theoretical basis for it, and usually improvements on it. Any significance test is only a slight variant of our multiple hypothesis testing procedures given in Chapter 4.

9.15 Halley's mortality table

An early example of the use of observed frequencies as probabilities, in a more useful and dignified context than gambling, and by a procedure that is so nearly correct that we could not improve on it appreciably today, was provided by the astronomer Edmund Halley (1656–1742) of 'Halley's Comet' fame. Interested in many things besides astronomy, he also prepared in 1693 the first modern mortality table. Let us dwell a moment on the details of this work because of its great historical interest.

The subject does not quite start with Halley, however. In England, due presumably to increasing population densities, various plagues were rampant from the 16th century up to the adoption of public sanitation policies and facilities in the mid-19th century. In London, starting intermittently in 1591, and continuously from 1604 for several decades, there were published weekly Bills of Mortality, which listed for each parish the number of births and deaths of males and females and the statistics compiled by the *Searchers*, a body of 'ancient

Matrons' who carried out the unpleasant task of examining corpses, and, from the physical evidence and any other information they were able to elicit by inquiry, judged as best as they could the cause of each death.

In 1662, John Graunt (1620–74) called attention to the fact that these Bills, in their totality, contained valuable demographic information that could be useful to governments and scholars for many other purposes besides judging the current state of public health (Graunt, 1662).[6] He aggregated the data for 1632 into a single more useful table and made the observation that, in sufficiently large pools of data on births, there are always slightly more boys than girls, which circumstance provoked many speculations and calculations by probabilists for the next 150 years. Graunt was not a scholar, but a self-educated shopkeeper. Nevertheless, his short work contained so much valuable good sense that it came to the attention of Charles II, who as a reward ordered the Royal Society (which he had founded shortly before) to admit Graunt as a Fellow.[7]

Edmund Halley was highly educated, mathematically competent (later succeeding Wallis (in 1703) as Savilian Professor of Mathematics at Oxford University and Flamsteed (in 1720) as Astronomer Royal and Director of the Greenwich Observatory), a personal friend of Isaac Newton and the one who had persuaded him to publish his *Principia* by dropping his own work to see it through publication and paying for it out of his own modest fortune. He was eminently in a position to do more with demographic data than was John Graunt.

In undertaking to determine the actual distribution of age in the population, Halley had extensive data on births and deaths from London and Dublin. But records of the age at death were often missing, and he perceived that London and Dublin were growing rapidly by in-migration, biasing the data with people dying there who were not born there. Those data were so contaminated with trend that he had no means of extracting the information he needed. So he found instead five years' data (1687–91) for a city with a stable population: Breslau in Silesia (today called Wroclaw, in what is now Poland). Silesians, more meticulous in record keeping and less inclined to migrate, generated better data for his purpose.

Of course, contemporary standards of nutrition, sanitation, and medical care in Breslau might differ from those in England. But in any event Halley produced a mortality table surely valid for Breslau and presumably not badly in error for England. We have converted it into a graph, with three emendations described below, and present it in Figure 9.1.

In the 17th century, even so learned a man as Halley did not have the habits of full, clear expression that we expect in scholarly works today. In reading his work we are exasperated

---

[6] It appears that this story may be repeated some 330 years later, in the recent realization that the records of credit card companies contain a wealth of economic data which have been sitting there unused for many years. For the largest such company (Citicorp), a record of 1% of the nation's retail sales comes into its computers every day. For predicting some economic trends and activity, this is far more detailed, reliable, and timely than the monthly government releases.

[7] Contrast this enlightened attitude and behavior with that of Oliver Cromwell shortly before, who, through his henchmen, did more wanton, malicious damage to Cambridge University than any other person in history. The writer lived for a year in the Second Court of St John's College, Cambridge, which Cromwell appropriated and put to use, not for scholarly pursuits, but as the stockade for holding his prisoners. Whatever one may think of the private escapades of Charles II, one must ask also: against what alternative do we judge him? Had the humorless fanatic Cromwell prevailed, there would have been no Royal Society, and no recognition for scholarly accomplishment in England; quite likely, the magnificent achievements of British science in the 19th century would not have happened. It is even problematical whether Cambridge and Oxford Universities would still exist today.

Table 9.1. *Halley's first table.*

Age	$d(y)/5$	Age	$d(y)/5$	Age	$d(y)/5$	Age	$d(y)/5$
0	348	28	8	⋮	10	90	1
⋮	198	⋮	7	63	12	91	1
7	11	35	7	⋮	9.5	98	0
8	11	36	8	70	14	99	0.5
9	6	⋮	9.5	71	9	100	3/5
⋮	5.5	42	8	72	11		
14	2	⋮	9	⋮	9.5		
⋮	3.5	45	7	77	6		
18	5	⋮	7	⋮	7		
⋮	6	49	10	81	3		
21	4.5	54	11	⋮	4		
⋮	6.5	55	9	84	2		
27	9	56	9	⋮	1		

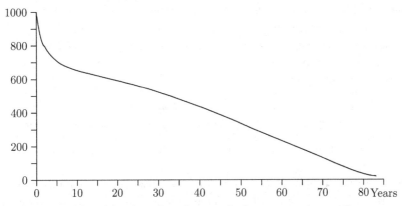

Fig. 9.1. $n(y)$: estimated number of persons in the age range $(y, y + 1)$ years.

at the ambiguities and omissions, which make it impossible to ascertain some important details about his data and procedure. We know that his data consisted of monthly records of the number of births and deaths and the age of each person at death. Unfortunately, he does not show us the original, unprocessed data, which would today be of far greater value to us than anything in his work, because, with modern probability theory and computers, we could easily process the data for ourselves, and extract much more information from them than Halley did.

Halley presents two tables derived from the data, giving respectively the estimated number $d(x)$ of annual deaths (total number/5) at each age of x years, Table 9.1 (but which inexplicably contains some entries that are not multiples of 1/5), and the estimated distribution $n(x)$ of population by age, Table 9.2. Thus, the first table is, crudely, something like

Table 9.2. *Halley's second table.*

Age	$n(y)$	Age	$n(y)$	Age	$n(y)$	Age	$n(y)$	Age	$n(y)$	Age	$n(y)$	Age	$n(y)$
1	1000	13	640	25	567	37	472	49	357	61	232	73	109
2	855	14	634	26	560	38	463	50	346	62	222	74	98
3	798	15	628	27	553	39	454	51	335	63	212	75	88
4	760	16	622	28	546	40	445	52	324	64	202	76	78
5	732	17	616	29	539	41	436	53	313	65	192	77	68
6	710	18	610	30	531	42	427	54	302	66	182	78	58
7	692	19	604	31	523	43	417	55	292	67	172	79	49
8	680	20	598	32	515	44	407	56	282	68	162	80	41
9	670	21	592	33	507	45	397	57	272	69	152	81	34
10	661	22	586	34	499	46	387	58	262	70	142	82	28
11	653	23	579	35	490	47	377	59	252	71	131	83	23
12	646	24	573	36	481	48	367	60	242	72	120	84	20

the negative derivative of the second. But, inexplicably, he omits the very young (< 7 yr) from the first table, and the very old (> 84 yr) from the second, thus withholding what are in many ways the most interesting parts, the regions of strong curvature of the graph.

Even so, if we knew the exact procedure by which Halley constructed the tables from the raw data, we might be able to reconstruct both tables in their entirety. But he gives absolutely no information about this, saying only,

From these Considerations I have formed the *adjoyned Table*, whose Uses are manifold, and give a more just *Idea* of the *State* and *Condition of Mankind*, than any thing yet extant that I know of.

But he fails to inform us what 'these Considerations' are, so we are reduced to conjecturing what he actually did.

Although we were unable to find any conjecture which is consistent with all the numerical values in Halley's tables, we can clarify things to some extent. Firstly, the actual number of deaths at each age in the first table naturally shows considerable 'statistical fluctuations' from one age to the next. Halley must have done some kind of smoothing of this, because the fluctuations do not show in the second table.

From other evidence in his article, we infer that he reasoned as follows. If the population distribution is stable (exactly the same next year as this year), then the difference $n(25) - n(26)$ between the number now alive at ages 25 and 26 must be equal to the number $d(25)$ now at age 25 who will die in the next year. Thus we would expect that the second table might be constructed by starting with the estimated number (1238) born each year as $n(0)$, and by recursion taking $n(x) = n(x - 1) - \overline{d}(x)$, where $\overline{d}(x)$ is the smoothed estimate of d. Finally, the total population of Breslau is estimated as $\sum_x n(x) = 34\,000$. But, although the later parts of Table 9.2 are well accounted for by this surmise, the early parts ($0 < x < 7$) do not fit it, and we have been unable to form even a conjecture about how he determined the first six entries of Table 9.2.

We have shifted the ages downward by one year in our graph because it appears that the common meanings of terms have changed in 300 years. Today, when we say colloquially that a boy is 'eight years old', we mean that his exact age x is in the range $(8 \leq x < 9)$; i.e., he is actually in his ninth year of life. But we can make sense out of Halley's numbers only if we assume that for him the phrase 'eight years current' meant in the eighth year of life; $7 < x \leq 8$. These points were noted also by Greenwood (1942), whose analysis confirms our conclusion about the meaning of 'age current'. However, our attempt to follow his reasoning beyond that point leaves us more confused than before. At this point we must give up, and simply accept Halley's judgment, whatever it was.

In Figure 9.1 we give Halley's second table as a graph of a shifted function $n(y)$. Thus, where Halley's table reads (25   567) we give it as $n(24) = 567$, which we interpret to mean an estimated 567 persons in the age range $(24 \leq x < 25)$. Thus, our $n(y)$ is what we believe to be Halley's estimated number of persons in the age range $(y, y + 1)$ years.

Thirdly, Halley's second table stops at the entry (84   20); yet the first table has data beyond that age, which he used in estimating the total population of Breslau. His first table indicates what we interpret as 19 deaths in the range $(85, 100)$ in the five years, including three at 'age current' 100. He estimated the total population in that age range as 107. We have converted this meager information, plus other comparisons of the two tables, into a smoothed extrapolation of Halley's second table (our entries $n(84), \ldots, n(99)$), which shows the necessary sharp curvature in the tail.

What strikes us first about this graph is the appalling infant mortality rate. Halley states elsewhere that only 56% of those born survived to the age of six (although this does not agree with his Table 9.2) and that 50% survive to age 17 (which does agree with the table). The second striking feature is the almost perfect linearity in the age range 35–80.

Halley notes various uses that can be made of his second table, including estimating the size of the army that the city could raise, and the values of annuities. Let us consider only one, the estimation of future life expectancy. We would think it reasonable to assign a probability that a person of age y will live to age z, as $p = n(z)/n(y)$, to sufficient accuracy.

Actually, Halley does not use the word 'probability' but instead refers to 'odds' in exactly the same way that we use it today: '. . . if the number of Persons of any *Age* remaining after one year, be divided by the difference between that and the number of the Age proposed, it shews the *odds* that there is, that a Person of that Age does not die in a *Year*.' Thus, Halley's odds on a person living m more years, given a present age of y years, is $O(m|y) = n(y + m)/[n(y) - n(y + m)] = p/(1 - p)$, in agreement with our calculation.

Another exasperating feature is that Halley pooled the data for males and females, and thus failed to exhibit their different mortality functions; lacking his raw data, we are unable to rectify this.

Let the things which exasperate us in Halley's work be a lesson for us today. The First Commandment of scientific data analysis publication ought to be: 'Thou shalt reveal thy full original data, unmutilated by any processing whatsoever.' Just as today we could do far more with Halley's raw data than he did, future readers may be able to do more with our raw data than we can, if only we will refrain from mutilating it according to our present purposes and prejudices. At the very least, they will approach our data with a different state of prior knowledge than ours, and we have seen how much this can affect the conclusions.

Exercise 9.3.   Suppose you had the same raw data as Halley. How would you process them today, taking full advantage of probability theory? How different would the actual conclusions be?

9.16 Comments

9.16.1 The irrationalists

Philosophers have argued over the nature of induction for centuries. Some, from David Hume (1711–76) in the mid-18th century to Karl Popper in the mid-20th (for example, Popper and Miller, 1983), have tried to deny the possibility of induction, although all scientific knowledge has been obtained by induction. D. Stove (1982) calls them and their colleagues 'the irrationalists' and tries to understand (1) how could such an absurd view ever have arisen?; and (2) by what linguistic practices do the irrationalists succeed in gaining an audience? However, we are not bothered by this situation because we are not convinced that much of an audience exists.

In denying the possibility of induction, Popper holds that theories can never attain a high probability. But this presupposes that the theory is being tested against an infinite number of alternatives. We would observe that the number of atoms in the known universe is finite; so also, therefore, is the amount of paper and ink available to write alternative theories. It is not the absolute status of an hypothesis embedded in the universe of all conceivable theories, but the plausibility of an hypothesis *relative to a definite set of specified alternatives*, that Bayesian inference determines.

As we showed in connection with multiple hypothesis testing in Chapter 4, Newton's theory in Chapter 5, and the above discussion of significance tests, an hypothesis can attain a very high or very low probability *within a class of well-defined alternatives*. Its probability within the class of all conceivable theories is neither large nor small; it is simply undefined because the class of all conceivable theories is undefined. In other words, Bayesian inference deals with determinate problems – not the undefined ones of Popper – and we would not have it otherwise.

The objection to induction is often stated in different terms. If a theory cannot attain a high absolute probability against all alternatives, then there is no way to prove that induction from it will be right. But that quite misses the point; it is not the function of induction to be 'right', and working scientists do not use it for that purpose (and could not if we wanted to). The functional use of induction in science is not to tell us what predictions must be true, but rather *what predictions are most strongly indicated by our present hypotheses and our present information*?

Put more carefully: What predictions are most strongly indicated by the information *that we have put into the calculation*? It is quite legitimate to do induction based on hypotheses that we do not believe, or even that we know to be false, to learn what their predictable consequences would be. Indeed, an experimenter seeking evidence for his favorite theory does not know what to look for unless he knows what predictions are made by some alternative theory. He must give temporary lip-service to the alternative in order to find out what it predicts, although he does not really believe it.

If predictions made by a theory are borne out by future observation, then we become more confident of the hypotheses that led to them; and if the predictions never fail in vast numbers of tests, we come eventually to call those hypotheses 'physical laws'. Successful induction is, of course, of great practical value in planning strategies for the future. But from successful induction we do not learn anything basically new; we only become more confident of what we knew already.

On the other hand, if the predictions prove to be wrong, then induction has served its real purpose; we have learned that our hypotheses are wrong or incomplete, and from the nature of the error we have a clue as to how they might be improved. So those who criticize induction on the grounds that it might not be right, could not possibly be more mistaken. As Harold Jeffreys explained long ago, induction is most valuable to a scientist just when it turns out to be wrong; only then do we get new fundamental knowledge.

Some striking case histories of induction in use are found in biology, where causal relations are often so complex and subtle that it is remarkable that it was possible to uncover them at all. For example, it became clear in the 20th century that new influenza pandemics were coming out of China; the worst ones acquired names like the Asian Flu (in 1957), the Hong Kong Flu (in 1968), and Beijing A (in 1993). It appears that the cause has been traced to the fact that Chinese farmers raise ducks and pigs side by side. Humans are not infected directly by viruses in ducks, even by handling them and eating them; but pigs can absorb duck viruses, transfer some of their genes to other viruses, and in this form pass them on to humans, where they take on a life of their own because they appear as something entirely new, for which the human immune system is unprepared.

An equally remarkable causal chain is in the role of the gooseberry as a host transmuting and transmitting the white pine blister rust disease. Many other examples of unraveling subtle cause–effect chains are found in the classic work of Louis Pasteur, and of modern medical researchers who continue to succeed in locating the specific genes responsible for various disorders.

We stress that all of these triumphant examples of highly important detective work were accomplished by qualitative plausible reasoning using the format defined by Pólya (1954). Modern Bayesian analysis is just the unique quantitative expression of this reasoning format, the inductive reasoning that Hume and Popper held to be impossible. It is true that this reasoning format does not guarantee that the conclusion *must* be correct; but then direct tests can confirm it or refute it. Without the preparatory inductive reasoning phase, one would not know which direct tests to try.

9.16.2 Superstitions

Another curious circumstance is that, although induction has proved a tricky thing to understand and justify logically, the human mind has a predilection for rampant, uncontrolled induction, and it requires much education to overcome this. As we noted briefly in Chapter 5, the reasoning of those without training in any mental discipline – who are therefore unfamiliar with either deductive logic or probability theory – is mostly unjustified induction.

In spite of modern science, general human comprehension of the world has progressed very little beyond the level of ancient superstitions. As we observe constantly in news commentaries and documentaries, the untrained mind never hesitates to interpret every observed correlation as a causal influence, and to predict its recurrence in the future. For one with no comprehension of what science is, it makes no difference whether that causation is or is not explainable rationally by a physical mechanism. Indeed, the very idea that a causal influence requires a physical mechanism to bring it about is quite foreign to the thinking of the uneducated; belief in supernatural influences makes such hypotheses, for them, unnecessary.[8]

Thus, the commentators for the very numerous television nature documentaries showing us the behavior of animals in the wild, never hesitate to see in every random mutation some teleological purpose; always, the environmental niche is there and the animal mutates, purposefully, in order to adapt to it. Every conformation of feather, beak, and claw is explained to us in terms of its *purpose*, but never suggesting how an unsubstantial purpose could bring about a physical change in the animal.[9]

It would seem that we have here a valuable opportunity to illustrate and explain evolution; yet the commentators (usually out-of-work actors) have no comprehension of the simple, easily understood cause-and-effect mechanism pointed out over 100 years ago by Charles Darwin. When we have the palpable evidence, and a simple explanation of it, before us, it is incredible that anybody could look to something supernatural, that nobody has ever observed, to explain it. But never does a commentator imagine that the mutation occurs first, and the resulting animal is obliged to seek a niche where it can survive and use its body structures as best it can in that environment. We see only the ones who were successful at this; the others are not around when the cameraman arrives, and their small numbers make it unlikely that a paleontologist will ever find evidence of them.[10] These documentaries always have very beautiful photography, and they deserve commentaries that make sense.

Indeed, there are powerful counter-examples to the theory that an animal adapts its body structure purposefully to its environment. In the Andes mountains there are woodpeckers where there are no trees. Evidently, they did not become woodpeckers by adapting their body

---

[8] In the meantime, progress in human knowledge continues to be made by those who, like modern biologists, do think in terms of physical mechanisms; as soon as that premise is abandoned, progress ceases, as we observe in modern quantum theory.

[9] But it is hard to believe that the ridiculous color patterns of the wood duck and the pileated woodpecker serve any survival purpose; what would the teleologists have to say about this? Our answer would be that, even without subsequent natural selection, divergent evolution can proceed by mutations that have nothing to do with survival. We noted some of this in Chapter 7, in connection with the work of Francis Galton.

[10] But a striking exception was found in the Burgess shale of the Canadian Rockies (Gould, 1989), in which beautifully preserved fossils of soft-bodied creatures contemporary with trilobites, which did not survive to leave any evolutionary lines, were found in such profusion that it radically revised our picture of life in the Cambrian.

structures to their environment; rather, they were woodpeckers first who, finding themselves through some accident in a strange environment, survived by putting their body structures to a different use. Indeed, the creatures arriving at any environmental niche are seldom perfectly adapted to it; often they are just barely well enough adapted to survive. But then, in this stressful situation, bad mutations are eliminated faster than usual, so natural selection operates faster than usual to make them better adapted.

10

Physics of 'random experiments'

I believe, for instance, that it would be very difficult to persuade an intelligent physicist that current statistical practice was sensible, but that there would be much less difficulty with an approach via likelihood and Bayes' theorem.

G. E. P. Box (1962)

As we have noted several times, the idea that probabilities are physically real things, based ultimately on observed frequencies of random variables, underlies most recent expositions of probability theory, which would seem to make it a branch of experimental science. At the end of Chapter 8 we saw some of the difficulties that this view leads us to; in some real physical experiments the distinction between random and nonrandom quantities is so obscure and artificial that you have to resort to black magic in order to force this distinction into the problem at all. But that discussion did not reach into the serious physics of the situation. In this chapter, we take time off for an interlude of physical considerations that show the fundamental difficulty with the notion of 'random' experiments.

10.1 An interesting correlation

There have always been dissenters from the 'frequentist' view who have maintained, with Laplace, that probability theory is properly regarded as the 'calculus of inductive reasoning', and is not fundamentally related to random experiments at all. A major purpose of the present work is to demonstrate that probability theory can deal, consistently and usefully, with far more than frequencies in random experiments. According to this view, consideration of random experiments is only one specialized *application* of probability theory, and not even the most important one; for probability theory as logic solves far more general problems of reasoning which have nothing to do with chance or randomness, but a great deal to do with the real world. In the present chapter we carry this further and show that 'frequentist' probability theory has major logical difficulties in dealing with the very random experiments for which it was invented.

One who studies the literature of these matters perceives that there is a strong correlation; those who have advocated the non-frequency view have tended to be physicists, while, up until very recently, mathematicians, statisticians, and philosophers almost invariably favored the frequentist view. Thus, it appears that the issue is not merely one of philosophy or mathematics; in some way not yet clear, it also involves physics.

The mathematician tends to think of a random experiment as an abstraction – really nothing more than a sequence of numbers. To define the 'nature' of the random experiment, he introduces statements – variously termed assumptions, postulates, or axioms – which specify the sample space and assert the existence, and certain other properties, of limiting frequencies. But, in the real world, a random experiment is not an abstraction whose properties can be defined at will. It is surely subject to the laws of physics; yet recognition of this is conspicuously missing from frequentist expositions of probability theory. Even the phrase 'laws of physics' is not to be found in them. However, defining a probability as a frequency is not merely an excuse for ignoring the laws of physics; it is more serious than that. We want to show that maintenance of a frequency interpretation to the exclusion of all others *requires* one to ignore virtually all the professional knowledge that scientists have about real phenomena. If the aim is to draw inferences about real phenomena, this is hardly the way to begin.

As soon as a specific random experiment is described, it is the nature of a physicist to start thinking, not about the abstract sample space thus defined, but about the physical mechanism of the phenomenon being observed. The question whether the usual postulates of probability theory are compatible with the known laws of physics is capable of logical analysis, with results that have a direct bearing on the question, not of the mathematical consistency of frequency and non-frequency theories of probability, but of their applicability in real situations. In our opening quotation, the statistician G. E. P. Box noted this; let us analyze his statement in the light both of history and of physics.

10.2 Historical background

As we know, probability theory started in consideration of gambling devices by Gerolamo Cardano in the 16th century, and by Pascal and Fermat in the 17th; but its development beyond that level, in the 18th and 19th centuries, was stimulated by applications in astronomy and physics, and was the work of people – James and Daniel Bernoulli, Laplace, Poisson, Legendre, Gauss, Boltzmann, Maxwell, Gibbs – most of whom we would describe today as mathematical physicists.

Reactions against Laplace had begun in the mid-19th century, when Cournot (1843), Ellis (1842, 1863), Boole (1854), and Venn (1866) – none of whom had any training in physics – were unable to comprehend Laplace's rationale and attacked what he did, simply ignoring all his successful results. In particular, John Venn, a philosopher without the tiniest fraction of Laplace's knowledge of either physics or mathematics, nevertheless considered himself competent to write scathing, sarcastic attacks on Laplace's work. In Chapter 16 we note his possible later influence on the young R. A. Fisher. Boole (1854, Chaps XX and XXI) shows repeatedly that he does not understand the function of Laplace's prior probabilities (to represent a *state of knowledge* rather than a physical fact). In other words, he too suffers from the mind projection fallacy. On p. 380 he rejects a uniform prior probability assignment as 'arbitrary', and explicitly *refuses* to examine its consequences; by which tactics he prevents himself from learning what Laplace was really doing and why.

Laplace was defended staunchly by the mathematician Augustus de Morgan (1838, 1847) and the physicist W. Stanley Jevons,[1] who understood Laplace's motivations and for whom his beautiful mathematics was a delight rather than a pain. Nevertheless, the attacks of Boole and Venn found a sympathetic hearing in England among non-physicists. Perhaps this was because biologists, whose training in physics and mathematics was for the most part not much better than Venn's, were trying to find empirical evidence for Darwin's theory and realized that it would be necessary to collect and analyze large masses of data in order to detect the small, slow trends that they visualized as the means by which evolution proceeds. Finding Laplace's mathematical works too much to digest, and since the profession of statistician did not yet exist, they would naturally welcome suggestions that they need not read Laplace after all.

In any event, a radical change took place at about the beginning of the 20th century when a new group of workers, not physicists, entered the field. They were concerned mostly with biological problems and with Venn's encouragement proceeded to reject virtually everything done by Laplace. To fill the vacuum, they sought to develop the field anew based on entirely different principles in which one assigned probabilities only to data and to nothing else. Indeed, this did simplify the mathematics at first, because many of the problems solvable by Laplace's methods now lay outside the gambit of their methods. As long as they considered only relatively simple problems (technically, problems with sufficient statistics but no nuisance parameters and no important prior information), the shortcoming was not troublesome. This extremely aggressive school soon dominated the field so completely that its methods have come to be known as 'orthodox' statistics, and the modern profession of statistician has evolved mostly out of this movement. Simultaneously with this development, the physicists – with Sir Harold Jeffreys as almost the sole exception – quietly retired from the field, and statistical analysis disappeared from the physics curriculum. This disappearance has been so complete that if today someone were to take a poll of physicists, we think that not one in 100 could identify such names as Fisher, Neyman or Wald, or such terms as maximum likelihood, confidence interval, analysis of variance.

This course of events – the leading role of physicists in development of the original Bayesian methods, and their later withdrawal from orthodox statistics – was no accident. As further evidence that there is some kind of basic conflict between orthodox statistical doctrine and physics, we may note that two of the most eloquent proponents of nonfrequency definitions in the early 20th century – Poincaré and Jeffreys – were mathematical physicists of the very highest competence, as was Laplace. Professor Box's statement thus has a clear basis in historical fact.

---

[1] Jevons did so many things that it is difficult to classify him by occupation. Zabell (1989), apparently guided by the title of one of his books (Jevons, 1874), describes Jevons as a logician and philosopher of science; from examination of his other works we are inclined to list him rather as a physicist who wrote extensively on economics.

But what is the nature of this conflict? What is there in the physicist's knowledge that leads him to reject the very thing that the others regard as conferring 'objectivity' on probability theory? To see where the difficulty lies, we examine a few simple random experiments from the physicist's viewpoint. The facts we want to point out are so elementary that one cannot believe they are really unknown to modern writers on probability theory. The continual appearance of new textbooks which ignore them merely illustrates what we physics teachers have always known: you can teach a student the laws of physics, but you cannot teach him the art of recognizing the *relevance* of this knowledge, much less the habit of actually *applying* it, in his everyday problems.

10.3 How to cheat at coin and die tossing

Cramér (1946) takes it as an axiom that 'any random variable has a unique probability distribution'. From the later context, it is clear that what he really means is that it has a unique *frequency* distribution. If one assumes that the number obtained by tossing a die is a random variable, this leads to the conclusion that the frequency with which a certain face comes up is a physical property of the die; just as much so as its mass, moment of inertia, or chemical composition. Thus, Cramér (1946, p. 154) states:

The numbers p_r should, in fact, be regarded as physical constants of the particular die that we are using, and the question as to their numerical values cannot be answered by the axioms of probability theory, any more than the size and the weight of the die are determined by the geometrical and mechanical axioms. However, experience shows that in a well-made die the frequency of any event r in a long series of throws usually approaches 1/6, and accordingly we shall often assume that all the p_r are equal to 1/6. . . .

To a physicist, this statement seems to show utter contempt for the known laws of mechanics. The results of tossing a die many times do *not* tell us any definite number characteristic only of the die. They tell us also something about how the die was tossed. If you toss 'loaded' dice in different ways, you can easily alter the relative frequencies of the faces. With only slightly more difficulty, you can still do this if your dice are perfectly 'honest'.

Although the principles will be just the same, it will be simpler to discuss a random experiment with only two possible outcomes per trial. Consider, therefore, a 'biased' coin, about which I. J. Good (1962) has remarked:

Most of us probably think about a biased coin as if it had a physical probability. Now whether it is defined in terms of frequency or just falls out of another type of theory, I think we do argue that way. I suspect that even the most extreme subjectivist such as de Finetti would have to agree that he did sometimes think that way, though he would perhaps avoid doing it in print.

We do not know de Finetti's private thoughts, but would observe that it is just the famous exchangeability theorem of de Finetti which shows us how to carry out a probability analysis of the biased coin *without* thinking in the manner suggested.

In any event, it is easy to show how a physicist would analyze the problem. Let us suppose that the center of gravity of this coin lies on its axis, but displaced a distance x from its geometrical center. If we agree that the result of tossing this coin is a 'random variable', then, according to the axiom stated by Cramér and hinted at by Good, there must exist a definite functional relationship between the frequency of heads and x:

$$p_H = f(x). \tag{10.1}$$

But this assertion goes far beyond the mathematician's traditional range of freedom to invent arbitrary axioms, and encroaches on the domain of physics; for the laws of mechanics are quite competent to tell us whether such a functional relationship does or does not exist.

The easiest game to analyze turns out to be just the one most often played to decide such practical matters as the starting side in a football game. Your opponent first calls 'heads' or 'tails' at will. You then toss the coin into the air, catch it in your hand, and, without looking at it, show it first to your opponent, who wins if he has called correctly. It is further agreed that a 'fair' toss is one in which the coin rises at least nine feet into the air, and thus spends at least 1.5 seconds in free flight.

The laws of mechanics now tell us the following. The ellipsoid of inertia of a thin disc is an oblate spheroid of eccentricity $1/\sqrt{2}$. The displacement x does not affect the symmetry of this ellipsoid, and, so according to the Poinsot construction, as found in textbooks on rigid dynamics (such as Routh, 1905, or Goldstein, 1980, Chap. 5), the polhodes remain circles concentric with the axis of the coin. In consequence, the character of the tumbling motion of the coin while in flight is exactly the same for a biased as an unbiased coin, except that for the biased one it is the center of gravity, rather than the geometrical center, which describes the parabolic 'free particle' trajectory.

An important feature of this tumbling motion is conservation of angular momentum; during its flight the angular momentum of the coin maintains a fixed direction in space (but the angular *velocity* does not; and so the tumbling may appear chaotic to the eye). Let us denote this fixed direction by the unit vector n; it can be any direction you choose, and it is determined by the particular kind of twist you give the coin at the instant of launching. Whether the coin is biased or not, it will show the same face throughout the motion if viewed from this direction (unless, of course, n is exactly perpendicular to the axis of the coin, in which case it shows no face at all).

Therefore, in order to know which face will be uppermost in your hand, you have only to carry out the following procedure. Denote by k a unit vector passing through the coin along its axis, with its point on the 'heads' side. Now toss the coin with a twist so that k and n make an acute angle, then catch it with your palm held flat, in a plane normal to n. On successive tosses, you can let the direction of n, the magnitude of the angular momentum, and the angle between n and k, vary widely; the tumbling motion will then appear entirely different to the eye on different tosses, and it would require almost superhuman powers of observation to discover your strategy.

Thus, anyone familiar with the law of conservation of angular momentum can, after some practice, cheat at the usual coin-toss game and call his shots with 100% accuracy. You can obtain any frequency of heads you want – *and the bias of the coin has no influence at all on the results*!

Of course, as soon as this secret is out, someone will object that the experiment analyzed is too 'simple'. In other words, those who have postulated a physical probability for the biased coin have, without stating so, really had in mind a more complicated experiment in which some kind of 'randomness' has more opportunity to make itself felt.

While accepting this criticism, we cannot suppress the obvious comment: scanning the literature of probability theory, isn't it curious that so many mathematicians, usually far more careful than physicists to list all the qualifications needed to make a statement correct, should have failed to see the need for any qualifications here? However, to be more constructive, we can just as well analyze a more complicated experiment.

Suppose that now, instead of catching the coin in our hands, we toss it onto a table, and let it spin and bounce in various ways until it comes to rest. Is this experiment sufficiently 'random' so that the true 'physical probability' will manifest itself? No doubt, the answer will be that it is not sufficiently random if the coin is merely tossed up six inches starting at the table level, but it will become a 'fair' experiment if we toss it up higher.

Exactly how high, then, must we toss it before the true physical probability can be measured? This is not an easy question to answer, and we make no attempt to answer it here. It would appear, however, that anyone who asserts the existence of a physical probability for the coin ought to be prepared to answer it; otherwise, it is hard to see what content the assertion has (that is, the assertion has the nature of theology rather than science; there is no way to confirm it or disprove it).

We do not deny that the bias of the coin will now have some influence on the frequency of heads; we claim only that the amount of that influence depends very much on how you toss the coin so that, again in this experiment, there is no definite number $p_H = f(x)$ describing a physical property of the coin. Indeed, even the direction of this influence can be reversed by different methods of tossing, as follows.

However high we toss the coin, we still have the law of conservation of angular momentum; and so we can toss it by *method A*: to ensure that heads will be uppermost when the coin first strikes the table, we have only to hold it heads up, and toss it so that the total angular momentum is directed vertically. Again, we can vary the magnitude of the angular momentum, and the angle between n and k, so that the motion appears quite different to the eye on different tosses, and it would require very close observation to notice that heads remains uppermost throughout the free flight. Although what happens after the coin strikes the table is complicated, the fact that heads is uppermost at first has a strong influence on the result, which is more pronounced for large angular momentum.

Many people have developed the knack of tossing a coin by *method B*: it goes through a phase of standing on edge and spinning rapidly about a vertical axis, before finally falling to one side or the other. If you toss the coin this way, the eccentric position of the center of gravity will have a dominating influence, and render it practically certain that it will fall always showing the same face. Ordinarily, one would suppose that the coin prefers to fall in the position which gives it the lowest center of gravity; i.e., if the center of gravity is displaced toward tails, then the coin should have a tendency to show heads. However, for an interesting mechanical reason, which we leave for you to work out from the principles of rigid dynamics, method *B* produces the opposite influence, the coin strongly preferring to fall so that its center of gravity is high.

On the other hand, the bias of the coin has a rather small influence in the opposite direction if we toss it by *method C*: the coin rotates about a horizontal axis which is perpendicular to the axis of the coin, and so bounces until it can no longer turn over.

In this experiment also, a person familiar with the laws of mechanics can toss a biased coin so that it will produce predominantly either heads or tails, at will. Furthermore, the effect of method *A* persists whether the coin is biased or not; and so one can even do this with a perfectly 'honest' coin. Finally, although we have been considering only coins, essentially the same mechanical considerations (with more complicated details) apply to the tossing of any other object, such as a die.

The writer has never thought of a biased coin 'as if it had a physical probability' because, being a professional physicist, I know that it does *not* have a physical probability. From the fact that we have seen a strong preponderance of heads, we cannot conclude legitimately that the coin is biased; it may be biased, or it may have been tossed in a way that systematically favors heads. Likewise, from the fact that we have seen equal numbers of heads and tails, we cannot conclude legitimately that the coin is 'honest'. It may be honest, or it may have been tossed in a way that nullifies the effect of its bias.

10.3.1 Experimental evidence

Since the conclusions just stated are in direct contradiction to what is postulated, almost universally, in expositions of probability theory, it is worth noting that we can verify them easily in a few minutes of experimentation in a kitchen. An excellent 'biased coin' is provided by the metal lid of a small pickle jar, of the type which is not knurled on the outside, and has the edge rolled inward rather than outward, so that the outside surface is accurately round and smooth, and so symmetrical that on an edge view one cannot tell which is the top side. Suspecting that many people not trained in physics simply would not believe the things just claimed without experimental proof, we have performed these experiments with a jar lid of diameter $d = 2\frac{5}{8}$ inches, height $h = \frac{3}{8}$ inch. Assuming a uniform thickness for the metal, the center of gravity should be displaced from the geometrical center by a distance $x = dh/(2d + 8h) = 0.120$ inches; and this was confirmed by hanging the lid by its edge and measuring the angle at which it comes to rest. Ordinarily, one expects this bias to make the lid prefer to fall bottom side (i.e., the inside) up; and so this side will be called 'heads'. The lid was tossed up about 6 feet, and fell onto a smooth linoleum floor. I allowed myself ten practice tosses by each of the three methods described, and then recorded the results of a number of tosses by: method *A* deliberately favoring heads, method *A* deliberately favoring tails, method *B*, and method *C*, as given in Table 10.1.

In method *A* the mode of tossing completely dominated the result (the effect of bias would, presumably, have been greater if the 'coin' were tossed onto a surface with a greater

Table 10.1. *Results of tossing a 'biased coin' in four different ways.*

Method	Number of tosses	Number of heads
A(H)	100	99
A(T)	50	0
B	100	0
C	100	54

coefficient of friction). In method B, the bias completely dominated the result (in about 30 of these tosses it looked for a while as if the result were going to be heads, as one might naively expect; but each time the 'coin' eventually righted itself and turned over, as predicted by the laws of rigid dynamics). In method C, there was no significant evidence for any effect of bias. The conclusions are pretty clear.

A holdout can always claim that tossing the coin in any of the four specific ways described is 'cheating', and that there exists a 'fair' way of tossing it, such that the 'true' physical probabilities of the coin will emerge from the experiment. But again, the person who asserts this should be prepared to define precisely what this fair method is, otherwise the assertion is without content. Presumably, a fair method of tossing ought to be some kind of random mixture of methods $A(H)$, $A(T)$, B, C, and others; but what is a 'fair' relative weighting to give them? It is difficult to see how one could define a 'fair' method of tossing except by the condition that it should result in a certain frequency of heads; and so we are involved in a circular argument.

This analysis can be carried much further, as we shall do below; but perhaps it is sufficiently clear already that analysis of coin and die tossing is not a problem of abstract statistics, in which one is free to introduce postulates about 'physical probabilities' which ignore the laws of physics. It is a problem of mechanics, highly complicated and irrelevant to probability theory except insofar as it forces us to think a little more carefully about how probability theory must be formulated if it is to be applicable to real situations. Performing a random experiment with a coin does not tell us what the physical probability for heads is; it may tell us something about the bias, but it also tells us something about how the coin is being tossed. Indeed, unless we know how it is being tossed, we cannot draw any reliable inferences about its bias from the experiment.

It may not, however, be clear from the above that conclusions of this type hold quite generally for random experiments, and in no way depend on the particular mechanical properties of coins and dice. In order to illustrate this, consider an entirely different kind of random experiment, as a physicist views it.

10.4 Bridge hands

Elsewhere we quote Professor William Feller's pronouncements on the use of Bayes' theorem in quality control testing (Chapter 17), on Laplace's rule of succession (Chapter 18), and on Daniel Bernoulli's conception of the utility function for decision theory (Chapter 13). He does not fail us here either; in his interesting textbook (Feller, 1950), he writes:

The number of possible distributions of cards in bridge is almost 10^{30}. Usually, we agree to consider them as equally probable. For a check of this convention more than 10^{30} experiments would be required – a billion of billion of years if every living person played one game every second, day and night.

Here again, we have the view that bridge hands possess 'physical probabilities,' that the uniform probability assignment is a 'convention', and that the ultimate criterion for its correctness must be observed frequencies in a random experiment.

The thing which is wrong here is that none of us – not even Feller – would be willing to use this criterion with a real deck of cards. Because, if we know that the deck is an honest one, our common sense tells us something which carries more weight than 10^{30} random experiments do. We would, in fact, be willing to accept the result of the random experiment *only if it agreed with our preconceived notion that all distributions are equally likely.*

To many, this last statement will seem like pure blasphemy – it stands in violent contradiction to what we have all been taught is the correct attitude toward probability theory. Yet, in order to see why it is true, we have only to imagine that those 10^{30} experiments *had* been performed, and the uniform distribution was not forthcoming. If all distributions of cards have equal frequencies, then any combination of two specified cards will appear together in a given hand, on the average, once in $(52 \times 51)/(13 \times 12) = 17$ deals. But suppose that the combination (jack of hearts – seven of clubs) appeared together in each hand three times as often as this. Would we then accept it as an established fact that there is something about the particular combination (jack of hearts – seven of clubs) that makes it inherently more likely than others?

We would not. We would reject the experiment and say that the cards had not been properly shuffled. But once again we are involved in a circular argument, because there is no way to define a 'proper' method of shuffling except by the condition that it should produce all distributions with equal frequency!

Any attempt to find such a definition involves one in even deeper logical difficulties; one dare not describe the procedure of shuffling in exact detail because that would destroy the 'randomness' and make the exact outcome predictable and always the same. In order to keep the experiment 'random', one must describe the procedure incompletely, so that the outcome will be different on different runs. But how could one prove that an incompletely defined procedure will produce all distributions with equal frequency? It seems to us that the attempt to uphold Feller's postulate of physical probabilities for bridge hands leads one into an outright logical contradiction.

Conventional teaching holds that probability assignments must be based fundamentally on frequencies; and that any other basis is at best suspect, at worst irrational with disastrous consequences. On the contrary, this example shows very clearly that *there is a principle for determining probability assignments which has nothing to do with frequencies, yet is so compelling that it takes precedence over any amount of frequency data.* If present teaching does not admit the existence of this principle, it is only because our intuition has run so far ahead of logical analysis – just as it does in elementary geometry – that we have never taken the trouble to present that logical analysis in a mathematically respectable form. But if we learn how to do this, we may expect to find that the mathematical formulation can be applied to a much wider class of problems, where our intuition alone would hardly suffice.

In carrying out a probability analysis of bridge hands, are we really concerned with physical probabilities, or with inductive reasoning? To help answer this, consider the following scenario. The date is 1956, when the writer met Willy Feller and had a discussion with him about these matters. Suppose I had told him that I have dealt at bridge 1000 times, shuffling 'fairly' each time; and that in every case the seven of clubs was in my own hand. What would his reaction be? He would, I think, mentally visualize the number

$$\left(\frac{1}{4}\right)^{1000} = 10^{-602}, \tag{10.2}$$

and conclude instantly that I have not told the truth; and no amount of persuasion on my part would shake that judgment. But what accounts for the strength of his belief? Obviously, it cannot be justified if our assignment of equal probabilities to all distributions of cards (therefore probability 1/4 for the seven of clubs to be in the dealer's hand) is merely a 'convention', subject to change in the light of experimental evidence; he rejects my reported experimental evidence, just as we did above. Even more obviously, he is not making use of any knowledge about the outcome of an experiment involving 10^{30} bridge hands.

Then *what is the extra evidence he has*, which his common sense tells him carries more weight than do any number of random experiments, but whose help he refuses to acknowledge in writing textbooks? In order to maintain the claim that probability theory is an experimental science, based fundamentally not on logical inference but on frequency in a random experiment, it is necessary to suppress some of the information which is available. This suppressed information, however, is just what enables our inferences to approach the certainty of deductive reasoning in this example and many others.

The suppressed evidence is, of course, simply our recognition of the *symmetry* of the situation. The only difference between a seven and an eight is that there is a different number printed on the face of the card. Our common sense tells us that where a card goes in shuffling depends only on the mechanical forces that are applied to it; and not on which number is printed on its face. If we observe any systematic tendency for one card to appear in the dealer's hand, which persists on indefinite repetitions of the experiment, we can conclude from this only that there is some systematic tendency in the procedure of shuffling, which alone determines the outcome of the experiment.

Once again, therefore, performing the experiment tells you nothing about the 'physical probabilities' of different bridge hands. It tells you something about how the cards are being shuffled. But the full power of symmetry as cogent evidence has not yet been revealed in this argument; we return to it presently.

10.5 General random experiments

In the face of all the foregoing arguments, one can still take the following position (as a member of the audience did after one of the writer's lectures): 'You have shown only that coins, dice, and cards represent exceptional cases, where physical considerations obviate the usual probability postulates; i.e., they are not really 'random experiments'. But that is of no importance because these devices are used only for illustrative purposes; in the more dignified random experiments which merit the serious attention of the scientist, there *is* a physical probability.'

To answer this we note two points. Firstly, we reiterate that when anyone asserts the existence of a physical probability in any experiment, then the onus is on him to define the exact circumstances in which this physical probability can be measured; otherwise the assertion is without content.

This point needs to be stressed: those who assert the existence of physical probabilities do so in the belief that this establishes for their position an 'objectivity' that those who speak only of a 'state of knowledge' lack. Yet to assert as fact something which cannot be either proved or disproved by observation of facts is the opposite of objectivity; it is to assert something that one could not possibly know to be true. Such an assertion is not even entitled to be called a description of a 'state of knowledge'.

Secondly, note that any specific experiment for which the existence of a physical probability is asserted is subject to physical analysis like the ones just given, which will lead eventually to an understanding of its mechanism. But as soon as this understanding is reached, then this new experiment will also appear as an exceptional case like the above ones, where physical considerations obviate the usual postulates of physical probabilities.

For, as soon as we have understood the mechanism of any experiment E, then there is logically no room for any postulate that various outcomes possess physical probabilities; the question: 'What are the probabilities of various outcomes $(O_1, O_2, \ldots)$?' then reduces immediately to the question: 'What are the probabilities of the corresponding initial conditions $(I_1, I_2, \ldots)$ that lead to these outcomes?'

We might suppose that the possible initial conditions $\{I_k\}$ of experiment E themselves possess physical probabilities. But then we are considering an antecedent random experiment E', which produces conditions I_k as its possible outcomes: $I_k = O'_k$. We can analyze the physical mechanism of E' and, as soon as this is understood, the question will revert to: 'What are the probabilities of the various initial conditions I'_k for experiment E'?'

Evidently, we are involved in an infinite regress $\{E, E', E'', \ldots\}$; the attempt to introduce a physical probability will be frustrated at every level where our knowledge of physical law permits us to analyze the mechanism involved. The notion of 'physical probability' must retreat continually from one level to the next, as knowledge advances.

We are, therefore, in a situation very much like the 'warfare between science and theology' of earlier times. For several centuries, theologians with no factual knowledge of astronomy, physics, biology, and geology nevertheless considered themselves competent to make dogmatic factual assertions which encroached on the domains of those fields – assertions which they were later forced to retract one by one in the face of advancing knowledge.

Clearly, probability theory ought to be formulated in a way that avoids factual assertions properly belonging to other fields, and which will later need to be retracted (as is now the case for many assertions in the literature concerning coins, dice, and cards). It appears to us that the only formulation which accomplishes this, and at the same time has the analytical power to deal with the current problems of science, is the one which was seen and expounded on intuitive grounds by Laplace and Jeffreys. Its validity is a question of logic, and does not depend on any physical assumptions.

As we saw in Chapter 2, a major contribution to that logic was made by R. T. Cox (1946, 1961), who showed that those intuitive grounds can be replaced by theorems. We think it is no accident that Richard Cox was also a physicist (Professor of Physics and Dean of the Graduate School at Johns Hopkins University), to whom the things we have pointed out here would be evident from the start.

The Laplace–Jeffreys–Cox formulation of probability theory does not require us to take one reluctant step after another down that infinite regress; it recognizes that anything which – like the child's spook – continually recedes from the light of detailed inspection can exist only in our imagination. Those who believe most strongly in physical probabilities, like those who believe in astrology, never seem to ask what would constitute a controlled experiment capable of confirming or disproving their belief.

Indeed, the examples of coins and cards should persuade us that such controlled experiments are, in principle, impossible. Performing any of the so-called random experiments will not tell us what the 'physical probabilities' are, because *there is no such thing as a 'physical probability'*; we might as well ask for a square circle. The experiment tells us, in a very crude and incomplete way, something about how the initial conditions are varying from one repetition to another.

A much more efficient way of obtaining this information would be to observe the initial conditions directly. However, in many cases this is beyond our present abilities; as in determining the safety and effectiveness of a new medicine. Here the only fully satisfactory approach would be to analyze the detailed sequence of chemical reactions that follow the taking of this medicine, in persons of every conceivable state of health. Having this analysis, one could then predict, for each individual patient, exactly what the effect of the medicine will be.

Such an analysis being entirely out of the question at present, the only feasible way of obtaining information about the effectiveness of a medicine is to perform a 'random' experiment. No two patients are in exactly the same state of health; and the unknown variations in this factor constitute the variable initial conditions of the experiment, while the sample space comprises the set of distinguishable reactions to the medicine. Our use of probability theory in this case is a standard example of inductive reasoning which amounts to the following.

If the initial conditions of the experiment (i.e., the physiological conditions of the patients who come to us) continue in the future to vary over the same unknown range as they have in the past, then the relative frequency of cures will, in the future, approximate those which we have observed in the past. In the absence of positive evidence giving a reason why there should be some change in the future, *and* indicating in which direction this change should go, we have no grounds for predicting any change in either direction, and so can only suppose that things will continue in more or less the same way. As we observe the relative frequencies of cures and side-effects to remain stable over longer and longer times, we become more and more confident about this conclusion. But this is only inductive reasoning – there is no deductive proof that frequencies in the future will not be entirely different from those in the past.

Suppose now that the eating habits or some other aspect of the lifestyle of the population starts to change. Then, the state of health of the incoming patients will vary over a different range than before, and the frequency of cures for the same treatment may start to drift up or down. Conceivably, monitoring this frequency could be a useful indicator that the habits of the population are changing, and this in turn could lead to new policies in medical procedures and public health education.

At this point, we see that the logic invoked here is virtually identical with that of industrial quality control, discussed in Chapter 4. But looking at it in this greater generality makes us see the role of induction in science in a very different way than has been imagined by some philosophers.

10.6 Induction revisited

As we noted in Chapter 9, some philosophers have rejected induction on the grounds that there is no way to prove that it is 'right' (theories can never attain a high probability); but this misses the point. The function of induction is to tell us not which predictions are right, but which predictions are indicated by our present knowledge. If the predictions succeed, then we are pleased and become more confident of our present knowledge; but we have not learned much.

The real role of induction in science was pointed out clearly by Harold Jeffreys (1931, Chap. 1) over 60 years ago; yet, to the best of our knowledge, no mathematician or philosopher has ever taken the slightest note of what he had to say:

A common argument for induction is that induction has always worked in the past and therefore may be expected to hold in the future. It has been objected that this is itself an inductive argument and cannot be used in support of induction. What is hardly ever mentioned is that induction has often failed in the past and that progress in science is very largely the consequence of direct attention to instances where the inductive method has led to incorrect predictions.

Put more strongly, it is only when our inductive inferences are wrong that we learn new things about the real world. For a scientist, therefore, the quickest path to discovery is to examine those situations where it appears most likely that induction from our present knowledge will fail. But those inferences must be our *best* inferences, which make full use of all the knowledge we have. One can always make inductive inferences that are wrong in a useless way, merely by ignoring cogent information.

Indeed, that is just what Popper did. His trying to interpret probability itself as expressing physical causation not only cripples the applications of probability theory in the way we saw in Chapter 3 (it would prevent us from getting about half of all conditional probabilities right because they express logical connections rather than causal physical ones), it leads one to conjure up imaginary causes while ignoring what was already known about the real physical causes at work. This can reduce our inferences to the level of pre-scientific, uneducated superstition, even when we have good data.

Why do physicists see this more readily than others? Because, having created this knowledge of physical law, we have a vested interest in it and want to see it preserved and used. Frequency or propensity interpretations start by throwing away practically all the professional knowledge that we have labored for centuries to get. Those who have not comprehended this are in no position to discourse to us on the philosophy of science or the proper methods of inference.

10.7 But what about quantum theory?

Those who cling to a belief in the existence of 'physical probabilities' may react to the above arguments by pointing to quantum theory, in which physical probabilities appear to express the most fundamental laws of physics. Therefore let us explain why this is another case of circular reasoning. We need to understand that present quantum theory uses entirely different standards of logic than does the rest of science.

In biology or medicine, if we note that an effect E (for example, muscle contraction, phototropism, digestion of protein) does not occur unless a condition C (nerve impulse, light, pepsin) is present, it seems natural to infer that C is a necessary causative agent for E. Most of what is known in all fields of science has resulted from following up this kind of reasoning. But suppose that condition C does not always lead to effect E; what further inferences should a scientist draw? At this point, the reasoning formats of biology and quantum theory diverge sharply.

In the biological sciences, one takes it for granted that in addition to C there must be some other causative factor F, not yet identified. One searches for it, tracking down the assumed cause by a process of elimination of possibilities that is sometimes extremely tedious. But persistence pays off; over and over again, medically important and intellectually impressive success has been achieved, the conjectured unknown causative factor being finally identified as a definite chemical compound. Most enzymes, vitamins, viruses, and other biologically active substances owe their discovery to this reasoning process.

In quantum theory, one does not reason in this way. Consider, for example, the photoelectric effect (we shine light on a metal surface and find that electrons are ejected from it). The experimental fact is that the electrons do not appear unless light is present. So light must be a causative factor. But light does not always produce ejected electrons; even though the light from a unimode laser is present with absolutely steady amplitude, the electrons appear only at particular times that are not determined by any known parameters of the light. Why then do we not draw the obvious inference, that in addition to the light there must be a second causative factor, still unidentified, and the physicist's job is to search for it?

What is done in quantum theory today is just the opposite; when no cause is apparent one simply postulates that no cause exists – ergo, the laws of physics are indeterministic and can be expressed only in probability form. The central dogma is that the light determines not whether a photoelectron will appear, but only the probability that it will appear. The mathematical formalism of present quantum theory – incomplete in the same way that our present knowledge is incomplete – does not even provide the vocabulary in which one could ask a question about the real cause of an event.

Biologists have a mechanistic picture of the world because, being trained to believe in causes, they continue to use the full power of their brains to search for them – and so they find them. Quantum physicists have only probability laws because for two generations we have been indoctrinated not to believe in causes – and so we have stopped looking for them. Indeed, any attempt to search for the causes of microphenomena is met with scorn and a charge of professional incompetence and 'obsolete mechanistic materialism'. Therefore, to explain the indeterminacy in current quantum theory we need not suppose there is any indeterminacy in Nature; the mental attitude of quantum physicists is already sufficient to guarantee it.[2]

This point also needs to be stressed, because most people who have not studied quantum theory on the full technical level are incredulous when told that it does not concern itself with causes; and, indeed, it does not even recognize the notion of 'physical reality'. The currently taught interpretation of the mathematics is due to Niels Bohr, who directed the Institute for Theoretical Physics in Copenhagen; therefore it has come to be called 'The Copenhagen interpretation'.

As Bohr stressed repeatedly in his writings and lectures, present quantum theory can answer only questions of the form: 'If this experiment is performed, what are the possible results and their probabilities?' It cannot, as a matter of principle, answer any question of the form: 'What is really happening when . . .?' Again, the mathematical formalism of present quantum theory, like Orwellian *newspeak*, does not even provide the vocabulary in which one could ask such a question. These points have been explained in some detail by Jaynes (1986d, 1989, 1990a, 1992a).

We suggest, then, that those who try to justify the concept of 'physical probability' by pointing to quantum theory are entrapped in circular reasoning, not basically different from that noted above with coins and bridge hands. Probabilities in present quantum theory express the incompleteness of human knowledge just as truly as did those in classical statistical mechanics; only its origin is different.

---

[2] Here, there is a striking similarity to the position of the parapsychologists Soal and Bateman (1954), discussed in Chapter 5. They suggest that to seek a physical explanation of parapsychological phenomena is a regression to the quaint and reprehensible materialism of Thomas Huxley. Our impression is that by 1954 the views of Huxley in biology were in a position of complete triumph over vitalism, supernaturalism, or any other anti-materialistic teachings; for example, the long mysterious immune mechanism was at last understood, and the mechanism of DNA replication had just been discovered. In both cases the phenomena could be described in 'mechanistic' terms so simple and straightforward – templates, geometrical fit, etc. – that they would be understood immediately in a machine shop.

In classical statistical mechanics, probability distributions represented our ignorance of the true microscopic coordinates – ignorance that was avoidable in principle but unavoidable in practice, but which did not prevent us from predicting reproducible phenomena, just because those phenomena are independent of the microscopic details.

In current quantum theory, probabilities express our own ignorance due to our failure to search for the real causes of physical phenomena; and, worse, our failure even to think seriously about the problem. This ignorance may be unavoidable in practice, but in our present state of knowledge we do not know whether it is unavoidable in principle; the 'central dogma' simply asserts this, and draws the conclusion that belief in causes, and searching for them, is philosophically naïve. If everybody accepted this and abided by it, no further advances in understanding of physical law would ever be made; indeed, no such advance has been made since the 1927 Solvay Congress in which this mentality became solidified into physics.[3] But it seems to us that this attitude places a premium on stupidity; to lack the ingenuity to think of a rational physical explanation is to support the supernatural view.

To many people, these ideas are almost impossible to comprehend because they are so radically different from what we have all been taught from childhood. Therefore, let us show how just the same situation could have happened in coin tossing, had classical physicists used the same standards of logic that are now used in quantum theory.

10.8 Mechanics under the clouds

We are fortunate that the principles of Newtonian mechanics could be developed and verified to great accuracy by studying astronomical phenomena, where friction and turbulence do not complicate what we see. But suppose the Earth were, like Venus, enclosed perpetually in thick clouds. The very existence of an external universe would be unknown for a long time, and to develop the laws of mechanics we would be dependent on the observations we could make locally.

Since tossing of small objects is nearly the first activity of every child, it would be observed very early that they do not always fall with the same side up, and that all one's efforts to control the outcome are in vain. The natural hypothesis would be that it is the volition of the object tossed, not the volition of the tosser, that determines the outcome; indeed, that is the hypothesis that small children make when questioned about this.

Then it would be a major discovery, once coins had been fabricated, that they tend to show both sides about equally often; and the equality appears to get better as the number of tosses increases. The equality of heads and tails would be seen as a fundamental law of physics; symmetric objects have a symmetric volition in falling (as, indeed, Cramér and Feller seem to have thought).

---

[3] Of course, physicists continued discovering new particles and calculation techniques – just as an astronomer can discover a new planet and a new algorithm to calculate its orbit, without any advance in his basic understanding of celestial mechanics.

With this beginning, we could develop the mathematical theory of object tossing, discovering the binomial distribution, the absence of time correlations, the limit theorems, the combinatorial frequency laws for tossing of several coins at once, the extension to more complicated symmetric objects like dice, etc. All the experimental confirmations of the theory would consist of more and more tossing experiments, measuring the frequencies in more and more elaborate scenarios. From such experiments, nothing would ever be found that called into question the existence of that volition of the object tossed; they only enable one to confirm that volition and measure it more and more accurately.

Then, suppose that someone was so foolish as to suggest that the motion of a tossed object is determined, not by its own volition, but by laws like those of Newtonian mechanics, governed by its initial position and velocity. He would be met with scorn and derision; for in all the existing experiments there is not the slightest evidence for any such influence. The Establishment would proclaim that, since all the observable facts are accounted for by the volition theory, it is philosophically naïve and a sign of professional incompetence to assume or search for anything deeper. In this respect, the elementary physics textbooks would read just like our present quantum theory textbooks.

Indeed, anyone trying to test the mechanical theory would have no success; however carefully he tossed the coin (not knowing what we know) it would persist in showing head and tails about equally often. To find any evidence for a causal instead of a statistical theory would require control over the initial conditions of launching, orders of magnitude more precise than anyone can achieve by hand tossing. We would continue almost indefinitely, satisfied with laws of physical probability and denying the existence of causes for individual tosses external to the object tossed – just as quantum theory does today – because those probability laws account correctly for everything that we can observe reproducibly with the technology we are using.

After thousands of years of triumph of the statistical theory, someone finally makes a machine which tosses coins in absolutely still air, with very precise control of the exact initial conditions. Magically, the coin starts giving unequal numbers of heads and tails; the frequency of heads is being controlled partially by the machine. With development of more and more precise machines, one finally reaches a degree of control where the outcome of the toss can be predicted with 100% accuracy. Belief in 'physical probabilities' expressing a volition of the coin is recognized finally as an unfounded superstition. The existence of an underlying mechanical theory is proved beyond question; and the long success of the previous statistical theory is seen as due only to the lack of control over the initial conditions of the tossing.

Because of recent spectacular advances in the technology of experimentation, with increasingly detailed control over the initial states of individual atoms (see, for example, Rempe, Walter and Klein, 1987), we think that the stage is going to be set, before very many more years have passed, for the same thing to happen in quantum theory; a century from now the true causes of microphenomena will be known to every schoolboy and, to paraphrase Seneca, they will be incredulous that such clear truths could have escaped us throughout the 20th (and into the 21st) century.

10.9 More on coins and symmetry

Now we go into a more careful, detailed discussion of some of these points, alluding to technical matters that must be explained more fully elsewhere. The rest of this chapter is not for the casual reader; only the one who wants a deeper understanding than is conveyed by the above simple scenarios. But many of the attacks on Laplace arise from failure to comprehend the following points.

The problems in which intuition compels us most strongly to a uniform probability assignment are not the ones in which we merely apply a principle of 'equal distribution of ignorance'. Thus, to explain the assignment of equal probabilities to heads and tails on the grounds that we 'saw no reason why either face should be more likely than the other', fails utterly to do justice to the reasoning involved. The point is that we have not merely 'equal ignorance'. We also have *positive knowledge of the symmetry* of the problem; and introspection will show that when this positive knowledge is lacking, so also is our intuitive compulsion toward a uniform distribution. In order to find a respectable mathematical formulation, we therefore need to find first a more respectable verbal formulation. We suggest that the following verbalization does do justice to the reasoning, and shows us how to generalize the principle.

I perceive here two different problems: having formulated one definite problem – call it P_1 – involving the coin, the operation which interchanges heads and tails transforms the problem into a different one – call it P_2. If I have positive knowledge of the symmetry of the coin, then I know that all relevant dynamical or statistical considerations, however complicated, are exactly the same in the two problems. Whatever state of knowledge I had in P_1, I must therefore have exactly the same state of knowledge in P_2, except for the interchange of heads and tails. Thus, whatever probability I assign to heads in P_1, consistency demands that I assign the *same* probability to tails in P_2.

Now, it might be quite reasonable to assign probability 2/3 to heads, 1/3 to tails in P_1; whereupon, from symmetry, it must be 2/3 to tails, 1/3 to heads in P_2. This might be the case, for example, if P_1 specified that the coin is to be held between the fingers heads up, and dropped just one inch onto a table. Thus, symmetry of the coin by no means compels us to assign equal probabilities to heads and tails; the question necessarily involves the other conditions of the problem.

But now suppose the statement of the problem is changed in just one respect; we are no longer told whether the coin is held initially with heads up or tails up. In this case, our intuition suddenly takes over with a compelling force, and tells us that we *must* assign equal probabilities to heads and tails; and, in fact, we must do this *regardless of what frequencies have been observed in previous repetitions of the experiment*.

The great power of symmetry arguments lies just in the fact that they are not deterred by any amount of complication in the details. The conservation laws of physics arise in this way; thus, conservation of angular momentum for an arbitrarily complicated system of particles is a simple consequence of the fact that the Lagrangian is invariant under space rotations. In current theoretical physics, almost the only known exact results in atomic and nuclear structure are those which we can deduce by symmetry arguments, using the methods of group theory.

These methods could be of the highest importance in probability theory also, if orthodox ideology did not forbid their use. For example, they enable us, in many cases, to extend the principle of indifference to find consistent prior probability assignments in a continuous parameter space Θ, where its use has always been considered ambiguous. The basic point is that a consistent principle for assigning prior probabilities must have the property that it assigns equivalent priors to represent equivalent states of knowledge.

The prior distribution must therefore be invariant under the symmetry group of the problem; and so the prior can be specified arbitrarily only in the so-called 'fundamental domain' of the group (Wigner, 1959). This is a subspace $\Theta_0 \subset \Theta$ such that (1) applying two different group elements $g_i \neq g_j$ to Θ_0, the subspaces $\Theta_i \equiv g_i, \Theta_0, \Theta_j \equiv g_j, \Theta_0$ are disjoint; and (2) carrying out all group operations on Θ_0 just generates the full hypothesis space: $\cup_j \Theta_j = \Theta$.

For example, let points in a plane be defined by their polar coordinates (r, α). If the group is the four-element one generated by a 90° rotation of the plane, then any sector 90° wide, such as $(\beta \leq \alpha < \beta + \pi/2)$, is a fundamental domain. Specifying the prior in any such sector, symmetry under the group then determines the prior everywhere in the plane.

If the group contains a continuous symmetry operation, the dimensionality of the fundamental domain is less than that of the parameter space; and so the probability density need be specified only on a set of points of measure zero, whereupon it is determined everywhere. If the number of continuous symmetry operations is equal to the dimensionality of the space Θ, the fundamental domain reduces to a single point, and the prior probability distribution is then uniquely determined by symmetry alone, just as it is in the case of an honest coin. Later we shall formalize and generalize these symmetry arguments.

There is still an important constructive point to be made about the power of symmetry arguments in probability theory. To see it, let us go back for a closer look at the coin-tossing problem. The laws of mechanics determine the motion of the coin, as describing a certain trajectory in a 12-dimensional phase space (three coordinates (q_1, q_2, q_3) of its center of mass, three Eulerian angles (q_4, q_5, q_6) specifying its orientation, and six associated momenta $(p_1, \ldots, p_6)$). The difficulty of predicting the outcome of a toss arises from the fact that very small changes in the location of the initial phase point can change the final results.

Imagine the possible initial phase points to be labeled H or T, according to the final results. Contiguous points labeled H comprise a set which is presumably twisted about in the 12-dimensional phase space in a very complicated, convoluted way, parallel to and separated by similar T-sets.

Consider now a region R of phase space, which represents the accuracy with which a human hand can control the initial phase point. Because of limited skill, we can be sure only that the initial point is somewhere in R, which has a phase volume

$$\Gamma(R) = \int_R dq_1 \cdots dq_6 dp_1 \cdots dp_6. \tag{10.3}$$

If the region R contains both H and T domains, we cannot predict the result of the toss. But what probability should we assign to heads? If we assign equal probability to equal phase volumes in R, this is evidently the fraction $p_H \equiv \Gamma(H)/\Gamma(R)$ of phase volume of R that is occupied by H domains. This phase volume Γ is the 'invariant measure' of phase space. The cogency of invariant measures for probability theory will be explained later; for now we note that the measure Γ is invariant under a large group of 'canonical' coordinate transformations, and also under the time development, according to the equations of motion. This is Liouville's theorem, fundamental to statistical mechanics; the exposition of Gibbs (1902) devotes the first three chapters to discussion of it, before introducing probabilities.

Now, if we have positive knowledge that the coin is perfectly 'honest', then it is clear that the fraction $\Gamma(H)/\Gamma(R)$ is very nearly 1/2, and becomes more accurately so as the size of the individual H and T domains become smaller compared with R. Because, for example, if we are launching the coin in a region R where the coin makes 50 complete revolutions while falling, then a 1% change in the initial angular velocity will just interchange heads and tails by the time the coin reaches the floor. Other things being equal (all dynamical properties of the coin involve heads and tails in the same manner), this should just reverse the final result.

A change in the initial 'orbital' velocity of the coin, which results in a 1% change in the time of flight, should also do this (strictly speaking, these conclusions are only approximate, but we expect them to be highly accurate, and to become more so if the changes become less than 1%). Thus, if all other initial phase coordinates remain fixed, and we vary only the initial angular velocity $\dot\theta$ and upward velocity $\dot z$, the H and T domains will spread into thin ribbons, like the stripes on a zebra. From symmetry, the width of adjacent ribbons must be very nearly equal.

This same 'parallel ribbon' shape of the H and T domains presumably holds also in the full phase space.[4] This is quite reminiscent of Gibbs' illustration of fine-grained and coarse-grained probability densities, in terms of the stirring of colored ink in water. On a sufficiently fine scale, every phase region is either H or T; the probability for heads is either zero or unity. But on the scale of sizes of the 'macroscopic' region R corresponding to ordinary skills, the probability density is the coarse-grained one, which from symmetry must be very nearly 1/2 if we know that the coin is honest.

What if we don't consider all equal phase volumes within R as equally likely? Well, it doesn't really matter if the H and T domains are sufficiently small. 'Almost any' probability density which is a smooth, continuous function within R, will give nearly equal weight to the H and T domains, and we will still have very nearly 1/2 for the probability for heads. This is an example of a general phenomenon, discussed by Poincaré, that, in cases where small

---

[4] Actually, if the coin is tossed onto a perfectly flat and homogeneous level floor, and is not only perfectly symmetrical under the reflection operation that interchanges heads and tails, but also perfectly round, the probability for heads is independent of five of the 12 coordinates, so we have this intricate structure only in a seven-dimensional space. Let the reader for whom this is a startling statement think about it hard, to see why symmetry makes five coordinates irrelevant (they are the two horizontal coordinates of its center of mass, the direction of its horizontal component of momentum, the Eulerian angle for rotation about a vertical axis, and the Eulerian angle for rotation about the axis of the coin).

changes in initial conditions produce big changes in the final results, our final probability assignments will be, for all practical purposes, independent of the initial ones.

As soon as we know that the coin has perfect dynamical symmetry between heads and tails – i.e., its Lagrangian function

$$L(q_1, \ldots, p_6) = (\text{kinetic energy}) - (\text{potential energy}) \qquad (10.4)$$

is invariant under the symmetry operation that interchanges heads and tails – then we know an exact result. No matter where in phase space the initial region R is located, for every H domain there is a T domain of equal size and identical shape, in which heads and tails are interchanged. Then if R is large enough to include both, we shall persist in assigning probability 1/2 to heads.

Now suppose the coin is biased. The above argument is lost to us, and we expect that the phase volumes of H and T domains within R are no longer equal. In this case, the 'frequentist' tells us that there still exists a definite 'objective' frequency of heads, $p_H \neq 1/2$, which is a measurable physical property of the coin. Let us understand clearly what this implies. *To assert that the frequency of heads is a physical property only of the coin, is equivalent to asserting that the ratio $v(H)/v(R)$ is independent of the location of region R.* If this were true, it would be an utterly unprecedented new theorem of mechanics, with important implications for physics which extend far beyond coin tossing.

Of course, no such thing is true. From the three specific methods of tossing the coin discussed in Section 10.3 which correspond to widely different locations of the region R, it is clear that the frequency of heads will depend very much on how the coin is tossed. Method A uses a region of phase space where the individual H and T domains are large compared with R, so human skill is able to control the result. Method B uses a region where, for a biased coin, the T domain is very much larger than either R or the H domain. Only method C uses a region where the H and T domains are small compared with R, making the result unpredictable from knowledge of R.

It would be interesting to know how to calculate the ratio $v(H)/v(R)$ as a function of the location of R from the laws of mechanics; but it appears to be a very difficult problem. Note, for example, that the coin cannot come to rest until its initial potential and kinetic energy have been either transferred to some other object or dissipated into heat by frictional forces; so all the details of how that happens must be taken into account. Of course, it would be quite feasible to do controlled experiments which measure this ratio in various regions of phase space. But it seems that the only person who would have any use for this information is a professional gambler.

Clearly, our reason for assigning probability 1/2 to heads when the coin is honest is not based merely on observed frequencies. How many of us can cite a single experiment in which the frequency 1/2 was established under conditions we would accept as significant? Yet none of us hesitates a second in choosing the number 1/2. Our real reason is simply common sense recognition of the *symmetry* of the situation. *Prior information which does not consist of frequencies is of decisive importance in determining probability assignments even in this simplest of all random experiments.*

Those who adhere publicly to a strict frequency interpretation of probability jump to such conclusions privately just as quickly and automatically as anyone else; but in so doing they have violated their basic premise that (probability) $\equiv$ (frequency); and so in trying to justify this choice they must suppress any mention of symmetry, and fall back on remarks about assumed frequencies in random experiments which have, in fact, never been performed.[5]

Here is an example of what one loses by so doing. From the result of tossing a die, we cannot tell whether it is symmetrical or not. But if we know, from direct physical measurements, that the die *is* perfectly symmetrical and we accept the laws of mechanics as correct, then it is no longer plausible inference, but deductive reasoning, that tells us this: *any nonuniformity in the frequencies of different faces is proof of a corresponding nonuniformity in the method of tossing.* The qualitative nature of the conclusions we can draw from the random experiment depend on whether we do or do not know that the die is symmetrical.

This reasoning power of arguments based on symmetry has led to great advances in physics for 60 years; as noted, it is not very exaggerated to say that the only known exact results in mathematical physics are the ones that can be deduced by the methods of group theory from symmetry considerations. Although this power is obvious once noted, and it is used intuitively by every worker in probability theory, it has not been widely recognized as a legitimate formal tool in probability theory.[6]

We have just seen that, in the simplest of random experiments, any attempt to define a probability merely as a frequency involves us in the most obvious logical difficulties as soon as we analyze the mechanism of the experiment. In many situations where we can recognize an element of symmetry, our intuition readily takes over and suggests an answer; and of course it is the same answer that our basic desideratum – that equivalent states of knowledge should be represented by equivalent probability assignments – requires for consistency.

But situations in which we have positive knowledge of symmetry are rather special ones among all those faced by the scientist. How can we carry out consistent inductive reasoning in situations where we do not perceive any clear element of symmetry? This is an open-ended problem because there is no end to the variety of different special circumstances that might arise. As we shall see, the principle of maximum entropy gives a useful and versatile tool for many such problems. But in order to give a start toward understanding this, let's go way back to the beginning and consider the tossing of the coin still another time, in a different way.

10.10 Independence of tosses

'When I toss a coin, the probability for heads is one-half.' What do we mean by this statement? Over the past two centuries, millions of words have been written about this simple question. A recent exchange (Edwards, 1991) shows that it is still enveloped in total

---

[5] Or rather, whenever anyone has tried to perform such experiments under sufficiently controlled conditions to be significant, the expected equality of frequencies is *not* observed. The famous experiments of Weldon (E. S. Pearson, 1967; K. Pearson, 1980) and Wolf (Czuber, 1908) are discussed elsewhere in this book.

[6] Indeed, L. J. Savage (1962, p. 102) rejects symmetry arguments, thereby putting his system of 'personalistic' probability in the position of recognizing the need for prior probabilities, but refusing to admit any formal principles for assigning them.

confusion in the minds of some. But, by and large, the issue is between the following two interpretations:

(A) 'The available information gives me no reason to expect heads rather than tails, or vice versa – I am completely unable to predict which it will be.'
(B) 'If I toss the coin a very large number of times, in the long run heads will occur about half the time – in other words, the frequency of heads will approach 1/2.'

We belabor still another time, what we have already stressed many times before. Statement (A) does not describe any property of the coin, but only the robot's *state of knowledge* (or, if you prefer, of ignorance). Statement (B) is, at least by implication, asserting something about the coin. Thus, (B) is a very much stronger statement than (A). Note, however, that (A) does not in any way contradict (B); on the contrary, (A) could be a consequence of (B). For if our robot were told that this coin has in the past given heads and tails with equal frequency, this would give it no help at all in predicting the result of the next toss.

Why, then, has interpretation (A) been almost universally rejected by writers on probability and statistics for two generations? There are, we think, two reasons for this. In the first place, there is a widespread belief that if probability theory is to be of any use in applications, we must be able to interpret our calculations in the strong sense of (B). But this is simply untrue, as we have demonstrated throughout the preceding eight chapters. We have seen examples of almost all known applications of frequentist probability theory, and many useful problems outside the scope of frequentist probability theory, which are nevertheless solved readily by probability theory as logic.

Secondly, it is another widely held misconception that the mathematical rules of probability theory (the 'laws of large numbers') would lead to (B) as a consequence of (A), and this seems to be 'getting something for nothing'. For, the fact that I know nothing about the coin is clearly not enough to make the coin give heads and tails equally often!

This misconception arises because of a failure to distinguish between the following two statements:

(C) 'Heads and tails are equally likely on a single toss.'
(D) 'If the coin is tossed N times, each of the 2^N conceivable outcomes is equally likely.'

To see the difference between statements (C) and (D), consider a case where it is known that the coin is biased, but not whether the bias favors heads or tails. Then (C) is applicable but (D) is not. For, on this state of knowledge, as was noted already by Laplace, the sequences HH and TT are each somewhat more likely than HT or TH. More generally, our common sense tells us that any unknown influence which favors heads on one toss will likely favor heads on the other toss. Unless our robot has positive knowledge (symmetry of both the coin and the method of tossing) which definitely rules out *all* such possibilities, (D) is not a correct description of his true state of knowledge; it assumes too much.

Statement (D) implies (C), but says a great deal more. Statement (C) says only, 'I do not know enough about the situation to give me any help in predicting the result of any throw', while (D) seems to be saying, 'I know that the coin is honest, *and* that it is being tossed in a way which favors neither face over the other, *and* that the method of tossing and the wear of the coin give no tendency for the result of one toss to influence the result of another'. But probability theory is subtle; in Chapter 9 we met the poorly informed robot, who makes statement (D) without having any of that information.

Mathematically, the laws of large numbers require much more than (C) for their derivation. Indeed, if we agree that tossing a coin generates an exchangeable sequence (i.e., the probability that N tosses will yield heads at n specified trials depends only on N and n, not on the order of heads and tails), then application of the de Finetti theorem, as in Chapter 9, shows that the weak law of large numbers holds *only* when (D) can be justified. In this case, it is almost correct to say that the probability assigned to heads is equal to the frequency with which the coin gives heads; because, for any $\epsilon \to 0$, the probability that the observed frequency $f = (n/N)$ lies in the interval $(1/2 \pm \epsilon)$ tends to unity as $N \to \infty$. Let us describe this by saying that there exists a *strong connection* between probability and frequency. We analyze this more deeply in Chapter 18.

In most recent treatments of probability theory, the writer is concerned with situations where a strong connection between probability and frequency is taken for granted – indeed, this is usually considered essential to the very notion of probability. Nevertheless, the existence of such a strong connection is clearly only an ideal limiting case, unlikely to be realized in any real application. For this reason, the laws of large numbers and limit theorems of probability theory can be grossly misleading to a scientist or engineer who naïvely supposes them to be experimental facts, and tries to interpret them literally in his problems. Here are two simple examples.

(1) Suppose there is some random experiment in which you assign a probability p for some particular outcome A. It is important to estimate accurately the fraction f of times A will be true in the next million trials. If you try to use the laws of large numbers, it will tell you various things about f; for example, that it is quite likely to differ from p by less than one-tenth of 1%, and enormously unlikely to differ from p by more than 1%. But now imagine that, in the first 100 trials, the observed frequency of A turned out to be entirely different from p. Would this lead you to suspect that something was wrong, and would you revise your probability assignment for the 101st trial? If it would, then your state of knowledge is different from that required for the validity of the law of large numbers. You are not sure of the independence of different trials, and/or you are not sure of the correctness of the numerical value of p. Your prediction of f for one million trials is probably no more reliable than for 100.

(2) The common sense of a good experimental scientist tells him the same thing without any probability theory. Suppose someone is measuring the velocity of light. After making allowances for the known systematic errors, he could calculate a probability distribution for the various other errors, based on the noise level in his electronics, vibration amplitudes, etc. At this point, a naïve application of the law of large numbers might lead him to think that he can add three significant figures to his measurement merely by repeating it one million times and averaging the results. But, of course, what he would actually do is to repeat some unknown systematic error one million times. It is idle to repeat a physical measurement an enormous number of times in the hope that 'good statistics' will average out your errors, because we cannot know the full systematic error. This is the old 'Emperor of China' fallacy, discussed in Chapter 8.

Indeed, unless we know that all sources of systematic error – recognized or unrecognized – contribute less than about one-third the total error, we cannot be sure that the average of one million measurements is any more reliable than the average of ten. Our time is much better spent in designing a new experiment which will give a lower probable error *per trial*. As Poincaré put it, 'The physicist is persuaded that one good measurement is worth many bad ones'. In other words, the common sense of a scientist tells him that the probabilities he assigns to various errors do not have a strong connection with frequencies, and that methods of inference which presuppose such a connection could be disastrously misleading in his problems.

Then, in advanced applications, it will behoove us to consider: How are our final conclusions altered if we depart from the universal custom of orthodox statistics, and relax the assumption of strong connections? Harold Jeffreys showed a very easy way to answer this, as we shall see later. As common sense tells us it must be, the ultimate accuracy of our conclusions is then determined not by anything in the data or in the orthodox picture of things, but rather by our own state of knowledge about the systematic errors. Of course, the orthodoxian will protest that, 'We understand this perfectly well; and in our analysis we assume that systematic errors have been located and eliminated'. But he does not tell us how to do this, or what to do if – as is the case in virtually every real experiment – they are unknown and so cannot be eliminated. Then all the usual 'asymptotic' rules are qualitatively wrong, and only probability theory as logic can give defensible conclusions.

10.11 The arrogance of the uninformed

Now we come to a very subtle and important point, which has caused trouble from the start in the use of probability theory. Many of the objections to Laplace's viewpoint which you find in the literature can be traced to the author's failure to recognize it. Suppose we do not know whether a coin is honest, and we fail to notice that this state of ignorance allows the possibility of unknown influences which would tend to favor the same face on all tosses. We say, 'Well, I don't see any reason why any one of the 2^N outcomes in N tosses should be more likely than any other, so I'll assign uniform probabilities by the principle of indifference'.

We would be led to statement (D) and the resulting strong connection between probability and frequency. But this is absurd – in this state of uncertainty, we could not possibly make reliable predictions of the frequency of heads. Statement (D), which is supposed to represent a great deal of positive knowledge about the coin and the method of tossing, can also result from *failure* to make proper use of all the available information! In other applications of mathematics, if we fail to use all of the relevant data of a problem, the result will not be that we get an incorrect answer. The result will be that we are unable to get any answer at all. But probability theory cannot have any such built-in safety device, because, in principle, the theory must be able to operate no matter what our incomplete information might be. If we fail to include all of the relevant data, or to take into account all the possibilities allowed by the data and prior information, probability theory will still give us a definite answer; and that answer will be the correct conclusion from the information that we actually gave the robot. But that answer may be in violent contradiction to our common sense judgments which did take everything into account, if only crudely. *The onus is always on the user to make sure that all the information, which his common sense tells him is relevant to the problem, is actually incorporated into the equations, and that the full extent of his ignorance is also properly represented.* If you fail to do this, then you should not blame Bayes and Laplace for your nonsensical answers.

We shall see examples of this kind of misuse of probability theory later, in the various objections to the rule of succession. It may seem paradoxical that a more careful analysis of a problem may lead to less certainty in prediction of the frequency of heads. However, look at it this way. It is commonplace that in all kinds of questions the fool feels a certainty that is denied to the wise man. The semiliterate on the next bar stool will tell you with absolute, arrogant assurance just how to solve all the world's problems; while the scholar who has spent a lifetime studying their causes is not at all sure how to do this. Indeed, we have seen just this phenomenon in Chapter 9, in the scenario of the poorly informed robot, who arrogantly asserts all the limit theorems of frequentist probability theory out of its ignorance rather than its knowledge.

In almost any example of inference, a more careful study of the situation, uncovering new facts, can lead us to feel either more certain or less certain about our conclusions, depending on what we have learned. New facts may support our previous conclusions, or they may refute them; we saw some of the subtleties of this in Chapter 5. If our mathematical model of reasoning failed to reproduce this phenomenon, it could not be an adequate 'calculus of inductive reasoning'.

Part 2

Advanced applications

11

Discrete prior probabilities: the entropy principle

At this point, we return to the job of designing the robot. We have part of its brain designed, and we have seen how it would reason in a few simple problems of hypothesis testing and estimation. In every problem it has solved thus far, the results have either amounted to the same thing as, or were usually demonstrably superior to, those offered in the 'orthodox' statistical literature. But it is still not a very versatile reasoning machine, because it has only one means by which it can translate raw information into numerical values of probabilities, the principle of indifference (2.95). Consistency requires it to recognize the relevance of prior information, and so in almost every problem it is faced at the onset with the problem of assigning initial probabilities, whether they are called technically prior probabilities or sampling probabilities. It can use indifference for this if it can break the situation up into mutually exclusive, exhaustive possibilities in such a way that no one of them is preferred to any other by the evidence. But often there will be prior information that does not change the set of possibilities but does give a reason for preferring one possibility to another. What do we do in this case?

Orthodoxy evades this problem by simply ignoring prior information for fixed parameters, and maintaining the fiction that sampling probabilities are known frequencies. Yet, in some 40 years of active work in this field, the writer has never seen a real problem in which one actually has prior information about sampling frequencies! In practice, sampling probabilities are always assigned from some standard theoretical model (binomial distribution, etc.) which starts from the principle of indifference. If the robot is to rise above such false pretenses, we must give it more principles for assigning initial probabilities by logical analysis of the prior information. In this chapter and the following one we introduce two new principles of this kind, each of which has an unlimited range of useful applications. But the field is open-ended in all directions; we expect that more principles will be found in the future, leading to a still wider range of applications.

11.1 A new kind of prior information

Imagine a class of problems in which the robot's prior information consists of average values of certain things. Suppose, for example, that statistics were collected in a recent earthquake and that, out of 100 windows broken, there were 976 pieces found. But we are not given the numbers 100 and 976; we are told only that 'the average window is broken into $\overline{m} = 9.76$ pieces'. Given only that information, what is the probability that a window would be broken into exactly m pieces? There is nothing in the theory so far that will answer that question.

As another example, suppose we have a table covered with black cloth, and some dice, but, for reasons that will be clear in a minute, they are black dice with white spots. A die is tossed onto the black table. Above there is a camera. Every time the die is tossed, we take a snapshot. The camera will record only the white spots. Now we don't change the film in between, so we end up with a multiple exposure; uniform blackening of the film after we have done this a few thousand times. From the known density of dots and the number of tosses, we can infer the average number of spots which were on top, but not the frequencies with which various faces came up. Suppose that the average number of spots turned out to be 4.5 instead of 3.5. Given only this information (i.e., not making use of anything else that you or I might know about dice except that they have six faces), what estimates should the robot make of the frequencies with which n spots came up? Supposing that successive tosses form an exchangeable sequence as defined in Chapter 3, what probability should it assign to the nth face coming up on the next toss?

As a third example, suppose that we have a string of $N = 1000$ cars, bumper to bumper, and that they occupy the full length of $L = 3$ miles. As they drive onto a rather large ferry boat, the distance that it sinks into the water determines their total weight W. But the numbers N, L, W are withheld from us; we are told only their average length L/N and average weight W/N. We can look up statistics from the manufacturers, and find out how long the Volkswagen is, how heavy it is, how long a Cadillac is, and how heavy it is, and so on, for all the other brands. From knowledge only of the average length and the average weight of these cars, what can we then infer about the proportion of cars of each make that were in the cluster?

If we knew the numbers N, L, W, then this could be solved by direct application of Bayes' theorem; without that information, we could still introduce the unknowns N, L, W as nuisance parameters and use Bayes' theorem, eliminating them at the end. We shall give an example of this procedure in the nonconglomerability problem in Chapter 15. However, the Bayesian solution would not really address our problem; it only transfers it to the problem of assigning priors to N, L, W, leaving us back in essentially the same situation; how do we assign informative probabilities?

Now, it is not at all obvious how our robot should handle problems of this sort. Actually, we have defined two different problems; *estimating* a frequency distribution, and *assigning* a probability distribution. But in an exchangeable sequence these are almost identical mathematically. So let's think about how we would want the robot to behave in this situation. Of course, we want it to take into account fully all the information it has, of whatever kind. But we would not want it to jump to conclusions that are not warranted by the evidence it has. We have seen that a uniform probability assignment represents a state of mind completely noncommittal with regard to all possibilities; it favors no one over any other, and thus leaves the entire decision to the subsequent data which the robot may receive. The knowledge of average values does give the robot a reason for preferring some possibilities to others, but we would like it to assign a probability distribution which is as uniform as it can be while agreeing with the available information. The most conservative, noncommittal distribution is the one which is as 'spread-out' as possible. In particular, the robot must not ignore any possibility – it must not assign zero probability to any situation unless its information really rules out that situation.

This sounds very much like defining a variational problem; the information available defines constraints fixing some properties of the initial probability distribution, but not all of them. The ambiguity remaining is to be resolved by the policy of honesty; frankly acknowledging the full extent of its ignorance by taking into account all possibilities allowed by its knowledge.[1] To cast it into mathematical form, the aim of avoiding unwarranted conclusions leads us to ask whether there is some reasonable numerical measure of how uniform a probability distribution is, which the robot could maximize subject to constraints which represent its available information. Let's approach this in the way most problems are solved: the time-honored method of trial and error. We just have to invent some measures of uncertainty, and put them to the test to see what they give us.

One measure of how broad an initial distribution is would be its variance. Would it make sense if the robot were to assign probabilities so as to maximize the variance subject to its information? Consider the distribution of maximum variance for a given $\overline{m}$, if the conceivable values of m are essentially unlimited, as in the broken window problem. Then the maximum variance solution would be the one where the robot assigns a very large probability for no breakage at all, and an enormously small probability of a window to be broken into billions and billions of pieces. You can get an arbitrarily high variance this way, while keeping the average at 9.76. In the dice problem, the solution with maximum variance would be to assign all the probability to the one and the six, in such a way that $p_1 + 6p_6 = 4.5$, or $p_1 = 0.3$, $p_6 = 0.7$. So that, evidently, is not the way we would want our robot to behave; it would be jumping to wildly unjustified conclusions, since nothing in its information says that it is impossible to have spots two through five up.

11.2 Minimum $\sum p_i^2$

Another kind of measure of how spread out a probability distribution is, which has been used a great deal in statistics, is the sum of the squares of the probabilities assigned to each of the possibilities. The distribution which minimizes this expression, subject to constraints represented by average values, might be a reasonable way for our robot to behave. Let's see what sort of a solution this would lead to. We want to make

$$\sum_m p_m^2 \tag{11.1}$$

---

[1] This is really an ancient principle of wisdom, recognized clearly already in such sources as Herodotus and the Old Testament.

a minimum, subject to the constraints that the sum of all p_m shall be unity and the average over the distribution is $\overline{m}$. A formal solution is obtained at once from the variational problem

$$\delta\left[\sum_m p_m^2 - \lambda \sum_m mp_m - \mu \sum_m p_m\right] = \sum_m (2p_m - \lambda m - \mu)\delta p_m = 0, \qquad (11.2)$$

where λ and μ are Lagrange multipliers. So p_m will be a linear function of m: $2p_m - \lambda m - \mu = 0$. Then μ and λ are found from

$$\sum_m p_m = 1, \qquad (11.3)$$

and

$$\sum_m mp_m = \overline{m}, \qquad (11.4)$$

where $\overline{m}$ is the average value of m, given to us in the statement of the problem.

Suppose that m can take on only the values 1, 2, and 3. Then the formal solution is

$$p_1 = \frac{4}{3} - \frac{\overline{m}}{2}, \qquad p_2 = \frac{1}{3}, \qquad p_3 = \frac{\overline{m}}{2} - \frac{2}{3}. \qquad (11.5)$$

This would be at least usable for some values of $\overline{m}$. But, in principle, $\overline{m}$ could be anywhere in $1 \leq \overline{m} \leq 3$, and p_1 becomes negative when $\overline{m} > 8/3 = 2.667$, while p_3 becomes negative when $\overline{m} < 4/3 = 1.333$. The formal solution for minimum $\sum p_i^2$ lacks the property of non-negativity. We might try to patch this up in an *ad hoc* way by replacing the negative values by zero and adjusting the other probabilities to keep the constraint satisfied. But then the robot is using different principles of reasoning in different ranges of $\overline{m}$; and it is still assigning zero probability to situations that are not ruled out by its information. This performance is not acceptable; it is an improvement over maximum variance, but the robot is still behaving inconsistently and jumping to unwarranted conclusions. We have taken the trouble to examine this criterion because some writers have rejected the entropy solution given next and suggested on intuitive grounds, without examining the actual results, that minimum $\sum p_i^2$ would be a more reasonable criterion.

But the idea behind the variational approach still looks like a good one. There should be some consistent measure of the uniformity, or 'amount of uncertainty', of a probability distribution which we can maximize, subject to constraints, and which will have the property that forces the robot to be completely honest about what it knows, and in particular it does not permit the robot to draw any conclusions unless those conclusions are really justified by the evidence it has.

11.3 Entropy: Shannon's theorem

At this stage, we turn to the most quoted theorem in Shannon's work on information theory (Shannon, 1948). If there exists a consistent measure of the 'amount of uncertainty' represented by a probability distribution, there are certain conditions it will have to satisfy. We shall state them in a way which will remind you of the arguments we gave in Chapter 2; in fact, this is really a continuation of the basic development of probability theory.

(1) We assume that some numerical measure $H_n(p_1, \ldots, p_n)$ exists; i.e., that it is possible to set up some kind of association between 'amount of uncertainty' and real numbers.
(2) We assume a continuity property: H_n is a continuous function of the p_i. Otherwise, an arbitrarily small change in the probability distribution would lead to a big change in the amount of uncertainty.
(3) We require that this measure should correspond qualitatively to common sense in that, when there are many possibilities, we are more uncertain than when there are few. This condition takes the form that in the case that the p_i are all equal, the quantity

$$h(n) = H_n \left(\frac{1}{n}, \frac{1}{n}, \ldots, \frac{1}{n} \right) \tag{11.6}$$

is a monotonic increasing function of n. This establishes the 'sense of direction'.
(4) We require that the measure H_n be consistent in the same sense as before; i.e., if there is more than one way of working out its value, we must get the same answer for every possible way.

Previously, our conditions of consistency took the form of the functional equations (2.13) and (2.45). Now we have instead a hierarchy of functional equations relating the different H_n to each other. Suppose the robot perceives two alternatives, to which it assigns probabilities p_1 and $q \equiv 1 - p_1$. Then the 'amount of uncertainty' represented by this distribution is $H_2(p_1, q)$. But now the robot learns that the second alternative really consists of two possibilities, and it assigns probabilities p_2, p_3 to them, satisfying $p_2 + p_3 = q$. What is now the robot's full uncertainty $H_3(p_1, p_2, p_3)$ as to all three possibilities? Well, the process of choosing one of the three can be broken down into two steps. Firstly, decide whether the first possibility is or is not true; the uncertainty removed by this decision is the original $H_2(p_1, q)$. Then, with probability q, the robot encounters an additional uncertainty as to events 2, 3, leading to

$$H_3(p_1, p_2, p_3) = H_2(p_1, q) + q H_2 \left(\frac{p_2}{q}, \frac{p_3}{q} \right) \tag{11.7}$$

as the condition that we shall obtain the same net uncertainty for either method of calculation. In general, a function H_n can be broken down in many different ways, relating it to the lower order functions by a large number of equations like this.

Note that (11.7) says rather more than our previous functional equations did. It says not only that the H_n are consistent in the aforementioned sense, but also that they are to be additive. So this is really an additional assumption which we should have included in our list.

Exercise 11.1. It seems intuitively that the most general condition of consistency would be a functional equation which is satisfied by any monotonic increasing function of H_n. But this is ambiguous unless we say something about how the monotonic functions for different n are to be related; is it possible to invoke the same function for all n? Carry out some new research in this field by investigating this matter; try either to find a possible form of the new functional equations, or to explain why this cannot be done.

At any rate, the next step is perfectly straightforward mathematics; let's see the full proof of Shannon's theorem, now dropping the unnecessary subscript on H_n.

We find the most general form of the composition law (11.7) for the case that there are n mutually exclusive propositions $(A_1, \ldots, A_n)$, to which we assign probabilities $(p_1, \ldots, p_n)$. Instead of giving the probabilities for the $(A_1, \ldots, A_n)$ directly, we might group the first k of them together as the proposition $(A_1 + A_2 + \cdots + A_k)$ and assign probability $w_1 = (p_1 + \cdots + p_k)$; then the next m propositions are grouped into $(A_{k+1} + \cdots + A_{k+m})$, to which we assign probability $w_2 = (p_{k+1} + \cdots + p_{k+m})$, etc. The amount of uncertainty as to the composite propositions is $H(w_1, \ldots, w_r)$.

Next we give the conditional probabilities $(p_1/w_1, \ldots, p_k/w_1)$ for the propositions $(A_1, \ldots, A_k)$, given that the composite proposition $(A_1 + \cdots + A_k)$ is true. The additional uncertainty, encountered with probability w_1, is then $H(p_1/w_1, \ldots, p_k/w_k)$. Carrying this out for the composite propositions $(A_{k+1} + \cdots + A_{k+m})$, etc., we arrive ultimately at the same state of knowledge as if the $(p_1, \ldots, p_n)$ had been given directly; so consistency requires that these calculations yield the same ultimate uncertainty, no matter how the choices were broken down. Thus we have

$$
H(p_1, \ldots, p_n) = H(w_1, \ldots, w_r) + w_1 H\left(\frac{p_1}{w_1}, \ldots, \frac{p_k}{w_1}\right)
$$
$$
+ w_2 H\left(\frac{p_{k+1}}{w_2}, \ldots, \frac{p_{k+m}}{w_2}\right) + \cdots,
$$
(11.8)

which is the general form of the functional equation (11.7). For example,

$$
H\left(\frac{1}{2}, \frac{1}{3}, \frac{1}{6}\right) = H\left(\frac{1}{2}, \frac{1}{2}\right) + \frac{1}{2} H\left(\frac{2}{3}, \frac{1}{3}\right).
$$
(11.9)

Since $H(p_1, \ldots, p_n)$ is to be continuous, it will suffice to determine it for all rational values

$$
p_i = \frac{n_i}{\sum n_i}
$$
(11.10)

with n_i integers. But then (11.8) determines the function H already in terms of the quantities $h(n) \equiv H(1/n, 1/n, \ldots, 1/n)$ which measure the 'amount of uncertainty' for the case of n equally likely alternatives. For we can regard a choice of one of the alternatives $(A_1, \ldots, A_n)$

as the first step in the choice of one of

$$\sum_{i=1}^{n} n_i \qquad (11.11)$$

equally likely alternatives in the manner just described, the second step of which is also a choice between n_i equally likely alternatives. As an example, with $n = 3$, we might choose $n_1 = 3$, $n_2 = 4$, $n_3 = 2$. For this case the composition law (11.8) becomes

$$h(9) = H\left(\frac{3}{9}, \frac{4}{9}, \frac{2}{9}\right) + \frac{3}{9}h(3) + \frac{4}{9}h(4) + \frac{2}{9}h(2). \qquad (11.12)$$

For a general choice of the n_i, (11.8) reduces to

$$h\left(\sum n_i\right) = H(p_1, \ldots, p_n) + \sum_i p_i h(n_i). \qquad (11.13)$$

Now we can choose all $n_i = m$; whereupon (11.13) collapses to

$$h(mn) = h(m) + h(n). \qquad (11.14)$$

Evidently, this is solved by setting

$$h(n) = K \log(n), \qquad (11.15)$$

where K is a constant. But is this solution unique? If m, n were continuous variables, this would be easy to answer; differentiate with respect to m, set $m = 1$, and integrate the resulting differential equation with the initial condition $h(1) = 0$ evident from (11.14), and you have proved that (11.15) is the only solution. But in our case, (11.14) need hold only for integer values of m, n; and this elevates the problem from a trivial one of analysis to an interesting little exercise in number theory.

Firstly, note that (11.15) is no longer unique; in fact, (11.14) has an infinite number of solutions for integer m, n. Each positive integer N has a unique decomposition into prime factors; and so, by repeated application of (11.14), we can express $h(N)$ in the form $\sum_i m_i h(q_i)$, where q_i are the prime numbers and m_i are the non-negative integers. Thus we can specify $h(q_i)$ *arbitrarily* for the prime numbers q_i, whereupon (11.14) is just sufficient to determine $h(N)$ for all positive integers.

To get any unique solution for $h(n)$, we have to add our qualitative requirement that $h(n)$ be monotonic increasing in n. To show this, note first that (11.14) may be extended by induction:

$$h(nmr \cdots) = h(n) + h(m) + h(r) + \cdots, \qquad (11.16)$$

and setting the factors equal in the kth order extension gives

$$h(n^k) = kh(n). \qquad (11.17)$$

Now let t, s be any two integers not less than 2. Then, for arbitrarily large n, we can find an integer m such that

$$\frac{m}{n} \leq \frac{\log(t)}{\log(s)} < \frac{m+1}{n}, \qquad \text{or} \qquad s^m \leq t^n < s^{m+1}. \tag{11.18}$$

Since h is monotonic increasing, $h(s^m) \leq h(t^n) \leq h(s^{m+1})$; or, from (11.17),

$$mh(s) \leq nh(t) \leq (m+1)h(s), \tag{11.19}$$

which can be written as

$$\frac{m}{n} \leq \frac{h(t)}{h(s)} \leq \frac{m+1}{n}. \tag{11.20}$$

Comparing (11.18) and (11.20), we see that

$$\left| \frac{h(t)}{h(s)} - \frac{\log(t)}{\log(s)} \right| \leq \frac{1}{n}, \qquad \text{or} \qquad \left| \frac{h(t)}{\log(t)} - \frac{h(s)}{\log(s)} \right| \leq \epsilon, \tag{11.21}$$

where

$$\epsilon \equiv \frac{h(s)}{n \log(t)} \tag{11.22}$$

is arbitrarily small. Thus $h(t)/\log(t)$ must be a constant, and the uniqueness of (11.15) is proved.

Now, different choices of K in (11.15) amount to the same thing as taking logarithms to different bases; so if we leave the base arbitrary for the moment, we can just as well write $h(n) = \log(n)$. Substituting this into (11.13), we have Shannon's theorem: The only function $H(p_1, \ldots, p_n)$ satisfying the conditions we have imposed on a reasonable measure of 'amount of uncertainty' is

$$H(p_1, \ldots, p_n) = -\sum_{i=1}^{n} p_i \log(p_i). \tag{11.23}$$

Accepting this interpretation, it follows that the distribution $(p_1, \ldots, p_n)$ which maximizes (11.23), subject to constraints imposed by the available information, will represent the 'most honest' description of what the robot knows about the propositions $(A_1, \ldots, A_n)$. The only arbitrariness is that we have the option of taking the logarithm to any base we please, corresponding to a multiplicative constant in H. This, of course, has no effect on the values of $(p_1, \ldots, p_n)$ which maximize H.

As in Chapter 2, we note the logic of what has and has not been proved. We have shown that use of the measure (11.23) is a *necessary* condition for consistency; but, in accordance with Gödel's theorem, one cannot prove that it actually is consistent unless we move out into some as yet unknown region beyond that used in our proof. From the above argument, given originally in Jaynes (1957a) and leaning heavily on Shannon, we conjectured that any other choice of 'information measure' will lead to inconsistencies if carried far enough; and a direct proof of this was found subsequently by Shore and Johnson (1980) using an argument entirely independent of ours. Many years of use of the maximum entropy principle (variously abbreviated to PME, MEM, MENT, MAXENT by various writers) has not revealed any inconsistency; and of course we do not believe that one will ever be found.

The function H is called the *entropy*, or, better, the *information entropy* of the distribution $\{p_i\}$. This is an unfortunate terminology, which now seems impossible to correct. We must warn at the outset that the major occupational disease of this field is a persistent failure to distinguish between the *information entropy*, which is a property of any probability distribution, and the *experimental entropy* of thermodynamics, which is instead a property of a thermodynamic state as defined, for example by such observed quantities as pressure, volume, temperature, magnetization, of some physical system. They should never have been called by the same name; the experimental entropy makes no reference to any probability distribution, and the information entropy makes no reference to thermodynamics.[2] Many textbooks and research papers are flawed fatally by the author's failure to distinguish between these entirely different things, and in consequence proving nonsense theorems.

We have seen the mathematical expression $\sum p \log(p)$ appearing incidentally in several previous chapters, generally in connection with the multinomial distribution; now it has acquired a new meaning as a fundamental measure of how uniform a probability distribution is.

Exercise 11.2. Prove that any change in the direction of equalizing two probabilities will increase the information entropy. That is, if $p_i < p_j$, then the change $p_i \rightarrow p_i + \epsilon$, $p_j \rightarrow p_j - \epsilon$, where ϵ is infinitesimal and positive, will increase $H(p_1, \ldots, p_n)$ by an amount proportional to ϵ. Applying this repeatedly, it follows that the maximum attainable entropy is one for which all the differences $|p_i - p_j|$ are as small as possible. This shows also that information entropy is a *global* property, not a local one; a difference $|p_i - p_j|$ has just as great an effect on entropy whether $|i - j|$ is 1 or 1000.

Although the above demonstration appears satisfactory mathematically, it is not yet in completely satisfactory form conceptually. The functional equation (11.7) does not seem quite so intuitively compelling as our previous ones did. In this case, the trouble is probably that we have not yet learned how to verbalize the argument leading to (11.7) in a fully convincing manner. Perhaps this will inspire others to try their hand at improving the verbiage that we used just before writing (11.7). Then it is comforting to know that there are several other possible arguments, like the aforementioned one of Shore and Johnson, which also lead uniquely to the same conclusion (11.23). We note another of them.

11.4 The Wallis derivation

This resulted from a suggestion made to the writer in 1962 by Graham Wallis (although the argument we give differs slightly from his). We are given information I, which is to be used

---

[2] But in case the problem happens to be one of thermodynamics, there is a relation between them, which we shall find presently.

in assigning probabilities $\{p_1, \ldots, p_m\}$ to m different possibilities. We have a total amount of probability

$$\sum_{i=1}^{m} p_i = 1 \tag{11.24}$$

to allocate among them. Now, in judging the reasonableness of any particular allocation, we are limited to consideration of I and the rules of probability theory; to call upon any other evidence would be to admit that we had not used all the available information in the first place.

The problem can also be stated as follows. Choose some integer $n \gg m$, and imagine that we have n little 'quanta' of probability, each of magnitude $\delta = n^{-1}$, to distribute in any way we see fit. In order to ensure that we have a 'fair' allocation, in the sense that none of the m possibilities shall knowingly be given either more or fewer of these quanta than it 'deserves', in the light of the information I, we might proceed as follows.

Suppose we were to scatter these quanta at random among the m choices – you can make this a blindfolded penny-pitching game into m equal boxes if you like. If we simply toss these 'quanta' of probability at random, so that each box has an equal probability of getting them, nobody can claim that any box is being unfairly favored over any other. If we do this, and the first box receives exactly n_1 quanta, and the second n_2, etc., we will say that the random experiment has generated the probability assignment

$$p_i = n_i \delta = \frac{n_i}{n}, \quad i = 1, 2, \ldots, m. \tag{11.25}$$

The probability that this will happen is the multinomial distribution

$$m^{-n} \frac{n!}{n_1! \cdots n_m!}. \tag{11.26}$$

Now imagine that a blindfolded friend repeatedly scatters the n quanta at random among the m boxes. Each time he does this we examine the resulting probability assignment. If it happens to conform to the information I, we accept it; otherwise we reject it and tell him to try again. We continue until some probability assignment $\{p_1, \ldots, p_m\}$ is accepted.

What is the most likely probability distribution to result from this game? From (11.26) it is the one which maximizes

$$W = \frac{n!}{n_1! \cdots n_m!} \tag{11.27}$$

subject to whatever constraints are imposed by the information I. We can refine this procedure by choosing smaller quanta; i.e. large n. In this limit we have, by the Stirling approximation,

$$\log(n!) = n \log(n) - n + \sqrt{2\pi n} + \frac{1}{12n} + O\left(\frac{1}{n^2}\right), \tag{11.28}$$

where $O(1/n^2)$ denotes terms that tend to zero as $n \to \infty$, as $(1/n^2)$ or faster. Using this result, and writing $n_i = np_i$, we find easily that as $n \to \infty$, $n_i \to \infty$, in such a way that $n_i/n \to p_i = \text{const.}$,

$$\frac{1}{n} \log(W) \to -\sum_{i=1}^{m} p_i \log(p_i) = H(p_1, \ldots, p_m), \tag{11.29}$$

and, so, the *most likely* probability assignment to result from this game is just the one that has maximum entropy subject to the given information I.

You might object that this game is still not entirely 'fair', because we have stopped at the first acceptable result without seeing what other acceptable ones might also have turned up. In order to remove this objection, we can consider all possible acceptable distributions and choose the average $\overline{p_i}$ of them. But here the 'laws of large numbers' come to our rescue. We leave it as an exercise for the reader to prove that in the limit of large n, *the overwhelming majority of all acceptable probability allocations that can be produced in this game are arbitrarily close to the maximum entropy distribution.*[3]

From a conceptual standpoint, the Wallis derivation is quite attractive. It is entirely independent of Shannon's functional equations (11.8); it does not require any postulates about connections between probability and frequency; nor does it suppose that the different possibilities $\{1, \ldots, m\}$ are themselves the result of any repeatable random experiment. Furthermore, it leads automatically to the prescription that H is to be *maximized* – and not treated in some other way – without the need for any quasi-philosophical interpretation of H in terms of such a vague notion as 'amount of uncertainty'. Anyone who accepts the proposed game as a fair way to allocate probabilities that are not determined by the prior information is thereby led inexorably to the maximum entropy principle.

Let us stress this point. It is a big mistake to try to read too much philosophical significance into theorems which lead to (11.23). In particular, the association of the word 'information' with entropy expressions seems in retrospect quite unfortunate, because it persists in carrying the wrong connotations to so many people. Shannon himself, with prophetic insight into the reception his work would get, tried to play it down by pointing out immediately after stating the theorem that it was in no way necessary for the theory to follow. By this he meant that the inequalities which H satisfies are already quite sufficient to justify its use; it does not really need the further support of the theorem, which deduces it from functional equations expressing intuitively the properties of 'amount of uncertainty'.

However, while granting that this is perfectly true, we would like now to show that *if we do* accept the expression for entropy, very literally, as *the* correct expression for the 'amount of uncertainty' represented by a probability distribution, this will lead us to a much more unified picture of probability theory in general. It will enable us to see that the principle of indifference, and many frequency connections of probability, are special cases

---

[3] This result is formalized more completely in the entropy concentration theorem given later.

of a single principle, and that statistical mechanics, communication theory, and a mass of other applications are all instances of a single method of reasoning.

11.5 An example

Let's test this principle by seeing how it would work on the example discussed above, in which m can take on only the values 1, 2, 3, and $\overline{m}$ is given. We can use our Lagrange multiplier argument again to solve this problem; as in (11.2),

$$\delta\left[H - \lambda \sum_{m=1}^{3} mp_m - \mu \sum_{m=1}^{3} p_m\right] = \sum_{m=1}^{3}\left[\frac{\partial H}{\partial p_m} - \lambda m - \mu\right]\delta p_m = 0. \tag{11.30}$$

Now,

$$\frac{\partial H}{\partial p_m} = -\log(p_m) - 1, \tag{11.31}$$

so our solution is

$$p_m = \exp\left\{-\lambda_0 - \lambda m\right\}, \tag{11.32}$$

where $\lambda_0 \equiv \mu + 1$.

So the distribution which has maximum entropy, subject to a given average value, will be in exponential form, and we have to fit the constants λ_0 and λ by forcing this to agree with the constraints that the sum of the p's must be one and the expectation value must be equal to the average $\overline{m}$ that we assigned. This is accomplished quite neatly if we define a function

$$Z(\lambda) \equiv \sum_{m=1}^{3} \exp\{-\lambda m\}, \tag{11.33}$$

which we called the *partition function* in Chapter 9. The equations (11.3) and (11.4) which fix our Lagrange multipliers take the form

$$\lambda_0 = \log Z(\lambda), \tag{11.34}$$

$$\overline{m} = -\frac{\partial \log Z(\lambda)}{\partial \lambda}. \tag{11.35}$$

We find that $p_1(\overline{m})$, $p_2(\overline{m})$, $p_3(\overline{m})$ are given in parametric form by

$$p_k = \frac{\exp\{-k\lambda\}}{\exp\{-\lambda\} + \exp\{-2\lambda\} + \exp\{-3\lambda\}}$$

$$= \frac{\exp\{(3-k)\lambda\}}{\exp\{2\lambda\} + \exp\{\lambda\} + 1}, \quad k = 1, 2, 3; \tag{11.36}$$

$$\overline{m} = \frac{\exp\{2\lambda\} + 2\exp\{\lambda\} + 3}{\exp\{2\lambda\} + \exp\{\lambda\} + 1}. \tag{11.37}$$

In a more complicated problem, we would just have to leave it in parametric form, but in this particular case we can eliminate the parameter λ algebraically, leading to the explicit solution

$$p_1 = \frac{3 - \overline{m} - p_2}{2},$$

$$p_2 = \frac{1}{3}\left[\sqrt{4 - 3(\overline{m} - 2)^2} - 1\right], \tag{11.38}$$

$$p_3 = \frac{\overline{m} - 1 - p_2}{2}.$$

As a function of $\overline{m}$, p_2 is the arc of an ellipse which comes in with unit slope at the end points. p_1 and p_3 are also arcs of ellipses, but slanted one way and the other.

We have finally arrived here at a solution which meets the objections we had to the first two criteria. The maximum entropy distribution (11.36) has automatically the property $p_k \geq 0$ because the logarithm has a singularity at zero which we could never get past. It has, furthermore, the property that it never allows the robot to assign zero probability to any possibility unless the evidence forces that probability to be zero.[4] The only place where a probability goes to zero is in the limit where $\overline{m}$ is exactly one or exactly three. But of course, in those limits, some probabilities did have to be zero by deductive reasoning, whatever principle we invoked.

11.6 Generalization: a more rigorous proof

The maximum entropy solution can be generalized in many ways. Suppose a variable x can take on n different discrete values $(x_1, \ldots, x_n)$, which correspond to the n different propositions $(A_1, \ldots, A_n)$; and that there are m different functions of x,

$$f_k(x), \quad 1 \leq k \leq m < n, \tag{11.39}$$

and that we want them to have expectations

$$\langle f_k(x) \rangle = F_k, \quad 1 \leq k \leq m, \tag{11.40}$$

where the $\{F_k\}$ are numbers given to us in the statement of the problem. What probabilities $(p_1, \ldots, p_n)$ will the robot assign to the possibilities $(x_1, \ldots, x_n)$? We shall have

$$F_k = \langle f_k(x) \rangle = \sum_{i=1}^{n} p_i f_k(x_i), \tag{11.41}$$

and, to find the set of p_i's which has maximum entropy subject to all these constraints simultaneously, we introduce as many Lagrange multipliers as there are

---

[4] This property was stressed by David Blackwell, who considered it the most fundamental requirement of a rational procedure for assigning probabilities.

constraints:

$$0 = \delta\left[H - (\lambda_0 - 1)\sum_i p_i - \sum_{j=1}^{m} \lambda_j \sum_i p_i f_j(x_i)\right]$$

$$= \sum_i \left[\frac{\partial H}{\partial p_i} - (\lambda_0 - 1) - \sum_{j=1}^{m} \lambda_j f_j(x_i)\right]\delta p_i. \tag{11.42}$$

So from (11.23) our solution is the following:

$$p_i = \exp\left\{-\lambda_0 - \sum_{j=1}^{m} \lambda_j f_j(x_i)\right\}, \tag{11.43}$$

as always, exponential in the constraints. The sum of all probabilities has to be unity, so

$$1 = \sum_i p_i = \exp\{-\lambda_0\}\sum_i \exp\left\{-\sum_{j=1}^{m} \lambda_j f_j(x_i)\right\}. \tag{11.44}$$

If we now define the partition function

$$Z(\lambda_1 \cdots \lambda_m) \equiv \sum_{i=1}^{n} \exp\left\{-\sum_{j=1}^{m} \lambda_j f_j(x_i)\right\}, \tag{11.45}$$

then (11.44) reduces to

$$\lambda_0 = \log Z(\lambda_1, \ldots, \lambda_m). \tag{11.46}$$

The average value F_k must be equal to the expected value of $f_x(x)$ over the probability distribution

$$F_k = \exp\{-\lambda_0\}\sum_i f_k(x_i)\exp\left\{-\sum_{j=1}^{m} \lambda_j f_j(x_i)\right\}, \tag{11.47}$$

or

$$F_k = -\frac{\partial \log Z(\lambda_1, \ldots, \lambda_m)}{\partial \lambda_k}. \tag{11.48}$$

The maximum value of the entropy is

$$H_{\max} = \left[-\sum_{i=1}^{n} p_i \log(p_i)\right]_{\max}, \tag{11.49}$$

and from (11.43) we find that

$$H_{\max} = \lambda_0 + \sum_{j=1}^{m} \lambda_j F_j. \tag{11.50}$$

Now, these results open up so many new applications that it is important to have as rigorous a proof as possible. But to solve a maximization problem by variational means, as we just did, is not 100% rigorous. Our Lagrange multiplier argument has the nice feature that it gives us the answer instantaneously. It has the bad feature that after we done it, we're not quite sure it *is* the answer. Suppose we wanted to locate the maximum of a function whose absolute maximum happened to occur at a cusp (discontinuity of slope) instead at a rounded top. Variational methods will locate some subsidiary rounded maxima, but they will not find the cusp. Even after we've proved that we have the highest value that can be reached by variational methods, it is possible that the function reaches a still higher value at some cusp that we can't locate by variational methods. There would always be a little grain of doubt remaining if we do only the variational problem.

So now we give an entirely different derivation which is strong just where the variational argument is weak. For this we need a lemma. Let p_i be any set of numbers which could be a possible probability distribution; in other words,

$$\sum_{i=1}^{n} p_i = 1, \qquad p_i \geq 0, \tag{11.51}$$

and let u_i be another possible probability distribution,

$$\sum_{i=1}^{n} u_i = 1, \qquad u_i \geq 0. \tag{11.52}$$

Now,

$$\log(x) \leq (x-1), \qquad 0 \leq x < \infty, \tag{11.53}$$

with equality if and only if $x = 1$. Therefore,

$$\sum_{i=1}^{n} p_i \log \left(\frac{u_i}{p_i} \right) \leq \sum_{i=1}^{n} p_i \left(\frac{u_i}{p_i} - 1 \right) = 0, \tag{11.54}$$

or

$$H(p_1, \ldots, p_n) \leq \sum_{i=1}^{n} p_i \log \left(\frac{1}{u_i} \right), \tag{11.55}$$

with equality if and only if $p_i = u_i$, $i = 1, \ldots, n$. This is the lemma we need.

Now we simply pull a distribution u_i out of the hat;

$$u_i \equiv \frac{1}{Z(\lambda_1, \ldots, \lambda_m)} \exp \left\{ -\sum_{j=1}^{m} \lambda_j f_j(x_i) \right\}, \tag{11.56}$$

where $Z(\lambda_1, \ldots, \lambda_m)$ is defined by (11.45). Never mind why we chose u_i this particular way; we'll see why in a minute. We can now write the inequality (11.55) as

$$H \leq \sum_{i=1}^{n} p_i \left[\log Z(\lambda_1, \ldots, \lambda_m) + \sum_{j=1}^{m} \lambda_j f_j(x_i) \right] \tag{11.57}$$

or

$$H \leq \log Z(\lambda_1, \ldots, \lambda_m) + \sum_{j=1}^{m} \lambda_j \langle f_j(x) \rangle. \tag{11.58}$$

Now let the p_i vary over the class of all possible probability distributions that satisfy the constraints (11.41). The right-hand side of (11.58) stays constant. Our lemma now says that H attains its absolute maximum $H_{\max}$, making (11.58) an equality, if and only if the p_i are chosen as the canonical distribution (11.56).

This is the rigorous proof, which is independent of the things that might happen if we try to do it as a variational problem. This argument is, as we see, strong just where the variational argument is weak. On the other hand, this argument is weak where the variational argument is strong, because we just had to pull the answer out of a hat in writing (11.56). We had to know the answer before we could prove it. If you have both arguments side by side, then you have the whole story.

11.7 Formal properties of maximum entropy distributions

Now we want to list the formal properties of the canonical distribution (11.56). This is a bad way to proceed in one sense because it all sounds very abstract and we don't see the connections to real problems. On the other hand, we get all the things we need a lot faster if we first become aware of all the formal properties that are in the theory; and then later go into specific physical problems and see that every one of these formal relations has many different useful meanings, depending on the particular problem.

The maximum attainable H that we can obtain by holding these averages fixed depends, of course, on the average values we specified,

$$H_{\max} = S(F_1, \ldots, F_m) = \log Z(\lambda_1, \ldots, \lambda_m) + \sum_{k=1}^{m} \lambda_k F_k. \tag{11.59}$$

We can regard H as a measure of the 'amount of the uncertainty' in any probability distribution. After we have maximized it, it becomes a function of the definite data of the problem $\{F_i\}$, and we'll call this maximum $S(F_1, \ldots, F_m)$ with a view to the original application in physics. It is still a measure of 'uncertainty', but it is uncertainty *when all the information we have consists of just these numbers*. It is completely 'objective' in the sense that it depends only on the *given data of the problem*, and not on anybody's personality or wishes.

If S is to be a function only of $(F_1, \ldots, F_m)$, then in (11.59) the $Z(\lambda_1, \ldots, \lambda_m)$ must also be thought of as functions of $(F_1, \ldots, F_m)$. At first, the λ's were just unspecified Lagrange multipliers, but eventually we will want to know what they are. If we choose different λ_i, we are writing down different probability distributions (11.56); and we saw in (11.48) that the averages over these distributions agree with the given averages F_k if

$$F_k = \langle f_k \rangle = -\frac{\partial \log Z(\lambda_1, \ldots, \lambda_m)}{\partial \lambda_k}, \qquad k = 1, 2, \ldots, m. \tag{11.60}$$

Equation (11.60) is a set of m simultaneous nonlinear equations which must be solved for the λ's in terms of the F_k. Generally, in a nontrivial problem, it is impractical to solve for the λ's explicitly (although there is a simple formal solution, (11.62), below). We leave the λ_k where they are, expressing things in parametric form. Actually, this isn't such a tragedy, because the λ's usually turn out to have such important physical meanings that we are quite happy to use them as the independent variables. However, if we can evaluate the function $S(F_1, \ldots, F_m)$ explicitly, then we *can* give the λ's as explicit functions of the $\{F_k\}$ as follows.

Suppose we make a small change in one of the F_k; how does this change the maximum attainable H? We have, from (11.59),

$$\frac{\partial S(F_1, \ldots, F_m)}{\partial F_k} = \sum_{j=1}^{m} \left[\frac{\partial \log Z(\lambda_1, \ldots, \lambda_m)}{\partial \lambda_j} \right] \left[\frac{\partial \lambda_j}{\partial F_k} \right] + \sum_{j=1}^{m} \frac{\partial \lambda_j}{\partial F_k} F_k + \lambda_k, \quad (11.61)$$

which, thanks to (11.60), collapses to

$$\lambda_k = \frac{\partial S(F_1, \ldots, F_m)}{\partial F_k}, \quad (11.62)$$

in which λ_k is given explicitly.

Compare this equation with (11.60); one gives F_k explicitly in terms of the λ_k, the other gives the λ_k explicitly in terms of the F_k. Specifying $\log Z(\lambda_1, \ldots, \lambda_m)$ or $S(F_1, \ldots, F_m)$ are equivalent in the sense that each gives full information about the probability distribution. The complete story is contained in either function, and in fact (11.59) is just the Legendre transformation that takes us from one representative function to the other.

We can derive some more interesting laws simply by differentiating either (11.60) or (11.62). If we differentiate (11.60) with respect to λ_j, we obtain

$$\frac{\partial F_k}{\partial \lambda_j} = \frac{\partial^2 \log Z(\lambda_1, \ldots, \lambda_m)}{\partial \lambda_j \partial \lambda_k} = \frac{\partial F_j}{\partial \lambda_k}, \quad (11.63)$$

because the second cross-derivatives of $\log Z(\lambda_1, \ldots, \lambda_m)$ are symmetric in j and k. So, here is a general reciprocity law which will hold in any problem we do by maximizing the entropy. Likewise, if we differentiate (11.62) a second time, we have

$$\frac{\partial \lambda_k}{\partial F_j} = \frac{\partial^2 S}{\partial F_j \partial F_k} = \frac{\partial \lambda_j}{\partial F_k}, \quad (11.64)$$

another reciprocity law, which is, however, not independent of (11.63), because, if we define the matrices $A_{jk} \equiv \partial \lambda_j / \partial F_k$, $B_{jk} \equiv \partial F_j / \partial \lambda_k$, we see easily that they are inverse matrices: $A = B^{-1}$, $B = A^{-1}$. These reciprocity laws might appear trivial from the ease with which we derived them here; but when we get around to applications we'll see that they have highly nontrivial and nonobvious physical meanings. In the past, some of them were found by tedious means that made them seem mysterious and arcane.

Now let's consider the possibility that one of the functions $f_k(x)$ contains a parameter α which can be varied. If you want to think of applications, you can say $f_k(x_i; \alpha)$ stands for the ith energy level of some system and α represents the volume of the system. The energy levels depend on the volume. Or, if it's a magnetic resonance system, you can say that $f_k(x_i)$ represents the energy of the ith stationary state of the spin system and α represents the magnetic field $\mathbf{H}$ applied to it. Often we want to make a prediction of how certain quantities change as we change α. We may want to calculate the pressure or the susceptibility. By the criterion of minimum mean-square error, the best estimate of the derivative would be the mean value over the probability distribution

$$\left\langle \frac{\partial f_k}{\partial \alpha} \right\rangle = \frac{1}{Z} \sum_i \exp\{-\lambda_1 f_1(x_i) - \cdots - \lambda_k f_k(x_i;\alpha) - \cdots - \lambda_m f_m(x_i)\} \frac{\partial f_k(x_i,\alpha)}{\partial \alpha},$$
(11.65)

which reduces to

$$\left\langle \frac{\partial f_k}{\partial \alpha} \right\rangle = -\frac{1}{\lambda_k} \frac{\partial \log Z(\lambda_1, \ldots, \lambda_m; \alpha)}{\partial \alpha}.$$
(11.66)

In this derivation, we supposed that α appeared in only one function, f_k. If the same parameter is in several different f_k, then we verify easily that this generalizes to

$$\sum_{k=1}^m \lambda_k \left\langle \frac{\partial f_k}{\partial \alpha} \right\rangle = -\frac{\partial \log Z(\lambda_1, \ldots, \lambda_m; \alpha)}{\partial \alpha}.$$
(11.67)

This general rule contains, among other things, the equation of state of any thermodynamic system.

When we add α to the problem, both $Z(\lambda_1, \ldots, \lambda_m; \alpha)$ and $S(F_1, \ldots, F_k; \alpha)$ become functions of α. If we differentiate $\log Z(\lambda_1, \ldots, \lambda_m; \alpha)$ or $S(F_1, \ldots, F_k; \alpha)$, we get the same thing:

$$\frac{\partial S(F_1, \ldots, F_k; \alpha)}{\partial \alpha} = -\sum_{k=1}^m \lambda_k \left\langle \frac{\partial f_k}{\partial \alpha} \right\rangle = \frac{\partial \log Z(\lambda_1, \ldots, \lambda_m; \alpha)}{\partial \alpha},$$
(11.68)

with one tricky point: in (11.68) we have to understand that in $\partial S(F_1, \ldots, F_m; \alpha)/\partial \alpha$ we are holding the F_k fixed, while in $\partial \log Z(\lambda_1, \ldots, \lambda_m; \alpha)/\partial \alpha$ we are holding the λ_k fixed. The equality of these derivatives then follows from the Legendre transformation (11.59). Evidently, if there are several different parameters $\{\alpha_1, \alpha_2, \ldots, \alpha_r\}$ in the problem, a relation of the form (11.68) will hold for each of them.

Now let's note some general 'fluctuation laws', or moment theorems. Firstly, a comment about notation: we were using the F_k and $\langle f_k \rangle$ to stand for the same *number*. They are equal because we specified that the expectation values $\{\langle f_1 \rangle, \ldots, \langle f_m \rangle\}$ are to be set equal to the given data $\{F_1, \ldots, F_m\}$. When we want to emphasize that these quantities are expectation values over the canonical distribution (11.56), we will use the notation $\langle f_k \rangle$. When we want to emphasize that they are the given data, we will call them F_k. At the moment, we want to do the former, and so the reciprocity law (11.63) can be written equally well as

$$\frac{\partial \langle f_k \rangle}{\partial \lambda_j} = \frac{\partial \langle f_j \rangle}{\partial \lambda_k} = \frac{\partial^2 \log Z(\lambda_1, \ldots, \lambda_m)}{\partial \lambda_j \partial \lambda_k}.$$
(11.69)

In varying the λ's here, we were changing from one canonical distribution (11.56) to a slightly different one in which the $\langle f_k \rangle$ are slightly different. Since the new distribution corresponding to $(\lambda_k + d\lambda_k)$ is still of canonical form, it is still a maximum entropy distribution corresponding to slightly different data $(F_k + dF_k)$. Thus we are comparing two slightly different maximum entropy problems. For later physical applications it will be important to recognize this in interpreting the reciprocity law (11.69).

Now we want to show that the quantities in (11.69) also have an important meaning with reference to a *single* maximum entropy problem. In the canonical distribution (11.56), how are the different quantities $f_k(x)$ correlated with each other? More specifically, how are departures from their mean values $\langle f_k \rangle$ correlated? The measure of this is the *covariance*, or second central moments, of the distribution:

$$\Big\langle \big(f_j - \langle f_j \rangle\big)\big(f_k - \langle f_k \rangle\big)\Big\rangle = \Big\langle f_j f_k - f_j \langle f_k \rangle - \langle f_j \rangle f_k + \langle f_j \rangle \langle f_k \rangle \Big\rangle$$

$$= \langle f_j f_k \rangle - \langle f_j \rangle \langle f_k \rangle.$$

(11.70)

If a value of f_k greater than the average $\langle f_k \rangle$ is likely to be accompanied by a value of f_j greater than its average $\langle f_j \rangle$, the covariance is positive; if they tend to fluctuate in opposite directions, it is negative; and if their variations are uncorrelated, the covariance is zero. If $j = k$, this reduces to the *variance*:

$$\big\langle (f_k - \langle f_k \rangle)^2 \big\rangle = \langle f_k^2 \rangle - \langle f_k \rangle^2 \geq 0.$$

(11.71)

To calculate these quantities directly from the canonical distribution (11.56), we can first find

$$\langle f_j f_k \rangle = \frac{1}{Z(\lambda_1, \ldots, \lambda_m)} \sum_{i=1}^{n} f_j(x_i) f_k(x_i) \exp\left\{ -\sum_{j=1}^{m} \lambda_j f_j(x_i) \right\}$$

$$= \frac{1}{Z(\lambda_1, \ldots, \lambda_m)} \sum_{i-1}^{n} \frac{\partial^2}{\partial \lambda_j \partial \lambda_k} \exp\left\{ -\sum_{j=1}^{m} \lambda_j f_j(x_i) \right\}$$

(11.72)

$$= \frac{1}{Z(\lambda_1, \ldots, \lambda_m)} \frac{\partial^2 Z(\lambda_1, \ldots, \lambda_m)}{\partial \lambda_j \partial \lambda_k}.$$

Then, using (11.60), the covariance becomes

$$\langle f_j f_k \rangle - \langle f_j \rangle \langle f_k \rangle = \frac{1}{Z} \frac{\partial^2 Z}{\partial \lambda_j \partial \lambda_k} - \frac{1}{Z^2} \frac{\partial Z}{\partial \lambda_j} \frac{\partial Z}{\partial \lambda_k} = \frac{\partial^2 \log Z}{\partial \lambda_j \partial \lambda_k}.$$

(11.73)

But this is just the quantity (11.69); therefore the reciprocity law takes on a bigger meaning,

$$\langle f_j f_k \rangle - \langle f_j \rangle \langle f_k \rangle = -\frac{\partial \langle f_j \rangle}{\partial \lambda_k} = -\frac{\partial \langle f_k \rangle}{\partial \lambda_j}.$$

(11.74)

The second derivatives of $\log Z(\lambda_1, \ldots, \lambda_m)$ which gave us the reciprocity law also give us the covariance of f_j and f_k in our distribution.

Note that (11.74) is in turn only a special case of a more general rule. Let $q(x)$ be any function; then the covariance with $f_k(x)$ is, as can be easily verified,

$$\langle qf_k \rangle - \langle q \rangle \langle f_k \rangle = -\frac{\partial \langle q \rangle}{\partial \lambda_k}. \tag{11.75}$$

Exercise 11.3. From comparing (11.60), (11.69) and (11.74), we might expect that still higher derivatives of $\log Z(\lambda_1, \ldots, \lambda_m)$ would correspond to higher central moments of the distribution (11.56). Check this conjecture by calculating the third and fourth central moments in terms of $\log Z(\lambda_1, \ldots, \lambda_m)$.
Hint: See Appendix C on the theory of cumulants.

For noncentral moments, it is customary to define a *moment generating function*

$$\Phi(\beta_1, \ldots, \beta_m) \equiv \left\langle \exp\left\{ \sum_{j=1}^{m} \beta_j f_j \right\} \right\rangle, \tag{11.76}$$

which evidently has the property

$$\langle f_i^{m_i} f_j^{m_j} \cdots \rangle = \left(\frac{\partial^{m_i}}{\partial \beta_i^{m_i}} \frac{\partial^{m_j}}{\partial \beta_j^{m_j}} \cdots \right) \Phi(\beta_1, \ldots, \beta_m) \Big|_{\beta_k = 0}. \tag{11.77}$$

However, we find from (11.76),

$$\Phi(\beta_1, \ldots, \beta_m) = \frac{Z([\lambda_1 - \beta_1], \ldots, [\lambda_m - \beta_m])}{Z(\lambda_1, \ldots, \lambda_m)}, \tag{11.78}$$

so that the partition function $Z(\lambda_1, \ldots, \lambda_m)$ serves this purpose; instead of (11.77) we may write equally well

$$\langle f_i^{m_i} f_j^{m_j} \cdots \rangle = \frac{1}{Z(\lambda_1, \ldots, \lambda_m)} \left(\frac{\partial^{m_i}}{\partial \lambda_i^{m_i}} \frac{\partial^{m_j}}{\partial \lambda_j^{m_j}} \cdots \right) Z(\lambda_1, \ldots, \lambda_m), \tag{11.79}$$

which is the generalization of (11.72).

Now, we might ask, what are the covariances of the derivatives of f_k with respect to a parameter α? Define

$$g_k \equiv \frac{\partial f_k}{\partial \alpha}; \tag{11.80}$$

then, for example, if f_k is the energy and α is the volume then $-g_k$ is the pressure. We easily verify another reciprocity relation:

$$\frac{\partial \langle g_j \rangle}{\partial \lambda_k} = -\left[\langle g_j f_k \rangle - \langle g_j \rangle \langle g_k \rangle \right] = \frac{\partial \langle g_k \rangle}{\partial \lambda_j} \tag{11.81}$$

analogous to (11.74). By a similar derivation, we find the identity

$$\sum_{j=1}^{m} \lambda_j \left[\langle g_j g_k \rangle - \langle g_j \rangle \langle g_k \rangle \right] = \left\langle \frac{\partial g_k}{\partial \alpha} \right\rangle - \frac{\partial \langle g_k \rangle}{\partial \alpha}. \tag{11.82}$$

We had found and used special cases of this for some time before realizing its generality.

Other derivatives of $\log Z(\lambda_1, \ldots, \lambda_m)$ are related to various moments of the f_k and their derivatives with respect to α. For example, closely related to (11.82) is

$$\frac{\partial^2 \log Z(\lambda_1, \ldots, \lambda_m)}{\partial \alpha^2} = \sum_{jk} \lambda_j \lambda_k \left[\langle g_j g_k \rangle - \langle g_j \rangle \langle g_k \rangle \right] - \sum_k \lambda_k \left\langle \frac{\partial g_k}{\partial \alpha} \right\rangle. \tag{11.83}$$

The cross-derivatives give us a simple and useful relation,

$$\frac{\partial^2 \log Z(\lambda_1, \ldots, \lambda_m)}{\partial \alpha \partial \lambda_k} = -\frac{\partial \langle f_k \rangle}{\partial \alpha} = \sum_j \lambda_j \left[\langle f_k g_j \rangle - \langle f_k \rangle \langle g_j \rangle \right] - \langle g_k \rangle, \tag{11.84}$$

which also follows from (11.69) and (11.75); by taking further derivatives, an infinite hierarchy of similar moment relations is obtained. As we will see later, the above theorems have, as special cases, many relations, such as the Einstein fluctuation laws for black-body radiation and for density of a gas or liquid, the Nyquist voltage fluctuations, or 'noise' generated by a reversible electric cell, etc.

It is evident that if several different parameters $\{\alpha_1, \ldots, \alpha_r\}$ are present, relations of the above form will hold for each of them; and new ones such as

$$\frac{\partial^2 \log Z(\lambda_1, \ldots, \lambda_m)}{\partial \alpha_1 \partial \alpha_2} = \sum_k \lambda_k \left\langle \frac{\partial^2 f_k}{\partial \alpha_1 \partial \alpha_2} \right\rangle - \sum_{kj} \lambda_j \lambda_k \left[\left\langle \frac{\partial f_k}{\partial \alpha_1} \frac{\partial f_j}{\partial \alpha_2} \right\rangle - \left\langle \frac{\partial f_k}{\partial \alpha_1} \right\rangle \left\langle \frac{\partial f_j}{\partial \alpha_2} \right\rangle \right] \tag{11.85}$$

will appear.

The relationship between $\log Z(\lambda_1, \ldots, \lambda_m; \alpha_1, \ldots, \alpha_r)$ and $S(\langle f_1 \rangle, \ldots, \langle f_m \rangle; \alpha_1, \ldots, \alpha_r)$ shows that they can all be stated also in terms of derivatives (i.e. variational properties) of S; see (11.59). In the case of S, however, there is a still more general and important variational property.

In (11.62) we supposed that the definitions of the functions $f_k(x)$ were fixed once and for all, the variation of $\langle f_k \rangle$ being due only to variations in the p_i. We now derive a more general variational statement in which both of these quantities are varied. Let $\delta f_k(x_i)$ be specified arbitrarily and independently for each value of k and i, let $\delta \langle f_k \rangle$ be specified independently of the $\delta f_k(x_i)$, and consider the resulting change from one maximum entropy distribution p_i to a slightly different one $p_i' = p_i + \delta p_i$, the variations δp_i and $\delta \lambda_k$ being determined in terms of $\delta f_k(x_i)$ and $\delta \langle f_k \rangle$ through the above equations. In other words, we are now considering two slightly different maximum entropy problems in which all conditions of the problem – including the definitions of the functions $f_k(x)$ on which it is based – are varied arbitrarily. The variation in $\log Z(\lambda_1, \ldots, \lambda_m)$ is

$$
\begin{aligned}
\delta \log Z(\lambda_1, \ldots, \lambda_m) &= \frac{1}{Z} \sum_{i=1}^{n} \left[\sum_{k=1}^{m} \left[-\lambda_k \delta f_k(x_i) - \delta\lambda_k f_k(x_i) \right] \exp\left\{ -\sum_{j=1}^{m} \lambda_j f_j(x_i) \right\} \right] \\
&= -\sum_{k=1}^{m} \left[\lambda_k \langle \delta f_k \rangle + \delta\lambda_k \langle f_k \rangle \right],
\end{aligned}
$$

$$(11.86)$$

and thus from the Legendre transformation (11.59)

$$
\delta S = -\sum_{k} \lambda_k \left[\delta \langle f_k \rangle - \langle \delta f_k \rangle \right], \quad \text{or} \quad \delta S = \sum_{k} \lambda_k \delta Q_k, \tag{11.87}
$$

where

$$
\delta Q_k \equiv \delta \langle f_k \rangle - \langle \delta f_k \rangle = \sum_{i=1}^{n} f_k(x_i) \delta p_i. \tag{11.88}
$$

This result, which generalizes (11.62), shows that the entropy S is stationary not only in the sense of the maximization property which led to the canonical distribution (11.56); it is also stationary with respect to small variations in the functions $f_k(x_i)$ if the p_i are held fixed.

As a special case of (11.87), suppose that the functions f_k contain parameters $\{\alpha_1, \ldots, \alpha_r\}$ as in (11.85), which generate the $\delta f_k(x_i)$ by

$$
\delta f_k(x_i, \alpha_j) = \sum_{j=1}^{r} \frac{\partial f_k(x_i, \alpha)}{\partial \alpha_j} \delta\alpha_j. \tag{11.89}
$$

While δQ_k is not in general the exact differential of any function $Q_k(\langle f_i \rangle; \alpha_j)$, (11.87) shows that λ_k is an integrating factor such that $\sum \lambda_k \delta Q_k$ is the exact differential of a 'state function' $S(\langle f_i \rangle; \alpha_j)$. At this point, perhaps all this is beginning to sound familiar to those who have studied thermodynamics. Finally, we leave it for you to prove from (11.87) that

$$
\sum_{k=1}^{m} \langle f_k \rangle \frac{\partial \lambda_k}{\partial \alpha} = 0, \tag{11.90}
$$

where $\langle f_1 \rangle, \ldots, \langle f_r \rangle$ are held constant in the differentiation.

Evidently, there's now a large new class of problems which we can ask the robot to do, which it can solve in rather a wholesale way. It first evaluates this partition function Z, or, better still, $\log Z$. Then, just by differentiating $\log Z$ with respect to all its arguments in every possible way, it obtains all sorts of predictions in the form of mean values over the maximum entropy distribution. This is quite a neat mathematical procedure, and, of course, you recognize what we have been doing here. These relations are all just the standard equations of statistical mechanics given to us by J. Willard Gibbs, but now in a disembodied form with all the physics removed.

Indeed, virtually all known thermodynamic relations, found over more than a century ago by the most diverse and difficult kinds of physical reasoning and experimentation, are now seen as special cases of simple mathematical identities of the maximum entropy formalism. This makes it clear that those relations are actually independent of any particular physical assumptions and are properties of extended logic in general, giving us a new insight into why the relations of thermodynamics are so general, independent of the properties of any particular substance. Gibbs' statistical mechanics is historically the oldest application of the principle of maximum entropy and is still the most used (although many of its users are still unaware of its generality).

The maximum entropy mathematical formalism has a mass of other applications outside of physics. In Chapter 14 we work out the full numerical solution to a nontrivial problem of inventory control, and in Chapter 22 we give a highly nontrivial analytical solution of a problem of optimal encoding in communication theory. In a sense, once we have understood the maximum entropy principle as explained in this chapter, most applications of probability theory are seen as invoking it to assign the initial probabilities – whether called technically prior probabilities or sampling probabilities. Whenever we assign uniform prior probabilities, we can say truthfully that we are applying maximum entropy (although in that case the result is so simple and intuitive that we do not need any of the above formalism). As we saw in Chapter 7, whenever we assign a Gaussian sampling distribution, this is the same as applying maximum entropy for given first and second moments. And we saw in Chapter 9 that, whenever we assign a binomial sampling distribution, this is mathematically equivalent to assigning the uniform maximum entropy distribution on a deeper hypothesis space.

11.8 Conceptual problems – frequency correspondence

The principle of maximum entropy is basically a simple and straightforward idea, and, in the case that the given information consists of average values, it leads, as we have just seen, to a surprisingly concise mathematical formalism, since essentially everything is known if we can evaluate a single function $\log Z(\lambda_1, \ldots, \lambda_m; \alpha_1, \ldots, \alpha_r)$. Nevertheless, it seems to generate some serious conceptual difficulties, particularly to people who have been trained to think of probability only in the frequency sense. Therefore, before turning to applications, we want to examine, and hopefully resolve, some of these difficulties. Here are some of the objections that have been raised against the principle of maximum entropy.

(A) If the only justification for the canonical distribution (11.56) is 'maximum uncertainty', that is a negative thing which can't possibly lead to any useful predictions; you can't get reliable results out of mere ignorance.

(B) The probabilities obtained by maximum entropy cannot be relevant to physical predictions because they have nothing to do with frequencies – there is absolutely no reason to suppose that distributions observed experimentally would agree with ones found by maximizing entropy.

(C) The principle is restricted to the case where the constraints are average values – but almost always the given data $\{F_1, \ldots, F_n\}$ are *not* averages over anything. They are definite measured numbers. When you set them equal to averages, $F_k = \langle f_k \rangle$, you are committing a logical contradiction, for the given data said that f_k had the value F_k; yet you immediately write down a probability distribution that assigns nonzero probabilities to values of $f_k \neq F_k$.

(D) The principle cannot lead to any definite physical results because different people have different
information, which would lead to different distributions – the results are basically arbitrary.

Objection (A) is, of course, nothing but a play on words. The 'uncertainty' was always there.
Our maximizing the entropy did not *create* any 'ignorance' or 'uncertainty'; it is rather the
means of determining quantitatively the full extent of the uncertainty already present. It is
failure to do this – and as a result using a distribution that implies more knowledge than we
really have – that would lead to unreliable conclusions.

Of course, the information put into the theory as constraints on our maximum entropy
distribution, may be so meager – the distribution is so weakly constrained from the uninformative uniform one – that no reliable predictions can be made from it. But in that case,
as we will see later, the theory automatically tells us this: if we emerge with a very broad
probability distribution for some quantity θ (such as pressure, magnetization, electric current density, rate of diffusion, etc.), that is the robot's way of telling us: 'You haven't given
me enough information to determine any definite prediction'. But if we get a very sharp
distribution for θ (for example – and typical of what does happen in many real problems –
if the theory says the odds on θ being in the interval $\theta_0(1 \pm 10^{-6})$ are greater than $10^{10} : 1$),
then the given information *was* sufficient to make a very definite prediction.

In both cases, and in the intermediate ones, the distribution for θ always tells us just what
conclusions we *are* entitled to draw about θ, on the basis of the information *which was put
into the equations*. If someone has additional cogent information, but fails to incorporate it
into his calculation, the result is not a failure, only a misuse, of the maximum entropy method.

To answer objection (B), we show that the situation is vastly more subtle than that.
The principle of maximum entropy has, fundamentally, nothing to do with any repeatable
'random experiment'. Some of the most important applications are to cases where the
probabilities p_i in (11.56) have no frequency connection – the x_i are simply an enumeration
of the *possibilities*, in the single situation being considered, as in the cars on the ferry
problem.

Nothing prevents us, however, from applying the principle of maximum entropy also to
cases where the x_i are generated by successive repetitions of some experiment as in the dice
problem; and, in this case, the question of the relationship between the maximum entropy
probability $p(x_i)$ and the frequency with which x_i is observed, is capable of mathematical
analysis. We demonstrate that (1) in this case the maximum entropy probabilities *do* have
a precise connection with frequencies; (2) in most real problems, however, this relation is
unnecessary for the usefulness of the method; and (3) in fact, the principle of maximum
entropy is most useful to us in just those cases where the observed frequencies do *not* agree
with the maximum entropy probabilities.

Suppose now that the value of x is determined by some random experiment; at each
repetition of the experiment, the final result is one of the values $x_i, i = 1, 2, \ldots, n$; in the
dice problem, $n = 6$. But now, instead of asking for the probability p_i, let's ask an entirely
different question: on the basis of the available information, what can we say about the
relative *frequencies* f_i with which the various x_i occur?

Let the experiment consist of N trials (we are particularly interested in the limit $N \to \infty$, because that is the situation contemplated in the usual frequency theory of probability), and let every conceivable sequence of results be analyzed. Each trial could give, independently, any one of the results $\{x_1, \ldots, x_n\}$, and so there are n^N conceivable outcomes of the whole experiment. But many of these will be incompatible with the given information. (Let's suppose again that this consists of average values of several functions $f_k(x)$, $k = 1, 2, \ldots, m$; in the end, it will be clear that the final conclusions are independent of whether it takes this form or some other.) We will, of course, assume that the result of the experiment agrees with this information – if it didn't, then the given information was false and we are doing the wrong problem. In the whole experiment, the results x_1 will be obtained n_1 times, x_2 will be obtained n_2 times, etc. Of course,

$$\sum_{i=1}^{n} n_i = N, \tag{11.91}$$

and if the specified mean values F_k given to us are in fact observed in the actual experiment, we have the additional relation

$$\sum_{i=1}^{n} n_i f_k(x_i) = N F_k, \qquad d = 1, 2, \ldots, m. \tag{11.92}$$

If $m < n - 1$, (11.91) and (11.92) are insufficient to determine the relative frequencies $f_i = n_i/N$. Nevertheless, we do have grounds for preferring some choices of the f_i to others. For, out of the original n^N conceivable outcomes, how many would lead to a given set of sample numbers $\{n_1, n_2, \ldots, n_n\}$? The answer is, of course, the multinomial coefficient

$$W = \frac{N!}{n_1! n_2! \cdots n_n!} = \frac{N!}{(N f_1)!(N f_2)! \cdots (N f_n)!}. \tag{11.93}$$

The set of frequencies $\{f_1, \ldots, f_n\}$ which can be realized in the greatest number of ways is therefore the one which maximizes W subject to the constraints (11.91), (11.92). Now we can equally well maximize any monotonic increasing function of W, in particular $N^{-1} \log(W)$; but as $N \to \infty$ we have, as we saw already in (11.29),

$$\frac{1}{N} \log(W) \to -\sum_{i=1}^{n} f_i \log(f_i) = H_f. \tag{11.94}$$

So you see that, in (11.91), (11.92) and (11.94) we have formulated exactly the same mathematical problem as in the maximum entropy derivation, so the two problems will have the same solution. This argument is mathematically reminiscent of the Wallis derivation given in Section 11.4; and the same result could have been found as well by direct application of Bayes' theorem, assigning uniform prior probabilities over all the n^N conceivable outcomes and passing to the limit $N \to \infty$.

You see also, in partial answer to objection (C), that this identity of the mathematical problems will persist whether or not the constraints take the form of mean values. If the given information does consist of mean values, then the mathematics is particularly neat, leading to the partition function, etc. But, for given information which places *any* definite kind of constraint on the problem, we have the same conclusion: the *probability* distribution which maximizes the entropy is numerically identical with the *frequency* distribution which can be realized in the greatest number of ways.

The maximum in W is, furthermore, enormously sharp. To show this, let $\{f_1, \ldots, f_n\}$ be the set of frequencies which maximizes W and has entropy H_f; and let $\{f'_1, \ldots, f'_n\}$ be any other set of possible frequencies (that is, a set which satisfies the constraints (11.91), (11.92) and has entropy $H_{f'} < H_f$). The ratio (number of ways in which f_i could be realized)/(number of ways in which f'_i could be realized) grows asymptotically, according to (11.94), as

$$\frac{W}{W'} \to \exp\{N(H_f - H_{f'})\} \tag{11.95}$$

and passes all bounds as $N \to \infty$. Therefore, the frequency distribution predicted by maximum entropy can be realized experimentally in *overwhelmingly* more ways than can any other that satisfies the same constraints.

We have here another precise and quite general connection between probability and frequency; it had nothing to do with the definition of probability, but emerged as a mathematical *consequence* of probability theory, interpreted as extended logic. Another kind of connection between probability and frequency, whose precise mathematical statement is different in form, but which has the same practical consequences, will appear in Chapter 12.

Turning to objection (C), our purpose in imposing constraints is to incorporate certain information into our probability distribution. Now, what does it mean to say that a probability distribution 'contains' some information? We take this as meaning that the information can be extracted from it by using the usual rule for estimating the expectation. Usually, the datum F_k is of unknown accuracy, and so using it to constrain only the $\langle F_k \rangle$ is just the process of being honest, leaving the width of the distribution for $f_k(x)$ to be determined by the range and density of the set of possibilities x_i. But if we do have independent information about the accuracy of F_1, that can be incorporated by adding a new constraint on $\langle f_1(x_i)^2 \rangle$; the formalism already allows for this. But this seldom makes any substantive difference in the final conclusions, because the variance of the maximum entropy distribution for $f_1(x)$ is usually small compared with any reasonable mean-square experimental error.

Now let's turn to objection (D) and analyze the situation with some care, because it is perhaps the most common of all of them. Does the above connection between probability and frequency justify our predicting that the maximum entropy distribution will in fact be observed as a frequency distribution in a real experiment? Clearly not, in the sense of deductive proof; for, just as objection (D) points out, we have to concede that different people may have different amounts of information, which will lead them to writing down different distributions, which make different predictions of observable facts, and they can't all be right. But this misses the point about what we are trying to do; let's look at it more closely.

Consider a specific case: Mr A imposes constraints on the mean values $\langle f_1(x) \rangle$, $\langle f_2(x) \rangle$ to agree with his data F_1, F_2. Mr B, better informed, imposes in addition a constraint on $\langle f_3(x) \rangle$ to agree with his extra datum F_3. Each sets up a maximum entropy distribution on the basis of his information. Since Mr B's entropy is maximized subject to one further constraint, we will have

$$S_B \leq S_A. \tag{11.96}$$

Suppose that Mr B's extra information was redundant, in the sense that it was only what Mr A would have predicted from his distribution. Now, Mr A has maximized his entropy with respect to all variations of the probability distribution which hold $\langle f_1 \rangle$, $\langle f_2 \rangle$ fixed at the specified values F_1, F_2. Therefore, he has *a fortiori* maximized it with respect to the smaller class of variations which also hold $\langle f_3 \rangle$ fixed at the value finally attained. Therefore Mr A's distribution also solves Mr B's problem in this case; $\lambda_3 = 0$, and Mr A and Mr B have identical probability distributions. In this case, and only in this case, we have equality in (11.96).

From this we learn two things. (1) Two people with different given information do not necessarily arrive at different maximum entropy distributions; this is the case only when Mr B's extra information was 'surprising' to Mr A. (2) In setting up a maximum entropy problem, it is not necessary to determine whether the different pieces of information used are independent: any redundant information will not be 'counted twice', but will drop out of the equations automatically. Indeed, this not only agrees with our basic desideratum that $AA = A$ in Boolean algebra; it would be true of any variational principle (imposing a new constraint cannot change the solution if the old solution already satisfied that constraint).

Now suppose the opposite extreme: Mr B's extra information was logically contradictory to what Mr A knows. For example, it might turn out that $f_3(x) = f_1(x) + 2f_2(x)$, but Mr B's data failed to satisfy $F_3 = F_1 + 2F_2$. Evidently, there is *no* probability distribution that fits Mr B's supposed data. How does our robot tell us this? Mathematically, you will then find that the equations

$$F_k = -\frac{\partial \log Z(\lambda_1, \lambda_2, \lambda_3)}{\partial \lambda_k} \tag{11.97}$$

have no simultaneous solution with real λ_k. In the example just mentioned,

$$Z(\lambda_1, \lambda_2, \lambda_3) = \sum_{i=1}^{n} \exp\{-\lambda_1 f_1(x_i) - \lambda_2 f_2(x_i) - \lambda_3 f_3(x_i)\}$$
$$= \sum_{i=1}^{n} \exp\{-(\lambda_1 + \lambda_3) f_1(x_i) - (\lambda_2 + 2\lambda_3) f_2(x_i)\} \tag{11.98}$$

and so

$$\frac{\partial Z(\lambda_1, \lambda_2, \lambda_3)}{\partial \lambda_3} = \frac{\partial Z(\lambda_1, \lambda_2, \lambda_3)}{\partial \lambda_1} + 2\frac{\partial Z(\lambda_1, \lambda_2, \lambda_3)}{\partial \lambda_2}, \tag{11.99}$$

and so (11.97) cannot have solutions for $\lambda_1, \lambda_2, \lambda_3$ unless $F_3 = F_1 + 2F_2$. So, when a new piece of information logically contradicts previous information, the principle of maximum entropy breaks down, as it should, refusing to give us any distribution at all.

The most interesting case is the intermediate one where Mr B's extra information was neither redundant nor contradictory. He then finds a maximum entropy distribution different from that of Mr A, and the inequality holds in (11.96), indicating that Mr B's extra information was 'useful' in further narrowing down the range of possibilities allowed by Mr A's information. The measure of this range is just W; and from (11.95) we have asymptotically

$$\frac{W_A}{W_B} \sim \exp\{N(S_A - S_B)\}. \tag{11.100}$$

For large N, even a slight decrease in the entropy leads to an enormous decrease in the number of possibilities.

Suppose now that we start performing the experiment with Mr A and Mr B watching. Since Mr A predicts a mean value $\langle f_3 \rangle$ different from the correct one known to Mr B, it is clear that the experimental distribution cannot agree in all respects with Mr A's prediction. We cannot be sure in advance that it will agree with Mr B's prediction either, for there may be still further constraints on $f_4(x)$, $f_5(x)$, ..., etc. operating in the experiment unknown to Mr B.

The property demonstrated above justifies the following weaker statement of frequency correspondence: If the information incorporated into the maximum entropy analysis includes all the constraints actually operating in the random experiment, then the distribution predicted by maximum entropy is overwhelmingly the most likely to be observed experimentally. Indeed, most frequency distributions observed in Nature are maximum entropy distributions, simply because they can be realized in so many more ways than can any other.

Conversely, suppose the experiment fails to confirm the maximum entropy prediction, and this disagreement persists indefinitely on repetition of the experiment. Then, since by hypothesis the data F_i were true if incomplete, we will conclude that the physical mechanism of the experiment must contain some additional constraint which was not taken into account in the maximum entropy calculation. The observed deviations then provide a clue as to the nature of this new constraint. In this way, Mr A can discover empirically that his information was incomplete.

In summary, the principle of maximum entropy is not an oracle telling which predictions *must* be right; it is a rule for inductive reasoning that tells us which predictions *are most strongly indicated by our present information*.

11.9 Comments

The little scenario just described in Section 11.8 is an accurate model of just what did happen in one of the most important applications of statistical analysis, carried out by J. Willard Gibbs. By the year 1901 it was known that, in classical statistical mechanics, use of the canonical ensemble (which Gibbs derived as the maximum entropy distribution over the classical state space, or phase volume, based on a specified mean value of the energy) failed to predict some thermodynamic properties (heat capacities, equation of state) correctly. Analysis of the data showed that the entropy of a real physical system was always less than the value predicted. At that time, therefore, Gibbs was in just the position of Mr *A* in the scenario, and the conclusion was that the microscopic laws of physics must involve some additional constraint not contained in the laws of classical mechanics.

But Gibbs died in 1903, and it was left to others to find the nature of this constraint; first by Planck, in the case of radiation, then by Einstein and Debye for solids, and finally by Bohr for isolated atoms. The constraint consisted in the discreteness of the possible energy values, thenceforth called energy levels. By 1927, the mathematical theory by which these could be calculated from first principles had been developed by Heisenberg and Schrödinger.

Thus, it is an historical fact that the first clues indicating the need for the quantum theory, and indicating some necessary features of the new theory, were uncovered by a seemingly 'unsuccessful' application of the principle of maximum entropy. We may expect that such things will happen again in the future, and this is the basis of the remark that the principle of maximum entropy is most useful to us in just those cases where it fails to predict the correct experimental facts. This illustrates the real nature, function, and value of inductive reasoning in science; an observation that was stressed also by Jeffreys (see 1957 edition of Jeffreys, 1931).

Gibbs (1902) wrote his probability density in phase space in the form

$$w(q_1, \ldots, q_n; p_1, \ldots, p_n) = \exp\{\eta(q_1, \ldots, q_n)\} \tag{11.101}$$

and called the function η the 'index of probability of phase'. He derived his canonical and grand canonical ensembles from constraints on average energy, and average energy and particle numbers, respectively, as (Gibbs, 1902, p. 143) 'the distribution in phase which without violating this condition gives the least value of the average index of probability of phase $\bar{\eta}$ ...'. This is, of course, just what we would describe today as maximizing the entropy subject to constraints.

Unfortunately, Gibbs' work was left unfinished due to failing health. He did not give any clear explanation, and we can only conjecture whether he possessed one, as to why this particular function is to be maximized in preference to all others. Consequently, his procedure appeared arbitrary to many, and for 60 years there was confusion and controversy over the justification for Gibbs' methods; they were rejected summarily by some writers on statistical mechanics, and treated with the greatest caution by others. Only with the work of Shannon (1948) could one see the way to new thinking on a fundamental level. These historical matters are discussed in more detail in Jaynes (1967) and Jaynes (1992b).

12

Ignorance priors and transformation groups

Ignorance is preferable to error and he is less remote from the truth who
believes nothing than he who believes what is wrong.

Thomas Jefferson (1781)

The problem of translating prior information uniquely into a prior probability assignment
represents the as yet unfinished half of probability theory, though the principle of maximum
entropy in the preceding chapter provides one important tool. It is unfinished because it
has been rejected for many decades by those who were unable to conceive of a probability
distribution as representing information; but, just because of that long neglect, many current
scientific, engineering, economic, and environmental problems are today calling out for new
solutions to this problem, without which important new applications cannot proceed.

12.1 What are we trying to do?

It is curious that, even when different workers are in substantially complete agreement on
what calculations should be done, they may have radically different views as to what we are
actually doing and why we are doing it. For example, there is a large Bayesian community,
whose members call themselves 'subjective Bayesians', who have settled into a position
intermediate between 'orthodox' statistics and the theory expounded here. Their members
have had, for the most part, standard orthodox training; but then they saw the absurdities
in it and defected from the orthodox philosophy, while retaining the habits of orthodox
terminology and notation.

These habits of expression put subjective Bayesians under a severe handicap. While
perceiving that probabilities cannot represent only frequencies, they still regard sampling
probabilities as representing frequencies of 'random variables'. But for them prior and posterior probabilities represent only private opinions, which are to be updated, in accordance
with de Finetti's principle of coherence. Fortunately, this leads to the Bayesian algorithm,
so we do the same calculations.

Subjective Bayesians face an awkward ambiguity at the beginning of a problem, when one
assigns prior probabilities. If these represent merely prior opinions, then they are basically
arbitrary and undefined; it seems that only private introspection could assign them, and

different people will make different assignments. Yet most subjective Bayesians continue to use a language which implies that there exists some unknown 'true' prior probability distribution in a real problem. In our view, problems of inference are ill-posed until we recognize three essential things.

(A) The prior probabilities represent our prior *information*, and are to be determined, not by introspection, but by *logical analysis* of that information.
(B) Since the final conclusions depend necessarily on both the prior information and the data, it follows that, in formulating a problem, one must specify the prior information to be used just as fully as one specifies the data.
(C) Our goal is that inferences are to be completely 'objective' in the sense that two persons with the same prior information must assign the same prior probabilities.

If one fails to specify the prior information, a problem of inference is just as ill-posed as if one had failed to specify the data. Indeed, since the time of Laplace, applications of probability theory have been hampered by difficulties in the treatment of prior information. In realistic problems of inference, it is typical that we have cogent prior information, highly relevant to the question being asked; to fail to take it into account is to commit the most obvious inconsistency of reasoning, and it may lead to absurd or dangerously misleading results.

Having specified the prior information, we then have the problem of translating that information into a specific prior probability assignment. It is this formal translation process that represents fully half of probability theory, as it is needed for real applications; yet it is entirely absent from orthodox statistics, and only dimly perceived in subjective Bayesian theory.

Just as zero is the natural starting point in adding a column of numbers, the natural starting point in translating a number of pieces of prior information is the state of complete ignorance. In the previous chapter we have seen that for discrete probabilities the principle of maximum entropy tells us, in agreement with our obvious intuition, that complete ignorance, but for specification of a finite set of possibilities, is represented by a uniform prior probability assignment. For continuous probabilities the problem is much more difficult, because intuition fails us and we must resort to formal desiderata and principles. In this chapter we examine the use of the mathematical tool of transformation groups for this purpose.

Some object to the very attempt to represent complete ignorance, on the grounds that a state of complete ignorance does not 'exist'. We would reply that a perfect triangle does not exist either; nevertheless, a surveyor who was ignorant of the properties of perfect triangles would not be competent to do his job. Complete ignorance is, for us, an ideal limiting case of real prior information, in exactly the same sense that a perfect triangle is an ideal limiting case of the real triangles made by surveyors. If we have not learned how to deal with complete ignorance, we are hardly in a position to solve a real problem.

The relatively simple problems examined up till now could be dealt with by reasonable common sense, which could see, nearly always, what the prior ought to be. When we advance to more complicated problems, a formal theory of how to find ignorance priors becomes more and more necessary. The principle of maximum entropy suffices in many cases, but other principles such as transformation groups, marginalization theory, and coding theory, should also be available in our toolbox. In this chapter we develop the method of transformation groups. Before beginning that development, we first, as a way of introduction, discuss the principle of maximum entropy for continuous distributions, and show how this naturally leads to the idea of assigning distributions to represent complete ignorance.

12.2 Ignorance priors

Thus far we have considered the principle of maximum entropy only for the, discrete case and have seen that, if the distribution sought can be regarded as having been produced by a random experiment, there is a correspondence property between probability and frequency, and the results are consistent with other principles of probability theory. However, nothing in the mathematics requires that any random experiment be in fact performed or conceivable; and so we interpret the principle in the broadest sense which gives it the widest range of applicability, i.e. whether or not any random experiment is involved, the maximum entropy distribution still represents the most 'honest' description of our state of knowledge.

In such applications, the principle is easy to apply and leads to the kind of results we should want and expect. For example, in Jaynes (1963a) a sequence of problems about decision making under uncertainty (essentially, of inventory control), of a type which arises constantly in practice, was analyzed. Here, the state of nature was not the result of any random experiment; there was no sampling distribution and no sample. Thus it might be thought to be a 'no data' decision problem, in the sense of Chernoff and Moses (1959). However, in successive stages of the sequence, there were available more and more pieces of prior information, and digesting them by maximum entropy led to a sequence of prior distributions in which the range of possibilities was successively narrowed down. They led to a sequence of decisions, each representing the rational one on the basis of the information available at that stage, which corresponds to intuitive common-sense judgments in the early stages where intuition was able to see the answer. It is difficult to see how this problem could have been treated at all without the use of the principle of maximum entropy, or some other device that turns out in the end to be equivalent to it.

In several years of routine application of this principle in problems of physics and engineering, we have yet to find a case involving a discrete prior where it fails to produce a useful and intuitively reasonable result. To the best of the author's knowledge, no other general method for setting up discrete priors has been proposed. It appears, then, that the principle of maximum entropy may prove to be the final solution to the problem of assigning discrete priors.

12.3 Continuous distributions

Use of the principle of maximum entropy in setting up continuous prior distributions, however, requires considerably more analysis because at first glance the results appear to depend on the choice of parameters. We do not refer here to the well-known fact that the quantity

$$H' = -\int dx \; p(x|I) \log[p(x|I)] \qquad (12.1)$$

lacks invariance under a change of variables $x \to y(x)$, for (12.1) is not the result of any derivation, and it turns out not to be the correct information measure for a continuous distribution. Shannon's theorem establishing (11.23) as an information measure goes through only for discrete distributions; to find the corresponding expression in the continuous case we can pass to the limit from a discrete distribution. The following argument can be made as rigorous as we please, but at considerable sacrifice of clarity.

In the discrete entropy expression

$$H_I^d = -\sum_{i=1}^{n} p_i \log[p_i], \qquad (12.2)$$

we suppose that the discrete points x_i, $i = 1, 2, \ldots, n$, become more and more numerous, in such a way that, in the limit $n \to \infty$,

$$\lim_{n\to\infty} \frac{1}{n} (\text{number of points in } a < x < b) = \int_a^b dx \; m(x). \qquad (12.3)$$

If this passage to the limit is sufficiently well-behaved, it will also be true that adjacent differences $(x_{i+1} - x_i)$ in the neighborhood of any particular value of x will tend to zero so that

$$\lim_{n\to\infty} [n(x_{i+1} - x_i)] = [m(x_i)]^{-1}. \qquad (12.4)$$

The discrete probability distribution p_i will go over into a continuous probability $p(x|I)$, according to the limiting form of

$$p_i = p(x_i|I)(x_{i+1} - x_i) \qquad (12.5)$$

or, from (12.4),

$$p_i \to p(x_i|I)[nm(x_i)]^{-1}. \qquad (12.6)$$

Consequently, the discrete entropy (12.2) goes over into the integral

$$H_I^d \to \int dx \; p(x|I) \log\left[\frac{p(x|I)}{nm(x)}\right]. \qquad (12.7)$$

In the limit, this contains an infinite term $\log(n)$; if we subtract this, the difference will, in the cases of interest, approach a definite limit, which we take as the continuous information measure:

$$H_I^c \equiv \lim_{n\to\infty} \left[H_I^d - \log(n)\right] = -\int dx \; p(x|I) \log\left[\frac{p(x|I)}{m(x)}\right]. \qquad (12.8)$$

The 'invariant measure' function, $m(x)$, is proportional to the limiting density of discrete points. (In all applications so far studied, $m(x)$ is a well-behaved continuous function, and so we continue to use the notion of Riemann integrals; we call $m(x)$ a 'measure' only to suggest the appropriate generalization, readily supplied if a practical problem should ever require it.) Since $p(x|I)$ and $m(x)$ transform in the same way under a change of variables, H_I^c is invariant.

We seek a probability density $p(x|I)$ which is normalized:

$$\int dx \, p(x|I) = 1 \tag{12.9}$$

(we understand the range of integration to be the full parameter space), and constrained by information fixing the mean values of m different functions $f_k(x)$:

$$F_k = \int dx \, p(x|I) f_k(x), \quad k = 1, 2, \ldots, m, \tag{12.10}$$

where the F_f are the given numerical values. Subject to these constraints, we are to maximize (12.8). The solution is again elementary:

$$p(x|I) = Z^{-1} m(x) \exp \{\lambda_1 f_1(x) + \cdots + \lambda_m f_m(x)\}, \tag{12.11}$$

with the partition function

$$Z(\lambda_1, \ldots, \lambda_m) \equiv \int dx \, m(x) \exp \{\lambda_1 f_1(x) + \cdots + \lambda_m f_m(x)\}, \tag{12.12}$$

and the Lagrange multipliers λ_k are determined by

$$F_k = -\frac{\partial \log Z(\lambda_1, \ldots, \lambda_m)}{\partial \lambda_k} \quad k = 1, \ldots, m. \tag{12.13}$$

Our 'best' estimate (by quadratic loss function) of any other quantity $q(x)$ is then

$$\langle q \rangle = \int dx \, q(x) p(x|I). \tag{12.14}$$

It is evident from these equations that when we use (12.8) rather than (12.1) as our information measure not only our final conclusions (12.14), but also the partition function and Lagrange multipliers are all invariant under a change of parameter $x \to y(x)$. In applications, these quantities acquire definite physical meanings.

There remains, however, a practical difficulty. If the parameter space is not the result of any obvious limiting process, what determines the proper measure $m(x)$? The conclusions, evidently, will depend on which measure we adopt. This is the shortcoming from which the maximum entropy principle has suffered until now, and which must be cleared up before we can regard it as a full solution to the prior probability problem.

Let us note the intuitive meaning of this measure. Consider the one-dimensional case, and suppose it is known that $a < x < b$ but we have no other prior information. Then there are no Lagrange multipliers λ_k, and (12.11) reduces to

$$p(x|I) = \left[\int_a^b dx \, m(x) \right]^{-1} m(x), \quad a < x < b. \tag{12.15}$$

Except for a constant factor, the measure $m(x)$ is also the prior distribution describing 'complete ignorance' of x. The ambiguity is, therefore, just the ancient one which has always plagued Bayesian statistics: how do we find the prior representing 'complete ignorance'? Once this problem is solved, the maximum entropy principle will lead to a definite, parameter-independent method of setting up prior distributions based on any testable prior information. Since this problem has been the subject of so much discussion and controversy for 200 years, we wish to state what appears to us a constructive attitude toward it.

To reject the question, as some have done, on the grounds that the state of complete ignorance does not 'exist' would be just as absurd as to reject Euclidean geometry on the grounds that a physical point does not exist. In the study of inductive inference, the notion of complete ignorance intrudes itself into the theory just as naturally and inevitably as the concept of zero in arithmetic.

If one rejects the consideration of complete ignorance on the grounds that the notion is vague and ill-defined, the reply is that the notion cannot be evaded in any full theory of inference. So if it is still ill-defined, then a major and immediate objective must be to find a precise definition which will agree with intuitive requirements and be of constructive use in a mathematical theory.

With this in mind, let us survey some previous thoughts on the problem. Bayes suggested, in one particular case, that we express complete ignorance by assigning a uniform prior probability density; the domain of useful applications of this rule is certainly not zero, for Laplace was led to some of the most important discoveries in celestial mechanics by using it in analysis of astronomical data. However, Bayes' rule has the obvious difficulty that it is not invariant under a change of parameters, and there seems to be no criterion for telling us which parameterization to use. (We note in passing that the notions of an unbiased estimator, an efficient estimator, and a shortest confidence interval are all subject to just the same ambiguity with equally serious consequences, and so orthodox statistics cannot claim to have solved this problem any better than Bayes did.)

Jeffreys (1931; 1939, 1957 edn) suggested that we assign a prior $d\sigma/\sigma$ to a continuous parameter σ known to be positive, on the grounds that we are then saying the same thing whether we use the parameter σ or σ^m. Such a desideratum is surely a step in the right direction; however, it cannot be extended to more general parameter changes. We do not want (and obviously cannot have) invariance of the form of the prior under all parameter changes; what we want is invariance of content, but the rules of probability theory already determine how the prior must transform, under any parameter change, so as to achieve this.

The real problem, therefore, must be stated rather differently. We suggest that the proper question to ask is: 'For which choice of parameters does a given form, such as that of Bayes or Jeffreys, apply?' Our parameter spaces seem to have a mollusk-like quality that prevents us from answering this, unless we can find a new principle that gives them a property of 'rigidity'.

Stated in this way, we recognize that problems of just this type have already appeared and have been solved in other branches of mathematics. In Riemannian geometry and general relativity theory, we allow arbitrary continuous coordinate transformations; yet the property of rigidity is maintained by the concept of the invariant line element, which enables us to make statements of definite geometrical and physical meaning independently of the choice of coordinates. In the theory of continuous groups, the group parameter space has just this mollusk-like quality until the introduction of invariant group measure by Harr (1933), Pontryagin (1946), and Wigner (1959). We seek to do something very similar to this for the parameter spaces of statistics.

The idea of utilizing groups of transformations in problems related to this was discussed by Poincaré (1912) and more recently by Hartigan (1964), Stone (1965) and Fraser (1966). In the following sections we give four examples of a different group theoretical method of reasoning developed largely by Wigner (1959) and Weyl (1961), which has met with great success in physical problems and seems uniquely adapted to our problem.

12.4 Transformation groups

The method of reasoning is best illustrated by some simple examples, the first of which also happens to be one of the most important in practice.

12.4.1 Location and scale parameters

We sample from a continuous two-parameter distribution

$$p(x|v\sigma)dx = \phi(x, v, \sigma)\,dx \qquad (12.16)$$

and consider problem A, as follows.

Problem A

Given a sample $\{x_1, \ldots, x_n\}$, estimate v and σ. The problem is indeterminate, both mathematically and conceptually, until we introduce a definite prior distribution

$$p(v\sigma|I)\,dv\,d\sigma = f(v, \sigma)\,dv\,d\sigma, \qquad (12.17)$$

but if we merely specify 'complete initial ignorance', this does not tell us which function $f(v, \sigma)$ to use.

Suppose we carry out a change of variables to the new quantities $\{x', v', \sigma'\}$ according to

$$v' = v + b$$
$$\sigma' = a\sigma \qquad (12.18)$$
$$x' - v' = a(x - v),$$

where $0 < a < \infty$, $-\infty < b < \infty$. The distribution (12.16) expressed in the new variables is

$$p(x'|v'\sigma')dx' = \psi(x', v', \sigma')dx' = \phi(x, v, \sigma)\,dx, \qquad (12.19)$$

or, from (12.18),

$$\psi(x', v', \sigma') = a^{-1}\phi(x, v, \sigma). \tag{12.20}$$

Likewise, the prior distribution is changed to $g(v', \sigma')$, where, from the Jacobian of the transformation (12.18),

$$g(v', \sigma') = a^{-1} f(v, \sigma). \tag{12.21}$$

The above relations will hold whatever the distributions $\phi(x, v, \sigma)$, $f(v, \sigma)$.

Now suppose the distribution (12.16) is invariant under the group of transformations (12.18), so that ψ and ϕ are the same *function:*

$$\psi(x, v, \sigma) = \phi(x, v, \sigma), \tag{12.22}$$

whatever the values of a, b. The condition for this invariance is that $\phi(x, v, \sigma)$ must satisfy the functional equation

$$\phi(x, v, \sigma) = a\phi(ax - av + v + b, v + b, a\sigma). \tag{12.23}$$

Differentiating with respect to a, b and solving the resulting differential equation, we find that the general solution of (12.23) is

$$\phi(x, v, \sigma) = \frac{1}{\sigma} h\left(\frac{x - v}{\sigma}\right), \tag{12.24}$$

where $h(q)$ is an arbitrary function. Thus, the usual definition of a location parameter v and a scale parameter σ is equivalent to specifying that the distribution shall be invariant under the group of transformations (12.18).

What do we mean by the statement that we are 'completely ignorant' of v and σ except for the knowledge that v is a location parameter and σ is a scale parameter? To answer this, we might reason as follows. If a change of scale can make the problem appear in any way different to us, then we were *not* completely ignorant; we must have had some kind of information about the absolute scale of the problem. Likewise, if a shift of location can make the problem appear in any way different, then we must have had some prior information about location. In other words, 'complete ignorance' of a location and a scale parameter is a state of knowledge such that *a change of scale and shift of location does not change that state of knowledge.* We shall presently have to state this more carefully, but first let us see its consequences. Consider, therefore, problem B.

Problem B

Given a sample $\{x'_1, \ldots, x'_n\}$, estimate v' and σ'. If we are 'completely ignorant' in the above sense, then we must consider A and B as entirely equivalent problems; they have identical sampling distributions, and our state of prior knowledge about v' and σ' in problem B is exactly the same as for v and σ in problem A.

Our basic desideratum now acquires a nontrivial content; for we have formulated two problems in which we have the same prior information. Consistency demands, therefore, that we assign the same prior probability distribution in them. Thus, f and g must be the same function:

$$f(v, \sigma) = g(v, \sigma) \tag{12.25}$$

whatever the values of (a, b). But the form of the prior distribution is now uniquely determined; for, combining (12.18), (12.21), and (12.25), we see that $f(v, \sigma)$ must satisfy the functional equation

$$f(v, \sigma) = af(v + b, a\sigma), \tag{12.26}$$

whose general solution is

$$f(v, \sigma) = \frac{\text{const.}}{\sigma} \tag{12.27}$$

which is the Jeffreys rule!

We must not jump to the conclusion that the prior (12.27) has been determined by the form (12.24) of the population. Indeed, it would be very disconcerting if the form of the prior were determined merely by the form of the population from which we are sampling; any principle which led to such a result would be suspect. Examination of the above reasoning shows, however, that the result (12.27) was uniquely determined by the *transformation group* (12.18), and not by the form of the distribution (12.24).

To illustrate this, note that there is more than one transformation group under which (12.24) is invariant. In the transformations (12.18) we carry out a change of scale by a factor a and a translation b. Denoting this operation by the symbol (a, b), we can carry out the transformation (a_1, b_1), then (a_2, b_2), and, from (12.18), obtain the composition law of group elements:

$$(a_2, b_2)(a_1, b_1) = (a_2 a_1, b_2 + b_1). \tag{12.28}$$

Thus the group (12.18) is Abelian, the direct product of two one-parameter groups. It has a faithful representation in terms of the matrices

$$\begin{pmatrix} a & 0 \\ 0 & \exp\{b\} \end{pmatrix}. \tag{12.29}$$

Now consider the group of transformations in which we first carry out a change of scale a on the quantities, and follow this by a translation b. This group is given by

$$\begin{aligned} v' &= av + b \\ \sigma' &= a\sigma \\ x' &= ax + b. \end{aligned} \tag{12.30}$$

These transformations have the composition law

$$(a_2, b_2)(a_1, b_2) = (a_2 a_1, a_2 b_1 + b_2), \tag{12.31}$$

and so the group (12.30) is non-Abelian; it has a faithful representation in terms of the matrices

$$\begin{pmatrix} a & b \\ 0 & 1 \end{pmatrix}, \tag{12.32}$$

which cannot be reduced to diagonal form. Therefore, (12.18) and (12.30) are entirely different groups.

If we specify the transformation group (12.30) instead of (12.18), (12.21) and (12.23) are modified to

$$g(v', \sigma') = a^{-2} f(v, \sigma), \tag{12.33}$$

and

$$\phi(x, v, \sigma) = a\phi(ax + b, av + b, a\sigma). \tag{12.34}$$

But we find that the general solution of (12.34) is also (12.24); and so both groups define location and scale parameters equally well. However, their consequences for the prior are different; for the functional equation (12.26) is modified to

$$f(v, \sigma) = a^2 f(av + b, a\sigma), \tag{12.35}$$

whose general solution is

$$f(v, \sigma) = \frac{\text{const.}}{\sigma^2}. \tag{12.36}$$

Thus, the state of knowledge which is invariant under the group (12.18) is *not* the same as that which is invariant under (12.30); and we see a new subtlety in the concept of 'complete ignorance'. In order to define it unambiguously, it is not enough to say merely, 'A change of scale and shift of location does not change that state of knowledge'. We must specify the precise manner in which these operations are to be carried out; i.e. *we must specify a definite group of transformations.*

We thus face the question: Which group, (12.18) or (12.30), really describes the prior information? The difficulty with (12.30) lies in the equations $x' = ax + b$, $v' = ax + b$; thus, the change of scale operation is to be carried out about two points denoted by $x = 0$, $v = 0$. But, if we are 'completely ignorant' about location, then the condition $x = 0$ has no particular meaning; what determines this fixed point about which the change of scale is to be carried out?

In every problem which I have been able to imagine, it is the group (12.18), and therefore the Jeffreys prior probability rule, which seems appropriate. Here the change of scale involves only the difference $\{x - v\}$; thus it is carried out about a point which is itself arbitrary, and so no 'fixed point' is defined by the group (12.18). However, it will be interesting to see whether others can produce examples in which the point $x = 0$ always has a special meaning, justifying the stronger prior (12.36).

To summarize: if we merely specify 'complete initial ignorance', we cannot hope to obtain any definite prior distribution, because such a statement is too vague to define any mathematically well-posed problem. We are defining this state of knowledge far more precisely if we can specify a set of operations which we recognize as transforming the problem into an equivalent one. Having found such a set of operations, the basic desideratum of consistency then places nontrivial restrictions on the form of the prior.

12.4.2 A Poisson rate

As another example, not very different mathematically but differently verbalized, consider a Poisson process. The probability that exactly n events will occur in a time interval t is

$$p(n|\lambda t) = \frac{(\lambda t)^n}{n!} \exp\{-\lambda t\}, \tag{12.37}$$

and by observing the number of events we wish to estimate the rate constant λ. We are initially completely ignorant of λ except for the knowledge that it is a rate constant of physical dimensions $(\text{seconds})^{-1}$, i.e. we are completely ignorant of the absolute time scale of the process.

Suppose, then, that two observers, Mr X and Mr X', whose watches run at different rates such that their measurements of a given interval are related by $t = qt'$, conduct this experiment. Since they are observing the same physical experiment, their rate constants must be related by $\lambda't' = \lambda t$, or $\lambda' = q\lambda$. They assign prior distributions

$$p(d\lambda|X) = f(\lambda)\, d\lambda, \tag{12.38}$$

$$p(d\lambda'|X') = g(\lambda')\, d\lambda', \tag{12.39}$$

and if these are mutually consistent (i.e. they have the same content), it must be that $f(\lambda)d\lambda = g(\lambda')d\lambda'$; or $f(\lambda) = qg(\lambda')$. But Mr X and Mr X' are both completely ignorant, and they are in the same state of knowledge, and so f and g must be the same function: $f(\lambda) = g(\lambda)$. Combining those relations gives the functional equation $f(\lambda) = qf(q\lambda)$ or

$$p(d\lambda|X) \sim \lambda^{-1}d\lambda. \tag{12.40}$$

To use any other prior than this will have the consequence that a change in the time scale will lead to a change in the form of the prior, which would imply a different state of prior knowledge; but if we are completely ignorant of the time scale, then all time scales should appear equivalent.

12.4.3 Unknown probability for success

As a third and less trivial example, where intuition did not anticipate the result, consider Bernoulli trials with an unknown probability for success. Here the probability for success is itself the parameter θ to be estimated. Given θ, the probability that we shall observe r successes in n trials is

$$p(r|n\theta) = \binom{n}{r} \theta^r (1 - \theta)^{n-r}, \tag{12.41}$$

and again the question is: What prior distribution $f(\theta)\,d\theta$ describes 'complete initial ignorance' of θ?

In discussing this problem, Laplace followed the example of Bayes and answered the question with the famous sentence: 'When the probability for a simple event is unknown, we may suppose all values between zero and one as equally likely.' In other words, Bayes and Laplace used the uniform prior $f_B(\theta) = 1$. However, Jeffreys (1939) and Carnap (1952) have noted that the resulting rule of succession does not seem to correspond well with the inductive reasoning which we all carry out intuitively. Jeffreys suggested that $f(\theta)$ ought to give greater weight to the end-points $\theta = (0, 1)$ if the theory is to account for the kind of inferences made by a scientist.

For example, in a chemical laboratory we find a jar containing an unknown and unlabeled compound. We are at first completely ignorant as to whether a small sample of this compound will dissolve in water or not. But, having observed that one small sample does dissolve, we infer immediately that all samples of this compound are water soluble, and although this conclusion does not carry quite the force of deductive proof, we feel strongly that the inference was justified. Yet the Bayes–Laplace rule leads to a negligibly small probability for this being true, and yields only a probability of 2/3 that the next sample tested will dissolve.

Now let us examine this problem from the standpoint of transformation groups. There is a conceptual difficulty here, since $f(\theta)\,d\theta$ is a 'probability for a probability'. However, it can be removed by carrying the notion of a split personality to extremes; instead of supposing that $f(\theta)$ describes the state of knowledge of any one person, imagine that we have a large population of individuals who hold varying beliefs about the probability for success, and that $f(\theta)$ describes the distribution of their beliefs. Is it possible that, although each individual holds a definite opinion, the population as a whole is completely ignorant of θ? What distribution $f(\theta)$ describes a population in a state of total confusion on the issue?

Since we are concerned with a consistent extension of probability theory, we must suppose that each individual reasons according to the mathematical rules (Bayes' theorem, etc.) of probability theory. The reason they hold different beliefs is, therefore, that they have been given different and conflicting information; one man has read the editorials of the *St Louis Post-Dispatch*, another the *Los Angeles Times*, one has read the *Daily Worker*, another the *National Review*, etc., and nothing in probability theory tells one to doubt the truth of what he has been told in the statement of the problem.

Now suppose that, before the experiment is performed, one more definite piece of evidence E is given simultaneously to all of them. Each individual will change his state of belief according to Bayes' theorem; Mr X, who had previously held the probability for success to be

$$\theta = p(S|X), \tag{12.42}$$

will change it to

$$\theta' = p(S|EX) = \frac{p(S|X)p(E|SX)}{p(E|SX)p(S|X) + p(E|FX)p(F|X)}, \tag{12.43}$$

where $p(F|X) = 1 - p(S|X)$ is his prior belief in probability for failure. This new evidence thus generates a mapping of the parameter space $0 \leq \theta \leq 1$ onto itself, given from (12.43) by

$$\theta' = \frac{a\theta}{1 - \theta + a\theta}, \tag{12.44}$$

where

$$a = \frac{p(E|SX)}{p(E|FX)}. \tag{12.45}$$

If the population as a whole can learn nothing from this new evidence, then it would seem reasonable to say that the population has been reduced, by conflicting propaganda, to a state of total confusion on the issue. We therefore define the state of 'total confusion' or 'complete ignorance' by the condition that, after the transformation (12.44), the number of individuals who hold beliefs in any given range $\theta_1 < \theta < \theta_2$ is the same as before.

The mathematical problem is again straightforward. The original distribution for beliefs $f(\theta)$ is shifted by the transformation (12.44) to a new distribution $g(\theta')$ with

$$f(\theta)\,d\theta = g(\theta')\,d\theta', \tag{12.46}$$

and if the population as a whole learned nothing, then f and g must be the same function:

$$f(\theta) = g(\theta). \tag{12.47}$$

Combining (12.44), (12.46), and (12.47), we find that $f(\theta)$ must satisfy the functional equation

$$af\left(\frac{a\theta}{1 - \theta - a\theta}\right) = (1 - \theta + a\theta)^2 f(\theta). \tag{12.48}$$

This may be solved directly by eliminating a between (12.44) and (12.48) or, in the more usual manner, by differentiating with respect to a and setting $a = 1$. This leads to the differential equation

$$\theta(1 - \theta)f'(\theta) = (2\theta - 1)f(\theta), \tag{12.49}$$

whose solution is

$$f(\theta) = \frac{\text{const.}}{\theta(1 - \theta)}, \tag{12.50}$$

which has the qualitative property anticipated by Jeffreys. Now that the imaginary population of individuals has served its purpose of revealing the transformation group (12.44) of the problem, let them coalesce again into a single mind (that of a statistician who wishes to estimate θ), and let us examine the consequences of using (12.50) as our prior distribution.

If we had observed r successes in n trials, then from (12.41) and (12.50) the posterior distribution for θ is (provided that $r \geq 1, n - r \geq 1$)

$$p(d\theta \mid rn) = \frac{(n-1)!}{(r-1)!(n-r-1)!} \theta^{r-1}(1-\theta)^{n-r-1} d\theta. \tag{12.51}$$

This distribution has expectation value and variance

$$\langle \theta \rangle = \frac{r}{n} = f, \tag{12.52}$$

$$\sigma^2 = \frac{f(1-f)}{n+1}. \tag{12.53}$$

Thus the 'best' estimate of the *probability* of success, by the criterion of quadratic loss function, is just equal to the observed *frequency* of success f; and this is also equal to the probability for success at the next trial, in agreement with the intuition of everybody who has studied Bernoulli trials. On the other hand, the Bayes–Laplace uniform prior would lead instead to the mean value $\langle \theta \rangle_B = (r+1)/(n+2)$ of the rule of succession, which has always seemed a bit peculiar.

For interval estimation, numerical analysis shows that the conclusions drawn from (12.51) are, for all practical purposes, the same as those based on confidence intervals (i.e. the shortest 90% confidence interval for θ is nearly equal to the shortest 90% posterior probability interval determined from (12.51)). If $r \gg 1$ and $(n - r) \gg 1$, the normal approximation to (12.51) will be valid, and the $100P\%$ posterior probability interval is simply $(f \pm q\sigma)$, where q is the $(1 + P)/2$ percentile of the normal distribution; for the 90%, 95%, and 99% levels, $q = 1.645, 1.960$, and 2.576, respectively. Under conditions where this normal approximation is valid, the difference between this result and the exact confidence interval is generally less than the difference between various published confidence interval tables, which have been calculated from different approximation schemes.

If $r = (n - r) = 1$, (12.51) reduces to $p(d\theta \mid r, n) = d\theta$, the uniform distribution which Bayes and Laplace took as their prior. Therefore, we can now interpret the Bayes–Laplace prior as describing not a state of complete ignorance, but the state of knowledge in which we have observed one success and one failure. It thus appears that the Bayes–Laplace choice will be the appropriate prior if the prior information assures us that it is physically possible for the experiment to yield either a success or a failure, while the distribution for complete ignorance (12.50) describes a 'pre-prior' state of knowledge in which we are not even sure of that.

If $r = 0$, or $r = n$, the derivation of (12.51) breaks down and the posterior distribution remains unnormalizable, proportional to $\theta^{-1}(1-\theta)^{n-1}$ or $\theta^{n-1}(1-\theta)^{-1}$, respectively. The weight is concentrated overwhelmingly on the value $\theta = 0$ or $\theta = 1$. The prior (12.50) thus accounts for the kind of inductive inference noted in the case of chemicals, which we all make intuitively. However, once we have seen at least one success and one failure, then we know that the experiment is a true binary one, in the sense of physical possibility, and from that point on all posterior distributions (12.51) remain normalized, permitting definite inferences about θ.

The transformation group method therefore yields a prior which appears to meet the common objections raised against the Laplace rule of succession; but we also see that whether (12.50) or the Bayes–Laplace prior is appropriate depends on the exact prior information available.

12.4.4 Bertrand's problem

Finally, we give an example where transformation groups may be used to find more informative priors. Bertrand's problem (Bertrand, 1889) was stated originally in terms of drawing a straight line 'at random' intersecting a circle. It will be helpful to think of this in a more concrete way; presumably, we do no violence to the problem (i.e. it is still just as 'random') if we suppose that we are tossing straws onto the circle, without specifying how they are tossed. We therefore formulate the problem as follows.

A long straw is tossed at random onto a circle; given that it falls so that it intersects the circle, what is the probability that the chord thus defined is longer than a side of the inscribed equilateral triangle? Since Bertrand proposed it in 1889, this problem has been cited to generations of students to demonstrate that Laplace's 'principle of indifference' contains logical inconsistencies. For there appear to be many ways of defining 'equally possible' situations, and they lead to different results. Three of these are: assign uniform probability density to (A) the linear distance between centers of chord and circle, (B) angles of intersections of the chord on the circumference, (C) the center of the chord over the interior area of the circle. These assignments lead to the results $p_A = 1/2$, $p_B = 1/3$, and $p_C = 1/4$, respectively.

Which solution is correct? Of the ten authors cited (Bertrand 1889; Borel 1909; Poincaré 1912; Uspensky 1937; Northrop 1944; von Mises 1957; Gnedenko 1962; Kendell and Moran 1963; Mosteller 1965), only Borel is willing to express a definite preference, although he does not support it by any proof. Von Mises takes the opposite extreme, declaring that such problems (including the similar Buffon needle problem) do not belong to the field of probability theory at all. The others, including Bertrand, take the intermediate position of saying simply that the problem has no definite solution because it is ill-posed, the phrase 'at random' being undefined.

In works on probability theory, this state of affairs has been interpreted, almost universally, as showing that the principle of indifference must be totally rejected. Usually, there is the further conclusion that the only valid basis for assigning probabilities is frequency in some random experiment. It would appear, then, that the only way of answering Bertrand's question is to perform the experiment.

But do we really believe that it is beyond our power to predict by 'pure thought' the result of such a simple experiment? The point at issue is far more important than merely resolving a geometric puzzle; for, as discussed further in the conclusion of this chapter, applications of probability theory to physical experiments usually lead to problems of just this type; i.e. they appear at first to be undetermined, allowing many different solutions with nothing to choose among them. For example, given the average particle density and total energy of a gas, predict its viscosity. The answer, evidently, depends on the exact spatial and velocity distributions of the molecules (in fact, it depends critically on position–velocity correlations), and nothing in the given data seems to tell us which distribution to assume. Yet physicists *have* made definite choices, guided by the principle of indifference, and they *have* led us to correct and nontrivial predictions of viscosity and many other physical phenomena.

Thus, while in some problems the principle of indifference has led us to paradoxes, in others it has produced some of the most important and successful applications of probability theory. To reject the principle without having anything better to put in its place would lead to consequences so unacceptable that for many years even those who profess the most faithful adherence to the strict frequency definition of probability have managed to overlook these logical difficulties in order to preserve some very useful solutions.

Evidently, we ought to examine the apparent paradoxes such as Bertrand's more closely; there is an important point to be learned about the application of probability theory to real physical situations.

It is evident that if the circle becomes sufficiently large, and the tosser sufficiently skilled, various results could be obtained at will. However, in the limit where the skill of the tosser must be described by a 'region of uncertainty' large compared with the circle, the distribution for chord lengths must surely go into one unique function obtainable by 'pure thought'. A viewpoint toward probability theory which cannot show us how to calculate this function from first principles, or even denies the possibility of doing this, would imply severe – and, to a physicist, intolerable – restrictions on the range of useful applications of probability theory.

An invariance argument was applied to problems of this type by Poincaré (1912), and cited more recently by Kendall and Moran (1963). In this treatment we consider straight lines drawn 'at random' in the xy plane. Each line is located by specifying two parameters (u, v) such that the equation of the line is $ux + vy = 1$, and one can ask: Which probability density $p(u, v)\, du\, dv$ has the property that it is invariant in *form* under the group of Euclidean transformations (rotations and translations) of the plane? This is a readily solvable problem (Kendall and Moran 1963), with the answer $p(u, v) = (u^2 + v^2)^{-3/2}$.

Yet evidently this has not seemed convincing; for later authors have ignored Poincaré's invariance argument, and have adhered to Bertrand's original judgment that the problem has no definite solution. This is understandable, for the statement of the problem does not specify that the distribution for straight lines is to have this invariance property, and we do not see any compelling reason to expect that a rain of straws produced in a real experiment would have it. To assume this would seem to be an intuitive judgment resting on no stronger grounds than the ones which led to the three different solutions above. All of this amounts to trying to guess what properties a 'random' rain of straws should have, by specifying the intuitively 'equally possible' events; and the fact remains that different intuitive judgments lead to different results.

The viewpoint just expressed, which is by far the most common in the literature, clearly represents one valid way of interpreting the problem. If we can find another viewpoint according to which such problems *do* have definite solutions, *and define the conditions under which these solutions are experimentally verifiable,* then, while it would perhaps be overstating the case to say that this new viewpoint is more 'correct' in principle than the conventional one, it will surely be more useful in practice.

We now suggest such a viewpoint, and we understand from the start that we are not concerned at this stage with *frequencies* of various events. We ask rather: Which probability distribution describes our *state of knowledge* when the only information available is that given in the above statement of the problem? Such a distribution must conform to the desideratum of consistency formulated in Chapter 1: in two problems where we have the same state of knowledge we must assign the same probabilities. The essential point is this: if we start with the assumption that Bertrand's problem has a definite solution *in spite of the many things left unspecified,* then the statement of the problem automatically implies certain invariance properties, which in no way depend on our intuitive judgments. After the solution is found, it may be used as a prior for Bayesian inference whether or not it has any correspondence with frequencies; any frequency connections that may emerge will be regarded as an additional bonus, which justify its use also for direct physical prediction.

Bertrand's problem has an obvious element of rotational symmetry, recognized in all the proposed solutions; however, this symmetry is irrelevant to the distribution for chord lengths. There are two other 'symmetries' which are highly relevant: neither Bertrand's original statement nor our restatement in terms of straws specified the exact size of the circle, or its exact location. If, therefore, the problem is to have any definite solution at all, it must be 'indifferent' to these circumstances; i.e. it must be unchanged by a small change in the size or position of the circle. This seemingly trivial statement, as we will see, fully determines the solution.

It would be possible to consider all these invariance requirements simultaneously by defining a four-parameter transformation group, whereupon the complete solution would appear suddenly, as if by magic. However, it will be more instructive to analyze the effects of these invariances separately, and see how each places its own restrictions on the form of the solution.

Rotational invariance

Let the circle have radius R. The position of the chord is determined by giving the polar coordinates (r, θ) of its center. We seek to answer a more detailed question than Bertrand's: What probability density $f(r, \theta)\mathrm{d}A = f(r, \theta)r\,\mathrm{d}r\,\mathrm{d}\theta$ should we assign over the interior area of the circle? The dependence on θ is actually irrelevant to Bertrand's question, since the distribution for chord lengths depends only on the radial distribution

$$g(r) = \int_0^{2\pi} d\theta \, f(r, \theta). \tag{12.54}$$

However, intuition suggests that $f(r, \theta)$ should be independent of θ, and the formal transformation group argument deals with the rotational symmetry as follows.

The starting point is the observation that the statement of the problem does not specify whether the observer is facing north or east; therefore, if there is a definite solution, it must not depend on the direction of the observer's line of sight. Suppose, therefore, that two different observers, Mr X and Mr Y, are watching this experiment. They view the experiment from different directions, their lines of sight making an angle α. Each uses a coordinate system oriented along his line of sight. Mr X assigns the probability density $f(r, \theta)$ in his coordinate system S; and Mr Y assigns $g(r, \theta)$ in his system S_α. Evidently, if they are describing the same situation, then it must be true that

$$f(r, \theta) = g(r, \theta - \alpha), \tag{12.55}$$

which expresses a simple change of variables, transforming a fixed distribution f to a new coordinate system; this relation will hold whether or not the problem has rotational symmetry.

But now we recognize that, because of the rotational symmetry, the problem appears exactly the same to Mr X in his coordinate system as it does to Mr Y in his. Since they are in the same state of knowledge, our desideratum of consistency demands that they assign the same probability distribution; and so f and g must be the same function:

$$f(r, \theta) = g(r, \theta). \tag{12.56}$$

These relations must hold for all α in $0 \leq \alpha \leq 2\pi$; and so the only possibility is $f(r, \theta) = f(r)$.

This formal argument may appear cumbersome when compared with our obvious flash of intuition; and of course it is, when applied to such a trivial problem. However, as Wigner (1931) and Weyl (1946) have shown in other physical problems, it is this cumbersome argument that generalizes at once to nontrivial cases where our intuition fails us. It always consists of two steps: we first find a transformation equation like (12.55) which shows how two problems are related to each other, irrespective of symmetry; then a symmetry relation like (12.56) which states that we have formulated two equivalent *problems*. Combining them leads in most cases to a functional equation which imposes some restriction on the form of the distribution.

Scale invariance

The problem is reduced, by rotational symmetry, to determining a function $f(r)$, normalized according to

$$\int_0^{2\pi} d\theta \int_0^R r\,dr\, f(r) = 1. \tag{12.57}$$

Again, we consider two different problems; concentric with a circle of radius R, there is a circle of radius aR, $0 < a \le 1$. Within the smaller circle there is a probability $h(r)r\,dr\,d\theta$ which answers the question: given that a straw intersects the smaller circle, what is the probability that the center of its chord lies in the area $dA = r\,dr\,d\theta$?

Any straw that intersects the small circle will also define a chord on the larger one; and so, within the small circle $f(r)$ must be proportional to $h(r)$. This proportionality is, of course, given by the standard formula for a conditional probability, which in this case takes the form

$$f(r) = 2\pi h(r) \int_0^{aR} r\,dr\, f(r) \qquad 0 < a \le 1, \quad 0 \le r \le aR. \tag{12.58}$$

This transformation equation will hold whether or not the problem has scale invariance.

But we now invoke scale invariance; to two different observers with different size eyeballs, the problems of the large and small circles would appear exactly the same. If there is any unique solution independent of the size of the circle, there must be another relationship between $f(r)$ and $h(r)$, which expresses the fact that one problem is merely a scaled-down version of the other. Two elements of area $r\,dr\,d\theta$ and $(ar)\,d(ar)\,d\theta$ are related to the large and small circles, respectively, in the same way; and so they must be assigned the same probabilities by the distributions $f(r)$ and $h(r)$, respectively:

$$h(ar)(ar)\,d(ar)\,d\theta = f(r)r\,dr\,d\theta, \tag{12.59}$$

or

$$a^2 h(ar) = f(r), \tag{12.60}$$

which is the symmetry equation. Combining (12.58) and (12.60), we see that invariance under change of scale requires that the probability density satisfy the functional equation

$$a^2 f(ar) = 2\pi f(r) \int_0^{aR} u\,du\, f(u) \qquad 0 < a \le 1, \quad 0 \le r \le R. \tag{12.61}$$

Differentiating with respect to a, setting $a = 1$, and solving the resulting differential equation, we find that the most general solution of (12.61) satisfying the normalization condition (12.57) is

$$f(r) = \frac{qr^{q-2}}{2\pi R^q}, \tag{12.62}$$

where q is a constant in the range $0 < q < \infty$, not further determined by scale invariance.

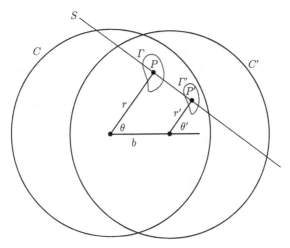

Fig. 12.1. A Straw S intersects two slightly displaced circles C and C'.

We note that the proposed solution B in the introduction has now been eliminated, for it corresponds to the choice $f(r) \sim 1/\sqrt{(R^2 - r^2)}$, which is not of the form (12.62). This means that if the intersections of chords on the circumference were distributed in angle uniformly and independently on one circle, this would not be true for a smaller circle inscribed in it; i.e. the probability assignment of B could be true for, at most, only one size of circle. However, solutions A and C are still compatible with scale invariance, corresponding to the choices $q = 1$ and $q = 2$, respectively.

Translational invariance

We now investigate the consequences of the fact that a given straw S can intersect two circles C, C' of the same radius R, but with a relative displacement b. Referring to Figure 12.1, the midpoint of the chord with respect to circle C is the point P, with coordinates (r, θ); while the same straw defines a midpoint of the chord with respect to C' at the point P' whose coordinates are (r', θ'). From Figure 12.1 the coordinate transformation $(r, \theta) \rightarrow (r', \theta')$ is given by

$$r' = |r - b\cos\theta|, \tag{12.63}$$

$$\theta' = \begin{cases} \theta & r > b\cos\theta \\ \theta + \pi & r < b\cos\theta. \end{cases} \tag{12.64}$$

As P varies over the region Γ, P' varies over Γ', and vice versa; thus the straws define a 1:1 mapping of Γ onto Γ'.

Now we note the translational symmetry; since the statement of the problem gave no information about the location of the circle, the problems of C and C' appear exactly the same to two slightly displaced observers O and O'. Our desideratum of consistency then demands that they assign probability densities in C and C', respectively, which have the same form (12.62) with the same value of q.

It is further necessary that these two observers assign equal probabilities to the regions Γ and Γ', respectively, since (a) they are probabilities of the same event, and (b) the probability that a straw which intersects one circle will also intersect the other, thus setting up this correspondence, is also the same in the two problems. Let us see whether these two requirements are compatible.

The probability that a chord intersecting C will have its midpoint in Γ is

$$\int_\Gamma r\,dr\,d\theta\,f(r) = \frac{q}{2\pi R^q}\int_\Gamma dr\,d\theta\,r^{q-1}. \tag{12.65}$$

The probability that a chord intersecting C' will have its midpoint in Γ' is

$$\frac{q}{2\pi R^q}\int_{\Gamma'} dr'\,d\theta'\,(r')^{q-1} = \frac{q}{2\pi R^q}\int_\Gamma dr\,d\theta\,|r - b\cos\theta|^{q-1}, \tag{12.66}$$

where we have transformed the integral back to the variables (r,θ) by use of (12.63) and (12.64), noting that the Jacobian is unity. Evidently, (12.65) and (12.66) will be equal for arbitrary Γ if and only if $q = 1$; and so our distribution $f(r)$ is now uniquely determined.

The proposed solution C in the introduction is thus eliminated for lack of translational invariance; a rain of straws which had the property assumed with respect to one circle could not have the same property with respect to a slightly displaced one.

We have found that the invariance requirements determine the probability density

$$f(r,\theta) = \frac{1}{2\pi Rr}, \quad 0 \le r \le R, \quad 0 \le \theta \le 2\pi, \tag{12.67}$$

corresponding to solution A in the introduction. It is interesting that this has a singularity at the center, the need for which can be understood as follows. The condition that the midpoint (r,θ) falls within a small region Δ imposes restrictions on the possible directions of the chord. But as Δ moves inward, as soon as it includes the center of the circle all angles are suddenly allowed. Thus there is an infinitely rapid change in the 'manifold of possibilities'.

Further analysis (almost obvious from contemplation of Figure 12.1) shows that the requirement of translational invariance is so stringent that it already determines the result (12.67) uniquely; thus the proposed solution B is incompatible with either scale or translational invariance, and in order to find (12.67) it was not really necessary to consider scale invariance. However, the solution (12.67) would in any event have to be tested for scale invariance, and if it failed to pass that test we would conclude that the problem as stated has *no* solution; i.e. although at first glance it appears underdetermined, it would have to be regarded, from the standpoint of transformation groups, as overdetermined. As luck would have it, these requirements *are* compatible; and so the problem has one unique solution.

The distribution for chord lengths follows at once from (12.67). A chord whose midpoint is at (r,θ) has a length $L = 2\sqrt{(R^2 - r^2)}$. In terms of the reduced chord lengths, $x \equiv L/2R$, we obtain the universal distribution law

$$p(x)\,\mathrm{d}x = \frac{x\,\mathrm{d}x}{\sqrt{(1 - x^2)}}, \qquad 0 \leq x \leq 1, \tag{12.68}$$

in agreement with Borel's conjecture (1909).

Frequency correspondence

From the manner of its derivation, the distribution (12.68) would appear to have only a subjective meaning; while it describes the only possible state of knowledge corresponding to a unique solution in view of the many things left unspecified in the statement of Bertrand's problem, we have as yet given no reason to suppose that it has any relation to frequencies observed in the actual experiment. In general, of course, no such claim can be made; the mere fact that my state of knowledge gives me no reason to prefer one event over another is not enough to make the events occur equally often! Indeed, it is clear that no 'pure thought' argument, whether based on transformation groups or any other principle, can predict with certainty what must happen in a real experiment. And we can easily imagine a very precise machine which tosses straws in such a way as to produce any distribution for chord lengths we please on a given circle.

Nevertheless, we are entitled to claim a definite frequency correspondence for the result (12.68). For there is one 'objective fact' which *has* been proved by the above derivation: any rain of straws which does *not* produce a frequency distribution agreeing with (12.68) will necessarily produce different distributions on different circles.

This is all we need in order to predict with confidence that the distribution (12.68) *will* be observed in any experiment where the 'region of uncertainty' is large compared with the circle. For, if we lack the skill to toss straws so that, with certainty, they intersect a given circle, then surely we lack *a fortiori* the skill consistently to produce different distributions on different circles *within* this region of uncertainty!

It is for this reason that distributions predicted by the method of transformation groups turn out to have a frequency correspondence after all. Strictly speaking, this result holds only in the limiting case of 'zero skill', but, as a moment's thought will show, the skill required to produce any appreciable deviation from (12.68) is so great that in practice it would be difficult to achieve even with a machine.

These conclusions seem to be in direct contradiction to those of von Mises (1957), who denied that such problems belong to the field of probability theory at all. It appears to us that if we were to adopt von Mises' philosophy of probability theory strictly and consistently, the range of legitimate physical applications of probability theory would be reduced almost to the vanishing point. Since we have made a definite, unequivocal prediction, this issue has now been removed from the realm of philosophy into that of verifiable fact. The predictive power of the transformation group method can be put to the test quite easily in this and other problems by performing the experiments.

The Bertrand experiment has, in fact, been performed by the writer and Dr Charles E. Tyler, tossing broom straws from a standing position onto a 5 inch diameter circle drawn on the floor. Grouping the range of chord lengths into ten categories, 128 successful tosses confirmed Eq. (12.68) with an embarrassingly low value of chi-squared. However, experimental results will no doubt be more convincing if reported by others.

12.5 Comments

Bertrand's problem has a greater importance than appears at first glance, because it is a simple crystallization of a deeper paradox which has permeated much of probability theory from its beginnings. In 'real' physical applications, when we try to formulate the problem of interest in probability terms, we find almost always that a statement emerges which, like Bertrand's, appears too vague to determine any definite solution, because apparently essential things are left unspecified.

We elaborate the example noted in the introduction of the preceding section. Given a gas of N molecules in a volume V, with known intermolecular forces, total energy E, predict its molecular velocity distribution, the pressure, distribution for pressure fluctuations, viscosity, thermal conductivity, and diffusion constant. Here again the viewpoint expressed by most writers on probability theory would lead one to conclude that the problem has no definite solution because it is ill-posed; the things specified are grossly inadequate to determine any unique probability distribution over microstates. If we reject the principle of indifference, and insist that the only valid basis for assigning probabilities is frequency in some random experiment, it would again appear that the only way of determining these quantities is to perform the experiment.

It is, however, a matter of record that over a century ago, without benefit of any frequency data on positions and velocities of molecules, James Clark Maxwell was able to predict all these quantities correctly by a 'pure thought' probability analysis, which amounted to recognizing the 'equally possible' cases. In the case of viscosity, the predicted dependence on density appeared at first to contradict common sense, casting doubt on Maxwell's analysis. But when the experiments were performed they confirmed Maxwell's prediction, leading to the first great triumph of kinetic theory. These are solid, positive accomplishments; and they cannot be made to appear otherwise merely by deploring Maxwell's use of the principle of indifference.

Likewise, we calculate the probability for obtaining various hands at poker; and we are so confident of the results that we are willing to risk money on bets which the calculations indicate are favorable to us. But underlying these calculations is the intuitive judgment that all distributions of cards are equally likely; and with a different judgment our calculations would give different results. Once again we are predicting definite, verifiable facts by 'pure thought' arguments based ultimately on recognizing the 'equally possible' cases; and yet present statistical doctrine, both orthodox and personalistic, denies that this is a valid basis for assigning probabilities!

The dilemma is thus apparent. On the one hand, one cannot deny the force of arguments which, by pointing to such things as Bertrand's paradox, demonstrate the ambiguities and dangers in the principle of indifference. On the other hand, it is equally undeniable that use of this principle has, over and over again, led to correct, nontrivial, and useful predictions. Thus it appears that, although we cannot wholly accept the principle of indifference, we cannot wholly reject it either; to do so would be to cast out some of the most important and successful applications of probability theory.

The transformation group method grew out of the writer's conviction that the principle of indifference has been unjustly maligned in the past; what it has needed was not blanket condemnation, but recognition of the proper way to apply it. We agree with most other writers on probability theory that it is dangerous to apply this principle at the level of indifference between *events,* because our intuition is a very unreliable guide in such matters, as Bertrand's paradox illustrates.

The principle of indifference may, in our view, be applied legitimately at the more abstract level of indifference between *problems;* because that is a matter that is definitely determined by the statement of a problem, independently of our intuition. Every circumstance left unspecified in the statement of a problem defines an invariance property which the solution must have if there is to be any definite solution at all. The transformation group, which expresses these invariances mathematically, imposes definite restrictions on the form of the solution, and in many cases fully determines it.

Of course, not all invariances are useful. For example, the statement of Bertrand's problem does not specify the time of day at which the straws are tossed, the color of the circle, the luminosity of Betelgeuse, or the number of oysters in Chesapeake Bay; from which we infer, correctly, that if the problem as stated is to have a unique solution, it must not depend on these circumstances. But this would not help us unless we had previously thought that these things might be germane.

Study of a number of cases makes it appear that the aforementioned dilemma can now be resolved as follows. We suggest that the cases in which the principle of indifference has been applied successfully in the past are just the ones in which the solution can be 'reverbalized' so that the actual calculations used are seen as an application of indifference between problems, rather than events.

The transformation group derivation of the Jeffreys prior enables us to see that prior in a new light. It has, perhaps, always been obvious that the real justification of the Jeffreys rule cannot lie merely in the fact that the parameter is positive. As a simple example, suppose that μ is known to be a location parameter; then both intuition and the preceding analysis agree that a uniform prior density is the proper way to express complete ignorance of μ. The relation $\mu = \theta - \theta^{-1}$ defines a 1:1 mapping of the region $(-\infty < \mu < \infty)$ onto the region $(0 < \theta < \infty)$; but the Jeffreys rule cannot apply to the parameter θ, consistency demanding that its prior density be taken proportional to $d\mu = (1 + \theta^{-2}) d\theta$. It appears that the fundamental justification of the Jeffreys rule is not merely that a parameter is positive, but that it is a *scale parameter.*

The fact that the distributions representing complete ignorance found by transformation groups cannot be normalized may be interpreted in two ways. One can say that it arises simply from the fact that our formulation of the notion of complete ignorance was an idealization that does not strictly apply in any realistic problem. A shift of location from a point in St Louis to a point in the Andromeda nebula, or a change of scale from the size of an atom to the size of our galaxy, does not transform any problem of earthly concern into a completely equivalent one. In practice we will always have some kind of prior knowledge about location and scale, and in consequence the group parameters (a, b) cannot vary over a truly infinite range. Therefore, the transformations (12.50) do not, strictly speaking, form a group. However, over the range which does express our prior ignorance, the above kind of arguments still apply. Within this range, the functional equations and the resulting form of the priors must still hold.

Our discussion of maximum entropy has shown a more constructive way of looking at this, however. Finding the distribution representing complete ignorance is only the first step in finding the prior for any realistic problem. The pre-prior distribution resulting from a transformation group does not strictly represent any realistic state of knowledge, but it does define the invariant measure for our parameter space, without which the problem of finding a realistic prior by maximum entropy is mathematically indeterminate.

13

Decision theory, historical background

'Your act was unwise,' I exclaimed 'as you see
  by the outcome.' He solemnly eyed me.
'When choosing the course of my action,' said he,
'I had not the outcome to guide me.'

Ambrose Bierce

In several previous discussions we inserted parenthetic remarks to the effect that 'there is still an essential point missing here, which will be supplied when we take up decision theory'. However, in postponing the topic until now, we have not deprived the reader of a needed technical tool, because the solution of the decision problem was, from our viewpoint, so immediate and intuitive that we did not need to invoke any underlying formal theory.

13.1 Inference vs. decision

The situation of appraising inference vs. decision arose as soon as we started applying probability theory to our first problem. When we illustrated the use of Bayes' theorem by sequential testing in Chapter 4, we noted that there is nothing in probability theory *per se* which could tell us where to put the critical levels at which the robot changes its decision: whether to accept the batch, reject it, or make another test. The location of these critical levels obviously depends in some way on value judgments as well as on probabilities; what are the consequences of making wrong decisions, and what are the costs of making further tests?

The same situation occurred in Chapter 6 when the robot was faced with the job of estimating a parameter. Probability theory determined only the robot's state of knowledge about the parameter; it did not tell the robot what estimate it should in fact make. We noted at that time that taking the mean value over the posterior pdf was the same as making that decision which minimizes the expected square of the error; but we noted also that in some cases we should really prefer the median.

Qualitatively and intuitively, these considerations are clear enough; but before we can claim to have a really complete design for our robot, we must clean up the logic of this, and show that our procedures were not just intuitive *ad hockeries*, but were optimal by some clearly defined criterion. Wald's decision theory aims to accomplish this.

A common feature of all the problems considered thus far was: probability theory alone can solve only the inference problem; i.e. it can give us only a probability distribution which represents the robot's final state of knowledge with all the available prior information and data taken into account. But in practice its job does not end at that point. *An essential thing which is still missing in our design of this robot is the rule by which it converts its final probability assignment into a definite course of action.* But for us, formal decision theory will only legitimize – not change – what our intuition has already told us to do.

Decision theory has for us a different kind of importance, in the light that it sheds on the centuries-old controversies about the foundations of probability theory. Decision theory can be derived equally well from either of two diametrically opposed views about what probability theory is, and it therefore forms a kind of bridge between them, and suggests that decision theory might help to resolve the controversy. We dwell here on the historical background of and relationship between the two approaches to the decision problem.

13.2 Daniel Bernoulli's suggestion

As one might expect from the way this situation appeared in the most elementary applications of probability theory, the relationship between the two approaches to decision theory is by no means a new problem. It was clearly recognized, and a definite solution offered for a certain class of problems, by Daniel Bernoulli (1738). In a crude form, the same principle had been seen even earlier, at the time when probability theory was concerned almost exclusively with problems of gambling. Although today it seems very hard to understand, the historical record shows clearly and repeatedly that the notion of 'expectation of profit' was very intuitive to the first workers in probability theory; even more intuitive than that of probability.

Consider each possibility, $i = 1, 2, \ldots n$, assign probabilities p_i to them, and also assign numbers M_i which represent the 'profit' we would obtain if the ith possibility should in fact turn out to be true. Then the expectation of profit is, in either of our standard notations,

$$E(M) = \langle M \rangle = \sum_{i=1}^{n} p_i M_i. \tag{13.1}$$

The prosperous merchants in 17th century Amsterdam bought and sold expectations as if they were tangible goods. It seemed obvious to many that a person acting in pure self-interest should always behave in such a way as to maximize his expected profit. This, however, led to some paradoxes (particularly that of the famous St Petersburg problem) which led Bernoulli to recognize that simple expectation of profit is not always a sensible criterion of action.

For example, suppose that your information leads you to assign probability 0.51 to heads in a certain slightly biased coin. Now you are given the choice of two actions: (1) to bet every cent you have at even money, on heads for the next toss of this coin; (2) not to bet at all. According to the criterion of expectation of profit, you should always choose to gamble when faced with this choice. Your expectation of profit, if you do not gamble, is zero; but if you do gamble, it is

$$\langle M \rangle = 0.51\, M_0 + 0.49\,(-M_0) = 0.02\, M_0 > 0, \qquad (13.2)$$

where M_0 is the amount you have now. Nevertheless it seemed obvious to Bernoulli, as it doubtless does also to the reader, that nobody in his right mind would really choose the first alternative. This means that our common sense, in some cases, rejects the criterion of maximizing expected profit.

Suppose that you are offered the following opportunity. You can bet any amount you want on the basis that, with probability $(1 - 10^{-6})$, you will lose your money; but with probability 10^{-6}, you will win $1\,000\,001$ times the amount you had wagered. Again, the criterion of maximizing expected profit says that you should bet all the money you have. Common sense rejects this solution even more forcefully.

Daniel Bernoulli proposed to resolve these paradoxes by recognition that the true value to a person, of receiving a certain amount of money, is not measured simply by the amount received; it depends also upon how much he has already. In other words, Bernoulli said that we should recognize that the mathematical expectation of profit is not the same thing as its 'moral expectation'. A modern economist is expressing the same idea when he speaks of the 'diminishing marginal utility of money'.

In the St Petersburg game we toss an honest coin until it comes up heads for the first time. The game is then terminated. If heads occurs for the first time at the nth throw, the player receives 2^n dollars. The question is: what is a 'fair' entrance fee for him to pay, for the privilege of playing this game? If we use the criterion that a fair game is one where the entrance fee is equal to the expectation of profit, you see what happens. The expectation is infinite:

$$\sum_{k=1}^{\infty}(2^{-k})(2^k) = \sum_{k=1}^{\infty} 1 = \infty. \qquad (13.3)$$

Nevertheless, it is clear again that no sane person would be willing to risk more than a very small amount on this game. We quote Laplace (1814, 1819) at this point:

Indeed, it is apparent that one franc has much greater value for him who possesses only 100 than for a millionaire. We ought then to distinguish the absolute value of the hoped-for benefit from its relative value. The latter is determined by the motives which make it desirable, whereas the first is independent of them. The general principle for assigning this relative value cannot be given, but here is one proposed by Daniel Bernoulli which will serve in many cases: The relative value of an infinitely small sum is equal to its absolute value divided by the total fortune of the person interested.

In other words, Bernoulli proposed that the 'moral value', or what the modern economist would call the 'utility' of an amount M of money, should be taken proportional to $\log(M)$. Laplace, in discussing the St Petersburg problem and this criterion, reports the following result without giving the calculation: a person whose total fortune is 200 francs ought not reasonably to stake more than 9 francs on the play of this game. Let us, 180 years later, check Laplace's calculation.

For a person whose initial 'fortune' is m francs, the fair fee $f(m)$ is determined by equating his present utility with his expected utility if he pays the fee and plays the game; i.e. $f(m)$ is the root of

$$\log(m) = \sum_{n=1}^{\infty} \frac{1}{2^n} \log(m - f + 2^n). \tag{13.4}$$

Computer evaluation gives $f(200) = 8.7204$; Laplace, without a computer, did his calculation very well. Likewise, $f(10^3) = 10.95$, $f(10^4) = 14.24$, $f(10^6) = 20.87$. Even a millionaire should not risk more than 21 francs on this dubious game.

It seems to us that this kind of numerical result is entirely reasonable. However, the logarithmic assignment of utility is not to be taken literally, either in the case of extremely small fortunes (as Laplace points out), or in the case of extremely large ones, as the following example of Savage (1954) shows.

Suppose your present fortune is $1 000 000; if your utility for money is proportional to the logarithm of the amount, you should be as willing as not to accept a wager in which, with probability one-half, you'll be left with only $1000, and with probability one-half you will be left with $1 000 000 000. Most of us would consider such a bet to be distinctly disadvantageous to a person with that initial fortune. This shows that our intuitive 'utility' for money must increase even less rapidly than the logarithm for extremely large values. Chernoff and Moses (1959) claim that it is bounded; this appears to us plausible theoretically, but not really demonstrated in the real world.

The gist of Daniel Bernoulli's suggestion was therefore that, in the gambler's problem of decision making under uncertainty, one should act so as to maximize the expected value, not necessarily of the profit itself, but of some function of the profit which he called the 'moral value'. In more modern terminology, the optimist will call this 'maximizing expected utility', while the pessimist will speak instead of 'minimizing expected loss', the loss function being taken as the negative of the utility function.

13.3 The rationale of insurance

Let us illustrate some of the above remarks briefly with the example of insurance, which is in some ways like the St Petersburg game. The following scenario is oversimplified in obvious ways; nevertheless, it makes some valid and important points. Insurance premiums are always set high enough to guarantee the insurance company a positive expectation of profit over all the contingencies covered in the contract, and every dollar the company earns is a dollar spent by a customer. Then why should anyone ever want to buy insurance?

The point is that the individual customer has a utility function for money that may be strongly curved over ranges of $1000; but the insurance company is so much larger that its utility for money is accurately linear over ranges of millions of dollars. Thus, let P be the premium for some proposed insurance contract, let $i = 1, \ldots, n$ enumerate the contingencies covered, the ith having probability w_i and cost to the insurance company, if

Table 13.1. *Expanded utility.*

	Buy	Don't buy
Company	$P - \sum w_i L_i$	0
Customer	$\log(M - P)$	$\sum w_i \log(M - L_i)$

it happens, of L_i. Let the prospective customer have Daniel Bernoulli's logarithmic utility for money and an initial amount M. Of course, by M we should understand his so-called 'net worth', not merely the amount of cash he has on hand. Then the expected utility for the insurance company and for the customer, if he does or does not buy the insurance, will be as given in Table 13.1. So if $\langle L \rangle < P$, the company wants to sell the insurance, and if $\langle \log(M - L) \rangle < \log(M - P)$ the customer wants to buy it. If the premium is in the range

$$\langle L \rangle < P < [M - \exp\langle \log(M - L) \rangle], \tag{13.5}$$

it will be advantageous for both to do business.

We leave it as an exercise for the reader to show from (13.5) that a poor man should buy insurance, but a rich man should not unless his assessment of expected loss $\langle L \rangle$ is much greater than the insurance company's. Indeed, if your present fortune is much greater than any likely loss, then your utility for money is nearly as linear as the insurance company's, in the region where it matters; and you may as well be your own insurance company.

Further insight into the rich man's psychology is had by noting that if $M \gg \langle L \rangle$ we may expand in powers of M^{-1};

$$M - \exp\langle \log(M - L) \rangle = \langle L \rangle + \frac{\text{var}(L)}{2M} + \cdots, \tag{13.6}$$

where $\text{var}(L) = \langle L^2 \rangle - \langle L \rangle^2$. Thus, a moderately rich man might be willing to buy insurance even if the premium is slightly larger than his expected loss, because this removes the uncertainty $\text{var}(L)$ about the actual loss which he would otherwise have to live with; we have an aversion not only to risk, but to uncertainty about it.

Further insight into the poor man's psychology is had by writing the right-hand side of (13.5) as

$$M - \exp\langle \log(M - L) \rangle = M - \prod_i \exp\{w_i \log(M - L_i)\}. \tag{13.7}$$

Let the L_i be enumerated so that $L_1 \geq L_2 \geq L_3 \cdots$, then this expression does not make sense unless $M > L_1$; but presumably it is not possible to have $M < L_1$, for one cannot lose more than he has. But if M approaches L_1, the last term becomes singular [$\exp\{-\infty\}$] and drops out. Equation (13.5) then reduces to $\langle L \rangle < P < M$; it appears that this unfortunate person should always buy insurance if he can, even if this leaves him as poor as if the worst possible contingency had happened to him!

Of course, this only illustrates that the logarithmic utility assignment is unrealistic for very small amounts. In fact, the utility is clearly bounded in that region also; he who possesses only one penny does not consider it a calamity to lose it. We may correct this by replacing $\log(M)$ by $\log(M + b)$, where b is an amount so small that we consider it practically worthless. This modifies our conclusion from (13.7) in a way that we leave for the reader to work out, and which may suggest a good choice for b.

13.4 Entropy and utility

The logarithmic assignment of utility is reasonable for many purposes, as long as it is not pushed to extremes. It is also, incidentally, closely connected with the notion of entropy, as shown by Bellman and Kalaba (1956, 1957). A gambler who receives partially reliable advance tips on a game acts (i.e. decides on which side and how much to bet) so as to maximize the expected logarithm of his fortune. Bellman and Kalaba show that (1) one can never go broke following this strategy, in contrast to the strategy of maximizing expected profit, where it is easily seen that with probability one this will happen eventually (the classical 'gambler's ruin' situation), and (2) the amount one can reasonably expect to win on any one game is clearly proportional to the amount M_0 he has to begin with, so, after n games, one could hope to have an amount $M = M_0 \exp\{\alpha n\}$. Evidently, to use the logarithmic utility function means that one acts so as to maximize the expectation of α.

Exercise 13.1. Show that the maximum attainable $\langle \alpha \rangle$ is just $(H_0 - H)$, where H is the entropy which describes the gambler's uncertainty as to the truth of his tips, and H_0 is the maximum possible entropy, if the tips were completely uninformative.

A similar result is derived below. This suggests that, with more development of the theory, entropy might have an important place in guiding the strategy of a businessman or stock market investor.

There is a more subtle use of these considerations; the possibility not only of maximizing our own utility, but of manipulating the utility considerations of others so as to induce them to behave as we wish. Competent administrators know, instinctively but qualitatively, how to offer rewards and punishments so as to keep their organizations running smoothly and on course. A much oversimplified but quantitative example of this follows.

13.5 The honest weatherman

The weatherman's prior information and data yield a probability $p = P(\text{rain}|\text{data}, I)$ that it will rain tomorrow. Then what probability q will he announce publicly, in his evening TV forecast? This depends on his perceived utility function. We suspect that weather forecasters systematically overstate the probability for bad weather, i.e. announce a value $q > p$, in the belief that they will incur more criticism from failing to predict a storm that arrives than from predicting one that fails to arrive.[1]

Nevertheless, we would prefer to be told the value p actually indicated by all the data at hand; indeed, if we were sure that we were being told this, we could not reasonably criticize the weatherman for his failures. Is it possible to give the weatherman a utility environment that will induce him always to tell the truth?

Suppose we write the weatherman's employment contract to stipulate that he will never be fired for making too many wrong predictions; but that, each day, when he announces a probability q of rain, his pay for that day will be $B \log(2q)$ if it actually rains the next day, and $B \log(2[1 - q])$ if it does not, where B is a base rate that does not matter for our present considerations, as long as it is high enough to make him want the job. Then the weatherman's expected pay for today, if he announces probability q, is

$$B[p \log(2q) + (1 - p) \log(2[1 - q])] = B[\log(2) + p \log(q) + (1 - p) \log(1 - q)]. \tag{13.8}$$

Taking the first and second derivatives, we find that this is a maximum when $q = p$.

Now any continuous utility function appears linear if we examine only a small segment of it. Thus, if the weatherman considers a single day's pay small enough so that his utility for it is linear in the amount, it will always be to his advantage to tell the truth. There exist combinations of rewards and utility functions for which, quite literally, honesty is the best policy.

More generally, let there be n possible events $(A_1, \ldots, A_n)$ for which the available prior information and data indicate the probabilities $(p_1, \ldots, p_n)$. But a predictor chooses to announce instead the probabilities $(q_1, \ldots, q_n)$. Let him be paid $B \log(nq_i)$ if the event A_i subsequently occurs; he is rewarded for placing a high probability on the true event. Then his expectation of pay is

$$B[\log(n) - I(q; p)], \tag{13.9}$$

where $I(q; p) \equiv \sum p_i \log(q_i)$ is essentially (to within an additive constant) the relative entropy of the distributions (today commonly called the Kullback–Leibler information (Kullback and Leibler, 1951), although its fundamental properties were proved and exploited already by Gibbs (1902, Chap. 11)). Then it will be to the weatherman's advantage to announce always $q_i = p_i$, and his maximum expectation of pay is

$$B[\log(n) - H(p_1, \ldots, p_n)], \tag{13.10}$$

where $H(p_i) = -\sum p_i \log(p_i)$ is the entropy that measures his uncertainty about the A_i. It is not only to his advantage to tell the truth; it is to his advantage to acquire the maximum possible amount of information so as to decrease that entropy.

---

[1] Evidence for this is seen in the fact that, in St Louis, we experience a predicted nonstorm almost every other week; but a nonpredicted storm is so rare that it is a major news item.

As a very real, concrete example, consider a drug company, which has only a finite amount of research and development facilities. We have two potential new drugs: drug A alleviates a disorder that afflicts 10^6 persons per year, while drug B would help only 1000 persons per year. Supposing equally good preliminary evidence for the efficacy and safety of the drugs, the company will naturally prefer to expend its development efforts on drug A; and for this decision we can predict confidently that it will come under attack from some misanthrope who charges it with being interested only in its own profits. Yet had he thought it through one more step, he might have perceived that this policy, while undeniably benefitting the company, also benefits a much larger proportion of society.

13.6 Reactions to Daniel Bernoulli and Laplace

The mathematically elementary – yet evidently important – nature of these results, might make one think that such things must have been not only perceived by many, but put to good use immediately, as soon as Daniel Bernoulli and Laplace had started this train of thought. Indeed, it seems in retrospect surprising that the notion of entropy was not discovered in this way, 100 years before Gibbs.

The actual course of history has been very different; for most of the 20th century the 'frequentist' school of thought either ignored the above line of reasoning or condemned it as metaphysical nonsense. In one of the best known books on probability theory (Feller, 1950, p. 199), Daniel Bernoulli's resolution of the St Petersburg paradox is rejected without even being described, except to assure the reader that he 'tried in vain to solve it by the concept of moral expectation'. Warren M. Hirsch, in a review of the book, amplified this as follows:

Various mystifying 'explanations' of this paradox had been offered in the past, involving, for example, the concept of moral expectation. These explanations are hardly understandable to the modern student of probability. Feller gives a straightforward mathematical argument which leads to the determination of finite entrance fee with which the St Petersburg game has all the properties of a fair game.

We have just seen how 'vain' and 'hardly understandable' Daniel Bernoulli's efforts were. Reading Feller, one finds that he 'resolved' the paradox merely by defining and analyzing a different game. He undertakes to explain the rationale of insurance in the same way; but, since he rejects Daniel Bernoulli's concept of a curved utility function, he concludes that insurance is always necessarily 'unfair' to the insured. These explanations are hardly understandable to the modern economist.

In the 1930s and 1940s, a form of decision rules, as an adjunct to hypothesis testing, was expounded by J. Neyman and E. S. Pearson. It enjoyed a period of popularity with electrical engineers (Middleton, 1960) and economists (Simon, 1977), but it is now obsolete because it lacks two fundamental features now recognized as essential to the problem. In Chapter 14 we give a simple example of the Neyman–Pearson procedure, which shows how it is related to others. In 1950, Abraham Wald gave a formulation that operates at a more fundamental level which makes it appear likely to have a permanent validity, as far as it goes, and gives a rather fundamental justification to Daniel Bernoulli's intuitive ideas. But these efforts were not appreciated in all quarters. Maurice Kendall (1963) wrote:

There has been a strong movement in the USA to regard inference as a branch of decision theory. Fisher would have maintained (and in my opinion rightly) that inference in science is not a matter of decision, and that, in any case, criteria for choice in decision based on pay-offs of one kind or another are not available. This, broadly speaking, is the English as against the American point of view. . . . I propound the thesis that some such difference of attitude is inevitable between countries where what a man does is more important than what he thinks, and those where what he thinks is more important than what he does.

We need not rely on second-hand sources for Fisher's attitude toward decision theory; as noted in Chapter 16, he was never at a loss to express himself on anything. In discussing significance tests, he writes (Fisher, 1956, p. 77):

. . . recently . . . a considerable body of doctrine has attempted to explain, or rather to reinterpret, these tests on the basis of quite a different acceptance procedure. The differences between these two situations seem to the author many and wide, and I do not think it would have been possible to overlook them had the authors of this reinterpretation had any real familiarity with work in the natural sciences, or consciousness of those features of an observational record which permit of an improved scientific understanding.

Then he identifies Neyman and Wald as the objects of his criticism.

Apparently, Kendall, appealing to motives usually disavowed by scholars, regarded decision theory as a defect of the American, as opposed to the British, character (although neither Neyman nor Wald was born or educated in America – they fled here from Europe). Fisher regarded it as an aberration of minds not versed in natural science (although the procedures were due originally to Daniel Bernoulli and Laplace, whose stature as natural scientists will easily bear comparison with Fisher's).

We agree with Kendall that the approach of Wald does indeed give the impression that inference is only a special case of decision; and we deplore this as much as he did. But we observe that in the original Bernoulli–Laplace formulation (and in ours), the clear distinction between these two functions is maintained, as it should be. But, while we perceive this necessary distinction between inference and decision, we perceive also that inference not followed by decision is largely idle, and no natural scientist worthy of the name would undertake the labor of conducting inference unless it served some purpose.

These quotations give an idea of the obstacles which the perfectly natural, and immensely useful, ideas of Daniel Bernoulli and Laplace had to overcome; 200 years later, anyone who suggested such things was still coming under attack from the entrenched 'orthodox' statistical establishment – and in a way that reflected no credit on the attackers. Let us now examine Wald's theory.

13.7 Wald's decision theory

Wald's formulation, in its initial stages, had no apparent connection with probability theory. We begin by imagining (i.e. enumerating) a set of possible 'states of nature', $\{\theta_1, \theta_2, \ldots, \theta_N\}$ whose number is always, in practice, finite, although it might be a useful limiting approximation to think of them as infinite or even as forming a continuum. In the quality control example of Chapter 4, the 'state of nature' was the unknown number of defectives in the batch.

There are certain illusions that tend to grow and propagate here. Let us dispel one by noting that, in enumerating the different states of nature, we are not describing any real (verifiable) property of nature – for one and only one of them is in fact true. The enumeration is only a means of describing a *state of knowledge* about the range of possibilities. Two persons, or robots, with different prior information may enumerate the θ_j differently without either being in error or inconsistent. One can only strive to do the best he can with the information he has, and we expect that the one with better information will naturally – and deservedly – make better decisions. This is not a paradox, but a platitude.

The next step in our theory is to make a similar enumeration of the decisions $\{D_1, D_2, \ldots, D_k\}$ that might be made. In the quality control example, there were three possible decisions at each stage:

$$D_1 \equiv \text{accept the batch,}$$
$$D_2 \equiv \text{reject the batch,}$$
$$D_3 \equiv \text{make another test.}$$

In the particle counter problem of Mr B in Chapter 6, where we were to estimate the number n_1 of particles passing through the counter in the first second, there were an infinite number of possible decisions:

$$D_i \equiv n_1 \text{ is estimated as equal to } 0, 1, 2, \ldots.$$

If we are to estimate the source strength, there are so many possible estimates that we thought of them as forming a continuum of possible decisions, even though in actual fact we can write down only a finite number of decimal digits.

This theory is clearly of no use unless by 'making a decision' we mean, 'deciding to act as if the decision were correct'. It is idle for the robot to 'decide' that $n_1 = 150$ is the best estimate unless we are then prepared to act on the assumption that $n_1 = 150$. Thus the enumeration of the D_i that we give the robot is a means of describing our knowledge as to what kinds of actions are *feasible*; it is idle and computationally wasteful to consider any decision which we know in advance corresponds to an impossible course of action.

There is another reason why a particular decision might be eliminated; even though D_1 is easy to carry out, we might know in advance that it would lead to intolerable consequences. An automobile driver can make a sharp turn at any time; but his common sense usually tells him not to. Here we see two more points: (1) there is a continuous gradation – the consequences of an action might be serious without being absolutely intolerable, and

(2) the consequences of an action will in general depend on what is the true state of nature – a sudden sharp turn does not always lead to disaster, and it may actually avert disaster.

This suggests a third necessary concept – the loss function $L(D_i, \theta_j)$, which is a set of numbers representing our judgment as to the 'loss' incurred by making decision D_i if θ_j should turn out to be the true state of nature. If the D_i and θ_j are both discrete, this is a loss matrix L_{ij}.

Quite a bit can be done with just the θ_j, D_i, L_{ij}, and there is a rather extensive literature dealing with criteria for making decisions with no more than this. In the early days of this theory the results were summarized in a very readable and entertaining form by Luce and Raiffa (1989), and in the aforementioned elementary textbook of Chernoff and Moses (1959), which we recommend as still very much worth reading today. This culminated in the more advanced work of Raiffa and Schlaifer (1961), which is still a standard reference work because of its great amount of useful mathematical material.

For a modern exposition with both the philosophy and the mathematics in more detail than we give here, see James Berger (1985). This is written from a Bayesian viewpoint almost identical to ours, and it takes up many technical circumstances important for inference but which are not, in our view, really part of decision theory.

The minimax criterion is: for each D_i find the maximum possible loss $M_i = \max_j (L_{ij})$; then choose that D_i for which M_i is a minimum. This would be a reasonable strategy if we regard Nature as an intelligent adversary who foresees our decision and deliberately chooses the state of nature so as to cause us the maximum frustration. In the theory of some games, this is not a completely unrealistic way of describing the situation, and consequently minimax strategies are of fundamental importance in game theory (von Neumann and Morgenstern, 1953).

In the decision problems of the scientist, engineer, or economist we have no intelligent adversary, and the minimax criterion is that of the long-faced pessimist who concentrates all his attention on the worst possible thing that could happen, and thereby misses out on the favorable opportunities.

Equally unreasonable from our standpoint is the starry-eyed optimist who believes that Nature is deliberately trying to help him, and so uses this 'minimin' criterion: for each D_i find the minimum possible loss $m_i = \min_j (L_{ij})$ and choose the D_i that makes m_i a minimum.

Evidently, a reasonable decision criterion for the scientist, engineer, or economist is in some sense intermediate between minimax and minimin, expressing our belief that Nature is neutral toward our goals. Many other criteria have been suggested, with such names as maximin utility (Wald), α-optimism–pessimism (Hurwicz), minimax regret (Savage), etc. The usual procedure, as described in detail by Luce and Raiffa, has been to analyze any proposed criterion to see whether it satisfies about a dozen qualitative common sense conditions such as

(1) *Transitivity*: If D_1 is preferred to D_2, and D_2 preferred to D_3, then D_1 should be preferred to D_3.
(2) *Strong domination*: If for all states of nature θ_j we have $L_{ij} < L_{kj}$, then D_i should always be preferred to D_k.

This kind of analysis, although straightforward, can become tedious. We do not follow it any further, because the final result is that there is only one class of decision criteria which passes all the tests, and this class is obtained more easily by a different line of reasoning.

A full decision theory, of course, cannot concern itself merely with the θ_j, D_i, L_{ij}. We also, in typical problems, have additional evidence E, which we recognize as relevant to the decision problem, and we have to learn how to incorporate E into the theory. In the quality control example of Chapter 4, E consisted of the results of the previous tests.

At this point, the decision theory of Wald takes a long, difficult, and, as we now realize, unnecessary mathematical detour. One defines a 'strategy' S, which is a set of rules of the form, 'If I receive new evidence E_i, then I will make decision D_k'. In principle, one first enumerates all conceivable strategies (whose number is, however, astronomical even in quite simple problems), and then eliminates the ones considered undesirable by the following criterion. Denote by

$$p(D_k|\theta_j S) = \sum_i p(D_k|E_i\theta_j S)\, p(E_i|\theta_j) \qquad (13.11)$$

the sampling probability that, if θ_j is the true state of nature, strategy S would lead us to make decision D_k, and define the *risk* presented by θ_j with strategy S as the expected loss over this distribution:

$$R_j(S) = \langle L \rangle_j = \sum_k p(D_k|\theta_j S) L_{kj}. \qquad (13.12)$$

Then a strategy S is called *admissible* if no other S' exists for which

$$R_j(S') \le R_j(S), \qquad \text{for all } j. \qquad (13.13)$$

If an S' exists for which the strict inequality holds for at least one θ_j, then S is termed *inadmissible*. The notions of risk and admissibility are evidently sampling theory criteria, not Bayesian, since they invoke only the sampling distribution. Wald, thinking in sampling theory terms, considered it obvious that the optimal strategy should be sought only within the class of admissible ones.

A principal object of Wald's theory is then to characterize the class of admissible strategies in mathematical terms, so that any such strategy can be found by carrying out a definite procedure. The fundamental theorem bearing on this is Wald's complete class theorem, which establishes a result shocking to sampling theorists (including Wald himself). Berger (1985, Chap. 8) discusses this in Wald's terminology. The term 'complete class' is defined in a rather awkward way (Berger, 1985, pp. 521–522). What Wald really wanted was just the set of all admissible rules, which Berger calls a 'minimal complete class'. From Wald's viewpoint it is a highly nontrivial mathematical problem to prove that such a class exists, and to find an algorithm by which any rule in the class can be constructed.

From our viewpoint, however, these are unnecessary complications, signifying only an inappropriate definition of the term 'admissible'. We shall return to this issue in Chapter 17 and come to a different conclusion: an 'inadmissible' decision may be overwhelmingly preferable to an 'admissible' one, because the criterion of admissibility ignores prior information – even information so cogent that, for example, in major medical, public health, or airline safety decisions, to ignore it would put lives in jeopardy and support a charge of criminal negligence.

The notion of admissibility is flawed in another respect. According to the above definition, an estimation rule which simply ignores the data and always estimates $\theta^* = 5$ is admissible if the point $\theta = 5$ is in the parameter space. In this case it is clear that almost any 'inadmissible' rule would be superior to the 'admissible' one.

This illustrates the folly of inventing noble-sounding names such as 'admissible' and 'unbiased' for principles that are far from noble; and not even fully rational. In the future we should profit from this lesson and take care that we describe technical conditions by names that are ethically and morally neutral, and so do not have false connotations which could mislead others for decades, as these have.

Since in real applications we do not want to – and could not – restrict ourselves to admissible rules anyway, we shall not follow this quite involved argument. We give a different line of reasoning which leads to the rules which are appropriate in the real world, while giving us a better understanding of the reason for them.

What makes a decision process difficult? Well, if we knew which state of nature was the correct one, there would be no problem at all; if θ_3 is the true state of nature, then the best decision D_i is the one which renders L_{i3} a minimum. In other words, once the loss function has been specified, our uncertainty as to the best decision arises solely from our uncertainty as to the state of nature. Whether the decision minimizing L_{i3} is or is not best depends on this: how strongly do we believe that θ_3 is the true state of nature? How plausible is θ_3?

To our robot it seems a trivial step – really only a rephrasing of the question – to ask next, 'Conditional on all the available evidence, what is the *probability* P_3 that θ_3 is the true state of nature?' Not so to the sampling theorist, who regards the word 'probability' as synonymous with 'long-run relative frequency in some random experiment'. On this definition it is meaningless to speak of the probability for θ_3, because the state of nature is not a 'random variable'. Thus, if we adhere consistently to the sampling theory view of probability, we shall conclude that probability theory cannot be applied to the decision problem, at least not in this direct way.

It was just this kind of reasoning which led statisticians, in the early part of the 20th century, to relegate problems of parameter estimation and hypothesis testing to a new field, statistical inference, which was regarded as distinct from probability theory, and based on entirely different principles. Let us examine a typical problem of this type from the sampling theory viewpoint, and see how introducing the notion of a loss function changes this conclusion.

13.8 Parameter estimation for minimum loss

There is some unknown parameter α, and we make n repeated observations of a quantity, obtaining an observed 'sample' $x \equiv \{x_1, \ldots, x_n\}$. We interpret the symbol x, without subscripts, as standing for a vector in an n-dimensional 'sample space', and suppose that the possible results x_i of individual observations are real numbers which we think of as continuously variable in some domain ($a \leq x_i \leq b$). From observation of the sample x, what can we say about the unknown parameter α? We have already studied such problems from the Bayesian 'probability theory as logic' viewpoint; now we consider them from the sampling theory viewpoint.

To state the problem more drastically, suppose that we are compelled to choose one specific numerical value as our 'best' estimate of α, on the basis of the observed sample x, and any other prior information we might have, and then to act as if this estimate were true. This is the decision situation which we all face daily, both in our professional capacity and in everyday life. The automobile driver approaching a blind intersection cannot know with certainty whether he will have enough time to cross it safely; but still he is compelled to make a decision based on what he can see, and act on it.

Now it is clear that in estimating α, the observed sample x is of no use to us unless we can see some kind of logical (not necessarily causal) connection between α and x. In other words, if we knew α, but not x, then the probabilities which we would assign to various observable samples must depend in some way on the value of α. If we consider the different observations as independent, as was almost always done in the sampling theory of parameter estimation, then the sampling density function factors:

$$f(x|\alpha) = f(x_1|\alpha) \cdots f(x_n|\alpha). \tag{13.14}$$

However, this very restrictive assumption is not necessary (and in fact does not lead to any formal simplification) in discussing the general principles of parameter estimation from the decision theory standpoint.

Let $\beta = \beta(x_1, \ldots, x_n)$ be an 'estimator,' i.e. any function of the data values, proposed as an estimate of α. Also, let $L(\alpha, \beta)$ be the 'loss' incurred by guessing the value β when α is in fact the true value. Then for any given estimator the risk is the 'pre-data' expected loss; i.e. the loss *for a person who already knows the true value of α but does not know what data will be observed*:

$$R_\alpha \equiv \int dx \, L(\alpha, \beta) f(x|\alpha). \tag{13.15}$$

By $\int dx \, (\ )$ we mean the n-fold integration

$$\int \cdots \int dx_1 \cdots dx_n (\ ). \tag{13.16}$$

We may interpret this notation as including both the continuous and discrete cases; in the latter, $f(x|\alpha)$ is a sum of delta-functions.

On the view of one who uses the frequency definition of probability, the above phrase 'for a person who already knows the true value of α' is misleading and unwanted. The notion of the probability for sample x *for a person with a certain state of knowledge* is entirely foreign to him; he regards $f(x|\alpha)$ not as a description of a mere state of knowledge about the sample, but as an objective statement of fact, giving the relative frequencies with which different samples would be observed 'in the long run'.

Unfortunately, to maintain this view strictly and consistently would reduce the legitimate applications of probability theory almost to zero; for one can (and most of us do) work in this field for a lifetime without ever encountering a real problem in which one actually has knowledge of the 'true' limiting frequencies for an infinite number of trials; how could one ever acquire such knowledge? Indeed, quite apart from probability theory, no scientist ever has sure knowledge of what is 'really true'; the only thing we can ever know with certainty is: *what is our state of knowledge?*

Then how could one ever assign a probability which he knew was equal to a limiting frequency in the real world? It seems to us that the belief that probabilities are realities existing in Nature is pure mind projection fallacy. True 'scientific objectivity' demands that we escape from this delusion and recognize that in conducting inference our equations are not describing reality; they are describing and processing our information about reality.

In any event, the 'frequentist' believes that R_α is not merely the 'expectation of loss' in the present situation, but is also, with probability one, the limit of the average of *actual* losses which would be incurred by using the estimator β an indefinitely large number of times; i.e. by drawing a sample of n observations repeatedly with a fixed value of α. Furthermore, the idea of finding the estimator which is 'best for the present specific sample' is quite foreign to his outlook; because he regards the notion of probability as referring to a collection of cases rather than a single case, he is forced to speak instead of finding that estimator 'which will prove best, on average, in the long run'.

On the frequentist view, therefore, it would appear that the best estimator will be the one that minimizes R_α. Is this a variational problem? A small change $\delta\beta(x)$ in the estimator changes the risk by

$$\delta R_\alpha = \int dx \, \frac{\partial L(\alpha, \beta)}{\partial \beta} f(x|\alpha) \delta\beta(x). \tag{13.17}$$

If we were to require this to vanish for all $\delta\beta(x)$, this would imply

$$\frac{\partial L}{\partial \beta} = 0, \qquad \text{all possible } \beta. \tag{13.18}$$

Thus the problem as stated has no truly stationary solution except in the trivial – and useless – case where the loss function is independent of the estimated value β; if there is any 'best' estimator by the criterion of minimum risk, it cannot be found by variational methods.

Nevertheless, we can get some understanding of what is happening by considering (13.15) for some specific choices of loss function. Suppose we take the quadratic loss function

$L(\alpha, \beta) = (\alpha - \beta)^2$. Then (13.15) reduces to

$$R_\alpha = \int dx \, (\alpha^2 - 2\alpha\beta + \beta^2) \, f(x|\alpha), \tag{13.19}$$

or

$$R_\alpha = (\alpha - \langle\beta\rangle)^2 + \text{var}(\beta), \tag{13.20}$$

where $\text{var}(\beta) \equiv \langle\beta^2\rangle - \langle\beta\rangle^2$ is the variance of the sampling pdf for β, and

$$\langle\beta^n\rangle \equiv \int dx \, [\beta(x)]^n \, f(x|\alpha) \tag{13.21}$$

is the nth moment of that pdf. The risk (13.20) is the sum of two positive terms, and a good estimator by the criterion of minimum risk has two properties:

(1) $\langle\beta\rangle = \alpha$,
(2) $\text{var}(\beta)$ is a minimum.

These are just the two conditions which sampling theory has considered most important. An estimator with property (1) is called *unbiased* (more generally, the function $b(\alpha) = \langle\beta\rangle - \alpha$ is called the *bias* of the estimator $\beta(x)$), and one with both properties (1) and (2) was called *efficient* by R. A. Fisher. Nowadays, it is often called an *unbiased minimum variance* (UMV) estimator.

In Chapter 17 we shall examine the relative importance of removing bias and minimizing variance, and derive the Cramér–Rao inequality which places a lower limit on the possible value of $\text{var}(\beta)$. For the present, our concern is only with the failure of (13.17) to provide any optimal estimator for a given loss function. This weakness of the sampling theory approach to parameter estimation, that it does not tell us how to find the best estimator, but only how to compare different guesses, can be overcome as follows: we give a simple substitute for Wald's complete class theorem.

13.9 Reformulation of the problem

It is easy to see why the criterion of minimum risk is bound to get us into trouble and is unable to furnish any general rule for constructing an estimator. The mathematical problem was: for given $L(\alpha, \beta)$ and $f(x|\alpha)$, what function $\beta(x_1, \ldots, x_n)$ will minimize R_α?

Although this is not a variational problem, it might have a unique solution; but the more fundamental difficulty is that the solution will still, in general, depend on α. Then the criterion of minimum risk leads to an impossible situation – even if we could solve the mathematical minimization problem and had before us the best estimator $\beta_\alpha(x_1, \ldots, x_n)$ for each value of α, we could use that result only if α were already known, in which case we would have no need to estimate. We were looking at the problem backwards!

This makes it clear how to correct the trouble. It is of no use to ask what estimator is 'best' for some particular value of α; the answer to that question is always, obviously, $\beta(x) = \alpha$, independent of the data. But the only reason for using an estimator is that α is unknown. The estimator must therefore be some compromise that allows for all possibilities within some prescribed range of α; within this range, it must do the best job of protecting against loss, no matter what the true value of α turns out to be.

Thus it is some weighted average of R_α,

$$\langle R \rangle = \int d\alpha \, g(\alpha) R_\alpha, \tag{13.22}$$

that we should really minimize, where the function $g(\alpha) \geq 0$ measures in some way the relative importance of minimizing R_α for the various possible values that α might turn out to have.

The mathematical character of the problem is completely changed by adopting (13.22) as our criterion; we now have a solvable variational problem with a unique, well-behaved, and useful solution. The first variation in $\langle R \rangle$ due to an arbitrary variation $\delta\beta(x_1, \ldots, x_n)$ in the estimator is

$$\delta\langle R \rangle = \int \cdots \int dx_1 \cdots dx_n \left\{ \int d\alpha \, g(\alpha) \frac{\partial L(\alpha, \beta)}{\partial \beta} f(x_1, \ldots, x_n | \alpha) \right\} \delta\beta(x_1, \ldots, x_n), \tag{13.23}$$

which vanishes independently of $\delta\beta$ if

$$\int d\alpha \, g(\alpha) \frac{\partial L(\alpha, \beta)}{\partial \beta} f(x_1, \ldots, x_n | \alpha) = 0 \tag{13.24}$$

for all possible samples $\{x_1, \ldots, x_n\}$.

Equation (13.24) is the fundamental integral equation which determines the 'best' estimator by our new criterion. Taking the second variation, we find that (13.24) yields a true minimum if

$$\int d\alpha \, g(\alpha) \frac{\partial^2 L}{\partial \beta^2} f(x_1, \ldots, x_n | \alpha) > 0. \tag{13.25}$$

Thus a sufficient condition for a minimum is simply $\partial^2 L / \partial \beta^2 \geq 0$, but this is stronger than necessary.

If we take the quadratic loss function $L(\alpha, \beta) = K(\alpha - \beta)^2$, (13.24) reduces to

$$\int d\alpha \, g(\alpha)(\alpha - \beta) f(x_1, \ldots, x_n | \alpha) = 0, \tag{13.26}$$

or the optimal estimator for quadratic loss is

$$\beta(x_1, \ldots, x_n) = \frac{\int d\alpha \, g(\alpha) \alpha \, f(x_1, \ldots, x_n | \alpha)}{\int d\alpha \, g(\alpha) f(x_1, \ldots, x_n | \alpha)}. \tag{13.27}$$

But this is just the mean value over the posterior pdf for α:

$$f(\alpha | x_1, \ldots, x_n, I) = \frac{g(\alpha) f(x_1, \ldots, x_n | \alpha)}{\int d\alpha \, g(\alpha) f(x_1, \ldots, x_n | \alpha)} \tag{13.28}$$

given by Bayes' theorem, if we interpret $g(\alpha)$ as a prior probability density! This argument shows, perhaps more clearly than any other we have given, why the *mathematical form* of Bayes' theorem intrudes itself inevitably into parameter estimation.

If we take as a loss function the absolute error, $L(\alpha, \beta) = |\alpha - \beta|$, then the integral (13.24) becomes

$$\int_{-\infty}^{\beta} d\alpha \, g(\alpha) f(x_1, \ldots, x_n|\alpha) = \int_{\beta}^{\infty} d\alpha \, g(\alpha) f(x_1, \ldots, x_n|\alpha), \qquad (13.29)$$

which states that $\beta(x_1 \ldots x_n)$ is to be taken as the *median* over the posterior pdf for α:

$$\int_{-\infty}^{\beta} d\alpha \, f(\alpha|x_1, \ldots, x_n, I) = \int_{\beta}^{\infty} d\alpha \, f(\alpha|x_1, \ldots, x_n, I) = \frac{1}{2}. \qquad (13.30)$$

Likewise, if we take a loss function $L(\alpha, \beta) = (\alpha - \beta)^4$, (13.24) leads to an estimator $\beta(x_1, \ldots, x_n)$, which is the real root of

$$f(\beta) = \beta^3 - 3\langle\alpha\rangle\beta^2 + 3\langle\alpha^2\rangle\beta - \langle\alpha^3\rangle = 0, \qquad (13.31)$$

where

$$\langle\alpha^n\rangle = \int d\alpha \, \alpha^n \, f(\alpha|x_1, \ldots, x_n I) \qquad (13.32)$$

is the nth moment of the posterior pdf for α. (That (13.31) has only one real root is seen on forming the discriminant; the condition $f'(\beta) \geq 0$ for all real β is just $(\langle\alpha^2\rangle - \langle\alpha^2\rangle) \geq 0$.)

If we take $L(\alpha, \beta) = |\alpha - \beta|^k$, and pass to the limit $k \to 0$, or if we just take

$$L(\alpha, \beta) = \begin{cases} 0 & \alpha = \beta \\ 1 & \text{otherwise,} \end{cases} \qquad (13.33)$$

(13.24) tells us that we should choose $\beta(x_1, \ldots, x_n)$ as the 'most probable value', or *mode* of the posterior pdf $f(\alpha|x_1, \ldots, x_n I)$. If $g(\alpha) = $ constant in the high-likelihood region, and is not much greater elsewhere, this is just the maximum-likelihood estimate advocated by Fisher.

In this result we see finally just what maximum likelihood accomplishes, and under what circumstances it is the appropriate method to use. The maximum-likelihood criterion is the one in which we care only about the chance of being exactly right; and, if we are wrong, we don't care how wrong we are. This is just the situation we have in shooting at a small target, where 'a miss is as good as a mile'. But it is clear that there are few other situations where this would be a rational way to behave; almost always, the amount of error is of some concern to us, and so maximum likelihood is not the best estimation criterion.

Note that in all these cases it was the posterior pdf, $f(\alpha|x_1, \ldots, x_n, I)$ that was involved. That this will always be the case is seen by noting that our 'fundamental integral equation' (13.24) is not so profound after all. It can be written equally well as

$$\frac{\partial}{\partial\beta} \int d\alpha \, g(\alpha) L(\alpha, \beta) f(x_1, \ldots, x_n|\alpha) = 0. \qquad (13.34)$$

But if we interpret $g(\alpha)$ as a prior probability density, this is just the statement that we are indeed to minimize the expectation of $L(\alpha, \beta)$: it is not the expectation over the sampling pdf for β; it is always the expectation over the Bayesian posterior pdf for α!

We have here an interesting case of 'chickens coming home to roost'. If a sampling theorist will think his estimation problems through to the end, he will find himself obliged to use the Bayesian mathematical algorithm, even if his ideology still leads him to reject the Bayesian rationale for it. But in arriving at these inevitable results, the Bayesian rationale has the advantages that (1) it leads us to this conclusion immediately; (2) it makes it obvious that its range of validity and usefulness is far greater than supposed by the sampling theorist. The Bayesian mathematical form is required for simple logical reasons, independently of all philosophical hangups over 'which quantities are random?' or the 'true meaning of probability'.

Wald's complete class theorem led him to essentially the same conclusion: if the θ_j are discrete and we agree not to include in our enumeration of states of nature any θ_j that is known to be impossible, then the class of admissible strategies is just the class of Bayes strategies (i.e. those that minimize expected loss over a posterior pdf). If the possible θ_j form a continuum, the admissible rules are the proper Bayesian ones; i.e. Bayes rules from proper (normalizable) prior probabilities. But few people have ever tried to follow his proof of this; Berger (1985) does not attempt to present it, but gives instead a number of isolated special results.

There is a great deal of mathematical nitpicking, also noted by Berger, over the exact situation when one tries to jump into an improper prior in infinite parameter spaces without considering any limit from a proper prior. But for us such questions are of no interest, because the concept of admissibility is itself flawed when stretched to such extreme cases. Because of its refusal to consider any prior information whatsoever, it must consider all points of an infinite domain equivalent; the resulting singular mathematics is only an artifact that corresponds to no singularity in the real problem, where prior information always excludes the region at infinity.

For a given sampling distribution and loss function, we are content to say simply that the defensible decision rules are the Bayes rules characterized by the different proper priors, and their well-behaved limits. This is the conclusion that was shocking to sampling theorists – including Wald himself, who had been one of the proponents of the von Mises' 'collective' theory of probability – and it was psychologically the main spark that touched off our present 'Bayesian revolution' in statistics. To his everlasting credit, Abraham Wald had the intellectual honesty to see the inevitable consequences of this result, and in his final work (Wald, 1950), he termed the admissible decision rules, 'Bayes strategies'.

13.10 Effect of varying loss functions

Since the new feature of the theory being expounded here lies only in the introduction of the loss function, it is important to understand how the final results depend on the loss functions by some numerical examples. Suppose that the prior information I and data D lead to the following posterior pdf for a parameter α:

$$f(\alpha|DI) = k \exp\{-k\alpha\}, \qquad 0 \le \alpha < \infty. \tag{13.35}$$

The nth moment of this pdf is

$$\langle \alpha^n \rangle = \int_0^\infty d\alpha\, \alpha^n f(\alpha|DI) = n!k^{-n}. \tag{13.36}$$

With loss function $(\alpha - \beta)^2$, the best estimator is the mean value

$$\beta = \langle \alpha \rangle = k^{-1}. \tag{13.37}$$

With the loss function $|\alpha - \beta|$, the best estimator is the median, determined by

$$\frac{1}{2} = \int_0^\beta d\alpha\, f(\alpha|DI) = 1 - \exp\{-k\beta\} \tag{13.38}$$

or

$$\beta = k^{-1} \log_e(2) = 0.693\langle \alpha \rangle. \tag{13.39}$$

To minimize $\langle(\alpha - \beta)^4\rangle$, we should choose β to satisfy (13.31), which becomes $y^3 - 3y^2 + 6y - 6 = 0$ with $y = k\beta$. The real root of this is at $y = 1.59$, so the optimal estimator is

$$\beta = 1.59\,\langle \alpha \rangle. \tag{13.40}$$

For the loss function $(\alpha - \beta)^{s+1}$, with s an odd integer, the fundamental equation (13.34) is

$$\int_0^\infty d\alpha\,(\alpha - \beta)^s \exp\{-k\alpha\} = 0, \tag{13.41}$$

which reduces to

$$\sum_{m=0}^s \frac{(-k\beta)^m}{m!} = 0. \tag{13.42}$$

The case $s = 3$ leads to (13.40), while in the case $s = 5$, loss function $(\alpha - \beta)^6$, we find

$$\beta = 2.025\,\langle \alpha \rangle. \tag{13.43}$$

As $s \to \infty$, β also increases without limit. But the maximum-likelihood estimate, which corresponds to the loss function $L(\alpha, \beta) = -\delta(\alpha - \beta)$, or equally well to

$$\lim_{k \to 0} |\alpha - \beta|^k, \tag{13.44}$$

is $\beta = 0$. These numerical examples merely illustrate what was already clear intuitively; when the posterior pdf is not sharply peaked, the best estimate of α depends very much on which particular loss function we use.

One might suppose that a loss function must always be a monotonically increasing function of the error $|\alpha - \beta|$. In general, of course, this will be the case; but nothing in this theory restricts us to such functions. You can think of some rather frustrating situations in which, if you are going to make an error, you would rather make a large one than a small one. William Tell was in just that fix. If you study our equations for this case, you will see that there is really no very satisfactory decision at all (i.e. no decision has small expected loss); and nothing can be done about it.

Note that the decision rule is invariant under any proper linear transformation of the loss function; i.e. if $L(D_i, \theta_j)$ is one loss function, then the new one,

$$L'(D_i, \theta_j) \equiv a + bL(D_i, \theta_j) \qquad \begin{cases} -\infty < a < \infty \\ 0 < b < \infty, \end{cases} \qquad (13.45)$$

will lead to the same decision, whatever the prior probabilities and data. Thus, in a binary decision problem, given the loss matrix

$$L_{ij} = \begin{pmatrix} 10 & 19 \\ 100 & 10 \end{pmatrix}, \qquad (13.46)$$

we can equally well use

$$L'_{ij} = \begin{pmatrix} 0 & 1 \\ 10 & 0 \end{pmatrix} \qquad (13.47)$$

corresponding to $a = -10/9$, $b = 1/9$. This may simplify the calculation of expected loss quite a bit.

13.11 General decision theory

In the foregoing, we examined decision theory only in terms of one particular application, parameter estimation. But we really have the whole story already; the criterion (13.34) for constructing the optimal estimator generalizes immediately to the criterion for finding the optimal decision of any kind. The final rules are simple; to solve the problem of inference, there are four steps.

(1) Enumerate the possible states of nature θ_j, discrete or continuous, as the case may be.
(2) Assign prior probabilities $p(\theta_j|I)$ which represent whatever prior information I you have about them.
(3) Assign sampling probabilities $p(E_i|\theta_j)$ which represent your prior knowledge about the mechanism of the measurement process yielding the possible data sets E_i.
(4) Digest any additional evidence $E = E_1 E_2 \cdots$ by application of Bayes' theorem, thus obtaining the posterior probabilities $p(\theta_j|EI)$.

That is the end of the inference problem, and $p(\theta_j|EI)$ expresses all the information about the θ_j that is contained in the prior information and data. To solve the problem of decision there are three more steps.

(5) Enumerate the possible decisions D_i.
(6) Assign the loss function $L(D_i, \theta_j)$ that tells what you want to accomplish.
(7) Make that decision D_i which minimizes the expected loss over the posterior probabilities for θ_j.

After all is said and done, the final rules of calculation to which the theorems of Cox, Wald, and Shannon lead are just the ones which had been given already by Laplace and Daniel Bernoulli in the 18th century on intuitive grounds, except that the entropy principle generalizes the principle of indifference in step (2).

Theoretically, these rules are now determined uniquely by elementary qualitative desiderata of rationality and consistency. Some protest that they do not have any prior probability or loss function. The theorem is that rationality and consistency require you to behave *as if* you had them; for every strategy that obeys the desiderata, there is a prior probability and loss function which would have led to that strategy; conversely, if a strategy is derived from a prior probability and loss function, it is guaranteed to obey the desiderata.

Pragmatically, these rules either include, or improve upon, practically all known statistical methods for hypothesis testing and point estimation of parameters. If you have mastered them, then you have just about the entire field at your fingertips. The outstanding thing about them is their intuitive appeal and simplicity – if we sweep aside all the polemics and false starts that have cluttered up this field in the past and consider only the constructive arguments that lead directly to these rules, it is clear that the underlying rationale could be developed fully in a one-semester undergraduate course.

However, in spite of the formal simplicity of the rules themselves, really facile application of them in nontrivial problems involves intricate mathematics, and fine subtleties of concept; so much so that several generations of workers in this field misapplied them and concluded that the rules were all wrong. So, we still need a good deal of leading by the hand in order to develop facility in using this theory. It is like learning how to play a musical instrument – anybody can make noise with it, but to play this instrument well requires years of practice.

13.12 Comments

13.12.1 'Objectivity' of decision theory

Decision theory occupies a unique position in discussion of the logical foundations of statistics, because, as we have seen in (13.24) and (13.34), its procedures can be derived from either of two diametrically opposed viewpoints about the nature of probability theory. While there appears to be universal agreement as to the actual procedures that should be followed, there remains a fundamental disagreement as to the underlying reason for them, having its origin in the old issue of frequency vs. nonfrequency definitions of probability.

From a pragmatic standpoint, such considerations may seem at first to be unimportant. However, in the attempt to apply decision theory methods in real problems one learns very quickly that these questions intrude in the initial stage of setting up the problem in mathematical terms. In particular, our judgment as to the generality and range of validity of decision theory depends on how these conceptual problems are resolved. Our aim is to expound the viewpoint according to which these methods have the greatest possible range of application.

Now, we find that the main source of controversy here is on the issue of prior probabilities; on the sampling theory viewpoint, if the problem involves use of Bayes' theorem then these methods are just not applicable unless the prior probabilities are known frequencies. But to maintain this position consistently would imply an enormous restriction on the range of legitimate applications; indeed, we doubt whether there has ever been a real problem in which the prior probabilities were, in fact, known frequencies. But can the mathematical form of our final equations shed any light on this issue?

Notice first that only the product $g(\alpha)L(\alpha, \beta)$ is involved in (13.24) or (13.34); thus we could interpret the problem in three different ways:

(1) prior probability $g(\alpha)$, loss function $L(\alpha, \beta) = (\alpha - \beta)^2$;
(2) uniform prior probability, loss function $L(\alpha, \beta) = g(\alpha)(\alpha - \beta)^2$;
(3) prior probability $h(\alpha)$, loss function $g(\alpha)(\alpha - \beta)^2 / h(\alpha)$;

but the optimal decision is just the same. This is equally true for any loss function.

We emphasize this rather trivial mathematical fact because of a curious psychological phenomenon. In expositions of decision theory written from the sampling theory viewpoint (for example, Chernoff and Moses, 1959), the writers are reluctant to introduce the notion of prior probability. They postpone it as long as possible, and finally give in only when the mathematics forces them to recognize that prior probabilities are the only basis for choice among the different admissible decision rules. Even then, they are so unhappy about the use of prior probabilities that they feel it necessary always to invent a situation – often highly artificial – which makes the prior probabilities appear to be frequencies; and they will not use this theory for any problem where they do not see how to do this.

But these same writers do not hesitate to pull a completely arbitrary loss function out of thin air without any basis at all, and proceed with the calculation! Our equations show that if the final decision depends strongly on which particular prior probability assignment we use, it is going to depend just as strongly on which particular loss function we use. If one worries about arbitrariness in the prior probabilities, then, in order to be consistent, one ought to worry just as much about arbitrariness in the loss functions. If one claims (as sampling theorists did for decades and as some still do) that uncertainty as to the proper choice of prior probabilities invalidates the Laplace–Bayes theory, then, in order to be consistent, one must claim also that uncertainty as to the proper choice of loss functions invalidates Wald's decision theory.

The reason for this strange lopsided attitude is closely connected with a certain philosophy variously called behavioristic, or positivistic, which wants us to restrict our statements and concepts to objectively verifiable things. Therefore the observable *decision* is the thing to emphasize, while the process of plausible reasoning and the judgment described by a prior probability must be deprecated and swept under the rug. But we see no need to do this, because it seems to us obvious that rational action can come only as the result of rational thought.

If we refuse to consider the problem of rational thought merely on the grounds that it is not 'objective', the result will not be that we obtain a more 'objective' theory of inference or decision. The result will be that we have lost the possibility of getting any satisfactory theory at all, because we have denied ourselves any way of describing what is actually going on in the decision process. And, of course, the loss function is just the expression of a purely subjective value judgment, which can in no way be considered any more 'objective' than the prior probabilities.

In fact, prior probabilities are usually far more 'objective' than loss functions, both in the mathematical theory and in the everyday decision problems of 'real life'. In the mathematical theory we have general formal principles – maximum entropy, transformation groups, marginalization – that remove the arbitrariness of prior probabilities for a large class of important problems, which includes most of those discussed in textbooks. But we have no such principles for determining loss/utility functions.

This is not to say that the problem has not been discussed; de Groot (1970) notes the very weak abstract conditions (transitivity of preferences, etc.) sufficient to guarantee existence of a utility function. Long ago, L. J. Savage considered construction of utility functions by introspection. This is described by Chernoff and Moses (1959): suppose there are two possible rewards r_1 and r_2; then for what reward r_3 would you be indifferent between (r_3 for sure) or (either r_1 or r_2 as decided by the flip of a coin)? Presumably, r_3 is somewhere between r_1 and r_2. If one makes enough such intuitive judgments and manages to correct all intransitivities, a crude utility function emerges. Berger (1985, Chap. 2) gives a scenario in which this happens.

This is hardly a practical procedure, however, much less a formal principle; the result is just as arbitrary as if one simply drew a curve freehand. Indeed, the latter is much easier and cannot get one into intransitivity difficulties. One can, of course, invent a crude prior in the same way, as L. J. Savage often demonstrated. Such constructions, if one can transfer them into a computer, will be better than nothing; but they are clearly desperation moves in lieu of a really satisfactory formal theory such as we have in the principles of maximum entropy and transformation groups for priors.

Noting that the decision depends only on the product of loss function and prior suggests what seems at first an attractive possibility; could we simplify the foundations of this theory so as to make it obvious that we need only a single function, not two? The writer pondered this for some time, but decided finally that this is not the right direction for future development, because (1) priors and loss functions have very different – almost opposite – roles to play, both in the mathematical theory and in 'real life', and (2) the theory of inference involving priors is more fundamental than that of loss functions; the latter would need to be developed much further before it would be fit to join with priors into a single mathematical quantity.

What determines the validity of this theory? We would say, unhesitatingly, 'logical consistency'. But there is a perennial fallacy of basing validity judgments on whether people actually reason in the way required by consistency arguments. The theory is held by some to be invalid if real people do not always reason this way. It seems to us that this is getting it exactly backward; the theory being developed is, just because of the consistency properties, the normative goal which people should strive to approach in the real world.

Some authors get into even stranger problems in approaching decision theory. L. J. Savage (1954) faces many inexplicable difficulties. He thinks (p. 16) that the proverbs 'Look before you leap' and 'You can cross that bridge when you come to it' are contradictory. We feel that we routinely obey both, and see no conflict between them. That is, we do not act without considering the likely consequences; but at the same time we do not waste time and effort planning for future contingencies that are very unlikely to happen.

The original formulation of Wald contemplates, following the orthodox line of thought, that *before seeing the data* one will plan in advance for every possible contingency and list the decision to be made after getting every conceivable data set. The problem with this is that the number of such data sets is usually astronomical; no worker has the computing facilities needed to do it. Yet Savage (1954) thinks that planning for every contingency in advance is the proper course for decision theory because orthodox practice is confined to a small class of artificially simple problems. We take exactly the opposite view: it is only by delaying a decision until we know the actual data that it is possible to deal with complex problems at all. The defensible inferences are the post-data inferences.

As Chernoff and Moses (1959) demonstrate very convincingly, the Bayesian formulation saves us from this; whatever data set is actually observed, we enter it into the computer program and it calculates the appropriate response *for that data set*. It is wasteful and irrelevant to calculate the response to any data set that is not observed. This is not a trivial point; at stake is many orders of magnitude in computation. So carrying this observation a bit further, we fill out our proverb list with 'Never make an irrevocable decision until you have to'.

13.12.2 Loss functions in human society

We note the sharp contrast between the roles of prior probabilities and loss functions in human relations. People with similar prior probabilities get along well together, because they have about the same general view of the world and philosophy of life. People with radically different prior probabilities cannot get along – this has been the root cause of all the religious wars and most of the political repressions throughout history.

Loss functions operate in just the opposite way. People with similar loss functions are after the same thing, and are in contention with each other. People with different loss functions get along well because each is willing to give something that the other wants. Amicable trade or business transactions, advantageous to all, are possible only between parties with very different loss functions. We illustrated this by the example of insurance above.

In 'real life' decision problems, each man knows, pretty well, what his prior probabilities are; and because his beliefs are based on all his past experience, they are not easily changed by one more experience, so they are fairly stable. But, in the heat of argument, he may lose sight of his loss function.

Thus the labor mediator must deal with parties with sharply opposing ideologies; policies considered good by one are considered evil by the other. The successful mediator realizes that mere talk will not alter prior beliefs; and so his role must be to turn the attention of both parties away from this area, and explain clearly to each what his loss function is. In this sense, we can claim that in real life decision problems, the loss function is often far more 'subjective' (in the sense of being less well-fixed in our minds) than the prior probabilities.

Indeed, failure to judge one's own loss function correctly is one of the major dangers that humans face. Having a little intelligence, one can invent myths out of his own imagination, and come to believe them. Worse, one person may persuade thousands of others to believe his private myths, as the sordid history of religious, political, and military disasters shows.

We think that these considerations have a bearing on other social problems. For example, some psychologists never tire of trying to explain criminal behavior in terms of early childhood experiences. It is conceivable that these may generate a certain general 'propensity' to crime; but the fact that the vast majority of people with the same experiences do not become criminals shows that a far more important and immediate cause must exist. Perhaps criminal behavior has a much simpler explanation: poor reasoning, leading to a wrongly perceived loss function. Whatever our early childhood experiences, law abiding citizens have just the same motivations as do criminals; all of us have felt the urge to commit robbery, assault, and murder. The difference is that the criminal does not think ahead far enough to appreciate the predictable consequences of his actions; we were not surprised to learn that most violent criminals have very low intelligence.

Inability to perceive one's own loss function can have disastrous personal consequences in other ways. Consider the case of Ramanujan, whom many would consider to be, in one particular area, the greatest mathematical genius who ever lived. His death at age 32 was probably the result of his own ridiculous dietary views. He refused to eat the food served in Hall at Trinity College, Cambridge (although it was undoubtedly more wholesome than any food he had ever eaten before coming to England) and tried to subsist on rotten fruit shipped from India without refrigeration.

A strikingly similar case is that of Kurt Gödel, whom many would consider the greatest – certainly the best known – of all logicians. He died of starvation in a hospital with the finest food facilities, because he became obsessed with the idea that the doctors were trying to poison him. It is curious that the greatest intellectual gifts sometimes carry with them the inability to perceive simple realities that would be obvious to a moron.

We stress that the real world is vastly more complicated than supposed in Wald's theory, and many real decision problems are not covered by it. For example, the state of nature tomorrow might be influenced by our decision today (as when one decides to get an education). Recognizing this is a step in the direction of game theory, or dynamic programming. But to treat such problems does not require any departure from the principles of probability theory as logic; only a generalization of what we did above.

Actually, human intuition, in making decisions with seemingly no rational basis, does surprisingly well; persons with no mathematical comprehension whatsoever may still make good decisions. However, 'intuition' may make use of facts and memories so deeply buried in the subconscious that one is not aware of them; but without mathematical understanding it can also fail disastrously. For example, attempts to apply probability theory and decision theory to strategy in athletic performance provide several amusing illustrations of the fallacies that one can produce by combining a little bit of mathematics with a great deal of superstitious folklore. The book of Machol, Ladany and Morrison (1976) is a good source for this.

13.12.3 A new look at the Jeffreys prior

Our noting that the optimal decision depends only on the product of prior probability and loss function sets off several other lines of thought. As we noted in Chapter 12, Jeffreys (1939) proposed that, in the case of a continuous parameter α known to be positive, we should express prior ignorance by assigning, not uniform prior density, but a prior density proportional to $(1/\alpha)$. The theoretical justification of this rule was long unclear, but it yields very sensible-looking results in practice, which led Jeffreys to adopt it as fundamental in his significance tests.

We learned that, in the case that α is a scale parameter, the Jeffreys prior is uniquely determined by invariance under the scale transformation group; but now we can see a quite different justification for it. If we use the absolute error loss function $|\beta - \alpha|$ when α is known to be positive, then to assign $g(\alpha) = $ constant in (13.24) and (13.34) amounts to saying that we demand an estimator which yields, as nearly as possible, a constant absolute accuracy for all values of α in $0 < \alpha < \infty$. That is clearly asking for too much in the case of large α; and we must pay the price in a poor estimate for small α. But the median of Jeffreys' posterior distribution is mathematically the same thing as the optimal estimator for uniform prior and loss function $|\beta - \alpha|/\alpha$; we ask for, as nearly as possible, a constant *percentage* accuracy over all values of α. This is, of course, what we do want in most cases where we know that $0 < \alpha < \infty$. Another reason for the superior performance of Jeffreys' rule is thus made apparent, if we reinterpret it as saying that the $(1/\alpha)$ factor is part of the loss function. This requires only that α be positive, not necessarily a scale parameter; just what Jeffreys originally stated.

13.12.4 Decision theory is not fundamental

What parts of the theories expounded here will be a permanent part of human thinking, what parts may evolve on into different forms in the future? We can only speculate, but it seems clear to the writer that there is something necessary and timeless in the methods of inference developed here; not only their compelling theoretical basis[2] explained in Chapters 1 and 2, but, equally well, the beautiful way they work out in practice in all the later chapters – always giving us the right answer to whatever question we ask of them, while orthodox methods yield sense and nonsense about equally often – convinces us that these methods cannot be altered in any substantive way in the future.

However, views as to the foundation of those methods may change; for example, instead of our desiderata of logical consistency, future workers may prefer desiderata of optimal information processing, as suggested by the work of Zellner (1988). Indeed, many advantages would result from more common recognition that inference has fundamentally nothing to do with 'randomness' or 'chance' but is concerned rather with optimal *processing of information*. We noted at the end of Chapter 2 how Gödel's theorem appears as a platitude rather than a paradox, as soon as we recognize the information processing aspect of mathematics.

---

[2] Of course, better proofs than those we were able to give in Chapter 2 will be found.

But we can feel no such certainty about the decision theory addendum to inference. In the first place, many present applications already require an extension to game theory, dynamic programming or beyond. The state of nature may be chosen by another person; or it may be influenced by our decision without the intervention of a conscious second agent. There may be more than two agents involved. They might be either adversaries or helpful friends. Those are more complicated situations than the ones we have considered here. We do not think such extensions appropriate to our present topic of scientific inference, because we do not think of ourselves as playing an adversary game against Nature. However, future scientists may find good reasons to consider the more general theory.

For all the reasons noted in this chapter, it now appears that from a fundamental standpoint loss functions are less firmly grounded than are prior probabilities. This is just the opposite of the view that propelled the Wald-inspired development of decision theory in the 1950s, when priors were regarded as vague and ill-defined, but nobody seemed to notice that loss functions are far more so. For reasons we cannot explain, loss functions appeared to workers at that time to be 'real' and definite, although no principles for determining them were ever given, beyond the truism that any function with a continuous derivative appears linear if we examine a sufficiently small piece of it.

In the meantime, there have been several advances in the technique for assigning priors by logical analysis of the prior information. But, to the best of our knowledge, we have as yet no formal principles at all for assigning numerical values to loss functions; not even when the criterion is purely economic, because the utility of money remains ill-defined.

13.12.5 Another dimension?

There is another respect in which loss functions are less firmly grounded than are prior probabilities. We consider it an important aspect of 'objectivity' in inference – almost a principle of morality – that we should not allow our opinions to be swayed by our desires; what we believe should be independent of what we want. But the converse need not be true; on introspection, we would probably agree that what we want depends very much on what we know, and we do not feel guilty of any inconsistency or irrationality on that account.[3]

Indeed, it is clear that the act of assigning a loss function is itself only a means of describing certain *prior information* about the phenomena of interest, which now notes not just their plausibilities, but also their consequences. Thus a change in prior information which affects the prior probabilities could very well induce a change in the loss function as well.

But then, having admitted this possibility, it appears that value judgments need not be introduced in the form of loss functions at all. Already at the end of Chapter 1 we noted the possibility of future 'multidimensional' models of human mental activity. In view of the above considerations, the doors now seem wide open for new developments in that direction;

---

[3] Quasimodo, condemned by an accident of Nature to be something intermediate between man and gargoyle, wished that he had been made a whole man. But, after learning about the behavior of men, he wished instead that he had been made a whole gargoyle: 'O, why was I not made of stone like these?'

representing a mental state about a proposition or action not by one coordinate (plausibility) as in present probability theory, but by two coordinates (plausibility and value). Thus, while the principles of 'one-dimensional' inference seem permanent, the future can still bring many kinds of change in the representation of value judgments, which need not resemble present decision theory at all. But this in turn reacts back on the question of foundations of probability theory.

Thomas Bayes (1763) thought it necessary to explain the notion of probability in terms of that of expectation;[4] and this persisted to modern times in the work of both Wald (1950) and de Finetti (1972, 1974b). At first glance, it appears that the work of de Finetti on foundations of probability theory could hardly be more different in outlook from Wald's decision theory; yet these two avenues to Bayesianity shared the common premise that value judgments are in some way primary to inference.

de Finetti would base probability theory on the notion of 'coherence', which means roughly that in betting one should behave as if he assigned probabilities to the events (dice tosses, etc.) being betted on; but those probabilities should be chosen so that he cannot be made a sure loser, whatever the final outcome of those events.

It has always seemed objectionable to some, including this writer, to base probability theory on such vulgar things as betting, expectation of profit, etc. We think that the principles of logic ought to be on a higher plane. But that was only an aesthetic feeling; now, in recognizing the indefinite and provisional nature of loss functions, we have a more cogent reason for not basing probability theory on decisions or betting. Any rules which were found to be coherent, but not consistent, would be unusable in practice because a well-posed question would have more than one 'right' answer with nothing to choose between them. This is, in our view, still another aspect of the superiority of Richard Cox's approach, which stresses logical consistency instead and, just for that reason, is more likely to have a lasting place in probability theory.

---

[4] The difficulty of reading Bayes today can be appreciated from the bewildering sentence in which he states this: 'The *probability of any event* is the ratio between the value at which an expectation depending on the happening of the event ought to be computed, and the value of the thing expected upon its happening.'

14

Simple applications of decision theory

We now examine in detail two of the simplest applications of the general decision theory just formulated, and compare the first with the older Neyman–Pearson procedure. The problem of detection of signals in noise is really the same as Laplace's old problem of detecting the presence of unknown systematic influences in celestial mechanics, and Shewhart's (1931) more recent problem of detecting a systematic drift in machine characteristics, in industrial quality control. Statisticians would call the procedure a 'significance test'. It is unfortunate that the basic identity of all these problems was not more widely recognized, because it forced workers in several different fields to rediscover the same things, with varying degrees of success, over and over again.

As is clear by now, all we really have to do to solve this problem is to take the principles of inference developed in Chapters 2 and 4, and supplement them with the loss function criterion for converting final probabilities into decisions (and, if needed, the maximum entropy principle for assigning priors). However, the literature of this field has been created largely from the standpoint of the original decision theory before this was realized. The existing literature therefore uses a different sort of vocabulary and set of concepts than we have been using up to now. Since it exists, we have no choice but to learn these terms and viewpoints if we want to read the literature of the field. This material appeared in the papers of Middleton and Van Meter (1955, 1956), and later in the monumental treatise of Middleton (1960), in an enormously expanded form where a beginner can get lost for months without ever finding the real underlying principles. So we need a very rapid, condensed review of the literature of the 1950s on these problems. To have a complete, self-contained summary, we repeat a little from previous chapters as a way of introducing this different language.

14.1 Definitions and preliminaries

We employ the notation:

$$p(A|B) = \text{ conditional probability for } A, \text{ given } B,$$
$$p(AB|CD) = \text{ joint conditional probability for } A \text{ and } B, \text{ given } C \text{ and } D, \text{ etc.} \tag{14.1}$$

For our purposes, everything follows from the product rule:

$$p(AB|C) = p(A|BC)p(B|C) = p(B|AC)p(A|C). \qquad (14.2)$$

If the propositions B and C are not mutually contradictory, this may be rearranged to give the rule of 'learning by experience', Bayes' theorem:

$$p(A|BC) = p(A|C)\frac{p(B|AC)}{p(B|C)} = p(A|B)\frac{p(C|AB)}{p(C|B)}. \qquad (14.3)$$

If there are several mutually exclusive and exhaustive propositions B_i, then, by summing (14.2) over them, we obtain the chain rule

$$p(A|C) = \sum_i p(A|B_iC)p(B_i|C) \qquad (14.4)$$

or, in a simple skeleton notation,

$$p(A|C) = \sum_B p(A|BC)p(B|C). \qquad (14.5)$$

Now let

$$X = \text{prior knowledge, of any kind whatsoever,}$$
$$S = \text{signal,}$$
$$N = \text{noise,} \qquad (14.6)$$
$$V = V(S, N) = \text{observed voltage,}$$
$$D = \text{decision about the nature of the signal.}$$

Thus we have

$$p(S|X) = \text{prior probability for the particular signal } S,$$
$$p(N|X) = W(N) = \text{prior probability for the particular sample of noise } N. \qquad (14.7)$$

We understand that the prior information X is always built into the right-hand side of all our probability symbols, whether or not we write it explicitly. Thus, in a linear system, $V = S + N$ and

$$p(V|S) \equiv p(V|SX) = W(V - S). \qquad (14.8)$$

The reader may be disturbed by the absence of density functions, dS, dN, etc., which might be expected in the case of continuous S, N. Note, however, that our equations are homogeneous in these quantities, so they cancel out anyway. We are trying only to convey the broad ideas, without bothering with fine details which would make the notation very intricate. Thus by $\sum_A$ we mean ordinary summation over some previously agreed set of possible values if A is discrete, integration with appropriate density functions if A is continuous.

A *decision rule* $p(D_i|V_j)$, or for brevity just $p(D|V)$, represents the process of drawing inferences about the signal from the observed voltage. If it is always made in a definite way, then $p(D|V)$ has only the values 0, 1 for any choice of D and V; however, we may also have a 'randomized' decision rule according to which $p(D|V)$ is a true probability distribution. Maintaining this more general view turns out to be a help in formulating the theory.

The essence of any decision rule, and in particular any one which can be built into automatic equipment, is that the decision must be made on the basis of V alone; V is, by definition, the quantity which contains all the information actually used (in addition to the ever-present X) in arriving at the decision. Thus, if $Y \neq D$ is any other proposition, we have

$$p(D|V) = p(D|VY). \tag{14.9}$$

The fact that Y is to be ignored in the presence of V might appear a departure from our previous exhortations that the robot is always to take into account all the relevant information it has. However, if we consider that the property (14.9) is a part of the prior information X there is no difficulty. To put it differently, (14.9) expresses the prior knowledge that there is a direct logical relation by which D is determined by V alone. If this relationship was a known law of physics, there would be nothing strange in (14.9). The only difference is that in the present case this relationship does not express any law of Nature, but rather our own design of the apparatus. Then Y is ignored not because the robot has relaxed its rules, but because our design makes Y irrelevant.

An equivalent statement is that the probability for reaching a decision D depends on any proposition Y only through the intermediate influence of Y on V:

$$p(D|Y) = \sum_V p(D|V)p(V|Y) \tag{14.10}$$

which is a kind of 'Huygens principle' for logic. To see the analogy, think of Y as a light source which cannot be seen from D, but it illuminates various points V. Then the resulting light arriving at D is the sum of the Huygens wavelets $p(D|V)$ with amplitudes $p(V|Y)$. The almost exact mathematical analogy between conditional information flow and the flow of light according to the Huygens principle of optics appears in statistical mechanics of irreversible processes.

14.2 Sufficiency and information

Equation (14.9) has interesting consequences; suppose we wish to judge the plausibility of some proposition Y, on the basis of knowledge of V and D. From the product rule (14.2),

$$p(DY|V) = p(Y|VD)p(D|V) = p(D|VY)p(Y|V) \tag{14.11}$$

and, using (14.9), this reduces to

$$p(Y|VD) = p(Y|V). \tag{14.12}$$

Thus, if V is known, knowledge of D is redundant and cannot help us in estimating any other quantity. The reverse is not true, however; we could equally well use (14.9) in another way:

$$p(VY|D) = p(Y|VD)p(V|D) = p(Y|D)p(V|YD). \tag{14.13}$$

Combining this with (14.12), there results the following theorem.

Theorem

Let D be a possible decision, given V. Then $p(V|D) \neq 0$, and

$$p(Y|V) = p(Y|D) \text{ if and only if } p(V|D) = p(V|YD). \tag{14.14}$$

In words: knowledge of D is as good as knowledge of V for judgments about Y if and only if Y is irrelevant for judgments about V, given D. Stated differently: in the 'environment' produced by knowledge of D, the probabilities for Y and V are independent, i.e.

$$p(YV|D) = p(Y|D)p(V|D). \tag{14.15}$$

In this case, in the literature of this field D is said to be a *sufficient statistic* for judgments about Y. We shall want to see whether this is in agreement with our earlier definitions of sufficiency, made from a quite different point of view in Chapter 8.

Evidently, a decision rule which makes D a sufficient statistic for judgments about the signal S is superior to one without this property, in that it tells us more about the signal. However, such a rule does not necessarily exist. Equation (14.15) is a very restrictive condition, since it must be satisfied for all values of Y, V, and all D for which $p(D|V) \neq 0$.

As you might guess from this, the concept of sufficiency is closely related to that of information. The above definition of sufficiency could be stated equally well as: D is a sufficient statistic for judgments about Y if it contains all the information about Y which V contains. Since D is determined from V, if it is not a sufficient statistic, it necessarily contains *less* information about Y than does V. In this statement, the term 'information' was used in a loose, intuitive sense; does it remain true if we adopt Shannon's measure of information?

Imagine that there are several mutually exclusive propositions Y_i, one of which must be true. For brevity we use, as above, the notation $\sum_Y f(Y) \equiv \sum_i f(Y_i)$. With a specific value of D given, the entropy which measures our information about the propositions Y_i is

$$H_D(Y) = -\sum_Y p(Y|D) \log[p(Y|D)], \tag{14.16}$$

and its expectation over all values of D is

$$\overline{H}_D(Y) = \sum_D p(D|X)H_D(Y). \tag{14.17}$$

If

$$\overline{H}_C(Y) < \overline{H}_D(Y), \tag{14.18}$$

we say colloquially that C contains, 'on the average', more information about Y than does D. Note, however, that it may be otherwise for specific values of C and D.

Acquisition of new information can never increase $\overline{H}$; let $\{Z_i\}$ be, for the moment, any set of propositions and form the expression

$$
\begin{aligned}
\overline{H}_V(Z) - \overline{H}_{DV}(Z) &= \sum_{DVZ} p(DV|X)p(Z|DV)\log[p(Z|DV)] \\
&\quad - \sum_{VZ} p(V|X)p(Z|V)\log[p(Z|V)] \\
&= \sum_{DVZ} p(DV|X)p(Z|DV)\log\left[\frac{p(Z|DV)}{p(Z|V)}\right].
\end{aligned}
\tag{14.19}
$$

Using the fact that on the positive real line $\log(x) \geq (1 - x^{-1})$, with equality if and only if $x = 1$, this becomes

$$
\overline{H}_V(Z) - \overline{H}_{DV}(Z) \geq \sum_{DVZ} p(DV|X)[p(Z|DV) - p(Z|V)] = 0.
\tag{14.20}
$$

Thus, $\overline{H}_{DV}(Z) \leq \overline{H}_V(Z)$, with equality if and only if (14.12) holds for all D, V and Z for which $p(DV|X) \neq 0$.

But now, since (14.20) holds regardless of the meaning of D and V, we can conclude equally well that, for all D, V, Z,

$$
\overline{H}_D(Y) \geq \overline{H}_{DV}(Z) \leq \overline{H}_V(Z).
\tag{14.21}
$$

Choosing $Z = Y$, we have in consequence of (14.12) $H_V(Y) = H_{DV}(Y)$, so that

$$
\overline{H}_V(Y) \leq \overline{H}_D(Y),
\tag{14.22}
$$

with equality if and only if (14.15) holds, i.e. if and only if D is a sufficient statistic as just defined. Thus, if by 'information' we mean minus the expectation of the entropy of Y over the prior distribution of D or V, zero information loss in going from V to D is equivalent to sufficiency of D. Note that inequality (14.20) holds only for the expections of $\overline{H}$, not for the H. Acquisition of a specific piece of information (that an event previously considered improbable had in fact occurred) may in some cases increase the entropy of Y. However, this is an improbable situation, and on the average the entropy can only be lowered by additional information. This shows again that the term 'information' is not a happy choice of words to describe entropy expressions. In spite of the entropy increases, the situation just described could hardly be called one of less *information* in the colloquial sense of that word; but rather one of less *certainty*.

14.3 Loss functions and criteria of optimum performance

In order to say that one decision rule is better than another, we need some specific criterion of what we want our detection system to accomplish. The criterion will vary with the application, and obviously no single decision rule can be best for all purposes. But our discussion in Chapter 13 will apply, almost unchanged, in this slightly different language. A very general type of criterion is obtained by assigning a *loss function* $L(D, S)$ which represents our judgment of how serious it is to make decision D when signal S is in fact present.

In the case where there are only two possible signals, $S_0 = 0$ (i.e. no signal), and $S_1 > 0$, and consequently two possible decisions D_0, D_1 about the signal, there are two types of error, the false alarm $A = (D_1, S_0)$ and the false rest $R = (D_0, S_1)$. In some applications, one type of error might be much more serious than the other.

Suppose that a false rest is considered ten times as serious as is a false alarm, while a correct decision of either type represents no 'loss'. We could then take $L(D_0, S_0) = L(D_1, S_1) = 0$, $L(D_0, S_1) = 10$, $L(D_1, S_0) = 1$. Whenever the possible signals and the possible decisions form discrete sets, the loss function becomes a *loss matrix*. In the above example,

$$L_{ij} = \begin{pmatrix} 0 & 10 \\ 1 & 0 \end{pmatrix}. \tag{14.23}$$

Instead of assigning arbitrarily a certain loss value to each possible type of detection error, we may consider *information loss* by the assignment $L(D, S) = -\log[p(S|D)]$. This is somewhat more difficult to manipulate, because now $L(D, S)$ depends on the decision rule. A decision rule which minimizes information loss is one which makes the decision in some sense as close as possible to being a sufficient statistic for judgments about the signal. In exactly what sense seems never to have been clarified. The *conditional loss* $L(S)$ is the expected loss incurred when the specific signal S is present:

$$L(S) = \sum_D L(D, S)p(D|S), \tag{14.24}$$

which may in turn be expressed in terms of the decision rule and the properties of the noise by using (14.10). What is often called colloquially the 'average loss' is the expectation of the conditional loss over all possible signals:

$$\langle L \rangle = \sum_S L(S)p(S|X). \tag{14.25}$$

Two different criteria of optimal performance now suggest themselves:

The minimax criterion. For a given decision rule $p(D|V)$, consider the conditional loss $L(S)$ for all possible signals, and let $[L(S)]_{max}$ be the maximum value attained by $L(S)$. We seek that decision rule for which $[L(S)]_{max}$ is as small as possible. As we noted in Chapter 13, this criterion concentrates attention on the worst possible case, regardless of the probability for occurrence of this case, and it is thus in a sense too conservative. However, it gives some the psychological comfort that it does not involve the prior probabilities for the different signals $p(S|X)$, and therefore it can be applied by persons who, under the handicap of orthodox training, have a mental hangup against prior probabilities.
The Bayes criterion. We seek that decision rule for which the expected loss $\langle L \rangle$ is minimized. In order to apply this, a prior distribution $p(S|X)$ must be available.

Other criteria were proposed before the days of Wald's decision theory. In the Neyman–Pearson theory, we fix the probability for occurrence of one type of error at some small value ϵ, and then minimize the probability δ of the other type of error subject to this constraint.[1] Arnold Siegert's 'ideal observer' minimizes the total probability for error $(\epsilon + \delta)$.

After having invented many different such *ad hoc* criteria from various viewpoints, and arguing their relative merits on philosophical grounds, the basic mathematical identity of all these criteria came as quite a surprise to the early workers in this field. We shall see below that all of them are special cases of the Bayes criterion, for particular prior probabilities.

Let us find the Bayes solution, as it was rationalized in decision theory. Substituting in succession (14.24), (14.10), and (14.9) into (14.25), we obtain for the expected loss

$$\langle L \rangle = \sum_{DV} \left[\sum_{S} L(D, S) p(VS|X) \right] p(D|V). \tag{14.26}$$

If $L(D, S)$ is a definite function independent of $p(D|V)$ (this assumption excludes for the moment the information loss function), there is no function $p(D|V)$ for which this expression is stationary in the sense of the calculus of variations. We then minimize $\langle L \rangle$ merely by choosing for each possible V that decision $D_1(V)$ for which the coefficient in (14.26)

$$K(D, V) \equiv \sum_{S} L(D_1, S) p(VS|X) \tag{14.27}$$

is a minimum. Thus, we adopt the decision rule

$$p(D|V) = \delta(D, D_1). \tag{14.28}$$

In general, there will be only one such D_1, and the best decision rule is nonrandom. However, in case of 'degeneracy', $K(D_1, V) = K(D_2, V)$, any randomized rule of the form

$$p(D|V) = a\delta(D, D_1) + b\delta(D, D_2), \quad a + b = 1, \tag{14.29}$$

is just as good by the criterion being used. This degeneracy occurs at 'threshold' values of V, where we change from one decision to another.

14.4 A discrete example

Consider the case already mentioned, where there are two possible signals, S_0 and S_1, and a loss matrix

$$L_{ij} = \begin{pmatrix} L_{00} & L_{01} \\ L_{10} & L_{11} \end{pmatrix} = \begin{pmatrix} 0 & L_r \\ L_a & 0 \end{pmatrix}, \tag{14.30}$$

---

[1] For example, we suspect that at an Early Warning Radar Installation, the primary constraint might be that the Commanding Officer shall not be roused out of bed by a false alarm more often than once per month, and, subject to that requirement, we minimize the probability for a false rest.

where L_a, L_r are the losses incurred by a false alarm and a false rest, respectively. Then

$$K(D_0, V) = L_{01} p(V S_1|X) = L_r p(V S_1|X),$$
$$K(D_1, V) = L_{10} p(V S_0|X) = L_a p(V S_0|X),$$
(14.31)

and the decision rule that minimizes $\langle L \rangle$ is

$$\text{choose } D_1 \quad \text{if } \frac{p(V S_1|X)}{p(V S_0|X)} > \frac{L_a}{L_r},$$

$$\text{choose } D_0 \quad \text{if } \frac{p(V S_1|X)}{p(V S_0|X)} < \frac{L_a}{L_r},$$
(14.32)

choose either at random in case of equality.

If the prior probabilities for a signal and no signal are

$$p(S_1|X) = p, \qquad p(S_0|X) = q = 1 - p,$$
(14.33)

respectively, the decision rule becomes

$$\text{choose } D_1 \text{ if } \quad \frac{p(V|S_1)}{p(V|S_0)} > \frac{q L_a}{p L_r}, \quad \text{etc.}$$
(14.34)

The left-hand side of (14.34) is a likelihood ratio, which depends only on the pdf assigned to the noise, and is the quantity which should be computed by the optimum receiver according to the Bayes criterion.

This same quantity is the essential one regardless of the assumed loss function and regardless of the probability for the occurrence of the signal; these affect only the threshold of detection. Furthermore, if the receiver merely computes this likelihood ratio and delivers it at the output without making any decision, it provides us with all the information we need to make optimum decisions in the Bayes sense. Note the generality of this result, which is important for applications; no assumptions were needed as to the type of signal, linearity of the system, or properties of the noise.

We now work out, for purposes of illustration, the decision rules and their degree of reliability, for several of the above criteria, in the simplest possible problem. We have a linear system in which the voltage is observed at a single instant. We are to decide whether a signal, which can have only amplitude S_1, is present in noise. We assign a Gaussian pdf for the noise with variance σ^2:

$$W(N) = \frac{1}{\sqrt{2\pi\sigma^2}} \exp\left\{ -\frac{N^2}{2\sigma^2} \right\}.$$
(14.35)

The likelihood ratio in (14.34) then becomes

$$\frac{p(V|S_1)}{p(V|S_0)} = \frac{W(V - S_1)}{W(V)} = \exp\left\{ \frac{2V S_1 - S_1^2}{2\sigma^2} \right\},$$
(14.36)

and, since this is a monotonic function of V, the Bayesian decision rule, V_b can be written as

$$\text{choose} \begin{pmatrix} D_1 \\ D_0 \end{pmatrix} \quad \text{when} \quad V \begin{pmatrix} > \\ < \end{pmatrix} V_b, \tag{14.37}$$

with

$$\frac{V_b}{\sigma} = \frac{1}{2s} \left[2 \log \left(\frac{q L_a}{p L_r} \right) + s^2 \right] = v_b, \tag{14.38}$$

in which

$$s \equiv \frac{S_1}{\sigma} \text{ is the voltage signal-to-noise ratio,} \tag{14.39a}$$

and

$$v \equiv \frac{V}{\sigma} \text{ is the normalized voltage.} \tag{14.39b}$$

Now we find the probability for a false rest:

$$\begin{aligned} p(R|X) &= p(D_0 S_1 | X) \\ &= p \sum_V p(D_0|V) p(V|S_1) \\ &= p \int_{-\infty}^{V_b} dV \, W(V - S_1) \\ &= p \Phi(v_b - s) \end{aligned} \tag{14.40}$$

and for a false alarm

$$\begin{aligned} p(A|X) &= p(D_1 S_0 | X) \\ &= q \sum_V p(D_1|V) p(V|S_0) \\ &= q \int_{V_b}^{\infty} dV \, W(V) \\ &= q [1 - \Phi(v_b)]. \end{aligned} \tag{14.41}$$

Here $\Phi(x)$ is the cumulative normal distribution function and, as shown in (7.2), it may be computed from an error function:

$$\Phi(x) = \frac{1}{\sqrt{2\pi}} \int_{-\infty}^{x} dt \, \exp\{-t^2/2\} = \frac{1}{2} [1 + \text{erf}(x)]. \tag{14.42}$$

For $x > 2$, a good approximation is

$$1 - \Phi(x) \approx \frac{\exp\{-x^2/2\}}{x\sqrt{2\pi}}. \tag{14.43}$$

As a numerical example, if $L_r = 10 L_a$, $q = 10p$, these expressions reduce to

$$p(A|X) = 10 p(R|X) = \frac{10}{11} \left[1 - \Phi \left(\frac{s}{2} \right) \right]. \tag{14.44}$$

The probability for a false alarm is less than 0.027, and for a false rest less than 0.0027 for $s > 4$. For $s > 6$, these numbers become 1.48×10^{-3}, 1.48×10^{-4}, respectively.

Let us see what the minimax criterion would give in this problem. The conditional losses are

$$L(S_0) = L_a \sum_V p(D_1|V)p(V|S_0) = L_a \int_{-\infty}^{\infty} dV \, p(D_1|V)W(V),$$

$$(14.45)$$

$$L(S_1) = L_r \sum_V p(D_0|V)p(V|S_1) = L_r \int_{-\infty}^{\infty} dV \, p(D_0|V)W(V - S_1).$$

Writing $f(V) \equiv p(D_1|V) = 1 - p(D_0|V)$, the only restriction on $f(V)$ is $0 \leq f(V) \leq 1$. Since L_a, L_r, and $W(V)$ are all positive, a change $\delta f(V)$ in the neighborhood of any given point V will always increase one of the quantities in (14.45) and decrease the other. Thus, when the maximum $L(S)$ has been made as small as possible, we will certainly have $L(S_0) = L(S_1)$, and the problem is thus to minimize $L(S_0)$ subject to this constraint.

Suppose that for some particular $p(S|X)$ the Bayes decision rule happened to give $L(S_0) = L(S_1)$. Then this particular solution must be identical with the minimax solution, for with the above constraint, $\langle L \rangle = [L(S)]_{\max}$, and, if the Bayes solution minimizes $\langle L \rangle$ with respect to all variations $\delta f(V)$ in the decision rule, it *a fortiori* minimizes it with respect to the smaller class of variations which keep $L(S_0) = L(S_1)$. Therefore the decision rule will have the same form as before: there is a minimax threshold V_m such that

$$f(V) = \begin{cases} 0 & V < V_m \\ 1 & V > V_m. \end{cases} \qquad (14.46)$$

Any change in V_m from the value which makes $L(S_0) = L(S_1)$ necessarily increases one or the other of these quantities. The equation determining V_m is therefore

$$L_a \int_{V_m}^{\infty} dV \, W(V) = L_r \int_{-\infty}^{V_m} dV \, W(V - S_1), \qquad (14.47)$$

or, in terms of normalized quantities,

$$L_a[1 - \Phi(v_m)] = L_r\Phi(v_m - s). \qquad (14.48)$$

Note that (14.40) and (14.41) give the conditional probabilities for a false rest and false alarm for any decision rule of type (14.46), regardless of whether the threshold was determined from (14.38) or not; for the arbitrary threshold V_0

$$p(R|S_1) = p(V < V_0|S_1) = \Phi(v_0 - s)$$

$$(14.49)$$

$$p(A|S_0) = p(V > V_0|S_0) = \frac{1}{2}[1 - \Phi(v_0)].$$

From (14.38) we see that there is always a particular ratio (p/q) which makes the Bayes threshold V_b equal to the minimax threshold V_m. For values of (p/q) other than this worst value, the Bayes criterion gives a lower expected loss than does the minimax, although one of the conditional losses $L(S_0)$, $L(S_1)$ will be greater than the minimax value.

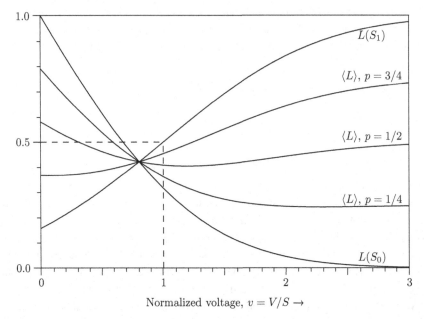

Fig. 14.1. Various risks as a function of voltage for $L_r = 1$, $L_a = 2$, $p = 1/4, 1/2, 3/4$.

These relations and several previous remarks are illustrated in Figure 14.1, in which we plot the conditional losses $L(S_0)$, $L(S_1)$ and the expected loss $\langle L \rangle$ as functions of the threshold V_0, for the case $L_a = (3/2)L_r$, $p = q = 1/2$. The minimax threshold is at the common crossing point of these curves, while the Bayes threshold occurs at the lowest point of the $\langle L \rangle$ curve.

One sees how the Bayes threshold moves as the ratio (p/q) is varied, and in particular that the value of (p/q) which makes $V_b = V_m$ also leads to the maximum value of the $\langle L \rangle_{min}$ obtained by the Bayes criterion. Thus we could also define a 'maximin' criterion; first find the Bayes decision rule which gives minimum $\langle L \rangle$ for a given $p(S|X)$, then vary the prior probability $p(S|X)$ until the maximum value of $\langle L \rangle_{min}$ is attained. The decision rule thus obtained is identical with the one resulting from the minimax criterion; this is the worst possible prior probability, in the sense that the most pessimistic rule is the best that can be done.

The Neyman–Pearson criterion is easily discussed in this example. Suppose the conditional probability for a false alarm $p(D_1|S_0)$ is held fixed at some value ϵ, and we wish to minimize the conditional probability $p(D_0|S_1)$ of a false rest, subject to this constraint. Now the Bayes criterion minimizes the expected loss

$$\langle L \rangle = pL_r p(D_0|S_1) + qL_a p(D_1|S_0) \tag{14.50}$$

with respect to any variation $\delta p(D|V)$ in the decision rule. In particular, therefore, it minimizes it with respect to the smaller class of variations which hold $p(D_1|S_0)$ constant at the value finally obtained. Thus it minimizes $p(D_0|S_1)$ with respect to these variations and solves the Neyman–Pearson problem; we need only choose the particular value of the ratio (qL_a/pL_r) which results in the assumed value of ϵ according to (14.38) and (14.41).

We find for the Neyman–Pearson threshold, from (14.49),

$$\Phi(v_{np}) = 1 - \epsilon, \qquad (14.51)$$

and the conditional probability for detection is

$$p(D_1|S_1) = 1 - p(D_0|S_1) = \Phi(s - v_{np}). \qquad (14.52)$$

If $\epsilon = 10^{-3}$, a detection probability of 99% or better is attained for $s > 6$.

It is important to note that these numerical examples depend critically on our noise pdf assignment. If we have prior information about the noise beyond its first and second moments, the noise pdf expressing this may not be Gaussian, and the actual situation may be either more or less favorable than indicated by the above relations.

It is well known that in one sense noise with a Gaussian frequency distribution is the worst possible kind; because of its maximum entropy properties, it can obscure a weak signal more completely than can any other noise of the same average power. On the other hand, Gaussian noise is a very favorable kind from which to extract a fairly strong signal, because the probability that the noise will exceed a few times the RMS value $\sigma = \sqrt{\langle N^2 \rangle}$ becomes vanishingly small. Consequently, the probability for making an incorrect decision on the presence or absence of a signal goes to zero very rapidly as the signal strength is increased. The high reliability of operation found above for $s > 6$ would not be found for noise possessing a frequency distribution with wider tails.

The type of noise frequency distribution to be expected in any particular case depends, of course, on the physical mechanism which gives rise to the noise. When the noise is the result of a large number of small, independent effects, the Landon derivation of Chapter 7 and the central limit theorem both tell us that a Gaussian frequency distribution for the total noise is by far the most likely to be found, regardless of the nature of the individual sources.

All of these apparently different decision criteria lead to a probability ratio test. In the case of a binary decision, it took the simple form (14.32). Of course, any decision process can be broken down into successive binary decisions, so this case really has the whole story in it. All the different criteria amounted, in the final analysis, only to different philosophies about how you choose the threshold value at which you change your decision.

14.5 How would our robot do it?

Now, let's see how this problem appears from the viewpoint of our robot. The rather long arguments we had to go through above (and even they are very highly condensed from the original literature) to obtain the result are due only to the orthodox view which insists on looking at the problem backwards, i.e. on concentrating attention on the final decision rather than on the inference process which logically has to precede it.

To the robot, if our job is to make the best possible decision as to whether the signal is present, the obvious thing we must do is to calculate the *probability* that the signal is present, conditional on all the evidence at hand. If there are only two possibilities, S_0, S_1, to be taken into account, then, after we have seen voltage V, the posterior odds on S_1 are, from (4.7),

$$O(S_1|VX) = O(S_1|X)\frac{p(V|S_1)}{p(V|S_0)}. \tag{14.53}$$

If we give the robot the loss function (14.30) and ask it to make the decision which minimizes the expected loss, it will evidently use the decision rule

$$\text{choose } D_1 \text{ if } O(S_1|V) = \frac{p(S_1|V)}{p(S_0|V)} > \frac{L_a}{L_r}, \tag{14.54}$$

etc. But from the product rule, $p(VS_1|X) = p(S_1|V)p(V|X)$, $p(VS_0|X) = p(S_0|V)$ $p(V|X)$, and (14.54) is identical with (14.32). So, just from looking at this problem the other way around, our robot obtains the same final result in just two lines!

You see that all this discussion of strategies, admissibility, conditional losses, etc. was unnecessary. Except for the introduction of the loss function at the end, there is nothing in the actual functional operation of Wald's decision theory that isn't contained already in basic *probability* theory, if we will only use it in the full generality given to it by Laplace and Jeffreys.

14.6 Historical remarks

This comparison shows why the development of decision theory has, more than any other single factor, touched off our 'Bayesian revolution' in statistical thought. For some 50 years, Harold Jeffreys tried valiantly to explain the great advantages of the Laplace methods to statisticians, and his efforts met only with a steady torrent of denials and ridicule. It was then a real irony that the work of one of the most respected of 'orthodox' statisticians (Abraham Wald), which was hailed, very properly, as a great advance in statistical practice, turned out to give, after very long and complicated arguments, exactly the same final results that the Laplace methods give you immediately. Wald showed in great generality what we have just illustrated by one simple example.

The only proper conclusion, as a few recognized at once, is that the supposed distinction between statistical inference and probability theory was entirely artificial – a tragic error of judgment which has wasted perhaps 1000 man-years of our best mathematical talent in the pursuit of false goals.

In the works cited, addressed to electrical engineers, the viewpoint of Middleton and van Meter was that of the Neyman–Pearson and Wald decision theories. At about the same time, Herbert Simon expounded the Neyman–Pearson viewpoint to economists. The writer collaborated with David Middleton for a short time while he was writing his large work, and tried to persuade him of the superiority of the straight Bayesian approach to decision theory. The success of the effort may be judged by comparing Middleton (1960, Chap. 18) – particularly its length – with our exposition deriving (14.54). It seems that persons with orthodox training had received such strong anti-Bayesian indoctrination that they were locked in an infinite regress situation; although they could not deny the results that Bayesians got on any specific problem, they could never believe that Bayesian methods would work on the next problem until that next solution was also presented to them.

14.6.1 The classical matched filter

A funny thing happened in the history of this subject. In the 1930s, electrical engineers knew nothing whatsoever about probability theory; they knew about signal to noise ratios. Receiver input circuits were designed for many years on the basis that signal to noise ratio was maximized by empirical trial and error. Then a general theoretical result was found: if you take the ratio of (peak signal)^2 to mean-square noise, and find, as a variational principle, the design of input stages of the receiver which will maximize it, this turned out to have an analytically neat and useful solution. It is now called the *classical matched filter*, and it has been discovered independently by dozens of people.

To the best of our knowledge, the first person to derive this matched filter solution was the late Professor W. W. Hansen of Stanford University. The writer was working with him, beginning in May 1942, on problems of radar detection. Shortly before then, Hansen had circulated a little memorandum dated 1941, in which he gave this solution for the design of the optimum response curve for the receiver first stage. Years later, I was thinking about an entirely different problem (an optimum antenna pattern for a radar system to maximize the ratio (signal)/(ground clutter response)), and when I finally got the solution, I recognized it as the same result that Bill Hansen had shown me many years before.

Throughout the 1950s, almost every time one opened a journal concerned with these problems, somebody else had a paper announcing the discovery of the same solution. The situation was satirized in a famous editorial by Peter Elias (1958), entitled 'Two famous papers'. He suggested that it was high time that people stopped rediscovering the easiest solution, and started to think about the many harder problems still in need of solution.

But also in the 1950s people became more sophisticated about the way they handled their detection problems, and they started using this wonderful new tool, statistical decision theory, to see if there were still better ways of handling these design problems. The strange thing happened that in the case of a linear system with Gaussian noise, the optimum solution which decision theory leads you to turns out to be exactly the same old classical matched filter! At first glance, it was very surprising that two approaches so entirely different conceptually should lead to the same solution. But note that our robot represents a viewpoint from which it is obvious that the two lines of argument would have to give the same result.

To our robot, it is obvious that the best analysis we can make of the problem will always be one in which we calculate the *probabilities* that the various signals are present by means of Bayes' theorem. But let us apply Bayes' theorem in the logarithmic form of Chapter 4. If we now let S_0 and S_1 stand for numerical values giving the amplitude of two possible signals as a function of V, the *evidence* for S_1 is increased by

$$\log\left[\frac{p(V|S_1)}{p(V|S_0)}\right] = \frac{(V-S_0)^2 - (V-S_1)^2}{2\langle\sigma^2\rangle} = \text{const.} + \frac{(S_1-S_0)}{\langle\sigma^2\rangle}V. \tag{14.55}$$

In the case of a linear system with Gaussian noise, the observed voltage is itself just a linear function of the posterior probability measured in decibels. So, they are essentially just two different ways of formulating the same problem. Without recognizing it, we had essentially solved this problem already in the Bayesian hypothesis testing discussion of Chapter 4.

In England, P. M. Woodword had perceived much of this correctly in the 1940s – but he was many years ahead of his time. Those with conventional statistical training were unable to see any merit in his work, and simply ignored it. His book (Woodword, 1953) is highly recommended reading; although it does not solve any of our current problems, its thinking is still in advance of some current literature and practice.

We have seen that the other non-Bayesian approaches to the theory all amounted to different philosophies of how you choose the threshold at which you change your decision. Because of the fact that they all lead to the same probability ratio test, they must necessarily all be derivable from Bayes' theorem.

The problem just examined by several different decision criteria is, of course, the simplest possible one. In a more realistic problem, we will observe the voltage $V(t)$ as a function of time, perhaps several voltages $V_1(t)$, $V_2(t)$, ... in several different channels. We may have many different possible signals $S_a(t)$, $S_b(t)$, ... to distinguish and correspondingly many possible decisions. We may need to decide not only *whether* a given signal is present, but also to make the best estimates of one or more signal parameters (such as intensity, starting time, frequency, phase, rate of frequency modulation, etc.). Therefore, just as in the problem of quality control discussed in Chapter 4, the details can become arbitrarily complicated. But these extensions are, from the Bayesian viewpoint, straightforward in that they require no new principles beyond those already given, only mathematical generalization.

We shall return to some of these more complicated problems of detection and filtering when we take up frequency/shape estimation; but for now let's look at another elementary kind of decision problem. In the ones just discussed, we needed Bayes' theorem, but not maximum entropy. Now we examine a kind of decision problem where we need maximum entropy, but not Bayes' theorem.

14.7 The widget problem

This problem was first propounded at a symposium held at Purdue University in November, 1960 – at which time, however, the full solution was not known. This was worked out later (Jaynes, 1963c), and some numerical approximations were improved in the computer work of Tribus and Fitts (1968).

The widget problem has proved to be interesting in more respects than originally realized. It is a decision problem in which there is no occasion to use Bayes' theorem, because no 'new' information is acquired. Thus it would be termed a 'no data' decision problem in the sense of Chernoff and Moses (1959). However, at successive stages of the problem we have more and more prior information; and digesting it by maximum entropy leads to a sequence of prior probability assignments, which lead to different decisions. Thus it is an example of the 'pure' use of maximum entropy, as in statistical mechanics. It is hard to see how the problem could be formulated mathematically at all without use of maximum entropy, or some other device (such as the method of Darwin and Fowler (Fowler, 1929) in statistical mechanics, or the 'method of the most probable distribution' dating back to Boltzmann (1871)) which turns out in the end to be mathematically equivalent to maximum entropy.

The problem is interesting also in that we can see a continuous gradation from decision problems so simple that common sense tells us the answer instantly, with no need for any mathematical theory, through problems more and more involved so that common sense has more and more difficulty in making a decision, until finally we reach a point where nobody has yet claimed to be able to see the right decision intuitively, and we require the mathematics to tell us what to do.

Finally, the widget problem turns out to be very close to an important real problem faced by oil prospectors. The details of the real problem are shrouded in proprietary caution; but it is not giving away any secrets to report that, a few years ago, the writer spent a week at the research laboratories of one of our large oil companies, lecturing for over 20 hours on the widget problem. We went through every part of the calculation in excruciating detail – with a room full of engineers armed with calculators, checking up on every stage of the numerical work.

Here is the problem: Mr A is in charge of a widget factory, which proudly advertises that it can make delivery in 24 hours on any size order. This, of course, is not really true, and Mr A's job is to protect, as best he can, the advertising manager's reputation for veracity. This means that each morning he must decide whether the day's run of 200 widgets will be painted red, yellow or green. (For complex technological reasons, not relevant to the present problem, only one color can be produced per day.) We follow his problem of decision through several stages of increasing knowledge.

Stage 1

When he arrives at work, Mr A checks with the stock room and finds that they now have in stock 100 red widgets, 150 yellow, and 50 green. His ignorance lies in the fact that he does not know how many orders for each type will come in during the day. Clearly, in this state of ignorance, Mr A will attach the highest significance to any tiny scrap of information about orders likely to come in today; and if no such scraps are to be had, we do not envy Mr A his job. Still, if a decision must be made here and now on no more information than this, his common sense will probably tell him that he had better build up that stock of green widgets.

Stage 2

Mr A, feeling the need for more information, calls up the front office and asks, 'Can you give me some idea of how many orders for red, yellow, and green widgets are likely to come

Table 14.1. *Summary of four stages of the widget problem.*

Stage	R	Y	G	Decision
1. In stock	100	150	50	G
2. Av. daily order total	50	100	10	Y
3. Av. individual order	75	10	20	R
4. Specific order			40	?

in today?' They reply, 'Well, we don't have the breakdown of what has been happening each day, and it would take us a week to compile that information from our files. But we do have a summary of the total sales last year. Over the last year, we sold a total of 13 000 red, 26 000 yellow, and 2600 green. Figuring 260 working days, this means that last year we sold an average of 50 red, 100 yellow, and 10 green each day.' If Mr A ponders this new information for a few seconds, I think he will change his mind, and decide to make yellow ones today.

Stage 3

The man in the front office calls Mr A back and says, 'It just occurred to me that we do have a little more information that might possibly help you. We have at hand not only the total number of widgets sold last year, but also the total number of orders we processed. Last year we got a total of 173 orders for red, 2600 for yellow, and 130 for green. This means that the customers who use red widgets order, on the average, $13\,000/173 = 75$ widgets per order, while the average order for yellow and green were $26\,000/2600 = 10$, and $2600/130 = 20$, respectively.' These new data do not change the expected daily demand; but, if Mr A is very shrewd and ponders it very hard, I think he may change his mind again, and decide to make red ones today.

Stage 4

Mr A is just about to give the order to make red widgets when the front office calls him again to say, 'We just got word that a messenger is on his way here with an emergency order for 40 green widgets.' Now, what should he do? Up to this point, Mr A's decision problem has been simple enough so that reasonably good common sense will tell him what to do. But now he is in trouble; qualitative common sense is just not powerful enough to solve his problem, and he needs a mathematical theory to determine a definite optimum decision.

Let's summarize all the above data in Table 14.1. In the final column, we give the decision that seemed intuitively to be the best one before we had worked out the mathematics. Do other people agree with this intuitive judgment? Professor Myron Tribus has put this to a test by giving talks about this problem, and taking votes from the audience before the solution is given. We quote his findings as given in Tribus and Fitts

(1968). They use D_1, D_2, D_3, D_4 to stand for the optimum decisions in stages 1, 2, 3, 4, respectively:

Before taking up the formal solution, it may be reported that Jaynes' widget problem has been presented to many gatherings of engineers who have been asked to vote on D_1, D_2, D_3, D_4. There is almost unanimous agreement about D_1. There is about 85% agreement on D_2. There is about 70% agreement on D_3, and almost no agreement on D_4. One conclusion stands out from these informal tests; the average engineer has remarkably good intuition in problems of this kind. The majority vote for D_1, D_2, and D_3 has always been in agreement with the formal mathematical solution. However, there has been almost universal disagreement over how to defend the intuitive solution. That is, while many engineers could agree on the best course of action, they were much less in agreement on *why* that course was the best one.

14.7.1 Solution for Stage 2

Now, how are we to set up this problem mathematically? In a real life situation, evidently, the problem would be a little more complicated than indicated so far, because what Mr A does today also affects how serious his problem will be tomorrow. That would get us into the subject of dynamic programming. But for now, just to keep the problem simple, we shall solve only the truncated problem in which he makes decisions on a day to day basis with no thought of tomorrow.

We have just to carry out the steps enumerated in Section 13.11. Since Stage 1 is almost too trivial to work with, consider the problem of Stage 2. Firstly, we define our underlying hypothesis space by enumerating the possible 'states of nature' θ_j that we will consider. These correspond to all possible order situations that could arise; if Mr A knew in advance exactly how many red, yellow, and green widgets would be ordered today, his decision problem would be trivial. Let $n_1 = 0, 1, 2, \ldots$ be the number of red widgets that will be ordered today, and similarly n_2, n_3 for yellow and green, respectively. Then, any conceivable order situation is given by specifying three non-negative integers $\{n_1, n_2, n_3\}$. Conversely, every ordered triple of non-negative integers represents a conceivable order situation.

Next, we are to assign prior probabilities $p(\theta_j|X) = p(n_1 n_2 n_3|X)$ to the states of nature, which maximize the entropy of the distribution subject to the constraints of our prior knowledge. We solved this problem in general in Chapter 11, Eqs. (11.39)–(11.50); and so we just have to translate the result into our present notation. The index i on x_i in Chapter 11 now corresponds to the three integers n_1, n_2, n_3; the function $f_k(x_i)$ also corresponds to the n_i, since the prior information at this stage will be used to fix the expectations $\langle n_1 \rangle, \langle n_2 \rangle, \langle n_3 \rangle$ of orders for red, yellow, and green widgets at 50, 100, 10, respectively. With three constraints we will have three Lagrange multipliers $\lambda_1, \lambda_2, \lambda_3$, and the partition function (11.45) becomes

$$Z(\lambda_1, \lambda_2, \lambda_3) = \sum_{n_1=0}^{\infty} \sum_{n_2=0}^{\infty} \sum_{n_3=0}^{\infty} \exp\{-\lambda_1 n_1 - \lambda_2 n_2 - \lambda_3 n_3\}$$

$$= \prod_{i=1}^{3} (1 - \exp\{-\lambda_i\})^{-1}. \tag{14.56}$$

The λ_i are determined from (11.46):

$$\langle n_i \rangle = -\frac{\partial \log(Z)}{\partial \lambda_i} = \frac{1}{\exp\{\lambda_i\} - 1}. \tag{14.57}$$

The maximum entropy probability assignment (11.41) for the states of nature $\theta_j = \{n_1 n_2 n_3\}$ therefore factors:

$$p(n_1 n_2 n_3) = p_1(n_1) p_2(n_2) p_3(n_3) \tag{14.58}$$

with

$$p_i(n_i) = (1 - \exp\{-\lambda_i\}) \exp\{-\lambda_i n_i\}, \quad n_i = 1, 2, 3 \ldots$$

$$\tag{14.59}$$

$$= \frac{1}{\langle n_i \rangle + 1} \left[\frac{\langle n_i \rangle}{\langle n_i \rangle + 1} \right]^{n_i}.$$

Thus, in Stage 2, Mr A's state of knowledge about today's orders is given by three exponential distributions:

$$p_1(n_1) = \frac{1}{51} \left(\frac{50}{51}\right)^{n_1}, \quad p_2(n_2) = \frac{1}{101} \left(\frac{100}{101}\right)^{n_2}, \quad p_3(n_3) = \frac{1}{11} \left(\frac{10}{11}\right)^{n_3}.$$

$$\tag{14.60}$$

Applications of Bayes' theorem to digest new evidence E is absent because there is no new evidence. Therefore, the decision must be made directly from the prior probabilities (14.60), as is always the case in statistical mechanics.

So, we now proceed to enumerate the possible decisions. These are $D_1 \equiv$ make red ones today, $D_2 \equiv$ make yellow ones, $D_3 \equiv$ make green ones, for which we are to introduce a loss function $L(D_i, \theta_j)$. Mr A's judgment is that there is no loss if all orders are filled today; otherwise, the loss will be proportional to – and in view of the invariance of the decision rule under proper linear transformations that we noted at the end of Chapter 13, we may as well take it equal to – the total number of unfilled orders.

The present stock of red, yellow, and green widgets is $S_1 = 100$, $S_2 = 150$, $S_3 = 50$, respectively. On decision D_1 (make red widgets) the available stock S_1 will be increased by the day's run of 200 widgets, and the loss will be

$$L(D_1; n_1, n_2, n_3) = R(n_1 - S_1 - 200) + R(n_2 - S_2) + R(n_3 - S_3), \tag{14.61}$$

where $R(x)$ is the ramp function

$$R(x) \equiv \begin{cases} x & x \geq 0 \\ 0 & x \leq 0. \end{cases} \tag{14.62}$$

Likewise, on decisions D_2, D_3, the loss will be

$$L(D_2; n_1, n_2, n_3) = R(n_1 - S_1) + R(n_2 - S_2 - 200) + R(n_3 - S_3), \tag{14.63}$$

$$L(D_3; n_1, n_2, n_3) = R(n_1 - S_1) + R(n_2 - S_2) + R(n_3 - S_3 - 200). \tag{14.64}$$

So, if decision D_1 is made, the expected loss will be

$$
\begin{aligned}
\langle L \rangle_1 &= \sum_{n_i} p(n_1 \, n_2 \, n_3) L(D_1; n_1, n_2, n_3) \\
&= \sum_{n_1=0}^{\infty} p_1(n_1) R(n_1 - S_1 - 200) \\
&+ \sum_{n_2=0}^{\infty} p_2(n_2) R(n_2 - S_2) \\
&+ \sum_{n_3=0}^{\infty} p_3(n_3) R(n_3 - S_3),
\end{aligned}
\tag{14.65}
$$

and similarly for D_2, D_3. The summations are elementary, giving

$$
\begin{aligned}
\langle L \rangle_1 &= \langle n_1 \rangle \exp\{-\lambda_1 (S_1 + 200)\} + \langle n_2 \rangle \exp\{-\lambda_2 S_2\} + \langle n_3 \rangle \exp\{-\lambda_3 S_3\}, \\
\langle L \rangle_2 &= \langle n_1 \rangle \exp\{-\lambda_1 S_1\} + \langle n_2 \rangle \exp\{-\lambda_2 (S_2 + 200)\} + \langle n_3 \rangle \exp\{-\lambda_3 S_3\}, \\
\langle L \rangle_3 &= \langle n_1 \rangle \exp\{-\lambda_1 S_1\} + \langle n_2 \rangle \exp\{-\lambda_2 S_2\} + \langle n_3 \rangle \exp\{-\lambda_3 (S_3 + 200)\},
\end{aligned}
\tag{14.66}
$$

or, inserting numerical values,

$$
\begin{aligned}
\langle L \rangle_1 &= 0.131 + 22.48 + 0.085 = 22.70, \\
\langle L \rangle_2 &= 6.902 + 3.073 + 0.085 = 10.6, \\
\langle L \rangle_3 &= 6.902 + 22.48 + 4 \times 10^{-10} = 29.38,
\end{aligned}
\tag{14.67}
$$

showing a strong preference for decision $D_2 \equiv$ 'make yellow ones today', as common sense had already anticipated.

Physicists will recognize that Stage 2 of Mr A's decision problem is mathematically the same as the theory of harmonic oscillators in quantum statistical mechanics. There is still another engineering application of the harmonic oscillator equations, in some problems of message encoding, to be noted when we take up communication theory. We are trying to emphasize the generality of this theory, which is mathematically quite old and well-known, but which has been applied in the past only in some specialized problems in thermodynamics. This general applicability can be seen only after we are emancipated from the orthodox view of probability.

14.7.2 Solution for Stage 3

In Stage 3 of Mr A's problem we have some additional pieces of information: the average *individual* orders for red, yellow, and green widgets. To take account of this new information, we need to go down into a deeper hypothesis space; set up a more detailed enumeration of the states of nature in which we take into account not only the total orders for each type, but also the breakdown into individual orders. We could have done this also in Stage 2, but since at that stage there was no information available bearing on this breakdown, it would have added nothing to the problem (the subtle difference that this makes after all will be noted later).

In Stage 3, a possible state of nature can be described as follows. We receive u_1 individual orders for one red widget each, u_2 orders for two red widgets each, ..., u_r individual orders for r red widgets each. Also, we receive v_y orders for y yellow widgets each, and w_g orders for g green widgets each. Thus, a state of nature is specified by an infinite number of non-negative integers

$$\theta = \{u_1 \cdots ; v_1 \cdots ; w_1 \cdots \}, \tag{14.68}$$

and conversely every such set of integers represents a conceivable state of nature, to which we assign a probability $p(u_1 \cdots ; v_1 \cdots ; w_1 \cdots)$.

Today's total demands for red, yellow, and green widgets are, respectively,

$$n_1 = \sum_{r=1}^{\infty} r u_r, \qquad n_2 = \sum_{y=1}^{\infty} y v_y, \qquad n_3 = \sum_{g=1}^{\infty} g w_g, \tag{14.69}$$

the expectations of which were given in Stage 2 as $\langle n_1 \rangle = 50$, $\langle n_2 \rangle = 100$, $\langle n_3 \rangle = 10$. The total number of individual orders for red, yellow, and green widgets are, respectively,

$$m_1 = \sum_{r=1}^{\infty} u_r, \qquad m_2 = \sum_{y=1}^{\infty} v_y, \qquad m_3 = \sum_{g=1}^{\infty} w_g, \tag{14.70}$$

and the new feature of Stage 3 is that $\langle m_1 \rangle$, $\langle m_2 \rangle$, $\langle m_3 \rangle$ are also known. For example, the statement that the average individual order for red widgets is 75 means that $\langle n_1 \rangle = 75 \langle m_1 \rangle$.

With six average values given, we will have six Lagrange multipliers $\{\lambda_1, \mu_1; \lambda_2, \mu_2; \lambda_3, \mu_3\}$. The maximum entropy probability assignment will have the form

$$p(u_1 \cdots ; v_1 \cdots ; w_1 \cdots) = \exp\{-\lambda_0 - \lambda_1 n_1 - \mu_1 m_1 - \lambda_2 n_2 - \mu_2 m_2 - \lambda_3 n_3 - \mu_3 m_3\}, \tag{14.71}$$

which factors:

$$p(u_1 \cdots ; v_1 \cdots ; w_1 \cdots) = p_1(u_1 \cdots) p_2(v_1 \cdots) p_3(w_1 \cdots). \tag{14.72}$$

The partition function also factors:

$$Z = Z_1(\lambda_1 \mu_1) Z_2(\lambda_2 \mu_2) Z(\lambda_3 \mu_3), \tag{14.73}$$

with

$$Z_1(\lambda_1 \mu_1) = \sum_{u_1=1}^{\infty} \sum_{u_2=1}^{\infty} \cdots \exp\{-\lambda_1(u_1 + 2u_2 + 3u_3 + \cdots) - \mu_1(u_1 + u_2 + u_3 + \cdots)\}$$

$$= \prod_{r=1}^{\infty} \frac{1}{1 - \exp\{-r\lambda_1 - \mu\}} \tag{14.74}$$

and similar expressions for Z_2, Z_3. To find λ_1, μ_1 we apply the general rule, Eq. (14.57):

$$\langle n_1 \rangle = \frac{\partial}{\partial \lambda_1} \sum_{r=1}^{\infty} \log(1 - \exp\{-r\lambda_1 - \mu_1\}) = \sum_{r=1}^{\infty} \frac{r}{\exp\{r\lambda_1 + \mu_1\} - 1}, \qquad (14.75)$$

$$\langle m_1 \rangle = \frac{\partial}{\partial \mu_1} \sum_{r=1}^{\infty} \log(1 - \exp\{-r\lambda_1 - \mu_1\}) = \sum_{r=1}^{\infty} \frac{1}{\exp\{r\lambda_1 + \mu_1\} - 1}. \qquad (14.76)$$

Combining with (14.69) and (14.70), we see that

$$\langle u_r \rangle = \frac{1}{\exp\{r\lambda_1 + \mu_1\} - 1}, \qquad (14.77)$$

and now the secret is out – Stage 3 of Mr A's decision problem is just the theory of the ideal Bose–Einstein gas in quantum statistical mechanics!

If we treat the ideal Bose–Einstein gas by the method of the Gibbs grand canonical ensemble, we obtain just these equations, in which the number r corresponds to the rth single-particle energy level, u_r to the number of particles in the rth state, and λ_1 and μ_1 to the temperature and chemical potential.

In the present problem it is clear that for all r, $\langle u_r \rangle \ll 1$, and that $\langle u_r \rangle$ cannot decrease appreciably below $\langle u_1 \rangle$ until r is of the order of 75, the average individual order. Therefore, μ_1 will be numerically large, and λ_1 numerically small, compared with unity. This means that the series (14.75), (14.76) converge very slowly and are useless for numerical work unless you write a computer program to do it. However, we can do it analytically if we transform them into rapidly converging sums as follows:

$$\sum_{r=1}^{\infty} \frac{1}{\exp\{\lambda r + \mu\} - 1} = \sum_{r=1}^{\infty} \sum_{n=1}^{\infty} \exp\{-n(\lambda r + \mu)\}$$
$$= \sum_{n=1}^{\infty} \frac{\exp\{-n\mu\}}{1 - \exp\{-n\lambda\}}. \qquad (14.78)$$

The first term is already an excellent approximation. Similarly,

$$\sum_{r=1}^{\infty} \frac{r}{\exp\{\lambda r + \mu\} - 1} = \sum_{n=1}^{\infty} \frac{\exp\{-n(\lambda r + \mu)\}}{(1 - \exp\{-n\lambda\})^2}, \qquad (14.79)$$

and so (14.75) and (14.76) become

$$\langle n_1 \rangle = \frac{\exp\{-\mu_1\}}{\lambda_1^2}, \qquad (14.80)$$

$$\langle m_1 \rangle = \frac{\exp\{-\mu_1\}}{\lambda_1}, \qquad (14.81)$$

$$\lambda_1 = \frac{\langle m_1 \rangle_1}{\langle n_1 \rangle} = \frac{1}{75} = 0.0133, \qquad (14.82)$$

$$\exp\{\mu_1\} = \frac{\langle n_1 \rangle_1}{\langle m_1 \rangle} = 112.5, \tag{14.83}$$

$$\mu_1 = 4.722. \tag{14.84}$$

Tribus and Fitts, evaluating the sums by computer, obtain $\lambda_1 = 0.0131$, $\mu_1 = 4.727$; so our approximations (14.80), (14.81) are very good, at least in the case of red widgets.

The probability that u_r has a particular value is, from (14.72) or (14.74),

$$p(u_r) = (1 - \exp\{-r\lambda_1 - \mu\}) \exp\{(-r\lambda_1 + \mu_1)u_r\}, \tag{14.85}$$

which has the mean value (14.77) and the variance

$$\mathrm{var}(u_r) = \langle u_r^2 \rangle - \langle u_r \rangle^2 = \frac{\exp\{r\lambda_1 + \mu_1\}}{\exp\{r\lambda_1 + \mu_1\} - 1}. \tag{14.86}$$

The total demand for red widgets

$$n_1 = \sum_{r=1}^{\infty} r u_r \tag{14.87}$$

is expressed as the sum of a large number of independent terms. The pdf for n_1 will have the mean value (14.80) and the variance

$$\mathrm{var}(n_1) = \sum_{r=1}^{\infty} r^2 \mathrm{var}(u_r) = \sum_{r=1}^{\infty} \frac{r^2 \exp\{r\lambda_1 + \mu_1\}}{(\exp\{r\lambda_1 + \mu_1\} - 1)^2}, \tag{14.88}$$

which we convert into the rapidly convergent sum

$$\sum_{r,n=1}^{\infty} n r^2 \exp\{-n(r\lambda + \mu)\} = \sum_{n=1}^{\infty} n \frac{\exp\{-n(\lambda + \mu)\} + \exp\{-n(2\lambda + \mu)\}}{(1 - \exp\{-n\lambda\})^3} \tag{14.89}$$

or, approximately,

$$\mathrm{var}(n_1) = \frac{2 \exp\{-\mu_1\}}{\lambda_1^3} = \frac{2}{\lambda_1} \langle n_1 \rangle. \tag{14.90}$$

At this point we can use some mathematical facts concerning the central limit theorem. Because n_1 is the sum of a large number of small terms to which we have assigned independent probabilities, our probability distribution for n_1 will be very nearly Gaussian:

$$p(n_1) \approx A \exp\left\{-\frac{\lambda_1(n_1 - \langle n_1 \rangle)^2}{4\langle n_1 \rangle}\right\} \tag{14.91}$$

for those values of n_1 which can arise in many different ways. For example, the case $n = 2$ can arise in only two ways: $u_1 = 2$, or $u_2 = 1$, all others u_k being zero. On the other hand, the case $n_1 = 150$ can arise in an enormous number of different ways, and the 'smoothing' mechanism of the central limit theorem can operate. Thus, Eq. (14.91) will be a good approximation for the large values of n_1 of interest to us, but not for small n_1.

The expected loss on the various decisions is, as we saw in (14.66), the sum of three terms arising from failure to meet orders for red, yellow, or green widgets, respectively. If we do not make red ones today, then the possibility of failing to meet orders for red widgets contributes to the expected loss the amount

$$\sum_{n_1=0}^{\infty} p(n_1) R(n_1 - S_1) \simeq \sqrt{\left[\frac{\lambda_1}{4\pi \langle n_1 \rangle}\right]} \int_{S_1}^{\infty} dn_1 \, (n_1 - S_1) \exp\left\{-\frac{\lambda_1(n_1 - \langle n_1 \rangle)^2}{4\langle n_1 \rangle}\right\}$$

$$= (\langle n_1 \rangle - S_1) \Phi\left[\alpha_1 \sqrt{2}(\langle n_1 \rangle - S_1)\right]$$

$$+ \frac{1}{2\alpha_1 \sqrt{\pi}} \exp\left\{-\alpha_1^2(\langle n_1 \rangle - S_1)^2\right\}, \qquad (14.92)$$

where $\alpha_1^2 = \lambda_1/4\langle n_1 \rangle$, and $\Phi(x)$ is the cumulative normal distribution function (14.42).

If we do decide to make red widgets today, the possibility of failing to meet red orders contributes to the expected loss the above expression (14.92) with S_1 replaced by $(S_1 + 200)$.

Similar equations hold for yellow and green widgets. Although the approximations we made are not equally good in all cases, let us use (14.92) for the partial losses and apply it three times with the given numerical values

$$\begin{array}{lll}
S_1 = 100, & S_2 = 150, & S_3 = 50, \\
\langle n_1 \rangle = 50, & \langle n_2 \rangle = 100, & \langle n_3 \rangle = 10, \\
\alpha_1 = 0.0082, & \alpha_2 = 0.0160, & \alpha_3 = 0.035.
\end{array} \qquad (14.93)$$

Doing the indicated calculations, we find that on the decisions D_1, D_2, D_3 the expected losses are

$$\begin{array}{l}
\langle L \rangle_1 = (0) + 2.86 + 0.18 = 3.04 \text{ unfilled orders} \\
\langle L \rangle_2 = 14.9 + (0) + 0.18 = 15.1 \text{ unfilled orders} \\
\langle L \rangle_3 = 14.9 + 2.86 + (0) = 17.8 \text{ unfilled orders}
\end{array} \qquad (14.94)$$

where (0) stands for a term orders of magnitude smaller than the others. The breakdown indicated is to be read as follows. If Decision D_1 (make red widgets) is made, there is negligible loss from the possibility of failing to meet red orders, while the possibility of failure with yellow orders contributes an expected loss of 2.86, and only 0.18 for green.

These results show the great preference for D_1 caused by the additional information about average individual orders, which had the intuitive effect of making the situation with respect to yellow widgets much safer than it seemed in Stage 2.

14.7.3 Solution for Stage 4

It is in the passage from Stage 3 to Stage 4 (where the new information consists of a specific order for 40 green widgets) that our common sense first fails us. Now both the red and green situations seem rather precarious, and our common sense lacks the 'resolving power' to tell which is the more serious. Strangely enough, this new knowledge, which makes the problem so hard for our common sense, causes no difficulty at all in the mathematics. The previous equations still apply, with the sole difference that the stock S_3 of green widgets is reduced from 50 to 10. We now have $(\langle n_3 \rangle - S_3) = 0$ so that (14.92) reduces to

$$\frac{1}{2\alpha_3\sqrt{\pi}} = 8.08,$$
(14.95)

and in place of (14.94) we have

$$
\begin{aligned}
\langle L \rangle_1 &= (0) + 2.86 + 8.08 = 10.9 \text{ unfilled orders} \\
\langle L \rangle_2 &= 14.9 + (0) + 8.08 = 23.0 \text{ unfilled orders} \\
\langle L \rangle_3 &= 14.9 + 2.86 + (0) = 17.8 \text{ unfilled orders.}
\end{aligned}
$$
(14.96)

So, Mr A should stick to his decision to make red widgets! Our common sense fails just because there is now so little difference between $\langle L \rangle_1$ and $\langle L \rangle_3$.

14.8 Comments

We have tried to show that use of probability theory in the sense of Laplace, with prior probabilities determined by the principle of maximum entropy, leads to a reasonable method of treating decision problems and to results in good correspondence with common sense. Mathematically, our equations are nothing but the Gibbs formalism in statistical mechanics, the only new feature being the recognition that the Gibbs methods are of far more general applicability than had been supposed.

The moral of this is simply that questions about 'interpretation of a formalism', which the positivist philosophy tends to reject as meaningless and useless, are, on the contrary, of central importance in scientific work. It is, of course, true that, in an application already established, a different interpretation of the equations cannot lead to any new numerical results. But our judgment as to the *range of validity* of a formalism can depend entirely on how we interpret it. The interpretation (probability) $\equiv$ (frequency) has led to a great and unnecessary restriction on the kinds of problem where probability theory can be applied. Today, the scientist, engineer, and economist face many problems which require the broader Laplace–Jeffreys interpretation.

15

Paradoxes of probability theory

I protest against the use of infinite magnitude as something accomplished, which is never permissible in mathematics. Infinity is merely a figure of speech, the true meaning being a limit.

C. F. Gauss

The term 'paradox' appears to have several different common meanings. Székely (1986) defines a paradox as anything which is true but surprising. By that definition, every scientific fact and every mathematical theorem qualifies as a paradox for someone. We use the term in almost the opposite sense; something which is absurd or logically contradictory, but which appears at first glance to be the result of sound reasoning. Not only in probability theory, but in all mathematics, it is the careless use of infinite sets, and of infinite and infinitesimal quantities, that generates most paradoxes.

In our usage, there is no sharp distinction between a paradox and an error. A paradox is simply an error out of control; i.e. one that has trapped so many unwary minds that it has gone public, become institutionalized in our literature, and taught as truth. It might seem incredible that such a thing could happen in an ostensibly mathematical field; yet we can understand the psychological mechanism behind it.

15.1 How do paradoxes survive and grow?

As we stress repeatedly, from a false proposition – or from a fallacious argument that leads to a false proposition – all propositions, true and false, may be deduced. But this is just the danger; if fallacious reasoning always led to absurd conclusions, it would be found out at once and corrected. But once an easy, shortcut mode of reasoning has led to a few correct results, almost everybody accepts it; those who try to warn against it are not listened to.

When a fallacy reaches this stage, it takes on a life of its own, and develops very effective defenses for self-preservation in the face of all criticisms. Mathematicians of the stature of Henri Poincaré and Hermann Weyl tried repeatedly to warn against the kind of reasoning used in infinite-set theory, with zero success. For details, see Appendix B and Kline (1980). The writer was also guilty of this failure to heed warnings for many years, until

451

absurd results that could no longer be ignored finally forced him to see the error in an easy mode of reasoning.

To remove a paradox from probability theory will require, at the very least, detailed analysis of the result and the reasoning that leads to it, showing that:

(1) the result is indeed absurd;
(2) the reasoning leading to it violates the rules of inference developed in Chapter 2;
(3) when one obeys those rules, the paradox disappears and we have a reasonable result.

There are too many paradoxes contaminating the current literature for us to analyze separately. Therefore we seek here to study a few representative examples in some depth, in the hope that the reader will then be on the alert for the kind of reasoning which leads to them.

15.2 Summing a series the easy way

As a kind of introduction to fallacious reasoning with infinite sets, we recall an old parlor game by which you can prove that any given infinite series $S = \sum_i a_i$ converges to any number x that your victim chooses. The sum of the first n terms is $s_n = a_1 + a_2 + \cdots + a_n$. Then, defining $s_0 \equiv 0$, we have

$$a_n = (s_n - x) - (s_{n-1} - x), \qquad 1 \le n < \infty, \tag{15.1}$$

so that the series becomes

$$\begin{aligned} S = (s_1 - x) + (s_2 - x) + (s_3 - x) + \cdots \\ - (s_0 - x) - (s_1 - x) - (s_2 - x) - \cdots. \end{aligned} \tag{15.2}$$

The terms $(s_1 - x), (s_2 - x), \ldots$ all cancel out, so the sum of the series is

$$S = -(s_0 - x) = x \qquad\qquad QED. \tag{15.3}$$

The reader for whom this reasoning appears at first glance to be valid has a great deal of company, and is urged to study this example carefully. Such fallacious arguments are avoided if we follow this advice, repeated from Chapter 2:

> *Apply the ordinary processes of arithmetic and analysis only to expressions with a finite number n of terms. Then after the calculation is done, observe how the resulting finite expressions behave as the parameter n increases indefinitely.*

Put more succinctly, passage to a limit should always be the last operation, not the first. In case of doubt, this is the only safe way to proceed. Our present theory of convergence of infinite series could never have been achieved if its founders – Abel, Cauchy, d'Alembert, Dirichlet, Gauss, Weierstrasz, and others – had not followed this advice meticulously. In pre-Bourbakist mathematics (such as Whittaker and Watson, 1927) this policy was considered so obvious that there was no need to stress it. The results thus obtained have never been found defective.

Had we followed this advice above, we would not have tried to cancel out an infinite number of terms in a single stroke; we would have found that at any finite nth stage, instead of the s_i cancelling out and one x remaining, the x values would have cancelled out and the last s remains, leading to the correct summation of the series.

Yet today, reasoning essentially equivalent to what we did in (15.2) is found repeatedly where infinite sets are used in probability theory. As an example, we examine another of the consequences of ignoring this advice, which has grown into far more than a parlor game.

15.3 Nonconglomerability

If $(C_1, \ldots, C_n)$ denote a finite set of mutually exclusive, exhaustive propositions on prior information I, then for any proposition A the sum and product rules of probability theory give

$$P(A|I) = \sum_{i=1}^{n} P(AC_i|I) = \sum_{i=1}^{n} P(A|C_i I)P(C_i|I) \qquad (15.4)$$

in which the prior probability $P(A|I)$ is written as a weighted average of the conditional probabilities $P(A|C_i I)$. Now, it is a very elementary theorem that a weighted average of a set of real numbers cannot lie outside the range spanned by those numbers; if

$$L \leq P(A|C_i I) \leq U, \qquad (1 \leq i \leq n) \qquad (15.5)$$

then necessarily

$$L \leq P(A|I) \leq U, \qquad (15.6)$$

a property which de Finetti (1972) called 'conglomerability' or, more precisely, 'conglomerability in the partition $\{C_i\}$', although it may seem too trivial to deserve a name. Obviously, nonconglomerability cannot arise from a correct application of the rules of probability theory on finite sets. It cannot, therefore, occur in an infinite set which is approached as a well-defined limit of a sequence of finite sets.

Yet nonconglomerability has become a minor industry, with a large and growing literature. There are writers who believe that it is a real phenomenon, and that they are proving theorems about the circumstances in which it occurs, which are important for the foundations of probability theory. Nonconglomerability has become, quite literally, institutionalized in our literature and taught as truth.

In spite of its mathematical triviality, then, we need to examine some cases where nonconglomerability has been claimed. Rather than trying to cite all of this vast literature, we draw upon a single reference (Kadane, Schervish and Seidenfeld, 1986), hereafter denoted by KSS, where several examples of nonconglomerability and some references to other work may be found.

Example 1: Rectangular array. Firstly, we note the typical way in which nonconglomerability is manufactured, and the illustrative example most often cited. We start from a two-dimensional ($M \times N$) set of probabilities:

$$p(i, j), \qquad 1 \leq i \leq M, \qquad 1 \leq j \leq N, \qquad (15.7)$$

and think of i plotted horizontally, j vertically, so that the sample space is a rectangular array of MN points in the first quadrant. It will suffice to take some prior information I for which these probabilities are uniform: $p(i, j) = (1/MN)$. Then the probability of the event $(A : i < j)$ is found by direct counting to be

$$P(A|I) = \begin{cases} (2N - M - 1)/2N & M \leq N \\ (N - 1)/2M & N \leq M. \end{cases} \tag{15.8}$$

Let us resolve this in the manner of (15.4), into probabilities conditional on the set of propositions $(C_1, \ldots, C_M)$, where C_i is the statement that we are on the ith column of the array: then $P(C_i|I) = (1/M)$, and

$$P(A|C_i I) = \begin{cases} (N - i)/N & 1 \leq i \leq M \leq N \\ (N - i)/N & 1 \leq i \leq N \leq M \\ 0 & N \leq i \leq M. \end{cases} \tag{15.9}$$

These conditional probabilities reach the upper and lower bounds

$$U = (N - 1)/N \qquad \text{all } M, N,$$

$$L = \begin{cases} 1 - R & M \leq N \\ 0 & N \leq M. \end{cases} \tag{15.10}$$

where R denotes the ratio $R = M/N$. Substituting (15.8) and (15.10) into (15.6), it is evident that the condition for conglomerability is always satisfied, as it must be, whatever the values of (M, N). How, then, can one possibly create a nonconglomerability out of this?

Just pass to the limit $M \to \infty$, $N \to \infty$, and ask for the probabilities $P(A|C_i I)$ for $i = 1, 2, \ldots$. But instead of examining the limiting form of (15.9), which gives the exact values for all (M, N), we try to evaluate these probabilities directly on the infinite set.

Then, it is argued that, for any given i, there are an infinite number of points where A is true and only a finite number where it is false. *Ergo*, the conditional probability $P(A|C_i I) = 1$ for all i; yet $P(A|I) < 1$. We see here the same kind of reasoning that we used in (15.2); we are trying to carry out very simple arithmetic operations (counting), but directly on an infinite set.

Now consider the set of propositions $(D_1, \ldots, D_N)$, where D_j is the statement that we are on the jth row of the array, counting from the bottom. Now, by the same argument, for any given j, there are an infinite number of points where A is false, and only a finite number where A is true. *Ergo*, the conditional probability $P(A|D_j I) = 0$ for all j; yet $P(A|I) > 0$. By this reasoning, we have produced two nonconglomerabilities, in opposite directions, from the same model (i.e. the same infinite set).

It is even more marvellous than that. In (15.8), it is true that if we pass to the limit holding i fixed, the conditional probability $P(A|C_i B)$ tends to one for all i; but if instead we hold $(N - i)$ fixed, it tends to zero for all i. Therefore, if we consider the cases $(i = 1, i = 2, \ldots)$ in increasing order, the probabilities $P(A|C_i B)$ appear to be one for all i. But it is equally valid to consider them in decreasing order $(i = N, i = N - 1, \ldots)$; then, by the same reasoning, they would appear to be zero for all i. (Note that we could redefine the labels by subtracting $N + 1$ from each one, thus numbering them $(i = -N, \ldots, i = -1)$ so that as $N \rightarrow \infty$ the upper indices stay fixed; this would have no effect on the validity of the reasoning.)

Thus, to produce two opposite nonconglomerabilities we need not introduce two different partitions $\{C_i\}$, $\{D_j\}$; they can be produced by two equally valid arguments from a single partition. What produces them is that one supposes the infinite limit already accomplished *before* doing the arithmetic, reversing the policy of Gauss which we recommended above. But if we follow that policy and do the arithmetic first, then an arbitrary redefinition of the labels $\{i\}$ has no effect; the counting for any N is the same.

Once one has understood the fallacy in (15.2), then whenever someone claims to have proved some result by carrying out arithmetic or analytical operations directly on an infinite set, it is hard to shake off a feeling that he could have proved the opposite just as easily and by an equally sound argument, had he wished to. Thus there is no reason to be surprised by what we have just found.

Suppose that instead we had done the calculation by obeying our rules strictly, doing first the arithmetic operations on finite sets to obtain the exact solution (15.8); then passing to the limit. However the infinite limit is approached, the conditional probabilities take on values in a wide interval whose lower bound is zero or $1 - R$, and whose upper bound tends to one. The condition (15.5) is always satisfied, and a nonconglomerability could never have been found.

The reasoning leading to this nonconglomerability contains another fallacy. Clearly, one cannot claim to have produced a nonconglomerability on the infinite set until the 'unconditional' probability $P(A|I)$ has also been calculated on that set, not merely bounded by a verbal argument. But as M and N increase, from (15.8) the limiting $P(A|I)$ depends only on the ratio $R = M/N$:

$$P(A|I) \rightarrow \begin{cases} 1 - R/2 & R \leq 1 \\ 1/(2R) & R \geq 1. \end{cases} \tag{15.11}$$

If we pass to the infinite limit without specifying the limiting ratio, the unconditional probability $P(A|I)$ becomes indeterminate; we can get any value in $[0, 1]$ depending on how the limit is approached. Put differently, the ratio R contains all the information relevant to the probability of A; yet it was thrown away in passing to the limit too soon. The unconditional probability $P(A|I)$ could not have been evaluated directly on the infinite set, any more than could the conditional probabilities.

Thus, nonconglomerability on a rectangular array, far from being a phenomenon of probability theory, is only an artifact of failure to obey the rules of probability theory as developed in Chapter 2. But from studying a single example we cannot see the common feature underlying all claims of nonconglomerability.

15.4 The tumbling tetrahedra

We now examine a claim that nonconglomerability can occur even in a one-dimensional infinite set $n \to \infty$ where there does not appear to be any limiting ratio like the above M/N to be ignored. Also we now consider a problem of inference, instead of the above sampling distribution example. The scenario (Stone, 1979) appears to be equivalent to the 'strong inconsistency' problem (Stone, 1976). We follow the KSS notation for the time being – until we see why we must not.

A regular tetrahedron with faces labeled e^+ (positron), e^- (electron), μ^+ (muon), μ^- (antimuon), is tossed repeatedly. A record is kept of the result of each toss, except that, whenever a record contains e^+ followed immediately by e^- (or e^- by e^+, or μ^+ by μ^-, or μ^- by μ^+), the particles annihilate each other, erasing that pair from the record. At some arbitrary point in the sequence, the player (who is ignorant of what has happened to date) calls for one more toss, and then is shown the final record $x \in X$, after which he must place bets on the truth of the proposition $A \equiv$ 'annihilation occurred at the final toss'. What probability $P(A|x)$ should he assign?

When we try to answer this by application of probability theory, we come up immediately against the difficulty that, in the problem as stated, the solution depends on a nuisance parameter, the unspecified length n of the original sequence of tosses. This was pointed out by Hill (1980), but KSS take no note of it. In fact, they do not mention n at all except by implication, in a passing remark that the die is 'rolled a very large number of times'. We infer that they meant the limit $n \to \infty$, from later phrases such as 'the countable set S' and 'every finite subset of S'.

In other words, once again an infinite set is supposed to be something already accomplished, and one is trying to find relations between probabilities by reasoning directly on the infinite set. Nonconglomerability enters through asking whether the prior probability $P(A)$ is conglomerable in the partition x, corresponding to the equation

$$P(A) = \sum_{x \in X} P(A|x)P(x). \tag{15.12}$$

KSS denote by $\theta \in S$ the record just before the final toss (thought of as a 'parameter' not known by the player), where S is the set of all possible such records, and conclude by verbal arguments that:

(a) $0 \leq p(A|\theta) \leq 1/4$,   all $\theta \in S$;
(b) $3/4 \leq p(A|x) \leq 1$,   all $x \in X$.

It appears that another violent nonconglomerability has been produced; for if $P(A)$ is conglomerable in the partition $\{x\}$ of final records, it must be true that $3/4 \leq P(A) \leq 1$, while if it is conglomerable in the partition $\{\theta\}$ of previous records, we require $0 \leq P(A) \leq 1/4$; it cannot be conglomerable in both. So where is the error this time?

We accept statement (a); indeed, given the independence of different tosses, knowing anything whatsoever about the earlier tosses gives us no information about the final one, so the uniform prior assignment $1/4$ for the four possible results of the final toss still holds. Therefore, $p(A|\theta) = 1/4$, except when the record θ is blank, in which case there is nothing to annihilate, and so $p(A|\theta) = 0$. But this argument does not hold for statement (b); since the result of the final toss affects the final record x, it follows that knowing x must give some information about the final toss, invalidating the uniform $1/4$ assignment.

Also, the argument that KSS gave for statement (b) supposed prior information different from that used for statement (a). This was concealed from view by the notation $p(A|\theta)$, $p(A|x)$ which fails to indicate prior information I. Let us repeat (15.12) with adequate notation:

$$P(A|I) = \sum_{x \in X} P(A|xI)P(x|I). \tag{15.13}$$

Now as I varies, all these quantities will in general vary. By 'conglomerability' we mean, of course, 'conglomerability *with some particular fixed prior information I*.' Recognizing this, we repeat statements (a) and (b) in a notation adequate to show this difference:

(a) $0 \le p(A|\theta I_a) \le 1/4$, $\theta \in S$;
(b) $3/4 \le p(A|x I_b) \le 1$, $x \in X$.

From reading KSS we find that prior information I_a, in effect, assigned uniform probabilities on the set T of 4^n possible outcomes of n tosses, as is appropriate for the case of 'independent repetitions of a random experiment' assumed in the statement of the problem. But I_b assigned uniform probabilities on the set S of different previous records θ. This is very different; an element of S (or X) may correspond to one element of T, or to many millions of elements of T, so a probability assignment uniform on the set of tosses is very nonuniform on the set of records. Therefore it is not evident whether there is any contradiction here; they are statements about two quite different problems.

Exercise 15.1. In $n = 40$ tosses there are $4^n = 1.21 \times 10^{24}$ possible sequences of results in the set T. Show that, if those tosses give the expected number $m = 10$ of annihilations leading to a record $x \in X$ of length 20, the specific record x corresponds to about 10^{14} elements of T. On the other hand, if there are no annihilations, the resulting record x of length 40 corresponds to only one element of T.

Perhaps this makes clearer the reason for our seemingly fanatical insistence on indicating the prior information I explicitly in every formal probability symbol $P(A|BI)$. Those who fail to do this may be able to get along without disaster for a while, judging the meaning of an equation from the surrounding context rather than from the equation as written. But eventually they are sure to find themselves writing nonsense, when they start inadvertently using probabilities conditional on different prior information in the same equation, or the same argument, and their notation conceals that fact. We shall see presently a more famous and more serious error (the marginalization paradox) caused by failure to indicate the fact that two probabilities are conditional on different prior information.

To show the crucial role that n plays in the problem, let I agree with I_a in assigning equal prior probabilities to each of the 4^n outcomes of n tosses. Then, if n is known, calculations of $p(A|nI)$, $p(x|nI)$, $p(A|nxI)$ are determinate combinatorial problems on finite sets (i.e. in each case there is one and only one correct answer), and the solutions obviously depend on n. So let us try to calculate $P(A|xI)$; denoting summation over all n in $(0 \leq n < \infty)$ by $\sum$, we have for the prior probabilities

$$p(A|I) = \sum p(An|I) = \sum p(A|nI)p(n|I)$$
$$p(x|I) = \sum p(xn|I) = \sum p(x|nI)p(n|I)$$
(15.14)

and for the conditional one

$$p(A|xI) = \sum p(A|nxI)p(n|xI) = \frac{\sum p(A|nxI)p(x|nI)p(n|I)}{\sum p(x|nI)p(n|I)},$$
(15.15)

where we expanded $p(n|xI)$ by Bayes' theorem. It is evident that the problem is indeterminate until the prior probabilities $p(n|I)$ are assigned. Quite generally, failure to specify the prior information makes a problem of inference just as ill-posed as does failure to specify the data.

Passage to infinite n then corresponds to taking the limit of prior probabilities $p(n|I)$ that are nonzero only for larger and larger n. Evidently, this can be done in many different ways, and the final results will depend on which limiting process we use unless $p(A|nI)$, $p(x|nI)$, $p(A|nxI)$ all approach limits independent of n.

The number of different possible records x is less than 4^n (asymptotically, about 3^n) because many different outcomes with annihilation may produce the same final record, as the above exercise shows. Therefore, for any $n < \infty$, there is a finite set X of different possible final records x, and *a fortiori* a finite set S of previous records θ, so the prior probability of final annihilation can be written in either of the forms:

$$p(A|nI) = \sum_{x \in X} p(A|xnI)p(x|nI) = \sum_{\theta \in S} p(A|\theta nI)p(\theta|nI),$$
(15.16)

and the general theorem on weighted averages guarantees that nonconglomerability cannot occur in either partition for any finite n, or for an infinite set generated as a well-behaved limit of a sequence of these finite sets.

A few things about the actual range of variability of the conditional probabilities $p(A|nxI)$ can be seen at once without any calculation. For any n, there are possible records of length n for which we know that no annihilation occurred; the lower bound is always reached for some x, and it is $p(A|nxI) = 0$, not $3/4$. The lower bound in statement (b) could never have been found for any prior information, had the infinite set been approached as a limit of a sequence of finite sets. Furthermore, for any even n there are possible records of length zero for which we know that the final toss was annihilated; the upper bound is always reached for some x, and it is $p(A|nxI) = 1$.

Likewise, for even n it is not possible for θ to be blank, so from (15.16) we have $p(A|nI) = p(A|\theta nI) = 1/4$ for all $\theta \in S$. Therefore, if n is even, there is no need to invoke even the weighted average theorem; there is no possibility for nonconglomerability in either the partition $\{x\}$ or $\{\theta\}$.

At this point it is clear that the issue of nonconglomerability is disposed of in the same way as in our first example; it is an artifact of trying to calculate probabilities directly on an infinite set without considering any limit from a finite set. Then it is not surprising that KSS never found any specific answer to their original question: 'What can we infer about final annihilation from the final record x?' But we would still like to see the answer (particularly since it reveals an even more startling feature of the problem).

15.5 Solution for a finite number of tosses

If n is known, we can get the exact analytical solution easily from valid application of our rules. It is a straightforward Bayesian inference in which we are asking only for the posterior probability for final annihilation A. But this enables us to simplify the problem; there is no need to draw inferences about every detail of the previous record θ.

If there is annihilation at the nth toss, then the length of the record decreases by one: $y(n) = y(n-1) - 1$. If there is no annihilation at the nth toss, the length increases by one: $y(n) = y(n-1) + 1$. The only exception is that $y(n)$ is not permitted to become negative; if $y(n-1) = 0$, then the nth toss cannot give annihilation. Therefore, since the available record x tells us the length $y(n)$ but not $y(n-1)$, any reasoning about final annihilation may be replaced immediately by reasoning about $\alpha \equiv y(n-1)$, which is the sole parameter needed in the problem.

Likewise, any permutations of the symbols $\{e^{\pm}, \mu^{\pm}\}$ in $x(n)$ which keep the same $y(n)$ will lead to just the same inferences about A. But then n and $y \equiv y(n)$ are sufficient statistics; all other details of the record x are irrelevant to the question being asked. Thus the scenario of the tetrahedrons is more complicated than it needs to be in order to define the mathematical problem (in fact, so complicated that it seems to have prevented recognition that it is a standard textbook random walk problem).

At each nth toss we have the sampling probability $1/4$ of annihilating, independently of what happened earlier (with a trivial exception if $y(n-1) = 0$). Therefore if we plot n horizontally, $y(n)$ vertically, we have the simplest random walk problem in one dimension, with a perfectly reflecting boundary on the horizontal axis $y = 0$. At each horizontal step, if $y > 0$ there is probability $3/4$ of moving up one unit, $1/4$ of moving down one unit; if $y = 0$, we can move only up. Starting with $y(0) = 0$, annihilation cannot occur on step 1, and immediately after the nth step, if there have been m annihilations, the length of the record is $y(n) = n - 2m$.

After the nth step we have a prior probability distribution for $y(n)$ to have the value i:

$$p_i^{(n)} \equiv p(i|nI), \qquad 0 \le i \le n, \qquad (15.17)$$

with the initial vector

$$p_i^{(0)} = \begin{pmatrix} 1 \\ 0 \\ 0 \\ \vdots \end{pmatrix}, \tag{15.18}$$

and successive distributions are connected by the Markov chain relation

$$p_i^{(n)} = \sum_{j=0}^{n-1} M_{ij} p_j^{(n-1)} \qquad \begin{array}{l} 0 \le i \le n \\ 1 \le n < \infty, \end{array} \tag{15.19}$$

with the transition matrix (number the rows and columns starting with zero)

$$M \equiv \begin{pmatrix} 0 & 1/4 & 0 & 0 & \dots \\ 1 & 0 & 1/4 & 0 & \dots \\ 0 & 3/4 & 0 & 1/4 & \dots \\ 0 & 0 & 3/4 & 0 & \dots \\ \vdots & \vdots & \vdots & \vdots & \ddots \end{pmatrix}. \tag{15.20}$$

The reflecting boundary at $y = 0$ is indicated by the element $M_{10} = 1$, which would be 3/4 without the reflection.

The matrix M is, in principle, infinite-dimensional, but for the nth step only the first $n + 1$ rows and columns are needed. The vector $p^{(n)}$ is also, in principle, infinite-dimensional, but $p_i^{(n)} = 0$ when $i > n$. Then the exact solution for the prior probabilities $p_i^{(n)}$ is the first column of M^n:

$$p_i^{(n)} = M_{i,0}^n \tag{15.21}$$

(note that this is intended to represent $(M^n)_{i,0}$, not $(M_{i,0})^n$).

Now let us see how this prior is to be used in our Bayesian inference problem. Denote the data and the hypothesis being tested by

$$D \equiv y(n) = i, \qquad H \equiv y(n-1) = \alpha, \tag{15.22}$$

which are the only parts of the data x and the parameter θ that are relevant to our problem. From the above their prior probabilities are

$$p(D|I) = M_{i,0}^n, \qquad p(H|I) = M_{\alpha,0}^{n-1}. \tag{15.23}$$

The sampling distribution is

$$p(D|HI) = \begin{cases} 3/4\, \delta(i, \alpha+1) + 1/4\, \delta(i, \alpha-1) & \alpha > 0 \\ \delta(i, 1) & \alpha = 0. \end{cases} \tag{15.24}$$

So, Bayes' theorem gives the posterior probability for α as

$$p(H|DI) = p(H|I)\frac{p(D|HI)}{p(D|I)} = \frac{M_{\alpha,0}^{n-1}}{M_{i,0}^n} \begin{cases} 3/4\,\delta(i, \alpha+1) + 1/4\,\delta(i, \alpha-1) & \alpha > 0 \\ \delta(i, 1) & \alpha = 0. \end{cases}$$

(15.25)

Now, final annihilation A occurs if and only if $\alpha = i+1$, so the exact solution for finite n is

$$p(A|DnI) = \frac{M_{i+1,0}^{n-1}}{4\,M_{i,0}^n},$$

(15.26)

in which $i = y(n)$ is a sufficient statistic. Another way of writing this is to note that the denominator of (15.26) is

$$4M_{i,0}^n = 4\sum_j M_{i,j}\,M_{j,0}^{n-1} = 3M_{i-1,0}^{n-1} + M_{i+1,0}^{n-1},$$

(15.27)

and so the posterior odds on A are

$$o(A|DnI) \equiv \frac{p(A|xnI)}{p(\overline{A}|xnI)} = \frac{1}{3}\frac{M_{i+1,0}^{n-1}}{M_{i-1,0}^{n-1}},$$

(15.28)

and it would appear, from their remarks, that the exact solution to the problem that KSS had in mind is the limit of (15.26) or (15.28) as $n \to \infty$.

This solution for finite n is complicated because of the reflecting boundary. Without it, the aforementioned matrix element $M_{1,0}$ would be 3/4 and the problem would reduce to the simplest of all random walk problems. That solution gives us a very good approximation to (15.26), which actually yields the exact solution to our problem in the limit. Let us examine this alternative formulation because its final result is very simple and the derivation is instructive about a point that is not evident from the above exact solution.

The problem where at each step there is probability p to move up one unit, $q = 1 - p$ to move down one unit, is defined by the recursion relation in which $f(i|n)$ is the probability to move a total distance i in n steps:

$$f(i|n+1) = pf(i-1|n) + qf(i+1|n).$$

(15.29)

With initial conditions $f(i|n = 0) = \delta(i, 0)$, the standard textbook solution is the binomial for r successes in n trials; $f_0(i|n) = b(r|np)$, with $r = (n+i)/2$. In our problem we know that on the first step we necessarily move up, $y(1) = 1$, so our initial conditions are $f(i|n = 1) = \delta(i, 1)$, and using the binomial recursion (15.29) after that the solution would be $f(i|n) = f_0(i-1|n-1) = b(r|n-1, p)$, with again $r = (n+i)/2$.

But with $p = 3/4$, this is not exactly the same as (15.19) because it neglects the reflecting boundary. If too many 'failures' (i.e. annihilations) occur early in the sequence, this could reduce the length of the record to zero, forcing the upward probability for the next step to be one rather than 3/4; and (15.19) is taking all that into account. Put differently, in the solution to (15.29), when n is small, some probability drifts into the region $y < 0$; but if $p = 3/4$ the amount is almost negligibly small, and it all returns eventually to $y > 0$.

When n is very large, the solution drifts arbitrarily far away from the reflecting boundary, putting practically all the probability into the region $(\hat{y} - \sqrt{n} < y < \hat{y} + \sqrt{n})$, where $\hat{y} \equiv (p - q)n = n/2$. So conclusions drawn from (15.29) become highly accurate (in the limit, exact).

The sampling distribution (15.24) is unchanged, but we need binomial approximations to the priors for i and α. The latter is the length of the record after $n - 1$ steps, or tosses. No annihilation is possible at the first toss, so after $n - 1$ tosses we know that there were $n - 2$ tosses at which annihilation could have occurred, with probability $1/4$ at each, so the prior probability for m annihilations in the first $n - 1$ tosses is the binomial $b(m|n - 2, 1/4)$:

$$f(m) \equiv p(m|n) = \binom{n-2}{m} \left(\frac{1}{4}\right)^m \left(\frac{3}{4}\right)^{n-2-m}, \qquad 0 \le m \le n - 2. \qquad (15.30)$$

Then the prior probability for α, replacing the numerator in (15.28), is

$$p(\alpha|n) = f\left(\frac{n - 1 - \alpha}{2}\right), \qquad (15.31)$$

from which we find the prior expectation $E(\alpha|I) = n/2$. Likewise in the denominator we want the prior for $y(n) = i$. This is just (15.31) with the replacements $n - 1 \to n, \alpha \to i$.

Given y, the possible values of α are $\alpha = y \pm 1$, so the posterior odds on final annihilation are, writing $m \equiv (n - y)/2$,

$$o = \frac{p(A|yn)}{p(\bar{A}|yn)} = \frac{p(\alpha = y + 1|yn)}{p(\alpha = y - 1|yn)} = \frac{(1/4)\binom{n-2}{m-1}(1/4)^{m-1}(3/4)^{n-1-m}}{(3/4)\binom{n-2}{m}(1/4)^m(3/4)^{n-2-m}}. \qquad (15.32)$$

But, at first sight astonishing, the factors $(1/4)$, $(3/4)$ cancel out, so the result depends only on the factorials:

$$o = \frac{m!\,(n - 2 - m)!}{(m - 1)!\,(n - 1 - m)!} = \frac{n - y}{n - 2 + y}, \qquad (15.33)$$

and the posterior probability of final annihilation reduces simply to

$$p(A|yn) = \frac{o}{1 + o} = \frac{n - y}{2(n - 1)}, \qquad (15.34)$$

which does not bear any resemblance to any of the solutions proposed by those who tried to solve the problem by reasoning directly on infinite sets. The sampling probabilities $p = 3/4$, $q = 1/4$, which figured so prominently in previous discussions, do not appear at all in the solution.

But now *think* about it. Given n and $y(n)$, we know that annihilation might have occurred in any of $n - 1$ tosses, but that in fact it did occur in exactly $(n - y)/2$ tosses. But we have no information about which tosses, so the posterior probability for annihilation at the final toss (or at any toss after the first) is, of course,

$$\frac{n - y}{2(n - 1)}. \qquad (15.35)$$

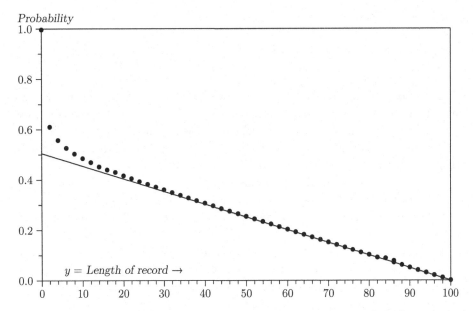

Fig. 15.1. Solution to the 'strong inconsistency' problem for $n = 100$ tosses. Solid line = approximation, Eq. (15.34); dots = exact solution, Eq. (15.26).

We derived (15.34) directly from the principles of probability theory by a rather long calculation; but with a modicum of intuitive understanding of the problem, we could have reasoned it out in our heads without any calculation at all!

In Figure 15.1 we compare the exact solution (15.26) with the asymptotic solution (15.34). The difference is negligible numerically when $n > 20$. So then, why did so many people think the answer should be 1/4? Perhaps it helps to note that the prior expectation for y is $E(y|I) = (n + 1)/2$, so the *predictive probability* of final annihilation is

$$p(A|nI) = \frac{n - E(y|I)}{2(n - 1)} = \frac{1}{4}. \tag{15.36}$$

The posterior probability of final annihilation is indeed 1/4, *if the observed record length y is the expected value*. If new information is only what we already expected, it does not change our estimates; it only makes us more confident of them. But if y is observed to be different from its prior expectation, this tells us the *actual number* of annihilations, and of course this information takes precedence over whatever initial probability assignments $(1/4, 3/4)$ we might have made. That is why they cancelled out in the posterior odds.[1] In spite of our initial surprise, then, Bayes' theorem is doing exactly the right thing here; and the exact solution of the problem originally posed is given also by

---

[1] This cancellation is the thing that is not evident at all in the exact solution (15.26), although it is still taking place out of sight.

the limit of (15.35) as $n \to \infty$:

$$p(A|xI) = \frac{1}{2}(1 - z) \tag{15.37}$$

where $z \equiv \lim y(n)/n$.

In summary, the common feature of these two claims of nonconglomerability is now apparent. In the first scenario, there was no mention of the existence of the finite numbers M, N whose ratio M/N is the crucial quantity on which the solution depends. In the second scenario, essentially the same thing was done; failure to introduce the length n of the sequence and, incredibly, even the length $y(n)$ of the observed record, likewise causes one to lose the crucial thing – in this case, the sufficient statistic y/n – on which the solution depends. In both cases, by supposing the infinite limit as something already accomplished at the start, *one is throwing away the very information required to find the solution.*

This has been a very long discussion, but it is hard to imagine a more instructive lesson in how and why one must carry out probability calculations where infinite sets are involved, or a more horrible example of what can happen if we fail to heed the advice of Gauss.

15.6 Finite vs. countable additivity

At this point, the reader will be puzzled and asking, 'Why should anybody care about nonconglomerability? What difference does it make?' Nonconglomerability is, indeed, of little interest in itself; it is only a kind of red herring that conceals the real issue. A follower of de Finetti would say that the underlying issue is the technical one of 'finite additivity'. To which we would reply that 'finite additivity' is also a red herring, because it is used for a purpose almost the opposite of what it sounds like.

In Chapter 2 we derived the sum rule (2.85) for mutually exclusive propositions: if as a statement of Boolean algebra, $A \equiv A_1 + A_2 + \cdots + A_n$ is a disjunction of a finite number of mutually exclusive propositions, then

$$p(A|C) = \sum_{i=1}^{n} p(A_i|C). \tag{15.38}$$

Then it is a trivial remark that our probabilities have 'finite additivity'. As $n \to \infty$ it seems rather innocuous to suppose that the sum rule goes in the limit into a sum over a countable number of terms, forming a convergent series; whereupon our probabilities would be called countably additive. Indeed (although we do not see how it could happen in a real problem), if this should ever fail to yield a convergent series we would conclude that the infinite limit does not make sense, and we would refuse to pass to the limit at all. In our formulation of probability theory, it is difficult to see how one could make any substantive issue out of this perfectly straightforward situation.

The conventional formulations, reversing our policy, suppose the infinite limit already accomplished at the beginning, before such questions as additivity are raised; and then are concerned with additivity over propositions about intervals on infinite sets. To quote Feller (1966, 1971 edn, p. 107):

Let F be a function assigning to each interval I a finite value $F\{I\}$. Such a function is called (finitely) *additive* if for every partition of an interval I into finitely many non-overlapping intervals $I_1 \cdots I_n$, $F\{I\} = F\{I_1\} + \cdots + F\{I_n\}$.

Then (p. 108) Feller gives an example showing why he wishes to replace finite additivity by countable additivity:

In R^1 put $F\{I\} = 0$ for any interval $I = (a, b)$ with $b < \infty$ and $F\{I\} = 1$ when $I = (a, \infty)$. This interval function is additive but weird because it violates the natural continuity requirement that $F\{(a, b)\}$ should tend to $F\{(a, \infty)\}$ as $b \to \infty$.

This last example shows the desirability of strengthening the requirement of finite additivity. We shall say that an interval function F is countably additive, or σ-additive, if for every partitioning of an interval I into countably many intervals $I_1, I_n, \ldots, F\{I\} = \sum F\{I_k\}$.

He then adds that the condition of countable additivity is 'manifestly violated' in the above weird example (let it be an exercise for the reader to explain clearly *why* this is manifest).

What is happening in that weird example? Surely, the weirdness does not lie in lack of continuity (since continuity is quite unnecessary in any event), but in something far worse. Supposing those intervals occupied by some variable x and the interval function $F\{I\}$ to be the probability $p(x \in I)$, one is assigning zero probability to any finite range of x, but unit probability to the infinite range. This is almost impossible to comprehend when we suppose the infinite interval already accomplished, but we can understand what is happening if we heed the advice of Gauss and think in terms of passage to a limit. Suppose we have a properly normalized pdf:

$$p(x|r) = \begin{cases} 1/r & 0 \le x < r \\ 0 & r \le x < \infty. \end{cases} \tag{15.39}$$

As long as $0 < r < \infty$, there is nothing strange, and we could describe this by an interval function

$$F(a, b) \equiv \int_a^b dx\, p(x|r) = \begin{cases} (b - a)/r & 0 \le a \le b \le r < \infty \\ (r - a)/r & 0 \le a \le r \le b < \infty \\ 0 & 0 \le r \le a \le b < \infty, \end{cases} \tag{15.40}$$

which is, rather trivially, countably additive and *a fortiori* finitely additive. As r increases, the density function becomes smaller and spread over a wider interval; but as long as $r < \infty$ we have a well-defined and nonparadoxical mathematical situation.

If we try to describe the limit of $p(x|r)$ as something already accomplished *before* discussing additivity, then we have created Feller's weird example. We are trying to make a probability density that is everywhere zero, but which integrates to unity. But *there is no such thing*, according not only to all the warnings of classical mathematicians from Gauss on, but according to our own elementary common sense.

Invoking finite additivity is a sneaky way of approaching the real issue. To see why the kind of additivity matters in the conventional formulation, let us note what happens when one carries out the order of operations corresponding to our advice above. We assign a continuous monotonic increasing cumulative probability function $G(x)$ on the real line, with the natural continuity property that

$$G(x) \rightarrow \begin{cases} 1 & x \rightarrow +\infty \\ 0 & x \rightarrow -\infty; \end{cases} \tag{15.41}$$

then, the interval function F for the interval $I = (a, b)$ may be taken as $F\{I\} = G(b) - G(a)$, and it is 'manifest' that this interval function is countably additive in the sense defined. That is, we can choose x_k satisfying $a < x_1 < x_2 < \cdots < b$ so as to break the interval (a, b) into as many nonoverlapping subintervals $\{I_0, I_1, \ldots, I_n\} = \{(a, x_1), (x_1, x_2), \ldots, (x_n, b)\}$ as we please, and it will be true that $F\{I\} = \sum F\{I_k\}$. If $G(x)$ is differentiable, then its derivative $f(x) \equiv G'(x)$ may be interpreted as a normalized probability density: $\int dx\, f(x) = 1$.

We see, finally, what the point of all this is: 'finite additivity' is a euphemism for 'reversing the proper order of approaching limits, and thereby getting into trouble with non-normalizable probability distributions'. Feller saw this instantly, warned the reader against it, and proceeded to develop his own theory in a way that avoids the many useless and unnecessary paradoxes that arise from it.[2]

As we saw in Chapter 6, passage to the limit $r \rightarrow \infty$ at the end of a calculation can yield useful results; some other probability derived from $p(x \mid r)$ might approach a definite, finite, and simple limiting value. We have now seen that trying to pass to the limit at the beginning of a calculation can generate nonsense because crucial information is lost before we have a chance to use it.

The real issue here is: do we admit such things as uniform probability distributions on infinite sets into probability theory as legitimate mathematical objects? Do we believe that an infinite number of zeroes can add up to one? In the strange language in which these things are discussed, to advocate 'finite additivity', as de Finetti and his followers do, is a devious way of answering 'yes' without seeming to do so. To advocate 'countable additivity', as Kolmogorov and Feller did, is an equally devious way to answer 'no' in the spirit of Gauss.

The terms are red herrings because 'finite additivity' sounds colloquially as if it were a cautious assumption, 'countable additivity' a bit more adventurous. de Finetti does indeed seem to think that finite additivity is the weaker assumption; and he rails against those who, as he sees it, are intellectually dishonest when they invoke countable additivity only for 'mathematical convenience', instead of for a compelling reason. As we see it, jumping directly into an infinite set at the very beginning of a problem is a vastly greater error of judgment, which has far worse consequences for probability theory; there is a little more than just 'mathematical convenience' at stake here.

---

[2] Since we disagree with Feller so often on conceptual issues, we are glad to be able to agree with him on nearly all technical ones. He was, after all, a very great contributor to the technical means for solving sampling theory problems, and practically everything he did is useful to us in our wider endeavors.

We noted the same psychological phenomenon in Chapter 3, when we introduced the binomial distribution for sampling with replacement; those who committed the sin of throwing away relevant information invented the term 'randomization' to conceal that fact and make it sound like they were doing something respectable. Those who commit the sin of doing reckless, irresponsible things with infinity often invoke the term 'finite additivity' to make it sound as if they are being *more* careful than others with their mathematics.

15.7 The Borel–Kolmogorov paradox

For the most part, the transition from discrete to continuous probabilities is uneventful, proceeding in the obvious way with no surprises. However, there is one tricky point concerning continuous densities that is not at all obvious, but can lead to erroneous calculations unless we understand it. The following example continues to trap many unwary minds.

Suppose I is prior information according to which (x, y) are assigned a bivariate normal pdf with variance unity and correlation coefficient ρ:

$$p(dx\,dy|I) = \frac{\sqrt{1-\rho^2}}{2\pi} \exp\left\{\frac{1}{2}(x^2 + y^2 - 2\rho x y)\right\} dx\,dy. \qquad (15.42)$$

We can integrate out either x or y to obtain the marginal pdfs (to prepare for integrating out x, write $x^2 + y^2 - 2\rho x y = (x - \rho y)^2 + (1 - \rho^2)\,y^2$, etc.):

$$p(dx|I) = \sqrt{\left(\frac{1-\rho^2}{2\pi}\right)} \exp\left\{-\frac{1}{2}(1 - \rho^2)x^2\right\} dx \qquad (15.43)$$

$$p(dy|I) = \sqrt{\left(\frac{1-\rho^2}{2\pi}\right)} \exp\left\{-\frac{1}{2}(1 - \rho^2)\,y^2\right\} dy. \qquad (15.44)$$

Thus far, all is routine. But now, what is the conditional pdf for x, given that $y = y_0$? We might think that we need only set $y = y_0$ in (15.42) and renormalize:

$$p(dx|y = y_0\,I) = A \exp\left\{-\frac{1}{2}(x^2 + y_0^2 - 2\rho x y_0)\right\} dx, \qquad (15.45)$$

where A is a normalizing constant. But there is no guarantee that this is valid, because we have obtained (15.45) by an intuitive *ad hoc* device; we did not derive it from (15.42) by applying the basic rules of probability theory, which we derived in Chapter 2 for the discrete case:

$$p(AB|X) = p(A|BX)p(B|X), \qquad (15.46)$$

from which a discrete conditional probability is given by the usual rule

$$p(A|BX) = \frac{p(AB|X)}{p(B|X)} \qquad (15.47)$$

often taken as the definition of a conditional probability. But we can do the calculation by strict application of our rules if we define the discrete propositions

$$A \equiv x \text{ in } dx$$
$$B \equiv y \text{ in } (y_0 < y < y_0 + dy). \tag{15.48}$$

Then we should write instead of (15.45), using (15.42) and (15.44),

$$p(A|BI) = p(dx|dy\,I) = \frac{p(dx\,dy|I)}{p(dy|I)} = \frac{1}{\sqrt{2\pi}} \exp\left\{-\frac{1}{2}(x - \rho y_0)^2\right\} dx. \tag{15.49}$$

Since dy cancels out, taking the limit $dy \to 0$ does nothing.

On working out the normalizing constant in (15.45), we find that (15.45) and (15.49) are in fact identical. So, why all this agony? Didn't the quick argument leading to (15.45) give us the right answer?

This is a good example of our opening remarks that a fallacious argument may lead to correct or incorrect results. The reasoning that led us to (15.45) happened to give a correct result here; but it can equally well yield any result we please instead of (15.45). It depends on the particular form in which you or I choose to write our equations. To show this, and therefore generate a paradox, suppose that we had used instead of (x, y) the variables (x, u), where

$$u \equiv \frac{y}{f(x)} \tag{15.50}$$

with $0 < f(x) < \infty$; for example, $f(x) = 1 + x^2$ or $f(x) = \cosh(x)$, etc. The Jacobian is

$$\frac{\partial(x, u)}{\partial(x, y)} = \left(\frac{\partial u}{\partial y}\right)_x = \frac{1}{f(x)} \tag{15.51}$$

so the pdf (15.42), expressed in the new variables, is

$$p(dx\,du|I) = \frac{\sqrt{1 - \rho^2}}{2\pi} \exp\left\{-\frac{1}{2}(x^2 + u^2 f^2(x) - 2\rho u f(x))\right\} f(x) dx\,du. \tag{15.52}$$

Again, we can integrate out u or x, leading to a marginal distribution $p(dx|I)$, which is easily seen to be identical with (15.43), and $p(du|I)$, which is found to be identical with (15.44) transformed to the variable u, as it should be; so far, so good.

But now, what is the conditional pdf for x, given that $u = 0$? If we follow the reasoning that led us to (15.45); i.e. simply set $u = 0$ in (15.52) and renormalize, we find

$$p(dx|u = 0\,I) = A \exp\left\{-\frac{1}{2}x^2\right\} f(x) dx. \tag{15.53}$$

Now from (15.50) the condition $u = 0$ is the same as $y = 0$; so it appears that this should be the same as (15.45) with $y_0 = 0$. But (15.53) differs from that by an extra factor $f(x)$, which could be arbitrary!

Many find this astonishing and unbelievable; they repeat over and over: 'But the condition $u = 0$ is *exactly the same condition* as $y = 0$; how can there be a different result?' We warned against this phenomenon briefly, and perhaps too cryptically, in Chapter 4; but there it did not actually cause an error because we had only one parameter in the problem. Now we need to examine it carefully to see the error and the solution.

We noted in Chapter 1 that we shall make no attempt to define any probability conditional on contradictory premises; there could be no unique solution to such a problem. We start each problem by defining a 'sample space' or 'hypothesis space' which sets forth the range of conditions we shall consider *in that problem*. In the present problem, our discrete hypotheses were of the form '$a \leq y \leq b$', placing y in an interval of positive measure $b - a$. Then what could we mean by the proposition '$y = 0$', which has measure zero? We could mean only the limit of some sequence of propositions referring to positive measure, such as

$$A_\epsilon \equiv |y| < \epsilon \qquad (15.54)$$

as $\epsilon \to 0$. The propositions A_ϵ confine the point (x, y) to successively narrower horizontal strips, but for any $\epsilon > 0$, A_ϵ is a discrete proposition with a definite positive probability, so by the product rule the conditional probability of any hypothesis $H \equiv$ 'x in dx',

$$p(H|A_\epsilon I) = \frac{p(H A_\epsilon | I)}{p(A_\epsilon | I)} \qquad (15.55)$$

is well-defined, and the limit of this as $\epsilon \to 0$ is also a well-defined quantity. Perhaps that limit is what one meant by $p(H | y = 0 \, I)$.[3]

But the proposition '$y = 0$' may be defined equally well as the limit of the sequence

$$B_\epsilon \equiv |y| < \epsilon |x| \qquad (15.56)$$

of successively thinner wedges, and $p(H | B_\epsilon I)$ is also unambiguously defined as in (15.55) for all $\epsilon > 0$. Although the sequences $\{A_\epsilon\}, \{B_\epsilon\}$ tend to the same limit $y = 0$, the conditional densities tend to different limits:

$$\begin{aligned} \lim_{\epsilon \to 0} p(H|A_\epsilon) &\propto g(x), \\ \lim_{\epsilon \to 0} p(H|B_\epsilon) &\propto |x| g(x), \end{aligned} \qquad (15.57)$$

and in place of $|x|$ we could put an arbitrary non-negative function $f(x)$. As we see from this, merely to specify '$y = 0$' without any qualifications is ambiguous; it tells us to pass to a measure-zero limit, but does not tell us which of any number of limits is intended.

We have here one more example showing why the rules of inference derived in Chapter 2 must be obeyed *strictly, in every detail*. Intuitive shortcuts have a potential for disaster, which is particularly dangerous just because of the fact that it strikes only intermittently. An intuitive *ad hockery* that violates those rules will probably lead to a correct result in some cases; but it will surely lead to disaster in others. Whenever we have a probability density on one space, and we wish to generate from it one on a subspace of measure zero, the only safe procedure is to pass to an explicitly defined limit by a process like (15.55). In general, the final result will and must depend on which limiting operation was specified.

---

[3] Note again what we belabor constantly: the rules of probability theory tell us unambiguously that it is the limit of the ratio, not the ratio of the limits, that is to be taken in (15.55). The former quantity remains finite and well-behaved in conditions where the latter does not exist.

This is extremely counter-intuitive at first hearing; yet it becomes obvious when the reason for it is understood.

A famous puzzle based on this paradox concerns passing from the surface of a sphere to a great circle on it. Given a uniform probability density over the surface area, what is the corresponding conditional density on any great circle? Intuitively, everyone says immediately that, from geometrical symmetry, it must be uniform also. But if we specify points by latitude $(-\pi/2 \le \theta \le \pi/2)$ and longitude $(-\pi < \phi \le \pi)$, we do not seem to get this result. If that great circle is the equator, defined by $|\theta| < \epsilon$ as $\epsilon \to 0$, we have the expected uniform distribution $p(\phi) = (2\pi)^{-1} \, (-\pi < \phi \le \pi)$. But if it is the meridian of Greenwich defined by $|\phi| < \epsilon$ as $\epsilon \to 0$, we have $p(\theta) = (1/2) \cos(\theta) \, (-\pi/2 \le \theta \le \pi/2)$, with the density reaching a maximum on the equator and zero at the poles.

Many quite futile arguments have raged – between otherwise competent probabilists – over which of these results is 'correct'. The writer has witnessed this more than once at professional meetings of scientists and statisticians. Nearly everybody feels that he knows perfectly well what a great circle is; so it is difficult to get people to see that the term 'great circle' is ambiguous until we specify what limiting operation is to produce it. The intuitive symmetry argument presupposes unconsciously the equatorial limit; yet one eating slices of an orange might presuppose the other.

15.8 The marginalization paradox

The tumbling tetrahedrons problem flared up into an even more spectacular case of probability theory gone crazy, with the work of Dawid, Stone, and Zidek (1973), hereafter denoted by DSZ, which for a time seemed to threaten the consistency of all probability theory. The marginalization paradox is more complicated than the ones discussed above, because it arises not from a single error, but from a combination of errors of logic and intuition, insidious because they happened to support each other. When first propounded it seems to have fooled every expert in the field, with the single exception of D. A. S. Fraser, who, as discussant of the DSZ paper, saw that the conclusions were erroneous and put his finger correctly on the cause of this; but he was not listened to.

The marginalization paradox also differs from the others in that it received the immediate, enthusiastic endorsement of the establishment, and therefore it has been able to do far more damage to the cause of scientific inference than any, other; yet, when properly understood, the phenomenon has useful applications in scientific inference. Marginalization as a potentially useful means of constructing uninformative priors is discussed incompletely in Jaynes (1980); this rather deep subject still has the status of ongoing research, in which the main theorems are probably not yet known.

In the present chapter we are concerned with the marginalization story only as a weird episode of history which accomplished one good thing by forcing Bayesians to revise some easy, shortcut inference procedures. We illustrate the original paradox by the scenario of DSZ, again following their notation until we see why we must not. It starts as a conventional, and seemingly harmless, nuisance parameter problem.

A conscientious Bayesian B_1 studies a problem with data $x \equiv (x_1, \ldots, x_n)$ and a multidimensional parameter θ which he partitions into two components, $\theta = (\eta, \zeta)$, being interested only in inferences about ζ. Thus his model is defined by some specified sampling distribution $p(x|\eta\zeta)$ supposed given in the statement of the problem, and η is a nuisance parameter to be integrated out. With a prior $\pi(\eta, \zeta)$, B_1 thus obtains the marginal posterior pdf for ζ:

$$p(\zeta|x) = \int d\eta\, p(\eta\zeta|x) = \frac{\int d\eta\, p(x|\eta\zeta)\pi(\eta, \zeta)}{\int d\zeta \int d\eta p(x|\eta\zeta)\pi(\eta, \zeta)}, \tag{15.58}$$

the standard result, which summarizes everything B_1 knows about ζ. The issue now turns on what class of priors $\pi(\eta, \zeta)$ we may assign for this purpose. Our answer is, of course:

> Any proper prior, or any limit of a sequence of such priors, such that the ratio of integrals in (15.58) converges to yield a proper posterior pdf for ζ, may be admitted into our theory as representing a conceivable state of prior knowledge about the parameters. Eq. (15.58) will then yield the correct conclusions that follow from that state of knowledge.

This need not be qualified by any special circumstances of the particular problem; we believe that this policy, followed strictly, cannot generate ambiguities or contradictions. But *failure* to follow it can lead to almost anything.

However, DSZ did not see it that way at all. They concentrate on a special circumstance, noting that in many cases the data x may be partitioned into two components: $x = (y, z)$ in such a way that 'the sampling distribution for z is independent of the nuisance parameter η', which property they write (DSZ, Eq. (1.2)) as

$$p(z|\eta\zeta) = \int dy\, p(yz|\eta\zeta) = p(z|\zeta), \tag{15.59}$$

which, by itself, would appear rather generally possible, but without any very deep significance. For example, if η is a location parameter, then any function $z(x)$ of the data that is invariant under rigid translations will have a sampling distribution independent of η. If η is a scale parameter, then any function $z(x)$ invariant under scale changes will have this property. If η is a rotation angle, then any component of the data that is invariant under those rotations will qualify. DSZ proceed to discover cases in which, when (15.59) holds and B_1 assigns an improper prior to η, he finds that his marginal posterior pdf for ζ 'is a function of z only', which, in view of (15.59), DSZ would presumably write as

$$p(\zeta|yz) = p(\zeta|z). \tag{15.60}$$

At this point there enters a lazy Bayesian B_2, who 'always arrives late on the scene of inference', and the combination of (15.59) and (15.60) sets off for him a curious train of thought. From (15.60) as written it appears that the component y of the data can be discarded as irrelevant to inferences about ζ. The appearance of (15.59) then suggests that η might also be removed from the model as irrelevant. So he proposes to simplify the calculation; his intuitive judgment is that, given (15.59) and (15.60), we should be able to derive the marginal pdf for ζ more easily by direct application of Bayes' theorem in a reduced model

$p(z|\zeta)$ in which (y, η) do not appear at all. Thus if B_2 assigns the prior $\pi(\zeta)$, he obtains the posterior distribution

$$p(\zeta|z) = \frac{p(z|\zeta)\pi(\zeta)}{\int d\zeta\ p(z|\zeta)\pi(\zeta)}. \tag{15.61}$$

But he finds to his dismay that he cannot reproduce B_1's result (15.58) whatever prior he assigns to ζ. What conclusions should we draw from this?

For DSZ, the reasoning of B_2 seemed compelling; on grounds of this intuitive 'reduction principle' they considered it obvious that B_1 and B_2 ought to get the same results, and therefore that one of them must be guilty of some transgression. They point the accusing finger at B_1 thus: 'B_2's intervention has revealed the paradoxical unBayesianity of B_1's posterior distribution for ζ.' They place the blame on his use of an improper prior for η.

For us, the situation appears very different; B_2's result was not derived by application of our rules. Eq. (15.61) was only an intuitive guess; as the reader may verify, it does not follow mathematically from (15.58), (15.59) and (15.60). Therefore, (15.61) *is not a valid application of probability theory to B_1's problem.* If intuition suggests otherwise, then that intuition needs educating – just as it did in the other paradoxes.

At this stage we are faced not just with one confusion, but with three. The notation used above conceals from view some crucial points:

(1) While the result (15.60) is 'a function of z only' in the sense that y does not appear explicitly in (15.60), it is a *different* function of z for different η-priors. That is, it is still a functional of the η-prior, as is clear from a glance at (15.58); through this dependence, probability theory is telling us that prior information about η still matters. As soon as we realize this, we see that B_2 comes to a different conclusion than B_1 not because B_1 is committing a transgression, but for just the opposite reason: B_1 is taking into account relevant prior information that B_2 is ignoring.

(2) But the real trouble starts farther back than that. We need to be aware that current orthodox notation has a more basic ambiguity that makes the meaning of (15.59) and (15.60) undefined, and this is corrected only by the notation introduced by Harold Jeffreys (1939) and expounded in our Chapter 2 and Appendix B. Thus, we understand that the symbol $p(yz|\eta\zeta)$ stands for the joint probability (density) for y, z conditional on *specific numerical values* for the two parameters η, ζ that are present in our model. But then what does $p(z|\zeta)$ stand for? Presumably this is not intended to say that η has no numerical value at all!

Indeed, if he wished to refer to a *different model* in which η is not present at all, the orthodoxian would use the same notation $p(z|\zeta)$. So it seems that, strictly speaking, we should always interpret the symbol $p(z|\zeta)$ as referring to that different model. But that is not the intention in (15.59); reference is being made to a model in which η is still present, but the probability for z is independent of its numerical value. It seems that the only way this could be expressed in orthodox notation is to rewrite (15.59) as

$$\frac{\partial}{\partial\eta}p(z|\eta\zeta) = 0. \tag{15.62}$$

(3) This ambiguity, and still another one, is present in (15.60); here the intention is only to indicate that $p(\zeta|yz)$ is independent of the numerical value of y; but the symbol $p(\zeta|z)$, strictly speaking, must be held to refer to a different model in which the datum y was not given at all. Now we have the additional ambiguity that any posterior probability depends necessarily on the prior

information; yet the notation in (15.60) makes no reference to any prior information.[4] We begin to see why the marginalization paradox was so confusing!

There is a better way of looking at this, which avoids all the above confusions while using the mathematics that was intended by DSZ; we may take a more charitable view of B_2 if we put these equations in a different scenario. The lazy Bayesian, B_2, was introduced as a fellow who invents a shortcut method that violates the rules of probability theory. But we may suppose equally well that, through no fault of his own, he is only an uninformed fellow who was given only the reduced model $p(z|\zeta)$ in which η is not present; and he is unaware of the existence of (η, y). Then (15.61) is a valid inference for the *different state of knowledge* that B_2 has; and it is valid whether or not the separation property (15.60) holds.[5] Although the equations are the same because we defined B_2's model by B_1's marginal sampling distribution $p(z|\zeta)$, this avoids much confusion; viewed in this way, B_1 and B_2 are both making valid inferences, but about two different problems.

Both of these new ambiguities arise from the fact that orthodox notation fails to indicate which model is being considered. But both are corrected by including the prior information symbol I, understood to be a proposition defined somewhere in the surrounding context, that includes full specification of the model. If we follow the example of Jeffreys and write the right-hand sides of (15.58) and (15.61) correctly as $p(\zeta|yzI_1)$ and $p(\zeta|zI_2)$, thereby making this difference in the problems clear, there can be no appearance of paradox. The prior information I_1 specifies the full sampling distribution $p(yz|\eta\zeta)$, while I_2 specifies a model only by $p(z|\zeta)$, which makes no reference to (η, y). That B_1 and B_2 came to different conclusions from different prior information is no more strange than if they had come to different conclusions from different data.

Exercise 15.2. Consider the intermediate case of a third Bayesian, B_3, who has the same prior information as B_1 about η, ζ but is not given the data component y. Then y never appears in B_3's equations at all; his model is the marginal sampling distribution $p(z|\eta\zeta I_3)$. Show that, nevertheless, if (15.59) still holds (in the interpretation intended, as indicated by (15.62)), then B_2 and B_3 are always in agreement, $p(\zeta|zI_3) = p(\zeta|zI_2)$, and that to prove this it is not necessary to appeal to (15.60). Merely withholding the datum y automatically makes any prior knowledge about η irrelevant to inference about ζ. Ponder this until you can explain in words why it is, after all, intuitively obvious.

---

[4] Yet, as we stress again, if you fail to specify the prior information, a problem of inference is just as ill-posed as if you had failed to specify the data. In practice, orthodoxy is able to function in spite of this in some problems, by the tacit assumption that an uninformative prior is to be used. Of course, the dedicated orthodoxian will deny vehemently that he is making any such assumption; nevertheless, it is a mathematical fact that his conclusions are what a Bayesian would obtain *from an uninformative prior*. This was demonstrated already by Jeffreys (1939).

[5] The fact that (15.60) is not essential to the problem was not yet clearly seen in Jaynes (1980); the marginalization problem was more subtle than any that Bayesians had faced up to that time. Because DSZ laid so much stress on (15.60), we followed them in concentrating on finding conditions for its validity. Today, with the benefit of hindsight, it is clear that there is in general no reason to expect (15.60) to hold, so it loses its supposed importance. This deeper understanding enables us to find useful solutions to current problems of inference far more subtle than marginalization, as demonstrated by Bretthorst (1988). But the secret of success here is, as always, simply: *absolutely strict adherence* to the rules of conduct derived in Chapter 2. As these paradoxes show, the slightest departure from them can generate gross absurdities.

15.8.1 On to greater disasters

Up to this point, we had only a misreading of equations through inadequate notation; but now a comedy of mutually reinforcing errors commenced. In support of their contention that B_1 is the guilty party, DSZ offered a proof that this paradox (i.e. the discrepancy in the results of B_1 and B_2) 'could not have arisen if B_1 had employed proper prior distributions'. Let us examine their proof of this, still using their notation. With a general joint proper prior $\pi(\eta, \zeta)$ the integrals in (15.58) are separately convergent and positive, so if we multiply through by the denominator, we are neither multiplying nor dividing by zero. Then

$$p(x|\eta\zeta) = p(yz|\eta\zeta) = p(y|z\eta\zeta)p(z|\eta\zeta) = p(y|z\eta\zeta)p(z|\zeta), \qquad (15.63)$$

where we used the product rule and (15.59). Then (15.58) becomes

$$p(\zeta|yz)\int d\zeta \int d\eta\, p(y|z\eta\zeta)p(z|\zeta)\pi(\eta, \zeta) = \int d\eta\, p(y|z\eta\zeta)p(z|\zeta)\pi(\eta, \zeta). \quad (15.64)$$

But now we assume that (15.60) still holds; because the integrals are absolutely convergent, we may integrate out y from both sides of (15.64), whereupon $\int d\eta\, \pi(\eta, \zeta) = \pi(\zeta)$ and (15.64) reduces to

$$p(\zeta|z)\int d\zeta\, p(z|\zeta)\pi(\zeta) = p(z|\zeta)\pi(\zeta), \qquad (15.65)$$

which is identical with (15.61). DSZ concluded that, if B_1 uses a proper prior, then B_1 and B_2 are necessarily in agreement – from which it would follow again, in agreement with their intuition, that the paradox must be caused by B_1's use of improper priors.

But this proof of (15.65) has used mutually contradictory assumptions. As Fraser recognized, if B_1 uses a proper prior, then (15.60) *cannot* be true and (15.65) does not follow; it is no accident that DSZ had found (15.60) only with improper priors. This is easiest to see in terms of a specific example, after which it will become obvious why it is true in general. In the following we use the full notation of Jeffreys so that we always distinguish between the two problems.

The change-point problem

Observations have been made of n successive, independent, positive real, 'exponentially distributed' quantities $\{x_1, \ldots, x_n\}$. It is known (definition of the model) that the first ζ of these have expectations $1/\eta$ and the remaining $(n - \zeta)$ have expectations $1/(c\eta)$, where c is known and $c \neq 1$, while η and ζ are unknown. From the data, we want to estimate at what point in the sequence the change occurred. The sampling density for $x \equiv (x_1, \ldots, x_n)$ is

$$p(x|\eta\zeta I_1) = c^{n-\zeta}\eta^n \exp\left\{-\eta\left(\sum_{i=1}^{\zeta} x_i + c\sum_{i=\zeta+1}^{n} x_i\right)\right\}, \qquad 1 \leq \zeta \leq n. \quad (15.66)$$

If $\zeta = n$, then there is no change, the last sum in (15.66) is absent, and c disappears from the model. Since η is a scale parameter, the sampling distribution for ratios of observations

$z_i \equiv x_i/x_1$ should be independent of η. Indeed, separating the data $x = (y, z)$ into $y \equiv x_1$, which sets the scale and the ratios $(z_2, \ldots, z_n)$, and noting that the volume element transforms as $dx_1 \cdots dx_n = y^{n-1}dy\,dz_2 \cdots dz_n$, we find that the joint sampling distribution for $z \equiv (z_2, \ldots, z_n)$ depends only on ζ:

$$p(z_2 \cdots z_n | \eta \zeta I_1) = \int_0^\infty dy\, c^{n-\zeta} \eta^n y^{n-1} \exp\{\eta y Q(\zeta, z)\} = \frac{c^{n-\zeta}(n-1)!}{Q(\zeta, z)^n} = p(z|\zeta I_1), \tag{15.67}$$

where $z_1 \equiv 1$ and

$$Q(\zeta, z) \equiv \sum_1^\zeta z_i + c \sum_{\zeta+1}^n z_i \tag{15.68}$$

is a function that is known from the data. Let B_1 choose a properly normalized discrete prior $\pi(\zeta)$ in $(1 \le \zeta \le n)$, and independently a prior $\pi(\eta)\,d\eta$ in $(0 < \eta < \infty)$. Then B_1's marginal posterior distribution for ζ is, from (15.66),

$$p(\zeta | yz I_1) \propto \pi(\zeta) c^{n-\zeta} \int_0^\infty d\eta\, \exp\{-\eta y Q\}\pi(\eta)\eta^n, \tag{15.69}$$

and, from (15.67), B_2's posterior distribution (15.61) for ζ is now

$$p(\zeta | z I_2) \propto \pi(\zeta) p(z|\zeta) = \frac{\pi(\zeta) c^{-\zeta}}{[Q(\zeta, z)]^n}, \tag{15.70}$$

which takes no note of $\pi(\eta)$. But, as expected from the above discussion, not only does B_1's knowledge about ζ depend on both y and z, it depends just as strongly on what prior $\pi(\eta)$ he assigned to the nuisance parameter.

On meditation, we see that a little common sense would have anticipated this result at once. If we know absolutely nothing about η except that it is positive, then the only evidence we can have about the change point ζ must come from noting the relative values of the x_i; for example, at which i does the ratio x_i/x_1 appear to change? On the other hand, suppose that we knew η exactly; then clearly not only the ratios x_i/x_1, but also the absolute values of the x_i, would be relevant to inference about ζ. Then, whether x_i is closer to $1/\eta$ or to $1/(c\eta)$ tells us something about whether $(i < \zeta)$ or $(i > \zeta)$ that the ratio x_i/x_1 does not tell us, this extra information would enable us to make better estimates of ζ. If we had only partial prior knowledge of η, then knowledge of the absolute values of the x_i would be less helpful, but still relevant, so, as Fraser noted, (15.60) could not be valid.

But now B_1 discovers that use of the improper prior

$$\pi(\eta) = \eta^{-k}, \qquad 0 < \eta < \infty, \tag{15.71}$$

where k is any real number for which the integral (15.69) converges, leads to the separation property (15.60), and to the posterior pdf

$$p(\zeta | z I_1) \propto \frac{\pi(\zeta) c^{-\zeta}}{[Q(\zeta, z)]^{n-k+1}}, \tag{15.72}$$

which still depends, through k, on the prior assigned to η. We see that for no prior $\pi(\zeta)$ can B_2 agree with B_1, except when $k = 1$, in which case B_2 and B_1 find themselves in agreement after all, and with the same prior $\pi(\zeta)$. But this result is not peculiar to the change-point model; it holds quite generally, as the following Exercise shows.

Exercise 15.3. Prove that the $k = 1$ prior is always uninformative in this sense whenever η is a scale parameter for y. That is, if the sampling distribution has the functional form

$$p(yz|\eta\zeta) = \eta^{-1} h(z, \zeta; y/\eta), \qquad (15.73)$$

then (15.59) follows at once, and B_1 and B_2 agree if and only if we use a prior $\pi(\eta) \propto \eta^{-1}$.

It seems to us that this is an eminently satisfactory result without any trace of paradox. For the case $k = 1$ is just the Jeffreys prior, which we have already seen to be 'completely uninformative' about any scale parameter η, by several different criteria. Then, of course, with this prior B_1 has no extra information after all, and should, indeed, find himself in agreement with B_2.

DSZ did not see it that way at all, and persisted in their intuitive judgment that there is a serious paradox and that B_1 was at fault for using an improper prior; so the story continues. DSZ proceed to exhibit many more examples in which this 'paradox' appears – invariably when an improper prior was used. The totality of all these demonstrations appeared to mount up into overwhelming evidence that to use any improper prior is to generate inconsistencies. But, in the belief that their proof of (15.65) had already dealt with it, they failed to examine what happens in those examples in the case of proper priors, and so they managed to get through a long string of examples without discovering the error in that proof.[6]

To correct this omission, and reveal the error in (15.65) clearly, we need only to examine any of the DSZ examples, to see what happens in the case of proper priors $\pi(\eta)$. In the change-point problem, whatever this prior, B_1's result (15.69) depends on y and z through a function of the product $yQ(\zeta, z)$. Then for what functions $f(yQ)$ will the separation property (15.60) hold? Evidently, the necessary and sufficient condition for this is that y and ζ appear in separate factors: in the case where the integrals in (15.58) converge, we

---

[6] Another reason for this was their tendency to write the priors in terms of the 'wrong' parameters. Usually, a model was defined initially with certain parameters α, β. The parameters η, ζ for which the relations (15.59), (15.60) held were certain functions of them: $\eta = \eta(\alpha, \beta)$, etc. But DSZ continued to write the priors in terms of α, β, which made it seem that the Jeffreys prior has no particular significance; a wide variety of different priors appeared to 'avoid the paradox' in various different problems. In Jaynes (1980) we showed that, had they transformed their parameters to the relevant ones η, ζ, they would have found in every such case except one that η was a scale parameter for y and the 'paradox' disappeared for and only for the Jeffreys prior $\pi(\eta)$. Thus Exercise 15.3 includes, in effect, all their examples except the infamous Example #5, which requires a separate treatment given below.

require the integral to have the functional form

$$\int_0^\infty d\eta \, \exp\{-\eta y Q\} \pi(\eta) \eta^n = f(yQ) = g(y, z) h(\zeta, z), \tag{15.74}$$

for then and only then will y cancel out upon normalization of $p(\zeta|yz)$. The answer is obvious: if a function of $[\log(y) + \log Q(\zeta)]$ has the form $[\log g(y) + \log h(\zeta)]$, the only possibility is a linear function: $\log f(yQ) = a[\log(y) + \log(Q)]$ or $f(yQ) = (yQ)^a$, where $a(z, n)$ may depend on z and n. But then, noting that the Laplace transform is uniquely invertible, and that

$$\int_0^\infty d\eta \, \exp\{-\eta y Q\} \eta^{a-1} = \frac{(a-1)!}{(yQ)^a}, \tag{15.75}$$

we see that, contrary to the assumption of DSZ, (15.60) *cannot hold unless the prior is of the improper form* $\pi(\eta) = \eta^{-k}, 0 < \eta < \infty$.

Exercise 15.4. Show that this result is also general; that is, not only in the change-point problem, but in any problem like that of Exercise 15.3 where η is a scale parameter for y, a prior of the form $\pi(\eta) = \eta^{-k}$ will lead to a factorization of the form $\int d\eta \, p(yz|\eta\zeta) \pi(\eta) = g(y, z) h(\zeta, z)$ for some functions g, h, whereupon (15.60) will hold. For this reason, the many later examples of DSZ are essentially repetitious; they are only making the same point over and over again.

Evidently, any value of k which makes the integral (15.74) converge will lead to a well-behaved posterior distribution for ζ; but a still wider class of values of k may do so if the improper prior is approached, as it should be, as the limit of a sequence of proper priors, as explained previously.

But use of a proper prior $\pi(\eta)$ necessarily means that the separation property (15.60) cannot hold. For example, choose the prior $\pi(\eta) \propto \eta^a \exp\{-b\eta\}$. Then (15.69) becomes

$$p(\zeta|yzI_1) \propto \frac{\pi(\zeta)c^{-\zeta}}{(b + yQ)^{n+a+1}}, \tag{15.76}$$

and as long as the prior is proper (that is, $b > 0$), the datum y cannot be disentangled, but remains relevant; and so (15.60) does not hold, as we expected from (15.75). The 'paradox' disappears, not because B_1 and B_2 agree, but because B_2 cannot invoke his 'reduction principle' at all. Indeed, in any of the DSZ examples, inserting any proper prior $\pi(\eta)$ for which we can do the integrals will yield an equally good counter-example to (15.65); how could this have gone undetected for years? We note some of the circumstances that led to this.

15.9 Discussion

Some have denied that there is any such thing as 'complete ignorance', much less any 'completely uninformative' prior. From their introductory remarks, it appears that to demonstrate this was the original goal of DSZ, and several discussants continued to emphasize the point in agreement with them. But their arguments were verbal, expressing only intuitive feelings; the mathematical facts confirm the sense of the idea of 'complete ignorance' after all. The Jeffreys prior is doing here what we should naturally suppose an uninformative prior ought to do, and it does this quite generally (whenever η is a scale parameter).

Technically, the concurrence of many different results like that of Exercise 15.3 shows us that the notion of complete ignorance is consistent and useful; the fact that the same Jeffreys prior emerges uniquely from many different and independent lines of reasoning shows how impossible it would be to modify it or abandon it. As is invariably the case in this field, past difficulties with the ideas of Jeffreys signified not any defects in his ideas, but only misapplications of probability theory by his critics.

Exercise 15.3 shows another sense in which our previous conclusion (that the prior $d\eta/\eta$ is uninformative about a scale parameter η) is quite literally true; not as an intuitive judgment, but now as a definite theorem that follows from the rules of probability theory. Of course, our ultimate goal is always to represent honestly the prior information that we actually have. But, both conceptually and mathematically, the notion of 'complete ignorance' is a valid and necessary part of this program, as the starting point from which all inference proceeds; just as the notion of zero is a necessary part of arithmetic.

In the discussion following the DSZ paper, nobody noticed that there was a counter-example to their proof of (15.65) already in plain sight in the DSZ article (their Example #5, where it is evident by inspection that B_1 and B_2 remain in disagreement for all priors, proper or improper), and only Fraser expressed any doubts about the DSZ conclusions. He noted that DSZ

... propose that the confusion can be avoided by a restriction to *proper* priors. This is a strange proposal as a resolution of the difficulties – for it means in the interesting cases that one cannot eliminate a variable, and hence cannot go to the marginal likelihood.

But it seems that these words were, like the prophecies of Nostradamus, too cryptic for anyone to understand until he had first located the error for himself. Fraser's point – and ours above – is that when B_1 uses a proper prior, then in general B_2's 'reduction principle' cannot be applied because (15.60) ceases to be true. In other words, when B_1 uses proper priors, this does not bring B_1 and B_2 into agreement. In (15.74) and (15.75) we have demonstrated that in the change-point problem, agreement of B_1 and B_2 *requires* that B_1 uses an improper prior; just the opposite of the DSZ conclusion.

It is evident, to one who has understood the above analysis, that the situation found in the change-point problem is actually quite general. For, if one knew both y and η, that information must be relevant to the inference about ζ unless the sampling distributions are completely independent; that is, unless $p(yz|\eta\zeta) = p(y|\eta)p(z|\zeta)$. Except in this trivial case, if one knows y, any partial information about η must still be relevant for inference about ζ or, similarly, if one knew η, any partial information about y would be relevant.

But common sense should have told us that any proper prior $\pi(\eta)$ on an infinite domain is necessarily informative about η, for it determines finite upper and lower bounds within which η is almost certain to lie. Seen in this way, Fraser's cryptic remark becomes obvious – and in full generality.

In any event, what happened was that nearly everybody accepted the DSZ conclusions uncritically, without careful examination of their argument. Anti-Bayesians, who very much wanted the DSZ conclusion to be true, seized upon it eagerly as sounding the death-knell of all Bayesianity. Under this pressure the prominent Bayesian D. V. Lindley broke down and confessed to sins of which he was not guilty, and the Royal Statistical Society bestowed a warm vote of thanks upon DSZ for this major contribution to our understanding of inference.

As a result, since 1973 a flood of articles has appeared, rejecting the use of improper priors under any and all circumstances, on the grounds that they have been proved by DSZ to generate inconsistencies. Incredibly, the fact that proper priors never 'correct' the supposed inconsistencies never came out in all this discussion. Thus the marginalization paradox became, like nonconglomerability, quite literally institutionalized in the literature of this field, and taught as truth. Scientific inference thus suffered a setback from which it will require decades to recover.

Nobody noted that this same 'paradox' had been found and interpreted correctly long before by Harold Jeffreys (1939, Sect. 3.8) in connection with estimating the correlation coefficient ρ in a bivariate normal distribution, in which the location parameters are the uninteresting nuisance parameters. He gives two examples of B_1's result, corresponding to different prior information about the nuisance parameters, in his equations (10) and (24), their difference indicating the effect of that prior information. Then he gives B_2's result in (28), the agreement with (24) indicating that a uniform prior for the location parameters is uninformative about ρ.

This was seen again independently by Geisser and Cornfield (1963) in connection with priors for multivariate normal distributions. They perceived that the difference between the results of B_1 and B_2, their equations (3.10) and (3.26), was not a paradox, because B_2's result was not a valid solution to the problem; they termed it, very properly, a 'pseudoposterior distribution'. DSZ refer to this work, but when faced with this discrepancy they still place more confidence in the 'reduction principle' than in the rules of probability theory.

In all these examples except one – that Example #5 again – an interesting phenomenon occurred. While the paradox was present for general improper priors in some infinite class C, there was always one particular improper prior in that class for which the paradox disappeared; B_1 and B_2 found themselves in agreement after all. DSZ noted this curious fact, but do not appear to have noticed its significance. We suggest that this was by far the most important fact uncovered in all the marginalization work.

Any prior $\pi(\eta)$ which leaves B_1 and B_2 in agreement must be *completely uninformative* about η (and, *a fortiori*, about ζ). This means that, far from casting doubt on the notion of complete ignorance, in the marginalization phenomena we have for the first time a purely *objective* definition of complete ignorance that springs directly out of the product and sum rules of probability theory without appeal to any other notions like entropy or group invariance.

This is, again, an eminently satisfactory result; but why does it seem not to be true in DSZ's Example #5? There is still something new and important to be learned here.

15.9.1 The DSZ Example #5

We have data $D = \{x_1, \ldots, x_n\}$ consisting of n observations from the standard normal sampling distribution $N(\mu, \sigma)$. With prior information I described by the proper prior pdf

$$p(d\mu d\sigma \,|\, I) = f(\mu, \sigma) d\mu d\sigma, \tag{15.77}$$

we have the usual joint posterior pdf for the parameters:

$$p(d\mu d\sigma \,|\, DI) = g(\mu, \sigma) d\mu d\sigma \tag{15.78}$$

with

$$g(\mu, \sigma) = \frac{f(\mu, \sigma) L(\mu, \sigma)}{\int d\mu \int d\sigma \; f(\mu, \sigma) L(\mu, \sigma)} \tag{15.79}$$

and the likelihood function

$$L(\mu, \sigma) = \sigma^{-n} \exp\left\{-\frac{n}{2\sigma^2}[s^2 + (\mu - \bar{x})^2]\right\}, \tag{15.80}$$

in which, as usual, $\bar{x} \equiv n^{-1} \sum x_i$ and $s^2 \equiv n^{-1} \sum (x_i - \bar{x})^2$ are the sufficient statistics. Although we suppose the prior $f(\mu, \sigma)$ normalizable, it need not be actually normalized in (15.79) because any normalization constant appears in both numerator and denominator, and cancels out.

As long as $s^2 > 0$, the likelihood is bounded throughout the region of integration $-\infty < \mu < \infty, 0 \leq \sigma < \infty$, and therefore with a proper prior the integral in (15.79) is guaranteed to converge, leading to a proper posterior pdf. Furthermore, if the prior has moments of order m, k,

$$\int_{-\infty}^{\infty} d\mu \int_{0}^{\infty} d\sigma \; \mu^m \sigma^k f(\mu, \sigma) < \infty, \tag{15.81}$$

the posterior distribution is guaranteed to have moments of higher order (in fact, all orders for μ and at least as high as order $k + n$ for σ). The solution is therefore very well-behaved mathematically.

But now we throw the proverbial monkey-wrench into this by declaring that we are interested only in the quantity

$$\zeta \equiv \frac{\mu}{\sigma}. \tag{15.82}$$

Making the change of variables $(\mu, \sigma) \to (\zeta, \sigma)$, the volume element transforms as $d\mu\,d\sigma = \sigma\,d\zeta\,d\sigma$, so writing $p(d\zeta|DI_1) = h_1(\zeta)d\zeta$, B_1's marginal posterior pdf is

$$h_1(\zeta) = \int_0^\infty \sigma\,d\sigma\; g(\sigma\zeta, \sigma), \tag{15.83}$$

and in view of the high moments of g there are no convergence problems here, as long as $n > 1$. Thus far, there is no hint of trouble.

Now we examine the solution for a specific proper prior that can approach an improper prior. Consider the conjugate prior probability element

$$f(\mu, \sigma)\,d\mu\,d\sigma \propto \sigma^{-\gamma-1}\exp\{-\beta/\sigma - \alpha\mu^2\}d\mu\,d\sigma, \tag{15.84}$$

which is proper when $(\alpha, \beta, \gamma) > 0$, and tends to the Jeffreys uninformative prior $d\mu\,d\sigma/\sigma$ as $(\alpha, \beta, \gamma) \to 0$. This leads to the joint posterior pdf, $p(d\mu\,d\sigma|DI) = g(\mu, \sigma)\,d\mu\,d\sigma$ with density function

$$g(\mu, \sigma) \propto \sigma^{-n-\gamma-1}\exp\left\{-\frac{\beta}{\sigma} - \alpha\mu^2 - \frac{n}{2\sigma^2}[s^2 + (\mu - \bar{x})^2]\right\}, \tag{15.85}$$

from which we are to calculate the marginal posterior pdf for ζ alone by the integration (15.83). The result depends on both sufficient statistics $(\bar{x}, s)$, but is most easily written in terms of a different set. The quantities R, r, where

$$R^2 \equiv n(\bar{x}^2 + s^2) = \sum x_i^2, \qquad r \equiv \frac{n\bar{x}}{R} = \frac{\sum x_i}{\sqrt{\sum x_i^2}}, \tag{15.86}$$

also form a set of jointly sufficient statistics, and from (15.85) and (15.83) we find the functional form $p(d\zeta|DI_1) = h_1(\zeta|r, R)d\zeta$, where

$$h_1(\zeta|r, R) \propto \exp\left\{-\frac{n\zeta^2}{2}\right\}\int_0^\infty d\omega\,\omega^{n+\gamma-1}\exp\left\{-\frac{1}{2}\omega^2 + r\zeta\omega - \beta R^{-1}\omega - \alpha\zeta^2 R^2\omega^{-2}\right\}. \tag{15.87}$$

As long as α or β is positive, the result depends on both sufficient statistics, as Fraser predicted; but, as α, β tend to zero and we approach an improper prior, the statistic R becomes less and less informative about ζ, and when α, β both vanish the dependence on R drops out altogether:

$$h_1(\zeta|r, R) \to h_1(\zeta|r) \propto \exp\left\{-\frac{n\zeta^2}{2}\right\}\int_0^\infty d\omega\,\omega^{n+\gamma-1}\exp\left\{-\frac{1}{2}\omega^2 + r\zeta\omega\right\}. \tag{15.88}$$

If then one were to look only at the limiting case $\alpha = \beta = 0$ and not at the limiting process, it might appear that just r alone is a sufficient statistic for ζ, as it did in (15.60). This supposition is encouraged by noting that the sampling distribution for r in turn depends only on ζ, not on μ and σ separately:

$$p(r|\mu\sigma) \propto (n - r^2)^{(n-3)/2}\int_0^\infty d\omega\,\omega^{n-1}\exp\left\{\frac{1}{2}\omega^2 + r\zeta\omega\right\}. \tag{15.89}$$

It might then seem that, in view of (15.88) and (15.89), we should be able to derive the same result by applying Bayes' theorem to the reduced sampling distribution (15.89). But one who supposes this finds, to his dismay, that (15.89) is not a factor of (15.88); that is, the ratio $h_1(\zeta|r)/p(r|\zeta)$ depends on r as well as ζ. The Jeffreys uninformative prior $\gamma = 0$ does indeed make the two integrals equal, but there remains an uncompensated factor with $(n - r^2)$, and so even the uninformative Jeffreys prior for (μ, σ) cannot bring about agreement of B_1 and B_2. There is no prior $p(\zeta|I_2)$ that can yield B_1's posterior distribution (15.88) from B_2's sampling distribution (15.89).

Since the paradox is still present for a proper prior, this is another counter-example to (15.65); but it has a deeper meaning for us. What is now the information being used by B_1 but ignored by B_2? It is not the prior probability for the nuisance parameter; the new feature is that in this model the mere qualitative fact of the existence of the nuisance parameter in the model *already constitutes prior information relevant to B_1's inference*, which B_2 is ignoring.

Recognizing this, we suddenly see the whole subject in a much broader light. We found above that (15.60) is not essential to the marginalization phenomenon; now we see that concentration on the nuisance parameter η is not an essential feature either! If there is any prior information whatsoever that is relevant to ζ, *whether or not it refers to η*, that B_1 is taking into account but B_2 is not, then we are in the same situation, and our two Bayesians come, necessarily, to different conclusions. In other words, DSZ considered only a very special case of the real phenomenon.

This situation is discussed in Jaynes (1980, following Eq. (79)), where the phenomenon is called 'ζ-overdetermination'. Reverting to our original notation in (15.58) and denoting B_1's prior information by I_1, it is shown that the general necessary and sufficient condition for agreement of B_1 and B_2 is that

$$\int d\eta \, p(y|z\eta\zeta I_1)\pi(\eta) = p(y|z\zeta I_1) \qquad (15.90)$$

shall be independent of ζ for all possible samples y, z. Denoting the parameter space and our partitioning into subspaces by $S_\theta = S_\zeta \otimes S_\eta$, we may write this as

$$\int_{S_\eta} d\eta \, p(yz|\eta\zeta)\pi(\eta) = p(y|zI_1)p(z|\zeta) \qquad \begin{cases} \zeta \in S_\zeta \\ \text{all } y, z \end{cases} \qquad (15.91)$$

or, more suggestively,

$$\int_{S_\eta} d\eta \, K(\zeta, \eta)\pi(\eta) = \lambda f(\zeta). \qquad (15.92)$$

This is a Fredholm integral equation in which the kernel is B_1's likelihood, $K(\zeta, \eta) = p(yz|\zeta\eta)$, the 'driving force' is B_2's likelihood $f(\zeta) = p(z|\zeta)$, and $\lambda(y, z) \equiv p(y|zI_1)$ is an unknown function to be determined from (15.92). But now we see the meaning of 'uninformative' much more deeply; for every different data set (y, z) there is a different integral equation. Therefore, for a single prior $\pi(\eta)$ to qualify as 'uninformative', it must satisfy many different (in general, an uncountable number) of these integral equations simultaneously.

At first glance, it seems almost beyond belief that any prior could do this; from a mathematical standpoint the condition seems hopelessly overdetermined, casting doubt on the notion of an uninformative prior. Yet we have many examples where such a prior does exist. In Jaynes (1980) we analyzed the structure of these integral equations in some detail, showing that the different status of Example #5 is due to the 'incompleteness' of the kernel.

More specifically, the set of all L^2 functions on S_ζ forms a Hilbert space H_ζ. For any specified data set $x = (y, z)$, as η ranges over S_η, the functions $K(\zeta, \eta)$, in their dependence on ζ, span a certain subspace $H'_\zeta(y, z) \in H_\zeta$. The kernel is said to be *complete* if $H'_\zeta = H_\zeta$. If it is incomplete, then if there is any data set (y, z) for which $f(\zeta)$ does not lie in H'_ζ, there can be no solution of (15.92). In such cases, the mere qualitative fact of the *existence* of the components (y, η) – irrespective of their numerical values – already constitutes prior information relevant to B_1's inference, because introducing them into the model restricts the space of B_1's possible likelihood functions (from different data sets y, z) from H_ζ to H'_ζ. In this case the shrinkage of H_ζ cannot be restored by any prior on S_η, and there is no possibility for agreement of B_1 and B_2.

In general, the point is that the integral equation for any one data set x imposes only very weak conditions on $\pi(\eta)$, determining its projection on only a tiny subspace $H(x) \in H_\zeta$. As we consider different data sets, the $H(x)$ are scattered about, like stars in the sky, within the full Hilbert space H_ζ. There is room for all of them, so the system of integral equations has nontrivial solutions after all.

15.9.2 Summary

Looking at the above equations with all this in mind, we now see that there was never any paradox or inconsistency after all; one should not have expected (15.88) to be derivable from (15.89) by Bayes' theorem because they are the posterior distribution and sampling distribution for two different problems, in which the model has different parameters. Eq. (15.88) is the correct marginal posterior pdf for ζ in a problem P_1 with two parameters (ζ, σ); but, although σ is integrated out to form the marginal pdf, the result still depends on what prior we have assigned to σ – as it should, since, if σ is known, it is highly relevant to the inference; if it is unknown, any partial prior information we have about it must still be relevant.

In contrast, (15.89) can be interpreted as a valid sampling distribution for a problem P_2 in which ζ is the only parameter present; the prior information does not even include the existence of the parameter σ which was integrated out in P_1. With a prior density $f_2(\zeta)$ it would yield a posterior pdf

$$h_2(\zeta) \propto f_2(\zeta) \int d\omega \, \omega^{n-1} \exp\left\{ -\frac{1}{2}\omega^2 + r\zeta\omega \right\} \qquad (15.93)$$

of a different functional form than (15.88). In view of the earlier work of Jeffreys and of Geisser and Cornfield, one could hardly claim that the situation was new and startling, much less paradoxical.

Forty years earlier, Harold Jeffreys was immune from such errors because (1) he perceived that the product and sum rules of probability theory are adequate to conduct inference and they take precedence over intuitive *ad hoc* devices like the reduction principle; (2) he had recognized from the start that all inferences are necessarily conditional not only on the data, but also on the prior information – therefore his formal probability symbols $P(A|BI)$ always indicated the prior information I, which included specification of the model.

Today, it seems to us incredible that anyone could have examined even one problem of inference without perceiving this necessary role of prior information; what kind of logic could they have been using? Nevertheless, those trained in the 'orthodox' tradition of probability theory did not recognize it. They did not have a term for prior information in their vocabulary, much less a symbol for it in their equations; and *a fortiori* no way of indicating when two probabilities are conditional on different prior information.[7] So they were helpless when prior information mattered.

15.10 A useful result after all?

In most paradoxes there is something of value to be salvaged from the debris, and we think (Jaynes, 1980) that the marginalization paradox may have made an important and useful contribution to the old problem of 'complete ignorance'. How is the notion to be defined, and how is one to construct priors expressing complete ignorance? We have discussed this from the standpoint of entropy and symmetry (transformation groups) in previous chapters; now marginalization suggests still another principle for constructing uninformative priors.

Many cases are known, of which we have seen examples in DSZ, where a problem has a parameter of interest ζ and an uninteresting nuisance parameter η. Then the marginal posterior pdf for ζ will depend on the prior assigned to η as well as on the sufficient statistics. Now for certain particular priors $p(\eta|I)$ one of the sufficient statistics may drop out of the marginal distribution $p(\zeta|DI)$, as R did in (15.88). It is at first glance surprising that the sampling distribution for the remaining sufficient statistics may in turn depend only on ζ as in (15.89).

Put differently, suppose a problem has a set of sufficient statistics (t_1, t_2) for the parameters (ζ, η). Now, if there is some function $r(t_1, t_2)$ whose sampling distribution depends only on ζ, so that $p(r|\zeta\eta I) = p(r|\zeta I)$, this defines a pseudoproblem with different prior information I_2, in which η is never present at all. Then there may be a prior $p(\eta|I)$ for which the posterior marginal distribution $p(\zeta|DI) = p(\zeta|rI)$ depends only on the component r of the sufficient

---

[7] Indeed, in the period 1930–1960 nearly all orthodoxians, under the influence of R. A. Fisher, scorned Jeffreys' work, and some took a militant stand against prior information, teaching their students that it is not only intellectually foolish, but also morally reprehensible – a deliberate breach of 'scientific objectivity' – to allow one's self to be influenced by prior information at all! This did little damage in the very simple problems considered in the orthodox literature, where there was no significant prior information anyway. And it did relatively little damage in physical science where prior information is important, because scientists ignored orthodox teaching and persisted in doing, qualitatively, the Bayesian reasoning using prior information that their own common sense told them was the right thing to do. But we think it was a disaster for fields such as econometrics and artificial intelligence, where adoption of the orthodox view of probability had the automatic consequence that the significant problems could not even be formulated, much less solved, because the orthodox view of probability theory does not recognize probability as expressing information at all.

statistic. This happened in the example studied above; but now, more may be true. It may be that for that prior on η the pseudoposterior pdf for ζ is identical with the marginal pdf in the original problem. If a prior brings about agreement between the marginal posterior and the pseudoposterior distributions, how should we interpret this?

Suppose we start from the pseudoproblem. It seems that if introducing a new parameter η and using the prior $p(\eta|I)$ makes no difference, then it has conveyed no *information* at all about ζ: that prior must express 'complete ignorance' of η in a rather fundamental sense. In all cases yet found the prior $p(\eta|I)$ which does this on an infinite domain is improper; this lends support to that conclusion because, as noted, our common sense should have told us that *any proper prior on an infinite domain is necessarily informative about η*; it places some finite limits on the range of values that η could reasonably have, whether we interpret 'reasonably' as 'with 99% probability' or 'with 99.9% probability' ... and so on.

Can this observation be extended to a general technique for constructing uninformative priors beyond the location and scale parameter cases? This is at present an ongoing research project rather than a finished part of probability theory, so we defer it for the future.

15.11 How to mass-produce paradoxes

Having examined a few paradoxes, we can recognize their common feature. Fundamentally, the procedural error was always failure to obey the product and sum rules of probability theory. Usually, the mechanism of this was careless handling of infinite sets and limits, sometimes accompanied by attempts to replace the rules of probability theory by intuitive *ad hoc* devices like B_2's 'reduction principle'. Indeed, paradoxes caused by careless dealing with infinite sets or limits can be mass-produced by the following simple procedure:

(1) Start from a mathematically well-defined situation, such as a finite set, a normalized probability distribution, or a convergent integral, where everything is well-behaved and there is no question about what is the correct solution.
(2) Pass to a limit – infinite magnitude, infinite set, zero measure, improper pdf, or some other kind – without specifying how the limit is approached.
(3) Ask a question whose answer depends on how the limit was approached.

This is guaranteed to produce a paradox in which a seemingly well-posed question has more than one seemingly right answer, with nothing to choose between them. The insidious thing about it is that, as long as we look only at the limit, and not the limiting process, the source of the error is concealed from view.

Thus, it is not surprising that those who persist in trying to evaluate probabilities directly on infinite sets have been able to study finite additivity and nonconglomerability for decades – and write dozens of papers of impressive scholarly appearance about it. Likewise, those who persist in trying to calculate probabilities conditional on propositions of probability zero, have before them an unlimited field of opportunities for scholarly looking research and publication – without hope of any meaningful or useful results.

In our opening quotation, Gauss had a situation much like this in mind. Whenever we find a belief that such infinite sets possess some kind of 'existence' and mathematical properties in their own right, independent of any such limiting process, we can expect to see paradoxes of the above type. But note that this does not in any way prohibit us from using infinite sets to define *propositions*. Thus the proposition

$$G \equiv 1 \leq x \leq 2 \tag{15.94}$$

invokes an uncountable set, but it is still a single discrete proposition, to which we may assign a probability $P(G|I)$ defined on a sample space of a finite number of such propositions without violating our 'probabilities on finite sets' policy. We are not assigning any probability directly on an infinite set.

But then if we replace the upper limit 2 by a variable quantity z, we may (and nearly always do) find that this defines a well-behaved function, $f(z) \equiv P(G|zI)$. In calculations, we are then free to make use of whatever analytic properties this function may have, as we noted in Chapter 6. Even if $f(z)$ is not an analytic function, we may be able to define other analytic functions from it, for example by integral transforms. In this way, we are able to deal with any real application that we have been able to imagine, by discrete algebraic or continuum analytical methods, without losing the protection of Cox's theorems.

15.12 Comments

In this chapter and Chapter 5, we have seen two different kinds of paradox. There are 'conceptually generated' ones, such as the Hempel paradox of Chapter 5, which arise from placing faulty intuition above the rules of probability theory, and 'mathematically generated' ones, such as nonconglomerability, which arise mostly out of careless use of infinite sets. Marginalization is an elaborate example of a compound paradox, generated by both conceptual errors and mathematical errors which happened to reinforce each other. It seems that nothing in the mathematics can protect us against conceptual errors, but we might ask whether there are better ways of protection against mathematical ones.

Back in Chapter 2 we saw that the rules of probability theory can be derived as necessary conditions for consistency, as expressed by Cox's functional equations. The proofs applied to finite-sets of propositions, but when the results of a finite-set calculation can be extended to an infinite set by a mathematically well-behaved passage to a limit, we also accept that limit.

It might be thought that it would be possible, and more elegant, to generalize Cox's proofs so that they would apply directly to infinite sets; and indeed that is what the writer believed and tried to carry out for many years. However, since at least the work of Bertrand (1889), the literature has been turning up paradoxes that result from attempts to apply the rules of probability theory directly and indiscriminately on infinite sets; we have just seen some representative examples and their consequences. Since in recent years there has been a sharp increase in this paradoxing, one must take a more cautious view of infinite sets.

Our conclusion – based on some 40 years of mathematical efforts and experience with real problems – is that, at least in probability theory, an infinite set should be thought of only as the limit of a specific (i.e. unambiguously specified) sequence of finite sets. Likewise, an improper pdf has meaning only as the limit of a well-defined sequence of proper pdfs. The mathematically generated paradoxes have been found only when we tried to depart from this policy by treating an infinite limit as something already accomplished, without regard to any limiting operation. Indeed, experience to date shows that almost any attempt to depart from our recommended 'finite-sets' policy has the potentiality for generating a paradox, in which two equally valid methods of reasoning lead us to contradictory results.

The paradoxes studied here stand as counter-examples to any hope that we can ever work with full freedom on infinite sets. Unfortunately, the Borel–Kolmogorov and marginalization paradoxes turn up so seldom as to encourage overconfidence in the inexperienced. As long as one works on problems where they do not cause trouble, the psychological phenomenon: 'You can't argue with success!', noted at the beginning of this Chapter, controls the situation. Our reply to this is, of course, 'You can and should argue with success that was obtained by fraudulent means'.

Mea culpa

For many years, the present writer was caught in this error just as badly as anybody else, because Bayesian calculations with improper priors continued to give just the reasonable and clearly correct results that common sense demanded. So warnings about improper priors went unheeded; just that psychological phenomenon. Finally, it was the marginalization paradox that forced recognition that we had only been lucky in our choice of problems. If we wish to consider an improper prior, the only correct way of doing it is to approach it as a well-defined limit of a sequence of proper priors. If the correct limiting procedure should yield an improper posterior pdf for some parameter α, then probability theory is telling us that the prior information and data are too meager to permit any inferences about α. Then the only remedy is to seek more data or more prior information; probability theory does not guarantee in advance that it will lead us to a useful answer to every conceivable question.

Generally, the posterior pdf is better behaved than the prior because of the extra information in the likelihood function, and the correct limiting procedure yields a useful posterior pdf that is analytically simpler than any from a proper prior. The most universally useful results of Bayesian analysis obtained in the past are of this type, because they tended to be rather simple problems, in which the data were indeed so much more informative than the prior information that an improper prior gave a reasonable approximation – good enough for all practical purposes – to the strictly correct results (the two results agreed typically to six or more significant figures).

In the future, however, we cannot expect this to continue because the field is turning to more complex problems in which the prior information is essential and the solution is found by computer. In these cases it would be quite wrong to think of passing to an improper prior. That would lead usually to computer crashes; and, even if a crash is avoided, the conclusions would still be, almost always, quantitatively wrong. But, since likelihood functions are bounded, the analytical solution with proper priors is always guaranteed to converge properly to finite results; therefore it is always possible to write a computer program in such a way (avoid underflow, etc.) that it cannot crash when given proper priors. So, even if the criticisms of improper priors on grounds of marginalization were unjustified, it remains true that in the future we shall be concerned necessarily with proper priors.

Note added

Preliminary versions of this chapter were made available to many interested persons, for comments and suggestions. Several have expressed, both privately and publicly, their appreciation for these clarifications of issues that have long been mysterious and confused, and even some compulsive nitpickers have failed to raise any objections. Only one source has exhibited that psychological phenomenon noted in Chapter 5 in connection with the Hempel paradox; someone asserts a principle that seems to him intuitively right, and when probability analysis reveals the error, instead of taking this opportunity to educate his intuition, he reacts by rejecting the probability analysis. For him, his intuitive *ad hoc* principle takes precedence over the rules of probability theory.

If the issue is only which is to take precedence, there does not seem to be any way to resolve it; if one is not convinced by Cox's theorems and our great deal of experience confirming what they tell us, then we shall just have to agree to disagree. But if the issue is one of mathematically demonstrable fact, then it can be resolved at once – in the minds of everyone except the one who proposed the principle. One can be so deeply committed to his position that mathematical proof to the contrary, and any number of counter-examples, carry no weight for him. That is just what has happened.

In the case of the tumbling tetrahedra problem, we pointed out the error in previous discussions, gave the exact solution (15.26) according to the rules of probability theory, an asymptotic approximation (15.18) to it, and after a little thought could see that the final result (15.34) was really obvious from the start. That should be enough; this is an issue of mathematically demonstrable fact, the mathematics is before us, and every reader can judge it for himself.

The case of the marginalization paradox is very similar. The real purpose of this note is to stress what is the issue here. DSZ (p. 194) purported to have proved that the disagreement of B_1 and B_2 'could not have arisen if B_1 had employed proper prior distributions'. We pointed out that their proof is based on mutually contradictory assumptions, and reinforced this by (1) pointing out that a counter-example to what they claimed to have proved was already present in plain sight in the original DSZ article (their Example #5, where it is evident by inspection that the disagreement is present for all priors, proper or improper), and (2) gave in (15.76) another counter-example, where B_1 uses a proper prior, but B_1 and B_2 still disagree because B_1's posterior distribution for ζ depends on the datum y and the prior $p(\eta|I_1)$, as common sense tells us it must when B_1 uses a proper – therefore informative – prior for η. In fact, we can leave it as an exercise for the reader to verify that every one of the DSZ examples is an equally good counter-example, if you look at what happens when B_1 uses a proper prior. Of course, B_1 and B_2 agree when B_1 uses a proper prior in the trivial case where we are concerned with two independent problems; then the sampling distribution factors in the form $p(yz|\eta\zeta) = p(y|\eta)p(z|\zeta)$ and the prior $p(\eta\zeta|I_1)$ also factors.

But if the basic theorem is invalid, then the entire marginalization tale collapses; if one applies the rules of probability theory correctly, as explained long ago by Harold Jeffreys, there is no paradox. Again, this is an issue of mathematically demonstrable fact for which we have given the relevant mathematics, so we see no reason to engage in continuing debate over it; every reader can judge it for himself.

16

Orthodox methods: historical background

> With all this confounded trafficking in hypotheses about invisible connec-
> tions with all manner of inconceivable properties, which have checked
> progress for so many years, I believe it to be most important to open
> people's eyes to the number of superfluous hypotheses they are making,
> and would rather exaggerate the opposite view, if need be, than proceed
> along these false lines.
>
> *H. von Helmholtz (1868)*

This chapter and Chapter 13 are concerned with the history of the subject rather than its present status. There is a complex and fascinating history before 1900, recounted by Stigler (1986c), but we are concerned now with more recent developments. In the period from about 1900 to 1970, one school of thought dominated the field so completely that it has come to be called 'orthodox statistics'. It is necessary for us to understand it, because it is what most working statisticians active today were taught, and its ideas are still being taught, and advocated vigorously, in many textbooks and universities.

In Chapter 17 we want to examine the 'orthodox' statistical practice thus developed and compare its technical performance with that of the 'probability as logic' approach expounded here. But first, to understand this weird course of events, we need to know something about the problems faced then, the sociology that evolved to deal with them, the roles and personalities of the principal figures, and the general attitude toward scientific inference that orthodoxy represents.

16.1 The early problems

The beginnings of scientific inference were laid in the 18th and 19th centuries out of the needs of astronomy and geodesy. The principal figures were Daniel Bernoulli, Laplace, Gauss, Legendre, Poisson and others, whom we would describe today as mathematical physicists. This reached its highest technical development in the hands of Laplace, and it was a 'Bayesian' theory.

Transitions in the dominant mode of thinking take place slowly over a few decades, the working lifetime of one generation. The beginning of the period we are concerned with,

1900, marks roughly the time when non-physicists moved in and proceeded to take over the field with quite different ideas. The end, 1970, marks roughly the time when those ideas in turn came under serious, concerted attack in our present 'Bayesian revolution'.

During this period, as we analyzed in Chapter 10, the non-physicists thought that probability theory was a physical theory of 'chance' or 'randomness', with no relation to logic, while 'statistical inference' was thought to be an entirely different field, based on entirely different principles. But, having abandoned the principles of probability theory, it seemed that they could not agree on what those new principles of inference were; or even on whether the reasoning of statistical inference was deductive or inductive.

The first problems, dating back to the 18th century, were of course of the very simplest kind, estimating one or more location parameters θ from data $D = \{x_1, \ldots, x_n\}$ with sampling distributions of the form $p(x|\theta) = f(x - \theta)$. However, in practice this was not a serious limitation, because even a pure scale parameter problem becomes approximately a location parameter one if the quantities involved are already known rather accurately, as is generally the case in astronomy and geodesy.

Thus, if the sampling distribution has the functional form $f(x/\sigma)$, and x and σ are already known to be about equal to x_0 and σ_0, we are really making inferences about the small corrections $q \equiv x - x_0$ and $\delta \equiv \sigma - \sigma_0$. Expanding in powers of δ and keeping only the linear term, we have

$$\frac{x}{\sigma} = \frac{x_0 + q}{\sigma_0 + \delta} = \frac{1}{\sigma_0}(x - \theta + \cdots), \tag{16.1}$$

where $\theta \equiv x_0 \delta / \sigma_0$. Thus we may define a new sampling distribution function

$$h(x - \theta) \propto f(x/\sigma), \tag{16.2}$$

and we are considering, at least approximately, a location parameter problem after all. In this way, almost any problem can be linearized into a location parameter one if the quantities involved are already known to fairly good accuracy. The 19th century astronomers took good advantage of this, as we should also.

Only toward the end of the 19th century did practice advance to the problem of estimating simultaneously both a location and scale parameter θ, σ from a sampling distribution of the form

$$p(x|\theta\sigma) = f\left(\frac{x - \theta}{\sigma}\right)\frac{1}{\sigma} \tag{16.3}$$

and to the marvellous developments by Galton (1886) associated with the bivariate Gaussian distribution, which we studied in Chapter 7. Virtually all of the development of orthodox statistics was concerned with these three problems or their reverbalizations in hypothesis testing form, and most of it only with the first. But even that seemingly trivial problem had the power to generate fundamental differences of opinion and fierce controversy over matters of principle.

16.2 Sociology of orthodox statistics

During the aforementioned period, the average worker in physics, chemistry, biology, medicine, or economics with a need to analyze data could hardly be expected to understand theoretical principles that did not exist, and so the approved methods of data analysis were conveyed to him in many different, unrelated *ad hoc* recipes in 'cookbooks' which, in effect, told one to 'Do this . . . then do that . . . and don't ask why.'

R. A. Fisher's *Statistical Methods for Research Workers* (1925) was the most influential of these cookbooks. In going through 13 editions in the period 1925–1960 it acquired such an authority over scientific practice that researchers in some fields such as medical testing found it impossible to get their work published if they failed to follow Fisher's recipes to the letter.

Fisher's recipes include maximum likelihood parameter estimation (MLE), analysis of variance (ANOVA), fiducial distributions, randomized design of experiments, and a great variety of significance tests, which make up the bulk of his book. The rival Neyman–Pearson school of thought offered unbiased estimators, confidence intervals, and hypothesis testing. The combined collection of the *ad hoc* recipes of the two schools came to be known as orthodox statistics, although arguments raged back and forth between them over fine details of their respective ideologies. It was just the absence of any unifying principles of inference that perpetuated this division; there was no criterion acceptable to all for resolving differences of opinion.

Whenever a real scientific problem arose that was not covered by the published recipes, the scientist was expected to consult a professional statistician for advice on how to analyze his data, and often on how to gather them as well. There developed a statistician–client relationship rather like the doctor–patient one, and for the same reason. If there are simple unifying principles (as there are today in the theory we are expounding), then it is easy to learn them and apply them to whatever problem one has; each scientist can become his own statistician. But in the absence of unifying principles, the collection of all the empirical, logically unrelated procedures that a data analyst might need, like the collection of all the logically unrelated medicines and treatments that a sick patient might need, was too large for anyone but a dedicated professional to learn.

Undoubtedly, this arrangement served a useful purpose at the time in bringing about a semblance of order into the way scientists analyzed and interpreted their data and published their conclusions. It was workable as long as scientific problems were simple enough so that the cookbook procedures could be applied and made some intuitive sense, even though they were not derived from any first principles. Then, had the proponents of orthodox methods behaved with the professional standards of a good doctor (who notes that some treatments have been found to be effective, but admits frankly that the real cause of a disorder is not known and welcomes further research to supply the missing knowledge) there could be no criticism of the arrangement.

That is not how they behaved, however; they adopted a militant attitude, each defending his own little bailiwick against intrusion and opposing every attempt to find the missing unifying principles of inference. R. A. Fisher (1956) and M. G. Kendall (1963) attacked Neyman and Wald for seeking unifying principles in decision theory. R. A. Fisher (in numerous articles, e.g. 1933), H. Cramér (1946), W. Feller (1950), J. Neyman (1952), R. von Mises (1957) – and even the putative Bayesian L. J. Savage (1954, 1981) – accused Laplace and Jeffreys of committing metaphysical nonsense for thinking that probability theory was an extension of logic, and seeking the unifying principles of inference on that basis. We are at a loss to explain how they could have felt such a certainty about this, since they were all quite competent mathematically and presumably understood perfectly well what does and what does not constitute a proof. Yet they did not examine the consistency of probability theory as logic, as R. T. Cox did; nor did they examine its qualitative correspondence with common sense, as Pólya did. They did not even deign to take note of how it works out in practice, as H. Jeffreys had shown so abundantly in works which were there for their inspection. In fact, they offered no demonstrative arguments or factual evidence at all in support of their position; they merely repeated ideological slogans about 'subjectivity' and 'objectivity' which were quite irrelevant to the issues of logical consistency and useful results.

We are equally helpless to explain why James Bernoulli and John Maynard Keynes (who expounded essentially the same views as did Laplace and Jeffreys) escaped scorn. Evidently, the course of events must have had something to do with personalities; let us examine a few of them.

16.3 Ronald Fisher, Harold Jeffreys, and Jerzy Neyman

Sir Ronald Aylmer Fisher (1890–1962) was by far the dominant personality in this field in the period 1925–1960. A personal account of his life is given by his daughter, Joan Fisher Box (1978). On the technical side, Fisher had a deep intuitive understanding and produced a steady stream of important research in genetics. Sir Harold Jeffreys (1891–1989), working in geophysics, wielded no such influence, and for most of his life found himself the object of scorn and derision from Fisher and his followers.

Fisher's early fame (1915–1925) rested on his mathematical ability: given data $D \equiv \{x_1, \ldots, x_n\}$ to which we assign a multivariate Gaussian sampling probability $p(D|\theta)$ with parameters $\theta \equiv \{\theta_1, \ldots, \theta_m\}$, how shall we best estimate those parameters from the data? Probability theory as logic considers it obvious that in any problem of inference we are always to calculate the probability of whatever is unknown and of interest, conditional on whatever is known and relevant; in this case, $p(\theta|DI)$.

But the orthodox view rejects this on the grounds that $p(\theta|DI)$ is meaningless because it is not a frequency; θ is not a 'random variable', only an unknown constant. Instead, we are to choose some function of the data $f(D)$ as our 'estimator' of θ. The merits of any proposed estimator are to be determined solely from its sampling distribution $p(f|\theta)$. The data are always supposed to be obtained by 'drawing from a population' urn-wise, and $p(f|\theta)$ is always supposed to be a limiting frequency in many repetitions of that draw.

A good estimator is one whose sampling distribution is strongly concentrated in a small neighborhood of the true value of θ.

But, as we noted in Chapter 13, orthodoxy, having no general theoretical principles for constructing the 'best' estimator, must in every new problem guess various functions $f(D)$ on grounds of intuitive judgment, and then test them by determining their sampling distributions, to see how concentrated they are near the true value. Thus, calculation of sampling distributions for estimators is the crucially important part of orthodox statistics; without it one has no grounds for choosing an estimator.

The sampling distribution for some complicated function of the data, such as the sample correlation coefficient, can become quite a difficult mathematical problem; but Fisher was very good at this, and found many of these sampling distributions for the first time. Technical details of these derivations, in more modern language and notation, may be found in Feinberg and Hinkley (1980).

Many writers have wondered how Fisher was able to acquire the multidimensional space intuition that enabled him to solve these problems. We would point out that, just before starting to produce those results, Fisher spent a year (1912–1913) as assistant to the theoretical physicist Sir James Jeans, who was then preparing the second edition of his book on kinetic theory and worked daily on calculations with high-dimensional multivariate Gaussian distributions (called Maxwellian velocity distributions).

But nobody seemed to notice that Jeffreys was able to bypass Fisher's calculations and derive those parameter estimates in a few lines of the most elementary algebra. For Jeffreys, using probability theory as logic, in the absence of any cogent and detailed prior information, the best estimators were always determined by the likelihood function, which can be written down at once, merely by inspection of $p(D|\theta)$. This automatically constructed the optimal estimator for him, with no need for intuitive judgment and without ever calculating a sampling distribution for an estimator. Fisher's difficult calculations calling for all that space intuition, although interesting as mathematical results in their own right, were quite unnecessary for the actual conduct of inference.

Fisher's later dominance of the field derives less from his technical work than from his flamboyant personal style and the worldly power that went with his official position, in charge of the work and destinies of many students and subordinates. For 14 years (1919–1933) he was at the Rothamsted agricultural research facility with an increasing number of assistants and visiting students, then holder of the Chair of Eugenics at University College, London, and finally in 1943 Balfour Professor of Genetics at Cambridge, where he also became President of Caius College. He was elected Fellow of the Royal Society in 1929, and was knighted in 1952.

Within his field of geophysics, Harold Jeffreys also showed an outstandingly high competence, was elected Fellow of the Royal Society in 1925, became Plumian Professor of Astronomy at Cambridge in 1946, and was knighted in 1953. The treatise on mathematical physics by Sir Harold and Lady Jeffreys (1946) was for many years the standard textbook in the field. But Jeffreys remained all his life as a Fellow of St John's College, Cambridge, working quietly and modestly, and hardly visible outside his field of geophysics; he had only one doctoral student in probability theory (V. S. Huzurbazar).

In sharp contrast, Fisher, possessed of a colossal, overbearing ego, thrashed about in the field, attacking the work of everyone else[1] with equal ferocity. Somehow, early in life, Fisher's mind became captured by the dogma that by 'probability' one is allowed to mean only limiting frequency in a random experiment. However, he usually stated this as the ratio of two infinite numbers rather than the limit of a ratio of finite numbers, and said that any other meaning is metaphysical nonsense, unworthy of a scientist. Conceivably, this view might have come from the philosopher John Venn, an earlier President of Caius College, Cambridge, where Fisher was an undergraduate from 1909 to 1912. In a very influential work, which went through three editions, Venn ridiculed Laplace's conception of probability theory as logic; and Fisher's early work sounds very much like this.

However, we see a weakening of resolve in Fisher's final book (1956), where he actually defends Laplace against the criticisms of Venn, and suggests that Venn did not understand mathematics well enough to comprehend what Laplace was saying. His criticisms of Jeffreys are now much toned down. Noting this, some have opined that, were Fisher alive today, he would be a Bayesian.[2]

In both science and art, every creative person must, at the beginning of his career, do battle with an establishment that, not comprehending the new ideas, is more intent on putting him down than understanding his message. Karl Pearson (1857–1936), as editor of *Biometrika*, performed that 'service' for Fisher in his early attempts at publication, and Fisher never forgave him for this. But curiously, in his last book, Fisher's attacks against Pearson are, if anything, more violent and personal than ever before. This is hard to understand, for by 1956 the battle was long since won; Pearson had been dead for 20 years, and it was universally recognized that in all their disputes Fisher had been in the right. Why should the bitterness remain 30 years after it had ceased to be relevant? This tells us much about Fisher's personality.

Fisher's articles are most easily found today in two 'collected works' (Fisher, 1950, 1974). The ones on the principles of inference have an interesting characteristic pattern. They start with a paragraph or two of polemical denunciation of Jeffreys' use of Bayes' theorem (at that time called *inverse probability*). Then he formulates a problem, sees the correct solution intuitively, and does the requisite calculations in a very efficient, competent way. But, just at the point where one more step of the logical argument would have forced him to see that he was only rediscovering, in his own way, the results of applying Bayes' theorem, the article comes to an abrupt end.

---

[1] For the record, we consider Fisher's criticisms of Karl Pearson on grounds of maximum likelihood vs. moment fitting and the proper number of degrees of freedom in chi-squared, and of Jerzy Neyman on grounds of confidence intervals, unbiased estimators, and the meaning of significance levels, to be justified on grounds of technical fact. It is perhaps a measure of Fisher's influence that the two disputes where we think that Fisher was in the wrong – the one with W. S. Gossett over randomization and the one with Jeffreys on the whole meaning and philosophy of inference – are still of serious concern today.

[2] But against this supposition is the fact that in the last year of his life Fisher published an article (Fisher, 1962) examining the possibilities of Bayesian methods, but *with the prior probabilities to be determined experimentally*! This shows that he never accepted – and probably never comprehended – the position of Jeffreys about the meaning and function of a prior probability.

Harold Jeffreys (1939) was able to derive all the same results far more easily, by direct use of probability theory as logic, and this automatically yielded additional information about the range of validity of the results and how to generalize them, that Fisher never did obtain. But whenever Jeffreys tried to point this out, he was buried under an avalanche of criticism which simply ignored his mathematical demonstrations and substantive results and attacked his ideology. His perceived sin was that he did not require a probability to be also a frequency, and so admitted the notion of probability of an hypothesis. Nobody seemed to perceive the fact that this broader conception of probability was just what was giving him those computational advantages.

Jerzy Neyman, whom we discussed in Chapter 14, also rejected Jeffreys' work on the same ideological grounds as did Fisher (but in turn had his own work rejected by Fisher). Neyman also directed scathing ridicule at Jeffreys, far beyond what would have been called for even if Neyman had been technically correct and Jeffreys wrong. For example, Neyman (1952, p. 11) becomes heated over a problem involving five balls in two urns, so simple that it would not be considered worthy of being an undergraduate homework problem today, in which Jeffreys (1939, Sect. 7.02) is clearly in the right.

In view of all this, it is pleasant to be able to record that, in the end, Harold Jeffreys outlived his critics, and the merit of his work, on both the theoretical and the pragmatic levels, was finally recognized. In the last years of his life he had the satisfaction of seeing Cambridge University – from the Cavendish Physics Laboratory to the north to the Molecular Biology Laboratory to the south – well populated with young scientists studying and applying his work and, with the new tool of computers, demonstrating its power for the current problems of science.

The exchanges between Fisher and Jeffreys over these issues in the British journals of the 1930s were recalled recently by S. Geisser (1980) and D. Lane (1980), with many interesting details. But we want to add some additional comments to theirs, because a fellow physicist is in a better position to appreciate Jeffreys' motivations, highly relevant for the applications we are concerned with today.

Firstly, we need to recognize that a large part of their differences arose from the fact that Fisher and Jeffreys were occupied with very different problems. Fisher studied biological problems, where one had no prior information and no guiding theory (this was long before the days of the DNA helix), and the data taking was very much like drawing from Bernoulli's urn. Jeffreys studied problems of geophysics, where one had a great deal of cogent prior information and a highly developed guiding theory (all of Newtonian mechanics giving the theory of elasticity and seismic wave propagation, plus the principles of physical chemistry and thermodynamics), and the data taking procedure had no resemblance to drawing from an urn. Fisher, in his cookbook (1925, Sect. 1) defines statistics as *the study of populations*; Jeffreys devotes virtually all of his analysis to problems of inference where there is no population.

Late in life, Jerzy Neyman was able to perceive this difference. His biographer, Constance Reid (1982, p. 229), quotes Neyman thus: 'The trouble is that what we statisticians call modern statistics was developed under strong pressure on the part of biologists. As a result, there is practically nothing done by us which is directly applicable to problems of astronomy.'

Fisher advanced, very aggressively, the opposite view: that the methods which were successful in his biological problems must be also the general basis of all scientific inference. What Fisher was never able to see is that, from Jeffreys' viewpoint, Fisher's biological problems were trivial, both mathematically and conceptually. In his early chapters, Jeffreys (1939) disposes of them in a few lines, obtaining Fisher's inference results far more easily than Fisher did, as the simplest possible applications of Bayes' theorem,[3] then goes on to more complex problems beyond the ambit of Fisher's methods. Jeffreys (1939, Chap. 7) then summarizes the comparisons with Fisher and Neyman in more general terms.

As science progressed to more and more complicated problems of inference, the short-comings of the orthodox methods became more and more troublesome. Fisher would have been nearly helpless, and Neyman completely helpless, in a problem with many nuisance parameters but no sufficient or ancillary statistics. Accordingly, neither ever attempted to deal with what is actually the most common problem of inference faced by experimental scientists: linear regression with both variables subject to unknown error. Generations of scientists in several different fields searched the statistical literature in vain for help on this; but for Bayesian methods (Zellner, 1971; Bretthorst, 1988) the nuisance parameters are only minor technical details that do not deter one from finding the straightforward and useful solutions. Scientists, engineers, biologists, and economists with good Bayesian training are now finding for themselves the correct solutions appropriate to their problems, which can adapt effortlessly to many different kinds of prior information, thus achieving a flexibility unknown in orthodox statistics.

However, we recognize Fisher's high competence in the problems which concerned him. An honest man can maintain an ideology only as long as he confines himself to problems where its shortcomings are not evident. Had Fisher tried more complex problems, we think that he would have perceived the superior power of Jeffreys' methods rather quickly; as we demonstrate in Chapters 13 and 17, the mathematics forces one to it, independently of all ideology. As noted, it may be that he started to see this toward the end of his life.

Secondly, we note the very different personalities and habits of scholarly conduct of the combatants. In any field, the most reliable and instantly recognizable sign of a fanatic is a lack of any sense of humor. Colleagues have reported their experiences at meetings, where Fisher could fly into a trembling rage over some harmless remark that others would only smile at. Even his disciples (for example, Kendall, 1963) noted that the character defects which he attributed to others were easily discernible in Fisher himself; as one put it, 'Whenever he paints a portrait, he paints a self-portrait'.

Harold Jeffreys maintained his composure, never took these disputes personally, and, even in his 90s, when the present writer knew him, it was a delight to converse with him

---

[3] Of course, Fisher's randomized planting methods – which we think to be not actually wrong, but hopelessly inefficient in information handling – were not reproduced by Jeffreys; nor would he wish to. It appears to be a quite general principle that, whenever there is a randomized way of doing something, then there is a nonrandomized way that delivers better performance but requires more thought. We illustrate this by example in Chapter 17 under 'The folly of randomization'.

because he still retained a wry, slightly mischievous, sense of humor. The greatest theoretical physicists of the 19th and 20th centuries, James Clerk Maxwell and Albert Einstein, showed just the same personality trait, as testified by many who knew them.

Needless to say (since Fisher's methods were mathematically only special cases of those of Jeffreys), Fisher was never able to exhibit a specific problem in which his methods gave a satisfactory result and Jeffreys' methods did not. Therefore we see in Fisher's words almost no pointing to actual results in real problems. Usually Fisher's words convey only a spluttering exasperation at the gross ideological errors of Jeffreys and his failure to repent. His few attempts to address technical details only reveal his own misunderstandings of Jeffreys.

For example, Jeffreys (1932) gave a beautiful derivation of the $d\sigma/\sigma$ prior for a scale parameter, which we referred to in Chapter 12. Given two observations x_1, x_2 from a Gaussian distribution, the predictive probability density for the third observation is

$$p(x_3|x_1x_2I) = \int d\mu \int d\sigma \; p(x_3|\mu\sigma I)p(\mu\sigma|x_1x_2I). \tag{16.4}$$

If initially σ is completely unknown, then our estimates of σ ought to follow the data difference $|x_2 - x_1|$, with the result that the predictive probability for the third observation to lie between them ought to be $1/3$, independently of x_1 and x_2 (with independent sampling, every permutation of the three observations has the same probability). He shows that this will be true only for the $d\sigma/\sigma$ prior.

But Fisher (1933), failing to grasp the concept of a predictive distribution, takes this to be a statement about the sampling distribution $p(x_3|\mu\sigma I)$, which is an entirely different thing; he jumps to the conclusion that Jeffreys is guilty of a ridiculous elementary error, and then launches into seven pages of polemical attacks on all of Jeffreys' work, which display in detail his own total lack of comprehension of what Jeffreys was doing. All readers who want to understand the conceptual hangups that delayed the progress of this field for decades should read this exchange very carefully.

But in Jeffreys' words there is no misunderstanding of Fisher, no heaping of scorn and no ideological sloganeering; only a bemused sense of humor at the whole business. The issue as Jeffreys saw it was not any error of Fisher's actual procedures on his particular biological problems, but the incompleteness of his methods for more general problems and the lack of any justification for his dogmatically asserted premises. In particular, that one must conjure up some hypothetical infinite population from which the data are drawn, and that every probability must have an objectively 'true' value, independently of human information; Jeffreys' whole objective was to use probability to *represent* human information. Furthermore, Jeffreys always made his point quite gently.

For example, Jeffreys (1939, p. 325), perceiving what we noted above, writes of Fisher that, 'In fact, in spite of his occasional denunciations of inverse probability, I think that he has succeeded better in making use of what it really says than many of its professed users have.' As another example, in one of the exchanges Jeffreys complained that Fisher had 'reduced his work to nonsense'. In reply, Fisher pounced upon this, and wrote, gleefully: 'I am not inclined to deny it.' Geisser (1980) concludes that Jeffreys came off second best here; we see instead Jeffreys smiling at the fact that Fisher was deflected from the issue and fell headlong into the little trap that Jeffreys had set for him.

Having said something of their differences, we should add that, as competent scientists, Fisher and Jeffreys were necessarily in close agreement on more basic things; in particular on the role of induction in science. Neyman, not a scientist but a mathematician, tried to claim that his methods were entirely deductive. For example, in Neyman (1952, p. 210), he states: '. . . in the ordinary procedure of statistical estimation there is no phase corresponding to the description of "inductive reasoning". . . . all the reasoning is deductive and leads to certain formulae and their properties.' But Neyman (1950) was willing to speak of inductive *behavior*.

Fisher and Jeffreys, aware that all scientific knowledge has been obtained by inductive *reasoning* from observed facts, naturally enough denied the claim of Neyman that inference does not use induction, and of the philosopher Karl Popper that induction was impossible. We discussed this claim at the end of Chapter 9. Jeffreys expressed himself on this more in private conversations (at one of which the writer was present) than in public utterances; Fisher publicly likened Popper's and Neyman's strictures to political thought-control. As he put it (Fisher, 1956, p. 7): 'To one brought up in the free intellectual atmosphere of an earlier time there is something rather horrifying in the ideological movement represented by the doctrine that reasoning, properly speaking, cannot be applied to empirical data to lead to inferences valid in the real world.'

Indeed, Fisher's and Jeffreys' reactions to Popper may be a repetition of what happened in the 18th century. Fisher (1956, p. 10), Stigler (1983), and Zabell (1989) present quite good evidence – which seems to us, in its totality, just short of proof – that Thomas Bayes had found his result as early as 1748, and the original motivation for this work was his annoyance at the claim of the 18th century philosopher David Hume of the impossibility of induction. We may conjecture that Bayes sought to give an explicit counter-example, but found it a bit more difficult than he had at first expected, and so delayed publishing it. This would give a neat and natural explanation of many otherwise puzzling facts.

16.4 Pre-data and post-data considerations

The basic pragmatic difference in the two approaches is in how they relate to the data; orthodox practice is limited at the outset to pre-data considerations. That is, it gives correct answers to questions of the form:

(A) Before you have seen the data, what data do you expect to get?
(B) If the as yet unknown data are used to estimate parameters by some known algorithm, how accurate do you expect the estimates to be?
(C) If the hypothesis being tested is in fact true, what is the probability that we shall get data indicating that it is true?

Of course, probability theory as logic automatically includes all sampling distribution calculations; so, in problems where such questions are the ones of interest, we shall do the same calculations and reach the same numerical conclusions, with at worst a verbal disagreement over terminology.

As we have stressed repeatedly, virtually all real problems of scientific inference are concerned with post-data questions:

(A′) After we have seen the data, do we have any reason to be surprised by them?
(B′) After we have seen the data, what parameter estimates can we now make, and what accuracy are we entitled to claim?
(C′) What is the probability *conditional on the data*, that the hypothesis is true?

Orthodoxy is prevented from dealing with post-data questions by its different philosophy. The basic tenet that determines the form of orthodox statistics is that the reason why inference is needed lies not in mere human ignorance of the true causes operative, but in a 'randomness' that is attributed instead to Nature herself; just what we call the 'mind projection fallacy'. This leads to the belief that probability statements can be made only about random variables and not about unknown fixed parameters. However, although the property of being 'random' is considered a real objective attribute of a variable, orthodoxy has never produced any definition of the term 'random variable' that could actually be used in practice to decide whether some specific quantity, such as the number of beans in a can, is or is not 'random'.

Therefore, although the question 'Which quantities are random?' is crucial to everything an orthodox statistician does, we are unable to explain how he actually decides this; we can only observe what decisions he makes. For some reason, data are always considered random, almost everything else is nonrandom; but to the best of our knowledge, there is no principle in orthodox statistics which would have enabled one to predict this choice. Indeed, in a real situation the data are usually the only things that *are* definite and known, and almost everything else in the problem is unknown and only conjectured; so the opposite choice would seem far more natural.

This orthodox choice has the consequence that orthodox theory does not admit the existence of prior or posterior probabilities for a fixed parameter or an hypothesis, because they are not considered random variables. We want, then, to examine how orthodoxy manages to pass off the answer to a pre-data question as if it were the answer to a post-data one. Mostly this is possible because of mathematical accidents, such as symmetry in parameter and estimator.

16.5 The sampling distribution for an estimator

We have noted why a major part of the orthodox literature is devoted, necessarily, to calculating, approximating, and comparing sampling pdfs for estimators; this is the only criterion orthodoxy has for judging estimators, and in a new problem one may need to find sampling distributions for a half-dozen different estimators before deciding which one is best.

The sampling pdf for an estimator does not have the same importance in Bayesian analysis, because we do have the needed theoretical principles; if an estimator has been derived from Bayes' theorem and a specified loss function, then we know from perfectly general theorems that it is the optimal estimator for the problem as defined, whatever its sampling distribution may be. In fact, the sampling pdf for an estimator plays no functional role in post-data inference, and so we have no reason to mention it at all, unless pre-data considerations are of some interest; for example, in planning an experiment and deciding what kind of data to take and when to stop.

In addition to this negative (nonfunctionality) reason, there is a stronger positive reason for diverting attention away from the sampling pdf for an estimator; it is not the proper *criterion* of the quality of an inference. Suppose a scientist is estimating a physical parameter α such as the mass of a planet. If the sampling pdf for the estimator is indeed equal to the long-run frequencies in many repetitions of the measurement, then its width would answer the pre-data question:

(Q1) How much would the estimate of α vary over the class of all data sets that we might conceivably get?

This is not the relevant question for the scientist, however. His concern is with the post-data one:

(Q2) How accurately is the value of α determined by the one data set D that we actually have?

According to probability theory as logic, the correct measure of this is the width of the posterior pdf for the parameter, not the sampling pdf for the estimator. Since this is a major bone of contention between the orthodox and Bayesian schools of thought, let us understand why they can sometimes be the same, with resulting confusion of pre-data and post-data considerations. In the next chapter, we shall see some of the horrors that can arise when they are not the same.

Historically, since the time of Laplace, scientific inference has been dominated overwhelmingly by the case of Gaussian sampling distributions which have the aforementioned symmetry. Suppose we have a data set $D = \{y_1, \ldots, y_n\}$ and a sampling distribution

$$p(D|\mu\sigma I) \propto \exp\left\{-\sum_i \frac{(y_i - \mu)^2}{2\sigma^2}\right\} \tag{16.5}$$

with σ known. Then the Bayesian posterior pdf for μ, with uniform prior, is

$$p(\mu|D\sigma I) \propto \exp\left\{-\frac{n(\mu - \bar{y})^2}{2\sigma^2}\right\}, \tag{16.6}$$

from which the post-data (mean $\pm$ standard deviation) estimate of μ is

$$(\mu)_{\text{est}} = \bar{y} \pm \frac{\sigma}{\sqrt{n}}, \tag{16.7}$$

which shows that the sample mean $\bar{y} \equiv n^{-1} \sum y_i$ is a sufficient statistic. Then, if the orthodoxian decided to use $\bar{y}$ as an estimator of μ, he would find its sampling distribution to be

$$p(\bar{y}|\mu\sigma I) \propto \exp\left\{-\frac{n(\bar{y}-\mu)^2}{2\sigma^2}\right\},\tag{16.8}$$

and this would lead him to make the pre-data estimate

$$(\bar{y})_{\text{est}} = \mu \pm \frac{\sigma}{\sqrt{n}}.\tag{16.9}$$

But although (16.7) and (16.9) have entirely different meanings conceptually, they are mathematically so nearly identical that the Bayesian and orthodoxian would make the same actual numerical estimate of μ and claim the same accuracy. In problems like this, which have sufficient statistics but no nuisance parameters, there is a mathematical symmetry (approximate or exact) which can make the answers to a pre-data question and a post-data question closely related if we have no very cogent prior information which would break that symmetry.

This accidental equivalence has produced a distorted picture of the field; the Gaussian case is the one in which orthodox methods do best – not only for the reasons explained in Chapter 7, but also because if there is no prior information the symmetry is exact, so pre-data and post-data results are numerically the same. On the basis of such limited evidence, orthodoxy tried to claim general validity for its methods. But had the early experience referred instead to Cauchy sampling distributions,

$$p(y|\mu) = \frac{1}{\pi}\left[\frac{1}{1+(y-\mu)^2}\right],\tag{16.10}$$

the distinction could never have been missed because the answers to the pre-data and post-data questions are so different that common sense would never have accepted the answer to one as the answer to the other. In this case, with an uninformative prior the Bayesian posterior pdf for μ is

$$p(\mu|DI) \propto \prod_{i=1}^{n}\frac{1}{1+(\mu-y_i)^2}\tag{16.11}$$

which is still straightforward, if analytically inconvenient. Numerically, the (posterior mean $\pm$ standard deviation) or (posterior median $\pm$ interquartile) estimates are readily found by computer, but there is no sufficient statistic and therefore no good analytical solution.

But orthodoxy has never found any satisfactory estimator at all for this problem! If we try again to use the sample mean $\bar{y}$ as an estimator, we find, to our dismay, that its sampling pdf is

$$p(\bar{y}|\mu I) \propto \frac{1}{1+(\bar{y}-\mu)^2},\tag{16.12}$$

which is identical with (16.10); the mean of any number of observations is, according to this orthodox criteria, no better than a single observation. Although Fisher noted that, for large samples, the sample median tends to be more strongly concentrated near the true μ than does the sample mean, this gives no reason to think that it is the *best* estimator by orthodox criteria, even in the limit of large samples, and the question remains open today.

We expect that both the Bayesian posterior mean and posterior median value estimators would prove to be considerably better, by orthodox criteria of performance, than any presently known orthodox estimator. Simple computer experiments would be able to confirm or refute this conjecture; we doubt whether they will be done, because the question is of no interest to a Bayesian, while a well-indoctrinated orthodoxian will never voluntarily examine any Bayesian result.[4]

16.6 Pro-causal and anti-causal bias

One criticism of orthodox methods that we shall find in the next chapter is not ideological, but that they have technical shortcomings (waste of information) which, in practice, all tend to bias our inferences in the same direction. The result is that, when we are testing for a new phenomenon, orthodoxy in effect considers it a calamity to give credence to a phenomenon that is not real, but is quite unconcerned about the consequences of failing to recognize a phenomenon that is real.

To be fair, at this point we should keep in mind the historical state of affairs, and the far worse practices that the early workers in this field had to counteract. As we noted in Chapter 5, the uneducated mind always sees a causal relationship – even where there is no conceivable physical mechanism for it – out of the most far-fetched coincidence.

Johannes Kepler (1571–1630) was obliged to waste much of his life casting horoscopes for his patron (and complained about it privately). No amount of evidence showing the futility of this seems to shake the belief in it; even today, more people make their living as astrologers than as astronomers.

In the 18th and 19th centuries, science was still awash with superstitious beliefs in causal influences that do not exist, and Laplace (1812) warned against this in terms that seem like platitudes today, although they made him enemies then. Our opening quotation from Helmholtz shows his exasperation at the fact that progress in physiology was made almost impossible by common belief in all kinds of causal influences for which there was no physical mechanism and no evidence. Louis Pasteur (1822–1895) spent much of his life trying to overcome the universal belief in spontaneous generation.

Although the state of public health was intolerable by present standards, hundreds of plants were credited with possessing miraculous medicinal properties; at the same time,

---

[4] For example, many years ago the writer attempted to publish an article demonstrating the superior performance of Bayesian estimation with a Cauchy distribution, in the small sample case which can be solved analytically – and had the work twice rejected. The referee accused me of unfair tactics for bringing up the matter of the Cauchy distribution at all, because '... it is well known that the Cauchy distribution is a pathological, exceptional case'. Thus did one orthodoxian protect the journal's readers from the unpleasant truth that Bayesian analysis does not break down on this problem. To the best of our knowledge, Bayesian analysis has no pathological, exceptional cases; a reasonable question always has a reasonable answer. Finally, after 13 years of struggling, we did manage to get that analysis published after all by sneaking it into a longer article (Jaynes, 1976).

tomatoes were believed to be poisonous. As late as 1910 it was still being reported as scientific fact that poison ivy plants emit an 'effluvium' which infects those who merely pass by them without actual contact, although the simplest controlled experiment would have disproved this at once.

Today, science has advanced far beyond this state of affairs, but common understanding has hardly progressed at all. On the package of a popular brand of rice, the cooking instructions tell us that we must use a closed vessel, because 'the steam does the cooking'. Since the steam does not come into contact with the rice, this seems to be on a par with the poison ivy myth. Surely, a controlled experiment would show that the *temperature of the water* does the cooking. But at least this myth does no harm.

Other spontaneously invented myths can do a great deal of harm. If we have a single unusually warm summer, we are besieged with dire warnings that the Earth will soon be too hot to support life. Next year we will have an unusually cold winter, and the same disaster-mongers will be right there shouting about the imminent ice age. Both times they will receive the most full and sympathetic coverage by the news media, who, with their short memory and in their belief that they are doing a public service, amplify 1000-fold the capacity of the disaster-monger to do mischief. They encourage ever more irresponsible disaster-mongering as the surest way to get free personal publicity.

In 1991 some persons without the slightest conception of what either electricity or cancer are, needed only to hint that the weak 60 Hz electric and magnetic fields around home wiring or power lines are causing cancer; and the news media gave it instant credence and full prime-time radio and television coverage, throwing the uneducated public into a panic. They set up picket lines and protest marches to prevent installation of power lines where they were needed. The right of the public to be protected against the fraud of false advertising is recognized by all; so when will we have the right to be free of the fraud of sensationally false and irresponsible news reporting?

To counter this universal tendency of the untrained mind to see causal relations and trends where none exist, responsible science *requires* a very skeptical attitude, which demands cogent evidence for an effect; particularly one which has captured the popular imagination. Thus we can easily understand and sympathize with the orthodox conservatism in accepting new effects.

There is another side to this; skepticism can be carried too far. The orthodox bias against a real effect does help to hold irresponsibility in check, but today it is also preventing recognition of effects that *are* real and important. The history of science offers many examples of important discoveries that had their origin in the perception of someone who saw a small unexpected thing in his data, that an orthodox significance test would have dismissed as a random error.[5] The discovery of argon by Lord Rayleigh and of cosmic rays by Victor Hess are examples that come to mind immediately. Of course, they did not jump to sweeping

---

[5] Jeffreys (1939, p. 321) notes that there has never been a time in the history of gravitational theory when an orthodox significance test, which takes no note of alternatives, would not have rejected Newton's law and left us with no law at all. Nevertheless, Newton's law did lead to constant improvements in the accuracy of our accounting of the motions of the moon and planets for centuries, and it was only when an alternative (Einstein's law) had been stated fully enough to make very accurate known predictions of its own that a rational person could have thought of abandoning Newton's law.

conclusions from a single observation, as do the disaster-mongers; rather, they used the single surprising observation to motivate a careful investigation that culminated in overwhelming evidence for the new phenomenon. It is fortunate that physicists and astronomers do not, in practice, use orthodox significance tests; their own innate common sense is a safer and more powerful reasoning tool.

In other fields we must wonder how many important discoveries, particularly in medicine, have been prevented by editorial policies which refuse to publish that necessary first evidence for some effect, because the one data set that the researcher was able to obtain did not quite achieve an arbitrarily imposed significance level in an orthodox test. This could well defeat the whole purpose of scientific publication; for the cumulative evidence of three or four such data sets might have yielded overwhelming evidence for the effect. Yet this evidence may never be found unless the first data set can manage to get published.

How can editors recognize that scientific discovery is not a one-step process, but a many-step one, without thereby releasing a new avalanche of irresponsible, sensational publicity seekers? The problem is genuinely difficult, and we do not pretend to know the full answer.

Throughout this work we note instructive case histories of science gone wrong, when orthodox statistics was used to support either an unreasonable belief, or more often an unreasonable disbelief, in some phenomenon. In every case, a Bayesian analysis – taking into account all the evidence, not just the evidence of one data set – would have led to far more defensible conclusions; so editorial policies that required Bayesian standards of reasoning would go a long way toward solving this problem.

This orthodox bias against an effect is seen in the fact that Feller and others heap ridicule on 'cycle hunters' as being irresponsible, seeing in phenomena, such as economic time series, weather, sunspot numbers and earthquakes, periodicities that are not there. It is conceivable that there may be instances of this; but those who make the charge do not document specific examples which we can verify, and so we do not know of any. In economics, belief in business cycles goes in and out of style cyclically. Those who, like the economist Arthur Burns, merely look at a plot of the data, see the cycles at once. Those who, like Fisher, Feller, and Tukey (Blackman and Tukey, 1958), use orthodox data analysis methods, do not find them. Those who, like Bretthorst (1988), use probability theory as logic are taking into account more evidence than either of the above groups, and may or may not find them. More generally, the reason why some orthodox skeptics do not see real effects is that they use methods of data analysis which not only ignore prior information, but also violate the likelihood principle, and therefore waste some of the information in the data. We demonstrate this in Chapter 17.

16.7 What is real, the probability or the phenomenon?

This orthodox reluctance to see causal effects, even when they are real, has another psychological danger because eventually it becomes extrapolated into a belief in the existence of 'stochastic processes' in which no causes at all are operative, and probability itself is the only real physical phenomenon. When the search for any causal relation whatever is deprecated and discouraged, scientific progress is brought to a standstill.

Belief in the existence of 'stochastic processes' in the real world; i.e. that the property of being 'stochastic' rather than 'deterministic' is a real physical property of a process, that exists independently of human information, is another example of the mind projection fallacy: attributing one's own ignorance to Nature instead. The current literature of probability theory is full of claims to the effect that a 'Gaussian random process' is fully determined by its first and second moments. If it were made clear that this is only the defining property for an abstract mathematical model, there could be no objection to this; but it is always presented in verbiage that implies that one is describing an objectively true property of a real physical process. To one who believes such a thing literally, there could be no motivation to investigate the causes more deeply than noting the first and second moments, and so the real processes at work might never be discovered.

This is not only irrational because one is throwing away the very information that is essential to understand the physical process; if carried into practice it can have disastrous consequences. Indeed, there is no such thing as a 'stochastic process' in the sense that the individual events have no specific causes. One who views human diseases or machine failures as 'stochastic processes', as described in some orthodox textbooks, would be led thereby to think that in gathering statistics about them he is measuring the one controlling factor – the physically real 'propensity' of a person to get a disease or a machine to fail – and that is the end of it.

Yet where our real interests are involved, such foolishness is usually displaced rather quickly. Every individual disease in every individual person has a definite cause; fortunately, Louis Pasteur understood this in the 19th century, and our medical researchers understand it today. In medicine one does not merely collect statistics about the incidence of diseases; there are large organized research efforts to find their specific causes in individual cases.

Likewise, every machine failure has a definite cause; after every airplane crash the Federal Aviation Officials arrive and, if necessary, spend months sifting through all the evidence trying to determine the exact cause. Only by this pursuit of each individual cause can the level of public health and the safety and reliability of our machines be improved.

16.8 Comments

One lesson from the considerations of this chapter is that a deep change in the sociology of science – the relationship between scientist and statistician – is now underway. This is being brought about by the coincidence of recent improvements in both theoretical understanding and computation facilities.

The scientist who has learned the simple, unified principles of inference expounded here would not consult a statistician for advice because he can now work out the details of his specific data analysis problem for himself, and if necessary write a new computer program, in less time than it would take to read about them in a book or hire it done by a statistician. He is also alert to the defects in orthodox methods, and will avoid all advice from a statistician who continues to recommend them. Each scientist involved in data analysis can be his own statistician.

Another important general conclusion is that in analyzing data – particularly when searching for new effects – scientists are obliged to find a very careful compromise between seeing too little and seeing too much. Only methods of inference which realize all the 'resolving power' possible, by taking careful account of all the relevant prior information, all the previously obtained data, and all the information in the likelihood function, can steer a safe course between these dangers and yield justifiable conclusions. Probability theory as logic automatically takes into account the full range of conditions consistent with our information (our basic desiderata require this); and so it cannot give us misleading conclusions unless we feed it false information or withhold true and relevant information from it.

For many years, orthodox methods of data analysis, through their failure to take into account all the relevant evidence, have been misleading us in ways that have increasingly serious economic and social consequences. Often, orthodox methods are unable to find significant evidence for effects so clear that they are obvious at once from a mere glance at the data. More rarely, from failure to note cogent prior information orthodox methods may hallucinate, seeing nonexistent effects. We document cases of both in this work, and see how in all cases Bayesian analysis would have avoided the difficulty automatically.

16.8.1 Communication difficulties

As an example of the difficulties that Bayesians have trying to communicate with those trained only in the sampling theory viewpoint, the writer once gave a talk in which he mentioned in passing a very elementary and well-known theorem: that the posterior expectation of a parameter is the estimator that minimizes the expected square of the error.[6]

A sampling theorist in the audience objected violently to this, for in his lexicon an 'expectation' and an 'estimator' were not only different things, but things of a totally different qualitative nature: an estimator is a function of the data, but an expectation is an average over all possible data, a function of the parameter. So when I said that the best estimator is the posterior expectation, it sounded to him like I had said that apples are oranges; he not only denied the theorem, but thought that I had taken leave of my senses, and was ignorant of the meaning of statistical terms.

How would you reply to this objection? The problem was that the term 'posterior expectation' was, for him, meaningless, because he denied the existence of any such thing as a posterior distribution, and so could not comprehend that a posterior expectation is indeed a function of the data and is therefore a possible estimator. So unless he can be shifted into a completely different mindset, the simple theorem will continue to seem like pure nonsense to him. How can one explain this without offending – and thereby completely losing contact with – the objector?

---

[6] Proof: let the data be $(x_1, \ldots, x_n)$, and $\theta^*(x_1, \ldots, x_n)$ any proposed estimator. Then the expected square of the error over the posterior pdf for θ is $\langle (\theta^* - \theta)^2 \rangle = (\theta^* - \langle \theta \rangle)^2 + (\langle \theta^2 \rangle - \langle \theta \rangle^2)$, which is minimized for the choice $\theta^* = \langle \theta \rangle$.

L. J. Savage (1954), noting these communication blocks, caused by seemingly irreconcilable differences in ideology growing into fundamental differences in terminology, wrote that '... there has seldom been such complete disagreement and breakdown of communication since the Tower of Babel'. A more complete discussion of past communication difficulties is given in Jaynes (1986a). Today, with plentiful, powerful, and cheap computation facilities, we can bypass this and settle these issues by demonstrating the facts of actual performance. One of the purposes of the present work is to explain how such demonstrations can be carried out.

In the 1930s and 1940s, there were not only communication blocks, but rampant statistical gamesmanship. Everybody wants to be seen as taking a public stance for virtue and against sin, so the frequentist statisticians adopted the simple device of inventing virtuous-sounding terms (like unbiased, efficient, uniformly most powerful, admissible, robust) to describe their own procedures, therefore almost forcing others to apply the sinful-sounding antonyms (like biased, inadmissible) to all other methods.

Those who played this game were, in the long run, only caught in their own trap; for all their favored methods were arbitrary *ad hockeries* not derived from any first principles. It developed – inevitably in view of Cox's theorems – that all of them had serious defects that are overcome only by the Bayesian methods that they rejected. It is now clear, as we demonstrate in Chapter 17, that a 'biased' estimate may be considerably closer to the truth than an 'unbiased' one, an 'inadmissible' procedure may be far superior to an 'admissible' one; and so on. Today those emotionally loaded terms are only retarding progress and doing a disservice to science.

17

Principles and pathology of orthodox statistics

> The development of our theory beyond this point, as a practical statisti-
> cal theory, involves ... all the complexities of the use, either of Bayes'
> law on the one hand, or of those terminological tricks in the theory of
> likelihood on the other, which seem to avoid the necessity for the use
> of Bayes' law, but which in reality transfer the responsibility for its use
> to the working statistician, or the person who ultimately employs his
> results.
>
> *Norbert Wiener (1948)*

To the best of our knowledge, Norbert Wiener never actually applied Bayes' theorem in a
published work; yet he perceived the logical necessity of its use as soon as one builds beyond
the sampling distributions involved in his own statistical work. In the present chapter we
examine some of the consequences of failing to use Bayesian methods in some very simple
problems, where the paradoxes of Chapter 15 never arise.

In Chapter 16 we noted that the orthodox objections to Bayesian methods were always philosophical or ideological in nature, never examining the actual results that they
give, and we expressed astonishment that mathematically competent persons would use
such arguments. In order to give a fair comparison, we need to adopt the opposite
tactic here, and concentrate on the demonstrable facts that orthodoxians never mention. Since Bayesian methods have been so egregiously misrepresented in the orthodox
literature throughout our lifetimes, we must lean over backwards to avoid misrepresenting orthodox methods now; whenever an orthodox method does yield a satisfactory result in some problem, we shall acknowledge that fact, and we shall not deplore its use
merely on ideological grounds. On the other hand, when a common orthodox procedure leads to a result that insults our intelligence, we shall not hesitate to complain
about it.

Our present goal is to understand the following. *In what circumstances, and in what
ways, do the orthodox results differ from the Bayesian results? What are the pragmatic
consequences of this in real applications?* The theorems of Richard Cox provide all the
ideology we need, and all of our pragmatic comparisons only confirm, in many different
contexts, what those theorems lead us to expect.

509

17.1 Information loss

It is not easy to cover all this ground, because orthodox statistics is not a coherent body of theory that could be confirmed or refuted by a single analysis. It is a loose collection of independent *ad hoc* devices, invented and advocated by many different people on many different intuitive grounds; and they are often in sharp disagreement with each other.

But one can see generally, once and for all, when and why orthodox methods, quite aside from their failure to use prior information, must also waste some of the information in the data. Consider estimation of a parameter θ from a data set $D \equiv \{x_1, \ldots, x_n\}$ represented by a point in R^n. Orthodoxy requires us to choose a single estimator $b(D) \equiv b(x_1, \ldots, x_n)$ *before we have seen the data*, and then use only $b(x)$ for the estimation. Now, specifying the observed numerical value of $b(x)$ locates the sample on a manifold (subspace of R^n) of dimension $(n - 1)$. Specifying the actual data set D tells us that, and also where on the manifold we are. If position on the manifold is irrelevant to θ, then $b(D)$ is a sufficient statistic for θ, and unless there are further circumstances, such as highly cogent prior information, the orthodox method will be satisfactory pragmatically whatever its proclaimed rationale. Otherwise, specifying D conveys additional information about θ that is not conveyed by specifying the statistic $b(D)$.

Put differently, given the actual data set D, all estimators that the orthodoxian might have chosen $\{b_1, b_2, \ldots\}$ are known, so Bayes' theorem has available for its use simultaneously all the information contained in the class of all possible estimators. If there is no sufficient statistic, it is able to choose the optimal estimator *for the present data set*.

If the estimator is not a sufficient statistic, its sampling distribution is irrelevant for us, because with different data sets we shall use different estimators. We saw this in some detail, from different viewpoints, in Chapters 8 and 13. The same considerations apply to hypothesis testing; the Bayesian procedure has available all the relevant information in the data, but an orthodox procedure based on a single statistic does not unless it is a sufficient statistic. If it is not sufficient, then we expect that the Bayesian procedure will be superior (in the sense of more accurate or more reliable) because it is extracting more information from the data. Once one understands this, it is easy to produce any number of examples which demonstrate it.

From the Neyman–Pearson camp of orthodoxy we have the devices of unbiased estimators, confidence intervals, and hypothesis tests which amount to a kind of decision theory. This line of thought was adopted more or less faithfully in the works of Herbert Simon (1977) in economics, Erich Lehmann (1986) in hypothesis testing, and David Middleton (1960) in electrical engineering.

From the Fisherian (sometimes called the piscatorial) camp, there are the principles of maximum likelihood, analysis of variance, randomization in design of experiments, and a mass of specialized 'tail area' significance tests. Fortunately, the underlying logic is the same in all such significance tests, so they need not be analyzed separately. Adoption of these methods has been almost mandatory in biology and medical testing. Also, Fisher advocated fiducial probability, which most statisticians rejected, and conditioning on ancillary statistics, which we discussed in Chapter 8, and showed that it is mathematically equivalent to applying Bayes' theorem without prior information.

17.2 Unbiased estimators

Given a sampling distribution $p(x|\alpha)$ with some parameter α and a data set comprising n observations $D \equiv \{x_1, \ldots, x_n\}$, there are various orthodox principles for estimating α, in the particular use of an unbiased estimator, and maximum likelihood. In the former we choose some function of the observations $\beta(D) = \beta(x_1, \ldots, x_n)$ as our 'estimator'. The Neyman–Pearson school holds that it should be 'unbiased', meaning that its expectation over the sampling distribution is equal to the true value of α:

$$\langle \beta \rangle = E(\beta) = \int dx_1 \cdots dx_n \, \beta(x_1, \ldots, x_n) p(x_1 \cdots x_n | \alpha) = \alpha. \qquad (17.1)$$

As noted in Chapter 13, Eq. (13.20), the expected square of the error, over the sampling distribution, is the sum of two positive terms,

$$\langle (\beta - \alpha)^2 \rangle = (\langle \beta \rangle - \alpha)^2 + \mathrm{var}(\beta), \qquad (17.2)$$

where what the orthodoxian calls the 'sampling variance of β' (more correctly, the variance *of the sampling distribution for* β) is $\mathrm{var}(\beta) = \langle \beta^2 \rangle - \langle \beta \rangle^2$. At present, we are not after mathematical pathology of the kind discussed in Chapter 15 and Appendix B, but rather *logical* pathology – due to conceptual errors in the basic formulation of a problem – which persists even when all the mathematics is well-behaved. So we suppose that the first two moments of that sampling distribution, $\langle \beta \rangle$, $\langle \beta^2 \rangle$, exist for all the estimators to be considered. If we introduce a fourth moment $\langle \beta^4 \rangle$, we are automatically supposing that it exists also; this is the general mathematical policy advocated in Appendix B. Then an unbiased estimator has, indeed, the merit that it makes one of the terms of (17.2) disappear. But it does not follow that this choice minimizes the expected square of the error; let us examine this more closely.

What is the relative importance of removing bias and minimizing the variance? From (17.2) it would appear that they are of equal importance; there is no advantage in decreasing one if in so doing we increase the other more than enough to compensate. Yet that is what the orthodox statistician usually does! As the most common specific example, Cramér (1946, p. 351) considers the problem of estimating the variance μ_2 of a sampling distribution $p(x_1|\mu_2)$:

$$\mu_2 = \langle x_1^2 \rangle - \langle x_1 \rangle^2 = \langle x_1^2 \rangle \qquad (17.3)$$

from n independent observations $\{x_1, \ldots, x_n\}$. We assume, in (17.3) and in what follows, that $\langle x_1 \rangle = 0$, since a trivial change of variables would in any event accomplish this. An elementary calculation shows that the sample variance (now correctly called the variance *of the sample* because it expresses the variability of the data within the sample, and does not make reference to any probability distribution)

$$m_2 \equiv \overline{x^2} - \overline{x}^2 = \frac{1}{n} \sum_{i=1}^{n} x_i^2 - \left[\frac{1}{n} \sum_{i=1}^{n} x_i \right]^2 \qquad (17.4)$$

has expectation, over the sampling distribution $p(x_1 \cdots x_n | \mu_2) = p(x_1 | \mu_2) \cdots p(x_n | \mu_2)$, of

$$\langle m_2 \rangle = \frac{n-1}{n} \mu_2 \qquad (17.5)$$

and thus, as an estimator of μ_2, it has a negative bias. So, goes the argument, we should correct this by using the unbiased estimator

$$M_2 \equiv \frac{n}{n-1} m_2. \qquad (17.6)$$

Indeed, this has seemed so imperative that in most of the orthodox literature, the term 'sample variance' is *defined* as M_2 rather than m_2.

Now, of course, the only thing that really matters here is the *total* error of our estimate; the particular way in which you or I separate error into two abstractions labeled 'bias' and 'variance' has nothing to do with the actual quality of the estimate. So, let's look at the full mean-square error criterion (17.2) with the choices $\beta = m_2$ and $\beta = M_2$. Replacement of m_2 by M_2 removes a term $(\langle m_2 \rangle - \mu_2)^2 = \mu_2^2/n^2$, but it also increases the term var(m_2) by a factor $[n/(n-1)]^2$, so it seems obvious that, at least for large n, this has made things worse instead of better. More specifically, suppose we replace m_2 by the estimator

$$\beta \equiv c m_2. \qquad (17.7)$$

What is the best choice of c by orthodox criteria? The expected quadratic loss (17.2) is now

$$\langle (c m_2 - \mu_2)^2 \rangle = c^2 \langle m_2^2 \rangle - 2c \langle m_2 \rangle \mu_2 + \mu_2^2$$
$$= \langle (m_2 - \mu_2)^2 \rangle - \langle m_2^2 \rangle (\hat{c} - 1)^2 + \langle m_2^2 \rangle (c - \hat{c})^2, \qquad (17.8)$$

where

$$\hat{c} \equiv \frac{\mu_2 \langle m_2 \rangle}{\langle m_2^2 \rangle}. \qquad (17.9)$$

Evidently, the best estimator in the class (17.7) is the one with $c = \hat{c}$, and the term $-\langle m_2^2 \rangle$ $(\hat{c} - 1)^2$ in (17.8) represents the decrease in mean-square error obtainable by using $\hat{\beta} \equiv \hat{c} m_2$ instead of m_2. Another short calculation shows that

$$\langle m_2^2 \rangle = n^{-3}(n-1)[(n^2 - 2n + 3)\mu_2^2 + (n-1)\mu_4], \qquad (17.10)$$

where

$$\mu_4 \equiv \langle (x_1 - \langle x_1 \rangle)^4 \rangle = \langle x_1^4 \rangle \qquad (17.11)$$

is the fourth central moment of $p(x_1|\mu_2)$. We must understand $n > 1$ in all this, for if $n = 1$, we have $m_2 = 0$; in sampling theory, a single observation gives no information about the variance μ_2.[1]

From (17.5) and (17.10) we then find that $\hat{c}$ depends on the second and fourth moments of the sampling distribution:

$$\hat{c} = \frac{n^2}{n^2 - 2n + 3 + (n-1)K}, \tag{17.12}$$

where $K \equiv \mu_4/\mu_2^2 \geq 1$. We see that $\hat{c}$ is a monotonic decreasing function of K; so, if $K \geq 2$, (17.12) shows that $\hat{c} < 1$ for all n, instead of removing the bias in (17.5) we should always *increase* it!

In the case of a Gaussian distribution, $p(x|\mu_2) \propto \exp\{-x^2/2\mu_2\}$, we find $K = 3$. We will seldom have $K < 3$, for that would imply that $p(x|\mu_2)$ cuts off even more rapidly than Gaussian for large x. If $K = 3$, (17.12) reduces to

$$\hat{c} = \frac{n}{n+1}, \tag{17.13}$$

which, by comparison with (17.6), says that rather than removing the bias we should approximately double it, in order to minimize the mean-square sampling error.

How much better is the estimator $\hat{\beta} = \hat{c}m_2$ than M_2? In the Gaussian case the mean-square error of the estimator $\hat{\beta}$ is

$$\langle(\hat{\beta} - \mu_2)^2\rangle = \frac{2\mu_2^2}{n+1}. \tag{17.14}$$

The unbiased estimator M_2 corresponds to the choice

$$c = \frac{n}{n-1} \tag{17.15}$$

and thus to the mean-square error

$$\langle(M_2 - \mu_2)^2\rangle = \mu_2^2\left[\frac{2}{n+1} + \frac{2}{n}\right], \tag{17.16}$$

which is over twice the amount incurred by use of $\hat{\beta}$.[2] Most sampling distributions that arise in practice, if not Gaussian, have wider tails than Gaussian, so that $K > 3$; in this case the difference will be even greater.

Up to this point, it may have seemed that we are quibbling over a very small thing – changes in the estimator of one or two parts out of n. But now we see that the difference between (17.14) and (17.16) is not at all trivial. For example, *with the unbiased estimator M_2 you will need $n = 203$ observations in order to get as small a mean-square sampling*

---

[1] In Bayesian theory a single observation could give information about μ_2 if μ_2 is correlated, in the joint prior probability $p(\mu_2\theta|I)$, with some other parameter θ in the problem about which a single observation does give information; that is, $p(\mu\theta|I) \neq p(\mu|I)p(\theta|I)$. This kind of indirect information transfer can be helpful in problems where we have cogent prior information but only sparse data.

[2] Editor's footnote: It appears that Jaynes inadvertently calculated this expectation using $\langle m_2^2\rangle(c - c^*)$ rather than $\langle M_2^2\rangle(c - c^2)$ and so arrived at (17.16) rather than $\langle(M_2 - \mu_2)^2\rangle = 2\mu_2^2/(n-1)$.

error as the biased estimator $\hat{\beta}$ gives you with only 100 observations. This is typical of the way orthodox methods waste information; in this example we have, in effect, thrown away half of our data whatever the value of n, and therefore wasted half the work expended in acquiring the data.

R. A. Fisher, who often thought in terms of information, perceived this long ago; but modern orthodox practitioners seem never to perceive it, because they continue to fantasize about frequencies, and do not think in terms of information at all.[3] There is a work on econometrics (Valavanis, 1959, p. 60) where the author attaches such great importance to removing bias that he advocates throwing away not just half the data but practically all of them, if necessary, to achieve this.

Why do orthodoxians put such exaggerated emphasis on bias? We suspect that the main reason is simply that they are caught in a psychosemantic trap of their own making. When we call the quantity $(\langle \beta \rangle - \alpha)$ the 'bias', that makes it sound like something awfully reprehensible, which we must get rid of at all costs. If it had been called instead the 'component of error orthogonal to the variance', as suggested by the Pythagorean form of (17.2), it would have been clear to all that these two contributions to the error are on an equal footing; it is folly to decrease one at the expense of increasing the other. This is just the price one pays for choosing a technical terminology that carries an emotional load, implying value judgments; orthodoxy falls constantly into this tactical error.

Chernoff and Moses (1959) give a more forceful example showing how an unbiased estimate may be far from what we want. A company is laying a telephone cable across San Francisco Bay. They cannot know in advance exactly how much cable will be needed, and so they must estimate. If they overestimate, the loss will be proportional to the amount of excess cable to be disposed of; but if they underestimate, and the cable end falls into the water, the result may be financial disaster. Use of an unbiased estimate here could only be described as foolhardy; this shows why a Wald-type decision theory is needed to fully express rational behavior.

Another reason for such an undue emphasis on bias is a belief that if we draw N successive samples of n observations each and calculate the estimators $\beta_1, \ldots, \beta_N$, the average $\overline{\beta} = N^{-1} \sum \beta_i$ of these estimates will converge in probability to $\langle \beta \rangle$ as $N \to \infty$, and thus an unbiased estimator will, on sufficiently prolonged sampling, give an arbitrarily accurate estimate of α. Such a belief is almost never justified, even for the fairly well-controlled measurements of the physicist or engineer, not only because of unknown systematic error, but because successive measurements lack the logical independence required for these limit theorems to apply.

In such uncontrolled situations as economics, the situation is far worse; there is, in principle, no such thing as 'asymptotic sampling properties' because the 'population' is always finite, and it changes uncontrollably in a finite time. The attempt to use only sampling

---

[3] Note that this difficulty does not arise in the Bayesian approach, in spite of a mathematical similarity. Again choosing any function $\beta(x_1, \ldots, x_n)$ of the data as an estimator, and letting the brackets $\langle \ \rangle$ stand now for expectations over the posterior pdf for α, we have the expected square of the error of $\langle (\beta - \alpha)^2 \rangle = (\beta - \langle \alpha \rangle)^2 + \text{var}(\alpha)$, rather like (17.2). But now changing the estimator β does not change $\text{var}(\alpha) = (\langle \alpha^2 \rangle - \langle \alpha \rangle^2)$, and so, by this criterion, the optimal estimator over the class of *all* estimators is always $\beta = \langle \alpha \rangle$.

distributions – always interpreted as limiting frequencies – in such a situation forces one to expend virtually all his efforts on irrelevant fantasies. What is relevant to inference is not any imagined (that is, nonobserved) frequencies, but *the actual state of knowledge that we have about the real situation*. To reject that state of knowledge – or any human information – on the grounds that it is 'subjective' is to destroy any possibility of finding useful results; for human information is all we have.[4]

Even if we accept these limit theorems uncritically, and believe faithfully that our sampling probabilities are also the limiting frequencies, unbiased estimators are not the only ones which approach perfect accuracy with indefinitely prolonged sampling. Many biased estimators approach the true value of α in this limit, *and do it more rapidly*. Our $\hat{\beta}$ is an example. Furthermore, asymptotic behavior of an estimator is not really relevant, because the real problem is always to do the best we can with a finite data set; therefore the important question is not *whether* an estimator tends to the true value, but *how rapidly* it does so.

Long ago, R. A. Fisher disposed of the unbiased estimate by a different argument that we noted in Chapter 6, Eq. (6.94). The criterion of bias is not really meaningful, because it is not invariant under a change of parameters; the square of an unbiased estimate of α is not an unbiased estimate of α^2. With higher powers α^k, the difference in conclusions can become arbitrarily large, and nothing in the formulation of a problem tells us which choice of k is 'right'. Thus, if you and I happen to choose k differently, the criterion of an unbiased estimate will lead us to different conclusions about α from the same data. However, many orthodoxians simply ignore these ambiguities (although they can hardly be unaware of them) and continue to use unbiased estimators whenever they can, aware that they are violating a rather basic principle of rationality, but unaware that they are also wasting information.[5]

Note, however, that, after all this argument, nothing in the above entitles us to conclude that $\hat{\beta}$ is the best estimator of μ_2 by the criterion of mean-square sampling error! We have considered only the restricted class of estimators (17.7) constructed by multiplying the sample variance (17.4) by some preassigned number; we can say only that $\hat{\beta}$ is the best one in that class. The question whether some other function of the sample values, not a multiple of (17.4), might be still better by the criterion of mean-square sampling error, remains completely open. That the orthodox approach to parameter estimation does not tell us how to find the best estimator, but only how to compare different intuitive guesses, was noted in Chapter 13 following Eq. (13.21); and we showed that the difficulty is overcome

---

[4] 'Objectivity' in inference consists, then, in carefully considering all the information we have about the real situation, and carefully avoiding fantasies about situations that do not actually exist. It seems to us that this should have been obvious to orthodoxians from the start, since it was obvious already to ancient writers such as Herodotus (*c* 500 BC) in his discussion of the policy decisions of the Persian kings.

[5] We noted in Chapter 6, Eqs. (6.94)–(6.98) that the Bayesian criterion of the posterior expectation has potentially the same ambiguity; different definitions of parameters will lead to different conclusions if we continue to use the criterion of posterior expectation after a parameter transformation. Curiously, this problem did not arise with Laplace's original criterion of posterior median and quartiles. But these were not entirely correct applications of Bayesian theory. When we completed the theoretical apparatus with the decision theory of Chapter 13, a transformation of parameters was accompanied by a corresponding transformation of the loss function, with the result that our final substantive conclusions are now invariant under arbitrary parameter redefinitions.

by a slight reformulation of the problem, which leads inexorably to the Bayesian algorithm as the one which accomplishes what we really want.

Exercise 17.1. Try to extend sampling theory to deal with the many questions left unanswered by the orthodox literature and the above discussion. Is there a general theory of optimal sampling theory estimators for finite samples? If so, does bias play any role in it? We know already, from the analysis in Chapter 13, that this cannot be a variational theory; but it seems conceivable that a theory somewhat like dynamic programming might exist. In particular, can you find an orthodox estimator that is better than $\hat{\beta}$ by the mean-square error criterion? Or can you prove that $\hat{\beta}$ cannot be improved upon within sampling theory?

In contrast to the difficulty of these questions in sampling theory, we have noted above and in Chapter 13 that the Bayesian procedure automatically constructs the optimal estimator for any data set and loss function, whether or not a sufficient statistic exists; and it leads at once to a simple variational proof of its optimality not within any restricted class, but with respect to *all* estimators. And it does this without making any reference to the notion of bias, which plays no role in Bayesian theory.

17.3 Pathology of an unbiased estimate

On closer examination, an even more disturbing feature of unbiased estimates appears. Consider the Poisson sampling distribution: the probability that, in one time unit, we observe n events, or 'counts', is

$$p(n|l) = \exp(-l)\frac{l^n}{n!}, \qquad n = 0, 1, 2, \ldots, \qquad (17.17)$$

in which the parameter l is the sampling expectation of n, $\langle n \rangle = l$. Then what function $f(n)$ gives an unbiased estimate of l? Evidently, the choice $f(n) = n$ will achieve this; to prove that it is unique, note that the requirement $\langle f(n) \rangle = l$ is

$$\sum_{n=0}^{\infty} \exp(-l)\frac{l^n}{n!}f(n) = l, \qquad (17.18)$$

and, from the formula for coefficients of a Taylor series, this requires

$$f(n) = \frac{d^n}{dl^n}\{l\exp(l)\}\Big|_{l=0} = n. \qquad (17.19)$$

A reasonable result. But suppose we want an unbiased estimator of some function $g(l)$; by the same reasoning, the unique solution is

$$f(n) = \frac{d^n}{dl^n}\{\exp(l)g(l)\}\Big|_{l=0}. \qquad (17.20)$$

Thus the only unbiased estimator of l^2 is

$$f(n) = \begin{cases} 0 & n = 0, 1 \\ n(n-1) & n > 1, \end{cases} \tag{17.21}$$

which is absurd for $n = 1$. Likewise, the only unbiased estimator of l^3 is absurd for $n = 1, 2$; and so on. Here the unbiased estimator does violence to elementary logic; if we observe $n = 2$, we are advised to estimate $l^3 = 0$; but if l^3 were zero, it would be impossible to observe $n = 2!$ The only unbiased estimator of $\exp\{-l\}$ is

$$f(n) = \begin{cases} 1 & n = 0 \\ 0 & n > 0, \end{cases} \tag{17.22}$$

which is absurd for all positive n. An unbiased estimator for $(1/l)$ does not exist; it is mathematically pathological. Unbiased estimators can stand in conflict with deductive logic not just for a few data sets, but for all data sets. And if they can generate such pathology even in such a simple problem as this, what horrors await us in more complicated problems?

The remedy

In contrast, with uniform prior the Bayesian posterior mean estimate of any function $g(l)$ is

$$\langle g(l) \rangle = \frac{1}{n!} \int_0^\infty dl \, \exp(-l) l^n g(l), \tag{17.23}$$

which is readily verified to be mathematically well-behaved and intuitively reasonable for all the above examples. The Bayes estimate of $(1/l)$ is just $(1/n)$; no pathology here. It is at first surprising that the Bayes estimate of $\exp\{-l\}$ is

$$f(n) = 2^{-(n+1)}. \tag{17.24}$$

Why would it not be just $\exp\{-n\}$? To see why, note the following points.

(1) The posterior distribution for l is skewed; the posterior probability that $l > n$ is

$$P(l > n) = \int_n^\infty dl \, \exp\{-l\}\frac{l^n}{n!} = \exp(-n) \sum_{m=0}^n \frac{n^m}{m!}. \tag{17.25}$$

This decreases monotonically from 1 at $n = 0$ to $1/2$ as $n \to \infty$. Thus, given n, the parameter l is always more likely to be greater than n than less.
(2) The posterior distribution for l is proportional to $\exp\{-l\}l^n$, which is concentrated mostly in the interval $(n \pm \sqrt{n})$. But $\exp\{-l\}$ is so rapidly varying that, in calculating its expectation, most of the contribution to the integral $\int dl \, \exp\{-2l\}l^n$ comes from the region $(n/2 \pm \sqrt{n}/2)$; so $\exp\{-n/2\}$ would be closer to the correct estimator than $\exp\{-n\}$. Both of these circumstances affect the numerical value, in such as way that (17.24) finally emerges as the balance between these opposing tendencies. This is still another example where Bayes' theorem detects a genuinely complicated

situation and automatically corrects for it, but in such a slick, efficient way that one is unaware of what is happening.

Exercise 17.2. Consider the truncated Poisson distribution:

$$p(n|l) = \left[\frac{1}{\exp(l) - 1}\right]\frac{l^n}{n!}, \qquad n = 1, 2, \ldots. \tag{17.26}$$

Show that the unbiased estimator of l is now absurd for $n = 1$, and the unbiased estimator of $\exp(-l)$ is absurd for all even n and queer for all odd n.

Many other examples are known in which the attempt to find unbiased estimates leads to similar pathologies; several were noted by the orthodoxians Kendall and Stuart (1961). But their anti-Bayesian indoctrination was so strong that they would not deign to examine the corresponding Bayesian results; and so they never did learn that in all their cases Bayesian methods overcome the difficulty easily.[6]

17.4 The fundamental inequality of the sampling variance

A famous inequality, variously associated with the names Cramér, Rao, Darmois, Frechét and others, finds a lower bound to the sampling variance that can be achieved for any estimator – or, indeed, any statistic – with a continuous sampling distribution. Although the result is nearly trivial mathematically, it is important because it is almost the only bit of connected theory that orthodoxy has to guide it. An extensive discussion with examples is given by Cramér (1946, Chap. 32). Denote a data set of n observations by $x \equiv \{x_1, \ldots, x_n\}$ and integration over the sample space by $\int dx \, (\ )$. With a sampling distribution $p(x|\alpha)$ containing a parameter α, let

$$u(x, \alpha) \equiv \frac{\partial \log p(x|\alpha)}{\partial \alpha}. \tag{17.27}$$

Mathematically, the result we seek is just the Schwartz inequality: given two functions $f(x)$, $g(x)$ defined on the sample space, write $(f, g) \equiv \int dx \, f(x)g(x)$. Then $(f, g)^2 \le (f, f)(g, g)$ with equality if and only if $f(x) = qg(x)$, where q is a constant independent of x, although it may depend on α.[7] Now make the choices

$$f(x) \equiv u(x, \alpha)\sqrt{p(x|\alpha)}, \qquad g(x) \equiv [\beta(x) - \langle\beta\rangle]\sqrt{p(x|\alpha)}. \tag{17.28}$$

We find that $(f, g) = \langle\beta u\rangle - \langle\beta\rangle\langle u\rangle = \langle\beta u\rangle$, since $\langle u\rangle = \int dx \, u(x, \alpha)p(x|\alpha) = \partial/\partial\alpha$ $[\int dx \, p(x|\alpha)] = 0$. Likewise, $(f, f) = \text{var}(u)$, and $(g, g) = \text{var}(\beta)$, so the Schwartz

---

[6] Maurice Kendall could have learned this in five minutes from Harold Jeffreys, whom he saw almost daily for years because they were both Fellows of St John's College, Cambridge, and ate at the same high table.

[7] Proof: $\int dx \, [f(x) - qg(x)]^2 \ge 0$ for all constants q, in particular for the value $q = (f, g)/(g, g)$ which minimizes the integral. Then we have equality if and only if $f(x) - qg(x) = 0$. Note that this remains true whatever the range of integration; it need not be the entire sample space.

inequality reduces to

$$\langle \beta u \rangle \le \sqrt{\mathrm{var}(\beta)\mathrm{var}(u)}. \tag{17.29}$$

But $\langle \beta u \rangle = \int dx\, \beta \partial p(x|a)/\partial \alpha = d\langle \beta \rangle / d\alpha = 1 + b'(\alpha)$, where $b(\alpha) \equiv (\langle \beta \rangle - \alpha)$ is the bias of the estimator. Thus the famous inequality sought is

$$\mathrm{var}(\beta) \ge \frac{[1 + b'(\alpha)]^2}{\int d\alpha\, (\partial \log p(x|\alpha)/\partial \alpha)^2 p(x|\alpha)}. \tag{17.30}$$

Now, substituting (17.27) into the necessary and sufficient condition for equality ($f = qg$) and making a change of parameters ($\alpha \to l$), where l is defined by $q(\alpha) = -\partial l/\partial \alpha$, we have

$$\frac{\partial \log p(x|\alpha)}{\partial \alpha} = -l'(\alpha)[\beta(x) - \langle \beta \rangle], \tag{17.31}$$

and, integrating over α, the condition for equality becomes

$$\log p(x|a) = -l(\alpha)\beta(x) + \int dl\, \langle \beta \rangle + \mathrm{const.} \tag{17.32}$$

To put this into more familiar notation, note that the integral in (17.32) is a function of α; let us call it $-\log Z(\alpha)$, defining the function $Z(\alpha)$. Likewise, the constant of integration in (17.32) is independent of α but may depend on x; so call it $\log m(x)$, defining the function $m(x)$. With these changes of notation, the necessary and sufficient condition for equality in (17.30) becomes

$$p(x|\alpha) = \frac{m(x)}{Z(l)} \exp\{-l(\alpha)\beta(x)\}. \tag{17.33}$$

But we recognize this as just the distribution that we found in Chapter 11, produced by the maximum entropy principle with a constraint fixing $\langle \beta(x) \rangle$. In (17.33) the denominator $Z(l)$ is evidently a normalizing constant, therefore equal to

$$Z(l) = \int dx\, m(x) \exp\{-l\beta(x)\}, \tag{17.34}$$

whereupon the constraint is just

$$\langle \beta \rangle = -\frac{\partial \log(Z)}{\partial l}, \tag{17.35}$$

which is identical with (11.60). This generalizes at once to the case where α, β are vectors of any dimensionality, the exponent becoming $\{-\sum l_i(\alpha)\, \beta_i(x)\}$ as in (11.43); so we are just rediscovering the maximum entropy formalism of Chapter 11!

These results teach us something very important about the basic unity and mutual consistency of several principles that had seemed, up till now, distinct from each other. We noted in Chapter 14 that the notion of sufficiency, which was always associated with the notion of information, is in fact definable in terms of Shannon's information measure of entropy. Long ago, the Pitman–Koopman theorem (Koopman, 1936; Pitman, 1936) proved that the condition for existence of a sufficient statistic is just that the sampling distribution be of the functional form (17.32). Therefore, if we use the maximum entropy principle to assign sampling distributions, this automatically generates the distributions with the most desirable properties from the standpoint of inference in either sampling theory (because the sampling variance of an estimator is then the minimum possible value) or Bayesian theory (because then in applying Bayes' theorem we need only calculate a single function of the data).

Indeed, if we think of a maximum entropy distribution as a sampling distribution parameterized by the Lagrange multipliers l_j, we find that the sufficient statistics are precisely the data images of the constraints that were used in defining that distribution. Thus, the maximum entropy distribution generated from the set of constraints fixing $\{\langle \beta_1(x) \rangle, \langle \beta_2(x) \rangle, \ldots, \langle \beta_k(x) \rangle\}$ as expectations over the probability distribution, has k sufficient statistics which are just $\{\beta_1(x), \ldots, \beta_k(x)\}$, in which now x is the observed data set. This is proved in Jaynes (1978, Eq. B82); we leave it as an exercise for the reader to reconstruct the proof.

If the sampling distribution does not have the form (17.33) or its generalization, there are two possibilities. Firstly, if the sampling distribution is continuous in α, then the lower bound (17.28) cannot be attained, and there seems to be no theory to determine the correct lower bound, much less to construct an estimator that achieves it. Then if $\langle \beta \rangle$ is unbiased, the ratio of the minimum possible variance, right-hand side of (17.30), to the actual var (β) was called by Fisher the *efficiency* of the estimator β, and an estimator with efficiency of one was called an *efficient estimator*. Nowadays it is usually called an 'unbiased minimum variance' (UMV) estimator.[8]

Secondly, if $p(x|\alpha)$ has discontinuities, Cramér (1946, p. 485) finds that there are estimators that actually achieve a lower variance than (17.28). But how is this possible, since the Schwartz inequality does not seem to admit to any exceptions? We consider this a mathematical error, for reasons explained in Appendix B (had Cramér approached a discontinuous function as the limit of a sequence of continuous ones, another term, a delta-function, would have appeared in the limit, which just accounts for the discrepancy and makes the inequality (17.30) correct whether $p(x|\alpha)$ is continuous or discontinuous). This is a typical case where failure to recognize the necessary role of delta-functions in analysis leads one into errors.

17.5 Periodicity: the weather in Central Park

A common problem, important in economics, meteorology, geophysics, astronomy and many other fields, is to decide whether certain data taken over time provide evidence for a periodic behavior. Any clearly discernible periodic component (in births, diseases, rainfall, temperature, business cycles, stock market, crop yields, incidence of earthquakes, brightness of a star) provides an evident basis for improved prediction of future behavior,

---

[8] Note that the notion of efficiency is even more parameter-dependent than that of an unbiased estimate; if an efficient estimator of α exists, then an efficient estimator of α^2 does not.

on the presumption (that is, inductive reasoning) that periodicities observed in the past are likely to continue in the future. But even apart from prediction, the principle for analyzing the data for evidence of periodicity in the past is still controversial: is it a problem of significance tests, or one of parameter estimation? Different schools of thought come to opposite conclusions from the same data.

Consider an example from the recent literature of orthodox reasoning and procedure here; this will also provide an easy introduction to Bayesian spectrum analysis. Bloomfield (1976, p. 110) gives a graph showing mean January temperatures observed over about 100 years in Central Park, New York. The presence of a periodicity of roughly 20 years with a peak-to-peak amplitude of about $4\,°F$ is perfectly evident to the eye, since the irregular 'noise' is only about $0.5\,°F$. Yet Bloomfield, applying an orthodox significance test introduced by Fisher, concludes that there is no significant evidence for any periodicity!

17.5.1 The folly of pre-filtering data

In trying to understand this we note first that the data of Bloomfield's graph have been 'pre-filtered' by taking a 10 year moving average. What effect does this have on the evidence for periodicity? Let the original raw data be $D = \{y_1, \ldots, y_n\}$ and consider the discrete Fourier transform

$$Y(\omega) \equiv \sum_{t=1}^{n} y_t \exp\{i\omega t\}. \tag{17.36}$$

This is well-defined for continuous values of ω and is periodic: $Y(\omega) = Y(\omega + 2\pi)$. Therefore there is no loss of information if we confine the frequency to $|\omega| < \pi$. But even that is more than necessary; the values of $Y(\omega)$ at any n consecutive and discrete 'Nyquist' frequencies[9]

$$\omega_k \equiv 2\pi k/n, \quad 0 \le k < n, \tag{17.37}$$

already contain all the information in the data, for by the orthogonality $n^{-1} \sum_k \exp\{i\omega_k(s - t)\} = \delta_{st}$, the data can be recovered from them by the Fourier inversion:

$$\frac{1}{n} \sum_{k=1}^{n} Y(\omega_k) \exp\{-i\omega_k t\} = y_t, \quad 1 \le t \le n. \tag{17.38}$$

Suppose the data were replaced with an m year moving average over past values, with weighting coefficient of w_s for lag s:

$$z_t \equiv \sum_{s=0}^{m-1} y_{t-s} w_s. \tag{17.39}$$

---

[9] Harry Nyquist was a mathematician at the Bell Telephone Laboratories who, in the 1920s, discovered a great deal of the fundamental physics and information theory involved in electrical communication. The work of Claude Shannon is a continuation, 20 years later, of some of Nyquist's pioneering work. All of it is still valid and indispensable in modern electronic technology. In Chapter 7 we have already considered the fundamental, irreducible 'Nyquist noise' in electrical circuits due to random thermal motion of electrons.

The new Fourier transform would be, after some algebra,[10]

$$Z(\omega) = \sum_{t=1}^{n} z_t \, \exp\{i\omega t\} = W(\omega)Y(\omega), \tag{17.40}$$

where

$$W(\omega) \equiv \sum_{s=0}^{m-1} w_s \exp\{i\omega s\} \tag{17.41}$$

is the Fourier transform of the weighting coefficients. This is just the convolution theorem of Fourier theory. Thus, taking any moving average of the data merely multiplies its Fourier transform by a known function. In particular, for uniform weighting

$$w_s = \frac{1}{m}, \qquad 0 \le s < m, \tag{17.42}$$

we have

$$W(\omega) = \frac{1}{m} \sum_{s=0}^{m-1} \exp\{-i\omega s\} = \exp\left\{-i\frac{\omega}{2}(m-1)\right\}\left[\frac{\sin(m\omega/2)}{m \sin(\omega/2)}\right]. \tag{17.43}$$

In the case $m = 10$ we find, for a 10 year and 20 year periodicity, respectively,

$$W(2\pi/10) = 0; \qquad W(2\pi/20) = 0.639\exp\{-9\pi i/20\}. \tag{17.44}$$

Thus, taking a 10 year moving average of any time series data represents an irreversible loss of information; it completely wipes out any evidence for a 10 year periodicity, and reduces the amplitude of a 20 year periodicity by a factor 0.639, while shifting its phase by $9\pi/20 = 1.41$ radian. In addition, the magnitude of $W(\omega)$ is decreasing at $\omega = 2\pi/20$ so the apparent frequency is shifted; the peak in $Z(\omega)$ occurs at a lower frequency than the true peak in $Y(\omega)$. We conclude that the original data had a periodicity of roughly 20 years with a peak-to-peak amplitude of about $4/0.639 = 6.3\,°F$, even more obvious to the eye and nearly 90 degrees out of phase with the periodicity visible in Bloomfield's graph; and the true frequency is somewhat higher than one would estimate from the graph. Taking the moving average has severely mutilated and distorted the information in the data.

At several places we warn against the common practice of pre-filtering data in this way before analyzing them. The only thing it can possibly accomplish is the cosmetic one of making the graph of the data look prettier to the eye. But if the data are to be analyzed

---

[10] At this point, many authors get involved in a semantic hangup over exactly what one means by the term 'm year moving average' for a series of finite length. If we have only y_t for $t > 0$, then it seems to many that the m year moving average (17.39) could start only at $t = m$. But then they find that their formulas are not exact, but require small 'end-effect' correction terms of order m/n. We avoid this by a slight change in definitions. Consider the original time series $\{y_t\}$ augmented by 'zero-padding'; we define $y_t \equiv 0$ when $t < 1$ or $t > n$, and likewise the weighting coefficients are defined to be zero when $s < 0$ or $s \ge m$. Then we may understand the above sums over t, s to be over $(-\infty, +\infty)$, and the first few terms $(z_1, \ldots, z_{m-1})$, although averages over m years of the padded data, are actually averages over less than m years of nonzero data. The differences are numerically negligible when $m \ll n$, but we gain the advantage that the simple formulas (17.36)–(17.42) with sums taken instead over $\pm\infty$ and t in (17.39) allowed to take all positive values, are all exact as they stand, without our having to bother with messy correction terms. Furthermore, it is evident that failure to do this means that some of the information in the first m and last m data values is lost. This particular definition of the term 'moving average' for a finite series (which was basically arbitrary anyway) is thus the one appropriate to the subject.

by a computer, this does not help in any way; it only throws away or distorts some of the information that the computer could have extracted from the original, unaltered data. It renders the filtered data completely useless for certain purposes. For all we know, there might have been a strong periodicity of about 10 years in the original data, corresponding to the well known 11 year periodicity in sunspot numbers; but, if so, taking a 10 year moving average has wiped out the evidence for it.

The periodogram of the data is then the power spectral density:

$$P(\omega) \equiv \frac{1}{n}|Y(\omega)|^2 = \frac{1}{n}\sum_{t,s} y_t y_s \exp\{i\omega(t-s)\}. \tag{17.45}$$

Note that $P(0) = (\sum y_t)^2/n = n\bar{y}^2$ determines the mean value of the data, while the average of the periodogram at the Nyquist frequencies is the mean-square value of the data:

$$P(\omega_k)_{\text{av}} = \frac{1}{n}\sum_{k=1}^{n} P(\omega_k) = \overline{y^2}. \tag{17.46}$$

Fisher's proposed test statistic for a periodicity is the ratio of peak/mean of the periodogram:

$$q = \frac{P(\omega_k)_{\text{max}}}{P(\omega_k)_{\text{av}}}, \tag{17.47}$$

and one computes its sampling distribution $p(q|H_0)$ conditional on the null hypothesis H_0 that the data are Gaussian white noise. Having observed the value q_0 from our data, we find the so-called 'P-value', which is the sampling probability, conditional on H_0, that chance alone would have produced a ratio as great or greater:

$$P \equiv p(q > q_0|H_0) = \int_{q_0}^{\infty} dq \, p(q|H_0), \tag{17.48}$$

and if $P > 0.05$ the evidence for periodicity is rejected as 'not significant at the 5% level'.[11]

This test looks only at probabilities conditional on the 'null hypothesis' that there is no periodic term. It takes no note of probabilities of the data conditional on the hypothesis that a periodicity is present; or on any prior information indicating whether it is reasonable to expect a periodicity! We commented on this kind of reasoning in Chapter 5; how can one test any hypothesis rationally if he fails to specify (1) the hypothesis to be tested; (2) the alternatives against which it is to be tested; and (3) the prior information that we bring to the problem? Until we have done that much, we have not asked any definite, well-posed question.

Equally puzzling, how can one expect to find evidence for a phenomenon that is real, if he starts with all the cards stacked overwhelmingly against it? The only hypothesis H_0 that this test considers is one which assumes that the totality of the data are part of a

---

[11] This is a typical orthodox 'tail area' significance test; we discussed such tests in Chapter 9, and noted that the orthodox chi-squared test has serious shortcomings, but there is a similar Bayesian ψ-test that is exact and is free of those difficulties. Many other Bayesian ψ-tests can be set up, which test some hypothesis H against a specified class C of alternatives. But now we note a different way of looking at this situation that is generally more useful: a significance test can often be replaced by a parameter estimation problem that is simpler and more informative.

'stationary Gaussian random process' *without* any periodic component. According to that H_0, the appearance of anything resembling a sine wave would be purely a matter of chance; even if the noise conspires, by chance, to resemble one cycle of a sine wave, it would still be only pure chance – equally unlikely according to the orthodox sampling distribution – that would make it resemble a second cycle of that wave; and so on.

In almost every application one can think of, our prior knowledge about the real world tells us that in speaking of 'periodicity' we have in mind some systematic physical influence that repeats itself; indeed, our interest in it *is due entirely to the fact that we expect it to repeat.*[12] Thus we expect to see some periodicity in the weather because we know that this is affected by periodic astronomical phenomena; the rotation of the Earth on its axis, its yearly orbital motion about the Sun, and the observed periodicity in sunspot numbers, which affect atmospheric conditions on the Earth. So the hypothesis H_1 that we want to test for is quite unrelated to the hypothesis H_0 that is used in Fisher's test.[13]

This is the kind of logic that underlies all orthodox significance tests. In order to argue for an hypothesis H_1 that some effect exists, one does it indirectly: invent a 'null hypothesis' H_0 that denies any such effect, then argue against H_0 in a way that makes no reference to H_1 at all (that is, using only probabilities conditional on H_0). To see how far this procedure takes us from elementary logic, suppose we decide that the effect exists; that is, we reject H_0. Surely, we must also reject probabilities conditional on H_0; but then what was the logical justification for the decision? Orthodox logic saws off its own limb.[14]

Harold Jeffreys (1939, p. 316) expressed his astonishment at such limb-sawing reasoning by looking at a different side of it: 'An hypothesis that may be true is rejected because it has failed to predict observable results that have not occurred. This seems a remarkable procedure. On the face of it, the evidence might more reasonably be taken as evidence for the hypothesis, not against it.'

Thus, if we say that there is a periodicity in temperature, we mean by this that there is some periodic physical influence at work, the nature of which may not be known with certainty, but about which we could make some reasonable conjectures. For example, the aforementioned periodicity in solar activity, already known to occur by the 11 year periodic variation in sunspot numbers (which many believe, with good reason, to be a rectified

---

[12] It is not necessary for successful prediction that the physical cause of the periodicity be actually understood; in ancient India records of eclipses were maintained carefully over centuries. From these observations they 'got the rhythm of it' and were able to predict future eclipses very accurately, although they had no conception of their causes.

[13] If an apparent periodicity were only a momentary artifact of the noise as supposed by H_0, we would not consider it a real periodicity at all, and would not want our statistical test to take any note of it. But, unfortunately, it is always possible for noise artifacts to appear momentarily real to any test one can devise. The remedy is to check whether the apparent effect is reproducible; a noise artifact will in all probability never occur again in the same way. A physicist can, almost always, use this remedy easily; an economist usually cannot.

[14] An historical study has suggested that the culprit who started this kind of reasoning was not a statistician, but the physicist Arthur Schuster (1897), who invented the periodogram for the purpose of refuting some claims of periodicity in earthquakes in Japan. Never thinking in terms of information, he achieved his preconceived goal by the simple device of analyzing the data in a way that *threw away* the information about that periodicity! But then this was taken up by many others, including Fisher, Feller, Blackman and Tukey, and Bloomfield. Nevertheless, we shall see that the periodogram does contain basic information that Schuster and his followers failed to recognize. They thought that the information was contained in the *sampling distribution* for the periodogram; whereas the analysis given here shows that it was actually contained in the *shape* of the periodogram.

22 year periodicity), would cause a periodic variation in the number of charged particles entering our atmosphere (the reality of this is shown by the observed periodic variations in the *aurora borealis*), varying the ion concentration and therefore the number of raindrop condensation centers. This would cause periodic variations in the cloud cover, and hence in the temperature and rainfall, which might be very different in different locations on the Earth because of prevailing atmospheric circulation patterns.

We do not mean to say that we firmly believe this mechanism to be the dominant one; only that it is a conceivable one, which does not violate any known laws of physics, but whose magnitude is difficult to estimate theoretically. But already, this prior information prepares us not to be surprised by a periodic variation in temperature in Central Park somewhat like that observed[15] and leads us to conjecture that the July temperatures might give even better evidence for periodicity.

Once a data set has given mild evidence for such a periodicity, its reality could be definitely confirmed or refuted by other observations, correlating other data (astronomical, atmospheric electricity, fish populations, etc.) with weather data at many different locations. A person trained only in orthodox statistics would not hesitate to consider all these phenomena 'independent'; a scientist with some prior knowledge of astrophysics and meteorology would not consider them independent at all.

If editors of scientific journals refuse to publish that first mild evidence on the grounds that it is not significant *in itself* by an orthodox significance test at the 5% level, the confirmatory observations will, in all probability, never be made; a potentially important discovery could be delayed by a century. Physicists and engineers have been largely spared from such fiascos because they hardly ever took orthodox teachings seriously anyway; but others working in economics, artificial intelligence, biology, or medical research who, in the past, allowed themselves to be cowed by Fisher's authority, have not been so fortunate.

Contrast our position just stated with that of Feller (II, p 76–77), who delivers another polemic against what he calls the 'old wrong way'. Suppose the data are expanded in sinusoids:

$$y_t = \sum_{j=1}^{n} (A_j \cos \omega_j t + B_j \sin \omega_j t). \tag{17.49}$$

We can always approximate y_t this way. Then it seems that A_j, B_j must be 'random variables' if the $\{y_t\}$ are. Feller warns us against that old wrong way: fit such a series to the data with well-chosen frequencies $\{\omega_1, \ldots, \omega_n\}$ and assume all A_j, $B_j \sim N(0, \sigma)$. If one of the $R_j^2 = A_j^2 + B_j^2$ is big, conclude that there is a true period. He writes of this:

For a time it was fashionable to introduce models of this form and to detect 'hidden periodicities' for sunspots, wheat prices, poetic creativity, etc. Such hidden periodicities used to be discovered as easily as witches in medieval times, but even strong faith must be fortified by a statistical

---

[15] One who was also aware of the roughly 20 year periodicity in crop yields, well known to Kansas wheat farmers for a century, would be even less surprised.

test. A particularly large amplitude R_j is observed; One wishes to prove that this cannot be due to chance and hence that ω_j is a true period. To test this conjecture one asks whether the large observed value of R is plausibly compatible with the hypothesis that all n components play the same role.

Apparently, Feller did not even believe in the sunspot periodicity, which no responsible scientist has doubted for over a century; the evidence for it is so overwhelming that nobody needs a 'statistical test' to see it. He states that the usual procedure was to assume the A_j, B_j iid normal $N(0, \sigma)$.[16] Then the R_j^2 are held to be independent with an exponential distribution with expectation $2\sigma^2$. 'If an observed value R_j^2 deviated 'significantly' from this predicted expectation it was customary to jump to the conclusion that the hypothesis of equal weights was untenable, and R_j represented a 'hidden periodicity'. At this point, Feller detects that we are using the wrong sampling distribution:

> The fallacy of this reasoning was exposed by R. A. Fisher who pointed out that the maximum among n independent observations does not obey the same probability distribution as each variable taken separately. The error of treating the worst case statistically as if it had been chosen at random is still common in medical statistics, but the reason for discussing the matter here is the surprising and amusing connection of Fisher's test of significance with covering theorems.

Feller then states that the quantities

$$V_j = \frac{R_j^2}{\sum R_i^2}, \qquad 1 \le j \le n, \tag{17.50}$$

are 'distributed' as the lengths of the n segments into which the interval $(0,1)$ is partitioned by a random distribution of $n - 1$ points. The probability that all $V_j < a$ is then given by a covering theorem noted by Feller.

Of course, our position is that both Feller's 'old wrong' and 'new right' sampling distributions are irrelevant to the inference; the two quantities that are relevant (the prior information that expresses our knowledge of the phenomenon and the likelihood function that expresses the evidence of the data) are not even mentioned.

In any event, the bottom line of this discussion is that Fisher's test fails to detect the perfectly evident 20 year periodicity in the New York Central Park January temperatures. But this is not the only case where simple visual examination of the data is a more powerful tool for inference than the principles taught in orthodox textbooks. Crow, Davis and Maxfield (1960) present applications of the orthodox F-test and t-test which we examine in Jaynes (1976) with the conclusions that (1) the eyeball is a more reliable indicator of an effect than an orthodox equal-tails test, and (2) the Bayesian test confirms quantitatively what the eyeball sees qualitatively. This is also relevant to the notions of domination and admissibility discussed elsewhere.

---

[16] The abbreviation 'iid' is orthodox jargon standing for 'independently and identically distributed'. For us, this is another form of the mind projection fallacy. In the real world, each individual coefficient A_j, B_j is a definite, fixed quantity that is known from the data; it is not 'distributed' at all!

17.6 A Bayesian analysis

Now we examine a Bayesian analysis of these same data, and for pedagogical reasons we want to explain its rationale in some detail. There may be various different Bayesian treatments of data for periodicity, corresponding to different information about the phenomenon, expressed by different choices of a model, and different prior information concerning the parameters in a model. Our Bayesian model is as follows. We consider it possible that the temperature data have a periodic component due to some systematic physical influence on the weather:

$$A \cos \omega t + B \sin \omega t, \tag{17.51}$$

where, as noted, we may suppose $|\omega| \leq \pi$ (with yearly data it does not make sense to consider periods shorter than a year). In addition the data are contaminated with variable components e_t that we call 'irregular' because we cannot control them or predict them and therefore cannot make allowance for them. This could be because we do not know their real causes or because, although we know the causes, we lack the data on initial conditions that would enable predictions.[17] Then, as explained in Chapter 7, it will almost always do justice to the real prior information that we have to assign a Gaussian sampling distribution with parameters (μ, σ) to the irregulars. There is hardly any real problem in which we would have the detailed prior information that would justify any more structured sampling distribution.

Thus μ is the 'nominal true mean temperature' not known in advance; we can estimate it from the data very easily (intuition can see already that the mean value of the data $\overline{y}$ is about as good an estimate of μ that we can make from the information we have); but it is not of present interest and so we treat it as a nuisance parameter. We do not know σ in advance either, although we can easily estimate it too from the data. But that is not our present interest, and so we shall let σ also be a nuisance parameter to be integrated out as explained in Chapter 7. Our model equation for the data is then

$$y_t = \mu + A \cos \omega t + B \sin \omega t + e_t, \qquad 1 \leq t \leq n, \tag{17.52}$$

and our sampling distribution for the irregular component is

$$p(e_1 \cdots e_n | \mu \sigma I) = \left(\frac{1}{2\pi\sigma^2}\right)^{n/2} \exp\left\{-\frac{1}{2\sigma^2}\sum_t e_t^2\right\}. \tag{17.53}$$

Then the sampling (density) distribution for the data is

$$p(y_1 \cdots y_n | \mu \sigma I) = \left(\frac{1}{2\pi\sigma^2}\right)^{n/2} \exp\left\{-\frac{Q}{2\sigma^2}\right\} \tag{17.54}$$

---

[17] In meteorology, although the principles of thermodynamics and hydrodynamics that determine the weather are well-understood, weather data taken on a 50 mile grid are grossly inadequate to predict the weather 24 hours in advance. Partial differential equations require an enormous amount of information on initial conditions to determine anything like a unique solution.

with the quadratic form

$$Q(A, B, \omega) \equiv \sum (y_t - \mu - A \cos \omega t - B \sin \omega t)^2 \qquad (17.55)$$

or

$$Q = n \left[\overline{y^2} - 2\overline{y}\mu + \mu^2 - 2A\overline{y_t \cos \omega t} - 2B\overline{y_t \sin \omega t} + 2\mu A\overline{\cos \omega t} \right.$$
$$\left. + 2\mu B\overline{\sin \omega t} + 2AB\overline{\cos \omega t \sin \omega t} + A^2\overline{\cos^2 \omega t} + B^2\overline{\sin^2 \omega t} \right], \qquad (17.56)$$

where all the overbar symbols denote sample averages over t. A great deal of detail has suddenly appeared that was not present in the orthodox treatment; but now all of this detail is actually *relevant to the inference.* In any nontrivial Bayesian solution we may encounter much analytical detail because every possible contingency allowed by our information is being taken into account (as is required by our basic desiderata in Chapters 1 and 2). Most of this detail is not perceived at all by orthodox principles, and it would be difficult to handle by paper-and-pencil calculation.

In practice, a Bayesian learns to recognize that much of this detail actually makes a negligible difference to the final conclusions, and so we can almost always make such good approximations that we can do the special calculation needed for our present purpose with pencil and paper after all. But, fortunately, masses of details are no deterrent to a computer, which can happily grind out the exact solution.[18] Now, in the present problem, (A, B, ω) are the interesting parameters that we want to estimate, while (μ, σ) are nuisance parameters to be eliminated. We see that of the nine sums in (17.56), four involve the data y_t; and since this is the only place where the data appear, these four sums are the jointly sufficient statistics for all the five parameters in the problem. The other five sums can be evaluated analytically once and for all, before we have the data.

Now, what is our prior information? Surely, we knew in advance that A, B must be less than 200 °F. If there were a temperature variation that large, New York City would not exist; there would have been a panic evacuation of that area long before, by anyone who happened to wander into it and survived long enough to escape. Thus the empirical fact that New York City *exists* is highly cogent information relevant to the question being asked; it is already sufficient to ensure proper priors for (A, B) in the Bayesian calculation. Also, we have no prior information about the phase $\theta = \tan^{-1}(B/A)$ of any periodicity, which we express by a uniform prior over θ.

We could cite various other bits of relevant prior information, but we know already from the results found in Chapter 6, Exercise 6.6, that, unless we have prior information that reduces the possible range to something like 30 °F, it will make a numerically negligible difference in the conclusions (a strictly nil difference if we report our conclusions only to three decimal digits). So let us see what Bayesian inference gives with just this. By an

---

[18] Indeed, the exact general solution is often easier to program than is any particular special case of it or approximation to it, because one need not go into the details that make the case special. And the program for the exact solution has the merit of being crash-proof if written to prevent underflow or overflow (for approximations will almost surely break down for some data sets, but the exact solution – with proper priors – must always exist for every possible data set.

argument essentially the same as the Herschel derivation of the Gaussian distribution in Chapter 7, we may assign a joint prior

$$p(AB|I) = \frac{1}{2\pi\delta^2} \exp\left\{-\frac{A^2 + B^2}{2\delta^2}\right\},$$ (17.57)

where δ is of the order of magnitude of $100\,°F$; we anticipate that its exact numerical value can have no visible effect on our conclusions (nevertheless, such a proper prior may be essential to prevent computer crashes).

Now the most general application of Bayes' theorem for this problem would proceed as follows. We first find the joint posterior distribution for all five parameters:

$$p(AB\omega\mu\sigma|DI) = p(AB\omega\mu\sigma|I)\frac{p(D|AB\omega\mu\sigma I)}{p(D|I)}.$$ (17.58)

Then integrate out the nuisance parameters:

$$p(AB\omega|DI) = \int d\mu \int d\sigma\, p(AB\omega\mu\sigma|DI).$$ (17.59)

But this is a far more general calculation than we need for present purposes; it is prepared to take into account arbitrary correlations in the prior probabilities. Indeed, we can always factor the prior thus:

$$p(AB\omega\mu\sigma|I) = p(AB\omega|I)p(\mu\sigma|AB\omega I);$$ (17.60)

and so the most general solution appears formally simpler:

$$p(AB\omega|DI) = Cp(AB\omega|I)L^*(A, B, \omega),$$ (17.61)

where C is a normalization constant, and L^* is the quasi-likelihood

$$L^*(A, B, \omega) \equiv \int d\mu \int d\sigma\, p(\mu\sigma|AB\omega I)p(D|AB\omega\mu\sigma I).$$ (17.62)

In (17.61) the nuisance parameters are already out of sight. But in our present problem, evidently knowledge of the parameters (A, B, ω) of the systematic periodicity would tell us nothing about the parameters (μ, σ) of the irregulars; so the prior for the latter is just

$$p(\mu\sigma|AB\omega I) = p(\mu\sigma|I),$$ (17.63)

so what is our prior information about (μ, σ)? Surely we knew also, for the same 'panic evacuation' reason, that neither of these parameters could be as large as $200\,°F$. And we know that σ could not be as small as $10^{-6}\,°F$, because, after all, our data are taken with a real thermometer, and no meteorologist's thermometer can be read to that accuracy (if it could, it would not give reproducible readings to that accuracy). We could just as well ignore that practical consideration and argue that σ could not be as small as $10^{-20}\,°F$ because the concept of temperature is not defined, in statistical mechanics, to that accuracy. Numerically, it will make no difference at all in our final conclusions, but it is still conceivable that a proper prior may be needed to avoid computer crashes in all contingencies. So, to be on the safe side, we assign the prior Gaussian in μ, because it is a location parameter, a truncated Jeffreys prior for σ, because we have seen in Chapter 12 that the Jeffreys prior is uniquely determined as the only completely uninformative prior for a scale parameter:

$$p(\mu\sigma|I) \propto \frac{1}{\sigma\sqrt{2\pi\alpha^2}} \exp\{-\mu^2/2\alpha^2\}, \qquad a \leq \sigma \leq b, \qquad (17.64)$$

in which α and b are also of the order of $100\,°\mathrm{F}$, while $a \simeq 10^{-6}$; we are only playing it extremely safe in the expectation that most of this care will prove in the end to have been unnecessary.

Our quasi-likelihood is then

$$L^*(A, B, \omega) = \int_{-\infty}^{\infty} d\mu \, \exp\{-\mu^2/2\alpha^2\} \int_a^b \frac{d\sigma}{\sigma^{n+1}} \exp\{-Q/2\sigma^2\}. \qquad (17.65)$$

But now it is evident that the finite limits on σ are unnecessary; for if $n > 0$ the integral over σ converges both at zero and infinity, and

$$\int_0^{\infty} \frac{d\sigma}{\sigma^{n+1}} \exp\{-Q/2\sigma^2\} = \frac{1}{2}\frac{(n/2-1)!}{(Q/2)^{n/2}}, \qquad (17.66)$$

and the integral of this over μ is also guaranteed to converge. For tactical reasons, let us do the integration over μ first. We begin by rewriting Q as

$$Q = n[s^2 + (\mu - \overline{d})^2]. \qquad (17.67)$$

Editor's Exercise 17.3(a) The equation for Q is formally identical to (7.29); however, as written, none of the quantities were defined by Jaynes. Show that s^2 may be written as

$$s^2 \equiv \overline{d^2} - \overline{d}^2, \qquad (17.68)$$

where $\overline{d}$ and $\overline{d^2}$ are the mean and mean-square of an effective data defined as

$$d_i = y_i - A\cos(\omega t_i) - B\sin(\omega t). \qquad (17.69)$$

(b) Evaluate the integral over u and σ to obtain the marginal $p(AB\omega|DI)$.

(c) Unfortunately, the $p(AB\omega|DI)$ does not summarize all of the information in the data concerning frequency estimation, to do that we need $p(\omega|DI)$; derive it in closed form.

(d) The posterior probability $p(\omega|DI)$ makes the implicit assumption that a resonance is present and so will estimate the frequency regardless of whether or not such a resonance exists. How would you use probability theory and the results derived so far to determine if a resonance is present?[19]

---

[19] For an example of such a signal detection statistic, see my article: Bretthorst, G. L. (1990), *J. Mag. Resonance* **88**, 571–595.

17.7 The folly of randomization

Many writers introduce randomized methods by the example of 'Monte Carlo integration'. Let a function $y = f(x)$ have its domain of existence in the unit square $0 \leq x, y \leq 1$; we wish to compute numerically the integral

$$\theta \equiv \int_0^1 dx \, f(x). \tag{17.70}$$

Perhaps this is too complicated analytically, or perhaps $f(x)$ was only empirically determined; we do not have it in analytical form. Then let us just choose n points at random (x, y) in the unit square and determine for each whether it lies below the graph of $f(x)$; that is, whether $y \leq f(x)$. Let the number of such points be r; then we estimate the integral as $(\theta)_{\text{est}} = r/n$ and as $n \to \infty$ we might expect this to approach the correct Riemannian integral; but how accurate is it? Always, one would suppose independent binomial sampling: the sampling distribution for r is taken to be

$$p(r|n\theta) = \binom{n}{r} \theta^r (1 - \theta)^{n-r} \tag{17.71}$$

which has (mean) $\pm$ (standard deviation) of

$$\theta \pm \sqrt{\frac{\theta(1 - \theta)}{n}}, \tag{17.72}$$

and if the width of the sampling distribution is held to indicate the accuracy of our estimate, one would think it reasonable to assign a probable error to $(\theta)_{\text{est}}$ given by

$$(\theta)_{\text{est}} = \frac{r}{n} \pm \sqrt{\frac{r(n - r)}{n^3}}. \tag{17.73}$$

For example, suppose the true θ is 1/2, and $n = 100$. Then, having observed $r = 43$ we would get the estimate of

$$(\theta)_{\text{est}} = 0.43 \pm \sqrt{\frac{0.43 \times 0.57}{n}} = 0.43 \pm 0.05, \tag{17.74}$$

or an accuracy of about 11.5%. But the trouble with such methods is that they improve only as $1/\sqrt{n}$.

Now let's take our n sampling points in a nonrandomized way on a uniform grid: divide the unit square into $\sqrt{n}$ steps each way, take one sampling point at each grid point, and again count how many (r) are below the curve. The maximum error we can make in each step is

$$[\text{error in determining } f(x)] \times [\text{width of step}] = \frac{1}{2\sqrt{n}} \times \frac{1}{\sqrt{n}} = \frac{1}{2n}. \tag{17.75}$$

Therefore the maximum possible error in the integral is

$$[\text{number of steps}] \times [\text{maximum error in each step}] = \frac{1}{2\sqrt{n}}. \tag{17.76}$$

So if $\theta \simeq 0.5$, the *probable error* in the Monte Carlo method is about equal to the *maximum possible error* in the uniform grid sampling method. But the probable error in the uniform grid method is much less than this: the central limit theorem tells us that, with a rectangular distribution of error probability in each step, the expected square of the error in determining $f(x)$ in that step is

$$\sqrt{n} \int_0^{\sqrt{n}} dx \left(x - \frac{1}{2\sqrt{n}} \right)^2 = \frac{1}{12n^2}. \tag{17.77}$$

If the errors in different steps are independent, the expected square of the total error is

$$\text{(mean square error per step)} \times \text{(number of steps)} = \frac{1}{12n^{3/2}}, \tag{17.78}$$

and the probable error in the integral is about

$$\pm \frac{1}{\sqrt{12}n^{3/4}}. \tag{17.79}$$

Thus, if $n = 100$ and $\theta \simeq 0.5$, the Monte Carlo method gives a probable error of about 0.05, the uniform grid sampling gives 0.00913, less than one-fifth as much. With $n = 1000$, the Monte Carlo probable error is 0.0158, the uniform grid probable error is 0.00162, about one-tenth as much. The uniform grid calculation at $n = 100$ points yields the same probable error as does the Monte Carlo method at $n = 3000$ points. This corresponds rather nicely to the italicized statement following (17.16). Another example is given by Royall and Cumberland (1981); this is particularly cogent because the authors are not Bayesian and did not start out with the intention of exposing the folly of randomization, but did so anyway.

17.8 Fisher: common sense at Rothamsted

From the study of several such examples, we propose as a general principle: *Whenever there is a randomized way of doing something, there is a nonrandomized way that yields better results from the same data, but requires more thinking.* Perhaps this principle does not have quite the status of a theorem, but we are confident that, whenever one is willing to do the required extra thinking, it will be confirmed.

17.8.1 The Bayesian safety device

We note that Bayesian methods are not only more powerful than orthodox ones; they are also safer (i.e. they have automatic built-in safety devices that prevent them from misleading us with the over-optimistic or over-pessimistic conclusions that orthodox methods can produce). It is important to understand why this is true. In parameter estimation, for example, whether or not there is a sufficient statistic, the log-likelihood function is

$$\log L(\alpha) = \sum_{i=1}^{n} \log p(x_i | \alpha) = n\overline{\log p(x_i | \alpha)}, \tag{17.80}$$

in which we see the average of the log-likelihoods over each individual data point. The log-likelihood is always spread out over the full range of variability of the data, so if we happen to get a very bad (spread out) data set, no good estimate is possible and Bayes' theorem warns us about this by returning a wide posterior distribution. With a location parameter $p(x|\alpha) = h(x - \alpha)$ and an uninformative prior, the width of the posterior distribution for α is essentially $(R + W)$

$$\text{(range of the data)} + \text{(width of individual likelihoods)}. \tag{17.81}$$

If we happen to get a very good (sharply concentrated) data set, a more accurate estimate of α is possible and Bayes' theorem takes advantage of this, returning a posterior distribution whose width approaches a lower bound determined by that of the single-point likelihood $L_i(\alpha) = p(x_i|\alpha)$ and the amount n of data.

In the orthodox method, the accuracy claim is essentially the width of the sampling distribution for whatever estimator β we have chosen to use. But this takes no note of the range of the data! Orthodox estimation based on a single statistic will claim just the same accuracy whether the data range is large or small. Far worse, that accuracy expresses entirely the variability of the estimator *over other data sets that we think might have been obtained but were not*. But again this concentrates attention on an irrelevancy, while ignoring what is relevant; unobserved data sets are only a figment of our imagination. Surely, if we are only imagining them, we are free to imagine anything we please. That is, given two proposed conjectures about unobserved data, what is the test by which we could decide which one is correct?

In spite of its mathematical triviality, we stress the fundamental importance of (17.80) for demonstrating the inner mechanism of Bayes' theorem. It clarifies several other questions often raised about Bayesian methods. We note one of the most important.

17.9 Missing data

This is a problem that does not exist for us; Bayesian methods work by the same algorithm whatever data we have. For example, in estimating a parameter θ from a data set $D \equiv \{x_i\}$, where the indices i refer to the times $\{t_i\}$ of observation and take on values in some set T, the data affect the result through the likelihood function $L(\theta)$, given by

$$\log L(\theta) = \sum_{i \in T} \log p(x_i|\theta), \tag{17.82}$$

where the sum is over whatever data values we have. The point is that, whether the times $\{t_i\}$ are consecutive and equally spaced, or completely irregular with large gaps, makes no difference; probability theory as logic tells us that (17.82) yields the optimal inference, which captures all the evidence in the data set that we happen to have. One can write a single computer program which, once and for all, accepts whatever data (that is, whatever set of numbers $\{x_i; t_i\}$) we give it, and proceeds to do the correct calculation for that data set.

In contrast, note what happens in orthodox statistics, where estimation is obliged to proceed through the sampling distribution of some 'statistic' $\theta^*(x_i)$. If any data are missing from the set T which was assumed in setting up the problem, one has two ways of dealing with this. Firstly, the theoretically correct procedure would recognize that this not only changes the sampling distribution for any statistic; it requires one to go back to the beginning and reconsider the whole problem. This can get us into a horrendous situation – every different kind of missing data or extra data can oblige us to define a new sample space, choose a new statistic θ^{**} and calculate a new sampling distribution $p(\theta^{**}|\theta)$.

Alternatively, one can invent a new *ad hockery* and try to estimate the missing data values from the ones we have, and use these as if they were real data. Obviously, this procedure is not only unjustified logically, it is highly ambiguous because that estimation could be made in many different ways. These difficulties are seen at first hand in Little and Rubin (1987).

The missing data problem was so cumbersome in orthodox statistics that some who saw the light and moved over into the Bayesian camp failed to perceive that they had left this problem behind them. Instead of applying such simple rules as (17.82) directly, which would have led them immediately to the correct solution whatever the data, they proceeded out of force of habit to follow the orthodox custom and invent new *ad hoc* devices like the one just noted, as 'corrections' to the Bayesian or maximum entropy methods, thus grotesquely mutilating them and getting a worse inference by a bigger computation. To those accustomed to orthodox difficulties, the power and simplicity of the Bayesian method in this application seems unbelievable; and one must think long and hard to understand how it is possible.

As a more general comment, there is a simple strategy that will serve in almost all of these Bayes/orthodox comparisons: 'magnification', as demonstrated for the chi-squared test in Chapter 9. When we find a quantitative difference in the orthodox and Bayesian conclusions, it may appear at first glance so small that our common sense is unable to judge the issue. But then we can usually find some extreme problem in which the small difference is magnified into a large one, or preferably to a qualitative difference in the conclusions. Our common sense will then tell us very clearly which procedure is giving reasonable results, and which one is not. Indeed, it is often possible to magnify to the point where one procedure is yielding an obvious violation of deductive reasoning or pathology like that noted above for an unbiased estimate. Now we examine another very important example where we can compare orthodox and Bayesian results by magnification.

17.10 Trend and seasonality in time series

The observed time series generated by the real world seldom appear to be 'stationary' but exhibit more complicated behavior. In most series, particularly demographic or economic data, trend is the most common form of nonstationarity. Many economic time series are so dominated by trend (due, for example, to steadily rising population, inflation, or techno-logical advances) that any attempt to study other regularities, such as cyclical fluctuations or settling back after response to a shock – and, particularly, correlations between different time series – can be more misleading than helpful until we have a safe way of dealing with trend.

The story is told – perhaps apocryphal – of a researcher who announced the discovery of a strong positive correlation between membership in the Church of England and incidence of suicide in England, and concluded that it would be safer to keep away from the Church. The true explanation was, of course, that the population of England was growing steadily, so membership in the Church, incidence of suicides – and almost any other demographic variable – were all growing together. False correlations of this type have led many to nonsensical conclusions because of the almost universal tendency to jump to the conclusion that correlation implies causation.

The problem of contaminated data has been with us from the very beginning. We noted in Chapter 9 how Edmund Halley (1693) was obliged to deal with it in compiling the first tables of mortality. The real key to dealing with it is recognition of the useful functional role of nuisance parameters in probability theory.

Likewise, today many time series are so dominated by cyclic fluctuations (seasonal effects in economic data, periodicity in weather, hum in electrical circuits, synchronized growth in bacteriology, vibrations in helicopter blades) that the attempt to extract an underlying 'signal', such as a long-term trend from a short run of data, is frustrated. We want to contrast how orthodox statistics and probability theory as logic deal with the problem of extracting the information one wants, in spite of such data contaminations.

17.10.1 Orthodox methods

The traditional procedures do not apply probability theory to this problem; instead, one resorts to inventing the same kind of intuitive *ad hoc* devices that we have noted so often before. The usual ones are called 'detrending' and 'seasonal adjustment' in the economic literature, 'filtering' in the electrical engineering literature. Like all such *ad hockeries* not derived from first principles, they capture enough of the truth to be usable in some problems; but they can generate disaster in others.

The almost universal detrending procedure in economics is to suppose the data (or the logarithm of the data) to be $y(t) = x(t) + Bt + e(t)$, composed additively of the component of interest $x(t)$, a linear 'trend' Bt, and a 'random error' or 'noise' $e(t)$. We estimate the trend component, subtract it from the data, and proceed to analyze the resulting 'detrended data' for other effects. However, many writers have noted that detrending may introduce spurious artifacts that distort the evidence for other effects. Detrending may even destroy the relevance of the data for our purposes, and we saw in the scenario of the weather in Central Park that filtering of data can also do this.

Similarly, the traditional way of dealing with seasonal effects is to produce 'seasonally adjusted' data, in which one subtracts an estimate of the seasonal component from the true data, then tries to analyze the adjusted data for other effects. Indeed, most of the economic time series data one can obtain have been rendered nearly useless because they have been seasonally adjusted in an irreversible way that has destroyed information which probability theory could have extracted from the raw data. We think it imperative that this be recognized, and that researchers be able to obtain the true data – free of detrending, seasonal adjustment, pre-filtering, smoothing, or any other destructive mutilation of the information in the data.

Electrical engineers would think instead in terms of Fourier analysis and resort to 'high-pass filters' and 'band-rejection filters' to deal with trend and seasonality. Again, the philosophy is to produce a new time series (the output of the filter) which represents in some sense an estimate of what the real series would be if the contaminating effect were absent. Then choice of the 'best' physically realizable filter is a difficult and basically indeterminate problem; fortunately, intuition has been able to invent filters good enough to be usable if one knows in advance what kind of contamination will occur.

17.10.2 The Bayesian method

The direct application of probability theory as logic leads us to an entirely different philosophy; always, the correct procedure is to calculate the probability of whatever is unknown and of interest, conditional on whatever is known. This means that we do not seek to remove the trend or seasonal component from the data: that is fundamentally impossible because there is no way to know the 'true' trend or seasonal term. Any assumption about them is necessarily in some degree arbitrary, and is therefore almost certain to inject false information into the detrended or seasonally adjusted series. Rather, we seek to remove the effect of trend or seasonality from our *final conclusions*, taking into account all the relevant information we have, while leaving the actual data intact. We develop the Bayesian procedure for this and compare it in detail with the conventional one.

Firstly, we analyze the simplest possible nontrivial model, which can be solved completely from either point of view and will enable us to understand the exact relationship between the two procedures. Having this understanding, the extension to the most complicated multivariate case will be an easy mathematical generalization – essentially, just promotion of numbers to matrices while retaining the same formal equations.

Suppose the model consists of only a single sinusoid and a linear trend: $y(t) = A \sin \omega t + Bt + e(t)$, where A is the amplitude of interest to be estimated and B is the unknown trend rate. If the data are monthly economic data and the sinusoid represents a yearly seasonal effect, then ω will be $2\pi/12 = 0.524$ months^{-1}. But, for example, if we are trying to detect a cycle with a period of 20 years, ω will be $0.524/20 = 0.00262$. Estimation of an unknown ω from such data is the very important problem of spectrum analysis, considered in the scenario of the weather in Central Park. But for the present we consider the case where ω is known (usually, because we know that the seasonality has a period of one year for unchanging astronomical reasons). Writing for brevity $s_t = s(t) \equiv \sin(\omega t)$, our model equation is then

$$y(t) = As(t) + Bt + e(t), \qquad (17.83)$$

and the available data $y \equiv (y_1, \ldots, y_N)$ are values of this at N equally spaced times $t = 1, 2, \ldots, N$. Assigning – for reasons already explained sufficiently – the noise components

e_i an independent Gaussian prior probability density function $e_t \sim N(0, \sigma)$ with variance σ^2, the sampling pdf for the data is

$$p(y|AB\sigma) = \left(\frac{1}{2\pi\sigma^2}\right)^{N/2} \exp\left\{-\frac{N}{2\sigma^2} Q(A, B)\right\}, \tag{17.84}$$

and, as in any Gaussian calculation, the first task is to rearrange the quadratic form

$$Q(A, B) \equiv \frac{1}{N} \sum_t (y_t - As_t - Bt)^2 = \overline{y^2} + A^2\,\overline{s^2} + B^2\overline{t^2} - 2A\overline{sy} - 2B\overline{ty} + 2AB\overline{st}, \tag{17.85}$$

where

$$\overline{y^2} \equiv \frac{1}{N} \sum_{t=1}^{N} y_t^2, \qquad \overline{sy} \equiv \frac{1}{N} \sum_{t=1}^{N} s_t y_t, \tag{17.86}$$

etc. denote averages over the data sample. Three of these averages, $(\overline{s^2}, \overline{t^2}, \overline{st})$ are determined by the 'design of the experiment' and can be known before one has the data. In fact, we have nearly

$$\overline{s^2} \simeq 1/2, \qquad \overline{t^2} \simeq N^2/3 \tag{17.87}$$

with errors of relative order $O(1/N)$. But $\overline{st}$ is highly variable; it is certainly less than $N/2$, since that could be achieved only if $s(t) = 1$ at every sampling point. Generally, $\overline{st}$ is much less than this, of the order $\overline{st} \simeq 1/\omega$ due to near cancellation of positive and negative terms.

The other three averages $(\overline{y^2}, \overline{sy}, \overline{ty})$ depend on the data, and, since they are the only terms containing the data, they are the jointly sufficient statistics for our problem, to be calculated as soon as one has the data.

Suppose that it is the seasonal amplitude A that we wish to estimate, while the trend rate B is the nuisance parameter that contaminates our data. We want to make its effects disappear, as far as is possible. We shall do this by finding the joint posterior distribution for A and B,

$$p(AB|DI), \tag{17.88}$$

and integrating out B to get the marginal posterior distribution for A,

$$p(A|DI) = \int dB\, p(AB|DI). \tag{17.89}$$

This is the quantity that tells us everything the data D and prior information I have to say about A, whatever the value of B; this would be called 'Bayesian detrending'. Conversely, if we wanted to estimate B, then A would be the nuisance parameter, and we would integrate it out of (17.88) to get the marginal posterior distribution $p(B|DI)$; and this would be called 'Bayesian seasonal adjustment'.

In the limit of diffuse priors for A and B (i.e. their prior densities do not vary appreciably over the region of high likelihood), the appropriate integration formula for (17.89) is

$$\int_{-\infty}^{\infty} dB \, \exp\left\{ -\frac{N \, Q(A, B)}{2\sigma^2} \right\}$$

$$= (\text{const.}) \times \exp\left\{ -\frac{N}{2\sigma^2} \left[\frac{\left(\overline{s^2}\right)\left(\overline{t^2}\right) - \left(\overline{st}\right)^2}{\overline{t^2}} \right] (A - \hat{A})^2 \right\}, \tag{17.90}$$

where

$$\hat{A} \equiv \frac{\left(\overline{t^2}\right)\left(\overline{sy}\right) - \left(\overline{st}\right)\left(\overline{ty}\right)}{\left(\overline{s^2}\right)\left(\overline{t^2}\right) - \left(\overline{st}\right)^2} \tag{17.91}$$

and the (const.) is independent of A. Thus the marginal posterior distribution for A is proportional to (17.90), and the Bayesian posterior (mean) $\pm$ (standard deviation) estimate of A, regardless of the value of B, is

$$(A)_{\text{est}} = \hat{A} \pm \sigma \sqrt{\frac{\overline{t^2}}{N\left[\left(\overline{s^2}\right)\left(\overline{t^2}\right) - \left(\overline{st}\right)^2\right]}} = \hat{A} \pm \frac{\sigma}{\sqrt{N\overline{s^2}(1 - r^2)}}, \tag{17.92}$$

where

$$r \equiv \frac{\overline{st}}{\sqrt{\overline{s^2}\overline{t^2}}} \tag{17.93}$$

is the correlation coefficient of s and t.

Some orthodox writers have railed against this process of integrating out nuisance parameters – in spite of the fact that it is uniquely determined by the rules of probability theory as the correct procedure – on the usual grounds that the probability of a parameter is meaningless because a parameter is not a 'random variable'. Even worse, in the integration we introduced a prior that they consider arbitrary (although for us it represents our real state of prior information which is clearly relevant to the inference, but is ignored by orthodoxy). But, independently of all such philosophical hangups, we can examine the facts of actual performance of the Bayesian and orthodox procedures.

The integration of a nuisance parameter may be related to the detrending procedure as follows. The joint posterior pdf may be factored into marginal and conditional pdfs in two different ways:

$$p(AB|DI) = p(A|DI)p(B|ADI), \tag{17.94}$$

or, equally well,

$$p(AB|DI) = p(A|BDI)p(B|DI). \tag{17.95}$$

From (17.94) we see that (17.89) follows at once, and from (17.95) we see that (17.89) can be written as

$$p(A|DI) = \int dB\, p(A|BDI)p(B|DI). \qquad (17.96)$$

Thus the marginal pdf for A is a weighted average of the conditional pdfs that we would have if B were known:

$$p(A|BDI). \qquad (17.97)$$

But if B is known, then (17.97), in its dependence on A, is just (17.84) with B held fixed. This is, from (17.85),

$$p(A|BDI) \propto \exp\left\{-\frac{N\overline{s^2}}{2\sigma^2}(A - A^*)^2\right\}, \qquad (17.98)$$

where

$$A^* \equiv \frac{\overline{sy} - B\overline{st}}{\overline{s^2}}. \qquad (17.99)$$

This is just the estimate that one would make by ordinary least squares fitting of $As(t)$ to the detrended data $y(t)_{det} \equiv y(t) - Bt$

$$A^* = \frac{(\overline{sy})_{det}}{\left(\overline{s^2}\right)}. \qquad (17.100)$$

That is, A^* is the estimate the orthodoxian would make if he estimated the trend rate to be B. Of course, if his estimate of B were exactly correct, then he would indeed find the best estimate possible; but any error in his estimate of the trend rate will bias his estimate of A.

The Bayesian estimate of A obtained from (17.96) does not assume any particular trend rate B; it is a weighted average over all possible values that the trend rate might have, weighted according to their respective posterior probabilities. Thus, if the trend rate is very well determined by the data, so that the probability $p(B|DI)$ in (17.96) has a single very sharp peak at $B = B^*$, then the Bayesian and orthodoxian will be in essential agreement on the estimate of A if the orthodoxian also happens to estimate B as B^*. If the trend rate is not well determined by the data, then the Bayes estimate is a more conservative one that takes into account all possible values that B might have, while the orthodox estimate can vary widely.

Although an orthodoxian might accept what we have done as mathematically consistent, this argument would not convince him of the superiority of the Bayesian estimate (implicitly based, as usual, on a quadratic loss function), because he judges estimates by a different criterion. So let us compare them more closely.

17.10.3 Comparison of Bayesian and orthodox estimates

Having found a Bayesian estimator, which theorems demonstrate to be optimal by the Bayesian criterion of performance, nothing prevents us from examining its performance from the orthodox sampling theory viewpoint and comparing it with orthodox estimates. We introduce a useful method of doing this, which makes it clear what the two methods are doing. Let A_0 and B_0 be the unknown true values of the parameters, and let us describe the situation as it would appear to one who already knew A_0 and B_0, but not what data we shall find. As he would know, but we would not, our data vector will in fact be

$$y_t = A_0 s_t + B_0 t + e_t, \qquad (17.101)$$

and we shall calculate the statistic

$$\overline{sy} = A_0 \overline{s^2} + B_0 \overline{st} + \overline{se} \qquad (17.102)$$

in which the first two terms are fixed (i.e. independent of the noise) and only the last varies with different noise samples. Similarly, he knows that we shall find the statistic

$$\overline{ty} = A_0 \overline{st} + B_0 \overline{t^2} + \overline{te}. \qquad (17.103)$$

Although $\overline{sy}$ and $\overline{ty}$ are known to us from the data, we cannot solve (17.102) and (17.103) for (A_0, B_0) because $\overline{se}$ and $\overline{te}$ are unknown. We are obliged to continue using probability theory to get the best possible estimates of A_0, B_0. Substituting (17.102) and (17.103) into (17.91), we find that the Bayes estimate that we shall get reduces to

$$(\hat{A})_{\text{Bayes}} = A_0 + \frac{\left(\overline{t^2}\right)\left(\overline{se}\right) - \left(\overline{st}\right)\left(\overline{te}\right)}{\left(\overline{s^2}\right)\left(\overline{t^2}\right) - \left(\overline{st}\right)^2}, \qquad (17.104)$$

which is independent of the true trend rate, B_0 having cancelled out. Therefore the Bayesian estimate does indeed eliminate the effect of trend entirely from our conclusions; one could hardly do so more completely than that. But the unknown error vector e must, necessarily, produce some error in our estimate of A, and (17.104) tells us exactly how much.

On the other hand, if the orthodoxian uses the conventional ordinary least squares estimator (17.100) from detrended data $[y_t - \hat{B}t]$ based on any estimate $\hat{B}$, he will find instead

$$(\hat{A})_{\text{orthodox}} = A_0 + \frac{\overline{se} + (B_0 - \hat{B})\overline{st}}{\overline{s^2}}, \qquad (17.105)$$

and any error in the trend rate estimate $\hat{B}$ contributes to the error in his estimate of the seasonal component. If, as is the usual practice, one uses the ordinary least squares estimate of the trend from the original data,

$$\hat{B} = \frac{(\overline{ty})}{(\overline{t^2})},$$
(17.106)

(17.105) becomes

$$(\hat{A})_{\text{orthodox}} = A_0 + \frac{(\overline{t^2})(\overline{se}) - (\overline{ty})(\overline{st}) + B_0(\overline{t^2})(\overline{st})}{(\overline{t^2})(\overline{s^2})} = (1 - r^2)A_0 + \frac{(\overline{se})}{(\overline{s^2})},$$
(17.107)

where we have again used (17.102) and (17.103). Thus (17.107) is also exactly independent of the true trend rate B_0. But orthodox teaching would hold that the estimator (17.107) has a negative bias, since $\overline{se}$ is, 'on the average', zero. One might wish to 'correct' for this by the same device as in (17.6): by multiplying by a suitable factor. But this is not obviously the best procedure. It is far from clear that the optimal estimator can be found merely by multiplying the ordinary least squares estimate by a constant.

Likewise, having recognized what he would consider a shortcoming of (17.107), and perceiving that the Bayesian result (17.104) has at least the merit (from his viewpoint) of being unbiased, it still would not follow that the Bayesian solution is the best possible one. Indeed, one who has absorbed a strong anti-Bayesian indoctrination would, we suspect, reject any such suggestion and would say that we should be able to correct the defects of (17.107) by a little more careful thinking about the problem from the orthodox viewpoint. Let us try.

17.10.4 An improved orthodox estimate

Starting back at the beginning of the problem, orthodox reasoning proceeded as follows. If one had in mind only the seasonal term and was not aware of trend, one would be led to estimate the cyclic amplitude as

$$\hat{A}^{(0)} = \frac{(\overline{sy})}{(\overline{s^2})},$$
(17.108)

the conventional regression solution. Many different lines of reasoning, including ordinary least squares fitting of the data to the sinusoid As_t, lead us to this result.

But then one realizes that (17.108) is not a very good estimate because it ignores the disturbing effect of trend. A better seasonal estimate could be made from the detrended data

$$(y_t)_{\text{det}} \equiv y_t - \hat{B}t,$$
(17.109)

where $\hat{B}$ is an estimate of the trend rate, and it seems natural to estimate it by the conventional regression rule

$$\hat{B}^{(0)} = \frac{\overline{(ty)}}{\overline{(t^2)}} \tag{17.110}$$

from ordinary least squares fitting of a straight line Bt to the data. Using the detrended data (17.109) in (17.108) yields the 'corrected' cyclic amplitude estimate

$$\hat{A}^{(1)} = \frac{\overline{sy} - \overline{st}\,\hat{B}^{(0)}}{\overline{s^2}} = \frac{\left(\overline{t^2}\right)\left(\overline{sy}\right) - \left(\overline{st}\right)\left(\overline{ty}\right)}{\left(\overline{t^2}\right)\left(\overline{s^2}\right)} \tag{17.111}$$

which is the conventional orthodox result for the problem.

But now we see that this is not the end of the story; for A and B enter into the model on just the same footing. If it is true that we should estimate the cyclic amplitude A from detrended data $y_t - \hat{B}^{(0)}t$, surely it is equally true that we should estimate the trend rate B from the decyclized data $y_t - \hat{A}^{(0)}s_t$. Thus a better estimate of trend than (17.110) would be

$$\hat{B}^{(1)} = \frac{\overline{(ty)} - \overline{(st)}\,\hat{A}^{(0)}}{\overline{(t^2)}} = \frac{\left(\overline{s^2}\right)\left(\overline{ty}\right) - \left(\overline{st}\right)\left(\overline{sy}\right)}{\left(\overline{t^2}\right)\left(\overline{s^2}\right)}, \tag{17.112}$$

where we used (17.108). But now, with this better estimate of trend, we can obtain a better estimate of the seasonal component than (17.111) by using (17.112):

$$\hat{A}^{(2)} = \frac{\overline{(sy)} - \overline{(st)}\,\hat{B}^{(1)}}{\left(\overline{s^2}\right)}. \tag{17.113}$$

This improved estimate of the seasonal amplitude will in turn enable us to achieve a still better estimate of trend

$$\hat{B}^{(2)} = \frac{\overline{(ty)} - \overline{(st)}\,\hat{A}^{(1)}}{\left(\overline{t^2}\right)} \tag{17.114}$$

... and so on, forever!

Therefore, the reasoning underlying the conventional detrending procedure, if applied consistently, does not stop at the conventional result (17.100). It leads us into an infinite sequence of back-and-forth revisions of our estimates, each set $[\hat{A}^{(n)}, \hat{B}^{(n)}]$ better than the previous $[\hat{A}^{(n-1)}, \hat{B}^{(n-1)}]$. Does this infinite sequence converge to a final 'best of all' set of estimates $[\hat{A}^{(\infty)}, \hat{B}^{(\infty)}]$? If so, this is surely the optimal way of dealing with a nuisance parameter from the orthodox viewpoint. But can we calculate these final optimal estimates directly without going through the infinite sequence of updatings?

To answer this define the (2×1) vector of nth order estimates:

$$V_n \equiv \begin{pmatrix} \hat{A}^{(n)} \\ \hat{B}^{(n)} \end{pmatrix}. \tag{17.115}$$

Then the general recursion relation is, as we see from (17.111)–(17.113),

$$V_{n+1} = V_0 + M V_n, \tag{17.116}$$

where the matrix M is

$$M = \begin{pmatrix} 0 & -\dfrac{\left(\overline{st}\right)}{\left(\overline{s^2}\right)} \\[3mm] \dfrac{\left(\overline{st}\right)}{\left(\overline{t^2}\right)} & 0 \end{pmatrix}. \tag{17.117}$$

The solution of (17.116) is

$$V_n = (1 + M + M^2 + \cdots + M^n) V_0. \tag{17.118}$$

By the Schwartz inequality, $\left(\overline{st}\right)^2 \le \left(\overline{s^2}\right)\left(\overline{t^2}\right)$, the eigenvalues of M are less than unity, so as $n \to \infty$ this infinite series sums to

$$V_\infty = (I - M)^{-1} V_0. \tag{17.119}$$

Now we find readily that

$$(I - M)^{-1} = \frac{1}{\left(\overline{s^2}\right)\left(\overline{t^2}\right) - \left(\overline{st}\right)^2} \begin{pmatrix} \left(\overline{t^2}\right)\left(\overline{s^2}\right) & -\left(\overline{t^2}\right)\left(\overline{st}\right) \\[2mm] -\left(\overline{s^2}\right)\left(\overline{st}\right) & \left(\overline{t^2}\right)\left(\overline{s^2}\right) \end{pmatrix}, \tag{17.120}$$

and so our final, best of all, estimate is

$$\hat{A}^{(\infty)} = \frac{\left(\overline{t^2}\right)\left(\overline{s^2}\right)\hat{A}^{(0)} - \left(\overline{t^2}\right)\left(\overline{st}\right)\hat{B}^{(0)}}{\left(\overline{s^2}\right)\left(\overline{t^2}\right) - \left(\overline{st}\right)^2} = \frac{\left(\overline{t^2}\right)\left(\overline{sy}\right) - \left(\overline{st}\right)\left(\overline{ty}\right)}{\left(\overline{s^2}\right)\left(\overline{t^2}\right) - \left(\overline{st}\right)^2}. \tag{17.121}$$

But this is precisely the Bayesian estimate that we calculated far more easily in (17.92)! Likewise, the final best possible orthodox estimate of trend rate is

$$\hat{B}^{(\infty)} = \frac{\left(\overline{s^2}\right)\left(\overline{ty}\right) - \left(\overline{ty}\right)\left(\overline{sy}\right)}{\left(\overline{s^2}\right)\left(\overline{t^2}\right) - \left(\overline{st}\right)^2}, \tag{17.122}$$

which is just the Bayesian estimate that we find by integrating out A as a nuisance parameter from (17.88).

This is another example of what we found in Chapter 13: if the orthodoxian will think his estimation problems through to the end, he will find himself obliged to use the Bayesian mathematical algorithm, even if his ideology still leads him to reject the Bayesian rationale for it. Independently of all philosophical hangups, this mathematical form is determined by elementary requirements of rationality and consistency.

Now we see the relationship between the orthodox and Bayesian procedures in an entirely different light. The Bayesian procedure of integrating out a nuisance parameter is summing an infinite series of mutual updatings for us, and in such a slick way that, to the best of our knowledge, no orthodox writer has yet realized that this is what is happening. What we have just found is not limited to trend and seasonal parameters: it will generalize effortlessly to far more complex problems.

As we noted before (Jaynes, 1976) in many other cases, it is a common phenomenon that orthodox results, when improved to the maximum possible extent, become mathematically equivalent to the results that Bayesian methods give us far more easily. Indeed, it is one of the problems we have that Bayesian and maximum entropy methods are so easy that orthodoxians accuse us of trying to get something for nothing.

Thus, in the long run, attempts to evade the use of Bayes' theorem do not lead to different final results; they only make us work an order of magnitude harder to get them.

17.10.5 The orthodox criterion of performance

In our endeavor to understand this situation fully, let us examine it from a different viewpoint. According to orthodox theory, the accuracy of an estimation procedure is to be judged by the sampling distribution of the estimator, while in Bayesian theory it should be judged from the posterior pdf for the parameter. Let us compare these. For the orthodox analysis, note that in both (17.104) and (17.107) the terms containing the noise vector e combine to make a linear combination of the form

$$\overline{ge} \equiv \frac{1}{N} \sum_{t=1}^{N} g_t e_t. \tag{17.123}$$

Then over the sampling pdf for the noise we have

$$E(\overline{ge}) = \frac{1}{N} \sum_t g_t E(e_t) = 0 \tag{17.124}$$

$$E[(\overline{ge})^2] = \frac{1}{N} \sum g_t g_{t'} E(e_t e_{t'}) = \overline{g^2}\sigma^2, \tag{17.125}$$

since $E(e_t e_{t'}) = \sigma^2 \delta(t, t')$. Thus, the sampling pdf would estimate this error term by (mean) $\pm$ (standard deviation):

$$(\overline{ge})_{\text{est}} = 0 \pm \sigma\sqrt{\overline{g^2}}. \tag{17.126}$$

For the Bayes estimator (17.104)

$$g_t = \frac{\left(\overline{t^2}\right)(s_t) - \left(\overline{st}\right)t}{\left(\overline{t^2}\right)\left(\overline{s^2}\right) - \left(\overline{st}\right)^2}, \tag{17.127}$$

and after some algebra we find

$$\overline{g^2} = \frac{\left(\overline{t^2}\right)\left[\left(\overline{s^2}\right)\left(\overline{t^2}\right) - \left(\overline{st}\right)^2\right]}{\left(\overline{s^2}\right)\left(1 - r^2\right)}, \tag{17.128}$$

where r is the correlation coefficient defined before. Thus the sampling distribution for the Bayes estimator (17.104) has (mean) ± (standard deviation) of

$$\tilde{A} \pm \sigma \sqrt{\frac{1 - r^2}{\hat{N} s^2}}, \tag{17.129}$$

while for the orthodox estimator this is

$$(1 - r^2)\tilde{A} \pm \sigma \sqrt{\frac{1 - r^2}{N\left(\overline{s^2}\right)}}. \tag{17.130}$$

17.11 The general case

Having shown the nature of the Bayesian results from several different viewpoints, we now generalize them to a fairly wide class of useful problems. We assume that the N data are not necessarily uniformly spaced in time, but are taken at times in some set $\{t : t_1, \ldots, t_N\}$ that the noise probability distribution, although Gaussian, is not necessarily stationary or white (uncorrelated) and that the prior probabilities for the parameters are not necessarily independent. It turns out that the computer programs to take all this into account are not appreciably more difficult to write, if the most general analytical formulas are in view when we write them.

So now we have the model

$$y_{t_i} = T(t_i) + F(t_i) + e(t_i), \quad 1 \le i \le N, \tag{17.131}$$

in which we may write $y_i \equiv y(t_i)$, etc., with data $D = (y_1, \ldots, y_N)$, where $T(t)$ is the trend function, not necessarily linear, $F(t)$ is the periodic seasonal function, not necessarily sinusoidal, and $e(t)$ is the irregular component. To define our matrices we suppose $T(t)$ expanded in some linearly independent basis functions $\Phi_k(t)$ (for example, Legendre polynomials):

$$T(t) = \sum \gamma_k \Phi_k(t). \tag{17.132}$$

Similarly, $F(t)$ is expanded in sinusoids:

$$F(t) = \sum [A_k \cos(kt) + B_k \sin(kt)]. \tag{17.133}$$

The joint likelihood of all the parameters is

$$L(\gamma, A, B, \sigma) = p(D|\gamma A B \sigma) = \left(\frac{1}{2\pi\sigma^2}\right)^{N/2} \exp\left\{\frac{1}{2\sigma^2}\sum_{i=1}^{N}[y_i - T(t_i) - F(t_i)]^2\right\}. \tag{17.134}$$

The quadratic form may be written as

$$Q(\alpha_k, \gamma_j) \equiv \sum_{i=1}^{N} \left[y_i - \sum_{j=1}^{r} \gamma_j T_j(t_i) - \sum_{k=1}^{m} \alpha_k F_k(t_i) \right]^2, \qquad (17.135)$$

where, in the seasonal adjustment problem, $m = 12$ and

$$\{\alpha_1, \ldots, \alpha_m\} = \{A_0, A_1, \ldots, A_6, B_1, B_2, \ldots, B_5\}. \qquad (17.136)$$

Likewise,

$$F_k(t) = \begin{cases} \cos(k\omega t) & 0 \le k \le 6 \\ \sin([k-6]\omega t) & 7 \le k \le 12. \end{cases} \qquad (17.137)$$

But if we combine α, γ into a single vector of dimension $n = m + r$:

$$q \equiv (\alpha_1, \ldots, \alpha_m, \gamma_1, \ldots, \gamma_r) \qquad (17.138)$$

and define the function

$$G_k(t) = \begin{cases} F_k(t) & 1 \le k \le m \\ T_k(t) & m+1 \le k \le n, \end{cases} \qquad (17.139)$$

then the model is in the more compact form

$$y(t) = \sum_{j=1}^{n} q_j G_j(t) + e(t), \qquad (17.140)$$

and the data vector is

$$y_i = \sum_{j=1}^{n} q_j G_j(t_i) + e(t_i), \qquad 1 \le i \le N \qquad (17.141)$$

or

$$y = qG + e. \qquad (17.142)$$

The 'noise' values $e = e(t_i)$ have the joint prior probability density

$$p(e_1 \cdots e_N) = \frac{\sqrt{\det K}}{(2\pi)^{N/2}} \exp\left\{ -\frac{1}{2} e^T K e \right\}, \qquad (17.143)$$

where K^{-1} is the $(N \times N)$ noise prior covariance matrix. For 'stationary white noise', it reduces to

$$K^{-1} = \sigma^2 \delta_{ij}, \qquad 1 \le i, j \le N. \qquad (17.144)$$

Given K and the parameters $\{q_j\}$, the sampling pdf for the data takes the form

$$p(y_1 \cdots y_N | qKI) = \frac{\sqrt{\det(K)}}{(2\pi)^{N/2}} \exp\left\{ -\frac{1}{2}(y - qG)^T K(y - qG) \right\}. \qquad (17.145)$$

Likewise, a very general form of joint prior pdf for the parameters is

$$p(q_1 \cdots q_n | I) = \frac{\sqrt{\det(L)}}{(2\pi)^{n/2}} \exp\left\{ -\frac{1}{2}(q - q_0)^T L(q - q_0) \right\},$$ (17.146)

where L^{-1} is the $(n \times n)$ prior covariance matrix and q_0 is the vector of prior estimates. Almost always we shall take L to be diagonal:

$$L_{ij} = \sigma_j^2 \delta_{ij}, \qquad 1 \le i, j \le n,$$ (17.147)

and q_0 to be zero. But the general formulas without these simplifying assumptions are readily found and programmed.

The joint posterior pdf for the parameters $\{q_j\}$ is then

$$p(q|yI) = \frac{\exp\{-Q/2\}}{\int dq_1 \cdots dq_n \, \exp\{-Q/2\}},$$ (17.148)

where Q is the quadratic form

$$Q \equiv (y - Gq)^T K(y - Gq) + (q - q_0)^T L(q - q_0),$$ (17.149)

which we may expand into eight terms:

$$Q = y^T K y - y^T K G q - q^T G^T K y + q^T G^T K G q + q^T L q - q^T L q_0 - q_0^T L q + q_0^T L q_0.$$ (17.150)

We want to bring out the dependence on q by writing this in the form

$$Q = (q - \hat{q})^T M(q - \hat{q}) + Q_0,$$ (17.151)

where Q_0 is independent of q. Writing this out and comparing with (17.150), we have

$$M = G^T K G + L,$$
$$M\hat{q} = G^T K y + L q_0,$$ (17.152)
$$\hat{q}^T M\hat{q} + Q_0 = y^T K y + q_0^T L q_0.$$

Thus M, $\hat{q}$, and Q_0 are uniquely determined, because the equality of (17.150) and (17.151) must be an identity in q:

$$\hat{q} = M^{-1}\left[G^T K y + L q_0 \right]$$ (17.153)

$$Q_0 = y^T K y + q_0^T L q_0 - \hat{q}^T M\hat{q}.$$ (17.154)

The denominator of (17.148) is then found using (17.151), with the final result

$$p(q_1 \cdots q_n | yKLI) = \frac{\sqrt{\det(M)}}{(2\pi)^{n/2}} \exp\left\{ -\frac{1}{2}(q - \hat{q})^T M(q - \hat{q}) \right\}.$$ (17.155)

The components $q_1, \ldots, q_m$ are the seasonal amplitudes we wish to estimate, while $(q_{m+1}, \ldots, q_n)$ are the trend nuisance parameters to be eliminated. From (17.155) the marginal pdf we want is

$$p(q_1 \cdots q_m \mid yKLI) = \int dq_{m+1} \cdots dq_n \, p(q_1 \cdots q_n \mid yKLI)$$

$$= \frac{\sqrt{\det(M)}}{(2\pi)^{n/2}} \frac{(2\pi)^{(n-m)/2}}{\sqrt{\det(\overline{W})}} \exp\left\{ -\frac{1}{2}(u - \hat{u})^T U (u - \hat{u}) \right\} \quad (17.156)$$

$$= \frac{\sqrt{\det(U)}}{(2\pi)^{m/2}} \exp\left\{ -\frac{1}{2}(u - \hat{u})^T U (u - \hat{u}) \right\}$$

where U, V, W, u are defined by ( ), ( ), ( ), ( ).

Editor's Exercise 17.4. Jaynes never defined U, V, W, and u. In (17.155) multiply out all of the terms in the exponent, obtain the appropriate sub-matrices, vectors, and scalars and then define each of these four quantities.

From the fact that the various probabilities are normalized, we see that

$$\det(M) = \det(W)\det(U), \quad (17.157)$$

a remarkable theorem not at all obvious from the definitions except in the case $V = 0$. This is another good example of the power of probabilistic reasoning to prove purely mathematical theorems.

Thus, the most general solution consists, computationally, of a string of elementary matrix operations and is readily programmed. To summarize the final computation rules:

K^{-1} is the $(N \times N)$ prior covariance matrix for the 'noise';
L^{-1} is the $(n \times n)$ prior covariance matrix for the parameters;
F is the $(N \times n)$ matrix of model functions.

Firstly, calculate the $(n \times n)$ matrix

$$M \equiv F^T K F + L \quad (17.158)$$

and decompose it into block form representing the interesting and uninteresting subspaces:

$$M = \begin{pmatrix} U_0 & V \\ V^T & W_0 \end{pmatrix}. \quad (17.159)$$

Then calculate the $(m \times m)$ and $(r \times r)$ renormalized matrices

$$U \equiv U_0 - VW_0^{-1}V^T \quad (17.160)$$

$$W \equiv W_0 - V^T U_0^{-1} V. \quad (17.161)$$

This much is determined by the definition of the model; the computer can work all this out in advance, before the data are known, and use the result on any number of data sets.

Now given y, the $(N \times 1)$ data vector and q_0, the $(n \times 1)$ vector of prior estimates, the computer should calculate the $(n \times 1)$ vector

$$\hat{q} = M^{-1}\left[F^T K y + L q_0\right] \tag{17.162}$$

of 'best' estimates of the parameters. Actually, the first m of them are the interesting ones wanted, and the remaining $r = n - m$ components are not needed unless one also wants an estimate of the trend function. Then we can use the following result.

The inverse M^{-1} can be written in the same block form as M:

$$M^{-1} = \begin{pmatrix} U^{-1} & -U_0 V W^{-1} \\ -W_0 V^T U^{-1} & W^{-1} \end{pmatrix}, \tag{17.163}$$

where, analogous to U,

$$W \equiv W_0 - V^T U_0^{-1} V. \tag{17.164}$$

Then F^T has the same block form with respect to its rows:

$$(F^T)_{ji} = \left[G_j(t_i)T_i(t_i)\right] \quad \begin{matrix} 1 \le j \le m \\ 1 \le i \le N \\ (m+1) \le K \le n, \end{matrix} \tag{17.165}$$

where $G_j(t)$ are the seasonal sinusoids and $T_k(t)$ the trend functions.

Almost always, $q_0 = 0$, and so the 'interesting' seasonal amplitudes are given by

$$\hat{q} = RKy, \tag{17.166}$$

where R is the reduced $(m \times N)$ matrix

$$R \equiv U^{-1}G - U_0^{-1}V W^{-1}T \tag{17.167}$$

and U^{-1} is the joint posterior covariance matrix for the interesting parameters $\{q_1, \ldots, q_m\}$. Note that R and U^{-1} are also determined by the model, so the computer can calculate them once and for all before the data are available.

Editor's Exercise 17.5. Jaynes never finished this section, so we can only speculate as to what he would have put in here. So let's speculate. Firstly, look at (17.156): this is the joint posterior probability for all of the seasonal amplitudes, but the amplitudes are not the same thing as the seasonal component itself. The seasonal component is given by

$$S(t) = \sum_{k=1}^{m} q_k G_k(t) \tag{17.168}$$

and is a continuous function of time. Can (17.156) and (17.168) be used to compute $p(S(t)|yKLI)$, the joint posterior probability for the seasonal? In other words, can a simple change of variables plus marginalization over the remaining q's be used to compute $p(S(t))|yKLI)$? If not, how would you compute this joint posterior probability?

17.12 Comments

Let us try to summarize and understand the underlying technical reasons for the facts noted in the preceding two chapters. Sampling theory methods of inference were satisfactory for the relatively simple problems considered by R. A. Fisher in the 1930s. These problems had the features of:

(a) few parameters;
(b) presence of sufficient statistics;
(c) no important prior information;
(d) no nuisance parameters.

When all these conditions are met, and we have a reasonably large amount of data (say, $n \geq 30$), orthodox methods become essentially equivalent to the Bayesian ones, and it will make no pragmatic difference which ideology we prefer. But today we are faced with important problems in which some or all of these conditions are violated. Only Bayesian methods have the analytical apparatus capable of dealing with such problems without sacrificing much of the relevant information available to us. Bayesian methods are more powerful; if there is no sufficient statistic, they extract more information from the data for reasons explained at the beginning of this chapter. Also, they take note of possibly highly important prior information, and deal easily with nuisance parameters, turning them into an important asset.

Today one wonders how it is possible that orthodox logic continues to be taught in some places year after year and praised as 'objective', while Bayesians are charged with 'subjectivity'. Orthodoxians, preoccupied with fantasies about nonexistent data sets and, in principle, unobservable limiting frequencies – while ignoring relevant prior information – are in no position to charge anybody with 'subjectivity'. If there is no sufficient statistic, the orthodox accuracy claim based on a single 'statistic' simply ignores not only the prior information, but also all the evidence in the data that is relevant to that accuracy: hardly an 'objective' procedure. If there are ancillary statistics and the orthodoxian follows Fisher by conditioning on them, he obtains just the estimate that Bayes' theorem based on a noninformative prior would have given him by a shorter calculation. Bayes' theorem would have given also a defensible accuracy claim.

We shall illustrate this in later chapters with several examples, including interval estimation, dealing with trend, linear regression, detection of cycles, and prediction of time series. In all these cases, 'orthodox' methods can miss important evidence in the data; but they can also yield conclusions not justified by the evidence because they ignore highly cogent prior information. No case of such failure of Bayesian methods has been found; indeed, the optimality theorems well known in the Bayesian literature lead one to expect this from the start. Psychologically, however, practical examples seem to have more convincing power than do optimality theorems.

Historically, scientific inference has been dominated overwhelmingly by the case of univariate or bivariate Gaussian sampling distributions. This has produced a distorted picture of the field: the Gaussian case is the one in which 'orthodox', or 'sampling theory' methods do best, and the difference between pre-data and post-data procedures is the least. On the basis of this limited evidence, orthodox theory (in the hands of Fisher) tried to claim general validity for its methods, and attacked Bayesian methods savagely without ever examining the results they give.

Even in the Gaussian case, there are important problems where sampling theory methods fail for technical reasons. An example is linear regression with both variables subject to error of unknown variance; indeed, this is perhaps the most common problem of inference faced by experimental scientists. Yet sampling theory is helpless to deal with it, because each new data point brings with it a new nuisance parameter. The orthodox statistical literature offers us no satisfactory way of dealing with this problem. See, for example, Kempthorne and Folks (1971), in which the (for them) necessity of deciding which quantities are 'random' and which are not, leads the authors to formulate *16* different linear regression models to describe what is only a single inference problem; then they find themselves helpless to deal with most of them, and give up with the statement that 'It is all very difficult.'

When we depart from the Gaussian case, we open up a Pandora's box of new anomalies, logical contradictions, absurd results, and technical difficulties beyond the means of sampling theory to handle. Several examples were noted already by the devout orthodoxians Kendall and Stuart (1961).

These examples show the fundamental error in supposing that the quality of an estimate can be judged merely from the sampling distribution of the estimator. This is true only in the simpler Gaussian cases for reasons of mathematical symmetry; in general, as Fisher noted, many different samples which all lead to the same estimator nevertheless determine the values of the parameters to very different accuracy because they have different configurations (ranges). But Fisher's remedy – conditioning on ancillary statistics – is seldom possible, and, when it is possible, we saw in Chapter 8 that it is mathematically equivalent to use of Bayes' theorem. In the case of the 'student' t-distribution this was shown already by Jeffreys in the 1930s. In Jaynes (1976) we demonstrate it in detail for the Cauchy distribution, which orthodoxy regards as 'pathological'.

What the orthodox literature invariably fails to recognize is that all of these difficulties are resolved effortlessly by the uniform application of the single Bayesian method. In fact, once the Bayesian analysis has shown us the correct answer, one can often study it, understand intuitively why it is right, and, with this deeper understanding, see how that answer might have been found by some *ad hoc* device acceptable to orthodoxy.

We shall illustrate this in later chapters by giving the solution to the aforementioned regression problem, and to some inference problems with the Cauchy sampling distribution. To the best of our knowledge, these solutions cannot be found in any of the orthodox statistical literature.

But we must note with sadness that, in much of the current Bayesian literature, very little of the orthodox baggage has been cast off. For example, it is rather typical to see a Bayesian article start with such phrases as: 'Let X be a random variable with density function $p(x|\theta)$, where the value of the parameter θ is unknown. Suppose this parametric family contains the true distribution of X ....' Or, one describes a uniform prior $p(\theta|I)$ by saying: 'θ is supposed uniformly distributed'. The analytical solutions thus obtained will doubtless be a valid Bayesian result; but one is still clinging to the orthodox fiction of 'random variables' and 'true distributions'. θ is simply an unknown constant; it is not 'distributed' at all. What is 'distributed' is our state of knowledge about θ: again there is that persistent mind projection fallacy that contaminates all of probability theory, leading inexperienced readers far astray as to what we are doing. Equally bad, those who commit this fallacy seem unaware that this is restricting the application to a small fraction of the real situations where the solution might be useful. In the vast majority of real applications there are no 'random variables' (What defines 'randomness'?) and no 'true distribution' (What defines it? What test could we apply to decide whether some proposed distribution is or is not the 'true' one?); yet probability theory as logic applies to all of them.

Unlike orthodox tests, Bayesian posterior probabilities or odds ratios can tell us quantitatively how strong the evidence is for some effect taking into account *all* the evidence at hand, not merely the evidence of one data set.

L. J. Savage (1962, pp. 63–67) gives, by a tortuously long, closely reasoned argument using only sampling probabilities, a rationale for the Bayesian algorithm. The Bayesian argument expounded here, which he rejects as a 'necessary' view, derives the same conclusion, in greater generality, directly from first principles.

These comparisons show that in order to deal successfully with current real problems, it is essential to jettison tradition and authority, which have retarded progress throughout this century. It is deplorable that orthodox methods and terminology continue to be taught at all to young statisticians, economists, biologists, psychologists, and medical researchers; this has done serious damage in these fields for decades.

Yet everywhere we look there are glimmerings of hope. In physics, Bretthorst (1988) has treated the analysis of magnetic resonance data, extracting by Bayesian methods far more information from the data than was possible with the previous *ad hoc* Fourier analysis. In econometrics Prof. Arnold Zellner is the founder of a large, active, and growing school of Bayesian analysis which has given rise to a vast literature. In medical diagnosis the great physician Sir William Osler (1849–1919) noted long ago that:[20] *Medicine is a science of uncertainty and an art of probability.* In recent years several people have started to take this remark seriously. Lee Lusted (1968) gives worked-out examples, with flow charts and source code, of the Bayesian computer diagnoses of six important medical conditions, as well as a great deal of qualitative wisdom in medical testing.[21] Peter Cheeseman (1988) has been developing expert systems for medical diagnosis based on Bayesian principles.

---

[20] Quoted by Bean (1950, p. 125).

[21] Lusted later founded the Society for Medical Decision Making in 1978, and served as the first editor of its journal. At the time of his death in February 1994, he was retired but still serving as Adjunct Professor at the Stanford University Medical School, advising medical students in problems of decision analysis.

18

The A_p distribution and rule of succession

Inside every Non-Bayesian, there is a Bayesian struggling to get out.

Dennis V. Lindley

Up to this point, we have given our robot fairly general principles by which it can convert information into numerical values of prior probabilities, and convert posterior probabilities into definite final decisions; so it is now able to solve lots of problems. But it still operates in a rather inefficient way in one respect. When we give it a new problem, it has to go back into its memory (this proposition that we have denoted by X or I, which represents everything it has ever learned). It must scan its entire memory archives for anything relevant to the problem before it can start working on it. As the robot grows older this gets to be a more and more time-consuming process.

Now, human brains don't do this. We have some machinery built into us which summarizes our past conclusions, and allows us to forget the details which led us to those conclusions. We want to see whether it is possible to give the robot a definite mechanism by which it can store general conclusions rather than isolated facts.

18.1 Memory storage for old robots

Note another thing, which we will see is closely related to this problem. Suppose you have a penny and you are allowed to examine it carefully, and convince yourself that it is an honest coin; i.e. accurately round, with head and tail, and a center of gravity where it ought to be. Then you're asked to assign a probability that this coin will come up heads on the first toss. I'm sure you'll say $1/2$. Now, suppose you are asked to assign a probability to the proposition that there was once life on Mars. Well, I don't know what your opinion is there, but on the basis of all the things that I have read on the subject, I would again say about $1/2$ for the probability. But, even though I have assigned the same 'external' probabilities to them, I have a very different 'internal' state of knowledge about those propositions.

To see this, imagine the effect of getting new information. Suppose we tossed the coin five times and it comes up tails every time. You ask me what's my probability for heads on the next throw; I'll still say $1/2$. But if you tell me one more fact about Mars, I'm ready to change my probability assignment completely. There is something which makes my state of belief very stable in the case of the penny, but very unstable in the case of Mars.[1]

This might seem to be a fatal objection to probability theory as logic. Perhaps we need to associate with a proposition not just a single number representing plausibility, but two numbers: one representing the plausibility, and the other how stable it is in the face of new evidence. And so, a kind of two-valued theory would be needed. In the early 1950s, the writer gave a talk at one of the Berkeley statistical symposiums, expounding this viewpoint.

But now, with more mature reflection we think that there is a mechanism by which our present theory automatically contains all these things. So far, all the propositions we have asked the robot to think about are 'Aristotelian' ones of two-valued logic: they had to be either true or false. Suppose we bring in new propositions of a different type. It doesn't make sense to say the proposition is either true or false, but still we are going to say that the robot associates a real number with it, which obeys the rules of probability theory. Now, these propositions are sometimes hard to state verbally; but we noticed before that if we give the probabilities conditional on X for all propositions that we are going to use in a given problem, we have told you everything about X which is relevant to that mathematical problem (although of course, not everything about its meaning and significance to us, that may make us interested in the problem). So, we introduce a new proposition A_p, defined by

$$P(A|A_pE) \equiv p, \tag{18.1}$$

where E is any additional evidence. If we had to render A_p as a verbal statement, it would come out something like this:

$$A_p \equiv \text{regardless of anything else you may have been told,}$$
$$\text{the probability of } A \text{ is } p. \tag{18.2}$$

Now, A_p is a strange proposition, but if we allow the robot to reason with propositions of this sort, Bayes' theorem guarantees that there's nothing to prevent it from getting an A_p worked over onto the left side in its probabilities: $P(A_p|E)$. What are we doing here? It seems almost as if we are talking about the 'probability of a probability'.

Pending a better understanding of what that means, let us adopt a cautious notation that will avoid giving possibly wrong impressions. We are not claiming that $P(A_p|E)$ is a 'real probability' in the sense that we have been using that term; it is only a number which is to obey the mathematical rules of probability theory. Perhaps its proper conceptual meaning will be clearer after getting a little experience using it. So let us refrain from using the prefix symbol p; to emphasize its more abstract nature, let us use the bare bracket symbol notation $(A_p|E)$ to denote such quantities, and call it simply 'the density for A_p, given E'.

We defined A_p by writing an equation. You ask what it means, and we reply by writing more equations. So let's write the equations: if X says nothing about A except that it is

---

[1] Note in passing a simple counter-example to a principle sometimes stated by philosophers, that theories cannot be proved true, only false. We seem to have just the opposite situation for the theory that there was once life on Mars. To prove it false, it would not suffice to dig up every square foot of the surface of Mars; to prove it true one needs only to find a single fossil.

possible for A to be true, and also possible for it to be false, then, as we saw in case of the 'completely ignorant population' in Chapter 12,

$$(A_p|X) = 1, \qquad 0 \le p \le 1. \tag{18.3}$$

The transformation group arguments of Chapter 12 apply to this problem. As soon as we have this, we can use Bayes' theorem to compute the density for A_p, conditional on the other things. In particular,

$$(A_p|EX) = (A_p|X)\frac{P(E|A_pX)}{P(E|X)} = \frac{P(E|A_p)}{P(E|X)}. \tag{18.4}$$

Now,

$$P(A|E) = \int_0^1 dp \, (AA_p|E). \tag{18.5}$$

The propositions A_p are mutually exclusive and exhaustive (in fact, every A_p flatly and dogmatically contradicts every other A_q), so we can do this. We're just going to apply all of our mathematical rules with total disregard of the fact that A_p is a funny kind of proposition. We believe that these rules form a consistent way of manipulating propositions. But now we recognize that consistency is a purely *structural* property of the rules, which could not depend on the particular semantic meaning you and I might attach to a proposition. So now we can blow up the integrand of (18.5) by the product rule:

$$P(A|E) = \int_0^1 dp \, P(A|A_pE)(A_p|E). \tag{18.6}$$

But from the definition (18.1) of A_p, the first factor is just p, and so

$$P(A|E) = \int_0^1 dp \, p \, (A_p|E). \tag{18.7}$$

The probability which our robot assigns to proposition A is just the *first moment* of the density for A_p. Therefore, the density for A_p should contain more information about the robot's state of mind concerning A, than just the probability for A. Our conjecture is that the introduction of propositions of this sort solves both of the problems mentioned, and also gives us a powerful analytical tool for calculating probabilities.

18.2 Relevance

To see why we propose our conjecture, let's note some lemmas about relevance. Suppose this evidence E consists of two parts, $E = E_aE_b$, where E_a is relevant to A and, given E_a, E_b is not relevant:

$$P(A|E) = P(A|E_aE_b) = P(A|E_a). \tag{18.8}$$

By Bayes' theorem, it follows that, given E_a, A must also be irrelevant to E_b, for

$$P(E_b|AE_a) = P(E_b|E_a)\frac{P(A|E_bE_a)}{P(A|E_a)} = P(E_b|E_a). \tag{18.9}$$

Let's call this property 'weak irrelevance'. Now, does this imply that E_b is irrelevant to A_p? Evidently not, for (18.8) says only that the first moments of $(A_p|E_a)$ and $(A_p|E_aE_b)$ are the same. But suppose that, for a given E_b, (18.8) holds independently of what E_a might be; call this 'strong irrelevance'. Then we have

$$P(A|E) = \int_0^1 dp\, p\, (A_p|E_aE_b) = \int_0^1 dp\, p\, (A_p|E_a). \tag{18.10}$$

But if this is to hold for all $(A_p|E_a)$, the integrands must be the same:

$$(A_p|E_aE_b) = (A_p|E_a), \tag{18.11}$$

and from Bayes' theorem it follows as in (18.9) that A_p is irrelevant to E_b:

$$P(E_b|A_pE_a) = P(E_b|E_a) \tag{18.12}$$

for all E_a.

Now, suppose our robot gets a new piece of evidence, F. How does this change its state of knowledge about A? We could expand directly by Bayes' theorem, which we have done before, but let's use our A_p this time:

$$P(A|EF) = \int_0^1 dp\, p\, (A_p|EF) = \int_0^1 dp\, p\, (A_p|E)\frac{P(F|A_pE)}{P(F|E)}. \tag{18.13}$$

In this likelihood ratio, any part of E that is irrelevant to A_p can be struck out; because, by Bayes' theorem, it is equal to

$$\frac{P(F|A_pE_aE_b)}{P(F|E_aE_b)} = \frac{P(F|A_pE_a)\left[\frac{P(E_b|FA_pE_a)}{P(E_b|A_pE_a)}\right]}{P(F|E_a)\left[\frac{P(E_b|FE_a)}{P(E_b|E_a)}\right]} = \frac{P(F|A_pE_a)}{P(F|E_a)}, \tag{18.14}$$

where we have used (18.12).

Now if E_a still contains a part irrelevant to A_p, we can repeat this process. Imagine this carried out as many times as possible; the part E_{aa} of E that is left contains nothing at all that is irrelevant to A_p. E_{aa} must then be some statement only about A. But then, by definition (18.1) of A_p, we see that A_p automatically cancels out E_{aa} in the numerator: $(F|A_pE_{aa}) = (F|A_p)$. And so we have (18.13) reduced to

$$P(A|EF) = \frac{1}{P(F|E_{aa})}\int_0^1 dp\, p\, (A_p|E)P(F|A_p). \tag{18.15}$$

The weak point in this argument is that we have not proved that it is always possible to resolve E into a completely relevant part and completely irrelevant part. However, it is easy to show that in many applications it *is* possible. So, let's just say that the following results apply to the case where the prior information is 'completely resolvable'. We have not shown that it is the most general case; but we do know that it is not an empty one.

18.3 A surprising consequence

Now, $(F|E_{aa})$ is a troublesome thing which we would like to eliminate. It's really just a normalizing factor, and we can eliminate it the way we did in Chapter 4: by calculating the odds on A instead of the probability. This is just

$$O(A|EF) = \frac{P(A|EF)}{P(\overline{A}|EF)} = \frac{\int_0^1 dp \, p \, (A_p|E)P(F|A_p)}{\int_0^1 dp(A_p|E)P(F|A_p)(1-p)}. \tag{18.16}$$

The significant thing here is that the proposition E, which for this problem represents our prior information, now appears only in the density $(A_p|E)$. This means that *the only property of E which the robot needs in order to reason out the effect of new information is this density* $(A_p|E)$. Everything that the robot has ever learned which is relevant to proposition A may consist of millions of isolated separate facts. But when it receives new information, it does not have to go back and search its entire memory for every little detail of its information relevant to A. Everything it needs in order to reason about A from that past experience is contained summarized in this one function, $(A_p|E)$.

So, for each proposition A about which it is to reason, the robot can store a density function $(A_p|E)$ like that in Figure 18.1. Whenever it receives new information F, it will be well advised to calculate $(A_p|EF)$, and then it can erase the previous $(A_p|E)$ and for the future store only $(A_p|EF)$. By this procedure, every detail of its previous experience is taken into account in future reasoning about A.

This suggests that in a machine which does inductive reasoning, the memory storage problem may be simpler than it is in a machine which does only deductive reasoning. This does not mean that the robot is able to throw away all of its past experience, because there is

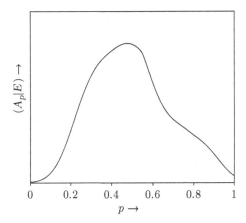

Fig. 18.1. An example A_p distribution.

always a possibility that some new proposition will come up which it has not had to reason about before. And, whenever this happens, then of course it *will* have to go back into its original archives and search for every scrap of information it has relevant to this proposition.

With a little introspection, we would all agree that this is just what goes on in our minds. If you are asked how plausible you regard some proposition, you don't go back and recall all the details of everything that you ever learned about this proposition. You recall your previous state of mind about it. How many of us can still remember the argument which first convinced us that $d \sin(x)/dx = \cos(x)$? But, unlike the robot, when you or I are confronted with some entirely new proposition Z, we do not have the ability to carry out a full archival search.

Let's look once more at (18.15). If the new information F is to make any appreciable change in the probability of A, we can see from this integral what has to happen. If the density $(A_p|E)$ was already very sharply peaked at one particular value of p, then $P(F|A_p)$ will have to be even more sharply peaked at some other value of p, if we are going to get any appreciable change in the probability. On the other hand, if the density $(A_p|E)$ is very broad, any small slope in $P(F|A_p)$ can make a big change in the probability which the robot assigns to A.

So, the stability of the robot's state of mind when it has evidence E is determined, essentially, by the *width* of the density $(A_p|E)$. There does not appear to be any single number which fully describes this stability. On the other hand, whenever it has accumulated enough evidence so that $(A_p|E)$ is fairly well peaked at some value of p, then the variance of that distribution becomes a pretty good measure of how stable the robot's state of mind is. The greater amount of previous information it has collected, the narrower its A_p-distribution will be, and therefore the harder it will be for any new evidence to change that state of mind.

Now we can see the difference between the penny and Mars. In the case of the penny, my $(A_p|E)$ density, based on my prior knowledge, is represented by a curve something like that shown in Figure 18.2(a). In the case of previous life on Mars, my state of knowledge is

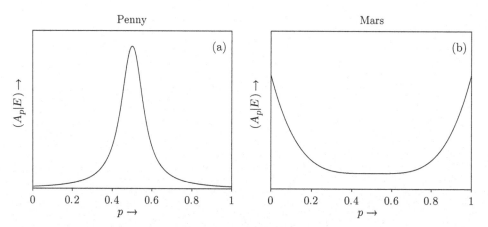

Fig. 18.2. Two A_p distributions having the same first moments, but representing very different states of knowledge.

described by an $(A_p|E)$ density something like that shown in Figure 18.2(b), qualitatively. The first moment is the same in the two cases, so I assign probability $1/2$ to either one; nevertheless, there's all the difference in the world between my state of knowledge about those two propositions, and this difference is represented in the $(A_p|E)$ densities.

Ideas very much like this have arisen in other contexts. While the writer was first speculating on these ideas, a newspaper story appeared entitled: 'Brain Stockpiles Man's Most Inner Thoughts'. It starts out:

Everything you have ever thought, done, or said – a complete record of every conscious moment – is logged in the comprehensive computer of your brain. You will never be able to recall more than the tiniest fraction of it to memory, but you'll never lose it either. These are the findings of Dr Wilder Penfield, Director of the Montreal Neurological Institute, and a leading Neurosurgeon. The brain's ability to store experiences, many lying below consciousness, has been recognized for some time, but the extent of this function is recorded by Dr Penfield.

Now, there are several examples given, of experiments on patients suffering from epilepsy. Stimulation of a definite location in the brain recalled a definite experience from the past, which the patients had not been able to recall to memory previously. Here are the concluding sentences of the article. Dr Penfield now says:

This is not memory as we usually use the word, although it may have a relation to it. No man can recall by voluntary effort such a wealth of detail. A man may learn a song so he can sing it perfectly, but he cannot recall in detail any one of the many times he heard it. Most things that a man is able to recall to memory are generalizations and summaries. If it were not so, we might find ourselves confused by too great a richness of detail.

This is exactly the hint we needed to form a clearer idea of what the A_p density means conceptually.

18.4 Outer and inner robots

We know from overwhelming evidence, of which the above is only a small part, that human brains have two different functions: a conscious mind and a subconscious one. They work together in some kind of cooperation. The subconscious mind is probably at work continually throughout life. It solves problems and communicates information to the conscious mind under circumstances not under our conscious control; everyone who has done original thinking about difficult problems has experienced this, and many (Henri Poincaré, Jacques Hadamard, Wm. Rowan Hamilton, Freeman Dyson) have recorded the experience for others to read. A communication from the subconscious mind appears to us as a sudden inspiration that seems to come out of nowhere when we are relaxed and not thinking consciously about the problem at all; instantly, we feel that we understand the problem that has perplexed us for weeks.[2]

---

[2] The writer has experienced this several times when, in unlikely situations like riding a tractor on his farm, he suddenly saw how to prove something long conjectured. But the inspiration does not come unless the conscious mind has prepared the way for it by intense concentration on the problem.

Now, if the human brain can operate on two different levels, so can our robot. Rather than trying to think of a 'probability of a probability', we may think of two different levels of reasoning: an 'outer robot' in contact with the external world and reasoning about it; and an 'inner robot' who observes the activity of the outer robot and thinks about it. The conventional probability formulas that we used before this chapter represent the reasoning of the outer robot; the A_p density represents the inner robot at work. But we would like our robot to have one advantage over the human brain. The outer robot should not be obliged as we are to wait for the inspiration from within; it should have the power to call at will upon the services of the inner robot.

Looking at the A_p distribution this way makes it much less puzzling conceptually. The outer robot, thinking about the real world, uses Aristotelian propositions referring to that world. The inner robot, thinking about the activities of the outer robot, uses propositions that are not Aristotelian in reference to the outer world, but they are still Aristotelian in its context, in reference to the thinking of the outer robot; so, of course, the same rules of probability theory will apply to them. The term 'probability of a probability' misses the point, since the two probabilities are at different levels.

Having had this much of a glimpse of things, our imagination races on far beyond it. The inner robot may prove to be more versatile than merely calculating and storing A_p densities; it may have functions that we have not yet imagined. Furthermore, could there be an 'inner inner' robot, twice removed from the real world, which thinks about the activity of the inner one? What prevents us from having a nested hierarchy of such robots, each inner to the next? Why not several parallel hierarchies, concerned with different contexts?

Questions like this may seem weird, until we note that just this same hierarchy has evolved already in the development of computers and computer programming methods. Our present microcomputers operate on three discernible hierarchical levels of activity, the inner 'BIOS' code which contacts the machine hardware directly, the 'COMMAND SHELL' which guards it from the outer world while sending information and instructions back and forth between them, and the outer level of human programmers who provide the 'high level' instructions representing the conscious ultimate purpose of the machine level activity. Furthermore, the development of 'massively parallel' computer architecture has been underway for several years.

In the evolution of computers this represented such a natural and inevitable division of labor that we should not be surprised to realize that a similar division of labor occurred in the evolution of the human brain. It has an inner 'BIOS' level which in some way exerts direct control over the body's biological hardware (such as rate of heartbeat and levels of hormone secretion), a 'COMMAND SHELL' which receives 'high level' instructions from the conscious mind and converts them into the finely detailed instructions needed to execute such complex activities as walking or playing a violin, without any need for the conscious mind to be aware of all those details. Then in some aspects of the present organization of the brain, not yet fully understood, we may be seeing some aspects of the future evolution of computers; in particular of our robot.

The idea of a nested hierarchy of robots, each thinking about propositions on a different level, is in some ways similar to Bertrand Russell's 'theory of types', which he introduced as a means of avoiding some paradoxes that arose in the first formulation of his *Principia Mathematica*. There may be a relationship between them; but these efforts at what Peano and Poincaré called 'logistic' made in the early 20th century, are now seen as so flawed and confused – with an unlimited proliferation of weird and self-contradictory definitions, yet with no recognition of the concept of information – that it seems safest to scrap this old work entirely and rebuild from the start using our present understanding of the role of information and our new respect for Kronecker's warnings, so appropriate in an age of computers, that *constructibility* is the first criterion for judging whether a newly defined set or other mathematical object makes any sense or can serve any useful purpose.

Our opening quotation from Dennis Lindley (made during a talk at a Bayesian seminar in the early 1980s) fits in nicely with these considerations and with our remarks in Chapter 5 about visual perception. There we noted that any reasoning format whose results conflict with Bayesian principles would place a creature at a decided survival disadvantage, so evolution by Darwinian natural selection would automatically produce brains which reason in the Bayesian format. But the outer brain can become corrupted by false indoctrination from contact with the outer world – even to the point of becoming anti-Bayesian – while the inner brain, protected from this, retains its pristine Bayesian purity. Thus, Lindley's remark, made as a kind of joke, may be quite literally true.

Here, however, we are treading on the boundaries of present knowledge, so the above material is necessarily a tentative, preliminary exploration of a possibly large new territory (call it wild speculation if you prefer), rather than expounding a well-established theory. With these cautions in mind, let us examine some concrete examples which follow from the above line of thought, but can also be justified independently.

18.5 An application

Now let's imagine that a 'random' experiment is being performed. From the results of the experiment in the past, we want to do the best job we can of predicting results in the future. To make the problem a definite one, introduce the propositions:

$X \equiv$ For each trial we admit two prior hypotheses: A true, and A false.

The underlying 'causal mechanism' is assumed the same at every trial. This means, for example, that (1) the probability assigned to A at the nth trial does not depend on n, and (2) evidence concerning the results of past trials retains its relevance for all time; thus for predicting the outcome of trial 100, knowledge of the result of trial 1 is just as relevant as is knowledge of the result of trial 99. There is no other prior evidence.

$N_n \equiv A$ true n times in N trials in the past.
$M_m \equiv A$ true m times in M trials in the future.

The verbal statement of X suffers from just the same ambiguities that we have found before, and which have caused so much trouble and controversy in the past. One of the important points we want to put across here is that we have not defined the prior information precisely until we have given, not just verbal statements, but equations, which show how we have translated them into mathematics by specifying the prior probabilities to be used. In the present problem, this more precise statement of X is, as before,

$$(A_p|X) = 1, \quad 0 \le p \le 1, \tag{18.17}$$

with the additional understanding (part of the prior information for this particular problem) that the *same* A_p distribution is to be used for calculations pertaining to all trials. What we are after is $P(M_m|N_n)$. Firstly, note that by many repetitions of our product and sum rules in the same way that we found Eq. (9.34), we have the binomial distributions

$$P(N_n|A_p) = \binom{N}{n} p^n (1-p)^{N-n},$$

$$\tag{18.18}$$

$$P(M_m|A_p) = \binom{M}{m} p^m (1-p)^{M-m},$$

and at this point we see that, although A_p sounds like an awfully dogmatic and indefensible statement to us the way we introduced it, this is actually the way in which probability *is* introduced in almost all present textbooks. One postulates that an event possesses some intrinsic, 'absolute' or 'physical' probability, whose numerical value we can never determine exactly. Nevertheless, no one questions that such an 'absolute' probability exists. Cramér (1946, p. 154), for example, takes it as his fundamental axiom. That is just as dogmatic a statement as our A_p; and we think it is, in fact, just our A_p. The equations we see in current textbooks are all like the two above; whenever p appears as a *given* number, an adequate notation would show that there is an A_p hiding invisibly in the right-hand side of the probability symbols.

Mathematically, the main functional differences between what we are doing here and what is done in current textbooks are: (1) we recognize the existence of that right-hand side of *all* probabilities, whether or not an A_p is hiding in them; and (2) thanks to Cox's theorems, we are not afraid to use Bayes' theorem to work any proposition – including A_p – back and forth from one side of our symbols to the other. In refusing to make free use of Bayes' theorem, orthodox writers are depriving themselves of the most powerful single principle in probability theory. When a problem of inference is studied long enough, sometimes through a string of *ad hockeries* for decades, one is always forced eventually to a conclusion that could have been derived in three lines from Bayes' theorem. But those cases refer to 'external' probabilities at the interface between the robot and the outside world; now we shall see that Bayes' theorem is equally powerful and indispensable for manipulating 'inner' probabilities.

Now we need to find the prior probability $P(N_n|X)$. This is determined already from $(A_p|X)$, for our trick of resolving a proposition into mutually exclusive alternatives gives us

$$P(N_n|X) = \int_0^1 \mathrm{d}p\,(N_n A_p|X) = \int_0^1 \mathrm{d}p\, P(N_n|A_p)(A_p|X) = \binom{N}{n} \int_0^1 \mathrm{d}p\, p^n (1-p)^{N-n}.$$

$$\tag{18.19}$$

The integral we have to evaluate is the complete Beta-function:

$$\int_0^1 dx\, x^r (1-x)^s = \frac{r!\,s!}{(r+s+1)!}. \tag{18.20}$$

Thus, we have

$$P(N_n|X) = \begin{cases} \dfrac{1}{N+1} & 0 \le n \le N \\[2mm] 0 & N < n, \end{cases} \tag{18.21}$$

i.e. just the uniform distribution of maximum entropy; $P(M_m|X)$ is found similarly. Now we can turn (18.18) around by Bayes' theorem:

$$(A_p|N_n) = (A_p|X)\frac{P(N_n|A_p)}{P(N_n|X)} = (N+1)P(N_n|A_p), \tag{18.22}$$

and so finally the desired probability is

$$P(M_m|N_n) = \int_0^1 dp\,(M_m A_p|N_n) = \int_0^1 dp\, P(M_m|A_p N_n)(A_p|N_n). \tag{18.23}$$

Since $P(M_m|A_p N_n) = P(M_m|A_p)$ by the definition of A_p, we have worked out everything in the integrand. Substituting into (18.23), we have again an Eulerian integral, and our result is

$$P(M_m|N_n) = \frac{\dbinom{n+m}{n}\dbinom{N+M-n-m}{N-n}}{\dbinom{N+M+1}{M}}. \tag{18.24}$$

Note that this is not the same as the hypergeometric distribution (3.22) of sampling theory. Let's look at this result first in the special case $M = m = 1$; it then reduces to the probability of A being true in the next trial, given that it has been true n time in the previous N trials. The result is

$$P(A|N_n) = \frac{n+1}{N+2}. \tag{18.25}$$

We recognize Laplace's rule of succession, which we found before and discussed briefly in terms of urn sampling in (6.29)–(6.46). Now we need to discuss it more carefully, in a wider context.

18.6 Laplace's rule of succession

This rule occupies a supreme position in probability theory; it has been easily the most misunderstood and misapplied rule in the theory, from the time Laplace first gave it in 1774. In almost any book on probability, this rule is mentioned very briefly, mainly in order to warn the reader not to use it. But we must take the trouble to understand it, because in our design of this robot Laplace's rule is, like Bayes' theorem, one of the most important constructive rules we have. It is a 'new' rule (i.e. a rule in addition to the principle of indifference and its generalization, maximum entropy) for converting raw information into numerical values of probabilities, and it gives us one of the most important connections between probability and frequency.

Poor old Laplace has been ridiculed for over a century because he illustrated use of this rule by calculating the probability that the sun will rise tomorrow, given that it has risen every day for the past 5000 years.[3] One obtains a rather large factor (odds of $5000 \times 365.2426 + 1 = 1\,826\,214 : 1$) in favor of the sun rising again tomorrow. With no exceptions at all as far as we are aware, modern writers on probability have considered this a pure absurdity. Even Keynes (1921) and Jeffreys (1939) find fault with the rule of succession.

We have to confess our inability to see anything at all absurd about the rule of succession. We recommend very strongly that you do a little independent literature searching, and read some of the objections various writers have to it. You will see that in every case the same thing has happened. Firstly, Laplace was quoted out of context, and secondly, in order to demonstrate the absurdity of the rule of succession, the author applies it to a case where it does not apply, because there is additional prior information which the rule of succession does not take into account.

But if you go back and read Laplace (1812) himself, you will see that in the very next sentence after this sunrise episode, he warns the reader against just this misunderstanding:

But this number is far greater for him who, seeing in the totality of phenomena the principle regulating the days and seasons, realizes that nothing at the present moment can arrest the course of it.

In this somewhat awkward phraseology he is pointing out to the reader that the rule of succession gives the probability based *only* on the information that the event occurred n times in N trials, and that our knowledge of celestial mechanics represents a great deal of additional information. Of course, if you have additional information beyond the numbers n and N, then you ought to take it into account. You are then considering a different problem, the rule of succession no longer applies, and you can reach an entirely different answer. Probability theory gives the results of consistent plausible reasoning on the basis of the information *which was put into it*.

It has to be admitted that, in mentioning the sunrise at all, Laplace made a very unfortunate choice of an example – because the rule of succession does not really apply to the sunrise, for just the reason that he points out. This choice has had a catastrophic effect on Laplace's reputation ever since. His statements make sense when the reader interprets 'probability', as Laplace did, as a means of representing a state of partial knowledge. But to those who thought of probability as a real physical phenomenon, existing independently of human knowledge, Laplace's position was quite incomprehensible; and so they jumped to the

---

[3] Some passages in the Bible led early theologians to conclude that the age of the world is about 5000 years. It seems that Laplace at first accepted this figure, as did everyone else. But it was during Laplace's lifetime that dinosaur remains were found almost under his feet (under the streets of Montmartre in Paris), and interpreted correctly by the anatomist Cuvier. Had he written this near the end of his life, we think that Laplace would have used a figure vastly greater than 5000 years.

conclusion that Laplace had committed a ludicrous error, without even bothering to read his full statement.

Here are some famous examples of the kind of objections to the rule of succession which may be found in the literature.

(1) Suppose the solidification of hydrogen to have been once accomplished. According to the rule of succession, the probability that it will solidify again if the experiment is repeated is 2/3. This does not in the least represent the state of belief of any scientist.

(2) A boy is 10 years old today. According to the rule of succession, he has the probability 11/12 of living one more year. The boy's grandfather is 70; according to this rule he has the probability 71/72 of living one more year. The rule violates qualitative common sense!

(3) Consider the case $N = n = 0$. It then says that any conjecture without verification has the probability 1/2. Thus there is probability 1/2 that there are exactly 137 elephants on Mars. Also there is probability 1/2 that there are 138 elephants on Mars. Therefore, it is certain that there are at least 137 elephants on Mars. But the rule says also that there is probability 1/2 that there are *no* elephants on Mars. The rule is logically self-contradictory!

The trouble with examples (1) and (2) is obvious in view of our earlier remarks; in each case, highly relevant prior information, known to all of us, was simply ignored, producing a flagrant misuse of the rule of succession. But let's look a little more closely at example (3). Wasn't the rule applied correctly here? We certainly can't claim that we had prior information about elephants on Mars which was ignored. Evidently, if the rule of succession is to survive example (3), there must be some very basic points about the use of probability theory which we need to emphasize.

Now, what do we mean when we say that there is 'no evidence' for a proposition? The question is not what you or I might mean colloquially by such a statement. The question is: *What does it mean to the robot?* What does it mean in terms of probability theory?

The prior information we used in derivation of the rule of succession was that the robot is told that there are only two possibilities: A is true, or A is false. Its entire 'universe of discourse' consists of only two propositions. In the case $N = 0$, we could solve the problem also by direct application of the principle of indifference, and this will of course give the same answer $P(A|X) = 1/2$, that we obtained from the rule of succession. But, just by noting this, we see what is wrong. Merely by admitting the possibility of one of three different propositions being true, instead of only one of two, we have already specified prior information different from that used in deriving the rule of succession.[4]

If the robot is told to consider 137 different ways in which A could be false, and only one way in which it could be true, and is given no other information, then its prior probability for A is 1/138, not 1/2. So, we see that the example of elephants on Mars was, again, a gross misapplication of the rule of succession.

---

[4] We see here only what should have been obvious: that our conclusions from some data can depend on the size of our hypothesis space. We saw a very similar thing in our study of the marginalization paradox in Chapter 15, in the discussion following Eq. (15.92), where we found that the size of a parameter space can affect our inferences. That is, introducing a new parameter can make a difference in our conclusions, even when we have no knowledge of its numerical value.

Moral

Probability theory, like any other mathematical theory, cannot give a definite answer unless we ask it a definite question. We should always start a problem with an explicit enumeration of the 'hypothesis space' consisting of the different propositions that we are going to consider in that problem. That is part of the 'boundary conditions' which must be specified before we have a well-posed mathematical problem. If we say, 'I don't know what the possible propositions are', that is mathematically equivalent to saying, 'I don't know what problem I want to solve'. The only answer the robot can give is: 'Come back and ask me again when you do know'.

18.7 Jeffreys' objection

As one would expect, the example used by Jeffreys (1939, p. 107) is more subtle. He writes:

I may have seen one in 1000 of the 'animals in feathers' in England; on Laplace's theory the probability of the propositions 'all animals with feathers have beaks' would be about 1/1000. This does not correspond to my state of belief, or anybody else's.

Now, while we agree with everything Jeffreys said, we must point out that he failed to add two important facts. Firstly, it is true that, on this evidence, P(all have beaks) $\approx 1/1000$ according to Laplace's rule. But also P(all but one have beaks) $\approx 1/1000$, P(all but two have beaks) $\approx 1/1000, \ldots$, etc. More specifically, if there are N feathered animals of which we have seen r (all with beaks), then rewriting (18.24) in this notation we see that P(all have beaks) $= P_0 = (r+1)/(N+1) \approx 1/1000$, while P (all but n have beaks) is

$$P_n = P_0 \frac{(N-r)!\,(N-n)!}{N!\,(N-n-r)!},\tag{18.26}$$

and the probability that there are n_0 or more without beaks is

$$\sum_{n=n_0}^{N} P_n = \frac{(N-r)!\,(N-n_0+1)!}{(N+1)!\,(N-n_0-r)!} \approx \exp\{-rn_0/N\}.\tag{18.27}$$

Thus if there are one million animals with feathers, of which we have seen 1000 (all with beaks), this leaves it an even bet that there are at least $1000\ln(2) = 693$ without beaks; and, of course, an even bet that the number is less than that. If the only relevant information one had was the aforementioned observation, we think that this *would* be just the proper and reasonable inference.

Secondly, Laplace's rule is not appropriate for this problem because we all have additional prior information that it does not take into account: hereditary stability of form, the fact that a beakless feathered animal would, if it existed, be such an interesting curiosity that we all should have heard of it even if we had not seen it (as has happened in the converse case of the duck-billed platypus), etc. To see fairly and in detail what Laplace's rule (18.24) says, we need to consider a problem where our prior information corresponds better to that supposed in its derivation.

18.8 Bass or carp?

A guide of unquestioned knowledge and veracity assures us that a certain lake contains only two species of fish: bass and carp. We catch ten and find them all to be carp – what is then our state of belief about the percentage of bass? Common sense tells us that, if the fish population were more than about 10% bass, then in ten catches we had a reasonably good chance of finding one; so our state of belief drops off rapidly above 10%. On the other hand, these data D provide no evidence against the hypothesis that the bass population is zero. So common sense without any calculation would lead us to conclude that the bass population is quite likely to be in the range, say, (0%, 15%), but intuition does not tell us quantitatively how likely this is.

What, then, does Laplace's rule say? Denoting the bass fraction by f, its posterior cumulative pdf is $P(f < f_0 | DX) = 1 - (1 - f_0)^{11}$. Thus we have a probability of $1 - (1 - 0.15)^{11} = 0.833$, or odds of 5:1, that the bass population is indeed below 15%. Likewise, the data yield a probability of $2/3$, or odds of 2:1, that the lake contains less than 9.5% bass, and odds of 10:1 that it is less than 19.6%, while the posterior median value is

$$f_{1/2} = 1 - \left(\frac{1}{2}\right)^{1/11} = 0.061, \tag{18.28}$$

or 6.1%; it is an even bet that the bass population is less than this. The interquartile range is $(f_{1/4}, f_{3/4}) = (2.6\%, 11.8\%)$; it is as likely to be within as outside that interval. The 'best' estimate of f by the criterion of minimum mean-square error is Laplace's posterior mean value (18.25): $\langle f \rangle = 1/12$, or 8.3%.

Suppose now that our 11th catch is a bass; how does this change our state of belief? Evidently, we shall revise our estimate of f upward, because the data now *do* provide evidence against the hypothesis that f is very small. Indeed, if the bass population were less than 5%, then we would be unlikely to find one in only 11 catches, so our state of belief drops off rapidly below 5%, but less rapidly than before above 10%.

Laplace's rule agrees, now saying that the best mean-square estimate is $\langle f \rangle = 2/13$, or 15.4%, and the posterior density is $P(df | DX) = 132 f(1 - f)^{10} df$. This yields a median value of 13.6%, raised very considerably because the new datum has effectively eliminated the possibility that the bass population might be below about 3%, which was just the most likely region before. The interquartile range is now (8.3%, 20.9%).

It appears to us that all these numbers correspond excellently to our common sense judgments. This, then, is the kind of problem to which Laplace's rule applies very realistically; i.e. there were known to be only two possibilities at each trial, and our prior knowledge gave no other information beyond assuring us that both were possible. Whenever the result of Laplace's rule of succession conflicts with our intuitive state of belief, we suggest that the reason is that our common sense is making use of additional prior information about the real world situation that is not used in the derivation of the rule of succession.

18.9 So where does this leave the rule?

Mathematically, the rule of succession is the solution to a certain problem of inference, defined by the prior probability and the data. The 200 year old hangup has been over the question: *what* prior information is being described by the uniform prior probability (18.3)? Laplace was not too clear about this – his discussion of it seemed to invoke the idea of a 'probability of a probability' which may appear to be metaphysical nonsense until one has the notion of an inner and outer robot – but his critics, instead of being constructive and trying to define the conceptual problem more clearly, seized upon this to denounce Laplace's whole approach to probability theory.

Of Laplace's critics, only Jeffreys (1939) and Fisher (1956) seem to have thought it through deeply enough to realize that the unclear definition of the prior information was the source of the difficulty; the others, following the example of Venn (1866), merely produce examples where common sense and Laplace's rule are in conflict, and, without making any attempt to understand the reason for it, reject the rule in any and all circumstances. As we noted in Chapter 16, Venn's criticisms were so unjust that even Fisher (1956) was impelled to come to Laplace's defense on this issue.

In this connection we have to remember that probability theory never solves problems of actual practice, because all such problems are infinitely complicated. We solve only idealizations of the real problem, and the solution is useful to the extent that the idealization is a good one. In the example of the solidification of hydrogen, the prior information, which our common sense uses so easily, is actually so complicated that nobody knows how to convert it into a prior probability assignment. There is no reason to doubt that probability theory is, in principle, competent to deal with such problems; but we have not yet learned how to translate them into mathematical language without oversimplifying rather drastically.

In summary, Laplace's rule of succession provides a definite, useful solution to a definite, real problem. Everybody denounces it as nonsense because it is not also the solution to some different problem. The case where the problem can be reasonably idealized to one with only two hypotheses to be considered, a belief in a constant 'causal mechanism', *and no other prior information*, is the only case where it applies. But we can, of course, generalize it to any number of hypotheses, as follows.

18.10 Generalization

We give the derivation in full detail, to present a mathematical technique of Laplace that is useful in many other problems. There are K different hypotheses, $\{A_1, A_2, \ldots, A_K\}$, a belief that the 'causal mechanism' is constant, and no other prior information. We perform a random experiment N times, and observe A_1 true n_1 times, A_2 true n_2 times, etc. Of course, $\sum_i n_i = N$. On the basis of this evidence, what is the probability that in the next $M = \sum_i m_i$ repetitions of the experiment, A_i will be true exactly m_i times? To find the probability $P(m_1 \cdots m_K | n_1, \ldots, n_K)$ that answers this, define the prior knowledge by a K-dimensional uniform prior A_p density:

$$(A_{p_1} \cdots A_{p_K} | X) = C\delta(p_1 + \cdots + p_K - 1), \quad p_i \geq 0. \tag{18.29}$$

To find the normalization constant C, we set

$$\int_0^\infty dp_1 \cdots dp_K \, (A_{p_1} \cdots A_{p_k}|X) = 1 = CI(1),$$ (18.30)

where

$$I(r) \equiv \int_0^\infty dp_1 \cdots dp_k \, \delta(p_1 + \cdots + p_K - r).$$ (18.31)

Direct evaluation of this would be rather messy, because all integrations after the first would be between limits that need to be worked out; so let's use the following trick. Firstly, take the Laplace transform of (18.31):

$$\int_0^\infty dr \, \exp\{-\alpha r\} I(r) = \int_0^\infty dp_1 \cdots dp_K \, \exp\{-\alpha(p_1 + \cdots + p_K)\} = \frac{1}{\alpha^K}.$$ (18.32)

Then, inverting the Laplace transform by Cauchy's theorem,

$$\begin{aligned} I(r) &= \frac{1}{2\pi i} \int_{-i\infty}^{+i\infty} d\alpha \, \frac{\exp\{\alpha r\}}{\alpha^K} \\ &= \frac{1}{(K-1)!} \frac{d^{K-1}}{d\alpha^{K-1}} \exp\{\alpha r\} \Big|_{\alpha=0} \\ &= \frac{r^{K-1}}{(K-1)!}, \end{aligned}$$ (18.33)

where, according to the standard theory of Laplace transforms, the path of integration passes to the right of the origin, and is closed by an infinite semicircle over the left half-plane, the integral over which is zero. Thus,

$$C = \frac{1}{I(1)} = (K-1)!.$$ (18.34)

By this device, we avoided having to consider complicated details about different ranges of integration over the different p_i, that would come up if we tried to evaluate (18.31) directly. The prior $P(n_1 \cdots n_K|X)$ is then, using the same trick,

$$\begin{aligned} P(n_1 \cdots n_K|X) &= \frac{N!}{n_1! \ldots n_K!} \int_0^\infty dp_1 \cdots \int_0^\infty dp_K \, p_1^{n_1} \cdots p_K^{n_K} (A_{p_1} \cdots A_{p_K}|X) \\ &= \frac{N!(K-1)!}{n_1! \cdots n_K!} J(1), \end{aligned}$$ (18.35)

where

$$J(r) \equiv \int_0^\infty dp_1 \cdots dp_K \, p_1^{n_1} \cdots p_K^{n_K} \delta(p_1 + \cdots + p_k - r),$$ (18.36)

which we evaluate as before by taking the Laplace transform:

$$\int_0^\infty dr\, e^{-\alpha r} J(r) = \int_0^\infty dp_1 \cdots dp_K\, p_1^{n_1} \cdots p_K^{n_K} \exp\{-\alpha(p_1 + \cdots + p_K)\}$$

$$= \prod_{i=1}^K \frac{n_i!}{\alpha^{n_i+1}}. \tag{18.37}$$

So, as in (18.33), we have

$$J(r) = \frac{n_1! \cdots n_K!}{2\pi i} \int_{-i\infty}^{+i\infty} d\alpha\, \frac{\exp\{\alpha r\}}{\alpha^{N+K}} = \frac{n_1! \cdots n_K!}{(N+K-1)!} r^{N+K-1} \tag{18.38}$$

and

$$P(n_1 \cdots n_k | X) = \frac{N!\,(K-1)!}{(N+K-1)!}, \quad n_i \geq 0, \quad n_1 + \cdots + n_K = N. \tag{18.39}$$

Therefore, by Bayes' theorem

$$(A_{p_1} \cdots A_{p_K} | n_1 \cdots n_K) = (A_{p_1} \cdots A_{p_K} | X) \frac{P(n_1 \cdots n_K | A_{p_1} \cdots A_{p_K})}{P(n_1 \cdots n_K | X)}$$

$$= \frac{(N+K-1)!}{n_1! \cdots n_K!} p_1^{n_1} \cdots p_K^{n_K} \delta(p_1 + \cdots + p_K - 1), \tag{18.40}$$

and finally

$$P(m_1 \cdots m_K | n_1 \cdots n_K)$$

$$= \int_0^\infty dp_1 \cdots dp_K\, P(m_1 \cdots m_K | A_{p_1} \cdots A_{p_K})(A_{p_1} \cdots A_{p_K} | n_1 \cdots n_K)$$

$$= \frac{M!}{m_1! \cdots m_K!} \frac{(N+K-1)!}{n_1! \cdots n_K!} \int_0^\infty dp_1 \cdots dp_K\, p_1^{n_1+m_1} \cdots p_K^{n_K+m_K}$$

$$\times \delta(p_1 + \cdots + p_K - 1). \tag{18.41}$$

The integral is the same as $J(1)$ except for the replacement $n_i \rightarrow n_i + m_i$. So, from (18.38),

$$P(m_1 \cdots m_K | n_1 \cdots n_K) = \frac{M!}{m_1! \cdots m_K!} \frac{(N+K-1)!}{n_1! \cdots n_K!} \frac{(n_1+m_1)! \cdots (n_K+m_K)!}{(N+M+K-1)!} \tag{18.42}$$

or, reorganizing into binomial coefficients, the generalization of (18.24) is

$$P(m_1 \cdots m_K | n_1 \cdots n_K) = \frac{\dbinom{n_1+m_1}{n_1} \cdots \dbinom{n_K+m_K}{n_K}}{\dbinom{N+M+K-1}{M}}. \tag{18.43}$$

In the case where we want just the probability that A_1 will be true on the next trial, we need this formula with $M = m_1 = 1$, all other $m_i = 0$. The result is the generalized rule of succession:

$$P(A_1|n_1NK) = \frac{n_1 + 1}{N + K}.$$ (18.44)

We see that, in the case $N = n_1 = 0$, this reduces to the answer provided by the principle of indifference, which it therefore contains as a special case. If K is a power of 2, this is the same as a method of inductive reasoning proposed by R. Carnap (1942), which he denotes $c^*(h, e)$ in his *Continuum of Inductive Methods*.

Use of the rule of succession in cases where N is very small is rather foolish, of course. Not really wrong; just foolish. Because, if we have no prior evidence about A, and we make such a small number of observations that we have practically no evidence, well, that's just not a very promising basis on which to do plausible reasoning. We can't expect to get anything useful out of it. We do, of course, obtain definite numerical values for the probabilities, but these values are very 'soft,' i.e. very unstable, because the A_p distribution is still very broad for small N. Our common sense tells us that the evidence N_n for small N provides no reliable basis for further predictions, and we'll see that this conclusion also follows as a consequence of the theory we are developing here.

The real reason for introducing the rule of succession lies in the cases where we *do* obtain a significant amount of information from the experiment; i.e. when N is a large number. In this case, fortunately, we can pretty much forget about these fine points concerning prior evidence. The particular initial assignment $(A_p|X)$ will no longer have much influence on the results, for the same reason as in the particle-counter problem of Chapter 6. This remains true for the generalized case leading to (18.43). You see from (18.44) that, as soon as the number of observations N is large compared with the number of hypotheses K, then the probability assigned to any particular hypothesis depends, for all practical purposes, only on what we have observed, and not on how many prior hypotheses there are. If you contemplate this for ten seconds, your common sense will tell you that the criterion $N \gg K$ is exactly the right one for this to be so.

In the literature starting with Venn (1866), those who issued polemical denunciations of Laplace's rule of succession have put themselves in an incredible situation. How is it possible for one human mind to reject Laplace's rule – and then advocate a frequency definition of probability? Anyone who assigns a probability to an event equal to its observed frequency in many trials is doing just what Laplace's rule tells him to do! The generalized rule (18.44) supplies an obviously needed refinement of this, small correction terms when the number of observations is not large compared with the number of propositions.

18.11 Confirmation and weight of evidence

A few new ideas – or rather, connections with familiar old ideas – are suggested by our calculations involving A_p. Although we shall not make any particular use of them, it seems worthwhile to point them out. We saw that the stability of a probability assignment in the face of new evidence is essentially determined by the width of the A_p distribution. If E is prior evidence and F is new evidence, then

$$P(A|EF) = \int_0^1 dp\, p\, (A_p|EF) = \frac{\int_0^1 dp\, p\, (A_p|F)(A_p|E)}{\int_0^1 dp\, (A_p|F)(A_p|E)}. \qquad (18.45)$$

We might say that F is *compatible* with E, as far as A is concerned, if having the new evidence, F, doesn't make any appreciable change in the probability of A;

$$P(A|EF) = P(A|E). \qquad (18.46)$$

The new evidence can make an enormous change in the distribution of A_p without changing the first moment. It might sharpen it up very much, or broaden it. We could become either more certain or more uncertain about A, but if F doesn't change the center of gravity of the A_p distribution, we still end up assigning the same probability to A.

Now, the stronger property: the new evidence F *confirms* the previous probability assignment, if F is compatible with it, and at the same time, gives us more confidence in it. In other words, we exclude one of these possibilities, and with new evidence F the A_p distribution narrows. Suppose F consists of performing some random experiment and observing the frequency with which A is true. In this case $F = N_n$, and our previous result, Eq. (18.22), gives

$$(A_p|N_n) = \frac{(N+1)!}{n!(N-n)!} p^n (1-p)^{N-n} \approx (\text{constant}) \cdot \exp\left\{-\frac{(p-f)^2}{2\sigma^2}\right\}, \qquad (18.47)$$

where

$$\sigma^2 = \frac{f(1-f)}{n}, \qquad (18.48)$$

and $f = (n/N)$ is the observed frequency of A. The approximation is found by expanding $\log(A_p|N_p)$ in a Taylor series about its peak value, and is valid when $n \gg 1$ and $(N-n) \gg 1$. If these conditions are satisfied, then $(A_p|N_n)$ is very nearly symmetric about its peak value. Then, if the observed frequency f is close to the prior probability $P(A|E)$, the new evidence N_n will not affect the first moment of the A_p distribution, but will sharpen it up, and that will constitute a confirmation as we have defined it.

This shows one more connection between probability and frequency. We defined the 'confirmation' of a probability assignment according to entirely different ideas than are usually used to define it. We define it in a way that agrees with our intuitive notion of confirmation of a previous state of mind. But it turned out that the *same* experimental evidence would constitute confirmation on either the frequency theory or our theory.

Now, from this we can see another useful notion, which we will call weight of evidence. Consider A_p, given two different pieces of evidence, E and F,

$$(A_p|EF) = (\text{constant}) \times (A_p|E)(A_p|F). \qquad (18.49)$$

If the distribution $(A_p|F)$ was very much sharper than the distribution $(A_p|E)$, then the product of the two would still have a peak at practically the value determined by F. In this case, we would say intuitively that the evidence F carries much greater 'weight' than the evidence E. If we have F, it doesn't really matter much whether we take E into account or not. On the other hand, if we don't have F, then whatever evidence E may represent will be extremely significant, because it will represent the best we are able to do. So, acquiring one piece of evidence which carries a great amount of weight can make it, for all practical purposes, unnecessary to continue keeping track of other pieces of evidence which carry only a small weight.

Of course, this is the way our minds operate. When we receive one very significant piece of evidence, we no longer pay so much attention to vague evidence. In so doing, we are not being very inconsistent, because it wouldn't make much difference anyway. So, our intuitive notion of weight of evidence is bound up with the sharpness of the A_p distribution. Evidence concerning A that we consider very significant is not necessarily evidence that makes a big change in the probability of A. It is evidence that makes a big change in our density for A_p. Seeing this, we can gain a little more insight into the principle of indifference and also make contact between this theory and Carnap's methods of inductive reasoning.

18.11.1 *Is indifference based on knowledge or ignorance?*

Before we can use the principle of indifference to assign numerical values of probabilities, there are two different conditions that must be satisfied: (1) we must be able to analyze the situation into mutually exclusive, exhaustive possibilities; (2) having done this, we must then find the available information gives us no reason to prefer any of the possibilities to any other. In practice, these conditions are hardly ever met unless there's some evident element of symmetry in the problem. But there are two entirely different ways in which condition (2) might be satisfied. It might be satisfied as a result of ignorance, or it might be satisfied as a result of positive knowledge about the situation. To illustrate this, let's suppose that a person who is known to be very dishonest is going to toss a coin, and that there are two people watching him. Mr A is allowed to examine the coin. He has all the facilities of the National Bureau of Standards at his disposal. He performs hundreds of experiments with scales and calipers, magnetometers and microscopes, X-rays and neutron beams, and so on. Finally, he is convinced that the coin *is* perfectly honest. Mr B is not allowed to do this. All he knows is that a coin is being tossed by a shady character. He suspects the coin is biased, but he has no idea in which direction. Condition (2) is satisfied equally well for both of them. Each would start out by assigning probability one-half to each face. The same probability assignment can describe a condition of complete ignorance or a condition of very great knowledge. This has seemed paradoxical for a long time. Why doesn't Mr A's extra knowledge make any difference? Well, of course, it *does* make a difference. It makes a very important difference, but one that doesn't show up until we start performing this experiment. The difference is not in the probability for A, but in the density for A_p.

Suppose the first toss is heads. To Mr B, that constitutes evidence that the coin is biased to favor heads. And so, on the next toss, he would assign new probabilities to take that into account. But to Mr A, the evidence that the coin is honest carries overwhelmingly greater weight than the evidence of one throw, and he'll continue to assign a probability of $1/2$.

You see what's going to happen. To Mr B, every toss of the coin represents new evidence about its bias. Every time it's tossed, he will revise his assignment for the next toss; but, after several tosses, his assignment will get more and more stable, and in the limit $n \to \infty$ they will tend to the observed frequency of heads. To observer A, the prior evidence of symmetry continues to carry greater weight than the evidence of almost any number of throws, and he persists in assigning the probability $1/2$. Each has done consistent plausible reasoning on the basis of the information available to him, and our theory accounts for the behavior of each.

If you assumed that Mr A had perfect knowledge of symmetry, you might conclude that his A_p distribution is a δ-function. In that case, his mind could never be changed by any amount of new data. Of course, that's a limiting case that's never reached in practice. Not even the Bureau of Standards can give us evidence that good.

18.12 Carnap's inductive methods

The philosopher Rudolph Carnap (1952) gives an infinite family of possible 'inductive methods' by which one can convert prior information and frequency data into a probability assignment and an estimate of frequencies for this future. His *ad hoc* principle (that is, a principle that is found from intuition rather than from the rules of probability theory) is that the final probability assignment $P(A|N_n X)$ should be a weighted average of the prior probability $P(A|X)$ and the observed frequency, $f = n/N$. Assigning a weight N to the 'empirical factor' f, and an arbitrary weight λ to the 'logical factor' $P(A|X)$ leads to the method which Carnap denotes by $c_\lambda(h, e)$. Introduction of the A_p distribution accounts for this in more detail; the theory developed here includes all of Carnap's methods as special cases corresponding to different prior densities $(A_p|X)$, and leads us to reinterpret λ as the weight of prior evidence. Thus, in the case of two hypotheses, the Carnap λ method is the one you can calculate from the prior density $(A_p|X) = (\text{constant}) \cdot [p(1-p)]^r$, with $2r = \lambda - 2$. The result is

$$P(A|N_n X) = \frac{2n + \lambda}{2N + 2\lambda} = \frac{(n+r)+1}{(N+2r)+2}. \tag{18.50}$$

Greater λ thus corresponds to a more sharply peaked $(A_p|X)$ density.

In our coin-tossing example, Mr A from the Bureau of Standards reasons according to a Carnap method with λ of the order of, perhaps, thousands; while Mr B, with much less prior knowledge about the coin, would use a λ of perhaps 5 or 6. (The case $\lambda = 2$, which gives Laplace's rule of succession, is much too broad to be realistic for coin tossing; for Mr B surely knows that the center of gravity of a coin can't be moved by more than half its thickness from the geometrical center. Actually, as we saw in Chapter 10, this analysis isn't always applicable to tossing of real coins, for reasons having to do with the laws of physics.)

From the second way we wrote Eq. (18.50), we see that the Carnap λ method corresponds to a weight of prior evidence which would be given by $(\lambda - 2)$ trials, in exactly half of which A was observed to be true. Can we understand why the weighting of prior evidence is $\lambda =$ (number of prior trials $+2$), while that of the new evidence N_p is only (number of new trials) $= N$? Well, look at it this way. The appearance of the $(+2)$ is the robot's way of telling us this: prior knowledge that is *possible* for A to be either true or false is equivalent to knowledge that A has been true at least once, and false at least once. This is hardly a derivation, but it makes reasonably good sense.

Let's pursue this line of reasoning a step further. We started with the statement X: it is *possible* for A to be either true or false at any trial. But that is still a somewhat vague statement. Suppose we interpret it as meaning that A has been observed true exactly once, and false exactly once. If we grant that this state of knowledge is correctly described by Laplace's assignment $(A_p|X) = 1$, then *what was the 'pre-prior' state of knowledge X_0 before we had the data X?* To answer this, we need only to apply Bayes' theorem backwards, as we did in the method of imaginary results in Chapter 5 and in urn sampling in Chapter 6. The result is: our 'pre-prior' A_p distribution must have been

$$(A_p|X_0)\,dp = \text{(constant)}\frac{dp}{p(1-p)}. \tag{18.51}$$

This is just the quasi-distribution representing 'complete ignorance', or the 'basic measure' of our parameter space, that we found by transformation groups in Chapter 12 and which Haldane (1932) had suggested long ago. So, here is another line of thought that could have led us to this measure. By the same line of thought we found the discrete version of (18.51) already in Chapter 6, Eq. (6.49).

It appears, then, that if we have definite prior evidence that it *is* possible for A to be either true or false on any one trial, then Laplace's rule $(A_p|X) = 1$ is the appropriate one to use. But if initially we are so completely uncertain that we're not even sure whether it is *possible* for A to be true on some trials and false on others, then we should use the prior (18.51).

How different are the numerical results which the pre-prior assignment (18.51) gives us? Repeating the derivation of (18.22) with this pre-prior assignment, we find that, provided n is not zero or N,

$$(A_p|N_n X_0) = \frac{(N-1)!}{(n-1)!\,(N-n-1)!}p^{n-1}(1-p)^{N-n-1}, \tag{18.52}$$

which leads, instead of to Laplace's rule of succession, to the mean-value estimate of p:

$$P(A|N_n X_0) = \int_0^1 dp\, p\,(A_p|N_n) = \frac{n}{N}, \tag{18.53}$$

equal to the observed frequency, and identical with the maximum likelihood estimate of p. Likewise, provided $0 < n < N$, we find instead of (18.24) the formula

$$P(M_m|N_n X_0) = \frac{\binom{m+n-1}{m}\binom{M-m+N-n-1}{M-m}}{\binom{N+M-1}{M}}. \tag{18.54}$$

All of these results correspond to having observed one less success and one less failure.

18.13 Probability and frequency in exchangeable sequences

We are now in a position to say quite a bit more about connections between probability and frequency. There are two main types of connections: (a) given an observed frequency in a random experiment, convert this information into a probability assignment; and (b) given a probability assignment, predict the frequency with which some condition will be realized. We have seen, in Chapters 11 and 12, how the principles of maximum entropy and transformation groups lead to probability assignments which, if the quantity of interest happens to be the result of some 'random experiment', correspond automatically to predicted frequencies, and thus solve problem (b) in some situations.

The rule of succession gives us the solution to problem (a) in a wide class of problems. If we have observed whether A was true in a very large number of trials, *and the only knowledge we have about A is the result of this random experiment, and the consistency of the 'causal mechanism'*, then it says that the probability we should assign to A at the next trial becomes practically equal to the observed frequency. Now, in fact, this is exactly what people who define probability in terms of frequency do: one postulates the existence of an unknown 'absolute' probability, whose numerical value is to be found by performing random experiments. Of course, you must perform a very large number of experiments. Then the observed frequency of A is taken as the estimate of the probability. As we saw earlier in this chapter, even the $+1$ and $+2$ in Laplace's formula turn up when the 'frequentist' refines his methods by taking the center of a confidence interval. So, I don't see how even the most ardent advocates of the frequency theory of probability can damn the rule of succession without thereby damning his own procedures; after all polemics, there remains the simple fact that in his own procedures, he is doing exactly what Laplace's rule of succession tells him to do. Indeed, to define probability in terms of frequency is equivalent to saying that the rule of succession is the *only* rule which can be used for converting observational data into probability assignments.

18.14 Prediction of frequencies

Now let's consider problem (b) in this situation: to reason from a probability to a frequency. This is simply a problem of parameter estimation, not different in principle from any other. Suppose that instead of asking for the probability that A will be true in the next trial, we wish to infer something about the relative frequency of A in an indefinitely large number of trials, on the basis of the evidence N_n. We must take the limit of (18.24) as $M \to \infty$, $m \to \infty$, in such a way that $(m/M) \to f$. Introducing the proposition

$$A_f = \text{the frequency of } A \text{ true in the indefinitely large number} \atop \text{of trials is } f, \tag{18.55}$$

we find in the limit that the probability density of A_f, given N_n, is

$$P(A_f|N_n) = \frac{(N+1)!}{N!(N-n)!} f^n (1-f)^{N-n}, \tag{18.56}$$

which is the same as our $(A_p|N_n)$ in (18.22), with f numerically equal to p. According to (18.55), the most probable frequency is equal to (n/N), the observed frequency in the past. But we have noted before that in parameter estimation (if you object to my calling f a 'parameter', then let's just call it 'prediction'), the most probable value is usually a poorer estimate than the mean value in the small sample case, where they can be appreciably different. The mean-value estimate of the frequency is

$$\overline{f} = \int_0^1 \mathrm{d} f \, f \, P(A_f|N_n) = \frac{n+1}{N+2}, \tag{18.57}$$

i.e. just the same as the value of $P(A|N_n)$, (18.25), given by Laplace's rule of succession. Thus, we can interpret the rule in either way; *the probability which Laplace's theory assigns to A at a single trial is numerically equal to the estimate of frequency which minimizes the expected square of the error.* You see how nicely this corresponds with the relationship between probability and frequency which we found in the maximum entropy and transformation group arguments.

Note also that the distribution $P(A_f|N_n)$ is quite broad for small N, confirming our expectation that no reliable predictions should be possible in this case. As a numerical example, if A has been observed true once in two trials, then $\overline{f} = P(A|N_n) = 1/2$, but according to (18.55) it is still an even bet that the true frequency f lies outside the interval $0.326 < f < 0.674$. With no evidence at all ($N = n = 0$), it would be an even bet that f lies outside the interval $0.25 < f < 0.75$. More generally, the variance of (18.55) is

$$\mathrm{var} P(A_f|N_n) = \overline{f^2} - \overline{f}^2 = \frac{\overline{f}(1-\overline{f})}{N+3}, \tag{18.58}$$

so that the expected error in the estimate (18.56) decreases like $1/\sqrt{N}$. More detailed conclusions about the reliability of predictions, which we could make from (18.56), are, for all practical purposes, identical with those the statistician would make by the method of confidence intervals.

All these results hold also for the generalized rule of succession. Taking the limit of (18.43) as $M \to \infty$, $m_i/M \to f_i$, we find the joint probability density function for A_i to occur with frequency f_i to be

$$P(f_1 \cdots f_k | n_1 \cdots n_k) = \frac{(N+K)!}{n_1! \cdots n_k!} (f_1^{n_1} \cdots f_k^{n_k}) \delta(f_1 + \cdots + f_k - 1). \tag{18.59}$$

The probability that the frequency f_1 will be in the range $\mathrm{d} f_1$ is found by integrating (18.59) over all values of $f_2, \ldots, f_k$ compatible with $f_1 \geq 0$, $(f_2 + \cdots + f_k) = 1 - f_1$. This can be carried out by application of Laplace transforms in a well-known way, and the result is

$$P(f_1 | n_1 \cdots n_k) = \frac{(N+K-1)!}{n_1!(N-n_1+K-2)!} f_1^{n_1} (1-f_1)^{N-n_1+K-2}, \tag{18.60}$$

from which we find the most probable value to be

$$(\hat{f}_1) = \frac{n_1}{N+K-2} \tag{18.61}$$

and the mean value to be

$$\overline{f_1} = \frac{n_1 + 1}{N + K},\qquad(18.62)$$

which is Laplace's rule of succession (18.44).

Another interesting result is found by taking the limit of $P(M_m|A_p)$ in (18.18) as $M \to \infty$, $(m/M) \to f$; we find

$$P(M_m|A_p) = \delta(f - p).\qquad(18.63)$$

Likewise, taking the limit of $(A_p|N_n)$ in (18.22) as $N \to \infty$, we find

$$(A_p|A_f) = \delta(f - p),\qquad(18.64)$$

which follows from (18.63) by application of Bayes' theorem. Therefore, if B is any proposition, we have, from our standard argument,

$$P(B|A_f) = \int_0^1 \mathrm{d}p\,(BA_p|A_f) = \int_0^1 \mathrm{d}p\,P(B|A_pA_f)(A_p|A_f)$$
$$= \int_0^1 \mathrm{d}p\,P(B|A_p)\delta(p - f).\qquad(18.65)$$

In the last step we used the property (18.1) that A_p automatically neutralizes any other statement about A. Thus, if f and p are numerically equal, we have $P(B|A_p) = P(B|A_p)$; A_p and A_f are *equivalent* statements in their implication for plausible reasoning.

To verify this equivalence in one case, note that in the limit $N \to \infty$, $(n/N) \to f$, $P(M_m|N_n)$ in Eq. (18.24) reduces to the binomial distribution $P(M_m|A_p)$ as given by (18.18). The generalized formula (18.43), in the corresponding limit, goes into the multinomial distribution,

$$P(m_1 \cdots m_k|f_1 \cdots f_k) = \frac{m!}{m_1! \cdots m_k!} f_1^{m_1} \cdots f_k^{m_k}.\qquad(18.66)$$

This equivalence shows why it is so easy to confuse the notion of probability and frequency, and why in many problems this confusion does no harm. Whenever the available information consists of observed frequencies in a large sample, and constancy of the 'causal mechanism', Laplace's theory becomes mathematically equivalent to the frequency theory. Most of the 'classical' problems of statistics (life insurance, etc.) are of just this type; and as long as one works only on such problems, all is well. The harm arises when we consider more general problems.

Today, physics and engineering offer many important applications for probability theory in which there is an absolutely essential part of the evidence which cannot be stated in terms of frequencies, and/or the quantities about which we need plausible inference have nothing to do with frequencies. The axiom (probability) $\equiv$ (frequency), if applied consistently, would prevent us from using probability theory in these problems.

18.15 One-dimensional neutron multiplication

Our discussion so far has been rather abstract; perhaps too much so. In order to make amends for this, I would like to show you a specific physical problem where these equations apply. This was first described in a short note by Bellman, Kalaba and Wing (1957) and further developed in a more recent book by Wing (1962). Neutrons are traveling in fissionable material, and we want to estimate how many new neutrons will be produced in the long run as a consequence of one incident trigger neutron. In order to have a tractable mathematical problem, we make some drastic simplifying assumptions as follows.

(a) The neutrons travel only in the $\pm x$ direction at a constant velocity.
(b) Each time a neutron, traveling either to the right or the left, initiates a fission reaction, the result is exactly two neutrons, one traveling to the right, one to the left. The net result is therefore that any neutron will from time to time emit a progeny neutron traveling in the opposite direction.
(c) The progeny neutrons are immediately able to produce still more progeny in the same manner.

We fire a single trigger neutron into a thickness x of fissionable material from the left, and the problem is to predict the number of neutrons that will emerge from the left and from the right, over all time, as a consequence. At least, that is what we would *like* to calculate. But, of course, the number of emerging neutrons is not *determined* by any of the given data, and so the best we can do is to calculate the *probability* that exactly n neutrons will be transmitted or reflected. We want to make a detailed comparison of the Laplace theory and the frequency theory of probability, as applied to the initial formulation of this problem. We are concerned mainly with the underlying rationale by which we relate probability theory to the physical model.

Many proponents of the frequency theory berate the Laplace theory on purely philosophical grounds that have nothing to do with its success or failure in applications. There is a more defensible position, held by some, who recognize that the present state of affairs gives them no reason for smugness, and a good reason for caution. While they believe that at present the frequency theory is superior, they also say, as one of my correspondents did to me, 'I will most cheerfully renounce the frequency theory for any theory that yields me a better understanding and a more efficient formalism.' The trouble is that the current statistical literature gives us no opportunity to see the Laplace theory in actual use so that valid comparisions could be made; and that is the situation we are trying to correct here.

18.15.1 The frequentist solution

Firstly, let us formulate the problem as it would be done using the frequency theory. Here is the way in which the 'frequentist' would reason:

The experimentalists have measured for us the relative frequency $p = a\Delta$ of fission in a very small thickness Δ of this material. This means that they have fired N trigger neutrons at a thin film of thickness Δ, and observed fission in n cases. Since N is finite, we cannot find the exact value of p from this, but it is approximately equal to the observed frequency (n/M). More precisely, we can find confidence limits for p. In similar situations, we can expect about $k\%$ of the time, the limits (Cramér, 1946, p. 515)

$$\frac{N}{N+\lambda^2}\left[\frac{2n+\lambda^2}{2N}\pm\lambda\sqrt{\frac{n(N-n)}{N^3}+\frac{\lambda^2}{4N^2}}\right] \tag{18.67}$$

will include the true value of p, where λ is the $(100-k)\%$ value of a normal deviate. For example, with $\lambda = \sqrt{2}$, the range

$$\frac{n+1}{N+2}\pm\frac{N}{N+2}\sqrt{\frac{2n(N-n)}{N^3}+\frac{1}{N^2}}=\frac{n+1}{N+2}\pm\sqrt{\frac{2n(N-n)}{N^3}} \tag{18.68}$$

will cover the correct p in about 84% of similar cases. [Again, there's that +1 and +2 of Laplace's rule of succession!] In general, the connection between λ and k is given by

$$\frac{1}{\sqrt{2\pi}}\int_{-\lambda}^{\lambda}dx\exp\left\{-\frac{x^2}{2}\right\}=\frac{k}{100}. \tag{18.69}$$

Equation (18.67) is an approximation valid when the numbers n and $(N-n)$ are sufficiently large; the exact confidence limits are difficult to express analytically, and for small N one should consult the graphs of E.S. Pearson and Clopper (1934). The number p is, of course, a definite, but imperfectly known, *physical constant* characteristic of the fissionable material.

Now, in order to calculate the relative frequency with which n neutrons will be reflected from a thickness x of this material, we have to make some additional assumptions. We assume that the probability of fission per unit length is always the same for each neutron independent of its history. Due to the complexity of the causes operating, it seems reasonable to assume this; but the real test of whether it is a valid assumption can come only from comparison of the final results of our calculation with experiment. This assumption means that the probabilities of fission in successive slabs of thickness Δ are independent, so that, for example, the probability that an incident neutron will undergo fission in the second slab of thickness Δ, but not in the first, is the product $p(1-p)$.

At this point, we turn to the mathematics and solve the problem by any one of several possible techniques, emerging with the relative frequencies $p_n(x)$, $q_n(x)$ for reflection or transmission of n neutrons, respectively. [Actually, the analytical solution has not yet been found, but Wing (1962) gives the results of numerical integration, which is equally good for our purposes.]

We now compare these predictions with experiment. When the first trigger neutron is fired into the thickness x, we observe r_1 neutrons reflected and t_1 neutrons transmitted. These data do not in any way affect the assignments $p_n(x)$ and $q_n(x)$, since the latter have no meaning in terms of a single experiment, but are predictions only of limiting frequencies for an indefinitely large number of experiments. We therefore must repeat the experiment many times, and record the numbers r_i, t_i for each experiment. If we find that the frequency of cases for which $r_i = n$ tends sufficiently close to $p_n(x)$ ('sufficiently close' being determined by certain significance tests such as chi-squared), then we conclude that the theory is satisfactory; or at least that it is not rejected by the data. If, however, the observed frequencies show a wide departure from $p_n(x)$, then we know that there is something wrong with our initial set of assumptions.

Now, of course, the theory is either right or wrong. If it is wrong, then in principle the entire theory is demolished, and we have to start all over again, trying to find the right theory. In practice, it may happen that only one minor feature of the theory has to be changed, so that most of the old calculations will be useful in the new theory.

18.15.2 The Laplace solution

Now let's state this same problem in terms of Laplace's theory. We regard it simply as an exercise in plausible reasoning, in which we make the best possible guesses as to the outcome of a *single* experiment, or of a finite number of them. We are not concerned with the prediction, or even the existence, of limiting frequencies; because any assertion about the outcome of an impossible experiment is obviously an empty statement, and cannot be relevant to any applications. We reason as follows.

The experimentalists have provided us with the evidence N_n, by firing N neutrons at a thin film of thickness Δ, and observing fission in n cases. Since by hypothesis the only prior knowledge was that a neutron either will or will not undergo fission, we have just the situation where Laplace's rule of succession applies and the probability, on this evidence, of fission for the $(N + 1)$th neutron in thickness Δ, is

$$p \equiv P(F_{N+1}|N_n) = \frac{n+1}{N+2}, \tag{18.70}$$

where

$$F_m \equiv \text{the } n\text{th neutron will undergo fission.} \tag{18.71}$$

Whether N is large or small, the question of the 'accuracy' of this probability does not arise – it is exact by definition. Of course, we will prefer to have as large a value of N as possible, since this increases the weight of the evidence N_n and makes the probability p not more *accurate*, but more *stable*. The probability p is manifestly *not* a physical property of the fissionable material, but is only a means of describing our state of knowledge about it, on the basis of the evidence N_n. If the preliminary experiment had yielded a different result N'_n, then we would of course assign a different probability p'; but the properties of the fissionable material would remain the same.

We now fire a neutron at a thickness $x = M_\delta$, and define the propositions

$F^n \equiv$ the neutron will cause fission in the nth slab of thickness Δ;

$f^n \equiv$ the neutron will not cause fission in the nth slab.

The probability of fission in slab 1 is then

$$p \equiv P(F_1|N_n) = \frac{n+1}{N+2}.$$ (18.72)

But now the probability that fission will occur in the second but not the first slab is *not* $p(1 - p)$ as in the first treatment. At this point we see one of the fundamental differences between the theories. From the product rule, we have

$$P(F^2 f^1|N_n) = P(F^2|f^1 N_n)P(f^1|N_n)$$

$$= \frac{n+1}{N+2}\left[1 - \frac{n+1}{N+2}\right]$$ (18.73)

$$= \frac{(n+1)(N-n+1)}{(N+2)(N+3)}.$$

The difference is that, in calculating the probability $P(F^2|f^1 N_n)$, we must take into account the evidence, f^1, that a neutron has passed through one more thickness Δ without fission. This amounts to one more experiment in addition to that leading to N_n. The evidence f^1 is fully as cogent as N_n, and it would be clearly inconsistent to take one into account and ignore the other. Continuing in this way, we find that the probability that the incident neutron will emit exactly m first-generation progeny in passing through thickness $M\Delta$ is just the expression

$$P(M_m|N_n) = \binom{M}{m}\frac{(n+m)!(N+1)!(N+M-n-m)!}{n!(N-n)!(N+M+1)!},$$ (18.74)

which we have derived before, Eq. (18.24). Now, if N is not a very large number, this may differ appreciably from the value

$$P(M_m|A_p) = \binom{M}{m}p^m(1-p)^{M-m},$$ (18.75)

which one obtains in the frequency approach. However, note again that, as the weight of evidence N_n increases, we find $(A_{p'}|N_n) \rightarrow \delta(p' - n/N)$, and

$$P(M_n|N_n) \rightarrow P(M_m|A_p)$$ (18.76)

in the limit $N \rightarrow \infty$, $(n/N) \rightarrow p$. The difference in the two results is negligible whenever $N \gg M$; i.e. when the weight of evidence N_n greatly exceeds M_m. Now let's study the difference between (18.74) and (18.75) more closely. From (18.74) we have, for the mean-value estimate of m, on the Laplace theory,

$$\overline{m} = M\frac{n+1}{N+2}.$$ (18.77)

To state the accuracy of this estimate, we can calculate the variance of the distribution (18.74). This is most easily done by using the representation (18.23):

$$\overline{m^2} = \sum_{m=0}^{M} m^2 \int_0^1 dp\, P(M_m|A_p)(A_p|N_n)$$

$$= \frac{(N+1)!}{n!(N-n)!} \int_0^1 dp\, \left[M_p + M(M-1)p^2\right] p^n (1-p)^{N-n} \qquad (18.78)$$

$$= M\frac{n+1}{N+2} + M(M-1)\frac{(n+1)(n+2)}{(N+2)(N+3)},$$

which gives the variance

$$V \equiv \overline{M^2} - \overline{m}^2 = M\left[\frac{N+M+2}{N+3}\right]\left[\frac{n+1}{N+2}\right]\left[n - \frac{n+1}{N+2}\right]; \qquad (18.79)$$

while, from (18.75), the frequency theory gives

$$\overline{m_0} = M_p \qquad (18.80)$$

$$V_0 \equiv \left[\overline{m^2} - \overline{m}^2\right]_0 = M_p(1-p). \qquad (18.81)$$

If the frequentist takes the center of the confidence interval (18.68) as his 'best' estimate of p, then he will take $p = (n+1)/(N+2)$ in these equations. So, we both obtain the same estimate, but the variance (18.79) is greater by the amount

$$V - V_0 = \frac{M-1}{N+3}M_p(1-p). \qquad (18.82)$$

Why this difference? Why is it that the Laplace theory seems to determine the value of m less precisely then the frequency theory? Well, appearances are deceptive here. The fact is that the Laplace theory determines the value of m *more* precisely than the frequency theory; the variance (18.81) is not the entire measure of the uncertainty as to m on the frequency theory, because there is still the uncertainty as to the 'true' value of p. According to (18.81), p is uncertain by about $\pm\sqrt{2p(1-p)/N}$, so the mean value (18.80) is uncertain by about

$$\pm\sqrt{\frac{2p(1-p)}{N}} \qquad (18.83)$$

in addition to the uncertainty represented by (18.81). If we suppose that the uncertainties (18.81) and (18.83) are independent, the total mean-square uncertainty as to the value of m on the frequency theory would be represented by the sum of (18.81) and

$$M^2\frac{2p(1-p)}{N}, \qquad (18.84)$$

which more than wipes out the difference (18.82). The factor of 2 in (18.84) would of course be changed somewhat by adopting a different confidence level, but no reasonable choice can change it very much.

In the frequency theory, the two uncertainties (18.81) and (18.84) appear as entirely separate effects which are determined by applying two different principles: one by conventional probability theory, the other by confidence intervals. In the Laplace theory no such distinction exists; both are given automatically by a *single* calculation. We found exactly this same situation in our particle-counter problem in Chapter 6, when we compared our robot's procedure with that of the orthodox statistician.

The mechanism by which the Laplace theory is able to do this is very interesting. It is just the difference already noted; in the derivation of (18.74), we are continually taking into account additional evidence accumulated in the new experiment, such as F^1 in (18.73). In the frequency theory, the uncertainty (18.83) in p arises because only a finite amount of data was provided by the preliminary experiment given N_n. It is just for that reason that new evidence, such as F^1, is still relevant. In giving a consistent treatment of *all* the evidence, the Laplace theory automatically includes the effect of the finiteness of the preliminary data, which the frequency theory is able to do only crudely by the introduction of confidence intervals. In the Laplace theory there is no need to decide on any arbitrary 'confidence level' because probability theory, when consistently applied to the *whole* problem, already tells us what weight should be given to the preliminary data N_n. What we get in return for this is not merely a more unified treatment; in yielding a smaller net uncertainty in m, the Laplace theory shows that the two sources of uncertainty (18.81) and (18.84) of the frequency theory are *not* independent: they have a small negative correlation, so that they tend to compensate each other. That is the reason for Laplace's smaller probable error. If you think about this very hard, you will be able to see intuitively why this negative correlation has to be there – I won't deprive you of the pleasure of figuring it out for yourself. All this subtlety is completely lost in the frequency theory.

'But,' someone will object, 'you are ignoring a very practical consideration which was the original reason for introducing confidence intervals. While I grant that *in principle* it is better to treat the whole problem in a single calculation, *in practice* we usually have to break it up into two different ones. After all, the preliminary data N_n were obtained by one group of people, who had to communicate their results to another group, who then carried out the second calculation applying these data. It is a practical necessity that the first group be able to state their conclusions in a way that tells honestly what they found, *and how reliable it was*. Their data can also be used in many other ways than in your second calculation, and the introduction of confidence intervals thus fills a very important practical need for communication between different workers.'

Of course, if you have followed everything so far, you know the answer to this. The memory storage problem was our original point of departure, and the problem just discussed is a specific example of just what we pointed out more abstractly in Eq. (18.16). You see from (18.23), and also in our derivation of (18.79), that the only property of the preliminary data which we needed in order to analyze the whole problem was the A_p distribution $(A_p|N_n)$

that resulted from the preliminary experiment. The principle of confidence intervals was introduced to fill a very practical need. But there was no need to introduce any new principle for this purpose; it is already contained in probability theory, which shows that the *exact* way of communicating what you have learned is not by specifying confidence intervals, but by specifying your final A_p distribution.

As a further point of comparison, note that in the Laplace theory there was no need to introduce any 'statistical assumption' about independence of events in successive slabs of thickness Δ. In fact, the theory told us, as in (18.73), that these probabilities are *not* independent when we have only a finite amount of preliminary data; and it was just this fact that enabled the Laplace theory to take account of the uncertainty which the frequency theory describes by means of confidence intervals.

This brings up a very fundamental point about probability theory, which the frequency theory fails to recognize, but which is essential for applications to both communication theory and statistical mechanics, as we will show later. What do we mean by saying that two events are 'independent'?

In the frequency theory, the only kind of independence recognized is *causal* independence; i.e. the fact that one event occurred does not in itself exert any physical influence on the occurrence of the other. Thus, in the coin-tossing example discussed in Chapter 6, the fact that the coin comes up heads on one toss of course doesn't *physically* affect the result of the next toss, and so on the frequency theory one would call the coin-tossing experiment a typical case of 'independent repetitions of a random experiment'; the probability of a heads at both tosses *must* be the product of separate probabilities. But then we lose any way of describing the difference between the reasoning of Mr A and Mr B!

In Laplace's theory, 'independence' means something entirely different, which we see from a glance at the product rule: $P(AB|C) = P(B|C)P(A|BC)$. Independence means that $P(A|BC) = P(A|C)$; i.e. *knowledge* that B is true does not affect the probability we assign to A. Thus, independence means not mere *causal* independence, but *logical* independence. Even though heads at one toss does not physically predispose the coin to give heads at the next, the *knowledge* that we got heads may have a very great influence on our predictions as to the next toss.

The importance of this is that the various limit theorems, which we will say more about later, require independence in their derivations. Consequently, even though there may be strict *causal* independence, if there is not also *logical* independence, these limit theorems will not hold. Writers of the frequency school of thought, who deny that probability theory has anything to do with inductive reasoning, recognize the existence only of *causal* connections, and, as a consequence, they have long been applying these limit theorems to physical and communications processes where, we claim, they are incorrect and completely misleading. This was noted long ago by Keynes (1921), who stressed exactly this same point.

I think these comparisons make it very clear that, at least in this kind of problem, the Laplace theory *does* provide the 'better understanding and more efficient formalism' that my colleague asked for.

18.16 The de Finetti theorem

So far we have considered the notion of an A_p distribution and derived a certain class of probability distributions from it, under the restriction that the *same* A_p distribution is to be used for all trials. Intuitively, this means that we have assumed the underlying 'mechanism' as constant, but unknown. It is clear that this is a very restrictive assumption, and the question arises: How general is the class of probability functions that we can obtain in this way? In order to state the problem clearly, let us define

$$ x_n \equiv \begin{cases} 1 & \text{if } A \text{ is true on the } n\text{th trial} \\ 0 & \text{if } A \text{ is false on the } n\text{th trial.} \end{cases} \tag{18.85} $$

Then a state of knowledge about N trials is described in the most general way by a probability function $P(x_1 \ldots x_N | N)$, which could, in principle, be defined arbitrarily (except for normalization) at each of the 2^N points.

We now ask: What is a necessary and sufficient condition on $P(x_1 \ldots x_N | N)$ for it to be derivable from an A_p distribution? What test could we apply to a given distribution $P(x_1 \ldots x_N | N)$ to tell whether it is included in our theory as given above? A necessary condition is clear from our previous equations: any distribution obtainable from an A_p distribution necessarily has the property that the probability that A is true in n *specified* trials, and false in the remaining $(N - n)$ trials, depends only on the numbers n and N; i.e. not on *which* trials in $1 \le n \le N$ were specified. If this is so, we say that $P(x_1 \ldots x_N | N)$ defines an *exchangeable sequence*.

An important theorem of de Finetti (1937) asserts that the converse is also true: *any exchangeable probability function $P(x_1 \ldots x_N | N)$ can be generated by an A_p distribution.* Thus there is a function $(A_p | X) = g(p)$ such that $g(p) \ge 0$, $\int_0^1 dp\, g(p) = 1$, and the probability that in N trials A is true in n specified trials and false in the remaining $(N - n)$ trials, is given by

$$ P(n|N) = \int_0^1 dp\, p^n (1 - p)^{N-n} g(p). \tag{18.86} $$

This can be proved as follows. Note that $p^n (1 - p)^{N-n}$ is a polynomial of degree N:

$$ p^n (1 - p)^{N-n} = p^n \sum_{m=0}^{N-n} \binom{N-m}{m} (-p)^m $$

$$ = \sum_{k=0}^{N} \alpha_k(N, n) p^k, \tag{18.87} $$

which defines $\alpha_k(N, n)$. Therefore, *if* (18.86) holds, we would have

$$ P(n|N) = \sum_{k=0}^{N} \alpha_k(N, n) \beta_k, \tag{18.88} $$

where

$$\beta_k = \int_0^1 dp\, p^k g(p) \tag{18.89}$$

is the kth moment of $g(p)$. Thus, specifying $\beta_0, \ldots, \beta_N$ is equivalent to specifying all the $P(n|N)$ for $n = 0, \ldots, N$. Conversely, for given N, specifying $P(n|N)$, $0 \le n \le N$, is equivalent to specifying $\{\beta_0, \ldots, \beta_N\}$. In fact, β_N is the probability that $x_1 = x_2 = \cdots = x_N = 1$, regardless of what happens in later trials, and its relation to $P(n|N)$ can be established directly without reference to any function $g(p)$.

So, the problem reduces to this: if the numbers $\beta_0, \ldots, \beta_N$ are specified, under what conditions does a function $g(p) \ge 0$ exist such that (18.89) holds? This is just the well-known *Hausdorff moment problem*, whose solution can be found in many places; for example Widder (1941, Chap. 3). Translated into our notation, the main theorem is as follows. A necessary and sufficient condition that a function $g(p) \ge 0$ exists satisfying (18.89) (and therefore also (18.86)) is that there exists a number B such that

$$\sum_{n=0}^N \binom{N}{n} P(n|N) \le B, \qquad N = 0, 1, \ldots . \tag{18.90}$$

But, from the interpretation of $P(n|N)$ as probabilities, we see that the equality sign always holds in (18.90) with $B = 1$, and the proof is completed.

Here is another way of looking at it, which might be made into a proof with a little more work, and perhaps discloses more clearly the intuitive reason for the de Finetti theorem, as well as showing immediately just how much we have said about $g(p)$ when we specify the $P(n|N)$. Imagine $g(p)$ expanded in the form

$$g(p) = \sum_{n=0}^{\infty} a_n \phi_n(p), \tag{18.91}$$

where $\phi_n(p)$ are the complete orthonormal set of polynomials in $0 \le p \le 1$, essentially the Legendre functions:

$$\phi_n(p) = \frac{\sqrt{2n+1}}{n!} \frac{d^n}{dp^n} [p(1-p)]^n \tag{18.92}$$

$$= (-1)^n \sqrt{2n+1}\, P_n(2p-1),$$

where $\phi_n(p)$ is a polynomial of degree n, and satisfies

$$\int_0^1 dp\, \phi_m(p) \phi_n(p) = \delta_{mn}. \tag{18.93}$$

If we substitute (18.93) into (18.86), only a finite number of terms will survive, because $\phi_k(p)$ is orthogonal to all polynomials of degree $N < k$. Then, it is easily seen that, for given N, specifying the values of $P(n|N)$, $0 \le n \le N$, is equivalent to specifying the first $(n+1)$ expansion coefficients $\{a_0, \ldots, a_N\}$. Thus, as $N \to \infty$, a function $g(p)$, defined by (18.91), becomes uniquely determined to the same extent that a Fourier series uniquely determines its generating function; i.e. 'almost everywhere'. The main trouble with this argument is that the condition $g(p) \geq 0$ is not so easily established from (18.91).

18.17 Comments

The de Finetti theorem is very important to us because it shows that the connection between probability and frequency which we have found in this chapter holds for a fairly wide class of probability functions $P(x_1, \ldots, x_N | N)$, namely the class of all exchangeable sequences. These results, of course, generalize immediately to the case where there are more than two possible outcomes at each trial.

Possibly even more important, however, is the light which the de Finetti theorem sheds on one of the oldest controversies in probability theory – Laplace's first derivation of the rule of succession. The idea of an A_p distribution is not, needless to say, our own invention. The way we have introduced it here is only our attempt to translate into modern language what we think Laplace was trying to say in that famous passage, 'When the probability of a simple event is unknown, we may suppose all possible values of this probability between 0 and 1 as equally likely.' This statement, which we interpret as saying that, with no prior evidence, $(A_p | X) = \text{const.}$, has been rejected as utter nonsense by virtually everyone who has written on probability theory in this century. And, of course, on any frequency definition of probability, Laplace's statement would have no justification at all. But on any theory it is conceptually difficult, since it seems to involve the idea of a 'probability of a probability', and the use of an A_p distribution in calculations has been largely avoided since the time of Laplace.

The de Finetti theorem puts some much more solid ground under these methods. Independently of all conceptual problems, it is a *mathematical theorem* that whenever you talk about a situation where the probability of a certain sequence of results depends only on the number of successes, not on the particular trials at which they occur, all your probability distributions can be generated from a single function $g(p)$, in just the way we have done here. The use of this generating function is, moreover, a very powerful technique mathematically, as you will quickly discover if you try to repeat some of the above derivations (for example, Eq. (18.24)) without using an A_p distribution. So, it doesn't matter what we might think about the A_p distribution conceptually; its validity as a mathematical tool for dealing with exchangeable sequences is a proven fact, standing beyond the reach of philosophical objections.

19

Physical measurements

We have seen, in Chapter 7, how the great mathematician Leonhard Euler was unable to solve the problem of estimating eight orbital parameters from 75 discrepant observations of the past positions of Jupiter and Saturn. Thinking in terms of deductive logic, he could not even conceive of the principles by which such a problem could be solved. But, 38 years later, Laplace, thinking in terms of probability theory as logic, was in possession of exactly the right principles to resolve the great inequality of Jupiter and Saturn. In this chapter we develop the solution as it would be done today by considering a simpler problem, estimating two parameters from three observations. But our general solution, in matrix notation, will include Laplace's automatically.

19.1 Reduction of equations of condition

Suppose we wish to determine the charge e and mass m of the electron. The Millikan oil-drop experiment measures e directly. The deflection of an electron beam in a known electromagnetic field measures the ratio e/m. The deflection of an electron toward a metal plate due to attraction of image charges measures e^2/m.

From the results of any two of these experiments we can calculate values of e and m. But all the measurements are subject to error, and the values of e and m obtained from different experiments will not agree. Yet each of the measurements does contain some information relevant to our question that is not contained in the others. How are we to process the data so as to make use of all the information available and get the best estimates of e and m? What is the probable error remaining? How much would the situation be improved by including still another experiment of given accuracy? Probability theory gives simple and elegant answers to these questions.

More specifically, suppose we have the results of these experiments:

(1) measures e with $\pm 2\%$ accuracy;
(2) measures (e/m) with $\pm 1\%$ accuracy;
(3) measures (e^2/m) with $\pm 5\%$ accuracy.

Supposing the values of e and m approximately known in advance to be $e \approx e_0$, $m \approx m_0$, the measurements are then linear functions of the corrections. Write the unknown true values of e and m as

$$e = e_0(1 + x_1),$$
$$m = m_0(1 + x_2); \tag{19.1}$$

then, x_1, x_2 are dimensionless corrections, small compared with unity, and our problem is to find the best estimates of x_1 and x_2. The results of the three measurements are three numbers M_1, M_2 and M_3, which we write as

$$M_1 = e_0(1 + y_1),$$

$$M_2 = \frac{e_0}{m_0}(1 + y_2), \tag{19.2}$$

$$M_3 = \frac{e_0^2}{m_0}(1 + y_3),$$

where the y_i are also small dimensionless numbers which are defined by (19.2) and are therefore known in terms of the old estimates e_0, m_0 and the new measurements M_1, M_2, M_3. On the other hand, the true values of e, e/m, e^2/m are expressible in terms of the x_j:

$$e = e_0(1 + x_1),$$

$$\frac{e}{m} = \frac{e_0(1 + x_1)}{m_0(1 + x_2)} = \frac{e_0}{m_0}(1 + x_1 - x_2 + \cdots), \tag{19.3}$$

$$\frac{e^2}{m} = \frac{e_0^2(1 + x_1)^2}{m_0(1 + x_2)} = \frac{e_0}{m_0}(1 + 2x_1 - x_2 + \cdots),$$

where higher order terms are considered negligible. Comparing (19.2) and (19.3), we see that if the measurements were exact we would have

$$y_1 = x_1,$$
$$y_2 = x_1 - x_2, \tag{19.4}$$
$$y_3 = 2x_1 - x_2.$$

But, taking into account the errors, the known y_i are related to the unknown x_j by

$$y_1 = a_{11}x_1 + a_{12}x_2 + \delta_1,$$
$$y_2 = a_{21}x_1 + a_{22}x_2 + \delta_2, \tag{19.5}$$
$$y_3 = a_{31}x_1 + a_{32}x_2 + \delta_3,$$

where the coefficients a_{ij} form a (3×2) matrix:

$$A = \begin{pmatrix} a_{11} & a_{12} \\ a_{21} & a_{22} \\ a_{31} & a_{32} \end{pmatrix} = \begin{pmatrix} 1 & 0 \\ 1 & -1 \\ 2 & -1 \end{pmatrix}, \tag{19.6}$$

and the δ_i are the unknown fractional errors of the three measurements. For example, the statement that $\delta_2 = -0.01$ means that the second measurement gave a result 1% too small.

More generally, we have n unknown quantities $\{x_1, \ldots, x_n\}$ to be estimated from N imperfect observations $\{y_1, \ldots, y_N\}$, and the N equations of condition,

$$y_i = \sum_{j=1}^{n} a_{ij} x_j + \delta_i, \qquad i = 1, 2, \ldots, N, \tag{19.7}$$

or, in matrix notation,

$$y = Ax + \delta, \tag{19.8}$$

where A is an $(N \times n)$ matrix. In the present discussion we suppose the problem 'overde-termined' in the sense that $N > n$. This condition defeated Euler (1749), who was facing the case $N = 75$, $n = 8$. But we keep in mind that the cases $N = n$ (ostensibly well-posed) and $N < n$ (underdetermined) can also arise in real problems, and it will be interesting to see what probability theory has to say about them.

In the early 19th century, it was common to reason as follows. It seems plausible that the best estimate of each x_j will be some linear combination of all the y_i, but if $N > n$ we cannot simply solve equation (19.8) for x, since A is not a square matrix and has no inverse. However, we can get a system of equations solvable for x if we take n linear combinations of the equations of condition; i.e. if we multiply (19.8) on the left by some $(n \times N)$ matrix B. Then the product BA exists and is a square $(n \times n)$ matrix. Choose B so that $(BA)^{-1}$ exists. Then the linear combinations are the n rows of

$$By = BAx + B\delta, \tag{19.9}$$

which has the unique solution

$$x = (BA)^{-1} B(y - \delta). \tag{19.10}$$

If the probabilities of various fractional errors δ_i are symmetric: $p(\delta_i) = p(-\delta_i)$ so that $\langle \delta_i \rangle = 0$, then corresponding to any given matrix B the 'best' estimate of x_j by almost any reasonable loss function criterion will be the jth row of

$$\hat{x} = (BA)^{-1} By, \tag{19.11}$$

but by making different choices of B (i.e. taking different linear combinations of the equations of condition) we get different estimates. In Euler's problem there were billions of possible choices. Which choice of B is best?

In the above we have merely restated, in modern notation but old language, the problem of 'reduction of equations of condition' described in Laplace's *Essai Philosophique* (1812). A popular criterion for solution was the principle of least squares: find that matrix B for which the sum of the squares of the errors in $\hat{x}_j$ is a minimum; or perhaps use a weighted sum. This problem can be solved directly; we shall find the same solution by different reasoning below.

19.2 Reformulation as a decision problem

We really solved this problem in Chapter 13, where we have seen in generality that the best estimate of any parameter, by the criterion of any loss function, is found by applying Bayes' theorem to find the probability, conditional on the data, that the parameter lies in various intervals, then making that estimate which minimizes the expected loss taken over the posterior probabilities.

Now in the original formulation of the problem, as given above, it was only a plausible conjecture that the best estimate of x_j is a linear combination of the y_i as in (19.11). The material in Chapter 13 shows us a much better way of formulating the problem, in which we don't have to depend on conjecture. Instead of trying to take linear combinations without knowing which combinations to take, we should apply Bayes' theorem directly to the equations of condition. Then, if the best estimates are indeed of the linear form (19.11), Bayes' theorem should not only tell us that fact, it will give us automatically the best choice of the matrix B and also tell us the accuracy of those estimates, which least squares does not give at all.

Let's do this calculation for the case that we assign independent Gaussian probabilities to the errors δ_i of the various measurements. From our discussion in Chapter 7 we expect this to be, nearly always, the best error law we can assign from the information we have. But in the orthodox literature one would not see it that way; instead one would argue that in most physical measurements the total error is the sum of contributions from many small, causally independent imperfections, and the central limit theorem would then lead us to a Gaussian *frequency distribution* of errors.[1] There is nothing wrong with that argument, except that it has been psychologically misleading to generations of workers, who concluded that if the frequency distribution of errors is not in fact Gaussian, then to assign a Gaussian probability distribution is to 'assume' something that is not true; and this will lead to some horrible kind of error in our final conclusions.

19.2.1 Sermon on Gaussian error distributions

The considerations of Chapter 7 reassure us that this danger is grossly exaggerated. The point is that, in probability theory as logic, the Gaussian probability assignment is not an *assumption* about the frequencies of the errors; it is a *description* of our state of knowledge about the errors. We hardly ever have prior knowledge about the errors beyond the general magnitude to be expected, which we can interpret reasonably as specifying the first two moments of the error distribution. This leads, by the principle of maximum entropy, to an independent Gaussian probability assignment as the one which agrees with that information without assuming anything else. The region Ω of reasonably probable noise vectors $(\delta_1, \ldots, \delta_N)$ or the region $Ax + \Omega$ of reasonably probable data vectors, is then as large as

---

[1] As noted in Chapter 14, this is subject to an important qualification: that in general the Gaussian approximation will be good only for those values of total error δ which can arise in many different ways by combination of the individual elementary errors. For unusually wide deviations we do not expect, and hardly ever observe, Gaussian frequencies.

it can be while agreeing with the second moment constraints. The frequency distribution of errors is almost always unknown before seeing the data; but even if it is far from Gaussian, the Gaussian probability assignment will still lead us to the best inferences possible from the information we have.

The privileged status of a Gaussian frequency distribution lies in a more subtle fact: acquisition of new information does not affect our inferences if that new information is only what we would have predicted from our old information. Thus, if we assigned Gaussian probabilities and then acquired new information that the true frequency distribution of errors is indeed Gaussian with the specified variance, *this would not help us* because it is only what we would have predicted. But if we had additional prior information about the specific way in which the error frequencies depart from Gaussian, that would be cogent new information constraining the possible error vectors to a smaller domain ($\Omega_1 \subset \Omega$). This would enable us to improve our parameter estimates over the ones to be obtained below, because data vectors in the complementary set ($\Omega - \Omega_1$), which were previously dismissed as noise, are now recognized as indicating a real 'signal'. Bayes' theorem does all this for us automatically.

Thus the covenant that we have with Nature is considerably more favorable than is supposed in orthodox teaching; for, given second moments, a non-Gaussian frequency distribution will not make our inferences worse, but *knowledge* of a non-Gaussian distribution would enable us to make them still better than the results to be found below.

Encouraged by the message of this sermon, we assign the probability for the errors $\{\delta_1, \ldots, \delta_N\}$ to lie in the intervals $\{d\delta_1, \ldots, d\delta_N\}$, respectively, as

$$p(\delta_1 \cdots \delta_N)d\delta_1 \cdots d\delta_N = (\text{const.}) \exp\left\{-\frac{1}{2}\sum_{i=1}^{N} w_i \delta_i^2\right\} d\delta_1 \cdots d\delta_N, \tag{19.12}$$

where the 'weight' w_i is the reciprocal variance of the error of the ith measurement. For example, the crude statement that the first measurement has $\pm 2\%$ accuracy now becomes the more precise statement that the first measurement has weight

$$w_1 = \frac{1}{\langle \delta_1^2 \rangle} = \frac{1}{(0.02)^2} = 2500. \tag{19.13}$$

For the time being, we suppose these weights known, as is generally the case with astronomical and other physical data. From (19.7) and (19.12) we have immediately the sampling probability density for obtaining measured values $\{y_1, \ldots, y_N\}$ given the true values $\{x_1, \ldots, x_N\}$:

$$p(y_1 \cdots y_N | x_1 \cdots x_n) = C_1 \exp\left\{-\frac{1}{2}\sum_{i=1}^{N} w_i \left[y_i - \sum_{j=1}^{n} a_{ij}x_j\right]^2\right\}, \tag{19.14}$$

where C_1 is independent of the y_i. According to Bayes' theorem, if we assign uniform prior probabilities to the x_j, then the posterior probability density for the x_j, given the actual measurements y_i, is of the form

$$p(x_1 \cdots x_n | y_1 \cdots y_N) = C_2 \exp \left\{ -\frac{1}{2} \sum_{i=1}^{N} w_i \left[y_i - \sum_{j=1}^{n} a_{ij} x_j \right]^2 \right\}, \qquad (19.15)$$

where now C_2 is independent of the x_j. Next, as in nearly all Gaussian calculations, we need to reorganize this quadratic form to bring out the dependence on the x_i. Expanding it, we have

$$\sum_{i=1}^{N} w_i \left(y_i - \sum_{j=1}^{n} a_{ij} x_j \right)^2 = \sum_{i=1}^{N} w_i \left\{ y_i^2 - 2 y_i \sum_{j=1}^{n} a_{ij} x_j + \sum_{j,k=1}^{n} a_{ij} a_{ik} x_j x_k \right\}$$
$$= \sum_{j,k=1}^{n} K_{jk} x_j x_k - 2 \sum_{j=1}^{n} L_j x_j + \sum_{i=1}^{N} w_i y_i^2, \qquad (19.16)$$

where

$$K_{jk} = \sum_{i=1}^{N} w_i a_{ij} a_{ik}, \qquad L_j = \sum_{i=1}^{N} w_i y_i a_{ij}, \qquad (19.17)$$

or, defining a diagonal 'weight' matrix $W_{ij} = w_i \delta_{ij}$, we have a matrix K and a vector L:

$$K = \tilde{A} W A, \qquad L = \tilde{A} W y, \qquad (19.18)$$

where $\tilde{A}$ is the transposed matrix. We want to write (19.15) in the form

$$p(x_1 \cdots x_n | y_1 \cdots y_N) = C_3 \exp \left\{ -\frac{1}{2} \sum_{j,k=1}^{n} K_{jk} (x_j - \hat{x}_j)(x_k - \hat{x}_k) \right\} \qquad (19.19)$$

whereupon the $\hat{x}_j$ will be the mean-value estimates desired. Comparing (19.16) and (19.19) we see that

$$\sum_{k=1}^{n} K_{jk} \hat{x}_k = L_j, \qquad (19.20)$$

so if K is nonsingular we can solve uniquely for $\hat{x}$.

19.3 The underdetermined case: K is singular

If we have fewer observations than parameters, $N < n$, then, from (19.17), K is still an $(n \times n)$ matrix, but it is at most of rank N, and so is necessarily singular. Then the trouble is not that (19.20) has no solution; but rather that it has an infinite number of them. The maximum likelihood is attained not at a point, but on an entire linear manifold of dimensionality $(n - N)$. Of course, maximum-likelihood solutions still exist, as is seen from the fact that, although $(\tilde{A} W A)^{-1}$ does not exist, $(A\tilde{A})^{-1}$ does, and so the parameter estimate

$$x^* = \tilde{A}(A\tilde{A})^{-1} y \qquad (19.21)$$

now makes the quadratic form in (19.15) vanish: $y = Ax^*$, achieving the maximum possible likelihood. This is called the canonical inverse solution, and the principle of maximum entropy may be used to calculate it. But the canonical inverse is far from unique, for we see from (19.8) that if we add to the estimate (19.21) any solution z of the homogeneous equation $Az = 0$, we have another estimate $x^* + z$ with just as high a likelihood; and there is a linear manifold Δ of such vectors $x^* + z$, of dimensionality $n - N$.

Exercise 19.1. Show that the canonical inverse solution (19.21) is also a least squares one, making $\sum (x_i^*)^2$ a minimum on the manifold Δ. Unfortunately, there seems to be no compelling reason why one should want the vector of estimates to have minimum length.

For a long time, no satisfactory way of dealing with such problems was recognized; yet we are not entirely helpless, for the data do restrict the possible values of the parameters $\{x_i\}$ to a 'feasible set' Δ satisfying (19.20). The data alone are incapable of picking out any unique point in this set; but the data may be supplemented with prior information which enables us to make a useful choice in spite of that. These are 'generalized inverse' problems, which are of current importance in many applications, such as image reconstruction. In fact, in the real world, generalized inverse problems probably make up the great majority, because the real world seldom favors us with all the information needed to make a well-posed problem. Yet useful solutions may be found in many cases by maximum entropy, which resolves the ambiguity in a way that is 'optimal' by several different criteria, as described in Chapters 11 and 20.

19.4 The overdetermined case: K can be made nonsingular

By its definition (19.17), K is an $(n \times n)$ matrix, and for all real $\{q_1, \ldots, q_n\}$ such that $\sum q_i^2 > 0$,

$$\sum_{j,k=1}^{n} K_{jk} q_j \, q_k = \sum_{i=1}^{N} w_i \left(\sum_{j=1}^{n} a_{ij} q_j \right)^2 \geq 0, \tag{19.22}$$

so if K is of rank n it is not only nonsingular, but positive definite. If $N \geq n$ this will be the case unless we have done something foolish in setting up the problem – including a useless observation or an irrelevant parameter.

In the first place, we suppose all the weights w_i to be positive: if any observation y_i has weight $w_i = 0$, then it is useless in our problem; i.e. it can convey no information about the parameters and we should not have included it in the data set at all. We can reduce N by one.

Secondly, if there is a nonzero vector q for which $\sum_j a_{ij} q_j$ is zero for all i, then in (19.7), for all c, the parameter sets $\{x_j\}$ and $\{x_j + cq_j\}$ would lead to identical data, and so could not be distinguished whatever the data. In other words, there is an irrelevant parameter in the problem which has nothing to do with the data; we can reduce n by one. Mathematically, this means that the columns of the matrix A are not linearly independent; then, if $q_k \neq 0$, we can remove the parameter x_k and the kth column of A with no essential change in the problem (i.e. no change in the information we get from it).

Removing irrelevant observations and parameters if necessary, and finally, the number of cogent observations is at least as great as the number of relevant parameters, then K is a positive definite matrix and (19.20) has a unique solution

$$\hat{x}_k = \sum_{j=1}^{n} (K^{-1})_{kj} L_j. \tag{19.23}$$

From (19.18), we can write the result as

$$\hat{x} = (\tilde{A}WA)^{-1}\tilde{A}Wy, \tag{19.24}$$

and, comparing with (19.11), we see that, in the Gaussian case with uniform prior probabilities, the best estimates are indeed linear combinations of the measurements, of the form (19.11), and the best choice of the matrix B is

$$B = \tilde{A}W, \tag{19.25}$$

a result perhaps first found by Gauss, and repeated in Laplace's *Essai Philosophique*. Let us evaluate this solution for our simple problem.

19.5 Numerical evaluation of the result

Applying the solution (19.24) to our problem of estimating e and m, the measurements of e, (e/m) and (e^2/m) were of 2%, 1% and 5% accuracy, respectively, and so

$$w_2 = \frac{1}{(0.01)^2} = 10\,000$$
$$w_3 = \frac{1}{(0.05)^2} = 400, \tag{19.26}$$

and we found $w_1 = 2500$ before. Thus we have

$$B = \tilde{A}W = \begin{pmatrix} 1 & 1 & 2 \\ 0 & -1 & -1 \end{pmatrix} \begin{pmatrix} w_1 & 0 & 0 \\ 0 & w_2 & 0 \\ 0 & 0 & w_3 \end{pmatrix} = \begin{pmatrix} w_1 & w_2 & 2w_3 \\ 0 & -w_2 & -w_3 \end{pmatrix}, \tag{19.27}$$

$$K = \tilde{A}WA = \begin{pmatrix} [w_1 + w_2 + 4w_3] & -[w_2 + 2w_3] \\ -[w_2 + 2w_3] & [w_2 + w_3] \end{pmatrix}, \tag{19.28}$$

$$K^{-1} = (\tilde{A}WA)^{-1} = \frac{1}{|K|} \begin{pmatrix} [w_2 + w_3] & [w_2 + 2w_3] \\ [w_2 + 2w_3] & [w_1 + w_2 + 4w_3] \end{pmatrix}, \tag{19.29}$$

where

$$|K| = \det(K) = w_1 w_2 + w_2 w_3 + w_3 w_1. \tag{19.30}$$

Thus, the final result is

$$(\tilde{A}WA)^{-1}\tilde{A}W = \frac{1}{|K|} \begin{pmatrix} w_1[w_2 + w_3] & -w_2 w_3 & w_2 w_3 \\ w_1[w_2 + 2w_3] & -w_2[w_1 + 2w_3] & w_3[w_2 - w_1] \end{pmatrix}, \tag{19.31}$$

and the best point estimates of x_1, x_2 are

$$\hat{x}_1 = \frac{w_1(w_2 + w_3)y_1 + w_2 w_3(y_3 - y_2)}{w_1 w_2 + w_2 w_3 + w_3 w_1},$$

$$\hat{x}_2 = \frac{w_1 w_2(y_1 - y_2) + w_2 w_3(y_3 - 2y_2) + w_3 w_1(2y_1 - y_3)}{w_1 w_2 + w_2 w_3 + w_3 w_1}. \tag{19.32}$$

Inserting the numerical values of w_1, w_2 and w_3, we have

$$\hat{x}_1 = \frac{13}{15}y_1 + \frac{2}{15}(y_2 - y_3),$$

$$\hat{x}_2 = \frac{5}{6}(y_1 - y_2) + \frac{2}{15}(y_3 - 2y_2) + \frac{1}{30}(2y_1 - y_3), \tag{19.33}$$

which exhibits the best estimates as weighted averages of the estimates taken from all possible pairs of experiments. Thus, y_1 is the estimate of x_1 obtained in the first experiment, which measures e directly. The second and third experiments combined yield an estimate of e given by $(e^2/m)(e/m)^{-1}$. Since

$$\frac{\frac{e_0^2}{m_0}(1 + y_3)}{\frac{e_0}{m_0}(1 + y_2)} \approx e_0(1 + y_3 - y_2), \tag{19.34}$$

$(y_3 - y_2)$ is the estimate of x_1 given by experiments 2 and 3. Equation (19.33) says that these two independent estimates of x_1 should be combined with weights 13/15 and 2/15. Likewise, $\hat{x}_2$ is given as a weighted average of three different (although not independent) estimates of x_2.

19.6 Accuracy of the estimates

From (19.19) we find the second central moments of $p(x_1 \cdots x_n | y_1 \cdots y_N)$:

$$\langle (x_j - \hat{x}_j)(x_k - \hat{x}_k) \rangle = \langle x_j x_k \rangle - \langle x_j \rangle \langle x_k \rangle = (K^{-1})_{jk}. \tag{19.35}$$

Thus, from the $(n \times n)$ inverse matrix

$$K^{-1} = (\tilde{A}WA)^{-1} \tag{19.36}$$

already found in our calculation of $\hat{x}_j$, we can also read off the probable errors, or, more conveniently, the standard deviations. From (19.29) we can state the results in the form (mean) $\pm$ (standard deviation) as

$$(x_j)_{\text{est}} = \hat{x}_j \pm \sqrt{(K^{-1})_{jj}}. \tag{19.37}$$

Equations (19.24) and (19.37) represent the general solution of the problem, which Euler needed. In the present case this is

$$(x_1)_{\text{est}} = \hat{x}_1 \pm \sqrt{\frac{w_2 + w_3}{w_1 w_2 + w_2 w_3 + w_3 w_1}},$$

$$(x_2)_{\text{est}} = \hat{x}_2 \pm \sqrt{\frac{w_1 + w_2 + 4w_3}{w_1 w_2 + w_2 w_3 + w_3 w_1}} \tag{19.38}$$

with numerical values

$$x_1 = \hat{x}_1 \pm 0.0186,$$
$$x_2 = \hat{x}_2 \pm 0.0216 \tag{19.39}$$

so that from the three measurements we obtain e with $\pm 1.86\%$ accuracy and m with $\pm 2.16\%$ accuracy.

How much did the rather poor measurement of (e^2/m), with only $\pm 5\%$ accuracy, help us? To answer this, note that in the absence of this experiment we would have arrived at conclusions given by (19.28), (19.29) and (19.32) in the limit $w_3 \to 0$. The results (also easily verified directly from the statement of the problem) are

$$\hat{x}_1 = y_1,$$
$$\hat{x}_2 = y_1 - y_2, \tag{19.40}$$

$$K^{-1} = \frac{1}{w_1 w_2} \begin{pmatrix} w_2 & w_2 \\ w_2 & [w_1 + w_2] \end{pmatrix}, \tag{19.41}$$

or, the (mean) $\pm$ (standard deviation) values are

$$x_1 = y_1 \pm \frac{1}{w_1} = y_1 \pm 0.020,$$

$$x_2 = y_1 - y_2 \pm \sqrt{\frac{w_1 + w_2}{w_1 w_2}} = y_1 - y_2 \pm 0.024. \tag{19.42}$$

As might have been anticipated by common sense, a low-accuracy measurement can add very little to the results of accurate measurements, and if the (e^2/m) measurement had been much worse than $\pm 5\%$ it would hardly be worthwhile to include it in our calculations. But suppose that an improved technique gives us an (e^2/m) measurement of $\pm 2\%$ accuracy. How much would this help? The answer is given by our previous formulas with $w_1 = w_3 = 2500$, $w_2 = 10\,000$. We find now that the mean-value estimates give much higher weight to the estimates using the (e^2/m) measurement:

$$\hat{x}_1 = 0.556y_1 + 0.444(y_3 - y_2),$$
$$\hat{x}_2 = 0.444(y_1 - y_2) + 0.444(y_3 - 2y_2) + 0.112(2y_1 - y_3),$$

(19.43)

which is to be compared with (19.33). The standard deviations are given by

$$x_1 = \hat{x}_1 \pm 0.0149,$$
$$x_2 = \hat{x}_2 \pm 0.020.$$

(19.44)

The accuracy of e (x_1) is improved roughly twice as much as that of m (x_2), since the improved measurement involves e^2, but only the first power of m.

Exercise 19.2. Write a computer program which solves this problem for general N and n, with $N \geq n$, and test it on the problem just solved. Estimate how long it would require for the compiled program to solve Euler's problem.

In the above we supposed the weights w_i known from prior information. If this is not the case, there are many different conceivable kinds of partial prior information about them, leading to many different possible prior probability assignments $p(w_1 \cdots w_n | I)$. This will make some minor quantitative changes in details, but no new difficulties of principle; only a straightforward mathematical generalization following the already established Bayesian principles.

19.7 Comments

19.7.1 A paradox

We can learn many more things from studying this problem. For example, let us note something which you will find astonishing at first. If you study (19.32), which gives the best estimate of m from the three measurements, you will see that y_3, the result of the (e^2/m) measurement, enters into the formula in a different way than y_1 and y_2. It appears once with a positive coefficient, and once with a negative one. If $w_1 = w_2$, these coefficients are equal, and (19.32) collapses to

$$\hat{x}_2 = y_1 - y_2.$$

(19.45)

Now, realize the full implications of this: *it says that the only reason we make use of the (e^2/m) measurement in estimating m is that the (e) measurement and the (e/m) measurement have different accuracy.* No matter how accurately we know (e^2/m), if the (e) and (e/m) measurements happen to have the same accuracy, however poor, then we should ignore the good measurement and base our estimate of m only on the (e) and (e/m) measurements!

We think that, on first hearing, your intuition will revolt against this conclusion, and your first reaction will be that there must be an error in (19.32). So, check the derivation at your leisure. This is a perfect example of the kind of result which probability theory gives us almost without effort, but which our unaided common sense might not notice in years of thinking about the problem. We won't deprive you of the pleasure of resolving this 'paradox' for yourself, and explaining to your friends how it can happen that consistent inductive reasoning may demand that you throw away your best measurement.

In Chapter 17, we complained about the fact that orthodox statisticians sometimes throw away relevant data in order to fit a problem to their preconceived model of 'independent random errors'. Are we now guilty of the same offense? No doubt, it looks very much that way! Yet we plead innocence: the numerical value of (e^2/m) is in fact *irrelevant* to inference about m, if we already have measurements of e and e/m of equal accuracy. To see this, suppose that we knew (e^2/m) exactly from the start. How would you make use of that information in this problem? If you try to do this, you will soon see why (e^2/m) is irrelevant. But to clinch matters, try the following exercise.

Exercise 19.3.   Consider a specific case: $w_1 = w_2 = 1$, $w_3 = 100$; the third measurement is ten times more accurate than the first two. But if the problem is such that the third measurement cancels out when we try to use all three as in (19.22), then it seems that the only way we could use the accurate third measurement is by discarding either the first or second. Show that, nevertheless, in this case the estimates made by (19.32) using only the first and second measurements are more accurate than those made by using the first and third; or the second and third. Now explain intuitively why this is as it should be; there is no paradox.

As another example, it is important that we understand the way our conclusions depend on our choice of loss functions and probability distributions for the errors δ_i. If we use instead of the Gaussian distribution (19.12) one with wider tails, such as the Cauchy distribution $p(\delta) \propto (1 + w\delta^2/2)^{-1}$, the posterior distribution $p(x_1x_2|y_1y_2y_3)$ may have more than one peak in the (x_1, x_2) plane. Then a quadratic loss function, or more generally any concave loss function (i.e. doubling the error more than doubles the loss) will lead one to make estimates of x_1 and x_2 which lie between the peaks, and are known to be very unlikely. With a convex loss function a different 'paradox' appears, in that the basic equation (19.26) for constructing the best estimator may have more than one solution, with nothing to tell us which one to use.

The appearance of these situations is the robot's way of telling us this: our state of knowledge about x_1 and x_2 is too complicated to be described adequately simply by giving best estimates and probable errors. The only honest way of describing what we know is to give the actual distribution $p(x_1x_2|y_1y_2y_3)$. This is one of the limitations of decision theory, which we need to understand in order to use it properly.

20

Model comparison

Entities are not to be multiplied without necessity.

William of Ockham, c 1330

We have seen in some detail how to conduct inferences – test hypotheses, estimate parameters, predict future observations – within the context of a preassigned model, representing some working hypothesis about the phenomenon being observed. But a scientist must also be concerned with a bigger problem: how to decide between different models when both seem able to account for the facts. Indeed, the progress of science requires comparison of different conceivable models; a false premise built into a model that is never questioned cannot be removed by any amount of new data.

Stated very broadly, the problem is hardly new; some 650 years ago the Franciscan Monk William of Ockham perceived the logical error in the mind projection fallacy.[1] This led him to teach that some religious issues might be settled by reason, but others only by faith. He removed the latter from his discourse, and concentrated on the areas where reason might be applied – just as Bayesians seek to do today when we discard orthodox mind projecting mythology (such as assertions of limiting frequencies in experiments that have never been performed), and concentrate on the things that are meaningful in the real world. His propositions 'amenable only to faith' correspond roughly to what we should call non-Aristotelian propositions. His famous epigram quoted above, generally called 'Ockham's razor', represents a good start on the principles of reasoning that he needed, and that we still need today. But it was also so subtle that only through modern Bayesian analysis has it been well understood.

Of course, from our present vantage point it is clear that this is really the same problem as that of compound hypothesis testing, considered already in Chapter 4. Here we need only generalize that treatment and work out further details. Then we are able to see conventional significance tests simply as model comparison in which we choose the best alternative in a prescribed *class* of alternative hypotheses.

---

[1] Ockham's position, stated in the language of his time, was that 'Reality exists solely in individual things, and universals are merely abstract signs.' Translated into 20th century language: the abstract creations of the mind are not realities in the external world. Unfortunately for him, some of the cherished 'realities' of contemporary orthodox theology were just the things to which he denied reality; so this got him into trouble with the Establishment. Evidently, Ockham was a forerunner of modern Bayesians, to whom all this sounds very familiar.

But some extra care is needed. As long as we work within a single model, normalization constants tend to cancel out and so need not be introduced at all in most cases. But when two different models appear in a single equation, the normalization constants do not cancel out, and it is imperative that all probabilities be correctly normalized.

20.1 Formulation of the problem

To see why the normalization constants no longer cancel, recall first what Bayes' theorem tells us about parameter estimation. A model M contains various parameters denoted collectively by θ. Given data D and prior information I, to estimate its parameters we first apply Bayes' theorem:

$$p(\theta|DMI) = p(\theta|MI)\frac{p(D|\theta MI)}{p(D|MI)}, \tag{20.1}$$

in which the presence of M on the right-hand side signifies that we are assuming the correctness of model M. The denominator serves as the normalizing constant:

$$p(D|MI) = \int d\theta\, p(D\theta|MI) = \int d\theta\, p(D|\theta MI)p(\theta|MI), \tag{20.2}$$

which we see is the prior expectation of the likelihood $L(\theta) = p(D|\theta MI)$; i.e. its expectation over our prior probability distribution $p(\theta|MI)$ for the parameters.

Now we move up to a higher level problem: to judge, in the light of the prior information and data, which of a given set of different models $\{M_1, \dots, M_r\}$ is most likely to be the correct one. Bayes' theorem gives the posterior probability for the jth model as

$$p(M_j|DI) = p(M_j|I)\frac{p(D|M_jI)}{p(D|I)}, \qquad 1 \le j \le r. \tag{20.3}$$

But we may eliminate the denominator $p(D|I)$ by calculating odds ratios as we did in Chapter 4. The posterior odds ratio for model M_j over M_k is

$$\frac{p(M_j|DI)}{p(M_k|DI)} = \frac{p(M_j|I)}{p(M_k|I)}\frac{p(D|M_jI)}{p(D|M_kI)}, \tag{20.4}$$

and we see that the same probability $p(D|M_jI)$ that appears in the single-model parameter estimation problem (20.1) only as a normalizing constant, now appears as the fundamental quantity determining the status of model M_j relative to any other.[2] The exact measure of what the data have to tell us about this is always the prior expectation of its likelihood function, over the prior probability $p(\theta_j|M_jI)$ for whatever parameters θ_j may be in that

---

[2] This logical structure is more general even than the Bayesian formalism; it persists in the pure maximum entropy formalism, where in statistical mechanics the relative probability P_j/P_k of two different phases, such as liquid and solid, is the ratio of their partition functions Z_j/Z_k, which are the normalization constants for the sub-problems of prediction within one phase. In Bayesian analysis, the data are indifferent between two models when their normalization constants become equal; in statistical mechanics the temperature of a phase transition is the one at which the two partition functions become equal. In Bayesian analysis we shall usually prefer to express (20.4) in log-odds form; in chemical thermodynamics it has been customary for a century to state the condition of indifference between phases as equality of the 'free energies' $F_j \propto \log(Z_j)$. This illustrates the basic unity of Bayesian and maximum entropy reasoning, in spite of their superficial differences arising from the different kind of information being processed.

model (they are generally different for different models). Probabilities must be correctly normalized here, otherwise we are violating our basic rules and the odds ratio in (20.4) is arbitrary.

Intuitively, the model favored by the data is the one that assigns the highest probability to the observed data, and therefore 'explains the data' best. This is just a repetition, at a higher level, of the likelihood principle for parameter estimation within a model.

But how can an Ockham principle emerge from this? The first difficulty is that the principle has never been stated in exact, well-defined terms. Later writers have tried, almost universally, to interpret our opening quotation as saying that the criterion of choice is the 'simplicity' of the competing models, although it is not clear that Ockham himself used that term. Perhaps we come closer to the notion of simplicity if we restate our opening quotation as: 'Do not introduce details that do not contribute to the quality of your inferences.' But centuries of discussion by philosophers brought no appreciable clarification of what is meant by 'simplicity'.[3] We think that concentration of attention exclusively on that undefined term has prevented understanding of the real point, which is merely that a model with unspecified parameters is a composite hypothesis, not a simple one; and it requires the kind of analysis given in Chapter 4 for composite hypotheses. Then some new features appear, arising from the different internal structures of the parameter spaces for the models considered.

20.2 The fair judge and the cruel realist

Now consider under what conditions we want this model comparison to take place. There are two possible positions. (1) We might adopt the posture of the scrupulously fair judge, who insists that fairness in comparing models requires that each is delivering the best performance of which it is capable, by giving each the best possible prior probability for its parameters (similarly, in Olympic games we would consider it unfair to judge two athletes by their performances when one of them is sick or injured; the fair judge wants to compare them when both are doing their absolute best). (2) We might consider it necessary to be cruel realists and judge each model taking into account the prior information we actually have pertaining to it; that is, we penalize a model if we do not have the best possible prior information about its parameters, although that is not really a fault of the model itself.

It develops that the Ockham factors express the position of the cruel realist; they are just the factors that convert the scrupulously fair comparison of the model itself – irrespective of the prior probability we are able to give it at the moment – into the comparisons of the cruel realist who insists on taking into account what is actually possible here and now. An athlete who is sick or injured merits our sympathy, but we cannot use him in the 'big game' tomorrow; likewise, a potentially superior model can be unusable if our prior information places its parameters far from their maximum-likelihood values. When real results are at stake, we are obliged to be cruel realists.

---

[3] For a time, the notion of simplicity was given up for dead, because of the seeming impossibility of defining it. The tedious details are recounted by Rosenkrantz (1977).

20.2.1 Parameters known in advance

To see this, suppose first that there is no internal parameter space; the parameters of a model are known exactly ($\theta = \theta'$) in advance. Then the model becomes, in effect, a simple hypothesis rather than a composite or compound one, and the simple form of Bayes' theorem applies. This amounts to assigning a prior $p(\theta_j | M_j I) = \delta(\theta_j - \theta'_j)$, whereupon (20.2) reduces to

$$p(D|M_j I) = p(D|\theta'_j M_j I) = L_j(\theta'_j), \qquad (20.5)$$

just the likelihood of θ'_j within the jth model. Evidently, the scrupulously fair judge will note that this is a maximum if θ'_j happens to be equal to the maximum-likelihood estimate $\hat{\theta}_j$ for that model and the data. Then his posterior odds ratio (20.4) reduces to

$$\frac{p(M_j|DI)}{p(M_k|DI)} = \frac{p(M_j|I)}{p(M_k|I)} \frac{(L_j)_{\max}}{(L_k)_{\max}}. \qquad (20.6)$$

But this extreme case, although fair in the aforementioned sense, may be very unrealistic; usually, the parameters are unknown, and in the problems 'amenable to reason', where useful inferences are possible, our prior information concerning the parameters must be good enough to allow useful inferences.

We have seen already in previous chapters that, if we have a reasonable amount of data, most models will give such sharply peaked likelihood functions that the prior is relatively unimportant for inferences about the *parameters*. But it is still important for inferences about the *models*, so Ockham factors defined by the priors remain important in model comparison. The simple biological problems studied by R. A. Fisher are generally of this type.

When prior information is important even for inferences about the parameters – whether from a loose model or sparse data – Ockham factors make a crucially important difference in our model comparisons. In the more complex problems studied by Harold Jeffreys and faced by modern scientists and economists, we ignore these factors at our peril.

20.2.2 Parameters unknown

Let a model M have parameters $\theta \equiv \{\theta_1, \ldots, \theta_m\}$. Then, comparing (20.4) and (20.6), we write

$$p(D|MI) = L_{\max} W, \qquad (20.7)$$

and this defines the Ockham factor W; it is just the amount by which the model M is penalized by our nonoptimal prior information. Written explicitly,

$$W \equiv \int d\theta \, \frac{L(\theta)}{L_{\max}} p(\theta | MI). \qquad (20.8)$$

If, as in Fisher's problems, the data are much more informative about θ than the prior information, then the likelihood function is sharply peaked and we could define a 'high-likelihood region' Ω' as the smallest subregion of the whole parameter space Ω that contains a specified amount (say, 95%) of the integrated likelihood. Then, most of the contribution to the integral (20.8) would come from the region Ω'. Better, the arbitrary number 0.95 can be done away with by defining first the volume $V(\Omega')$ by the condition that the integrated likelihood is just

$$\int d\theta \, L(\theta) = L_{\max} V(\Omega'). \tag{20.9}$$

Then Ω' is defined as the region of volume $V(\Omega')$ that contains the maximum possible amount of integrated likelihood; that is, within Ω' the likelihood is everywhere greater than some threshold value L_0 that is reached on the boundary of Ω'.

If the prior density $p(\theta|MI)$ is so broad that it is essentially constant over this high-likelihood region Ω' surrounding the maximum-likelihood point, (20.8) reduces to

$$W \simeq V(\Omega') p(\hat{\theta}|MI), \tag{20.10}$$

so in this case the Ockham factor is essentially just the *amount of prior probability* contained in the high-likelihood region Ω' picked out by the data.

In any case, our fundamental model comparison rule (20.4) becomes

$$\frac{p(M_j|DI)}{p(M_k|DI)} = \frac{p(M_j|I)}{p(M_k|I)} \frac{(L_j)_{\max}}{(L_k)_{\max}} \frac{W_j}{W_k}, \tag{20.11}$$

in which we see revealed, by comparison with (20.6), the net Ockham factor (W_j/W_k) arising from the internal parameter spaces of the models. In (20.11), the likelihood factor depends only on the data and the models. If two different models achieve the same likelihoods $(L_j)_{\max}$, then they are potentially capable of accounting for the data equally well, and in orthodox theory it would seem that we have no basis for choice between them. Yet Bayes' theorem tells us that there is an another quality to be considered: the prior information, which is ignored by orthodox theory, may still give strong grounds for preference of one model over the other. Indeed, the Ockham factor in (20.11) may be so strong that it reverses the likelihood judgment in (20.6).

20.3 But where is the idea of simplicity?

The relation (20.11) has much meaning that unaided intuition could not (or at least, did not) see. If the data are highly informative compared with the prior information, then the relative merit of two models is determined by two factors:

(1) how high a likelihood can be attained on their respective parameter spaces Ω_j, Ω_k;
(2) how much prior probability is concentrated in their respective high-likelihood regions Ω'_j, Ω'_k?

But neither of these seems concerned with the intuitive notion of simplicity (which seems for most of us to refer to the number of different assumptions that are made – for example, the number of different parameters that are introduced – in defining a model).

To understand this, let us ask: 'How do we all decide these things intuitively?' Having observed some facts, what is the real criterion that leads us to prefer one explanation of them over another? Suppose that two explanations, A and B, could account for some proven historical facts equally well. But A makes four assumptions, each of which seems to us already highly plausible, while B makes only two assumptions, but they seem strained, farfetched, and highly unlikely to be true. Every historian finds himself in situations like this, and he does not hesitate to opt for explanation A, although B is intuitively simpler. Thus our intuition asks, fundamentally, not how *simple* the hypotheses are, but rather how *plausible* they are.

Of course, there is a loose connection between plausibility and simplicity, because the more complicated a set of possible hypotheses, the larger the manifold of conceivable alternatives to some particular hypothesis, and so the smaller must be the prior probability of any particular hypothesis in the set.

Now we see why 'simplicity' could never be given a satisfactory definition (that is, a definition that accounted in a satisfactory way for these inferences); it was a poorly chosen word, directing one's attention away from an essential component of the inference. But from centuries of unquestioned acceptance, the idea of 'simplicity' became implanted with such an unshakeable mindset that some workers, even after applying Bayes' theorem where the contrary fact stares you in the face, continued doggedly trying to interpret the Bayesian analysis in terms of simplicity.[4]

Generations of writers opined vaguely that 'simple hypotheses are more plausible' without giving any logical reason for it. We suggest that this should be turned around: we should say rather that 'more plausible hypotheses tend to be simpler'. An hypothesis that we consider simpler is one that has fewer equally plausible alternatives.

None of this could be comprehended at all within the confines of orthodox statistical theory, whose ideology did not allow the concept of a probability for a model or for a fixed but unknown parameter, because they were not considered 'random variables'. Orthodoxy tried to compare models entirely in terms of their different sampling distributions, which took no note of *either* the simplicity of the model *or* the prior information! But it was unable to do even that, because then all the parameters within a model became nuisance parameters, and that same ideology denied one any way to deal with them.[5] Thus, orthodox statistics was a total failure on this problem – it did not provide even the vocabulary in which the problem could be stated – and this held up progress for most of the 20th century.

It is remarkable that, although the point at issue is trivial mathematically, generations of mathematically competent people failed to see it because of that conceptual mindset. But once the point is seen, it seems intuitively obvious and one cannot comprehend how anyone could ever have imagined that 'simplicity' alone was the criterion for judging models. This just reminds us again that the human brain is an imperfect reasoning device; although it is quite good at drawing reasonable conclusions, it often fails to give a convincing

---

[4] Indeed, one author, for whom Ockham's razor was *by definition* concerned with simplicity, rejected Bayesian analysis because of its failure to exhibit that error!

[5] This and other criticisms of orthodox hypothesis testing theory were made long ago by Pratt (1961).

rationale for those conclusions. For this we really do need the help of probability theory as logic.

Of course, Bayes' theorem does recognize simplicity as one component of the inference. But by what mechanism does this happen? Although Bayes' theorem always gives us the correct answer to whatever question we ask of it, it often does this in such a slick, efficient way that we are left bewildered and not quite understanding how it happened. The present problem is a good example of this, so let us try to understand the situation better intuitively.

Denote by M_n a model for which $\theta = \{\theta_1, \ldots, \theta_n\}$ is n-dimensional, ranging over a parameter space Ω_n. Now introduce a new model M_{n+1} by adding a new parameter θ_{n+1} and going to a new parameter space Ω_{n+1}, in such a way that $\theta_{n+1} = 0$ represents the old model M_n. We shall presently give an explicit calculation with this scenario, but first let us think about it in general terms.

On the subspace Ω_n the likelihood is unchanged by this change of model: $p(D|\theta M_{n+1}I) = p(D|\theta M_n I)$. But the prior probability $p(\theta|M_{n+1}I)$ must now be spread over a larger parameter space than before and will, in general, assign a lower probability to a neighborhood Ω' of a point in Ω_n than did the old model.

For a reasonably informative experiment, we expect that the likelihood will be rather strongly concentrated in small subregions $\Omega'_n \in \Omega_n$ and $\Omega'_{n+1} \in \Omega_{n+1}$. Therefore, if with M_{n+1} the maximum-likelihood point occurs at or near $\theta_{n+1} = 0$, Ω'_{n+1} will be assigned less prior probability than is Ω'_n with model M_n, and we have $p(D|M_n I) > p(D|M_{n+1}I)$; the likelihood ratio generated by the data will favor M_n over M_{n+1}. This is the Ockham phenomenon.

Thus, if the old model is already flexible enough to account well for the data, then as a general rule Bayes' theorem will, like Ockham, tell us to prefer the old model. It is intuitively simpler if by 'simpler' we mean a model that occupies a smaller volume of parameter space, and thus *restricts us to a smaller range of possible sampling distributions*. Generally, the inequality will go the other way only if the maximum-likelihood point is far from $\theta_{n+1} = 0$ (i.e. a significance test would indicate a need for the new parameter), because then the likelihood will be so much smaller on Ω'_n than on Ω'_{n+1} that it more than compensates for the lower prior probability of the latter; as noted, Ockham would not disagree.

But intuition does not tell us at all, quantitatively, how great this discrepancy in likelihoods must be in order to bring us to the point of indifference between the models. Furthermore, having seen this mechanism, it is easy to invent cases (for example, if the introduction of the new parameter is accompanied by a redistribution of prior probability on the old subspace S_n) in which Bayes' theorem may contradict Ockham because it is taking into account further circumstances undreamt of in Ockham's philosophy. So we need specific calculations to make these things quantitative.

20.4 An example: linear response models

Now we give a simple analysis that illustrates the above conclusions and allows us to calculate definite numerical values for the likelihood and Ockham factors. We have a common scenario: a data set $D \equiv \{(x_1, y_1), \ldots, (x_n, y_n)\}$ consisting of measured values of (x, y) in n pairs of observations. We may think of x as the 'cause' and y as the 'effect', although this is not required. For the general relations below, the 'independent variables' x_i need not be uniformly spaced or even monotonic increasing in the index i. From these data and any prior information we have, we are to decide between two conceivable models for the process generating the data. For model M_1 the responses are, but for irregular measurement errors e_i, linear in the cause:

$$M_1: \quad y_i = \alpha x_i + e_i, \qquad 1 \le i \le n, \tag{20.12}$$

while for model M_2 there is also a quadratic term:

$$M_2: \quad y_i = \alpha x_i + \beta x_i^2 + e_i, \tag{20.13}$$

which represents, if β is negative, an incipient saturation or stabilizing effect (if β is positive, an incipient instability). We may think, for concreteness, of x_i as the dose of some medicine given to the ith patient, y_i as the resulting increase in blood pressure. Then we are trying to decide whether the response to this medicine is linear or quadratic in the dosage. But this mathematical model applies equally well to many different scenarios.[6] Whichever model is correct, we assume that the x_i are measured with negligible error, but the errors of measurement of y_i are supposed to be the same for either model, so we assign a joint sampling distribution to them:

$$p(e_1 \cdots e_n | I) = \prod_{i=1}^{n} \frac{1}{\sqrt{2\pi\sigma^2}} \exp\left\{-\frac{e_i^2}{2\sigma^2}\right\} = \left(\frac{w}{2\pi}\right)^{n/2} \exp\left\{-\frac{w}{2}\sum_i e_i^2\right\} \tag{20.14}$$

where $w \equiv 1/\sigma^2$ is the 'weight' parameter, more convenient in calculations than σ^2. This simple scenario has the merit that all calculations can be done exactly with pencil and paper, so that the final result can be subjected to arbitrary extreme conditions and will remain correct, and we can see which limiting operations are and are not well-behaved.

20.4.1 Digression: the old sermon still another time

Again, we belabor the meaning of this, as discussed in Chapter 7. In orthodox statistics, a sampling distribution is always referred to as if it represented an 'objectively real' fact, the frequency distribution of the errors. But we doubt whether anybody has ever seen a real problem in which one had prior knowledge of any such frequency distribution, or indeed prior knowledge that any limiting frequency distribution exists.

---

[6] For example, x_i might be the amount of ozone in the air in the ith year, y_i the average temperature in that year. Or, x_i may be the amount of some food additive ingested by the ith Canadian rat, y_i the amount of cancer tissue that rat developed. Or, x_i may be the amount of acid rain falling on Northern Germany in the ith year, y_i the number of pine trees that died in that year; and so on. In other words, we are now in the realm of what were called 'linear response models' in the Preface, and the results of these calculations have a direct bearing on many currently controversial health and environmental issues. Of course, most real problems will require more sophisticated models than we are considering now, but having seen this simple calculation it will be clear how to generalize it in many different ways.

How could one ever acquire information about the long-run results of an experiment that has never been performed? That is part of the mind projecting mythology that we discard.

We recognize, then, that assigning this sampling distribution is only a means of describing our own *prior state of knowledge* about the measurement errors. The parameter σ indicates the general magnitude of the errors that we expect. The prior information I might, for example, be the variability observed in past examples of such data; or in a physics experiment it might not be the result of any observations, but rather obtained from the principles of statistical mechanics, indicating the level of Nyquist noise for the known temperature of the apparatus.

In particular, the absence of correlations in (20.14) is not an assertion that no correlations exist in the real data; it is only a recognition that we have no prior knowledge of such correlations, and therefore to suppose correlations of either sign is as likely to hurt as to help the accuracy of our inferences. In one sense, by being non-committal about it, we are only being honest and frankly acknowledging our ignorance. But in another sense, we are taking the safest, most conservative, course; using a sampling distribution which will yield reasonable results *whether or not correlations actually exist.* But if we knew of any such correlations, we would be able to make still better inferences (although not much better) by use of a sampling distribution which contains them.

The reason for this is that correlations in a sampling distribution tell the robot that some regions of the vector sample space are more likely than others, even though they have the same mean-square error magnitudes $\overline{e^2}$; then some details of the data that it would otherwise have to dismiss as noise can be recognized as providing further evidence about systematic effects in the model.

We return to the problem. The sampling distribution for model M_1 is

$$M_1: \qquad p(D|\alpha M_1) = \left(\frac{w}{2\pi}\right)^{n/2} \exp\left\{-\frac{nw}{2}Q_1(\alpha)\right\} \qquad (20.15)$$

with the quadratic form

$$Q_1(\alpha) \equiv \frac{1}{n}\sum_{i=1}^{n}(y_i - \alpha x_i)^2 = \overline{y^2} - 2\alpha\overline{xy} + \alpha^2\overline{x^2}, \qquad (20.16)$$

where the bars denote averages. The maximum-likelihood estimate of α is then found from $\partial Q_1/\partial\alpha = 0$, or,

$$\alpha = \hat{\alpha} \equiv \frac{\overline{xy}}{\overline{x}}, \qquad (20.17)$$

which in this case is also called the 'ordinary least squares' estimate. Supposing the weight w is known; the likelihood (20.15) for model M_1 is then

$$L_1(\alpha) = \left(\frac{w}{2\pi}\right)^{n/2} \exp\left\{-\frac{nw}{2}\left[\overline{y^2} + \overline{x^2}(\alpha - \hat{\alpha})^2 - \overline{x^2}\hat{\alpha}^2\right]\right\} \qquad (20.18)$$

in which we could discard any factor independent of α, but that will disappear presently of its own accord, in (20.23). If we were using this to estimate α from the data alone, our result would be

$$(\alpha)_{\text{est}} = \hat{\alpha} \pm \frac{1}{\sqrt{nw\overline{x^2}}} = \hat{\alpha} \pm \frac{1}{\sqrt{n}} \frac{\sigma}{x_{\text{rms}}} = \hat{\alpha} \pm \delta\alpha, \qquad (20.19)$$

where $x_{\text{rms}} = \sqrt{\overline{x^2}}$ is the root-mean-square value of the x_i. Thus the volume (in this case, width) of the high-likelihood region Ω' may be taken as roughly $V(\Omega') = 2(\delta\alpha)$.

Now, using (20.17), the 'global' sampling distribution for model M_1 in (20.3) contains two factors:

$$p(D|M_1 I) = \int d\alpha \, p(D|\alpha M_1) p(\alpha|M_1 I) = L_{\text{max}}(M_1) W_1, \qquad (20.20)$$

where

$$L_{\text{max}}(M_1) = L_1(\hat{\alpha}). \qquad (20.21)$$

The Ockham factor for model M_1 is therefore

$$W_1 = \int d\alpha \, \frac{L_1(\alpha)}{L_1(\hat{\alpha})} p(\alpha|M_1 I), \qquad (20.22)$$

and we find for the likelihood ratio

$$\frac{L_1(\alpha)}{L_1(\hat{\alpha})} = \exp\left[-\frac{nw\overline{x^2}}{2}(\alpha - \hat{\alpha})^2 \right]. \qquad (20.23)$$

This makes it evident that $W_1 \leq 1$, since the likelihood ratio cannot exceed unity and the prior is normalized.

Now we must assign a prior for α. Usually, we will have some reason, such as previous experience with such problems, for guessing a value of the general order of magnitude of some quantity α_0, but we are not at all confident of the accuracy of that guess, except to think that $|\alpha - \alpha_0|$ cannot be enormously large (else there would be such a catastrophe that we would not be concerned with this problem); but we would seldom have any more specific prior information about it. We can indicate this by assigning the normalized prior density

$$p(\alpha|M_1 I) = \sqrt{\frac{w_0}{2\pi}} \exp\left\{ -\frac{w_0}{2}(\alpha - \alpha_0)^2 \right\}, \qquad (20.24)$$

which says that we think it unlikely that $|\alpha - \alpha_0|$ is much greater than $\sigma_0 = 1/\sqrt{w_0}$. From both the central limit theorem as discussed in Chapter 7 and the maximum entropy principle as discussed in Chapter 11, this Gaussian functional form of prior is preferred in principle over all others as representing the actual state of knowledge that we have in virtually all real problems. Then it is fortunate that this form also enables us to do the integration (20.22) exactly, with the result

$$W_1 = \sqrt{\frac{w_0}{n\overline{wx^2} + w_0}} \exp\left\{-\frac{n\overline{wx^2}w_0}{2(n\overline{wx^2} + w_0)}(\hat{\alpha} - \alpha_0)^2\right\}. \qquad (20.25)$$

Rewriting this in terms of the half-width $\delta\alpha = 1/\sqrt{n\overline{wx^2}}$ of the high-likelihood region and the half-width $\sigma_0 = 1/\sqrt{w_0}$ of the prior for α, it becomes

$$W_1 = \frac{1}{\sqrt{1 + (\sigma_0/\delta\alpha)^2}} \exp\left\{-\frac{(\hat{\alpha} - \alpha_0)^2}{1 + (\sigma_0/\delta\alpha)^2}\right\}. \qquad (20.26)$$

This has several limiting forms. If the prior estimate α_0 is exactly equal to the ordinary least squares estimate $\hat{\alpha}$, it reduces to

$$W_1 = \frac{1}{\sqrt{1 + (\sigma_0/\delta\alpha)^2}}. \qquad (20.27)$$

Then, if $\sigma_0 \gg \delta\alpha$, we have

$$W_1 \simeq \frac{\delta\alpha}{\sigma_0}, \qquad (20.28)$$

which is indeed just the amount of prior probability contained in the high-likelihood region. In this case, the Ockham factor is the ratio by which the parameter space is contracted by the information in the data, which expresses how much the vagueness of our prior information deteriorates the performance of model M_1, by placing prior probability outside its high-likelihood region. If the prior estimate α_0 differs from the ordinary least squares estimate $\hat{\alpha}$ by less than σ_0, this remains approximately correct.

If in (20.27), $\sigma_0 \to 0$, we have $W_1 \to 1$, the maximum possible value; if the prior information already told us exactly the ordinary least squares estimate from the data, with zero error tolerance, model W_1 is not penalized at all. But in all other cases there is some penalty. For example, if $|\alpha_0 - \hat{\alpha}| \gg \sigma_0$, then the evidence of the data strongly contradicts the prior information, and the model is severely penalized.

For model M_2 the sampling distribution is still given by (20.15), but now with the quadratic form

$$Q_2(\alpha, \beta) \equiv \frac{1}{n}\sum(y_i - \alpha x_i - \beta x_i^2)^2 = \overline{y^2} + \alpha^2\overline{x^2} + \beta^2\overline{x^4} - 2\alpha\overline{xy} - 2\beta\overline{x^2y} + 2\alpha\beta\overline{x^3}, \qquad (20.29)$$

and the maximum-likelihood estimates $(\hat{\alpha}, \hat{\beta})$ are now the roots of the simultaneous equations $\partial Q_2/\partial\alpha = 0$, $\partial Q_2/\partial\beta = 0$, or

$$\begin{aligned}\overline{x^2}\hat{\alpha} + \overline{x^3}\hat{\beta} &= \overline{xy} \\ \overline{x^3}\hat{\alpha} + \overline{x^4}\hat{\beta} &= \overline{x^2y}, \end{aligned} \qquad (20.30)$$

of which the solution is

$$\hat{\alpha} = \frac{\overline{(x^4)}\overline{(xy)} - \overline{(x^3)}\overline{(x^2y)}}{\overline{(x^2)}\overline{(x^4)} - \overline{(x^3)}^2}, \qquad \hat{\beta} = \frac{\overline{(x^2)}\overline{(x^2y)} - \overline{(x^3)}\overline{(xy)}}{\overline{(x^2)}\overline{(x^4)} - \overline{(x^3)}^2}, \qquad (20.31)$$

and we note that, as $\overline{x^3} \to 0$, these relax into estimates

$$\hat{\alpha} \to \frac{\overline{xy}}{\overline{x^2}}, \qquad \hat{\beta} \to \frac{\overline{x^2y}}{\overline{x^4}}, \qquad (20.32)$$

where $\hat{\alpha}$ is the ordinary least squares estimate found using model M_1 (20.17). Now, as in (20.22), the Ockham factor for model M_2 is

$$W_2 = \int d\alpha \int d\beta \, \frac{L_2(\alpha, \beta)}{L_2(\hat{\alpha}, \hat{\beta})} p(\alpha\beta|M_2 I), \qquad (20.33)$$

and, after some rather tedious algebra, we find that the likelihood ratio just constructs a familiar quadratic form:

$$\frac{L_2(\alpha, \beta)}{L_2(\hat{\alpha}, \hat{\beta})} = \exp\left\{-\frac{nw}{2}Q(\alpha, \beta)\right\}, \qquad (20.34)$$

where

$$\begin{aligned} Q(\alpha, \beta) &\equiv Q_2(\alpha, \beta) - Q_2(\hat{\alpha}, \hat{\beta}) \\ &= \overline{x^2}(\alpha - \hat{\alpha})^2 + 2\overline{x^3}(\alpha - \hat{\alpha})(\beta - \hat{\beta}) + \overline{x^4}(\beta - \hat{\beta})^2. \end{aligned} \qquad (20.35)$$

Now we assign a joint prior

$$p(\alpha\beta|M_2 I) = \sqrt{\frac{w_0}{2\pi}} \exp\left\{-\frac{w_0}{2}(\alpha - \alpha_0)^2\right\} \sqrt{\frac{w_1}{2\pi}} \exp\left\{-\frac{w_1}{2}(\beta - \beta_0)^2\right\} \qquad (20.36)$$

in which w_0, a_0 are the same as in (20.24), so that the marginal prior for α is the same in the two models (otherwise we would be changing two different circumstances instead of one in going from M_1 to M_2, which would make the results very hard to interpret):

$$p(\alpha|M_1 I) = p(\alpha|M_2 I). \qquad (20.37)$$

The Ockham factor for model M_2 is then

$$W_2 = \frac{\sqrt{w_0 w_1}}{2\pi} \int d\alpha \int d\beta \exp\left\{-\frac{1}{2}\left[nwQ(\alpha, \beta) + w_0(\alpha - \alpha_0)^2 + w_1(\beta - \beta_0)^2\right]\right\}, \qquad (20.38)$$

and again this integration can be carried out exactly, with the result

$$W_2 = \sqrt{\frac{w_0 w_1}{(w_0 + n\overline{wx^2})(w_1 + n\overline{wx^4})}}\, \exp\{x\}. \tag{20.39}$$

Editor's Exercise 20.1. As written, the denominator in (20.39) is correct only if the condition $\overline{x^3} \to 0$ is used. Using this simplifying assumption, derive W_2 and define x.

The net Ockham factor in favor of M_1 over M_2 is computed from (20.27) and (20.39):

$$\frac{W_1}{W_2} = \frac{1/\sqrt{1 + (\sigma_0/\delta\alpha)^2}}{\sqrt{(w_0 w_1)/(w_0 + n\overline{wx^2})(w_1 + n\overline{wx^4})}\, \exp\{x\}}. \tag{20.40}$$

Editor's Exercise 20.2. Rewrite (20.40) in terms of the half-widths: $\delta\alpha = 1/\sqrt{n\overline{wx^2}}$, $\sigma_0 = 1/\sqrt{w_0}$, $\delta\beta = 1/\sqrt{n\overline{wx^4}}$, and $\sigma_1 = 1/\sqrt{w_1}$. Under what conditions will model M_2 will be favored over M_1?

20.5 Comments

Actual scientific practice does not really obey Ockham's razor, either in its previous 'simplicity' form or in our revised 'plausibility' form. As so many of us have deplored, the attractive new hypothesis or model, which accounts for the facts in such a neat, plausible way that you want to believe it at once, is usually pooh-poohed by the official Establishment in favor of some drab, complicated, uninteresting one; or, if necessary, in favor of no alternative at all. The progress of science is carried forward mostly by the few fundamental dissenting innovators, such as Copernicus, Galileo, Newton, Laplace, Darwin, Mendel, Pasteur, Boltzmann, Einstein, Wegener, Jeffreys – all of whom had to undergo this initial rejection and attack. In the cases of Galileo, Laplace, and Darwin, these attacks continued for more than a century after their deaths. This is not because their new hypotheses were faulty – quite the contrary – but because this is part of the sociology of science (and, indeed, of all scholarship). In any field, the Establishment is seldom in pursuit of the truth, because it is composed of those who sincerely believe that they are already in possession of it.

Progress is delayed also by another aspect of this. Scholars who failed to heed the teachings of William of Ockham about issues amenable to reason and issues amenable only to faith, were – and still are – doomed to a lifetime of generating nonsense. We note the most common form this nonsense has taken in the past.

20.5.1 *Final causes*

It seems that every discussion of scientific inference must deal, sooner or later, with the issue of belief or disbelief in final causes. Expressed views range all the way from Jacques Monod (1970) forbidding us even to mention purpose in the Universe, to the religious fundamentalist who insists that it is evil not to believe in such a purpose. We are astonished by the dogmatic, emotional intensity with which opposite views are proclaimed, by persons who do not have a shred of supporting factual evidence for their positions.

But almost everyone who has discussed this has supposed that by a 'final cause' one means some supernatural force that suspends natural law and takes over control of events (that is, alters positions and velocities of molecules in a way inconsistent with the equations of motion) in order to ensure that some desired final condition is attained. In our view, almost all past discussions have been flawed by failure to recognize that operation of a final cause does not imply controlling molecular details.

When the author of a textbook says: 'My purpose in writing this book was to …', he is disclosing that there was a true 'final cause' governing many activities of writer, pen, secretary, word processor, extending usually over several years. When a chemist imposes conditions on his system which forces it to have a certain volume and temperature, he is just as truly the wielder of a final cause dictating the final thermodynamic state that he wished it to have. A bricklayer and a cook are likewise engaged in the art of invoking final causes for definite purposes. But – and this is the point almost always missed – these final causes are *macroscopic*; they do not determine any particular 'molecular' details. In all cases, had those fine details been different in any one of billions of ways, the final cause would have been satisfied just as well.

The final cause may then be said to possess an entropy, indicating the number of microscopic ways in which its purpose can be realized; and the larger that entropy, the greater is the probability that it will be realized. Thus the principle of maximum entropy applies also here.

In other words, while the idea of a microscopic final cause runs counter to all the instincts of a scientist, a macroscopic final cause is a perfectly familiar and real phenomenon, which we all invoke daily. We can hardly deny the existence of purpose in the Universe when virtually everything we do is done with some definite purpose in mind. Indeed, anybody who fails to pursue some definite long-run purpose in the conduct of his life is dismissed as an idler by his colleagues. Obviously, this is just a familiar fact with no religious connotations – and no anti-religious ones. Every scientist believes in macroscopic final causes without thereby believing in supernatural contravention of the laws of physics. The wielder of the final cause is not suspending physical law; he is merely choosing the Hamiltonian with which some system evolves *according to physical law*. To fail to see this is to generate the most fantastic, mystical nonsense.

21

Outliers and robustness

Probably everybody who has been involved in quantitative measurements has found himself in the following situation. You are trying to measure some quantity θ (which might be, for example, the right ascension of Sirius, the mass of a π-meson, the velocity of seismic waves at a depth of 100 km, the melting point of a new organic compound, the elasticity of consumer demand for apples, etc.). But the apparatus or the data taking procedure is always imperfect and so, having made n independent measurements of θ, you have n different results $(x_1, \ldots, x_n)$. How are you to report what you now know about θ? More specifically, what 'best' estimate should you announce, and what accuracy are you entitled to claim?

If these n data values were closely clustered together making a reasonably smooth, single-peaked histogram, you would accept the solutions given in the previous chapters, and might feel that the problem of drawing conclusions from good data is not very difficult, even without any probability theory. But your data are not nicely clustered: one value, x_j, lies far away from the nice cluster made by the other $(n-1)$ values. How are you to deal with this outlier? What effect does it have on the conclusions that you entitled to draw about θ?

We have seen in Chapters 4 and 5 how the appearance of astonishing, unexpected data may cause the resurrection of dead hypotheses; it appears that something like that may be at work here. In fact, any surprisingly ugly looking data with unexpected features might raise the question in your mind. Here we consider only the special case of outliers, leaving it as an exercise for the reader to work out the corresponding theory for other kinds of unexpected structure.

21.1 The experimenter's dilemma

The problem of outliers in data has been a topic of lively discussion since the 18th century, when it arose in astronomy, geodesy, calorimetry, and doubtless many other measurements. Let us interpret 'apparatus' broadly as meaning any method for acquiring data. On the philosophical side, two opposite views have been expressed repeatedly.

(I) Something must have gone wrong with the apparatus; the outlier is not part of the good data and we must throw it out to avoid getting erroneous conclusions.

(II) No! It is dishonest to throw away any part of your data merely because it was unexpected. That outlier may well be the most significant datum you have, and it must be taken into account in your data analysis, otherwise you are 'fudging' the data arbitrarily and you can make no pretense of scientific objectivity.

From these statements we can understand why the issue can arouse controversy that is very hard to resolve. Not only has an element of righteous ethical fervor crept in; it is also clear that both positions do contain elements of truth. How can they be reconciled?

On the pragmatic side, several arbitrary *ad hoc* recipes were invented (such as the Chauvenet criterion found in the astronomy textbooks of a century ago) to decide when to reject an outlier. It is curious that the arbitrary criteria for rejection (two standard deviations, etc.) seem to have taken no note of the following, which we think is essential for any rational approach to the problem.

Pondering the two statements above, we see that they reflect different *prior information* about the apparatus. This is the crucial factor – ignored in all the aforementioned criteria. To take it into account properly requires, not still more *ad hockeries*, but straightforward probability analysis.

Position (I) seems reasonable if we know that the means of gathering data is unreliable, and it is indeed likely to break down without warning and give erroneous data. If we already expect this, then the appearance of a wild outlier seems far more likely to be due to 'apparatus failure' than to a real effect.[1]

Position (II), on the other hand, is the reasonable stance for one who has absolute confidence in his apparatus: he is sure that his voltmeter always gives readings reliable to $\pm 0.5\%$, and could not be in error by 5%; or that his telescope was aimed within 10 arc seconds of the recorded direction, and cannot be off by 1 degree. Then the appearance of an outlier must be accepted as a significant event, however unexpected; to ignore it could be to miss an important discovery.

But (I) and (II) are extreme positions, and the real experimenter is almost always in some intermediate situation. Presumably, if he knew that his apparatus was very unreliable, he would prefer not to take data with it at all; but in a field like biology or economics one may be obliged to use whatever 'apparatus' Nature has provided. On the other hand, few scientists – even in the best laboratories of the National Bureau of Standards – are ever so confident of their apparatus that they will affirm dogmatically that it *cannot* go awry.

One would like to see the estimate in the form of an unequivocal statement like $(\theta)_{est} = A \pm B$, where A, B are two definite numbers, presumably two functions of the data $D \equiv (x_1, \ldots, x_n)$. But what two functions? When the data are closely clustered together, it is surely a reasonable guess to take $A = \bar{x} \equiv n^{-1} \sum x_i$, the sample mean, as the estimated value. The observed scatter of the data values x_i indicates the *reproducibility* of the measurements, and one might suppose that this indicates also their *accuracy*. If so, it might seem reasonable to calculate the mean-square deviation from the mean, or sample

[1] We saw another example of this phenomenon in Chapter 6, in the discussion following Eq. (6.97). There probability theory told us that, if large fluctuations in counting rate are to be expected as an artifact of the apparatus, then the observed fluctuations become less cogent for estimating changes in beam intensity.

variance: $s^2 \equiv n^{-1} \sum (x_i - \bar{x})^2 = \overline{x^2} - \bar{x}^2$ and choose $B = s$, the sample standard deviation. A more educated intuition familiar with elementary results of probability theory can improve on this by taking $B = s/\sqrt{n}$, and even if it is not shown to be optimal by any clearly stated criterion, the conclusion

$$(\theta)_{\text{est}} = \bar{x} \pm \frac{s}{\sqrt{n}} \tag{21.1}$$

would not be criticized as wildly unreasonable in location or accuracy.

Exercise 21.1. We have seen in Chapter 7 that, under rather general conditions, a Gaussian sampling distribution, $p(x|\theta, \sigma) \propto \exp\{-(x - \theta)^2/2\sigma^2\}$, leads us to take our point estimate of θ as the data mean $\bar{x}$. Show that any sampling distribution with a rounded top (that is, $p(x|\theta) = a_0 - a_1(x - \theta)^2 + \cdots$) will lead us to the same mean-value estimate in the limit where the data are closely clustered.

If the data are not closely clustered, the above discussion seems to consider only two possible actions: keep the outlier and give it full credence; or throw it out altogether. Is there a more defensible intermediate position?

21.2 Robustness

Another viewpoint toward such problems has arisen recently, represented by Huber (1981) and noted briefly in Chapter 6. It still seeks to deal with them by intuitive *ad hoc* procedures that do not take explicit note of prior information or probability theory; but it does look for an intermediate position. One seeks data analysis methods that are *robust*, which means insensitive to the exact sampling distribution of errors or, as it is often stated, insensitive to the model, or are *resistant*, meaning that large errors in a small proportion of the data do not greatly affect the conclusions.

The general idea, stated vaguely, is that theoretical 'optimality', in the sense we have used it in previous chapters, is not always a good criterion in practice. Often we are unsure of the correct model; then a method which is useful for a variety of different models, even though not optimal for any, may be preferable to one that is exactly suited to one specific model, but misleading for others.

Evidently, there could be some merit in this view; but the 'robustnik/exploratory' school of thought, represented by Tukey and Mosteller (for example, Tukey, 1977), carries this to the point of deprecating all considerations of optimality. However, attempts to define this position less vaguely become troublesome. Given data D and any two estimators $f(D)$, $g(D)$ of some parameter, is there any explicit definition of the term 'robust' or 'resistant' which would make it meaningful to say that one is 'more robust' or 'more resistant' than the other? If so, then within a given set S of feasible estimators there is necessarily an 'optimally robust' one $a(D)$ and an 'optimally resistant' one $b(D)$, not necessarily unique.[2]

The point we make here is that if any intuitive property, such as robustness, is held to be desirable, then as soon as this property is defined with enough precision to allow transitive comparisons, an optimality principle follows inexorably. So one cannot consistently advocate any well-defined inference property and at the same time reject optimality principles.

Equally troublesome is the fact that robust/resistant qualities – however defined – must be bought at a price: the price of poorer performance when the model is correct. Indeed, this performance can be very much poorer; for it is clear that the most robust procedure of all – the 'optimal' procedure if one asks only for robustness – is the one that ignores the model, the data, and the prior information altogether, and considers all parameters zero, all hypotheses false! There must be, inevitably, some trade-off between the conflicting requirements of robustness and accuracy.[3] Advocates of robust/resistance have an obligation to show us just what trade-off, i.e. how much deterioration of performance, they are asking us to accept.

In estimating a location parameter, for example, the sample median M is often cited as a more robust estimator than the sample mean. But here it is obvious that this 'robustness' is bought at the price of insensitivity to much of the relevant information in the data. Many different data sets all have the same median; the values above or below the sample median may be moved about arbitrarily without affecting the estimate. Yet those data values surely contain information highly relevant to the question being asked, and all this is lost. We would have thought that the whole purpose of data analysis is to extract all the information we can from the data.

Thus, while we agree that robust/resistant properties may be desirable in some cases, we think it important to emphasize their cost in performance. In the literature, *ad hoc* procedures have been advocated on no more grounds than that they are 'robust' or 'resistant', with no mention of the quality of the inference they deliver, much less any comparison of performance with alternative methods; yet alternative methods such as Bayesian ones are criticized on grounds of lack of robustness, without any supporting factual evidence.

Those who criticize Bayesian methods on such grounds are simply revealing that they do not understand how to use Bayesian methods. We wish to show that Bayesian data analysis, properly applied, automatically delivers robustness and resistance whenever those qualities are desirable; in fact, it does so in a way that agrees qualitatively with what previous *ad hoc* intuitive procedures have done, but improves on them quantitatively, because Bayesian methods never throw away relevant information. In other words, present robust methods are, like the other orthodox methods, only intuitive approximations to what a full Bayesian analysis gives automatically.

Indeed, this situation is very much like that noted in Section 5.6, where we discussed horse racing and weather forecasting. The new information – there called the data – was not known

---

[2] The term 'robust/resistant estimator' was coined by John W. Tukey; the present writer suggested to him that this must mean, literally, 'an estimator which resists being robust', but he denied it.

[3] In parameter estimation, the orthodox criterion of admissibility suffers from just the same defect; a procedure which ignores the data and always estimates $\theta^* = 5$ is admissible if the point $\theta = 5$ is in the parameter space; yet it is clear that almost any 'inadmissible' estimation rule would be superior to this 'admissible' one.

with certainty to be true, and we saw how Bayesian analysis takes that into account automatically. Here it is the model – part of the prior information – that is in doubt, but that makes no difference in principle because the 'data' and the 'prior information' are just two components of our total evidence which enter into probability theory in the same way. In the present case, a detailed Bayesian analysis reveals some very interesting and unexpected insight.

Reasoning that is unresponsive to changes in the model must be also in some way unresponsive to changes in the data. Is this what we really want? We think the answer is: Sometimes: that is, in problems where we are unsure of the model *but nevertheless sure of the meaning of the parameters in it*. But if we are sure of the model, then robust/resistance is the last thing we want in our data analysis procedure; it would waste data by throwing away cogent information.

Again, we must take explicit note of the prior information before the issue can be judged. As demonstrated below, if we choose our sampling distribution to represent properly our prior knowledge of the phenomenon that is generating the data, Bayesian analysis gives us automatically both robustness/resistant qualities when we are unsure of the model, and optimal performance when we are sure of it.

We may, however, make one concession. Intuitive devices of the Tukey genre can take into account, after a fashion, all kinds of special, one-time transitory contingencies that would be difficult – and not even desirable – to build into a model. A formal probability model ought to describe nontransitory situations that deserve more careful treatment and recording for future use. As a mathematician once put it: 'A *method* is a *device* that you use twice.'

But this one-time intuition is necessarily also a one-man operation, because it offers no rationale or criterion of optimality for what it does so that others could judge its suitability. If your one-time intuition differs from mine, then, without a normative theory of rational inference, we are at an impasse that cannot be resolved. But a logically consistent 'normative theory of rational inference' means necessarily (because of the theorems of Cox) a Bayesian theory.

Let us examine first the Bayesian treatment of the most common situation, in which the data are classified into only two categories: good and bad.

21.3 The two-model model

We have a 'good' sampling distribution

$$G(x|\theta) \tag{21.2}$$

with a parameter θ that we want to estimate. Data drawn urn-wise from $G(x|\theta)$ are called 'good' data. But there is also a 'bad' sampling distribution

$$B(x|\eta), \tag{21.3}$$

possibly containing an uninteresting parameter η. Data from $B(x|\eta)$ are called 'bad' data; they appear to be useless or worse for estimating θ, since their probability of occurring has nothing to do with θ. Our data set consists of n observations

$$D = (x_1, \ldots, x_n).   \tag{21.4}$$

But the trouble is that some of these data are good and some are bad, and we do not know which is which (however, we may be able to make guesses: an obvious outlier – far out in the tails of $G(x|\theta)$ – or any datum in a region of x where $G(x|\theta) \ll B(x|\eta)$ comes under suspicion of being bad).

In various real problems we may, however, have some prior information about the process that determines whether a given datum will be good or bad. Various probability assignments for the good/bad selection process may express that information. For example, we may define

$$q_i \equiv \begin{cases} 1 \text{ if the ith datum is good} \\ 0 \text{ if it is bad,} \end{cases}   \tag{21.5}$$

and then assign joint prior probabilities

$$p(q_1 \cdots q_n | I)   \tag{21.6}$$

to the 2^n conceivable sequences of good and bad.

21.4 Exchangeable selection

Consider the most common case, where our information about the good/bad selection process can be represented by assigning an exchangeable prior. That is, the probability of any sequence of n good/bad observations depends only on the numbers r, $(n - r)$ of good and bad ones, respectively, and not on the particular trials at which they occur. Then the distribution (21.6) is invariant under permutations of the q_i, and by the de Finetti representation theorem (Chapter 18), it is determined by a single generating function $g(u)$:

$$p(q_1 \cdots q_n | I) = \int_0^1 du \, u^r \, (1 - u)^{n-r} \, g(u).   \tag{21.7}$$

It is much like flipping a coin with unknown bias where, instead of 'good' and 'bad', we say 'heads' and 'tails'. There is a parameter u such that if u were known we would say that any given datum x may, with probability u, have come from the good distribution; or with probability $(1 - u)$ from the bad one. Thus, u measures the 'purity' of our data; the closer to unity the better. But u is unknown, and $g(u)$ may, for present purposes, be thought of as its prior probability density (as was, indeed, done already by Laplace; further technical details about this representation are given in Chapter 18). Thus, our sampling distribution may be written as a probability mixture of the good and bad distributions:

$$p(x|\theta, \eta, u) = uG(x|\theta) + (1 - u)B(x|\eta) , \qquad 0 \leq u \leq 1.   \tag{21.8}$$

This is just a particular form of the general parameter estimation model, in which θ is the parameter of interest, while (η, u) are nuisance parameters; it requires no new principles beyond those expounded in Chapter 6.

Indeed, the model (21.8) contains the usual binary hypothesis testing problem as a special case, where it is known initially that all the observations are coming from G, or they are all coming from B, but we do not know which. That is, the prior density for u is concentrated on the points $u = 0, u = 1$:

$$p(u|I) = p_0 \, \delta(1 - u) + p_1 \, \delta(u), \tag{21.9}$$

where $p_0 = p(H_0|I)$, $p_1 = 1 - p_0 = p(H_1|I)$ are the prior probabilities for the two hypotheses:

$$H_0 \equiv \text{all the data come from the distribution } G(x|\theta), \\ H_1 \equiv \text{all the data come from the distribution } B(x|\eta). \tag{21.10}$$

Because of their internal parameters, they are composite hypotheses; the Bayesian analysis of this case was noted briefly in Chapter 4. Of course, the logic of what we are doing here does not depend on value judgments like 'good' and 'bad'.

Now consider u unknown and the problem to be that of estimating θ. A full nontrivial Bayesian solution tends to become intricate, since Bayes' theorem relentlessly seeks out and exposes every factor that has the slightest relevance to the question being asked. But often much of that detail contributes little to the final conclusions sought (which might be only the first few moments, or percentiles, of a posterior distribution). Then we are in a position to seek useful approximate algorithms that are 'good enough' without losing essential information or wasting computation on nonessentials. Such rules might conceivably be ones that intuition had already suggested, but, because they are good mathematical approximations to the full optimal solution, they may also be far superior to any of the intuitive devices that were invented without taking any note of probability theory; it depends on how good that intuition was.

Our problem of outliers is a good example of these remarks. If the good sampling density $G(x|\theta)$ is very small for $|x| > 1$, while the bad one $B(x|\eta)$ has long tails extending to $|x| \gg 1$, then any datum y for which $|y| > 1$ comes under suspicion of coming from the bad distribution, and intuitively one feels that we ought to 'hedge our bets' a little by giving it, in some sense, a little less credence in our estimate of θ. Put more specifically, if the validity of a datum is suspect, then intuition suggests that our conclusions ought to be less sensitive to its exact value. But then we have just about stated the condition of robustness (only now, this reasoning gives it a rationale that was previously missing). As $|x| \to \infty$ it is practically certain to be bad, and intuition probably tells us that we ought to disregard it altogether.

Such intuitive judgments were long since noted by Tukey and others, leading to such devices as the 'redescending psi function', which achieve robust/resistant performance by modifying the data analysis algorithms in this way. These works typically either do not deign to note even the existence of Bayesian methods, or contain harsh criticism of

Bayesian methods, expressing a belief that they are not robust/resistant and that the intuitive algorithms are correcting this defect – but never offering any factual evidence in support of this position.

In the following we break decades of precedent *actually examining* a Bayesian calculation of outlier effects, so that one can see – perhaps for the first time – what Bayesianity has to say about the issue, and thus give that missing factual evidence.

21.5 The general Bayesian solution

Firstly, we give the Bayesian solution based on the model (21.8) in full generality, then we study some special cases. Let $p(\theta\eta u|I)$ be the joint prior density for the parameters. Their joint posterior density given the data D is

$$f(\theta, \eta, u|DI) = Af(\theta, \eta, u|I)L(\theta, \eta, u), \tag{21.11}$$

where A is a normalizing constant, and, from (21.8),

$$L(\theta, \eta, u) = \prod_{i=1}^{n}\left[uG(x_i|\theta) + (1-u)B(x_i|\eta)\right] \tag{21.12}$$

is their joint likelihood. The marginal posterior density for θ is

$$p(\theta|DI) = \int\int d\eta du f(\theta, \eta, u|DI). \tag{21.13}$$

To write (21.12) more explicitly, factor the prior density:

$$f(\theta, \eta, u|I) = h(\eta, u|\theta, I)f(\theta|I), \tag{21.14}$$

where $f(\theta|I)$ is the prior density for θ, and $h(\eta, u|\theta, I)$ is the joint prior for (η, u), given θ. Then the marginal posterior density for θ, which contains all the information that the data and the prior information have to give us about θ, is

$$f(\theta|D, I) = \frac{f(\theta|I)\overline{L}(\theta)}{\int d\theta f(\theta|I)\overline{L}(\theta)}, \tag{21.15}$$

where we have introduced the quasi-likelihood

$$\overline{L}(\theta) \equiv \int\int d\eta du\, L(\theta, \eta, u)h(\eta, u|\theta, I). \tag{21.16}$$

Inserting (21.12) into (21.16) and expanding, we have

$$\overline{L}(\theta) = \int\int d\eta du\, h(\eta, u|\theta, I)\Big[u^n L(\theta) + u^{n-1}(1-u)\sum_{j=1}^{n}B(x_j|\eta)L_j(\theta)$$

$$+ u^{n-2}(1-u)^2\sum_{j<k}B(x_j|\eta)B(x_k|\eta)L_{jk}(\theta) + \cdots \tag{21.17}$$

$$+ (1-u)^n B(x_1|\eta)\cdots B(x_n|\eta)\Big],$$

in which

$$L(\theta) \equiv \prod_{i=1}^{n} G(x_i|\theta)$$

$$L_j(\theta) \equiv \prod_{i\neq j} G(x_i|\theta) \tag{21.18}$$

$$L_{jk}(\theta) \equiv \prod_{i\neq j,k} G(x_i|\theta)\dots \quad \text{etc.}$$

are a sequence of likelihood functions for the good distribution in which we use all the data, all except the datum x_j, all except x_j and x_k, and so on. To interpret the lengthy expression (21.17), note that the coefficient of $L(\theta)$,

$$\int_0^1 du \int d\eta\, h(\eta, u|\theta, I) u^n = \int du\, u^n h(u|\theta, I), \tag{21.19}$$

is the probability, conditional on θ and the prior information, that all the data $\{x_1, \dots, x_n\}$ are good. This is represented in the Laplace–de Finetti form (21.7) in which the generating function $g(u)$ is the prior density $h(u|\theta, I)$ for u, conditional on θ. Of course, in most real problems this would be independent of θ (which is presumably some parameter referring to an entirely different context than u); but preserving generality for the time being will help to bring out some interesting points later.

Likewise, the coefficient of $L_j(\theta)$ in (21.17) is

$$\int du\, u^{n-1}(1-u) \int d\eta\, B(x_j|\eta)h(\eta, u|\theta, I). \tag{21.20}$$

Now the factor

$$d\eta \int du\, u^{n-1}(1-u)h(\eta, u|\theta I) \tag{21.21}$$

is the joint probability density, given I and θ, that any specified datum x_j is bad, that the $(n-1)$ others are good, and that η lies in $(\eta, \eta + d\eta)$. Therefore the coefficient (21.20) is the probability, given I and θ, that the jth datum would be bad and would have the value x_j, and the other data would be good. Continuing in this way, we see that, to put in it words, our quasi-likelihood is:

$$\overline{L}(\theta) = \text{prob(all the data are good)} \times \text{(likelihood using all the data)}$$
$$+ \sum_j \text{prob(only } x_j \text{ bad)} \times \text{(likelihood using all data except } x_j)$$
$$+ \sum_{j,k} \text{prob(only } x_j, x_k \text{ bad)} \times \text{(likelihood using all except } x_j, x_k) \tag{21.22}$$
$$+ \cdots$$
$$+ \sum_j \text{prob(only } x_j \text{ good)} \times \text{(likelihood using only the datum } x_j)$$
$$+ \text{prob(all the data are bad)}.$$

In shorter words: the quasi-likelihood $\overline{L}(\theta)$ is a weighted average of the likelihoods for the good distribution $G(x|\theta)$ resulting from every possible assumption about which data are good, and which are bad, weighted according to the prior probabilities of those assumptions. We see how every detail of our prior knowledge about how the data are being generated is captured in the Bayesian solution.

This result has such wide scope that it would require a large volume to examine all its implications and useful special cases. But let us note how the simplest ones compare with our intuition.

21.6 Pure outliers

Suppose the good distribution is concentrated in a finite interval

$$G(x|\theta) = 0, \qquad |x| > 1, \tag{21.23}$$

while the bad distribution is positive in a wider interval which includes this. Then any datum x for which $|x| > 1$ is known with certainty to be an outlier, i.e. to be bad. If $|x| < 1$, we cannot tell with certainty whether it is good or bad. In this situation our intuition tells us quite forcefully: Any datum that is known to be bad is just not part of the data relevant to estimation of θ and we shouldn't be considering it at all. So just throw it out and base our estimate on the remaining data.

According to Bayes' theorem this is almost right. Suppose we find $x_j = 1.432$, $x_k = 2.176$, and all the other x's less than unity. Then, scanning (21.24) it is seen that only one term will survive:

$$\overline{L}(\theta) = \int du \int d\eta \, h(\eta, u|\theta, I) B(x_j|\eta) B(x_k|\eta) L_{jk}(\theta) = C_{jk}(\theta) L_{jk}(\theta). \tag{21.24}$$

As discussed above, the factor C_{jk} is almost always independent of θ, and since constant factors are irrelevant in a likelihood, our quasi-likelihood in (21.15) reduces to just the one obtained by throwing away the outliers, in agreement with that intuition.

But it is conceivable that in rare cases $C_{jk}(\theta)$ might, after all, depend on θ; and Bayes' theorem tells us that such a circumstance would make a difference. Pondering this, we see that the result was to be expected if only we had thought more deeply. For if the probability of obtaining two outliers with values x_j, x_k depends on θ, then the fact that we got those particular outliers is in itself evidence relevant to inference about θ.

Thus, even in this trivial case Bayes' theorem tells us something that unaided intuition did not see: even when some data are known to be outliers, their values might still, in principle, be relevant to estimation of θ. This is an example of what we meant in saying that Bayes' theorem relentlessly seeks out and exposes every factor that has any relevance at all to the question being asked.

In the more usual situations, Bayes' theorem tells us that whenever any datum is known to be an outlier, then we should simply throw it out, if the probability of getting that particular outlier is independent of θ. For, quite generally, a datum x_i can be known with certainty to be an outlier only if $G(x_i|\theta) = 0$ for all θ; but in that case every likelihood in (21.24) that contains x_i will be zero, and our posterior distribution for θ will be the same as if the datum x_i had never been observed.

21.7 One receding datum

Now suppose the parameter of interest is a location parameter, and we have a sample of ten observations. But one datum x_j moves away from the cluster of the others, eventually receding out 100 standard deviations of the good distribution G. How will our estimate of θ follow it? The answer depends on which model we specify.

Consider the usual model in which the sampling distribution is taken to be simply $G(x|\theta)$ with no mention of any other 'bad' distribution. If G is Gaussian, $x \sim N(\theta, \sigma)$, and our prior for θ is wide (say $> 1000\,\sigma$), then the Bayesian estimate for quadratic loss function will remain equal to the sample average, and our far-out datum will pull the estimate about ten standard deviations away from the average indicated by the other nine data values. This is presumably the reason why Bayesian methods are sometimes charged with failure to be robust/resistant.

However, that is the result only for the assumed model, which in effect proclaims dogmatically: I know in advance that $u = 1$; all the data will come from G, and I am so certain of this that no evidence from the data could change my mind. If one actually had this much prior knowledge, then that far-out datum would be highly significant; to reject it as an 'outlier' would be to ignore cogent evidence, perhaps the most cogent piece of evidence that the data provide. Indeed, it is a platitude that important scientific discoveries have resulted from an experimenter having that much confidence in his apparatus, so that surprising new data were believed; and not merely rejected as 'accidental' outliers.

If, nevertheless, our intuition tells us with overwhelming force that the deviant datum *should* be thrown out, then it must be that we do not really believe that $u = 1$ strongly enough to adhere to it in the face of the evidence of the surprising datum. A Bayesian may correct this by use of the more realistic model (21.8). Then the proper criticism of the first procedure is not of Bayesian *methods*, but rather of the saddling of Bayesian methodology with an inflexible, dogmatic model which denies the possibility of outliers. We saw in Section 4.4 on multiple hypothesis testing just how much difference it can make when we permit the robot to become skeptical about an overly simple model.

Bayesian methods have inherent in them all the desirable robust/resistant qualities, and they will exhibit these qualities automatically, whenever they *are* desirable – if a sufficiently flexible model permits them to do so. But neither Bayesian nor any other methods can give sensible results if we put absurd restrictions on them. There is a moral in this, extending to all of probability theory. In other areas of applied mathematics, failure to notice some feature (like the possibility of the bad distribution B) means only that it will not be taken into account. In probability theory, failure to notice some feature may be tantamount to making irrational assumptions about it.

Then why is it that Bayesian methods have been criticized more than orthodox ones on this issue? For the same reason that city B may appear in the statistics to have a higher crime rate than city A, when the fact is that city B has a lower crime rate, but more efficient means for detecting crime. Errors undetected are errors uncriticized.

Like any other problem in this field, this can be further generalized and extended endlessly, to a three-model model, putting parameters in (21.6), etc. But our model is already general enough to include both the problem of outliers and conventional hypothesis testing theory; and a great deal can be learned from a few of its simple special cases.

22

Introduction to communication theory

We noted in Chapter 11 that one of the motivations behind this work was the attempt to see Gibbsian statistical mechanics and Shannon's communication theory as examples of the same line of reasoning. A generalized form of statistical mechanics appeared as soon as we introduced the notion of entropy, and we ought now to be in a position to treat communication theory in a similar way.

One difference is that in statistical mechanics the prior information has nothing to do with frequencies (it consists of measured values of macroscopic quantities such as pressure), and so we have little temptation to commit errors. But in communication theory the prior information consists, typically, of frequencies; this makes the probability–frequency conceptual pitfalls much more acute. For this reason it seemed best to take up communication theory only after we had seen the general connections between probability and frequency, in a variety of conceptually simpler applications.

22.1 Origins of the theory

Firstly, the difficult matter of giving credit where credit is due. All major advances in understanding have their precursors, whose full significance is never recognized at the time. Relativity theory had them in the work of Mach, Fitzgerald, Lorentz and Poincaré, to mention only the most obvious examples. Communication theory had many precursors, in the work of Gibbs, Nyquist, Hartley, Szilard, von Neumann, and Wiener. But there is no denying that the work of Shannon (1948) represents the arrival of the main signal, just as did Einstein's of 1905. In both cases ideas which had long been, so to speak, 'in the air' in a vague form, are grasped and put into sharp focus.

Shannon's papers were so full of important new concepts and results that they exercised not only a stimulating effect, but also a paralyzing effect. During the first few years after their appearance it was common to hear the opinion expressed, rather sadly, that Shannon had anticipated and solved all the problems of the field, and left nothing else for others to do.

The post-Shannon developments, with few exceptions, can be classed into efforts in two entirely different directions. On the applications side, we have the expansionists (who try to apply Shannon's ideas to other fields, as we do here), the entropy calculator (who works out the entropy of a television signal, the French language, a chromosome, or almost anything else you can imagine; and then finds that nobody knows what to do with it), and the universalist (who assures us that Shannon's work will revolutionize all intellectual activity; but is unable to offer a specific example of anything that has been changed by it).

We should not be overly critical of these efforts because, as J. R. Pierce has remarked, it is very hard to tell at first which ones make sense, which are pure nonsense, and which are the beginning of something that will in time make sense. The writer's efforts have received all three classifications from various quarters. We expect that, eventually, the ideas introduced by Shannon will be indispensable to the linguist, the geneticist, the television engineer, the neurologist, the economist. But we share with many others a feeling of disappointment that 40 years of effort along these lines has led to so little in the way of really useful advances in these fields.

During this time there has been an overabundance of vague philosophy, and of abstract mathematics; but, outside of coding theory, a rather embarrassing shortage of examples where specific real problems have been solved by using this theory. We believe that the reason for this is that conceptual misunderstandings, almost all of which amount to the mind projection fallacy, have prevented workers from asking the right questions. In order to apply communication theory to other problems than coding, the first and hardest step is to state precisely *what is the specific problem that we want to solve?*

In almost diametric opposition to the above efforts, as far as aim was concerned, were the mathematicians, who viewed communication theory simply as a branch of pure mathematics. Characteristic of this school was a belief that, before introducing a continuous probability distribution, you have to talk about set theory, Borel fields, measure theory, the Lebesgue–Stieltjes integral, and the Radon–Nikodym theorem. The important thing was to make the theorems rigorous *by the criteria of rigor then fashionable among mathematicians*, even if in so doing we limit their scope for applications. The book on information theory by A. I. Khinchin (1957) can serve as a typical example of the style prevalent in this literature.

Here again, severe criticism of these efforts is not called for. Of course, we want our principles to be subjected to the closest scrutiny the human mind can bring to bear on them; if important applications exist, the need for this is so much the greater. However, the present work is not addressed to mathematicians, but to persons concerned with real applications. So we shall dwell on this side of the story only to the extent of pointing out that the rigorized theorems are not the ones relevant to problems of the real world. Typically, they refer *only* to situations that do not exist (such as infinitely long messages), and as a result they degenerate into 'nonsense theorems' which assign probability one to an impossible event, and therefore zero to all possible events. We have no way of using such results, because our probabilities are always conditional on our knowledge of the real world. Now let's turn to some of the specific things in Shannon's papers.

22.2 The noiseless channel

We deal with the transmission of information from some sender to some receiver. We shall speak of them in anthropomorphic terms, such as 'the man at the receiving end', although either or both might actually be machines, as in telemetry or remote control systems. Transmission takes place via some *channel*, which might be a telephone or telegraph circuit, a microwave link, a frequency band assigned by the Federal Communication Commission (FCC), the German language, the postman, the neighborhood gossip, or a chromosome. If, after having received a message, the receiver can always determine with certainty which message was intended by the sender, we say that the channel is *noiseless*.

It was recognized very early in the game, particularly by Nyquist and Hartley, that the capability of a channel is not described by any property of the specific message it sends, but rather by what it *could have* sent. The usefulness of a channel lies in its readiness to transmit any one of a large class of messages, which the sender can choose at will.

In a noiseless channel, the obvious measure of this ability is simply the maximum number, $W(t)$, of distinguishable (at the destination) messages which the channel is capable of transmitting in a time t. In all cases of interest to us, this number goes eventually into an exponential increase for sufficiently large t: $W(t) \propto \exp\{Ct\}$, so the measure of channel performance which is independent of any particular time interval is the coefficient C of this increase. We define the *channel capacity* as

$$C \equiv \lim_{t \to \infty} \left[\frac{1}{t} \log[W(t)] \right]. \qquad (22.1)$$

The units in which C is measured will depend on which base we choose for our logarithms. Usually one takes base 2, in which C is given in 'bits per second', one bit being the amount of information contained in a single binary (yes–no) decision. For easy interpretation of numerical values, the bit is by far the best unit to use; but in formal operations it may be easier to use the base e of natural logarithms. Our channel capacities are then measured in natural units, or 'nits per second'. To convert, note that 1 bit $= \ln(2) = 0.69315$ nits, or 1 nit $= 1.4427$ bits.

The capacity of a noiseless channel is a definite number, characteristic of the channel, which has nothing to do with human information. Thus, if a noiseless channel can transmit n symbols per second, chosen in any order from an alphabet of a letters, we have $W(t) = a^{nt}$, or $C = n \log_2(a)$ bits/s $= n \log_e(a)$ nits/s. Any constraint on the possible sequences of letters can only lower this number. For example, if the alphabet is $A_1, A_2, \ldots, A_a$, and it is required that in a long message of $N = nt$ symbols the letter A_i must occur with relative frequency f_i, then the number of possible messages in time t is only

$$W(t) = \frac{N!}{(Nf_1)! \cdots (Nf_a)!}, \qquad (22.2)$$

and from Stirling's approximation, as we found in Chapter 11,

$$C = -n \sum_i f_i \log(f_i) \quad \text{nits/s}. \qquad (22.3)$$

This attains its maximum value, equal to the previous $C = n \log(a)$, in the case of equal frequencies, $f_i = 1/a$. Thus we have the interesting result that a constraint requiring all letters to occur with equal frequencies does not decrease channel capacity at all. It does, of course, decrease the number $W(t)$ by an enormous factor; but the decrease in $\log(W)$ is what matters, and this grows less rapidly than t, so it makes no difference in the limit. In view of the entropy concentration theorem of Chapter 11, this can be understood in another way: the vast majority of all *possible* messages are ones in which the letter frequencies are nearly equal.

Suppose now that symbol A_i has transmission time t_i, but there is no other constraint on the allowable sequences of letters. What is the channel capacity? Well, consider first the case of messages in which letter A_i occurs n_i times, $i = 1, 2, \ldots, a$. The number of such messages is

$$W(n_1, \ldots, n_a) = \frac{N!}{n_1! \cdots n_a!},$$ (22.4)

where

$$N = \sum_{i=1}^{a} n_i.$$ (22.5)

The total number of different messages that could have been transmitted in time t is then

$$W(t) = \sum_{n_i} W(n_1, \ldots, n_a),$$ (22.6)

where we sum over all choices of $(n_1, \ldots, n_a)$ compatible with $N_i \geq 0$ and

$$\sum_{i=1}^{a} n_i t_i \leq t.$$ (22.7)

The number $K(t)$ of terms in the sum (22.6) satisfies $K(t) \leq (Bt)^a$ for some $B < \infty$. This is seen most easily by imagining the n_i as coordinates in an a-dimensional space and noting the geometrical meaning of $K(t)$ as the volume of a simplex.

Exact evaluation of (22.6) would be quite an unpleasant job. But it's only the limiting value that we care about right now, and we can get out of the hard work by the following trick. Note that $W(t)$ cannot be less than the greatest term $W_m = W_{\max}(n_1, \ldots, n_a)$ in (22.6) nor greater than $W_m K(t)$:

$$\log(W_m) \leq \log[W(t)] \leq \log(W_m) + a \log(Bt),$$ (22.8)

and so we have

$$C \equiv \lim_{t \to \infty} \frac{1}{t} \log[W(t)] = \lim_{t \to \infty} \frac{1}{t} \log[W_m];$$ (22.9)

i.e. to find the channel capacity, it is sufficient to maximize $\log W(n_1, \ldots, n_a)$ subject to the constraint (22.7). This rather surprising fact can be understood as follows. The logarithm of $W(t)$ is given, crudely, by $\log[W(t)] = \log(W_{\max}) + \log$ [number of reasonably large terms in (22.6)]. Even though the number of large terms tends to infinity as t^a, this is not rapid enough to make any difference in comparison with the exponential increase of $W_{\max}$. As explained by Schrödinger (1948), this same mathematical fact is the reason why, in statistical mechanics, the Darwin–Fowler method and the method of most probable distribution lead to the same results in the limit of large systems.

We can solve the problem of maximizing $\log W(n_1, \ldots, n_a)$ by the same Lagrange multiplier argument used in Chapter 11. The problem is not quite the same, however, because now N is also to be varied in finding the maximum. Using the Stirling approximation, which is valid for large n_i, we have

$$\log W(n_1, \ldots, n_a) \approx N \log(N) - \sum_{i=1}^{a} n_i \log(n_i). \tag{22.10}$$

The variational problem, with λ a Lagrangian multiplier, is

$$\delta[\log(W) + \lambda \sum n_i t_i] = 0, \tag{22.11}$$

but since $\delta N = \sum \delta n_i$ we have

$$\delta \log(W) = \delta N \log(N) - \delta N - \sum_i (\delta n_i \log(n_i) - \delta n_i)$$

$$= -\sum \delta n_i \log(n_i/N). \tag{22.12}$$

Therefore (22.11) reduces to

$$\sum_{i=1}^{a} [\log(n_i/N) + \lambda t_i] \delta n_i = 0 \tag{22.13}$$

with the solution

$$n_i = N \exp\{-\lambda t_i\}. \tag{22.14}$$

To fix the value of λ, we require

$$N = \sum n_i = N \sum \exp\{-\lambda t_i\}. \tag{22.15}$$

With this choice of n_i, we find

$$\frac{1}{t} \log(W_m) = -\frac{1}{t} \log(n_i/N) = \frac{1}{t} \sum n_i (\lambda t_i). \tag{22.16}$$

In the limit, $t^{-1} \sum n_i t_i \to 1$, and so

$$C = \lim_{t \to \infty} \frac{1}{t} \log[W(t)] = \lambda. \tag{22.17}$$

Our final result can be stated very simply:

> To calculate the capacity of a noiseless channel in which symbol A_i has transmission time t_i and which has no other constraints on the possible messages, define the partition function $Z(\lambda) \equiv \sum_i \exp\{-\lambda t_i\}$. Then the channel capacity C is the real root of

$$Z(\lambda) = 1. \tag{22.18}$$

You see already a very strong resemblance to the reasoning and the formalism of statistical mechanics, in spite of the fact that we have not yet said anything about probability.

From (22.15) we see that $W(n_1, \ldots, n_a)$ is maximized when the relative frequency of symbol A_i is given by the canonical distribution

$$f_i = \frac{n_i}{N} = \exp\{-\lambda t_i\} = \exp\{-C t_i\}. \tag{22.19}$$

Some have concluded from this that the channel is being 'used most efficiently' when we have encoded our messages so that (22.19) holds. But that would be quite mistaken because, of course, in time t the channel will actually transmit one message and only one; and this remains true regardless of what relative frequencies we use. Equation (22.19) tells us only that – in accordance with the entropy concentration theorem – the overwhelming majority of all possible messages that the channel *could have* transmitted in time t are ones where the relative frequencies are canonical.

On the other hand, we have a generalization of the remark following (22.3): if we impose an additional constraint requiring that the relative frequencies are given by (22.19), which might be regarded as defining a new channel, the channel capacity would not be decreased. But any constraint requiring that all possible messages have letter frequencies different from (22.19) will decrease channel capacity.

There are many other ways of interpreting these equations. For example, in our above arguments we supposed that the total time of transmission is fixed and we wanted to maximize the number W of possible messages which the sender can choose. In a practical communication system, the situation is usually the other way around: we know in advance the extent of choice which we demand in the messages which might be sent over the channel, so that W is fixed. We then ask for the condition that the total transmission time of the message be minimized subject to a fixed W.

It is well known that variational problems can be transformed into several different forms, the same mathematical result giving the solution to many different problems. A circle has maximum area for a given perimeter; but also it has minimum perimeter for a given area. In statistical mechanics, the canonical distribution can be characterized as one with maximum entropy for a given expectation of energy; or equally well as the one with minimum expectation of energy for a given entropy. Similarly, the channel capacity found from (22.18) gives the maximum attainable W for a given transmission time, or equally well the minimum attainable transmission time for a fixed W.

As another extension of the meaning of these equations, note that we need not interpret the quantity t_i as a time; it can stand equally well for the 'cost', as measured by any criterion, of transmitting the ith symbol. Perhaps the total length of time the channel is in operation is of no importance, because the apparatus has to sit there in readiness whether it is being used or not. The real criterion might be, for example, the amount of energy that a space probe must dissipate in transmitting a message back to Earth. In this case, we could define t_i as the energy required to transmit the ith symbol. The channel capacity given by (22.18) would then be measured, not in bits per second but in bits per joule, and its reciprocal is equal to the minimum attainable number of joules needed per bit of transmitted information.

A more complicated type of noiseless channel, also considered by Shannon, is one where the channel has a memory; it may be in any one of a set of 'states' $\{S_1, \ldots, S_k\}$ and the possible future symbols, or their transmission times, depend on the present state. For example, suppose that if the channel is in state S_i, it can transmit symbol A_n, which leaves the channel in state S_j, the corresponding transmission time being t_{inj}. Surprisingly, the calculation of the channel capacity in this case is quite easy.

Let $W_i(t)$ be the total number of different messages the channel can transmit in time t, starting from state S_i. Breaking down $W_i(t)$ into several terms according to the first symbol transmitted, we have the same difference equation that we used to introduce the partition function in Chapter 8:

$$W_i(t) = \sum_{jn} W_j(t - t_{inj}), \qquad (22.20)$$

where the sum is over all possible sequences $S_i \rightarrow A_n \rightarrow S_j$. As before, this is a linear difference equation with constant coefficients, so its asymptotic solution must be an exponential function:

$$W_i(t) \approx B_i \exp\{Ct\}, \qquad (22.21)$$

and from the definition (22.1) it is clear that, for finite k, the coefficient C is the channel capacity. Substituting (22.21) into (22.20), we obtain

$$B_i = \sum_{j=1}^{k} Z_{ij}(C) B_j, \qquad (22.22)$$

where

$$Z_{ij}(\lambda) = \sum_{n} \exp\{-\lambda t_{inj}\} \qquad (22.23)$$

is the 'partition matrix'. Compare this argument with our first derivation of a partition function in Chapter 8. If the sequence $S_i \rightarrow A_n \rightarrow S_j$ is impossible, we set $t_{inj} = \infty$. By this device we can understand the sum in (22.23) as extending over all symbols in the alphabet.

Equation (22.22) says that the matrix Z_{ij} has an eigenvalue equal to unity. Thus, the channel capacity is simply the greatest real root of $D(\lambda) = 0$, where

$$D(\lambda) \equiv \det[Z_{ij}(\lambda) - \delta_{ij}]. \qquad (22.24)$$

This is one of the prettiest results given by Shannon. In the case of a single state, $k = 1$, it reduces to the previous rule, (22.18).

The problems solved above are, of course, only especially simple ones. By inventing channels with more complicated types of constraints on the allowable sequences (i.e. with a long memory), we can generate mathematical problems as involved as we please. But it would still be just mathematics – as long as the channel is noiseless, there would be no difficulties of principle. In each case we simply have to count up the possibilities and apply the definition (22.1). For some weird channels, we might find that the limit therein does not exist, in which case we cannot speak of a channel capacity, but have to characterize the channel simply by giving the function $W(t)$.

22.3 The information source

When we take the next step and consider the information source feeding our channel, fundamentally new problems arise. There are mathematical problems aplenty, but there are also more basic conceptual problems which have to be considered before we can state which mathematical problems are the significant ones.

It was Professor Norbert Wiener who first suggested the enormously fruitful idea of representing an information source in probability terms. He applied this to some problems of filter design. This work was an essential step in developing a way of thinking which led to communication theory.

It is perhaps difficult nowadays for us to realize what a big step this was. Previously, communication engineers had considered an information source simply as a man with a message to send; for their purposes an information source could be characterized simply by describing that message. But Wiener suggested instead that an information source be characterized by giving the probabilities p_i that it will emit various messages M_i. Already we see the conceptual difficulties faced by a frequency theory of probability – the man at the sending end presumably knows perfectly well which message he is going to send. What, then, could we possibly mean by speaking of the *probability* that he will send something? There is nothing analogous to 'chance' operating here.

By the probability p_i of a message, do we mean the *frequency* with which he sends that particular message? The question is absurd – a sane man sends a given message at most once, and most messages never. Do we mean the frequency with which the message M_i occurs in some imaginary 'ensemble' of communication acts? Well, it's all right to state it that way if you want to, but it doesn't answer the question. It merely leads us to restate the question as: What defines that ensemble? How is it to be set up? Calling it by a different name doesn't help us. *What* information is that entropy $H = -\sum p_i \log(p_i)$ really measuring?

We take a halting first step toward answering this if we suppose that Shannon's H measures not the information of the sender, but the ignorance of the receiver, that is removed by receipt of the message. Indeed, most later commentators make this interpretation. Yet, on second thought, this does not make sense either; for Shannon proceeds to develop theorems relating H to the channel capacity C required to transmit the messages M_i. But how well a channel can transmit messages obviously depends on properties of the channel and the messages; and not at all on the state of ignorance of the receiver! You see the conceptual mess that the field has been in for 40 years.

Right at this point we have to state clearly *what the specific problem is that we want solved*. A probability distribution is a means of describing a state of knowledge. But *whose* state of knowledge do we want to talk about? Evidently, not the man at the sending end or the one at the receiving end; and Shannon offers us no explicit help on this. But implicitly, the answer seems to be clear; in view of the theorems Shannon gives, he cannot be describing the 'general philosophy' of communication between sender and receiver, as so many have supposed. He is thinking of the theory as something of practical value to an engineer whose job is to design the technical equipment in the communication system. In other words, *the state of knowledge Shannon is describing is that of the communication engineer when he designs the equipment*. It is *his* ignorance about the messages to be sent that is measured by H.

Although this viewpoint would seem perfectly natural for an engineer employed by the Bell Telephone Laboratories, as Shannon was at the time, you will not find it actually expressed in his words, or in the later literature based on the viewpoint which sees no distinction between probability and frequency. For on the frequentist view, the notion of a probability *for a person with a certain state of knowledge* simply doesn't exist, because probability is thought to be a real physical phenomenon which exists independently of human information. But the problem of choosing some probability distribution to represent the information source still does exist; it cannot be evaded. It is now clear that the whole content of the theory depends on how we do this.

We have already emphasized several times that in probability theory we never solve an actual problem of practice. We solve only some abstract mathematical model of the real problem. Setting up this model requires not only mathematical ability, but also a great deal of practical judgment. If our model does not correspond well to the actual situation, then our theorems, however rigorous the mathematicians may have made them, can be more misleading than helpful. This is so with a vengeance in communication theory, because not only the quantitative details, but even the qualitative nature of the theorems that can be proved, depend on which probability model we use to represent an information source.

The purpose of this probability model is to describe the communication engineer's *prior knowledge* about what messages his communication system may be called upon to send. In principle, this prior knowledge could be of any sort; in particular, nothing prevents it from being semantic in nature. For example, he might know in advance that the channel will be used only to transmit stock market quotations, not quotations from the Bible, or obscene limericks. That is a perfectly valid kind of prior information, which would have definite implications for the probabilities p_i by restricting the sample space in definite, specific ways, although they might be hard to state in general mathematical terms.

We stress this point because some critics harp away incessantly on the theme that information theory does not consider semantic meaning, and hold this to be a basic defect of our whole philosophy. They could not be more mistaken: the issue of semantic meaning is not a philosophical one but a technical one. The only reason why we do not consider semantic meaning is that we do not know how to do it *as a general procedure*, although we could certainly do it 'by hand' in the context of a specific, finite set of possible messages. Probably all of us have tried to restore some corrupted text by drawing upon our perception of its semantic meaning; but how do you teach a computer to do this?

So let us assure those critics: if you will show us *a definite, usable algorithm* for assessing semantic meaning, we are most eager to incorporate this too into information theory. In fact, our present inability to do this is a serious handicap in many applications, from image restoration, to pattern recognition, to artificial intelligence. We need your constructive help, not your criticisms.

But in traditional Shannon-type communication theory the only kind of prior knowledge considered is 'statistical' because this is amenable to mathematical treatment at once. That is, it consists of frequencies of letters, or combinations of letters, which have been observed in *past* samples of similar messages. Then a typical practical problem – indeed, the actual problem of writers of those popular text compression computer programs – is to design encoding systems which will transmit binary digits representing English text, reliably and at the maximum possible rate, given an available channel with known properties. This would be also the actual problem of designers of computer hardware such as disk drives and modems, if they became a little more sophisticated. The designer will then, according to the usual viewpoint, need accurate data giving the correct frequencies of English text. Let's think about that a little.

22.4 Does the English language have statistical properties?

Suppose we try to characterize the English language, for purposes of communication theory, by specifying the relative frequencies of various letters, or combinations of letters. Now we all know that there is a great deal of truth in statements such as 'the letter E occurs more frequently than the letter Z'. Long before the days of communication theory, many people made obvious common-sense use of this knowledge. One of the earliest examples is the design of the Morse telegraphic code, in which the most frequently used letters are represented by the shortest codes – the exact prototype of what Shannon formalized and made precise a century later.

The design of our standard typewriter keyboard makes considerable use of knowledge of letter frequencies. This knowledge was used in a much more direct and drastic way by Ottmar Mergenthaler, whose immortal phrase

$$\text{ETAOIN SHRDLU} \tag{22.25}$$

was a common sight in the newspapers many years ago when Linotype machines first came into use (an inexperienced operator, who allowed his fingers to brush lightly across the keys, automatically set this in type). But already we are getting into trouble, because there does not seem to be complete agreement even as to the relative order of the 12 most common letters in English, let alone the numerical values of their relative frequencies. For example, according to Pratt (1942) the above phrase should read

$$\text{ETANOR ISHDLF} \tag{22.26}$$

while Tribus (1961) gives it as

ETOANI RSHDLC. (22.27)

As we go into the less frequently used letters, the situation becomes still more chaotic.

Of course, we readily see the reason for these differences. People who have obtained different values for the relative frequencies of letters in English have consulted different samples of English text. It is obvious enough that the last volume of an encyclopedia will have a higher relative frequency for the letter Z than the first volume. The word frequencies would be very different in a textbook on organic chemistry, a treatise on the history of Egypt, and a modern American novel. The writing of educated people would reveal systematic differences in word frequencies from the writing of people who had never gone beyond grade school. Even within a much narrower field, we would expect to find significant differences in letter and word frequencies in the writings of James Michener and Ernest Hemingway. The letter frequencies in the transcript of a tape recording of a lecture will probably be noticeably different from those one would produce if the lecturer sat down and wrote out the lecture verbatim.

The fact that statistical properties of a language vary with the author and circumstances of writing is so clear that it has become a useful research tool. A doctoral thesis in classics submitted to Columbia University by James T. McDonough[1] contains a computer-run statistical analysis of Homer's *Iliad*. Classicists have long debated whether all parts of the *Iliad* were written by the same man, and indeed whether Homer is an actual historical person. The analysis showed stylistic patterns consistent throughout the work. For example, 40.4% of the 15 693 lines end on a word with one short syllable followed by two long ones, and a word of this structure never once appears in the middle of a line. Such consistency in a thing which is not a characteristic property of the Greek language seems rather strong evidence that the *Iliad* was written by a single person in a relatively short period of time, and it was not, as had been supposed by some 19th century classicists, the result of an evolutionary process over several centuries.

Of course, the evolutionary theory is not demolished by this evidence alone. If the *Iliad* was sung, we must suppose that the music had the very monotonous rhythmic pattern of primitive music, which persisted to a large extent as late as Bach and Haydn. Characteristic word patterns may have been forced on the writers, by the nature of the music.

Archaeologists tell us that the siege of Troy, described in the *Iliad*, is not a myth but an historical fact which occurred about 1200 BC, some four centuries before Homer. The decipherment of Minoan Linear B script by Michel Ventris in 1952 (Ventris and Chadwick, 1956; Chadwick, 1958; Ventris, 1988) established that Greek existed already as a spoken language in the Aegean area several centuries before the siege of Troy; but the introduction of the Phoenician alphabet, which made possible a written Greek language in the modern sense, occurred at only about the time of Homer.

---

[1] 'The structural metrics of the *Iliad*', Ph. D., 1966, Columbia University.

The considerations of the preceding two paragraphs still suggest an evolutionary development. It is clear that the question is very complex and far from settled; but we find it fascinating that a statistical analysis of word and syllable frequencies, representing evidence which has been there in the *Iliad* for some 28 centuries for anyone who had the wit to extract it, is finally recognized as having a definite bearing on the problem.

Well, to get back to communication theory, the point we are making is simply this: it is utterly wrong to say that there exists one and only one 'true' set of letter or word frequencies for English text. If we use a mathematical model which presupposes the existence of such uniquely defined frequencies, we might easily end up proving things which, while perfectly valid as mathematical theorems, are worse than useless to an engineer who is faced with the job of actually designing a communication system to transmit English text most efficiently.

But suppose our engineer does have extensive frequency data, and no other prior knowledge. How is he to make use of this in describing the information source? Many of the standard results of communication theory can, from the viewpoint we are advocating, be seen as simple examples of maximum entropy inference, i.e. as examples of the same kind of reasoning as in statistical mechanics.

22.5 Optimum encoding: letter frequencies known

Suppose our alphabet consists of different symbols $A_1, A_2, \ldots, A_a$, and we denote a general symbol by A_i, A_j, etc. Any message of N symbols then has the form $A_{i_1} A_{i_2} \cdots A_{i_N}$. We denote this message by M, which is a shorthand expression for the set of indices: $M = \{i_1 i_2 \cdots i_N\}$. The number of conceivable messages is a^N. By $\sum_M$ we mean a sum over all of them. Also, define

$$N_j(M) \equiv \text{number of times the letter } A_j \text{ appears in message } M,$$
$$N_{ij}(M) \equiv \text{number of times the digram } A_i A_j \text{ appears in } M, \tag{22.28}$$

and so on.

Consider first an engineer E_1, who has a set of numbers $(f_1, \ldots, f_a)$ giving the relative frequencies of the letters A_j, as observed in past samples of messages, but has no other prior knowledge. What communication system represents rational design on the basis of this much information, and what channel capacity does E_1 require in order to transmit messages at a given rate of n symbols per second?

To answer this, we need the probability distribution $p(M)$ which E_1 assigns to the various conceivable messages. Now, E_1 has no deductive proof that the letter frequencies in the future messages will be equal to the f_i observed in the past. On the other hand, his state of knowledge affords no grounds for supposing that the frequency of A_i will be greater than f_i rather than less, or vice versa. So he is going to suppose that frequencies in the future will be more or less the same as in the past, but he is not going to be too dogmatic about it. He can do this by requiring of the distribution $p(M)$ only that it yields *expected* frequencies equal to the known past ones. Put differently, if we say that our distribution $p(M)$ 'contains' certain information, we mean that that information can be extracted back out of it by the usual rule of estimation. In other words, E_1 will impose the constraints

$$\langle N_i \rangle = \sum_M N_i(M)p(M) = Nf_i, \quad i = 1, 2, \ldots, a. \tag{22.29}$$

Of course, $p(M)$ is not uniquely determined by these constraints, and so E_1 must at this point make a free choice of some distribution.

We emphasize again that it makes no sense to say there exists any 'physical' or 'objective' probability distribution $p(M)$ for this problem. This becomes especially clear if we suppose that only a single message is ever going to be sent over the communication system, but we still want it to be transmitted as quickly and reliably as possible, whatever that message turns out to be (perhaps we know that the system will be destroyed by impact on Ganymede immediately afterward); thus there is no conceivable way in which $p(M)$ could be measured as a frequency. But this would in no way affect the problem of engineering design which we are considering.

In choosing a distribution $p(M)$, it would be perfectly possible for E_1 to assume some message structure involving more than single letters. For example, he might suppose that the digram $A_1 A_2$ is twice as likely as $A_2 A_3$. But from the standpoint of E_1 this could not be justified, for *as far as he knows*, a design based on any such assumption is as likely to hurt as to help. From E_1's standpoint, rational conservative design consists just in carefully *avoiding* any such assumptions. This means, in short, that E_1 should choose the distribution $p(M)$ by maximum entropy consistent with (22.29).

All the formalism of the maximum entropy inference developed in Chapter 11 now becomes available to E_1. His distribution $p(M)$ will have the form

$$\log p(M) + \lambda_0 + \lambda_1 N_1(M) + \lambda_2 N_2 + \cdots + \lambda_a N_a(M) = 0, \tag{22.30}$$

and, in order to evaluate the Lagrangian multipliers λ_i, he will use the partition function

$$Z(\lambda_1, \ldots, \lambda_a) = \sum_M \exp\{-\lambda_1 N_1(M) - \cdots - \lambda_a N_a(M)\} = z^N, \tag{22.31}$$

where

$$z \equiv \exp\{-\lambda_1\} + \cdots + \exp\{-\lambda_a\}. \tag{22.32}$$

From (22.29) and the general relation

$$\langle N_i \rangle = -\frac{\partial}{\partial \lambda_i} \log Z(\lambda_1, \ldots, \lambda_a), \tag{22.33}$$

we find

$$\lambda_i = -\log(z f_i), \quad 1 \le i \le a, \tag{22.34}$$

and, substituting back into (22.30), we find the distribution which describes E_1's state of knowledge is just the multinomial distribution,

$$p(M) = f_1^{N_1} f_2^{N_2} \cdots f_a^{N_a}, \tag{22.35}$$

which is a special case of an exchangeable sequence; the probability of any particular message depends only on how many times the letters $A_1, A_2, \ldots$ appear, not on their order. The result (22.35) is correctly normalized, $\sum_M p(M) = 1$, as we see from the fact that the number of different messages possible for specified N_i is just the multinomial coefficient

$$\frac{N!}{N_1! \cdots N_a!}. \tag{22.36}$$

The entropy per symbol of the distribution (22.35) is

$$H_1 = -\frac{1}{N} \sum_M p(M) \log p(M) = \frac{\log(Z)}{N} + \sum_{i=1}^{a} \lambda_i f_i = -\sum_{i=1}^{a} f_i \log(f_i). \tag{22.37}$$

Having found the assignment $p(M)$, E_1 can encode into binary digits in the most efficient way by a method found independently by Shannon (1948, Sec. 9) and R. M. Fano. Arrange the messages in order of decreasing probability, and by a cut separate them into two classes so the total probability of all messages to the left of the cut is as nearly as possible equal to the probability of the messages on the right. If a given message falls in the left class, the first binary digit in its code is 0; if in the right, 1. By a similar division of these classes into subclasses with as nearly as possible a total probability of 1/4, we determine the second binary digit, etc. It is left for you to prove that (1) the expected number of binary digits required to transmit a symbol is equal to H_1, when expressed in bits, and (2) in order to transmit at a rate of n of the original message symbols per second, E_1 requires a channel capacity $C \geq nH_1$, a result first given by Shannon.

The preceding mathematical steps are so well-known that they might be called trivial. However, the rationale which we have given them differs essentially from that of conventional treatments, and in that difference lies the main point of this section. Conventionally, one would use the frequency definition of probability, and say that E_1's probability assignment $p(M)$ is the one resulting from the *assumption* that there are no intersymbol influences. Such a manner of speaking carries a connotation that the assumption might or might not be correct, and the implication that its correctness must be demonstrated if the resulting design is to be justified; i.e. that the resulting encoding rules might not be satisfactory if there are in fact intersymbol influences unknown to E_1.

On the other hand, we contend that the probability assignment (22.30) is not an assumption at all, but the opposite. Equation (22.30) represents, in a certain naïve sense which we shall come back to later, the complete *absence* of any assumption on the part of E_1, beyond specification of expected single-letter frequencies, and it is uniquely determined by that property. Because of this, the design based on (22.30) is the safest one possible on E_1's state of knowledge.

By that we mean the following. If, in fact, strong intersymbol correlations *do* exist unknown to E_1 (for example, Q is always followed by U), his encoding system will still be able to handle the messages perfectly well, whatever the nature of those correlations. This is what we mean by saying that the present design is the most conservative one; that it assumes *nothing* about correlations does not mean that it assumes *no* correlations and will be in trouble if correlations are in fact present. On the contrary, it means that it is prepared in advance *for whatever kind of correlations might exist*; they will not cause any deterioration in performance. We stress this point because it was not noted by Shannon, and it does not seem to be comprehended in the more recent literature.

But if E_1 had been given this additional information about some particular kind of correlations, he could have used it to arrive at a new encoding system which would be still more efficient (i.e. would require a smaller channel capacity), *as long as messages with only the specified type of correlation were transmitted*. But if the type of correlations in the messages were suddenly to change, this new encoding system would likely become worse than the one just found.

22.6 Better encoding from knowledge of digram frequencies

Here is a rather long mathematical derivation which has, however, useful applications outside the particular problem at hand. Consider a second engineer, E_2. He has a set of numbers f_{ij}, $1 \leq i \leq a$, $1 \leq j \leq a$, which represent the expected relative frequencies of the digrams $A_i A_j$. E_2 will assign message probabilities $p(M)$ so as to agree with his state of knowledge,

$$\langle N_{ij} \rangle = \sum_M N_{ij}(M)p(M) = (N-1)f_{ij}, \tag{22.38}$$

and, in order to avoid any further assumptions which are as likely to hurt as to help *as far as he knows*, he will determine the probability distribution over messages $p(M)$ which has maximum entropy subject to these constraints. The problem is solved if he can evaluate the partition function

$$Z(\lambda_{ij}) = \sum_M \exp\left\{ -\sum_{ij=1}^a \lambda_{ij} N_{ij}(M) \right\}. \tag{22.39}$$

This can be done by solving the combinatorial problem of the number of different messages with given $\{N_{ij}\}$, or by observing that (22.39) can be written in the form of a matrix product:

$$Z = \sum_{ij=1}^a (Q^{N-1})_{ij}, \tag{22.40}$$

where the matrix Q is defined by

$$Q_{ij} \equiv \exp\{-\lambda_{ij}\}. \tag{22.41}$$

The result can be simplified formally if we suppose that the message $A_{i_1} \ldots A_{i_N}$ is always terminated by repetition of the first symbol A_{i_1}, so that it becomes $A_{i_1} \ldots A_{i_N} A_{i_1}$. The digram $A_{i_N} A_{i_1}$ is added to the message and an extra factor $\exp\{-\lambda_{ij}\}$ appears in (22.39). The modified partition function then becomes a trace:

$$Z' = \text{Tr}(Q^N) = \sum_{k=1}^{a} q_k^N, \tag{22.42}$$

where the q_k are the roots of $|Q_{ij} - q\delta_{ij}| = 0$. This simplification would be termed 'use of periodic boundary conditions' by the physicist. Clearly, the modification leads to no difference in the limit of long messages; as $N \to \infty$,

$$\lim \frac{1}{N} \log(Z) = \lim \frac{1}{N} \log(Z') = \log(q_{max}), \tag{22.43}$$

where q_{max} is the greatest eigenvalue of Q. The probability of a particular message is now a special case of (22.40):

$$p(M) = \frac{1}{Z} \exp\left\{-\sum \lambda_{ij} N_{ij}(M)\right\}, \tag{22.44}$$

which yields the entropy as a special case of (22.42):

$$S = -\sum_M p(M) \log p(M) = \log(Z) + \sum_{ij} \lambda_{ij} \langle N_{ij} \rangle. \tag{22.45}$$

In view of (22.38) and (22.43), E_2's entropy per symbol reduces, in the limit $N \to \infty$, to

$$H_2 = \frac{S}{N} = \log(q_{max}) + \sum_{ij} \lambda_{ij} f_{ij}, \tag{22.46}$$

or, since $\sum_{ij} f_{ij} = 1$, we can write (22.46) as

$$H_2 = \sum_{ij} f_{ij}(\log[q_{max}] + \lambda_{ij}) = \sum_{ij} f_{ij} \log\left(\frac{q_{max}}{Q_{ij}}\right). \tag{22.47}$$

Thus, to calculate the entropy we do not need q_{max} as a function of the λ_{ij} (which would be impractical analytically for $a > 3$), but we need find only the ratio q_{max}/Q_{ij} as a function of the f_{ij}. To do this, we first introduce the characteristic polynomial of the matrix Q:

$$D(q) \equiv \det(Q_{ij} - q\delta_{ij}) \tag{22.48}$$

and note, for later purposes, some well-known properties of determinants. The first is

$$D(q)\delta_{ik} = \sum_{j=1}^{a} M_{ij}(Q_{kj} - q\delta_{kj}) = \sum_j M_{ij} Q_{kj} - q M_{ik} \tag{22.49}$$

and, similarly,

$$D(q)\delta_{ik} = \sum_j M_{ji} Q_{jk} - q M_{ki}, \tag{22.50}$$

in which M_{ij} is the cofactor of $(Q_{ij} - q\delta_{ij})$ in the determinant $D(q)$; i.e. $(-)^{i+j} M_{ij}$ is the determinant of the matrix formed by striking out the ith row and jth column of the matrix $(Q_{kj} - q\delta_{kj})$. If q is any eigenvalue of Q, the expression (22.49) vanishes for all choices of i and k.

The second identity applies only when q is an eigenvalue of Q. In this case, all minors of the matrix M are known to vanish. In particular, the second order minors are

$$M_{ik} M_{jl} - M_{il} M_{jk} = 0, \quad \text{if } D(q) = 0. \tag{22.51}$$

This implies that the ratios (M_{ik}/M_{jk}) and (M_{ki}/M_{kj}) are independent of k; i.e. that M_{ij} must have the form

$$M_{ij} = a_i b_j, \quad \text{if } D(q) = 0. \tag{22.52}$$

Substitution into (22.49) and (22.52) then shows that the quantities b_j form the *right eigen-vectors* of Q, while a_i is a *left eigenvector*:

$$\sum_j Q_{kj} b_j = q b_k, \quad \text{if } D(q) = 0 \tag{22.53}$$

$$\sum_i a_i Q_{ik} = a_k q, \quad \text{if } D(q) = 0. \tag{22.54}$$

Suppose now that any eigenvalue q of Q is expressed as an explicit function $q(\lambda_{11}, \lambda_{12}, \ldots, \lambda_{aa})$ of the Lagrangian multipliers λ_{ij}. Then, varying a particular λ_{kl} while keeping the other λ_{ij} fixed, q will vary so as to keep $D(q)$ identically zero. By the rule for differentiating the determinant (22.48), this gives

$$\frac{dD}{d\lambda_{kl}} = \frac{\partial D}{\partial \lambda_{kl}} + \frac{\partial D}{\partial q}\frac{\partial q}{\partial \lambda_{kl}} = -M_{kl} Q_{kl} - \frac{\partial q}{\partial \lambda_{kl}} \text{Tr}(M) = 0. \tag{22.55}$$

Using this relation, the condition (22.38) fixing the Lagrangian multipliers λ_{ij} in terms of the prescribed digram frequencies f_{ij}, become

$$f_{ij} = -\frac{\partial}{\partial \lambda_{ij}} \log(q_{\max}) = \frac{M_{ij} Q_{ij}}{q_{\max} \text{Tr}(M)}. \tag{22.56}$$

The single-letter frequencies are proportional to the diagonal elements of M:

$$f_i = \sum_{j=1}^{a} f_{ij} = \frac{M_{ii}}{\text{Tr}(M)}, \tag{22.57}$$

where we have used the fact that (22.49) vanishes for $q = q_{\max}$, $i = k$. Thus, from (22.56) and (22.57), the ratio needed in computing the entropy per symbol is

$$\frac{Q_{ij}}{q_{\max}} = \frac{f_{ij}}{f_i}\frac{M_{ii}}{M_{ij}} = \frac{f_{ij}}{f_i}\frac{b_i}{b_j}, \tag{22.58}$$

where we have used (22.52). Substituting this into (22.47), we find that the terms involving b_i and b_j cancel out, and E_2's entropy per symbol is just

$$H_2 = -\sum_{ij} f_{ij} \log\left(\frac{f_{ij}}{f_i}\right) = -\sum_{ij} f_{ij} \log(f_{ij}) + \sum_i f_i \log(f_i). \tag{22.59}$$

This is never greater than E_1's H_1, for, from (22.42) and (22.59),

$$H_2 - H_1 = \sum_{ij} f_{ij} \log\left(\frac{f_i f_j}{f_{ij}}\right) \le \sum_{ij} f_{ij} \left[\frac{f_i f_j}{f_{ij}} - 1\right] = 0, \tag{22.60}$$

where we used the fact that $\log(x) \le x - 1$ in $0 \le x < \infty$, with equality if and only if $x = 1$. Therefore,

$$H_2 \le H_1, \tag{22.61}$$

with equality if and only if $f_{ij} = f_i f_j$, in which case E_2's extra information was only what E_1 would have inferred. To see this, note that in the message $M = \{i_1 \ldots i_N\}$, the number of times the digram $A_i A_j$ occurs is

$$N_{ij}(M) = \delta(i, i_1)\delta(j, i_2) + \delta(i, i_2)\delta(i, i_3) + \cdots + \delta(i, i_{N-1})\delta(j, i_N), \tag{22.62}$$

and so, if we ask E_1 to estimate the frequency of digram $A_i A_j$ by the criterion of minimizing the expected square of the error, he will make the estimate

$$\langle f_{ij}\rangle = \frac{\langle N_{ij}\rangle}{N-1} = \frac{1}{N-1}\sum_M p(M)N_{ij}(M) = f_i f_j, \tag{22.63}$$

using for $p(M)$ the distribution (22.40) of E_1. In fact, the solutions found by E_1 and E_2 are identical if $f_{ij} = f_i f_j$, for then we have, from (22.56), (22.57) and (22.52),

$$Q_{ij} = \exp\{-\lambda_{ij}\} = q_{max}\sqrt{f_i f_j}. \tag{22.64}$$

Using (22.43), (22.62) and (22.64), we find that E_2's distribution (22.44) reduces to (22.40). This is a rather nontrivial example of what we noted in Chapter 11, Eq. (11.93).

22.7 Relation to a stochastic model

The quantities introduced above acquire a deeper meaning in terms of the following problem. Suppose that part of the message has been received, what can E_2 then say about the remainder of the message? This is answered by recalling our product rule

$$p(AB|I) = p(A|BI)p(B|I) \tag{22.65}$$

or by noting that the conditional probability of A, given B, is

$$p(A|BI) = \frac{p(AB|I)}{p(B|I)}, \tag{22.66}$$

a relation which in conventional theory, which never mentions prior information I, is taken as the *definition* of a conditional probability (i.e. the ratio of two 'absolute' probabilities). In our case, let I stand for the general statement of the problem leading to the solution (22.44), and let

$$B \equiv \text{the first } (m-1) \text{ symbols are } \{i_1 i_2 \ldots i_{m-1}\}, \tag{22.67}$$

$$A \equiv \text{the remainder of the message is } \{i_m \ldots i_N\}. \tag{22.68}$$

Then $p(AB|I)$ is the same as $p(M)$ in (22.44). Using (22.62), this reduces to

$$p(AB|I) = p(i_1 \ldots i_N|I) = Z^{-1} Q_{i_1 i_2} Q_{i_2 i_3} \cdots Q_{i_{N-1} i_N}, \tag{22.69}$$

and in

$$p(B|I) = \sum_{i_m=1}^{a} \cdots \sum_{i_N=1}^{a} p(i_1 \ldots i_N|I) \tag{22.70}$$

the sum generates a power of the matrix Q, just as in the partition function (22.40). Writing, for brevity, $i_{m-1} = i, i_m = j, i_N = k$, and

$$R \equiv \frac{1}{Z} Q_{i_1 i_2} \cdots Q_{i_{m-2} i_{m-1}}, \tag{22.71}$$

we have

$$p(B|I) = R \sum_{k=1}^{a} (Q^{N+m+1})_{ik} = R \sum_{jk=1}^{a} Q_{ij} (Q^{N-m})_{jk} \tag{22.72}$$

and so

$$p(A|BI) = \frac{Q_{ij} Q_{i_m i_{m+1}} \cdots Q_{i_{N-1} i_N}}{\sum_{k=1}^{a} (Q^{N-m+1})_{ik}} \tag{22.73}$$

since all the Q contained in R cancel out, we see that the probabilities for the remainder $\{i_m \ldots i_N\}$ of the message depend only on the immediately preceding symbol A_i, and not on any other details of B. This property defines a *generalized Markov chain*. There is a huge literature dealing with this; it is perhaps the most thoroughly worked out branch of probability theory, and we used a rudimentary form of it in calculating the conditional sampling distributions in Chapter 3. The basic tool, from which essentially all else follows, is the matrix p_{ij} of 'elementary transition probabilities'. This is the probability $p_{ij} = p(A_j|A_i I)$ that the next symbol will be A_j, given that the last one was A_i. Summing (22.73) over $i_{m+1} \ldots i_N$, we find that, for a chain of length N, the transition probabilities are

$$p_{ij}^{(N)} = p(A_j|A_i I) = \frac{Q_{ij} - T_j}{\sum_k Q_{ik} T_k}, \tag{22.74}$$

where

$$T_j \equiv \sum_{k=1}^{a} (Q^{N-m})_{jk}. \tag{22.75}$$

The fact that T_j depends on N and m is an interesting feature. Usually, one considers from the start a chain indefinitely prolonged, and so it is only the limit of (22.74) for $N \to \infty$ that is ever considered. This example shows that prior knowledge of the length of the chain can affect the transition probabilities; however, the limiting case is clearly of greatest interest.

To find this limit we need a little more matrix theory. The equation $D(q) = \det(Q_{ij} - q\delta_{ij}) = 0$ has roots $(q_1, q_2, \ldots, q_a)$, not necessarily all different, or real. Label them so that $|q_1| \geq |q_q| \geq \cdots \geq |q_a|$. There exists a nonsingular matrix A such that AQA^{-1} takes the canonical 'superdiagonal' form:

$$AQA^{-1} = \overline{Q} = \begin{pmatrix} C_1 & 0 & 0 & \cdots \\ 0 & C_2 & 0 & \cdots \\ 0 & 0 & C_3 & \cdots \\ \vdots & \vdots & \vdots & C_m \end{pmatrix}, \tag{22.76}$$

where the C_i are submatrices which can have either the forms

$$C_i = \begin{pmatrix} q_i & 1 & 0 & 0 & \cdots \\ 0 & q_i & 1 & 0 & \cdots \\ 0 & 0 & q_i & 1 & \cdots \\ 0 & 0 & 0 & q_i & 1 \\ \vdots & \vdots & \vdots & 0 & q_i \end{pmatrix} \quad \text{or} \quad C_i = \begin{pmatrix} q_i & & & \\ & q_i & & \\ & & \ddots & \\ & & & q_i \end{pmatrix}. \tag{22.77}$$

The result of raising Q to the nth power is

$$Q^n = A\overline{Q}^n A^{-1}, \tag{22.78}$$

and, as $n \to \infty$, the elements of $\overline{Q}^n$ arising from the greatest eigenvalue $q_{\max} = q_1$ become arbitrarily large compared with all others. If q_1 is nondegenerate, so that it appears only in the first row and column of $\overline{Q}$, we have

$$\lim_{N \to \infty} \left[\frac{T_j}{q_1^{N-m}} \right] = A_{j1} \sum_{k=1}^{a} (A^{-1})_{1k}, \tag{22.79}$$

$$\lim_{N \to \infty} \left[\frac{T_j}{\sum_k Q_{ik}T_k} \right] = \frac{A_{j1}}{q_1 A_{i1}}, \tag{22.80}$$

and the limiting transition probabilities are

$$p_{ij}^{(\infty)} = \frac{Q_{ij} A_{j1}}{q_1 A_{i1}} = \frac{Q_{ij} M_{ij}}{q_1 M_{ii}}, \tag{22.81}$$

where we have used the fact that the elements A_{j1} $(j = 1, 2, \ldots, a)$ from an eigenvector of Q with eigenvalue $q_1 = q_{max}$, so that, referring to (22.52), $A_{j1} = K b_j$ where K is some constant. Using (22.56) and (22.57), we have, finally,

$$p_{ij}^{(\infty)} = \frac{f_{ij}}{f_i}. \tag{22.82}$$

From this long calculation we learn many things. In the first place, for a sequence of finite length (the only kind that actually exists), the exact solution has intricate fine details that depend on the length. This, of course, could not be learned by those who try to jump directly into an infinite set at the beginning of a problem. Secondly, it is interesting that standard matrix theory was adequate to solve the problem completely. Finally, in the limit of infinitely long sequences, the exact solution of the maximum entropy problem does indeed go into the familiar Markov chain theory. This gives us a deeper insight into the basis of, and possible limitations on, Markov chain analysis.

Exercise 22.1. The exact meaning of this last statement might be unclear; in a classical Markov chain the transition probabilities two steps down the chain would be given by the square of the one-step matrix p_{ij}, three steps by the cube of that matrix, and so on. But our solution determines those multistep probabilities by summing (22.73) over the appropriate indices, which is not obviously the same thing. Investigate this and determine whether the maximum entropy multistep probabilities are the same as the classical Markov ones, or whether they become the same in some limit.

We see that the maximum entropy principle suffices to determine explicit solutions to problems of optimal encoding for noiseless channels. Of course, as we consider more complicated constraints (trigram frequencies, etc.), pencil and paper methods of solution will become impossibly difficult (there is no 'standard matrix theory' for them), and to the best of our knowledge we must resort to computers.

Now, Shannon's ostensibly strongest theorem concerns the limit as $n \to \infty$ of the problem with n-gram frequencies given; his $H \equiv \lim H_n$ is held to be the 'true' entropy of the English language, which determines the 'true' minimum channel capacity required to transmit it. We do not question this as a valid mathematical theorem, but from our discussion above it is clear that such a theorem can have no relevance to the real world, because there is no such thing as a 'true' n-gram frequency for English, even when $n = 1$.

Indeed, even if such frequencies did exist, think for a moment about how one would determine them. Even if we do not distinguish between capital and small letters and include no decimal digits or punctuation marks in our alphabet, there are $26^{10} = 1.41 \times 10^{14}$ ten-grams whose frequencies are to be measured and recorded. To store them all on paper at 1000 entries per sheet would require a stack of paper about 7000 miles high.

22.8 The noisy channel

Let us examine the simplest nontrivial case, where the noise acts independently (without memory) on each separate letter transmitted. Suppose that each letter has independently the probability ϵ of being transmitted incorrectly. Then in a message of N letters the probability that there are r errors is the binomial

$$p(r) = \binom{N}{r} \epsilon^r (1 - \epsilon)^{N-r} \tag{22.83}$$

and the expected number of errors is $\langle r \rangle = N\epsilon$. Then, if $N\epsilon << 1$, we might consider the communication system satisfactory for most purposes. However, it may be essential that the message be transmitted without any error at all (as in sending a computer code instruction to a satellite in orbit). The field of fancy error-correcting codes has a large literature and much sophisticated theory; but a very popular and simple procedure is the checksum.

Suppose, as is usually the case in computer practice, that our 'alphabet' consists of $2^8 = 256$ different characters sent as eight-bit binary numbers, called 'bytes'. At the end of the message one transmits one more byte, which is numerically the sum (mod 256) of the N previous ones. The receiver recalculates this sum from the first N bytes received, and compares it with the transmitted checksum. If they agree, then it is virtually certain that the transmission was error-free (if there is an error, then there must be at least two errors which just happened to cancel each other out in the checksum, and the probability of this is astronomically small, far less than ϵ). If they disagree, then it is certain that there was a transmission error, so the receiver sends back a 'please repeat' signal to the transmitter, and the process is repeated until error-free transmission is achieved.

Let us see just how good the checksum procedure is according to probability theory. Write, for brevity,

$$q \equiv (1 - \epsilon)^{N+1}. \tag{22.84}$$

Then to achieve error-free transmission, there is

probability q that it will require $(N + 1)$ symbols transmitted;
probability $(1 - q)q$ that $2(N + 1)$ symbols will be required;
probability $(1 - q)^2 q$ that $3(N + 1)$ symbols will be required;

and so on.

The expected length of transmission to achieve error-free operation is then the sum

$$\langle L \rangle = (N + 1)q[1 + 2(1 - q) + 3(1 - q)^2 + 4(1 - q)^3 + \cdots]. \tag{22.85}$$

Since $|1 - q| < 1$, the series converges to $1/q^2$, and so

$$\langle L \rangle = \frac{N + 1}{(1 - \epsilon)^{N+1}} \simeq N \exp\{N\epsilon\}, \tag{22.86}$$

the approximation holding reasonably well if $N >> 1$. But if the message is so long that $N\epsilon >> 1$, this procedure fails; there is almost no chance that we could transmit it without error in any feasible time.

But now an ingenious device comes to the rescue, and shows how much a little probability theory can help us to achieve exactitude. Let us break the long message into m shorter blocks of length $n = N/m$, and transmit each block with its own checksum. From (22.86) the expected total transmission length is now

$$\langle L \rangle = m \frac{n+1}{(1-\epsilon)^{n+1}} = N \frac{n+1}{n(1-\epsilon)^{n+1}}. \tag{22.87}$$

It is evident that if the blocks are too long, then we shall have to repeat too many of them; if they are too short, then we shall waste transmission time sending many unnecessary checksums. Thus there should be an optimal block length which minimizes (22.87). Providentially, this turns out to be independent of N; varying n, (22.87) reaches a minimum when

$$1 + n(n+1)\log(1-\epsilon) = 0, \qquad \text{or} \qquad (1-\epsilon)^{n+1} = \exp\{-1/n\}. \tag{22.88}$$

For all practical purposes, then, the optimal block length is

$$(n)_{\text{opt}} = \frac{1}{\sqrt{\epsilon}}, \tag{22.89}$$

and the minimum achievable expected length is

$$\langle L \rangle_{\min} = N \left(\frac{n+1}{n} \right) \exp \left\{ \frac{1}{n} \right\} \simeq N(1 + 2\sqrt{\epsilon}). \tag{22.90}$$

By breaking a long message into blocks, we have made an enormous improvement. If $\epsilon \simeq 10^{-4}$, then it would be impractical to send an error-free message of length $N = 100\,000$ bytes in a single block; for one expects about ten errors in each transmission. The expected transmission length would be about $22\,000N$ bytes, signifying that we would have to repeat the message, on the average, about $22\,000$ times before achieving one error-free result. But the optimal block length is about $n \simeq 100$, and by using this the expected length is reduced to $\langle L \rangle = 1.020N$. This signifies that we are sending 1000 blocks, of which each has one extra byte (which accounts for the factor $(n+1)/n \simeq 1 + \sqrt{\epsilon}$) and about ten will probably need to be repeated (which corresponds to the factor $\exp\{1/n\} \simeq 1 + \sqrt{\epsilon}$). But the minimum in (22.87) is very broad; if $40 \leq n \leq 250$, we have $\langle L \rangle \leq 1.030N$. If $\epsilon = 10^{-6}$, then the block technique allows us to transmit error-free messages of any length with virtually no penalty in transmission time ($\langle L \rangle \simeq 1.002N$ if n is anywhere near 1000).

To the best of our knowledge, the block technique is an intuitive *ad hockery*, not derived uniquely from any optimality criterion; yet it is so simple to use and comes so close to the best that could ever be hoped for ($\langle L \rangle = N$), that there is hardly any incentive to seek anything better.

In the early days of microcomputers, messages were sent to and from disks in block lengths of 128 or 256 bytes, which would be optimal if the error probability for each byte were of the order $\epsilon \simeq 10^{-5}$. At the time of writing (1991) they are being sent instead in blocks of 1024 to 4096 bytes, suggesting that disk reading and writing is now reliable to error probabilities of the order of 10^{-8} or better. Of course, it is conservative design to use block lengths somewhat shorter than the above optimal value, to hedge against deterioration in performance as the equipment wears out and the error rate increases.

But let us note a point of philosophy; in this discussion, have we abandoned our stance of probability theory as logic, and reverted to frequency definitions? Not at all! It is perfectly true that *if* the error probability ϵ is indeed an 'objectively real' frequency of errors measured in some class of repetitions of all this, then our $\langle L \rangle_{\min}$ is equally well the objectively real minimum achievable average transmission length *over that same class of repetitions*.

But there are few cases where this is really known to be true; such experiments are costly in time and resources. In the real world, they are never completed before the design becomes frozen and the manufactured product is delivered to the customers. Indeed, reliability experiments on highly reliable systems can never be really completed at all, because, in the time it requires to do them, our state of knowledge and technical capabilities will change, making the original purpose of the test irrelevant.

Our present point is that probability theory as logic works as well, in the following sense, whether our probabilities are or are not known to be real frequencies. As we saw in Chapter 8, it is an elementary derivable consequence of probability theory as logic that our probabilities are the best estimates of those frequencies that we can make on the information we have.

Then, whatever the evidence on which that probability assignment ϵ was based, the above equations still describe the most rational design *that could have been made, here and now, on the information we had*. As noted, this remains true even if we know in advance that only a single message is ever going to be sent over our communication system. Thus, probability theory as logic has a wider range of applications, even in situations where one sometimes pretends that he is using a frequency definition for psychological reasons.

Appendix A

Other approaches to probability theory

Needless to say, the way we developed probability theory in Chapter 2 is not the only way it could have been done. The particular conditions we used might have been chosen differently, and there are several other approaches based on entirely different notions.

As an example of the former, many qualitative statements seem so obvious that one might think of taking them as basic axioms or desiderata, instead of the ones we did use. Thus if A implies B, then for all C we should expect intuitively to have $P(A|C) \leq P(B|C)$. Of course, our rules do have this property, for the product rule is

$$P(AB|C) = P(B|AC)P(A|C) = P(A|BC)P(B|C). \tag{A.1}$$

But if A implies B, then $P(B|AC) = 1$ and $P(A|BC) \leq 1$, so the product rule reduces to the intuitive statement. It may well be that a different choice of axioms would have simplified the derivations of Chapter 2. However, that was not the criterion we used. We chose the ones that appeared to us the most primitive and most difficult to quarrel with, in the belief that the resulting theory would be seen thereby to have the greatest possible generality and range of application.

Now we examine briefly some other approaches that have been advocated in the past.

A.1 The Kolmogorov system of probability

In our comments at the end of Chapter 2, we noted the Venn diagram and the relation to set theory that it suggests, which became the basis of the Kolmogorov approach to probability theory. This approach could hardly be more different from ours in general viewpoint and motivation; yet the final results are identical in several respects.

The Kolmogorov system of probability (henceforth denoted by KSP) is a game played on a sample space Ω of elementary propositions ω_i (or 'events'; it does not matter what we call them at this level of abstraction). We may think of them as corresponding roughly to the individual points of the Venn diagram, although of course the abstract definition makes no such reference.

Then there is a field F consisting of certain selected subsets f_j of Ω, corresponding roughly to our propositions $A, \ldots, B, \ldots$ represented by areas of the Venn diagram (although, again, the abstract definition allows sets which need not correspond to areas). F is to have basically three properties:

(I) Ω is in F;
(II) F is a σ-field, meaning that if f_j is in F, then its complement with respect to Ω, $\bar{f}_j = \Omega - f_j$, is also in F;
(III) F is closed under countable unions, meaning that, if countably many f_j are in F, their union is also in F.

Finally, there is to be a probability measure P on F, with the properties of:

(1) normalization: $P(\Omega) = 1$;
(2) non-negativity: $P(f_i) \geq 0$ for all f_i in F;
(3) additivity: if $\{f_1 \cdots f_n\}$ are disjoint elements of F (i.e. they have no points ω_i in common, then
     $P(f) = \sum_i P(f_j)$, where $f = \cup_j f_j$ is their union;
(4) continuity at zero: if a sequence $f_1 \supseteq f_2 \supseteq f_3 \supseteq \ldots$ tends to the empty set, then $P(f_j) \to 0$.

There is nothing surprising in these axioms; they seem to be familiar echoes of just what we found in Chapter 2, except that they state analogous properties of sets rather than propositions.

We are with Kolmogorov in spirit when he wants F to be a σ-field, for any proposition that can be affirmed can also be denied; the operation NOT was also one of our primitive ones. Indeed, we went further by including the operation AND. Then it was a pleasant surprise that (AND, NOT), which are an adequate set for deductive logic, turn out to be also adequate for our extended logic (i.e. given a set of propositions $\{A_1, \ldots, A_n\}$ to be considered, our rules generate a system of inference that is formally complete in the sense that it is adequate to assign consistent probabilities to all propositions in the Boolean algebra generated from $\{A_1, \ldots, A_n\}$.)

Kolmogorov's closure under countable unions is also implied by this requirement, in the following sense. Working fundamentally with finite sets, we are content fundamentally with finite unions; yet a well-behaved limit to an infinite set may be a convenient simplification, removing intricate but irrelevant details from a finite-set calculation. On the infinite sets thus generated, our finite unions go into countable unions.

But, as noted in Chapter 2, a proposition A referring to the real world cannot always be viewed as a disjunction of elementary propositions ω_i from any set Ω that has meaning in the context of our problem; and its denial $\overline{A}$ may be even harder to interpret as set complementation. The attempt to replace logical operations on the propositions $A, B, \ldots$ by set operations on the set Ω does not change the abstract structure of the theory, but it makes it less general in respects that can matter in applications. Therefore we have sought to formulate probability theory in the wider sense of an extension of Aristotelian logic.

Finally, the properties (1)–(4) of the probability measure P were stated by Kolmogorov as seemingly arbitrary axioms; and KSP has been criticized for that arbitrariness. But we recognize them as statements, in the context of sets, of just the four properties that we derived in Chapter 2 from requirements of consistency. For example, the need for non-negativity is apparent from (2.35). Additivity also seems arbitrary when stated merely as an axiom; but in (2.85) we have derived it as necessary for consistency.

Many writers have thought that normalization is merely an arbitrary convention, but (2.31) shows that if certainty is not represented by $p = 1$, then we must restate the sum and product rules, or we shall have an inconsistency. For example, if we choose the convention that $p = 100$ is to represent certainty, then these rules take the form

$$p(A|B) + p(\overline{A}|B) = 100, \qquad p(A|BC)p(B|C) = 100p(AB|C). \tag{A.2}$$

More generally, by any change of variables $u = f(p)$ with some monotonic function $f(p)$, we can represent probability on a different scale than the one adopted; but then consistency will require that the product and sum rules also be modified in form, so that the content of our theory is not changed. For example, with the change of variables $u = \log[p/(1 - p)]$ the sum rule takes the equally simple form

$$u(A|B) + u(\overline{A}|B) = 0, \tag{A.3}$$

while the product rule becomes quite complicated. The substantive result is not that one is obliged to use any particular scale; but rather that a theory of probability whose *content* differs from one in

which there is a single scale that is normalized, non-negative, and additive, will contain
inconsistencies.

This should answer an objection sometimes raised (Fine, 1973, p. 65) that Kolmogorov's scale
was arbitrary. In 1973, such a charge might have seemed reasonable, calling for further
investigation. That further investigation has been made; since we now know that, in fact, he made
the only choices that will pass all our tests for consistency, the charge now seems to us unjustified.

We do not know how Kolmogorov was able to see the need for his axiom (4) of continuity at zero;
but our approach, in effect, derives it from a simple requirement of consistency. Firstly, let us dispel
a misunderstanding. Statement (4) in terms of sets seems to imply that an infinite sequence of
subsets is given. But its translation into a statement about propositions does not require that we
assign probabilities to an infinite number of propositions. What is essential is that we have an
infinite sequence of different *states of knowledge*, which may be about a single proposition, but
which tends to impossibility. Since Kolmogorov's sets are not associated with any such idea as a
'state of knowledge', there seems to be no way to say this in the context of sets; but in the context of
propositions it is easy.

We note this to emphasize that it would be a serious error to suppose that we can dispense with
this axiom merely by limiting our discourse to a finite set of propositions. The resulting theory
would have an arbitrary character which allows one to commit all kinds of inconsistencies.

In our system, 'continuity at zero' takes the following form: given a sequence of probabilities
$p(A)_1, p(A)_2, \ldots$ that tend to certainty, the probabilities $p(\overline{A})_1, p(\overline{A})_2, \ldots$ assigned to the denial
must tend to zero. Indeed, as we noted in (2.48), the functional equation that $S(x)$ satisfies ties
values at different x together so strongly that the exact way in which $S(x)$ tends to zero as $x \to 1$ is
the crucial thing that determines the function $S(x)$ over its entire range ($0 \le x \le 1$), and therefore
determines the additivity property (2.58). Thus, from our viewpoint, Kolmogorov's axioms (3) and
(4) appear to be closely related; it is not obvious whether they are logically independent.

For all practical purposes, then, our system will agree with KSP if we are applying it in the
set-theory context. But in more general applications, although we have a field of discourse F and
probability measure P on F with the same properties, we do not need, and do not always have, any
set Ω of elementary propositions into which the elements of F can be resolved. Of course, in many
of our applications such a set Ω will be present: for example, in equilibrium statistical mechanics the
elements ω_i of Ω can be identified with the stationary 'global' quantum states of a system, which
comprise a countable set. In these cases, there will be essentially complete agreement in the abstract
formulation, although we carry out practical calculations with more freedom in one respect – and
more inhibition in another – for reasons noted below.

Our approach supports KSP in another way also. KSP has been criticized as lacking connection to
the real world; it has seemed to some that its axioms are deficient because they contain no statement
to the effect that the measure P is to be interpreted as a frequency in a random experiment.[1] But,
from our viewpoint, this appears as a merit rather than a defect; to require that we invoke some
random experiment before using probability theory would have imposed an intolerable and arbitrary
restriction on the scope of the theory, making it inapplicable to most of the problems that we
propose to solve by extended logic.

Even when random experiments are involved in the real problem, propositions specifying
frequencies are properly considered, not as determinations of the measure P, but as elements of the
field F. In both Kolmogorov's system and ours, such propositions are not the tools for making
inferences, but the things about which inferences are being made.

There are some important differences, however, between these two systems of probability theory.
In the first place, in KSP attention is concentrated almost exclusively on the notion of additive

---

[1] Indeed, de Finetti (1972, p. 89) argues that Kolmogorov's system *cannot* be interpreted in terms of limits of frequencies.

measure. The Kolmogorov axioms make no reference to the notion of conditional probability; indeed, KSP finds this an awkward notion, really unwanted, and mentions it only reluctantly, as a seeming afterthought.[2] Although Kolmogorov has a section entitled 'Bayes' theorem', most of his followers ignore it. In contrast, we considered it obvious from the start that all probabilities referring to the real world are necessarily conditional on the information at hand. In Chapter 2 the product rule, with conditional probability and Bayes' theorem as immediate consequences, appeared in our system even before additivity.

Our derivation showed that, from the standpoint of logic, the product rule (and therefore Bayes' theorem) expresses simply the associativity and commutativity of Boolean algebra. This is what gives us that greater freedom of action in calculations, leading in later chapters to the unrestricted use of Bayes' theorem, in which we have complete freedom to move propositions back and forth between the left and right sides of our probability symbols in any way permitted by the product and sum rules. This is a superb computational device – and by far the most powerful tool of scientific inference – yet it is completely missing from expositions of probability theory based on the KSP work (which do not associate probability theory with information or inference at all).

In return for this freedom, we impose on ourselves an inhibition not present in KSP. Having been burned by de Finetti and his followers, we are wary of infinite sets, and approach them only cautiously, after ascertaining that in our problem there is a well-defined and well-behaved limiting process that will not lead us into paradoxes and will serve a useful purpose.

In principle, we start always by enumerating some finite set of propositions $A, B, \ldots$ to be considered. Our field of discourse F is then also finite, consisting of these and – automatically – all propositions that can be 'built up' from them by conjunction, disjunction, and negation. We have no need or wish to 'tear down' by resolving them into a disjunction of finer propositions, much less carrying this to infinite limits, except when this can be a useful calculational device due to the structure of a particular problem.

We have three reasons for taking this stance. The first was illustrated in Chapter 8 by the scenario of Sam's broken thermometer, where we saw that beyond a certain point this finer and finer resolving serves no purpose. Secondly, in Chapter 15 we saw some of the paradoxes that await those who jump directly into infinite sets without considering any limiting process from a finite set. But even here, when we considered the so-called 'Borel–Kolmogorov paradox', we found ourselves in agreement with Kolmogorov's resolution of it, and thus in disagreement with some of his critics. One must approach infinite sets carefully; but once in an uncountable set, one must then approach sets of measure zero just as carefully.

A third reason is that a different resolution often appears more useful to us. Instead of resolving a proposition A into the disjunction $A = B_1 + B_2 + \cdots$ of 'smaller' propositions and applying the sum rule, one can equally well resolve it into a conjunction $A = C_1 C_2 \cdots$ of 'larger' propositions and apply the product rule. This may be interpreted, in terms of sets, very simply. To specify the geographical layout of a country, there are two possible methods: (1) specify the points that are in it; (2) specify its boundary. Method (1) is the Venn–Kolmogorov viewpoint; but method (2) appears to us equally fundamental and often simpler and more directly related to the information we have in a real problem. In a Venn diagram the boundary of set A is composed of segments of the boundaries of $C_1, C_2, \ldots$, just as that of a country is composed of segments of rivers, coastlines, and adjacent countries.

These methods are not in conflict; rather, in each problem we may choose the one appropriate to the job before us. But in most of our problems method (2) is the natural one. A physical theory is always stated as a conjunction of hypotheses, not a disjunction; likewise a mathematical theory is defined by the set of axioms underlying it, which is always stated as a conjunction of elementary

---

[2] In the Kolmogorov system, conditional probability is such a foreign element that an entire book has been written (Rao, 1993) trying to explain the idea of conditional probability by giving it a separate axiomatic approach!

axioms. To express the foundations of any theory as disjunctions would be almost impossible; so we must demand this freedom of choice.

In summary, we see no substantive conflict between our system of probability and Kolmogorov's as far as it goes; rather, we have sought a deeper conceptual foundation which allows it to be extended to a wider class of applications, required by current problems of science.

The theory expounded here is still far from its final, complete form, however. In its present state of development, there are many situations where the robot does not know what to do with its information. For example, suppose it is told that 'Jones was very pleased at the suggestion that θ might be greater than 100'. By what principles is it to translate this into a probability statement about θ?

You and I, however, can make some use of that information to modify our opinions about θ (upward or downward according to our opinions about Jones). Indeed we can use almost any kind of prior information, and perhaps draw a free-hand curve which indicates roughly how it affects our probability distribution $p(\theta)$. In other words, *our brains are in possession of more principles than the robot's for converting raw information, semiquantitatively, into something which the computer can use.* This is the main reason why we are convinced that there must be more principles like maximum entropy and transformation groups, waiting to be discovered by someone. Each such discovery will open up a new area of useful applications of this theory.

A.2 The de Finetti system of probability

There is today an active school of thought, most of whose members call themselves 'Bayesians', but who are actually followers of Bruno de Finetti and concerned with matters that Bayes never dreamt of. In 1937, de Finetti published a work which expressed a philosophy somewhat like ours and contained not only his marvellous and indispensable exchangeability theorem, but also sought to establish the foundations of probability theory itself on the notion of 'coherence'. This means, roughly speaking, that one should assign and manipulate probabilities so that one cannot be made a sure loser in betting based on them. He appears to derive the rules of probability theory very easily from this premise (de Finetti, 1937).

Since 1937, de Finetti has published many more works on this topic (de Finetti, 1958, 1974a,b), as cited in our general references. Note particularly the large work published in English translation (de Finetti, 1974b). Some have thought that we should have followed de Finetti's coherence principle in the present work. Certainly, that would have shortened our derivations. However, we think that coherence is an unsatisfactory basis in three respects. The first is admittedly only aesthetic; it seems to us inelegant to base the principles of logic on such a vulgar thing as expectation of profit.

The second reason is strategic. If probabilities are thought to be defined basically in terms of betting preferences, then for assigning probabilities one's attention is focused on how to elicit the personal probabilities of different people. In our view, that is a worthy endeavor, but one that belongs to the field of psychology rather than probability theory; our robot does not have any betting preferences. When we apply probability theory as the normative extension of logic, our concern is not with the personal probabilities that different people might happen to have, but with the probabilities that they 'ought to' have, in view of their information – just as James Clerk Maxwell noted in our opening quotation for Chapter 1.

In other words, at the beginning of a problem our concern is not with anybody's personal opinions, but with specifying the *prior information* on which our robot's opinions are to be based, in the context of the current problem. The principles for assigning prior probabilities consistently by logical analysis of that prior information are for us an essential part of probability theory. Such considerations are almost entirely absent from expositions of probability theory based on the de Finetti approach (although of course it does not forbid us to consider such problems).

The third reason is thoroughly pragmatic: if any rules were found to possess the property of coherence in the sense of de Finetti, but not the property of consistency in the sense of Cox, they would be clearly unacceptable – indeed, functionally unusable – as rules for logical inference. There would be no 'right way' to do any calculation, and no 'right answer' to any question. Then there would be small comfort in the thought that all those different answers were at least 'coherent'.

To the best of our knowledge, de Finetti does not mention consistency as a desideratum, or test for it. Yet it is consistency – not merely coherence – that is essential here, and we find that, when our rules have been made to satisfy the consistency requirements, then they have automatically (and trivially) the property of coherence.

Another point was noted in our Preface. Like Kolmogorov, de Finetti is occupied mostly with probabilities defined directly on arbitrary uncountable sets; but he views additivity differently, and is led to such anomalies as an unlimited sequence of layers, like an onion, of different orders of zero probabilities that add up to one, etc. It is the followers of de Finetti who have perpetrated most of the infinite-set paradoxing that has forced us to turn to (and, like Helmholtz in Chapter 16, exaggerate if necessary) the opposite 'finite-sets' policy in order to avoid them. This line of thought continues, with technical details, in Chapter 15.

A.3 Comparative probability

In our Comments Section 1.8 at the end of Chapter 1, we noted a possible objection to our first desideratum:

(I) Degrees of plausibility are to be represented by real numbers.

Why must one do this? Our pragmatic reason was that we do not see how our robot's brain can function by carrying out definite physical operations – mechanical or electronic – unless, at some point, degrees of plausibility are associated with some definite physical quantity.

We recognize that this ignores some aesthetic considerations; for example, the geometry of Euclid derives its elegance in large part from the fact that it is not concerned with numerical values, but with recognizing qualitative conditions of equivalence or similarity. We had this very much in mind when choosing all our other axioms, being careful to ensure that 'consistency' and 'correspondence with common sense' expressed qualitative rather than quantitative properties.

Of course, our one pragmatic argument carries no weight for those concerned with abstract axiomatics rather than making something work, so let us consider the alternatives. If one wishes to pick away at our desideratum I, it can be dissected into simpler axioms. In the following, read '$(A|C) > (B|C)$' not as a numerical comparison, but simply as the verbal statement, 'given C, A is more plausible than B', etc. Then desideratum I may be replaced by two more elementary ones:

(Ia) Transitivity. If $(A|X) \geq (B|X)$ and $(B|X) \geq (C|X)$ then $(A|X) \geq (C|X)$.
(Ib) Universal comparability. Given propositions A, B, C, then one of the relations
  $(A|C) > (B|C)$, $(A|C) = (B|C)$, $(A|C) < (B|C)$ must hold.

To see this, note that, if we postulate both transitivity and universal comparability, then within any finite set of propositions, we can always set up a representation by real numbers (in fact, by rational numbers) that obeys all the ordering relations. For, suppose we have a set $\{A_1, \ldots, A_n\}$ of propositions with such numerical measures $\{x_1, \ldots, x_n\}$. Adding a new proposition A_{n+1}, the transitivity and universal comparability ensure that it fits into a unique place in those ordering relations. But since between any two rational numbers one can always find another rational number, we can always assign a number x_{n+1} to it so that all the ordering relations of the new set $\{A_1, \ldots, A_{n+1}\}$ are also obeyed by our rational numbers $\{x_1, \ldots, x_{n+1}\}$.

At this point, therefore, it is all over with any comparative theory which embodies both transitivity and universal comparability. Once the existence of a representation by real numbers is established, then Cox's theorems take over and force that theory to be identical with the theory of inference that we derived in Chapter 2. That is, either there is some monotonic function of the x_i that obeys the standard product and sum rules of probability theory; or else we can exhibit inconsistencies in the rules of the comparative theory.

Some systems of comparative probability theory have both of these axioms; then they have assumed everything needed to guarantee the equivalence to the standard numerical valued theory. This being the case, it would seem foolish to refuse to use the great convenience of the numerical representation. But now, can one drop transitivity or universal comparability and get an acceptable extension of logic with different content than ours?

No comparative probability theory is going to get far if it violates transitivity. Nobody would wish to – or be able to – use it, because we would be trapped in endless loops of circular reasoning. So transitivity is surely going to be one of the axioms of a comparative probability theory; discovery of an intransitivity would be grounds for immediate rejection of any system.

To many, universal comparability does not seem a compelling desideratum. By dropping it we could create a 'lattice' theory, so called because we can represent propositions by points, relations of comparability by lines connecting them in various ways. Then it is conceivable that A and C can be compared, and B and C can be compared; but A is not comparable to B. One might contemplate a situation in which $(A|D) < (C|D)$ and $(B|D) < (C|D)$; but neither $(A|D) < (B|D)$ nor $(A|D) \geq (B|D)$ could be established. This allows a looser structure which cannot be represented faithfully by assigning a single real number to each proposition (although it can be so represented by a lattice of vectors); any attempt to introduce a single-valued numerical representation would generate false comparisons not present in the system.

Much effort has been expended on attempts to develop such looser forms of probability theory in which one does not represent degrees of plausibility by real numbers, but admits only qualitative ordering relations of the form $(A|C) \geq (B|C)$, and attempts to deduce the existence of a (not necessarily unique) additive measure $p(A|B)$ with the property (2.85). The work of L. J. Savage (1954) is perhaps the best known example. A summary of such attempts is given by T. L. Fine (1973). These efforts appear to have been motivated only by an aesthetic feeling – that universal comparability is a stronger axiom than we need – rather than from the hope that any particular pragmatic advantage would be gained by dropping it. However, a restriction appeared in comparative probability theory, robbing it of its initial appeal.

Ordering relations may not be assigned arbitrarily because it must always be possible to extend the field of discourse, by adding more propositions and ordering relations, without generating contradictions. If adding a new ordering relation created an intransitive loop, it would be necessary to modify some ordering relations to restore transitivity. But such extensions may be carried out indefinitely, and, when a set of propositions with transitive ordering relations becomes, in a certain sense, 'everywhere dense' on the path from impossibility to certainty, consistency will require that the theory then approach the conventional numerical-valued probability theory expounded here.

In retrospect (i.e. in view of Cox's consistency theorems) this is hardly surprising; a comparative probability theory whose results conflict with those of our numerical probability theory necessarily contains within it either overtly visible inconsistencies or the seeds of inconsistencies which will become visible when one tries to extend the field of discourse.

Furthermore, it appears to us that any computer designed to carry out the operations of a comparative probability theory must at some stage represent the ordering relations as inequalities of real numbers. So attempts to evade numerical representation not only offer no pragmatic advantage, they are futile. Thus in the end the study of comparative probability theories serves only to show us still another aspect of the superiority of the Cox approach that we follow here.

A.4 Holdouts against universal comparability

The arguments in the preceding sections, however, do not quite close the subject, because some of the criticisms of probability theory as logic are from writers who have considered it absurd to suppose that all propositions can be compared. This view seems to arise from two different beliefs: (1) human brains cannot do this; and (2) they think that they have produced examples where it is fundamentally impossible to compare all propositions.

Argument (1) carries no weight for us; in our view, human brains do many absurd things while failing to do many sensible things. Our purpose in developing a formal theory of inference is not to imitate them, but to correct them. We agree that human brains have difficulty in comparing, and reasoning from, propositions that refer to different contexts. But we would observe also that the ability to do this improves with education.

For example, is it more likely that (A) Tokyo will have a severe earthquake on June 1, 2230; or that (B) Norway will have an unusually good fish catch on that day? To most people, the contexts of propositions A and B seem so different that we do not see how to answer this. But with a little education in geophysics and astrophysics, one realizes that the moon could well affect both phenomena, by causing phase-locked periodic variations in the amplitudes of both the tides and stresses in the Earth's crust. Recognition of a possible common physical cause at work makes the propositions seem comparable after all.

The second objection to universal comparability noted above appears to be a misunderstanding of our present theory, but one which does point to cases in which universal comparability would indeed be fundamentally impossible. These are the cases where we are trying to classify propositions with respect to more than one attribute, as in the conceivable multidimensional models of mental activity noted at the end of Chapter 1. All of the alleged counter-examples to comparability that we have seen prove on examination to be of this type.

For example, a mineralogist may classify a collection of rocks with respect to two qualities, such as density and hardness. If, within a certain subclass of them, density alone varies, then obviously there are transitive comparability relations that can be represented faithfully by real numbers d. If in another subclass hardness alone varies, there is a similar comparability representable by real numbers h. But if we classify rocks by both simultaneously, it requires two real numbers (d, h) to represent them; any attempt to arrange them in a unique one-dimensional order would be arbitrary.

The arbitrariness could be removed if we also introduced some new value judgment or 'objective function' $f(d, h)$ that tells us by relations such as $f(d_1, h_1) = f(d_2, h_2)$ how to trade off a change $\Delta d = d_2 - d_1$ in d against a change $\Delta h = h_2 - h_1$ in h. But then we are classifying the rocks with respect to only one attribute, namely f, and universal comparability is again possible.

In the theory of probability developed here we are, by definition, classifying propositions according to only one attribute, which we call intuitively 'degree of plausibility'. Once this is understood, we think that the *possibility* of representation by real numbers need never be questioned, and the *desirability* of doing this is attested to by all the nice results and useful applications of the theory.

Nevertheless, the general idea of a comparative probability theory might be useful to us in two respects. Firstly, for many purposes one has no need for precisely defined numerical probabilities; any values that preserve ordering relations within a small set of propositions may be adequate for our purpose. For example, if it is required only to choose between two competing hypotheses, or two feasible actions, a wide range of numerical probability values must all lead to the same final choice. Then the precise position within that range is irrelevant, and to determine it would be wasted computational effort. Something much like a comparative probability theory would then appear, not as a generalization of numerical probability theory, but as a simple, useful approximation to it.

Secondly, the above observation about Tokyo and Norway suggests a possible legitimate application for a lattice theory of probability. If our brains do not have automatically the property of universal comparability, then perhaps a lattice theory might come much closer than the Laplace–Bayes theory to describing the way we actually think. What are some of the properties that can be anticipated of a lattice theory?

A.5 Speculations about lattice theories

One evident property is that we could do plausible reasoning only in certain 'domains' consisting of sets of comparable propositions. We would not have any idea how to reason in cases involving a jump across widely separated parts of the lattice; unless we perceive some logical relationship between propositions, we have no criterion for comparing their plausibilities. Our scale of plausibility might be wildly different on different parts of the lattice, and we would have no way of knowing this until we had learned to increase the degree of comparability.

Indeed, the human brain does not start out as an efficient reasoning machine, plausible or deductive. This is something which we require years to learn, and a person who is an expert in one field of knowledge may do only rather poor plausible reasoning in another.[3] What is happening in the brain during this learning process?

Education could be defined as the process of becoming aware of more and more propositions, and of more and more logical relationships between them. Then it seems natural to conjecture that a small child reasons on a lattice of very open structure: large parts of it are not interconnected at all. For example, the association of historical events with a time sequence is not automatic; the writer has had the experience of seeing a child, who knew about ancient Egypt and had studied pictures of the treasures from the tomb of Tutankhamen, nevertheless coming home from school with a puzzled expression and asking: 'Was Abraham Lincoln the first person?'

It had been explained to him that the Egyptian artifacts were over 3000 years old, and that Abraham Lincoln was alive 120 years ago; but the meaning of those statements had not registered in his mind. This makes us wonder whether there may be primitive cultures in which the adults have no conception of time as something extending beyond their own lives. If so, that fact might not have been discovered by anthropologists, just because it was so unexpected that they would not have raised the question.

As learning proceeds, the lattice develops more and more points (propositions) and interconnecting lines (relations of comparability), some of which will need to be modified for consistency in the light of later knowledge. By developing a lattice with denser and denser structure, *one is making his scale of plausibilities more rigidly defined.*

No adult ever comes anywhere near to the degree of education where he would perceive relationships between all possible propositions, but he can approach this condition with some narrow field of specialization. Within this field, there would be a 'quasi-universal comparability', and his plausible reasoning within this field would approximate that given by the Laplace–Bayes theory.

A brain might develop several isolated regions where the lattice was locally quite dense; for example, one might be very well-informed about both biochemistry and musicology. Then for reasoning within each separate region, the Laplace–Bayes theory would be well-approximated, but there would still be no way of relating different regions to each other.

---

[3] The biologist James D. Watson has remarked before TV cameras that professional physicists can be 'rather stupid' when they have to think about biology. We do not deny this, although we wonder how far he would have got in finding the DNA structure without the help of the physicists Rosalind Franklin, to acquire the data for him, and Francis Crick, to explain to him how to interpret it.

Then what would be the limiting case as the lattice becomes everywhere dense with truly universal comparability? Evidently, the lattice would then collapse into a line, and some unique association of *all* plausibilities with real numbers would then be possible. Thus, *the Laplace–Bayes theory does not describe the inductive reasoning of actual human brains; it describes the ideal limiting case of an 'infinitely educated' brain.* No wonder that we fail to see how to use it in all problems!

This speculation may easily turn out to be nothing but science fiction; yet we feel that it must contain at least a little bit of truth. As in all really fundamental questions, we must leave the final decision to the future.

Appendix B

Mathematical formalities and style

We collect here a brief account of the various mathematical conventions used throughout this work, and discuss some basic mathematical issues that arise in probability theory. Careless notation has led to so many erroneous results in the recent literature that we need to find rules of notation and terminology that make it as difficult as possible to commit such errors.

A mathematical notation, like a language, is not an end in itself but only a communication device. Its purpose is best served if the notation, like the language, is allowed to evolve with use. This evolution usually takes the form of abbreviations for whatever expressions recur often, and reducing the number of symbols when their meaning can be read from the context.

But a living, changing language still needs a kind of safe harbor in the form of a fixed set of rules of grammar and orthography, hidden away in a dictionary for use when ambiguities threaten. Likewise, probability theory needs a fixed set of normative rules on which we can fall back in case of doubt. We state here our formal rules of notation and logical hierarchy; all chapters from Chapter 3 on start with these standard forms, and evolve from them. A notation which is so convenient that it is almost a necessity in one chapter might be only confusing in the next; so each separate topic must be allowed its own independent evolution from the standard beginning.

B.1 Notation and logical hierarchy

In our formal probability symbols (those with a capital P)

$$P(A|B) \tag{B.1}$$

the entries A, B always stand for *propositions*, with a sufficiently clear meaning (at least to us) that we are willing to use them as elements of Aristotelian logic, obeying a Boolean algebra. Thus $P(A|B)$ does not denote a 'function' in the usual sense.

We repeat the warning that a probability symbol is undefined and meaningless if the conditioning statement B happens to have zero probability in the context of our problem (for example, if $B = CD$, but $P(C|D) = 0$). Failure to recognize this can lead to erroneous calculations – just as inadvertently dividing by an expression that happens to have the value zero can invalidate all subsequent results.

To preserve the purity of our probability symbols (B.1) we must have also other symbols for probabilities. Thus, if proposition A has the meaning

$$A \equiv \text{the variable } q \text{ has the particular value } q' , \tag{B.2}$$

there is a tendency to write, instead of $P(A|B)$,

$$P(q'|B). \tag{B.3}$$

But q' is not a proposition, and so the writer evidently intends the symbol (B.3) to stand now for an ordinary mathematical function of the variable q'. In our system this is illegitimate, and so, when an ordinary mathematical function is intended, we shall take the precaution of inventing a different functional symbol such as $f(\quad | \quad)$, writing (B.3) instead as

$$f(q'|B). \tag{B.4}$$

Now the distinction between (B.3) and (B.4) may appear to some readers as pedantic nitpicking; so why do we insist on it? Many years ago, the present writer would also have dismissed this point as too trivial to deserve mention; but later experience has brought to light cases where failure to maintain the distinction in clear sight has tricked writers into erroneous calculations and conclusions. The amount of time and effort this has wasted – and which is still being wasted in this field – justifies our taking protective measures against it.

The point is that a proposition A is a verbal statement that may indeed specify the value of some variable q; but it generally contains qualifying clauses also:

$$A \equiv \text{the variable } q \text{ has the value } q' \text{ if the proposition } B \text{ is true.} \tag{B.5}$$

If we try to take the short-cut of replacing A by q' in the probability symbol, we lose sight of the qualification. Later in the calculation, the same variable q may appear in a proposition A_1 with a different qualification B_1; and again one may be tempted to replace A_1 by q' in the probability symbol. Still later in the calculation the same probability symbol will appear with two different meanings, and one is tricked into supposing that they represent the same quantity.

This is what happened in the famous 'marginalization paradox', in which the same probability symbol was used to denote probabilities conditional on two different pieces of prior information, with bizarre consequences described in Jaynes (1980) and in Chapter 15. This confusion is still causing trouble in probability theory, for those who have not yet understood it.

However, we are not fanatics about this. In cases so simple that there is very little danger of error anyway, we allow a compromise and follow the custom of most writers, even though it is not a strictly consistent notation. In probability symbols with a small p, we shall allow the arguments to be either propositions or numbers, in any mix: thus, if A is a proposition and q a number, the equation

$$p(A|B) = p(q|B) \tag{B.6}$$

is permitted; but with the warning that when small p symbols are used, the reader must judge their meaning from the context, and there is a possibility of error from failure to read them correctly.

A common and useful custom is to use Greek letters to denote parameters in a probability distribution, the corresponding Latin letters for the corresponding functions of the data. For example, one may denote a probability average (the mean of a probability distribution) by $\mu = \langle x \rangle = E(x)$, and then the average over the data would be $m = \bar{x} = n^{-1} \sum x_i$. We shall adhere to this except when it would be confusing because of a conflict with some other long established usage.

B.2 Our 'cautious approach' policy

The derivation of the rules of probability theory from simple desiderata of rationality and consistency in Chapter 2 applied to discrete, finite sets of propositions. Finite sets are therefore our safe harbor, where Cox's theorems apply and nobody has ever been able to produce an inconsistency from application of the sum and product rules. Likewise, in elementary arithmetic finite sets are the safe harbor in which nobody has been able to produce an inconsistency from applying the rules of addition and multiplication.

As soon as we try to extend probability theory to infinite sets, we are faced with the need to exercise the same kind of mathematical caution that one needs in proceeding from finite arithmetic expressions to infinite series. The 'parlor game' at the beginning of Chapter 15 illustrates how easy it is to commit errors by supposing that the operations of elementary arithmetic and analysis, that are always safe on finite sets, may be carried out also on infinite sets.

In probability theory, it appears that the only safe procedure known at present is to derive our results first by strict application of the rules of probability theory on finite sets of propositions; then, after the finite-set result is before us, observe how it behaves as the number of propositions increases indefinitely. There are, essentially, three possibilities:

(1) It tends smoothly to a finite limit, some terms just becoming smaller and dropping out, leaving behind a simpler analytical expression.
(2) It blows up, i.e. becomes infinite in the limit.
(3) It remains bounded, but oscillates or fluctuates forever, never tending to any definite limit.

In case (1) we say that the limit is 'well-behaved' and accept the limit as the correct solution on the infinite set. In cases (2) and (3) the limit is ill-behaved and cannot be considered a valid solution to the problem. Then we refuse to pass to the limit at all.

This is the 'look before you leap' policy: in principle, we pass to a limit only after verifying that the limit is well-behaved. Of course, in practice this does not mean that we conduct such a test anew on every problem; most situations arise repeatedly, and rules of conduct for the standard situations can be set down once and for all. But in case of doubt, we have no choice but to carry out this test.

In cases where the limit is well-behaved, it may be possible to get the correct answer by operating directly on the infinite set, but one cannot count on it. If the limit is not well-behaved, then any attempt to solve the problem directly on the infinite set would have led to nonsense, *the cause of which cannot be seen if one looks only at the limit, and not the limiting process.* The paradoxes noted in Chapter 15 illustrate some of the horrors that have resulted from carelessness in this regard.

B.3 Willy Feller on measure theory

In contrast to our policy, many expositions of probability theory begin at the outset to try to assign probabilities on infinite sets, both countable or uncountable. Those who use measure theory are, in effect, supposing the passage to an infinite set already accomplished before introducing probabilities. For example, Feller advocates this policy and uses it throughout his second volume (Feller, 1966).

In discussing this issue, Feller (1966) notes that specialists in various applications sometimes 'deny the need for measure theory because they are unacquainted with problems of other types and with situations where vague reasoning did lead to wrong results'. If Feller knew of any case where such a thing has happened, this would surely have been the place to cite it – yet he does not. Therefore we remain, just as he says, unacquainted with instances where wrong results could be attributed to failure to use measure theory.

But, as noted particularly in Chapter 15, there are many documentable cases where careless use of infinite sets has led to absurdities. We know of no case where our 'cautious approach' policy leads to inconsistency or error; or fails to yield a result that is reasonable.

We do not use the notation of measure theory because it presupposes the passage to an infinite limit already carried out at the beginning of a derivation – in defiance of the advice of Gauss, quoted at the start of Chapter 15. But in our calculations we often pass to an infinite limit at the end of a derivation; then we are in effect using 'Lebesgue measure' directly in its original meaning. We think that failure to use current measure theory notation is not 'vague reasoning'; quite the opposite. It is a matter of doing things in the proper order.

Feller does acknowledge, albeit grudgingly, the validity of our position. While he considers passage to a well-defined limit from a finite set unnecessary, he concedes that it is 'logically impeccable' and has 'the merit of a good exercise for beginners'. That is enough for us; for in this field we are all beginners. Perhaps the beginners who have the most to learn are those who now decline to practice this very instructive exercise.

We note also that measure theory is not always applicable, because not all sets that arise in real problems are measurable. For example, in many applications we want to assign probabilities to functions that we know in advance are continuous. But Mark Kac (1956) notes that the class of continuous functions is not measurable; its inner measure is zero, its outer measure one.[1] Being a mathematician, he was willing to sacrifice some aspects of the real world in order to conform to his preconception that his sets should be measurable. So to get a measurable class of functions he enlarges it to include the everywhere discontinuous functions. But then the resulting measure is concentrated 'almost entirely' on just the class of functions that, for physical reasons, we need to exclude most strongly from our set! So, while Kac gets a solution that is satisfactory to him, it is not always the solution to a real problem.

Our value judgment is just the opposite; being concerned with the real world, we are willing to sacrifice preconceptions about measurable classes in order to preserve the aspects of the real world that are important in our problem. In this case, a form of our cautious approach policy will always be able to bypass measure theory in order to get the useful results we seek; for example, (1) expand the continuous functions in a finite-number n of orthogonal functions, (2) assign probabilities to the expansion coefficients in a finite-dimensional space R_n; (3) do the probability calculation; (4) pass to the limit $n \to \infty$ at the end. In a real problem we find that increasing n beyond a certain value makes a numerically negligible change in our conclusions (i.e. if we are calculating to a finite number of decimal places, a strictly nil change). So we need never depart from finite sets after all.[2] Useful results, in various applications from statistical mechanics to radar detection, are found in this way.

It appears to us that most – perhaps all – of the paradoxes of infinite sets that arise in calculations are caused by the persistent tendency to pass to infinite limits too soon. Usually, this means that crucially important information is lost before we have a chance to use it; the case of nonconglomerabilty in Chapter 15 is a good example. In any event, whatever the cause and the cure, our position is that the paradoxes of infinite sets belong to the field of infinite-set theory, and have no place in probability theory. Our self-imposed inhibition of considering only finite sets and their well-behaved limits enables us to avoid all of the useless and unnecessary paradoxing that has appeared in the recent statistical literature. From this experience, we conjecture that perhaps all correct results in probability theory are either combinatorial theorems on finite sets or well-behaved limits of them.

But on this issue, too, we are not fanatics. We recognize that the language of set and measure theory was a useful development in *terminology*, in some cases enabling one to state mathematical propositions with a generality and conciseness that is quite lacking in 19th century mathematics. Therefore we are happy to use that language whenever it contributes to our goal, and we could hardly get along without an occasional 'almost everywhere' or 'of measure zero' phrase. However, when we use a bit of measure theory, it is never in the thought that this makes the argument more rigorous; but only a recognition of the compactness of that language.

Of course, we stand ready and willing to use set and measure theory – just as we stand ready and willing to use number theory, projective geometry, group theory, topology, or any other part of mathematics – wherever this should prove helpful for the technique of finding a result or for

[1] A continuous function is defined everywhere by specifying it at each rational point, whose number is countable. Thus the class of continuous functions is very much smaller than the class of everywhere discontinuous functions.

[2] But, even in the limit, the number of expansion coefficients is only countable, corresponding nicely to the property of continuous functions noted in footnote 1.

understanding it. But we see no reason why we must state every proposition in set/measure theory terminology and notation in cases where plain English is clearer and, as far as we can see, not only more efficient for our purposes but actually safer.

Indeed, an insistence that all of mathematics be stated in that language all of the time can place unnecessary burdens on a theory, particularly one intended for application in the real world. It can also degenerate into an affectation, used only linguistically rather than functionally. To give every old, familiar notion a new, impressive name and symbol unknown to Gauss and Cauchy has nothing to do with rigor. It is, more often than not, a form of gamesmanship whose real purpose is to conceal the Mickey Mouse triviality of what is being done. One would blush to state it in plain English.

B.4   Kronecker vs. Weierstrasz

At this point, a question will surely be in the reader's mind. After our emphasis on the safety of finite sets, it might appear that all of analysis, which seems to do everything on uncountable sets, is suspect. Let us explain why this is not the case, and why we do place full confidence in the analysis of Cauchy and Weierstrasz.[3]

In the late 19th century, both Karl Weierstrasz (1815–1897) and Leopold Kronecker (1823–1891) were at the University of Berlin,[4] lecturing on mathematics. A difference developed between them, which has been greatly exaggerated by later commentators, and it is only in the past few years that the real truth about their relationship has started to emerge.

Briefly, Weierstrasz was concerned with perfecting the tools of analysis – particularly power series expansions – with the specific case of elliptic functions in mind as an application. Kronecker was more concerned with the foundations of mathematics in number theory, and questioned the validity of reasoning that does not start back at the integers. On a superficial view, this might seem to deny us all the beautiful results of analysis. Even Morris Kline (1980) gives the impression that Kronecker's asceticism denies us some of the important advances in modern mathematics. But the record has been distorted.

For example, Bell (1937, p. 568) paints a picture of Weierstrasz as the great analyst, putting the final finishing touches on the work of Cauchy, and Kronecker as a mere gadfly, attacking the validity of everything he did without making any positive contribution. It is true that Kronecker annoyed Weierstrasz on at least one occasion, documented in Weierstrasz's correspondence; yet there was not really much conflict in their principles. To understand their positions, we just need a better witness than Eric Temple Bell, and fortunately we have two of them: Henri Poincaré and Harold M. Edwards.

When Weierstrasz died in 1897, Poincaré (1899) wrote a summary of his mathematical work, in which he pointed out that: '. . . all the equations which are the object of analysis and which deal with continuous magnitudes are nothing but symbols, replacing an infinite collection of inequalities relating whole numbers.' In the words of H. M. Edwards (1989), '. . . *both* Weierstrasz and Kronecker based their mathematics entirely on the whole numbers, so that all their work shared in the certitude of arithmetic.' Edwards notes also that several reactionary views commonly attributed to Kronecker are hearsay, for which no support can be found in Kronecker's own words.

For example, Bell (1937, p. 568) tells us, without any supporting documentation, that Kronecker, on hearing of Lindemann's proof that π is transcendental, asked of what use that could be, '. . . since

---

[3] Indeed, the writer's first love in mathematics was not probability theory, but the use of Cauchy's complex integration to solve systems of differential equations and boundary conditions, choosing the integrand to satisfy the differential equation, and then the contour of integration to satisfy the boundary conditions. Three generations of theoretical physicists have exploited this method enthusiastically; it is great fun to teach.

[4] More specifically, Weierstrasz was there from 1856–1897 and Kronecker from 1861–1891. E. T. Bell (1937) gives a portrait of the young Weierstrasz and a photograph of the old Kronecker; H. M. Edwards (1989) gives photographs of the old Weierstrasz and the young Kronecker.

irrational numbers do not exist?' The documentable fact is that Kronecker's own work on number theory (Kronecker, 1901, p. 4) describes the formula of Leibniz:

$$\frac{\pi}{4} = 1 - \frac{1}{3} + \frac{1}{5} - \frac{1}{7} + \cdots \tag{B.7}$$

as 'one of the most beautiful arithmetic properties of the odd integers, namely that of determining this geometrical irrational number.' Evidently, Kronecker considered irrational numbers as possessing at least enough 'existence' to allow them to be precisely defined. It is true that he did not consider irrationals to be a necessary part of the foundations; indeed, how could he, or anybody else, think that, in view of relations like the above one, which allow irrationals to be defined entirely in terms of integers? Curiously, Weierstrasz also defined irrationals from the integers in just the same way; so where was the difference between them?

The difference between Kronecker and Weierstrasz was aesthetic rather than substantive: Kronecker wants to keep first principles (the origin in the integers) constantly in view, while Weierstrasz, having made a new construction, is willing to forget the steps by which it was made, and use it as an element in its own right for further construction. Put in modern computer terminology, Weierstrasz did not deny Kronecker's 'machine language' basis of all mathematics, but wanted to develop analysis in a higher level language. Edwards points out that Kronecker's principles, '... in his mind and in fact, were no different from the principles of his predecessors, from Archimedes to Gauss.'

Thanks largely to the historical research of H. M. Edwards, the truth is emerging and Kronecker is being vindicated and rehabilitated. Perhaps Kronecker was overzealous, and perhaps he misunderstood the position of Weierstrasz; but events since then suggest that he was not zealous enough in his own cause. His failure to respond to Georg Cantor (1845–1918) seems unfortunate, but easy to understand.

To Kronecker, Cantor's ideas were so *outré* that they had nothing to do with mathematics, and there was no reason for a mathematician to take any note of them. If the editors of the mathematical journals made the mistake of publishing such stuff, that was their problem, not his. But the messages that Kronecker did communicate contained some very important truth; in particular, he complained that much of set theory was fantasy because it was not algorithmic (i.e. it contained no rule by which one could construct a given element or decide, in a finite number of operations, whether a given element did or did not belong to a given set). Today, with our computer mentalities, this seems such an obvious platitude that it is hard to imagine anyone ignoring it, much less denying it; yet that is just what happened. We think that, had mathematicians paid more attention to this warning of Kronecker, mathematics might be in a more healthy state today.

B.5 What is a legitimate mathematical function?

Much of the difference between current pure and applied mathematics lies in their different conceptions of the notion of a 'function'. Historically, one started with the well-behaved analytic entire functions like $f(x) = x^2$ or $f(x) = \sin x$. Then these 'good functions' were generalized, but in two different ways. In pure mathematics, the idea was generalized in such a way that set theory notions remained valid; first to piecewise continuous functions, then to quite arbitrary rules by which, given a number x, one can define another number f. Then, perceiving that a function or its argument need not be limited to real or complex numbers, this was generalized further to an arbitrary mapping of one set X onto another set F, the elements of which could be almost anything.

In applied mathematics, the notion of a function was generalized in a very different way, so that the useful *analytical operations* that we perform on functions remain valid. Perhaps the most important hint was provided by the operation of the Fourier transform. This is still a mapping, but at the higher level of mapping one function $f(x)$ onto another $F(k)$. This mapping was defined by the integrals

$$F(k) = \int dx e^{ikx} f(x), \qquad f(x) = \frac{1}{2\pi} \int dx e^{-ikx} F(k). \qquad \text{(B.8)}$$

If we indicate this Fourier transform pair symbolically as

$$\left[f(x) \leftrightarrow F(k) \right] \qquad \text{(B.9)}$$

we find the interesting properties that under translation, convolution, and differentiation,

$$\left[f(x - a) \leftrightarrow e^{ika} F(k) \right] \qquad \text{(B.10)}$$

$$\left[\int dy f(x - y)g(y) \leftrightarrow F(k)G(k) \right] \qquad \text{(B.11)}$$

$$\left[f'(x) \leftrightarrow ikF(k) \right], \qquad \left[-ix f(x) \leftrightarrow F'(k) \right]. \qquad \text{(B.12)}$$

In other words, analytical operations on one function correspond to algebraic operations on the other.

In practice, these are very useful properties. Thus, to solve a linear differential equation, or difference equation, or integral equation of convolution form $[\int dy K(x - y)f(y) = \lambda g(x)]$, or, indeed, a linear equation which contains all three of these operations, one may take its Fourier transform, which converts it into an algebraic equation for $F(k)$. If this can be solved directly for $F(k)$, then taking the inverse Fourier transform yields the solution $f(y)$ of the original equation. Thus the Fourier transform mapping reduces the solution of linear analytical equations to that of ordinary algebraic equations. In the early 20th century, the theoretical physicist Arnold Sommerfeld in Munich became a great artist in the technique of evaluating these solutions by fancy contour integrals, and some of the greatest of the next generation learned this from him. Today, physicists and engineers could hardly survive without it.

This procedure seemed to apply only to a limited class of functions. In the Dirichlet form of Fourier theory, one shows that, if $f(x)$ is absolutely integrable, then the integral (B.8) surely converges to a well-behaved continuous function $F(k)$ on the real axis, and all is well. If $f(x)$ also vanishes for negative x, then $F(k)$ is analytic and bounded in one-half of the complex plane, and all is even better. But if $f(x)$ is absolutely integrable, then $f'(x)$ or $f''(x)$ may not be; and there is some doubt whether the useful properties are still valid. In the early work on Fourier transforms, such as Titchmarsh (1937), virtually all one's attention was concentrated on the theory of convergence of the integrals, and any function for which the integral did not converge was held not to possess a Fourier transform. This placed an intolerable restriction on the range of useful applications of Fourier theory.

Then a more sophisticated view emerged in theoretical physics. One realized that the usefulness of the Fourier transform lies, not in convergence of any integral, but in the above properties (B.10)–(B.12) of the mapping. Therefore, as long as our functions are sufficiently well-behaved so that the operations in (B.10)–(B.12) make sense, then, if by any means we can define the mapping such that those properties are preserved, then the customary use of Fourier transforms to solve linear integrodifferential equations will be perfectly rigorous, and *it does not make the slightest difference* whether the integrals (B.8) or the analogous Fourier series do or do not converge. A divergent Fourier series is still a unique ordered sequence of numbers, conveying all the needed information (i.e. it is uniquely determined by, and uniquely determines, its Fourier transform). It was only an historical accident that this mapping was first discovered through series and integral representations, which exist only in special cases.

B.5.1 Delta-functions

Although its beginnings can be traced back to Duhamel and Green in the 19th century, this movement is commonly held to start with P. A. M. Dirac, who in the 1920s invented the notation of the delta-function $\delta(x - y)$ generalizing Kronecker's δ_{ij}, and showed how to use it to good advantage in applications. It is the 'Fourier transform of a constant' in the sense that as $F(k) \to 1$, we have $f(x) \to \delta(x)$. Mathematicians thinking in terms of the set theory definition of a 'function' were horrified and held this to be nonrigorous on the grounds that delta-functions do not 'exist'. But that was only because of their inappropriate definition of the term 'function'. A delta-function is not a mapping of any set onto any other. Laurent Schwartz (1950) tried to make the notion of a delta-function rigorous, but from our point of view awkwardly, because he persisted in defining the term 'function' in a way inappropriate to analysis.

Perceiving this, G. Temple (1955) and M. J. Lighthill (1957) showed how to remove the awkwardness simply by adopting a definition of functions as meaning 'good' functions and limits of sequences of good functions (thus, in our system, a discontinuous function is *defined* as the limit of a sequence of continuous functions). For this, there is almost no need to mention such things as open and closed sets. Lighthill saw that this definition of 'function' is the one appropriate to Fourier theory. It is now clear that it is also the one appropriate to probability theory and to all of analysis; with it our theorems become simpler and more general, without a long list of exceptions and special cases. For example, any Fourier series may now be differentiated term by term any number of times and the result, whether convergent or not, identifies (by 1:1 correspondence) a unique function in our sense of the word. Physicists had seen this intuitively and used it correctly long before the work of Schwartz, Temple, and Lighthill.

Lighthill produced a very thin book (1957) on the new form of Fourier analysis, which included a table of Fourier transforms in which every entry is a function which was held formerly not to possess a Fourier transform. Yet that table is a gold mine for the useful solution of linear integro-differential equations. In a famous review of Lighthill's book, the theoretical physicist Freeman J. Dyson (1958), a former student of the Cambridge mathematician G. H. Hardy, stated that Lighthill's book '... lays Hardy's work in ruins, and Hardy would have enjoyed it more than anybody.' Throughout the present work, we take Lighthill's approach for granted and assume that the reader is familiar with it.[5]

B.5.2 Nondifferentiable functions

The issue of nondifferentiable functions arises from time to time in probability theory. In particular, when one solves a functional equation such as those studied in Chapter 2, to assume differentiability is to have a horde of compulsive mathematical nitpickers descend upon one, with claims that we are excluding a large class of potentially important solutions. However, we noted that this is not the case; Aczel demonstrated that Cox's functional equations can all be solved without assuming differentiability (at the cost of much longer derivations) and with just the same solutions that we found above.

Let us take a closer look at the notion of nondifferentiable functions in general. This was not well-received at first by mathematicians. Charles Hermite wrote to Stieltjès: 'I turn away in horror from this awful plague of functions which have no derivatives.' The one generally blamed for this

---

[5] Lighthill defines the term 'good function' in a different way than we did above, which seems to us unnecessarily restrictive in its behavior at infinity. Apparently, this was because he did not like integrals over finite domains, whereas we do like them. But Lighthill's definition is more general than ours in that a 'good function' need not be analytic; however, this generality seems to us unnecessary because we have never seen a real problem in which the underlying good functions could not be chosen to be analytic.

plague was Henri Lebesgue (1875–1941), although Weierstrasz had noted them before him. The Weierstrasz nondifferentiable function is

$$f(x) \equiv \sum_{n=0}^{\infty} a^n \cos(m^n x), \tag{B.13}$$

where $(0 < a < 1)$ and m is a positive odd integer. It is an ordinary Fourier series with period 2π, since m^n is always an integer. Furthermore, the series is uniformly convergent for all real x (since it must converge at least as well as does $\sum a^n$), so it defines a continuous function. But if $ma > 1$, term-by-term differentiation yields a badly divergent series, whose coefficients grow exponentially in n. The proof that the derivative

$$f'(x) \equiv \lim_{h \to 0} \frac{f(x+h) - f(x)}{h} \tag{B.14}$$

then does not exist for any x is rather tedious.[6] Weierstrasz's function is, in fact, the limit of a sequence of good functions (the partial sums S_k of the first k terms), but it is not a very well-behaved limit, and such functions are of no apparent use to us because they fail to satisfy condition (B.12). Nevertheless, functions like this do arise in applications; for example, in Chapter 7 our attempt to solve the integral equation (7.49) by Fourier transform methods ran up against this difficulty if the kernel was too broad. Then our conclusion was that the integral equation does not have any usable solution unless the kernel $\phi(x - y)$ is at least as sharp as the 'driving force' $f(x)$.

B.5.3 Bogus nondifferentiable functions

The case most often cited as an example of a nondifferentiable function is derived from a sequence $f_n(x)$, each of which is a string of isosceles right triangles whose hypotenuses lie on the real axis and have length $1/n$. As $n \to \infty$, the triangles shrink to zero size. For any finite n, the slope of $f_n(x)$ is ± 1 almost everywhere. Then what happens as $n \to \infty$? The limit $f_\infty(x)$ is often cited carelessly as a nondifferentiable function. Now it is clear that the limit of the derivative, $f_n'(x)$, does not exist; but it is the derivative of the limit that is in question here, $f_\infty(x) \equiv 0$, and this is certainly differentiable. Any number of such sequences $f_n(x)$ with discontinuous slope on a finer and finer scale may be defined. The error of calling the resulting limit $f_\infty(x)$ nondifferentiable, on the grounds that the limit of the derivative does not exist, is common in the literature. In many cases, the limit of such a sequence of bad functions is actually a well-behaved function (although awkwardly defined), and we have no reason to exclude it from our system.

Lebesgue defended himself against his critics thus: 'If one wished always to limit himself to the consideration of well-behaved functions, it would be necessary to renounce the solution of many problems which were proposed long ago and in simple terms.' The present writer is unable to cite any specific problem which was thus solved; but we can borrow Lebesgue's argument to defend our own position.

To reject limits of sequences of good functions is to renounce the solution of many current real problems. Those limits can and do serve many useful purposes, which much current mathematical education and practice still tries to stamp out. Indeed, the refusal to admit delta-functions as legitimate mathematical objects has led mathematicians into error. For example, H. Cramér (1946, Chap. 32) gives an inequality, which we derived in Chapter 17, placing a lower limit to the variance

---

[6] See Hardy (1911). Titchmarsh (1939, pp. 350–353) gives only a shorter proof valid when $ma > 1 + 3\pi/2$. Some authors state that $f(x)$ is nondifferentiable only in this case; but, to the best of our knowledge, nobody has ever claimed that Hardy's proof contains an error.

of the sampling distribution for a parameter estimator θ^*:

$$\text{var}(\theta^*) \geq \frac{(1 + \mathrm{d}b/\mathrm{d}\theta)^2}{n \int \mathrm{d}x \, (\partial \log(f)/\partial\theta)^2 \, f(x|\theta)}, \tag{B.15}$$

where we have made n observations from a sampling distribution $f(x|\theta)$, and $b(\theta^*) \equiv E(\theta^* - \theta)$ is the bias of the estimator.

Then Cramér notes that, if $f(x|\theta)$ has discontinuities, then 'the conditions for the regular case are usually not satisfied. In such cases it is often possible to find unbiased estimates of 'abnormally high' precision, i.e. such that the variance is smaller than the lower limit [(B.15)] for regular estimates.' How could he have reached such a remarkable conclusion, since (B.15) is only the Schwartz inequality, which does not seem to admit of exceptions? We find that he has used the set-theory definition of a function, and concluded that the derivative $\partial \log(f)/\partial\theta$ does not exist at points of discontinuity. So he takes the integral in (B.15) only over those regions where $f(x|\theta)$ is continuous.

But the definition of a discontinuous function which is appropriate in analysis is our limit of a sequence of continuous functions. As we approach that limit, the derivative develops a higher and sharper spike. However close we are to that limit, the spike is part of the correct derivative of the function, and its contribution must be included in the exact integral. Thus the derivative of a discontinuous function $g(x)$ necessarily contains a delta-function $[g(y+) - g(y-)] \delta(x - y)$ at points y of discontinuity, *whose contribution is always present in the differentiated Fourier series for $g(x)$, and must be included in order to get the correct physical solution.* Had Cramér included this term, (B.15) would have reduced in the limit to $\text{var}(\theta^*) \geq 0$; hardly a useful statement, but at least there would have been no anomaly and no seeming violation of the Schwartz inequality.

In a similar way, the solution of an integral equation with finite limits, of the form

$$\int_a^b \mathrm{d}y \, K(x, y) f(y) = \lambda g(x), \tag{B.16}$$

generally involves delta-functions like $\delta(y - a)$ or $\delta'(y - b)$ at the end-points, and so those who do not believe in delta-functions consider such integral equations as not having solutions. But in real physical problems, exactly such integral equations occur repeatedly, and again the delta-functions must be included in order to get the correct physical solution. Some examples are given by D. Middleton (1960); they are virtually ubiquitous in the prediction of irreversible processes in statistical mechanics. It is astonishing that so few non-physicists have yet perceived this need to include delta-functions, but we think it only illustrates what we have observed independently; those who think of fundamentals in terms of set theory fail to see its limitations because they almost never get around to useful, substantive calculations.

So, bogus nondifferentiable functions are manufactured as limits of sequences of rows of tinier and tinier triangles, and this is accepted without complaint. Those who do this while looking askance at delta-functions are in the position of admitting limits of sequences of bad functions as legitimate mathematical objects, while refusing to admit limits of sequences of good functions! This seems to us a sick policy, for delta-functions serve many essential purposes in real, substantive calculations, but we are unable to conceive of any useful purpose that could be served by a nondifferentiable function. It seems that their only use is to provide trouble-makers with artificially contrived counter-examples to almost any sensible and useful mathematical statement one could make. Henri Poincaré (1909) noted this in his characteristically terse way:

> In the old days when people invented a new function they had some useful purpose in mind: now they invent them deliberately just to invalidate our ancestors' reasoning, and that is all they are ever going to get out of them.

We would point out that those trouble-makers did not, after all, invalidate our ancestors' reasoning; their pathology appeared only because they adopted, surreptitiously, a different definition of the term 'function' than our ancestors used. Had this been pointed out, it would have been clear that there was no need to modify our ancestors' conclusions.

Today, this fad of artificially contrived mathematical pathology seems nearly to have run its course, and for just the reason that Poincaré foresaw; nothing useful can be done with it. While we still see exhortations not to assume differentiability of an unknown function, it is difficult to find even one specific example of a nondifferentiable function appearing – much less actually being used for anything – in the recent literature. One must go back to old works like Titchmarsh (1939) to see them at all.

Note, therefore, that we stamp out this plague too, simply by our defining the term 'function' in the way appropriate to our subject. The definition of a mathematical concept that is 'appropriate' to some field is the one that allows its theorems to have the greatest range of validity and useful applications, without the need for a long list of exceptions, special cases, and other anomalies. In our work the term 'function' includes good functions and well-behaved limits of sequences of good functions; but not nondifferentiable functions. We do not deny the existence of other definitions which do include nondifferentiable functions, any more than we deny the existence of fluorescent purple hair dye in England; in both cases, we simply have no use for them.[7]

B.6 Counting infinite sets?

It is well known that Lewis Carroll's children's books were really expositions of the principles of logic, conveyed by the device of stating the opposite in a form that would appear ludicrous even to small children. One of his poems ends thus:

> He thought he saw an Argument that proved he was the Pope:
> He looked again and found it was a Bar of Mottled Soap.
> 'A fact so dread,' he faintly said, 'Extinguishes all hope!'

Indeed, many of the arguments seriously proposed in probability theory are seen, on second glance, to be nothing but mottled soap. The idea was appropriated in a famous anecdote[8] about the Cambridge mathematician G. H. Hardy; J. E. McTaggart expressed doubt that from a false proposition all propositions can be deduced, by challenging him thus: 'Given $2 + 2 = 5$: prove that I am the Pope.' Whereupon Hardy replied: 'Subtract 3 from each side and we have $1 = 2$. Now we agree that the Pope and you are two; therefore the Pope and you are one!' But that was only a play on words; infinite-set theory gives us a superior grade of mottled soap, with which we can prove McTaggart's papacy much more convincingly.

We start from the premise that two sets have the same number of elements if they can be put into 1:1 correspondence with each other. Then by the association $(n \leftrightarrow 2n)$, $n = 1, 2, \ldots$, we can put the positive integers into 1:1 correspondence with the positive even integers. And by the association $(2n \leftrightarrow 2n - 1)$, $n = 1, 2, \ldots$, we can, equally well, put the positive even integers into 1:1 correspondence with the positive odd integers; so by such logic it seems that we would be driven to conclude that:

(A) (number of integers) = (number of even integers);

(B) (number of even integers) = (number of odd integers);

(C) (number of integers) = $2 \times$ (number of even integers);

---

[7] On a different topic, in Chapter 17 (footnote 9 on p. 521) we follow the same policy by defining the term 'moving average' for a finite time series in such a way that our theorems are all exact, without any need for messy 'end effect' corrections. Of course, it then develops that this is the definition most directly useful in applications and that conserves information which would otherwise be lost.

[8] Cited by Jeffreys (1931; 1957 edn, p. 18).

and from (A) and (C) it follows that $1 = 2$. The reasoning here is not very different from that in Eqs. (15.2)–(15.3).

Our view is that the 'set of all integers' is undefined except as a limit of finite sets, and if it is approached in that way, by introducing the explicit limiting process, no contradiction can be produced whatever limiting process we choose, even though the limiting ratio of (number of even integers)/(number of integers) can be made to be any x we please in $0 \leq x \leq 1$. That is, the limit of (number of odd integers)/(number of integers) will be $(1 - x)$, and our counting will remain consistent in the limit.

For example, every integer is included once and only once in the sequence $\{1, 3, 2, 5, 7, 4, \ldots\}$, in which we take alternately two odd and one even. Then counting elements only in the finite sets consisting of the first n elements of this sequence, and passing to the limit $n \rightarrow \infty$ after doing the counting, we would find in place of the inconsistent statements (A), (B), (C) above, the consistent set

(A') (number of integers) $= 3 \times$ (number of even integers);
(B') (number of even integers) $= 1/2 \times$ (number of odd integers);
(C') (number of integers) $=$ (number of even integers) $+$ (number of odd integers).

These ideas are not as new as one might think. Galileo (1638), in his *Dialogues Concerning Two New Sciences*, notes two curious facts. On the one hand, each integer has one and only one square, and no two of them have the same square; from which it would seem that the number of integers and the number of squares must be the same. On the other hand, it is evident that there are many integers (in a certain sense, the 'great majority' of them) which are not squares. From this he draws the eminently sensible conclusion:

This is one of the difficulties which arise when we attempt, with our finite minds, to discuss the infinite, assigning to it those properties which we give to the finite and limited; but this I think is wrong, for we cannot speak of infinite quantities as being the one greater or less than or equal to another.

Hermann Weyl, 300 years later, expressed almost exactly the same judgment, as noted below. See, for example, Weyl (1949).

B.7 The Hausdorff sphere paradox and mathematical diseases

The inconsistent statements above are structurally almost identical with the Hausdorff paradox concerning congruent sets on a sphere, except for the promotion up to uncountable sets (here X, Y, Z are disjoint sets which nearly cover the sphere, and X is congruent to Y, in the sense that a rotation of the sphere makes X coincide with Y, and likewise Y is congruent to Z. But what is extraordinary is the claim that X is also congruent to the union of Y and Z, even though $Y \neq Z$). We are, like Poincaré and Weyl, puzzled by how mathematicians can accept and publish such results; why do they not see in this a blatant contradiction which invalidates the reasoning they are using?

Nevertheless, L. J. Savage (1962) accepted this antinomy as literal fact and, applying it to probability theory, said that someone may be so rash as to blurt out that he considers congruent sets on the sphere equally probable; but the Hausdorff result shows that his beliefs cannot actually have that property. The present writer, pondering this, has been forced to the opposite conclusion: my belief in the existence of a state of knowledge which considers congruent sets on a sphere equally probable, is vastly stronger than my belief in the soundness of the reasoning which led to the Hausdorff result.

Presumably, the Hausdorff sphere paradox and the Russell Barber paradox have similar explanations: one is defining weird sets with self-contradictory properties, so, of course, from that mess it will be possible to deduce any absurd proposition we please. Hausdorff entitled his work '*Mengenlehre*', and Poincaré made the famous quip that 'Future generations will regard *Mengenlehre* as a disease from which one has recovered.' But he would be appalled to see this recovery not yet achieved 80 years later; nevertheless, Poincaré's views are alive and well today among users of applied mathematics.

For example, in 1983 the writer heard a talk by a very prominent statistician, reporting on an historical investigation. He remarked: 'I was surprised to learn that, before the days of Bourbaki, the French actually produced some useful mathematics.' More recently, the Nobel Laureate theoretical physicist Murray Gell-Mann (1992), discussed this situation. He opined that there is still much in modern mathematics of value to physics, and the divergence of pure mathematics from science is in part only an illusion produced by the obscurantist language of Bourbakists and their reluctance to write up any non-trivial example in explicit detail. He concludes: 'Pure mathematics and science are finally being reunited and, mercifully, the Bourbaki plague is dying out.'

We wish we could feel that optimistic. In our view, this plague is far more serious than mere obscure language; it infects the substantive content of pure mathematics. A sane person can have no confidence in any of it; rules of conduct must be found which prevent the appearance of these ridiculous paradoxes, and then our mathematics textbooks must be rewritten. As is well known, Russell's theory of types can dispose of a few paradoxes, but far from all of them. We fear that, even with the best of good will on both sides, it will require at least another generation to bring about the reconciliation of pure mathematics and science. For now, it is the responsibility of those who specialize in infinite-set theory to put their own house in order before trying to export their product to other fields. Until this is accomplished, those of us who work in probability theory or any other area of applied mathematics have a right to demand that this disease, for which we are not responsible, be quarantined and kept out of our field.

In this view, too, we are not alone; and indeed have the support of many non-Bourbakist mathematicians. In our Preface we quoted Morris Kline (1980) on the dangers of allowing infinite-set theory to get a foothold in applied mathematics. He in turn (on his p. 237) quotes Hermann Weyl. Both Brouwer and Weyl noted that classical logic had been developed for application to finite sets. The attempt to apply classical logic, without justification, to infinite sets is, in Weyl's words: '. . . the Fall and original sin of set theory, for which it is justly punished by the antinomies. It is not that such contradictions showed up that is surprising, but that they showed up at such a late stage of the game.'

But there is a simple explanation for this late appearance, noted with examples in Chapter 15: if an erroneous argument leads to an absurd result immediately, it will be abandoned and we shall never hear about it. If it yields a reasonable result on the first two or three tries, then there is some range of problems where it will succeed. One will continue using it, but at first conservatively – on problems that are quite similar, so it is likely to continue giving reasonable results. Only later, when one becomes over-confident and tries to extend the application to different kinds of problems, do the contradictions appear.[9]

Just the same phenomenon occurred in orthodox statistics, where the *ad hoc* inventions such as confidence intervals yielded acceptable results for a long time because they were used at first only on simple problems which were free of nuisance parameters, but sufficient statistics existed, and there was no very important prior information. Nobody took any note of the fact that the numerical

---

[9] The writer knew Hermann Weyl, took his course in group theory at Princeton, and admired him as the final authority on both group theory and variational principles for general relativity. But the Bourbakist mathematicians at Princeton sneered at him and called him 'Holy Hermann' behind his back, because of his Biblical exhortations to virtue like the one just quoted. They would have been better advised to listen to him.

results were then the same as the Bayesian posterior probability intervals at the same level (based on the uninformative priors given by Jeffreys). Confidence intervals were widely held, by mathematicians such as Neyman, Cramér and Wilks, to be great advances over Bayesian methods, until their contradictions began to appear when one tried to apply them to more general problems.[10] Finally, we were able to show (Jaynes, 1976) that confidence intervals are satisfactory as inferences *only* in those special cases where they happen to agree with the Bayesian intervals after all.

Kline (1980, p. 285) also quotes J. Willard Gibbs on this subject: 'The pure mathematician can do what he pleases, but the applied mathematician must be at least partially sane.' In any event, no sane person would try to use such anomalies as the Hausdorff sphere paradox in a real application.

Finally, we offer a few more general comments on mathematical style.

B.8 What am I supposed to publish?

L. J. Savage (1962) asked this question to express his bemusement at the fact that, no matter what topic he chose to discuss, and no matter what style of writing he chose to adopt, he was sure to be criticized for not making a different choice. In this he was not alone. We would like to plead for a little more tolerance of our individual differences.

If anyone wants to concentrate his attention on infinite sets, measure theory, and mathematical pathology in general, he has every right to do so. And he need not justify this by pointing to useful applications or apologize for the lack of them; as was noted long ago, abstract mathematics is worth knowing for its own sake.

But others in turn have equal rights. If we choose to concentrate on those aspects of mathematics which *are* useful in real problems and which enable us to carry out the important substantive calculations correctly – but which the mathematical pathologists never get around to – we feel free to do so without apology.

Ultimately, the mathematical level and depth of this work were chosen with the aim of making it possible for all readers to extract what they want from it. Since those who approach a work with the sole purpose of finding fault with its style of presentation will always be able to do so no matter how it is presented, our aim was to ensure that those who approach it with sincere desire to understand its *content* will also be able to do so. Thus we try to give cogent reasons why the ideas we advocate are 'obvious', while those we deplore are not, when this can be done briefly enough not to interrupt the line of argument. This inevitably leaves some lacunae, in part filled in by the Comments sections at the end of most chapters.

In this connection, the question of what is and is not 'obvious' is a matter of gamesmanship that is played in two opposite directions. On the one hand, the standard way of introducing notions that do not stand up to critical examination – or to deprecate those that stand up too well to be safely opposed – is to call them 'obvious'. On the other hand, to express grave doubts about simple matters that *are* obvious is the equally standard technique for imputing to one's self deep critical faculties not possessed by others. We try to steer a middle course between these, but like Savage do so in the knowledge that, whatever our choice, it will receive opposite criticisms from the two types of gamesman.

But we avoid one common error: nothing could be more pathetically mistaken than the prefatory claim of one author in this field that mathematical rigor 'guarantees the correctness of the results'. On the contrary, much experience teaches us that the more one concentrates on the appearance of

---

[10] Confidence intervals are always correct as statements about sampling properties of estimators; yet they can be absurd as statements of inference about the values of parameters. For example, the entire confidence interval may lie in a region of the parameter space which we know, by deductive reasoning from the data, to be impossible.

mathematical rigor, the less attention one pays to the validity of the premises in the real world, and the more likely one is to reach final conclusions that are absurdly wrong in the real world.

B.9 Mathematical courtesy

A few years ago the writer attended a seminar talk by a young mathematician who had just received his Ph.D. degree and, we understood, had a marvellous new limit theorem of probability theory. He started to define the sets he proposed to use, but three blackboards were not enough for them, and he never got through the list. At the end of the hour, having to give up the room, we walked out in puzzlement, not knowing even the statement of his theorem.

A '19th century mathematician' like Poincaré would have been into the meat of the calculation within a few minutes and would have completed the proof and pointed out its consequences in time for discussion.

The young man is not to be blamed; he was only doing what he had been taught a '20th century mathematician' must do. Although he has perhaps now learned to plan his talks a little better, he is surely still wasting much of his own time and that of others in reciting all the preliminary incantations that are demanded in 20th century mathematics before one is allowed to proceed to the actual problem. He is a victim of what we consider to be, not higher standards of rigor, but studied mathematical discourtesy.

Nowadays, if you introduce a variable x without repeating the incantation that it is in some set or 'space' X, you are accused of dealing with an undefined problem. If you differentiate a function $f(x)$ without first having stated that it is differentiable, you are accused of lack of rigor. If you note that your function $f(x)$ has some special property natural to the application, you are accused of lack of generality. In other words, every statement you make will receive the discourteous interpretation.

Obviously, mathematical results cannot be communicated without some decent standards of precision in our statements. But a fanatical insistence on one particular form of precision and generality can be carried so far that it defeats its own purpose; 20th century mathematics often degenerates into an idle adversary game instead of a communication process.

The fanatic is not trying to understand your substantive message at all, but only trying to find fault with your style of presentation. He will strive to read nonsense into what you are saying, if he can possibly find any way of doing so. In self-defense, writers are obliged to concentrate their attention on every tiny, irrelevant, nitpicking detail of how things are said rather on what is said. The length grows; the content shrinks.

Mathematical communication would be much more efficient and pleasant if we adopted a different attitude. For one who makes the courteous interpretation of what others write, the fact that x is introduced as a variable *already implies* that there is some set X of possible values. Why should it be necessary to repeat that incantation every time a variable is introduced, thus using up two symbols where one would do? (Indeed, the range of values is usually indicated more clearly at the point where it matters, by adding conditions such as $(0 < x < 1)$ after an equation.)

For a courteous reader, the fact that a writer differentiates $f(x)$ twice already implies that he considers it twice differentiable; why should he be required to say everything twice? If he proves proposition A in enough generality to cover his application, why should he be obliged to use additional space for irrelevancies about the most general possible conditions under which A would be true?

A scourge as annoying as the fanatic is his cousin, the compulsive mathematical nitpicker. We expect that an author will define his technical terms, and then use them in a way consistent with his definitions. But if any other author has ever used the term with a slightly different shade of meaning, the nitpicker will be right there accusing you of inconsistent terminology. The writer has been subjected to this many times; and colleagues report the same experience.

Nineteenth century mathematicians were not being nonrigorous by their style; they merely, as a matter of course, extended simple civilized courtesy to others, and expected to receive it in return. This will lead one to try to read sense into what others write, if it can possibly be done in view of the whole context; not to pervert our reading of every mathematical work into a witch-hunt for deviations from the Official Style.

Therefore, sympathizing with the young man's plight but not intending to be enslaved like him, we issue the following:

Emancipation Proclamation

Every variable x that we introduce is understood to have some set X of possible values. Every function $f(x)$ that we introduce is understood to be sufficiently well-behaved so that what we do with it makes sense. We undertake to make every proof general enough to cover the application we make of it. It is an assigned homework problem for the reader who is interested in the question to find the most general conditions under which the result would hold.

We could convert many 19th century mathematical works to 20th century standards by making a rubber stamp containing this Proclamation, with perhaps another sentence using the terms 'sigma-algebra, Borel field, Radon–Nikodym derivative', and stamping it on the first page.

Modern writers could shorten their works substantially, with improved readability and no decrease in content, by including such a Proclamation in the copyright message, and writing thereafter in 19th century style. Perhaps some publishers, seeing these words, may demand that they do this for economic reasons; it would be a service to science.

In this appendix we have presented many short quotations without the references. Supporting documentation and many further interesting details may be found in Bell (1937), Félix (1960), Kline (1980), and Rowe and McCleary (1989).

Appendix C
Convolutions and cumulants

Firstly we note some general mathematical facts which have nothing to do with probability theory. Given a set of real functions $f_1(x)$, $f_2(x)$, ..., $f_n(x)$ defined on the real line and not necessarily non-negative, suppose that their integrals (zeroth moments) and their first, second, and third moments exist:

$$Z_i \equiv \int_{-\infty}^{\infty} dx \, f_i(x) < \infty, \qquad S_i \equiv \int_{-\infty}^{\infty} dx \, x^2 f_i(x) < \infty,$$

$$F_i \equiv \int_{-\infty}^{\infty} dx \, x f_i(x) < \infty \qquad T_i \equiv \int_{-\infty}^{\infty} dx \, x^3 f_i(x) < \infty. \tag{C.1}$$

The convolution of f_1 and f_2 is defined by

$$h(x) \equiv \int_{-\infty}^{\infty} dy \, f_1(y) f_2(x - y) \tag{C.2}$$

or, in condensed notation, $h = f_1 * f_2$. Convolution is associative: $(f_1 * f_2) * f_3 = f_1 * (f_2 * f_3)$, so we can write a multiple convolution as $(h = f_1 * f_2 * f_3 * \cdots * f_n)$ without ambiguity. What happens to the moments under this operation? The zeroth moment of $h(x)$ is

$$Z_h = \int_{-\infty}^{\infty} dx \int_{-\infty}^{\infty} dy \, f_1(y) \, f_2(x - y) = \int dy \, f_1(y) Z_2 = Z_1 Z_2. \tag{C.3}$$

Therefore, if $Z_i \neq 0$ we can multiply $f_i(x)$ by some constant factor which makes $Z_i = 1$, and this property will be preserved under convolution. In the following we assume that this has been done for all i. Then the first moment of the convolution is

$$F_h = \int_{-\infty}^{\infty} dx \int_{-\infty}^{\infty} dy \, f_1(y) x \, f_2(x - y) = \int dy \, f_1(y) \int_{-\infty}^{\infty} dq \, (y + q) f_2(q)$$

$$= \int_{-\infty}^{\infty} dy \, f_1(y) [y Z_2 + F_2] = F_1 Z_2 + Z_1 F_2 \tag{C.4}$$

so the first moments are additive under convolution:

$$F_h = F_1 + F_2. \tag{C.5}$$

For the second moment, we have by a similar argument

$$S_h = \int dy \, f_1(y) \int dq \, (y^2 + 2yq + q^2) f_2(q) = S_1 Z_2 + 2 F_1 F_2 + Z_1 S_2 \tag{C.6}$$

or

$$S_h = S_1 + 2 F_1 F_2 + S_2. \tag{C.7}$$

Subtracting the square of (C.5), the cross-product term cancels out, and we see that there is another quantity additive under convolution:

$$[S_h - (F_h)^2] = [S_1 - (F_1)^2] + [S_2 - (F_2)^2].$$ (C.8)

Proceeding to the third moment, we find

$$T_h = T_1 Z_2 + 3S_1 F_2 + 3F_1 S_2 + Z_1 T_2,$$ (C.9)

and after some algebra (subtracting off functions of (C.5) and (C.7)), we can confirm that there is a third quantity, namely

$$T_h - 3 S_h F_h + 2 (F_h)^3$$ (C.10)

that is additive under convolution.

This generalizes at once to any number of such functions: let $h(x) \equiv f_1 * f_2 * f_3 * \cdots * f_n$. Then we have found the additive quantities

$$F_h = \sum_{i=1}^{n} F_i$$

$$S_h - F_h^2 = \sum_{i=1}^{n} (S_i - F_i^2)$$ (C.11)

$$T_h - 3S_h F_h + 2F_h^3 = \sum_{i=1}^{n} (T_i - 3S_i F_i + 2F_i^3).$$

These quantities, which 'accumulate' additively under convolution, are called the *cumulants*; we have developed them in this way to emphasize that the notion has nothing, fundamentally, to do with probability.

At this point, we define the nth cumulant as the nth moment, with 'correction terms' from lower moments, so chosen as to make the result additive under convolution. Then two questions call out for solution: (1) do such correction terms always exist; and (2) if so, how do we find a general algorithm to construct them?

To answer these questions, we need a more powerful mathematical method. Introduce the Fourier transform of $f_i(x)$:

$$\mathcal{F}_i(\alpha) \equiv \int_{-\infty}^{\infty} dx\, f_i(x) e^{i\alpha x} \qquad f_i(x) = \frac{1}{2\pi} \int_{-\infty}^{\infty} d\alpha\, \mathcal{F}_i(\alpha) e^{-i\alpha x}.$$ (C.12)

Under convolution it behaves very simply:

$$\mathcal{H}(\alpha) = \int_{-\infty}^{\infty} dx\, h(x) e^{i\alpha x} = \int dy\, f_1(y) \int dx\, e^{i\alpha x} f_2(x - y)$$

$$= \int dy\, f_1(y) \int dq\, e^{i\alpha(y+q)} f_2(q)$$ (C.13)

$$= \mathcal{F}_1(\alpha) \mathcal{F}_2(\alpha).$$

In other words, $\log \mathcal{F}(\alpha)$ is additive under convolutions. This function has some remarkable properties in connection with the notion of the 'cepstrum' discussed later. For now, examine the power series expansions of $\mathcal{F}(\alpha)$ and $\log \mathcal{F}(\alpha)$. The first is

$$\mathcal{F}(\alpha) = M_0 + M_1(i\alpha) + M_2 \frac{(i\alpha)^2}{2!} + M_3 \frac{(i\alpha)^3}{3!} + \cdots$$ (C.14)

with the coefficients

$$M_n = \frac{1}{i^n} \frac{d^n \mathcal{F}(\alpha)}{d\alpha^n}\bigg|_{\alpha=0} = \int_{-\infty}^{\infty} dx\, x^n f(x), \tag{C.15}$$

which are just the nth moments of $f(x)$; if $f(x)$ has moments up to order N, then $\mathcal{F}(\alpha)$ is differentiable N times at the origin. There is a similar expansion for $\log \mathcal{F}(\alpha)$:

$$\log \mathcal{F}(\alpha) = C_0 + C_1(i\alpha) + C_2 \frac{(i\alpha)^2}{2!} + C_3 \frac{(i\alpha)^3}{3!} + \cdots. \tag{C.16}$$

Evidently, all its coefficients

$$C_n = \frac{1}{i^n} \frac{d^n}{d\alpha^n} \log \mathcal{F}(\alpha)\bigg|_{\alpha=0} \tag{C.17}$$

are additive under convolution, and are therefore cumulants. In fact, the term 'cumulant' is often defined by this relation, rather than by the property of accumulating. The first few are

$$C_0 = \log \mathcal{F}(0) = \log \int dx\, f(x) = \log(Z), \tag{C.18}$$

$$C_1 = \frac{1}{i} \frac{\int dx\, i x f(x)}{\int dx\, f(x)} = \frac{\mathcal{F}}{Z}, \tag{C.19}$$

$$C_2 = \frac{d^2}{d(i\alpha)^2} \log \mathcal{F}(\alpha) = \frac{d}{d(i\alpha)} \frac{\int dx\, x f(x) e^{i\alpha x}}{\int dx\, f(x) e^{i\alpha x}} = \frac{\int f \int \int x^2 f - (\int xf)^2}{(\int f)^2}, \tag{C.20}$$

or

$$C_2 = \frac{S}{Z} - \left(\frac{\mathcal{F}}{Z}\right)^2, \tag{C.21}$$

which we recognize as just the cumulants found directly above; likewise, after some tedious calculation, C_3 and C_4 prove to be equal to the third and fourth cumulants (C.10). Have we then found in (C.17) all the cumulants of a function, or are there still more that cannot be found in this way? We would argue that if all the C_i exist (i.e. $f(x)$ has moments of all orders, so $\mathcal{F}(\alpha)$ is an entire function) then the C_i uniquely determine $\mathcal{F}(\alpha)$ and therefore $f(x)$, so they must include all the algebraically independent cumulants; any others must be linear functions of the C_i. But if $f(x)$ does not have moments of all orders, the answer is not obvious, and further investigation is needed.

C.1 Relation of cumulants and moments

While adhering to our convention $Z = 1$, let us go to a more general notation for the nth moment of a function:

$$M_n \equiv \int_{-\infty}^{\infty} dx\, x^n f(x) = \frac{d^n}{d(i\alpha)^n} \int dx\, f(x) e^{i\alpha x}\bigg|_{\alpha=o} = i^{-n} \mathcal{F}^{(n)}(0), \qquad n = 0, 1, 2, \ldots. \tag{C.22}$$

It is often convenient to use also the notation

$$M_n = \overline{x^n}, \tag{C.23}$$

indicating an average of x^n with respect to the function $f(x)$. We stress that these are not in general probability averages; we are indicating some general algebraic relations in which $f(x)$ need not be non-negative. For probability averages we always reserve the notation $\langle x \rangle$ or $E(x)$.

If a function $f(x)$ has moments of all orders, then its Fourier transform has a power series expansion

$$\mathcal{F}(\alpha) = \sum_{n=0}^{\infty} M_n (i\alpha)^n. \tag{C.24}$$

Evidently, the first cumulant is the same as the first moment:

$$C_1 = M_1 = \overline{x}, \tag{C.25}$$

while for the second cumulant we have $C_2 = M_2 - M_1^2$; but this is

$$C_2 = \int dx\, [x - M_1]^2 f(x) = \overline{(x - \overline{x})^2} = \overline{x^2} - \overline{x}^2, \tag{C.26}$$

the second moment of x about its mean value, called the *second central moment* of $f(x)$. Likewise, the third central moment is

$$\int dx\, (x - \overline{x})^3 f(x) = \int dx\, [x^3 - 3\overline{x}x^2 + 3\overline{x}^2 x - \overline{x}^3] f(x), \tag{C.27}$$

but this is just the third cumulant (C.11):

$$C_3 = M_3 - 3M_1 M_2 + 2M_1^3, \tag{C.28}$$

and at this point we might conjecture that all the cumulants are just the corresponding central moments. However, this turns out not to be the case: we find that the fourth central moment is

$$\overline{(x - \overline{x})^4} = M_4 - 4M_3 M_1 + 6M_2 M_1^2 - 3M_1^4, \tag{C.29}$$

but the fourth cumulant is

$$C_4 = M_4 - 4M_3 M_1 - 3M_2^2 + 12M_2 M_1^2 - 6M_1^4. \tag{C.30}$$

So they are related by

$$\overline{(x - \overline{x})^4} = C_4 + 3C_2^2. \tag{C.31}$$

Thus the fourth central moment is not a cumulant; it is not a linear function of cumulants. However, we have found it true that, for $n = 1,\ 2,\ 3,\ 4$ the moments up to order n and the cumulants up to order n uniquely determine each other; we leave it for the reader to see, from examination of the above relations, whether this is or is not true for all n.

If our functions $f(x)$ are probability densities, many useful approximations are written most efficiently in terms of the first few terms of a cumulant expansion.

C.2 Examples

What are the cumulants of a Gaussian distribution? Let

$$f(x) = \frac{1}{\sqrt{2\pi\sigma^2}} \exp\left\{\frac{[x - \mu]^2}{2\sigma^2}\right\}. \tag{C.32}$$

Then we find the Fourier transform

$$\mathcal{F}(\alpha) = \exp\{i\alpha\mu - \alpha^2\sigma^2/2\} \tag{C.33}$$

so that

$$\log \mathcal{F}(\alpha) = i\alpha\mu - \alpha^2\sigma^2/2 \tag{C.34}$$

and so

$$C_0 = 0, \qquad C_1 = \mu, \qquad C_2 = \sigma^2, \tag{C.35}$$

and all higher C_n are zero. A Gaussian distribution is characterized by the fact that is has only two nontrivial cumulants, the mean and the variance.

References

Abel, N. H. (1826), 'Untersuchung der Functionen zweier unabhängig veränderlichen Gröszen x und y, wie $f(x, y)$, welche die Eigenschaft haben, dasz $f[z, f(x, y)]$ eine symmetrische Function von z, x und y ist.', *J. Reine u. angew. Math. (Crelle's Journal)*, **1**, 11–15.
> First known instance of the associativity functional equation.

Aczél, J. (1966), *Lectures on Functional Equations and their Applications*, Academic Press, New York.

Aczél, J. (1987), *A Short Course on Functional Equations*, D. Reidel, Dordrecht.

Aitken, G. A. (1892), *The Life and Works of John Arbuthnot*, Clarendon Press, Oxford.

Akaike, H. (1980), 'The interpretation of improper prior distributions as limits of data dependent proper prior distributions', *J. Roy. Stat. Soc.* **B42**, 46–52.
> A failed attempt to deal with the marginalization paradox, which never perceives that the paradox is just as present for proper priors as for improper ones, as discussed in Chapter 15.

Andrews, D. R. & Mallows, C. L. (1974), 'Scale mixtures of normal distributions', *J. Roy. Stat. Soc.* **B 36**, 99–102.

Anscombe, F. J. (1963), 'Sequential medical trials', *JASA* **58**, 365.
> Declares that the sequential analysis of Armitage (1960) is 'a hoax'.

Arbuthnot, J. (1710), 'An argument for Divine Providence', *Phil. Trans. Roy. Soc.* **27**, pp. 186–190.
> Reprinted in Kendall & Plackett (1977), Vol. 2, pp. 30–34. First known example of rejection of a statistical hypothesis on grounds of the improbability of the data, discussed in Chapter 5. John Arbuthnot (1667–1735) was physician to Queen Anne and a prolific writer on many topics. Biographical information on Arbuthnot is given in Aitken (1892).

Armitage, P. (1960), *Sequential Medical Trials*, Thomas, Springfield, Illinois; 2nd edn, Blackwell, Oxford (1975).
> One of the main origins of the 'optional stopping' controversy. See L. J. Savage (1962) and Anscombe (1963) for extensive discussions.

Barnard, G. A. (1947), 'Significance test for 2×2 tables', *Biometrika* **34**, 123–137.

Barnard, G. A. (1983), 'Pivotal inference and the conditional view of robustness (why have we for so long managed with normality assumptions?)', in Box, G. E. P., Leonard, T. & Wu, C-F., eds. (1983), *Scientific Inference, Data Analysis, and Robustness*, Academic Press Inc., Orlando, FL.
> Expresses somewhat the same surprise at the success of the normal distribution as Augustus de Morgan did 145 years earlier. We try to explain this in Chapter 7.

Bayes, Rev. T. (1763), 'An essay toward solving a problem in the doctrine of chances', *Phil. Trans. Roy. Soc.*, pp. 370–418.
> Photographic reproduction in Molina (1963). Reprint with biographical note by G. A. Barnard in *Biometrika* **45**, 293–313 (1958) and in Pearson & Kendall (1970). The date is slightly ambiguous; this work was read before the Royal Society in 1763, but not actually published until 1764. Further biographical information on Thomas Bayes (1702–61) is given by Holland (1962). Stigler (1983) and

Zabell (1989) present evidence that Bayes may have found his result and communicated it to friends as early as 1748, in response to a challenge from David Hume (1711–76).

Bean, W. B. (1950), *Aphorisms from the Bedside Teaching and Writings of Sir William Osler, (1849–1919)*. Henry Schumann, New York.

Osler perceived the reasoning format of medical diagnosis in a form essentially identical with that given later by George Pólya. This was taken up by L. Lusted (1968) as the basis for Bayesian medical diagnosis computer programs.

Bell, E. T. (1937) *Men and Mathematics*, Dover Publications, Inc., New York.

One needs to read this collection of biographical sketches because no substitute for it seems to exist; but let the reader be aware that Eric Temple Bell was also a well-known science fiction writer (under the pseudonym of John Taine), and this talent was not lost here. We can probably trust the accuracy of the names, dates, and documentable historical facts cited. But the interpretive statements tell us very little about the matter under discussion; they tell us a great deal about the fantasies and socio-political views of the writer, and the level of his comprehension of technical facts. For example (p. 167) he *endorses*, on the grounds of 'social justice' the beheading of Lavoisier, the father of modern chemical nomenclature. He makes blatantly false accusations against Laplace, and, equally falsely, portrays Boole as a saint who could do no wrong. He displays (p. 256) a ridiculous misconception of the nature of Einstein's work, getting the sequence of facts backward, and tells us (p. 459) that Archimedes never cared for applications of mathematics!

Bellman, R. & Kalaba, R. (1956), 'On the principle of invariant imbedding and propagation through inhomogeneous media', *Proc. Natl Acad. Sci. USA* **42**, 629–632.

Bellman, R. & Kalaba, R. (1957), 'On the role of dynamic programming in statistical communication theory', *IRE Trans.* **PGIT-1**, 197.

Bellman, R., Kalaba, R. & Wing, G. M. (1957), 'On the principle of invariant imbedding and one-dimensional neutron multiplication', *Proc. Natl Acad. Sci.* **43**, 517–520.

Berger, J. O. (1985), *Statistical Decision Theory and Bayesian Analysis*, Springer-Verlag, New York.

Berger, J. O. & Wolpert, R. L. (1988), *The Likelihood Principle*, 2nd edn, Institute of Mathematical Statistics, Hayward, CA.

Bernardo, J. M. (1980), 'A Bayesian analysis of classical hypothesis testing', in Bernardo, J. M. *et al.*, eds., *Bayesian Statistics*, University Press, Valencia, Spain, pp. 606–47.

With discussion.

Bernoulli, D. (1738), 'Specimen theoriae novae de mensura sortis', in Bernoulli, D. (1730–1), *Comment Acad. Sci. Imp. Petropolitanae*, **V**, pp. 175–192; English translation by Sommer, L. (1954) *Econometrica* **22**, 23–36.

Bernoulli, D. (1777), *Mem. St Petersburg Acad., Acta Acad. Petrop.*, pp. 3–33; English translation, 'The most probable choice between several discrepant observations', *Biometrika*, **45**, 1–18 (1961).

Reprinted in Pearson & Kendall (1970), pp. 157–167. Questions taking the mean of several observations; discussed in Chapter 7.

Bernoulli, J. (1713), *Ars Conjectandi*, Thurnisiorum, Basel.

Reprinted in *Die Werke von Jakob Bernoulli*, Vol. 3, Birkhaeuser, Basel (1975), pp. 107–286. First modern limit theorem. English translation of part IV (with the limit theorem) by Bing Sung, Harvard University Dept of Statistics Technical Report #2, 1966.

Bertrand, J. L. (1889), *Calcul des Probabilités*, Gauthier-Villars, Paris; 2nd edn, 1907; reprinted (1972) by Chelsea Publishing Co., New York.

This work is usually cited only for the 'Bertrand Paradox' which appears on pp. 4–5; however, it is full of neat, concise mathematics as well as good conceptual insight, both of which are often superior to the presentations in recent works. Bertrand understood clearly how much our conclusions from given data must depend on prior information, an understanding that was lost in the later 'orthodox' literature. We quote him on this at the end of Chapter 6. However, we disapprove of his criticism of the Herschel–Maxwell derivation of the Gaussian distribution that we give in Chapter 7; what he saw as a defect is what we consider its greatest merit, and a forerunner of Einstein's reasoning. Well worth knowing and reviving today.

Birnbaum, A. (1962), 'On the foundation of statistical inference (with discussion)', *J. Am. Stat. Assoc.* **57**, 269–326.

The first proof of the 'likelihood principle' to be accepted by anti-Bayesians.

Blackman, R. B. & Tukey, J. W. (1958), *The Measurement of Power Spectra*, Dover Publications, Inc., New York.
  Periodograms mutilated by *ad hoc* smoothing, which wipes out much of the useful information in them. We warn against this in several places.

Bloomfield, P. (1976), *Fourier Analysis of Time Series: An Introduction*, J. Wiley & Sons, New York.
  Blackman–Tukey methods carried to absurd extremes (some 50 db below the noise level!) on alleged astronomical data on variable stars from Whittaker & Robinson (1924), which Bloomfield fails to recognize as faked (he sees no difficulty in the implied claim that an unidentified observatory had clear skies on 600 successive midnights and is untroubled by the fact that the authors give no reference to the source of the alleged data). As a result, his periodogram is giving zero information about variable stars; its top displays the two sine waves put into the simulated data, its bottom reveals only the spectrum of the digitizing errors. A potent demonstration of the folly of blind, unthinking application of a statistical procedure where it does not apply. A Bayesian would not be able to make such errors, because he would be obliged to think about his prior information concerning the phenomenon and the data-taking procedure.

Boltzmann, L. W. (1871), *Wiener Berichte* **63**, pp. 397–418, 679–711, 712–732.
  First appearance of '$p \log(p)$' type entropy expressions in these three articles.

Boole, G. (1854), *An Investigation of The Laws of Thought*, Macmillan, London; reprinted by Dover Publications, Inc., New York (1958).

Boole, G. (1857), 'On the application of the theory of probabilities to the question of the combination of testimonies or judgments', *Edinburgh Phil. Trans.* **v**, xxi.

Borel, E. (1909), *Élements de la Théorie des Probabilités*, Hermann et Fils, Paris.
  Discusses the Bertrand paradox at some length and conjectures the correct solution, later found by group invariance arguments.

Borel, E. (1924), 'A propos d'un traitè de probabilitiès', *Rev. Philos.* **98**, 321–336.
  Review of Keynes (1921). Reprinted in Kyburg & Smokler (1981). Borel, like Bertrand (1889), understood very well how strongly probabilities must depend on our state of prior knowledge. It is a pity that neither undertook to demonstrate the important consequences of this in realistic applications; they might have averted 50 years of false teaching by others.

Boring, E. G. (1955), 'The present status of parapsychology', *Am. Sci.*, **43**, 108–16.
  Concludes that the curious phenomenon to be studied is the behavior of parapsychologists. Points out that, having observed any fact, attempts to prove that no natural explanation of it exists are logically impossible; one canot prove a universal negative (quantum theorists who deny the existence of causal explanations please take note).

Bortkiewicz, L. V. (1898), *Das Gesetz der Kleinen Zahlen*, Teubner, Leipzig.
  Contains his famous fitting of the Poisson distribution to the number of German soldiers killed by the kick of a horse in successive years.

Bortkiewicz, L. V. (1913), *Die radioaktive Strahlung als Gegenstand Warscheinlichkeitstheoretischer Untersuchungen*, B. G. Teubner, Leipzig.

Box, G. E. P. (1962), 'On the foundations of statistical inference: discussion', *J. Am. Stat. Assoc.* **57**, (298), 311.

Box, G. E. P. & Tiao, G. C. (1973), *Bayesian Inference in Statistical Analysis*, Addison-Wesley, Reading, MA.
  G. E. P. Box is, like L. J. Savage, a curious anomaly in this field; he was an assistant to R. A. Fisher and married his daughter, but became a Bayesian in issues of inference while remaining a Fisherian in matters of significance tests, which he held to be outside the ambit of Bayesian methods. In Jaynes (1985) we argue that, on the contrary, any rational significance test *requires* the full Bayesian apparatus.

Box, J. F. (1978), *R. A. Fisher: The Life of a Scientist*, Wiley, New York.
  Joan Fisher Box, being the youngest daughter of R. A. Fisher, gives many personal anecdotes that nobody else could know, interspersed with accounts of the problems he worked on.

Bredin, J.-D. (1986), *The Affair: The Case of Alfred Dreyfus*, G. Braziller, New York.
  A famous example of the psychological divergence phenomenon discussed in Chapter 5.

Bretthorst, G. L. (1988), *Bayesian Spectrum Analysis and Parameter Estimation*, Lecture Notes in Statistics, Vol. 48, Springer-Verlag, Berlin.
  A revised and expanded version of his 1987 Ph.D. Thesis. Important new developments in detrending in econometrics, and spectrum analysis of nuclear magnetic resonance data.

Buck, B. & Macaulay, V. A., eds. (1991), *Maximum Entropy in Action*, Clarendon Press, Oxford.
  Eight lectures given at Oxford University, covering introductory notions and applications in magnetic
  resonance, spectroscopy, plasma physics, X-ray crystallography, and thermodynamics. The best source to
  date for an introduction elementary enough to be useful to beginners; yet proceeding to enough technical
  detail to be useful to practicing scientists. Be warned that what is called 'maximum entropy' is in places
  distorted by *ad hoc* devices such as 'windowing' or 'prefiltering' the data – a practice that we condemn
  as destructive of some of the information in the data. Probability theory, correctly applied, is quite capable
  of extracting all the relevant information from the raw, unmutilated data and does best, with the least total
  computation, when it is allowed to do so freely.
Cantril, H. (1950), *The 'Why' of Man's Experience*, Macmillan, New York.
Carnap, R. (1952), *The Continuum of Inductive Methods*, University of Chicago Press.
Chadwick, J. (1958), *The Decipherment of Linear B*, Cambridge University Press.
Cheeseman, P. (1988), 'An inquiry into computer understanding', *Comput. Intell.* **4**, 58–66.
  See also the following 76 pages of discussion. This attempt to explain Bayesian principles to the artificial
  intelligence community ran into incredible opposition from discussants who had no comprehension of
  what the author was doing. The situation is desribed in Jaynes (1990b).
Chernoff, H. & Moses, L. E. (1959), *Elementary Decision Theory*, J. Wiley & Sons, Inc.,
  New York.
  When first issued, this work was described as 'the only textbook on statistics that is not 20 years behind the
  times'. It is now about 40 years behind the times because the authors could not accept the notion of a
  probability that is not a frequency, and so did not appreciate the fact that a straight Bayesian approach leads
  to all the same results with an order of magnitude less formal machinery. Still, it is an interesting and
  entertaining exposition of Wald's original ideas, far easier to read than Wald (1950).
Copi, I. M. (1994), *Introduction to Logic*, 9th edn, Macmillan, New York.
Cournot, A. A. (1843), *Exposition de la Theorie des Chances et des Probabilités*, L. Hachette, Paris.
  Reprinted (1984) in *Oeuvres Complètes*, J. Vrin, Paris. One of the first of the attacks against Laplace,
  which were carried on by Ellis, Boole, Venn, E. T. Bell, and others to this day.
Cox, R. T. (1946), 'Probability, frequency, and reasonable expectation', *Am. J. Phys.* **14**, 1–13.
  In our view, this article was the most important advance in the conceptual (as opposed to the purely
  mathematical) formulation of probability theory since Laplace.
Cox, R. T. (1961), *The Algebra of Probable Inference*, Johns Hopkins University Press, Baltimore
  MD.
  An extension of Cox (1946), with additional results and more discussion. Reviewed by E. T. Jaynes, *Am.
  J. Phys.* **31**, 66 (1963).
Craig, J. (1699), *Theologiae Christianae Principia Mathematica*, Timothy Child, London.
  Reprinted with commentary by Daniel Titus, Leipzig (1755). See also Stigler (1986a) for more comments.
Cramér, H. (1946), *Mathematical Methods of Statistics*, Princeton University Press.
  This marks the heyday of supreme confidence in confidence intervals over Bayesian methods, asserted as
  usual on purely ideological grounds, taking no note of the actual results that the two methods yield.
  Comments on it are in Appendix B and Jaynes (1976, 1986a).
Crick, F. (1988), *What Mad Pursuit*, Basic Books, Inc., New York.
  A reminiscence of Crick's life and work, full of important observations and advice about the conduct of
  science in general, and fascinating technical details about his decisively important work in biology – most
  of which occurred several years after the famous Crick–Watson discovery of the DNA structure. Almost
  equally important, this is an antidote to Watson (1968; see this bibliography); we have here the other side
  of the DNA double helix story as Crick recorded it in 1974, with a different recollection of events. From
  our viewpoint, this work is valuable as a case history of important scientific discoveries made without help
  of probabilistic inference in our mathematical form, but – at least in Crick's mind – obeying its principles
  strictly, in the qualitative form given by Pólya. We wish that theoretical physicists reasoned as well.
Crow, E. L., Davis, F. A. & Maxfield, M. W. (1960), *Statistics Manual*, Dover Publications, Inc.,
  New York.
  Has many useful tables and graphs, but expounds straight orthodox methods, never thinking in terms of
  information content, and therefore never perceiving their weakness in extracting information from the data.
  We have some fun with it in Jaynes (1976).

Dawid, A. P., Stone, M. & Zidek, J. V. (1973), 'Marginalization paradoxes in Bayesian and structural inference', *J. Roy Stat. Soc.* **B35**, 189–233.
Discussed in Chapter 15.

Dawkins, R. (1987), *The Blind Watchmaker*, W. W. Norton & Co., New York.
An answer to the unceasing attacks on Darwin's theory, by religious fundamentalists who do not understand what Darwin's theory is. Richard Dawkins, Professor of Zoology at Oxford University, goes patiently into much detail to explain, as did Charles Darwin 120 years earlier, why the facts of Nature can be accounted for as the operation of natural law, with no need to invoke teleological purpose; and we agree entirely. Unfortunately, Dawkins' enthusiasm seem to outrun his logic; on the cover he claims that it also explains a very different thing: 'Why the evidence of evolution reveals a universe without design.' We do not see how any evidence could possibly do this; elementary logic warns us of the folly of trying to prove a universal negative.

Dawkins' struggle against fundamentalist religion has continued; in 1993 the Starbridge Lectureship of Theology and Natural Science was established in the Faculty of Divinity of Cambridge University. Dawkins wrote in the national press to deplore this and to stress the vacuity of theology contrasted with the value of science. This prompted the Cambridge Nobel Laureate chemist Max Perutz to issue an unperceptive rejoinder, saying: 'Science teaches us the laws of nature, but religion commands us how we should live.... Dr Dawkins does a disservice to the public perception of scientists by picturing them as the demolition squad of religious beliefs.' It appears to us that Dawkins was deploring arbitrary systems of theology, rather than ethical teachings; these are very different things. Our own views on this are given at the end of Chapter 20.

de Finetti, B. (1937), 'La prevision: ses lois logiques, ses sources subjectives', *Ann. Inst. H. Poincaré*, **7**, 1–68.
English translation: 'Prevision, its logical laws, its subjective sources', in Kyburg & Smokler (1981).

de Finetti, B. (1972), *Probability, Induction and Statistics*, John Wiley & Sons, New York.

de Finetti, B. (1974a), 'Bayesianism', *Intern. Stat. Rev.* **42**, 117–130.

de Finetti, B. (1974b), *Theory of Probability*, 2 vols., J. Wiley & Sons, Inc., New York.
Adrian Smith's English translation could not hide the wit and humor of this work. Bruno de Finetti was having great fun writing it; but he could scarcely write two sentences without injecting some parenthetic remark about a different topic, that suddenly popped into his mind, and this is followed faithfully in the translation. Full of interesting information that all serious students of the field ought to know; but impossible to summarize because of its chaotic disorganization. Discussion of any one topic may be scattered over a half-dozen different chapters without cross-references, so one may as well read the pages at random.

de Groot, M. H. (1970), *Optimal Statistical Decisions*, McGraw-Hill Book Co., New York.

de Groot, M. H. (1975), *Probability and Statistics*, Addison-Wesley Publishing Co., Reading, MA; 2nd edn. (1986).
This textbook is full of useful results, but represents an intermediate transitional phase between orthodox statistics and modern Bayesian inference. Morrie de Groot (1931–1989), a Ph.D. student of the transitional Bayesian L. J. Savage, saw clearly the technical superiority of Bayesian methods and was a regular attendant and speaker at our twice-yearly NSF-NBER Bayesian Seminars; but he still retained the terminology, notation, and general absolutist mindset of orthodoxy. Thus he still speaks of 'true probabilities' and 'estimated probabilities' as if the former had a real existence, and distinguishes sharply between 'probability theory' and 'statistical inference' as if they were different topics. This does not prevent him from obtaining the standard useful results, often by continuing the orthodox habit of inventing *ad hoc* devices instead of application of the rules of probability theory. (Our present relativist theory recognizes that there is no such thing as an 'absolute' probability, because all probabilities express, and are necessarily conditional on, the user's state of information. This makes the general principles applicable uniformly to all problems of inference, with no need for *ad hockeries*.) A biography and bibliography of Morris de Groot may be found in *Statistical Science*, vol **6**, pp. 4–14 (1991).

de Groot, M. H. & Goel, P. (1980), 'Only normal distributions have linear posterior expectations in linear regression', *J. Am. Stat. Assoc.* **75**, 895–900.
Still another connection of the kind first found by Gauss (1809) and discussed in Chapter 7.

de Moivre, A. (1718), *The Doctrine of Chances: or, A Method of Calculating the Probability of Events in Play*, W. Pearson, London; 2nd edn, Woodfall, London (1738); 3rd edn, Millar, London (1756); reprinted by Chelsea Publishing Co., New York (1967).

de Moivre, A. (1733), *Approximatio ad Summam Terminorum Binomii* $(a + b)^n$ *in Seriem expansi*;
    Photographic reproduction in Archibald, R. C., *Isis* **8**, 671–683, (1926).

de Morgan, A. (1838), *An Essay on Probabilities*, Longman & Co., London.

de Morgan, A. (1847), *Formal Logic: or the Calculus of Inference Necessary and Probable*, Taylor
    & Watton, London.
    An enthusiastic exposition of Laplace's views.

de Morgan, A. (1872), *A Budget of Paradoxes*, 2 Vols, de Morgan, S., ed., London; 2nd edn,
    Smith, D. E., ed. (1915).
    Reprinted in one volume by Dover Publications, Inc. (1954). Augustus de Morgan (1806–1871) was a
    mathematician and logician, at University College, London, from 1828 to 1866. He collected notes
    concerning not only logic, but anomalies of logic; the latter are preserved in this delightful account of the
    activities of circle-squarers, anti-Copernicans, anti-Newtonians, religious fanatics, numerologists, and
    other demented souls that abounded in 19th century England. It gives a vivid picture of the difficulties that
    serious scholars had to overcome in order to make any forward progress in science. An inexhaustible
    supply of amusing anecdotes.

de Morgan, S. (1882), *Memoir of Augustus de Morgan*, Longman, Green, London.
    Further biographical and anecdotal material on de Morgan.

Devinatz, A. (1968), *Advanced Calculus*, Holt, Rinehart and Winston, New York.

Dyson, F. J. (1958), 'Review of Lighthill (1957)', *Phys. Today* **11**, 28.

Edwards, A. W. F. (1974), 'The history of likelihood', *Int. Stat. Rev.* **42**, 9–15.

Edwards, A. W. F. (1991), *Nature* Aug. 1, p. 386.

Edwards, H. M. (1989), 'Kronecker's philosophical views', in Rowe, D. E. & McCleary, J., eds.,
    The History of Modern Mathematics, Vol. 1, pp. 67–77.

Einstein, A. (1905a), 'On the electrodynamics of moving bodies', *Ann. Phys. Leipzig* **17**,
    891–921.

Einstein, A. (1905b), 'Does the inertia of a body depend on its energy contend?', *Ann. Phys. Leipzig*
    18, 639–641.

Elias, P. (1958), 'Two famous papers', Editorial, *IEEE Trans.* **IT-4**, p. 99.

Ellis, R. L. (1842), 'On the foundations of the theory of probability', *Camb. Phil. Soc.*, vol. viii.
    Ellis was the British counterpart of Cournot, in starting the anti-Laplace movement which set scientific
    inference back a century.

Ellis, R. L. (1863), *The Mathematical and Other Writings of Robert Leslie Ellis M. A.*, Wm. Walton,
    ed., Deighton, Bell, Cambridge.

Erickson, G. J. & Smith, C. R. (1988), eds., *Maximum-Entropy and Bayesian Methods in Science
    and Engineering, Vol. 1, Foundations; Vol. 2, Applications*, Kluwer Academic Publishers,
    Dordrecht, Holland.

Euler, L. (1749), *Recherches sur la Question des Inégalitiés du Mouvement de Saturne et de Jupiter,
    Sujet Proposé pour le Prix de l'Anneé 1748 pas l'Académie Royale des Sciences de Paris*;
    reprinted in *Leonhardi Euleri, Opera Omnia*, ser. 2, Vol. 25, Turici, Basel, (1960).
    Euler gave up at the problem of estimating eight unknown parameters from 75 discrepant observations, but
    won the prize anyway.

Feinberg, S. E. & Hinkley, D. V. (1980), *R. A. Fisher: An Appreciation*, 2nd edn, Lecture Notes in
    Statistics #1, Springer-Verlag, Berlin.
    This is the second printing of the work, which appeared originally in 1979. A valuable source, if it is
    regarded as an historical document rather than an account of present statistical principles. Rich in technical
    details of Fisher's most important derivations, it gives a large bibliography of his works, including four
    books and 294 published articles. But in its adulation of Fisher it fails repeatedly to note something that
    was already well established in 1979: the simpler and unified methods of Jeffreys, which Fisher rejected
    vehemently, actually accomplished everything that Fisher's methods did, with the same or better results,
    and almost always more easily. In addition, they deal easily with technical difficulties (such as nuisance
    parameters or lack of sufficient statistics) which Fisher was never able to overcome. Thus this work tends
    also to perpetuate harmful myths.

Félix, L. (1960), *The Modern Aspect of Mathematics*, Basic Books, Inc., New York.
    A Bourbakist view; for the contrary view, see Kline (1980).

Feller, W. (1950), *An Introduction to Probability Theory and its Applications*, Vol. 1, J. Wiley & Sons, New York; 2nd edn, 1957; 3rd edn, 1968.

Feller, W. (1966), *An Introduction to Probability Theory and its Applications, Volume 2*, J. Wiley & Sons, New York.

See also the second edition, 1971.

Fine, T. L. (1973), *Theories of Probability*, Academic Press, New York.

Fisher, R. A. (1922), 'On mathematical foundations of theoretical statistics', *Phil. Trans. Roy. Soc. Lond. Ser. A*, **222**, 309–368.

Introduction of the term 'sufficient statistic'.

Fisher, R. A. (1925), *Statistical Methods for Research Workers*, Oliver & Boyd, Edinburgh.

Twelve later editions followed.

Fisher, R. A. (1930b), *The Genetical Theory of Natural Selection*, Oxford University Press; 2nd (rev) edn, Dover Publications, Inc., New York (1958).

Here Fisher shows that Mendelian genetics is not in conflict with Darwinian evolution theory, as Mendelians supposed in the early 20th century; on the contrary, the 'particulate' or 'discrete' nature of Mendelian inheritance clears up some outstanding difficulties with Darwin's theory, resulting from the assumption of blending inheritance, which most biologists – including Darwin himself – took for granted in the 1860s. Recall that Mendel's work, with its lore of dominant and recessive genes, etc., was later than Darwin's; but Darwin (1809–1882) never knew of it, and it was not generally known until after 1900. The reinterpretation of Darwin's theory in these terms, by Fisher and others, is now known as neo-Darwinism. By the time of Fisher's second (1958) edition, the existence of mutations caused by radioactivity was well established, those caused by failures of DNA replication had become highly plausible, and genetic recombination (which had been suggested by August Weismann as early as 1886) was recognized as still another mechanism to provide the individual variations on which natural selection feeds, but whose origin was puzzling to Darwin. So Fisher added many new paragraphs, in smaller type, pointing out this newer understanding and its implications; how Darwin would have enjoyed seeing these beautiful solutions to his problems! Fisher's real, permanent contributions to science are in works like this, not in his statistical teachings, which were an advance in the 1920s, but have been a retarding force since the 1939 work of Jeffreys.

Fisher, R. A. (1933), 'Probability, likelihood and quantity of information in the logic of uncertain inference', *Proc. Roy. Soc.* **146**, 1–8.

A famous attempt to demolish Jeffreys' work, which we discuss in Chapter 16.

Fisher, R. A. (1934), 'Two new properties of mathematical likelihood', *Proc. Roy. Soc. London A* **144**, 285–307.

Fisher, R. A. (1950), *Contributions to Mathematical Statistics*, W. A. Shewhart, ed., J. Wiley & Sons, Inc., New York.

A collection of his best known early papers.

Fisher, R. A. (1956), *Statistical Methods and Scientific Inference*, Oliver & Boyd, London; second revised edition, Hafner Publishing Co., New York (1959).

Fisher's final book on statistics, in which he tries to sum up his views of the logical nature of uncertain inference. One discerns a considerable shift of position from his earlier works – he even admits, occasionally, that he had been wrong before. He is now more sympathetic toward the role of prior information, saying that recognizable subsets should be taken into account and that prior ignorance is essential for the validity of fiducial estimation. He shows his old power of intuitive insight in his neat explanation of Gödel's theorem, but also some apparent lapses of memory and numerical errors. Every serious student of the subject should read this work slowly and carefully at least twice, because the depth of thinking is so great that Fisher's meaning will not be grasped fully on a single reading. Also, Fisher goes into several specialized topics that we do not discuss in the present work.

Fisher, R. A. (1962), 'Some examples of Bayes' method of the experimental determination of probability a priori', *J. Roy. Stat. Soc.* **B 24**, 118–124.

Fisher, R. A. (1974), *Collected Papers of R. A. Fisher*, J. H. Bennett, ed., University of Adelaide.

Fowler, R. H. (1929), *Statistical Mechanics; The Theory of the Properties of Matter in Equilibrium*, Cambridge University Press.

Fraser, D. A. S. (1966), 'On fiducial inference', *Ann. Math. Stat.* **32**, 661–676.

Fraser, D. A. S. (1980), 'Comments on a paper by B. Hill', in J. M. Bernardo *et al.*, eds., *Bayesian Statistics*, University Press, Valencia, Spain, pp. 56–58.

Claims to have a counter-example to the likelihood principle. But it is the same as the tumbling tetrahedrons problem solved by Bayesian methods in Chapter 15; the correct solution to that problem was not known in 1980.

Galilei, Galileo (1638), *Dialogues Concerning Two New Sciences*; English translation by Henry Crew & Alfonso de Salvio, MacMillan Company, London (1914).

Paperback reprint by Dover Publishing Co., New York, undated (*c* 1960).

Galton, F. (1886), 'Family likeness in stature', *Proc. Roy. Soc. Lond.* **40**, 42–73.

Galton, F. (1908), *Memories of My Life*, Methuen, London.

More biographical and technical details are in Pearson (1914–1930).

Gardner, M. (1957), *Fads and Fallacies in the Name of Science*, Dover Publications, Inc., New York.

A kind of 20th century sequel to de Morgan (1872), with attention directed more to fakers in science than to their colleagues in mathematics. Here we meet both the sincere but tragically misguided souls, and the deliberate frauds out to make a dishonest dollar from the gullible.

Gardner, M. (1981), *Science – Good, Bad, and Bogus*, Paperbound edition (1989), Prometheus Books, Buffalo NY; paperbound edn (1989).

A sequel to Gardner (1957), with a sobering message that everyone ought to note. Particular details on several recent trends: the Creationist who utilizes TV to carry attacks on Darwin's theory to millions, while grossly misrepresenting what Darwin's theory is; the ESP advocate who invades scientific meetings to try to invoke quantum theory in his support, although he has no comprehension of what quantum theory is; the Gee Whiz publicist who turns every tiny advance in knowledge (artificial intelligence, chaos, catastrophe theory, fractals) into a revolutionary crusader cult; the professional disaster-monger who seeks personal publicity through inventing ever more ridiculous dangers out of every activity of Man; and, most frightening of all, the eagerness with which the news media give instant support and free publicity to all this. Today, our airwaves are saturated with bogus science and medieval superstitions belittling and misrepresenting real, responsible science. In the Introduction, Gardner documents the indignant refusal of network executives to correct this, on grounds of its profitability. Then at what point does persistent, deliberate abuse of freedom of speech for profit become a clear and present danger to society? See also Rothman (1989) and Huber (1992).

Gauss, K. F. (1809), *Theoria Motus Corporum Celestium*, Perthes, Hamburg; English translation, *Theory of the Motion of the Heavenly Bodies Moving About the Sun in Conic Sections*, Dover Publications, Inc., New York (1963).

Gauss, K. F. (1823), *Theoria Combinationis Observationum Erroribus Minimis Obnoxiae*, Dieterich, Göttingen; *Supplementum* (1826).

Geisser, S. & Cornfield, J. (1963), 'Posterior distribution for multivariate normal parameters', *J. Roy. Stat. Soc.* **B25**, 368–376.

Gives the correct treatment of a problem which was later corrupted into the marginalization paradox, as explained in Chapter 15, and more fully in Jaynes (1983, pp. 337–339, 374).

Geisser, S. (1980), 'The contributions of Sir Harold Jeffreys to Bayesian inference', in Zellner, A., ed., *Bayesian Analysis in Econometrics and Statistics*, North-Holland Publishing Co., Amsterdam, pp. 13–20.

Gell-Mann, M. (1992) 'Nature conformable to Herself', *Bull. Santa Fe Inst.* **7**, 7–10.

Some comments on the relationship between mathematics and physics; this Nobel Laureate theoretical physicist is, like us, happy that the 'plague of Bourbakism' is finally disappearing, raising the hope that mathematics and theoretical physics may become once more mutually helpful partners instead of adversaries.

Gibbs, J. W. (1875), 'On the equilibrium of heterogeneous substances'; reprinted in *The Scientific Papers of J. Willard Gibbs*, Vol. I, Longmans, Green & Co. (1906), and Dover Publications, Inc., New York (1961).

Gibbs, J. W. (1902), *Elementary Principles in Statistical Mechanics*, Yale University Press, New Haven, Connecticut.

Reprinted in *The Collected Works of J. Willard Gibbs*, Vol. 2, by Longmans, Green & Co. (1928), and by Dover Publications, Inc., New York (1960).

Glymour, C. (1985), 'Independence assumptions and Bayesian updating', *Artificial Intell.* **25**, 25–99.

Gnedenko, B. V. (1962), *The Theory of Probability,* Chelsea Publ. Co., New York, pp. 40–41.

Gödel, K. (1931), 'Über formal unendscheidbare Sätze der Principia Mathematica und verwandter Systeme I', *Monatshefte für Math. & Phys.*, **38**, 173–198.
> English translation, 'On formally undecidable propositions of Principia Mathematica and related systems', Basic Books, Inc., New York (1962); reprinted by Dover Publications, Inc., New York (1992).

Goldman, S. (1953), *Information Theory*, Prentice-Hall, Inc., New York.
> We would like to put in a friendly plug for this work, even though it has a weird reputation in the field. The author, in recounting the work of Norbert Wiener and Claude Shannon, explains it for the benefit of beginners much more clearly than Wiener did, and somewhat more clearly than Shannon. Its weirdness is the result of two unfortunate accidents: (1) a misspelled word in the title of Chapter 1 escaped both the author and the publisher, providing material for dozens of cruel jokes circulating in the 1950s; (2) on p. 295 there is a photograph of Gibbs, with the caption: 'J. Willard Gibbs (1839–1903), whose ergodic hypothesis is the forerunner of fundamental ideas in information theory.' Since Gibbs never mentioned ergodicity, this is a source of more jokes. However, the author is guilty only of trusting the veracity of Wiener (1948).

Goldstein, H. (1980), *Classical Mechanics*, Addison-Wesley, Reading, MA.

Good, I. J. (1950), *Probability and the Weighing of Evidence*, C. Griffin & Co., London.
> A work whose importance is out of all proportion to its small size. Still required reading for every student of scientific inference, and can be read in one evening.

Good, I. J. (1962), *The Scientist Speculates*, Heinemann Basic Books, New York.

Good, I. J. (1967), 'The white shoe is a red herring', *Brit. J. Phil. Sci.* **17**, 322.
> Reprinted in Good (1983). Points out the error in the Hempel paradox.

Good, I. J. (1980), 'The contributions of Jeffreys to Bayesian statistics', in A. Zellner, ed., *Bayesian Analysis in Econometrics and Statistics*, North-Holland Pub. Co., Amsterdam.

Good, I. J. (1983), *Good Thinking*, University of Minnesota Press.
> Reprints of 23 articles, scattered over many topics and many years, plus a long bibliography of other works. There are about 2000 short articles like these by Good, found throughout the statistical and philosophical literature starting in 1940. Workers in the field generally granted that every idea in modern statistics can be found expressed by him in one or more of these articles; but their sheer number made it impossible to find or cite them, and most are only one or two pages long, dashed off in an hour and never developed further. So, for many years, whatever one did in Bayesian statistics, one just conceded priority to Jack Good by default, without attempting the literature search for the relevant article, which would have required days. Finally, this book provided a bibliography of most of the first 1517 of these articles (presumably in the order of their writing, which is not the order of publication) with a long index, so it is now possible to give proper acknowledgment of his works up to 1983. Be sure to read Chap. 15, where he points out specific, quantitative errors in Karl Popper's work and demonstrates that Bayesian methods, which Popper rejects, actually correct those errors.

Gould, S. J. (1989), *Wonderful Life: The Burgess Shale and the Nature of History*, W. W. Norton & Co., New York.
> A tiny region in the Canadian Rockies had exactly the right geological history so that soft-bodied animals were preserved almost perfectly. As a result, we now know that the variety of life existing in early Cambrian time was vastly greater than had been supposed; this has profound implications for our view of evolution. Gould seems fanatical in his insistence that 'evolution' is not synonomous with 'progress'. Of course, anyone familiar with the principles of physics and chemistry will agree at once that a process that proceeded in one direction can also proceed in the opposite one. Nevertheless, it seems to us that at least 99% of observed evolutionary change *has in fact* been in the direction of progress (more competent, adaptable creatures). We also think that Darwinian theory, properly stated in terms conforming to present basic knowledge and present Bayesian principles of reasoning, predicts just this.

Graunt, J. (1662), *Natural and Political Observations Made Upon the Bills of Mortality*, Roycroft, London.
> Reprinted in Newman, J. R., ed. (1956), *The World of Mathematics*, Vol. 3, pp. 1420–1435, Simon & Schuster, New York. First recognition of the useful facts that can be inferred from records of births and deaths; the beginning of sociological inference, as distinguished from the mere collection of statistics. This work is sometimes attributed instead to William Petty; for details, see Greenwood (1942).

Gray, C. G. & Gubbins, K. E. (1984), *Theory of Molecular Fluids*, Oxford University Press.

Greenwood, Major (1942), 'Medical statistics from Graunt to Farr', *Biometrika*, **32**, 203–225.
> Part 2 of a three-part work. A lengthy but confusingly disorganized account of John Graunt (1620–74), William Petty (1623–87), and Edmund Halley (1656–1742) in the matter of the first mortality tables. Petty (a friend of Graunt and one of those restless but undisciplined minds, which dabbles for a short time in practically everything but never really masters anything) attempted to make a survey of Ireland many years before Halley, but did not reason carefully enough to produce a meaningful result. Greenwood ends in utter confusion over whether Petty is or is not the real author of Graunt's book, apparently unaware that Petty's connection is that he edited the fifth (posthumous) edition of Graunt's work; and it was Petty's edition that Halley referred to and saw how to correct. All this had been explained long before, with amusing sarcasm, by Augustus de Morgan (1872, Vol. I, pp. 113–115).

Grosser, M. (1979), *The Discovery of Neptune*, Dover Publications, Inc., New York.

Haldane, J. B. S. (1932), *Proc. Camb. Phil. Soc.* **28**, 58.
> Improper priors advocated and used. Harold Jeffreys (1939) acknowledges this as the source of some of his own ideas on them.

Halley, E. (1693), 'An estimate of the degrees of mortality of mankind', *Phil. Trans. Roy. Soc.* **17**, 596–610, 654–656.
> Reprinted in Newman, J. R., ed., *The World of Mathematics*, Simon & Schuster, New York (1956), Vol. 3, pp. 1436–1447. First mortality table, based on records of births and deaths in Breslau, 1687–1691. See also Greenwood (1942).

Hamilton, A. G. (1988), *Logic for Mathematicians*, revised 2nd edn, Cambridge University Press.

Hansel, C. E. M. (1980), *ESP and Parapsychology – A Critical Re-evaluation*, Prometheus Books, Buffalo, NY, Chap. 12.

Hardy, G. H. (1911), 'Theorems connected with MacLearin's test for the convergence of series', *Proc. Lond. Math. Soc.* **9**, (2), 126–144.

Harr, A. (1933), 'Der Massbegriff in der Theorie der Kontinuierlichen Gruppen', *Ann. Math. Stat.* **34**, 147–169.

Hartigan, J. (1964), 'Invariant prior distributons', *Ann. Math. Stat.* **35**, 836–845.

Hedges, L. V. & Olkin, I. (1985), *Statistical Methods for Meta-Analysis*, Academic Press, Inc., Orlando, FL.

Helmholtz, H. von (1868), 'Ueber discontinuirliche flussigkeitsbewegungen', Monatsberichte d. Konigl. akademie der wissenschaften zu Berlin.

Hempel, C. G. (1967), *Brit. J. Phil. Sci.* **18**, 239–240.
> Reply to Good (1967).

Herschel, J. (1850), *Edinburgh Rev.* **92**, 14.
> A two-dimensional 'Maxwellian velocity distribution' before Maxwell (1860).

Hill, B. M. (1980), 'On some statistical paradoxes and nonconglomerability', in Bernardo, J. M. *et al.*, eds., *Bayesian Statistics*, University Press, Valencia.

Holland, J. D. (1962), 'The Reverend Thomas Bayes F.R.S. (1702–1761)', *J. Roy. Stat. Soc. (A)*, **125**, 451–461.

Howson, C. & Urbach, P. (1989), *Scientific Reasoning: The Bayesian Approach*, Open Court Publishing Co., La Salle, Illinois.
> A curiously outdated work, which might have served a useful purpose 60 years earlier. Mostly a rehash of all the false starts of philosophers in the past, while offering no new insight into them and ignoring the modern developments by scientists, engineers, and economists which have made them obsolete. What little positive Bayesian material there is represents a level of understanding that Harold Jeffreys had surpassed 50 years earlier, minus the mathematics needed to apply it. They persist in the pre-Jeffreys notation, which fails to indicate the prior information in a probability symbol, take no note of nuisance parameters, and solve no problems.

Howson, C. & Urbach, P. (1991), 'Bayesian reasoning in science', *Nature* **350**, 371–374.
> An advertisement for Howson and Urbach (1989), with the same shortcomings. Since they expound Bayesian principles as they existed 60 years earlier, it is appropriate that Anthony Edwards responded (Nature **352**, 386–387) with the standard counter-arguments given by his teacher, R. A. Fisher, 60 years earlier. But to those actively engaged in actually *using* Bayesian methods in the real problems of science today, this exchange seems like arguing over two different systems of epicycles.

Hoyt, W. G. (1980), *Planets X and Pluto*, University of Arizona Press, Tucson.

Huber, P. J. (1981), *Robust Statistics*, J. Wiley & Sons, Inc., New York.

Huber, P. J. (1992), *Galileo's Revenge: Junk Science in the Courtroom*, Basic Books, Inc., NY.
Documents the devastating effects now being produced by charlatans and crackpots posing as scientists.
They are paid to give 'expert' testimony that claims all sorts of weird causal relations that do not exist, in
support of lawsuits that waste billions of dollars for consumers and businesses. The phenomena of
pro-causal and anti-causal bias are discussed in Chapters 5, 16 and 17. At present we seem to have no
effective way to counteract this; as noted by Gardner (1981), the news media will always raise a great wind
of publicity, giving support and encouragement to the charlatans, while denying responsible scientists a
hearing to present the real facts. It appears that the issue of what is and what is not valid scientific inference
must soon move out of academia and become a matter of legislation – a prospect even more frightening
than the present abuses.

Hume, D. (1739), *A Treatise of Human Nature*, London; as revised by P. H. Nidditch, Clarendon
Press, Oxford (1978).

Hume, D. (1777), *An Inquiry Concerning Human Understanding*, Clarendon Press, Oxford.

Jaynes, E. T. (1956), 'Probability theory in science and engineering', no. 4 in *Colloquium Lectures
in Pure and Applied Science*, Socony-Mobil Oil Co., USA.

Jaynes, E. T. (1957a), 'Information theory and statistical mechanics', *Phys. Rev.* **106**, 620–630; **108**,
pp. 171–190.
Reprinted in Jaynes (1983).

Jaynes, E. T. (1957b), 'How does the brain do plausible reasoning?', Stanford University Microwave
Laboratory Report 421.
Reprinted in Erickson & Smith (1988), Vol. 1, pp. 1–23.

Jaynes, E. T. (1963a), 'New engineering applications of information theory', in Bogdanoff, J. L. &
Kozin, F., eds., *Engineering Uses of Random Function Theory and Probability*, J. Wiley &
Sons, Inc., NY, pp. 163–203.

Jaynes, E. T. (1963b), 'Information theory and statistical mechanics', in K. W. Ford, ed., *Statistical
Physics*, W. A. Benjamin, Inc., New York, pp. 181–218.
Reprinted in Jaynes (1983).

Jaynes, E. T. (1963c), 'Comments on an article by Ulric Neisser', *Science* **140**, 216.
An exchange of views on the interaction of men and machines.

Jaynes, E. T. (1965), 'Gibbs vs. Boltzmann entropies', *Am. J. Phys.* **33**, 391.
Reprinted in Jaynes (1983).

Jaynes, E. T. (1967), 'Foundations of probability theory and statistical mechanics', in Bunge, M.
(ed.), *Delaware Seminar in Foundations of Physics*, Springer-Verlag, Berlin.
Reprinted in Jaynes (1983).

Jaynes, E. T. (1968), 'Prior probabilities', *IEEE Trans. Systems Science and Cybernetics*, **SSC-4**,
227–241.
Reprinted in Tummala, V. M. Rao and Henshaw, R. C., eds., *Concepts and Applications of Modern
Decision Models*, Michigan State University Business Studies Series (1976); and in Jaynes (1983).

Jaynes, E. T. (1976), 'Confidence intervals vs Bayesian intervals', in W. L. Harper & C. A. Hooker,
eds., *Foundations of Probability Theory, Statistical Inference, and Statistical Theories of
Science*, vol. II, Reidel Publishing Co., Dordrecht, Holland, pp. 175–257.
Reprinted in Jaynes (1983).

Jaynes, E. T. (1978) 'Where do we stand on maximum entropy?', in Levine, R. D. and Tribus, M.,
eds., *The Maximum Entropy Formalism*, M.I.T. Press, Cambridge MA, pp. 15–118.
Reprinted in Jaynes (1983).

Jaynes, E. T. (1980), 'Marginalization and prior probabilities', in Zellner, A., ed., *Bayesian Analysis
in Econometrics and Statistics*, North-Holland Publishing Co., Amsterdam.
Reprinted in Jaynes (1983).

Jaynes, E. T. (1983), in Rosenkrantz, R. D., ed., *Papers on Probability, Statistics, and Statistical
Physics*, D. Reidel Publishing Co., Dordrecht, Holland; second paperbound edition, Kluwer
Academic Publishers (1989).
Reprints of 13 papers dated 1957–1980.

Jaynes, E. T. (1984), 'The intuitive inadequacy of classical statistics', *Epistemologia, Special Issue on Probability, Statistics, and Inductive Logic* **VII**, 43–73.
With discussion.

Jaynes, E. T. (1985), 'Highly informative priors', in Bernardo, *et al.*, eds., *Bayesian Statistics 2*, Elsevier Science Publishers, North-Holland, pp. 329–360.
With discussion. An historical survey, followed by a worked-out example (seasonal adjustment in econometrics) showing how much prior information can affect our final conclusions in a way that cannot even be stated in the language of orthodox statistical theory, because it does not admit the concept of correlations in a posterior distribution function.

Jaynes, E. T. (1986a), 'Bayesian methods: general background', in Justice, J. H., ed., *Maximum Entropy and Bayesian Methods in Geophysical Inverse Problems*, Cambridge University Press.
A general, non-technical introductory tutorial for beginners, intended to explain the terminology and viewpoint, and warn of common pitfalls of misunderstanding and communication difficulties.

Jaynes, E. T. (1986b), 'Monkeys, kangaroos, and *N*', in Justice, J. H., ed., *Maximum Entropy and Bayesian Methods in Geophysical Inverse Problems*, Cambridge University Press.
Preliminary exploration of deeper hypothesis spaces in image reconstruction, with Dirichlet priors.

Jaynes, E. T. (1986c), 'Predictive statistical mechanics', in Moore, G. T. and Scully, M. O., eds., *Frontiers of Nonequilibrium Statistical Physics*, Plenum Press, NY, pp. 33–55.

Jaynes, E. T. (1987), 'Bayesian spectrum and chirp analysis', in Smith, R. C. & Erickson, G. J., eds., *Maximum Entropy and Bayesian Spectral Analysis and Estimation Problems*, D. Reidel, Dordrecht-Holland, pp. 1–37.
A reply to Tukey (1984; see the bibliography), carried much further by Bretthorst (1988).

Jaynes, E. T. (1989), 'Clearing up mysteries – the original goal', in Skilling, J., ed., *Maximum Entropy and Bayesian Methods*, Kluwer Publishing Co., Holland, pp. 1–27.
This contains what we think is the first application of Bayes' theorem to kinetic theory, the first recognition of hidden assumptions in Bell's theorem, and the first quantitative application of the second law of thermodynamics to biology.

Jaynes, E. T. (1990a), 'Probability in quantum theory', in Zurek, W. H., ed. *Complexity, Entropy and the Physics of Information*, Addison-Wesley Pub. Co., Redwood City, CA, pp. 381–404.
Use of probability theory as logic makes the meaning of quantum theory appear very different, and hints at possible future resolution of its conceptual difficulties.

Jaynes, E. T. (1990b), 'Probability theory as logic', in Fougère, P., ed., *Proceedings of the Ninth Annual Workshop on Maximum Entropy and Bayesian Methods*, Kluwer Publishers, Holland.
Shows by a nontrivial example that conditional probabilities need not express any causal influence of the Popper type – a fact highly relevant to the hidden assumptions in the Bell theorem, discussed in Jaynes (1989).

Jaynes, E. T. (1992), 'Commentary on two articles by C. A. Los', *Computers & Math. Appl.* **24**, 267–273.
This astonishing economist condemns not only our Bayesian analysis, but virtually every useful thing ever done in data analysis, going back to Gauss.

Jefferson, T. (1781), *Notes on Virginia*; reprinted in Koch, A. & Peden, W. (eds.), *The Life and Selected Writings of Thomas Jefferson*, The Modern Library, New York (1944).
Reprinted 1972 by Random House, Inc.

Jeffrey, R. C. (1983), *The Logic of Decision*, 2nd edn, University of Chicago Press.
Attempts to modify Bayes' theorem in an *ad hoc* way; as discussed in Chapter 5, this necessarily violates one of our desiderata.

Jeffreys, H. (1931), *Scientific Inference*, Cambridge University Press; later editions, 1937, 1957, 1973.
Be sure to read his introductory section with a Galilean dialogue showing how induction is actually used in science.

Jeffreys, H. (1932), 'On the theory of errors and least squares', *Proc. Roy. Soc.* **138**, 48–55.
A beautiful derivation of the $d\sigma/\sigma$ prior expressing complete ignorance of a scale parameter, fiercely attacked by Fisher (1933) and discussed in Chapters 7 and 16.

Jeffreys, H. (1939), *Theory of Probability*, Clarendon Press, Oxford; later editions, 1948, 1961, 1967, 1988.
Appreciated in our Preface.

Jeffreys, H. & Jeffreys, Lady Bertha Swirles (1946), *Methods of Mathematical Physics*, Cambridge University Press.

Jevons, W. S. (1874), *The Principles of Science: A Treatise on Logic and Scientific Method*, 2 vols., Macmillan, London.
Reprinted by Dover Publications, Inc., NY (1958). Jevons was a student of de Morgan, and also expounds the Laplacean viewpoint; for this both Jevons and de Morgan came under attack from Venn and others in what Zabell (1989) calls 'The Great Jevonian Controversy'.

Johnson, R. W. (1985), 'Independence and Bayesian updating methods', U. S. Naval Research Laboratory Memorandum Report 5689, November 1985.

Johnson, W. E. (1924), *Logic, Part III: The Logical Foundations of Science*, Cambridge University Press; reprinted by Dover Publications, Inc., NY (1964).

Johnson, W. E. (1932), 'Probability, the deduction and induction problem', *Mind* **44**, 409–413.

Justice, J. H. (1986), ed., *Maximum Entropy and Bayesian Methods in Geophysical Inverse Problems*, Cambridge University Press.
Proceedings of the fourth annual 'Maximum Entropy Workshop', held in Calgary in August 1984.

Kac, M. (1956), 'Some stochastic problems in physics and mathematics', Field Research Laboratory. Magnolia Petroleum Co., Dallas, Colloquium Lectures in Pure and Applied Science no. 2.

Kadane, J. B., Schervish, M. J. & Seidenfeld, T. (1986), 'Statistical implications of finitely additive probability', in Goel, P. K. & Zellner, A., eds., *Bayesian Inference and Decision Techniques*, Elsevier Science Publishers, Amsterdam.
The 'KSS' work discussed at length in Chapter 15.

Kahneman, D. & Tversky, A. (1972), 'Subjective probability: a judgment of representativeness', *Cog. Psychol.* **3**, 430–454.
See also Tversky & Kahneman (1981).

Kempthorne, O. & Folks, L. (1971), *Probability, Statistics and Data Analysis*, Iowa State University Press.

Kendall, M. G. (1963), 'Ronald Aylmer Fisher, 1890–1962', *Biometrika* **50**, 1–15.
Reprinted in Pearson and Kendall (1970). Like the previous reference, this tells us more about the author than the subject.

Kendall, M. G. & Moran, P. A. P. (1963), *Geometrical Probability*, Griffin, London.
Much useful mathematical material, all of which is readily adapted to Bayesian pursuits.

Kendall, M. G. & Plackett, R. L. (1977), *Studies in the History of Statistics and Probability*, 2 vols, Griffin, London.

Kendall, M. G. & Stuart, A. (1961), *The Advanced Theory of Statistics*: *Volume 2, Inference and Relationship*, Hafner Publishing Co., New York.
This represents the beginning of the end for the confidence interval; whereas the authors continued to endorse it on grounds of 'objectivity', they noted so many resulting absurdities that readers of this work were afraid to use confidence intervals thereafter. In Jaynes (1976) we explained the source of the difficulty and showed that these absurd results are corrected automatically by use of Bayesian methods.

Kendall, M. G. & Stuart, A. (1977), *The Advanced Theory of Statistics: Volume 1, Distribution Theory*, Macmillan, New York.

Kennard, E. H. (1938), *Kinetic Theory of Gases*, McGraw-Hill Book Co., NY.

Keynes, J. M. (1921), *A Treatise on Probability*, Macmillan, London; reprinted by Harper & Row, New York (1962).
The first clear explanation of the distinction between logical independence and causal independence. Important today because it served historically as the inspiration for the work of R. T. Cox. For an interesting review of Keynes, see Borel (1924).

Keynes, J. M. (1936), *Allgemeine theorie der beschuftigung, des Zinses und des Geldes*, von Duncker & Humblot, Munchen, Leipzig.

Khinchin, A. I. (1957), *Mathematical Foundations of Information Theory*, Dover Publications, Inc., New York.

> Attempts of a mathematician to 'rigorize' Shannon's work. But we do not think it was in need of this. In any event, when one tries to work directly on infinite sets from the beginning, the resulting theorems just do not refer to anything in the real world. Khinchin was probably careful enough to avoid actual error and thus produced theorems valid in his imaginary world; but we note in Chapter 15 some of the horrors that have been produced by others who tried to do mathematics this way.

Kline, M. (1980), *Mathematics: The Loss of Certainty*, Oxford University Press.

> A fairly complete history, recalling hundreds of interesting anecdotes, but expressing views very different from the Bourbakist ones of Félix (1960).

Kolmogorov, A. N. (1933), *Grundbegriffe der Wahrscheinlichkeitsrechnung*, Ergebnisse der Math. (2), Berlin; English translation, *Foundations of the Theory of Probability*, Chelsea Publishing Co., New York (1950).

> Described in Appendix A.

Koopman, B. O. (1936), 'On distributions admitting a sufficient statistic', *Trans. Am. Math. Soc.* **39**, 399–509.

> Proof that the NASC for existence of a sufficient statistic is that the sampling distribution have the exponential form, later recognized as identical with what maximum entropy generates automatically. Simultaneous with Pitman (1936).

Kronecker, L. (1901), *Vorlesungen über Zahlentheorie*, Teubner, Leipzig; republished by Springer-Verlag, 1978.

Kullback, S. & Leibler, R. A. (1951), 'On information and sufficiency', *Ann. Math. Stat.* **22**, 79–86.

Kurtz, P. (1985), *A Skeptic's Handbook of Parapsychology*, Prometheus Books, Buffalo, NY.

> Several chapters have relevant material; see particularly Chap. 11 by Betty Markwick.

Kyburg, H. E. & Smokler, H. E. (1981) *Studies in Subjective Probability*, 2nd edn, J. Wiley & Sons, Inc., New York.

Lancaster, H. O. (1969), *The Chi-squared Distribution*, J. Wiley & Sons, Inc., New York.

Lane, D. A. (1980), 'Fisher, Jeffreys, and the nature of probability', in Fienberg, S., *et al.*, eds., *R. A. Fisher, an Appreciation*, Springer-Verlag, New York, pp. 148–160.

Laplace, P. S. (1774), 'Mémoire sur la probabilité des causes par les évènemens', *Mem. Acad. Roy. Sci.* **6**, 621–656.

> Reprinted in Laplace (1878–1912), vol. 8, pp. 27–65; English translation by Stigler (1986b).

Laplace, P. S. (1781), 'Memoire sur les probabilités', *Mem. Acad. Roy. Sci. (Paris)*.

> Reprinted in Laplace (1878–1912), vol. 9, pp. 384–485. An early exposition of the properties of the 'Gaussian' distribution. Suggests that it is so important that it should be tabulated.

Laplace, P. S. (1810), 'Mémoire sur les approximations des formules qui sont fonctions de très grands nombres et sur leur application aux probabilités,' *Mem. Acad. Sci. Paris*, 1809, pp. 353–415, 559–565.

> Reprinted in Laplace (1878–1912), vol. 12, pp. 301–353. A massive compendium of the origin, properties, and uses of the Gaussian distribution.

Laplace, P. S. (1812), *Théorie Analytique des Probabilités*, 2 vols., Courcier Imprimeur, Paris; 3rd edition with supplements, 1820.

> Reprinted in Laplace (1878–1912), Vol. 7. Reprints of this rare but very important work and the following one are available from: Editions Culture et Civilisation, 115 Ave. Gabriel Lebron, 1160 Brussels, Belgium.

Laplace, P. S. (1814, 1819), *Essai Philosophique sur les Probabilités*, Courcier Imprimeur, Paris.

> Reprinted in *Oeuvres Complètes de Laplace*, Vol. 7, Gauthier-Villars, Paris, 1886; available from Editions Culture et Civilisation, 115 Ave. Gabriel Lebron, 1160 Brussels, Belgium. English translation by F. W. Truscott & F. L. Emory, Dover Publications, Inc., New York (1951). Be warned that this 'translation' is little more than a literal *transcription*, which distorts Laplace's meaning on many points. It is essential to check the original French version before accepting any interpretive statement in this work.

Laplace, P. S. (1878–1912), *Oeuvres Complètes*, 14 vols., Gauthier-Villars, Paris.

Lee, Y. W. (1960), *Statistical Theory of Communication*, J. Wiley & Sons, New York.

> The usable but watered-down pedagogical work that grew out of Wiener (1949). Masses of well-explained examples, but none of the mathematical techniques such as the Paley–Wiener factorization or functional

integration over Wiener measure, that were used in the original. Greatly extends the folklore about Gibbs that started in Wiener (1948). Reviewed by E. T. Jaynes, *Am. J. Phys.* **29**, 276 (1961).

Lehmann, E. L. (1959), *Testing Statistical Hypotheses*, Wiley, New York; 2nd edn, 1986.

Lighthill, M. J. (1957), *Introduction to Fourier Analysis and Generalised Functions*, Cambridge University Press.

Required reading for all who have been taught to mistrust delta-functions. See the review by Dyson (1958). Lighthill and Dyson were classmates in G. H. Hardy's famous course in 'pure mathematics' at Cambridge University, at a time when Fourier analysis was mostly preoccupied with convergence theory, as in Titchmarsh (1937). Now with a redefinition of the term 'function' as explained in our Appendix B, all that becomes nearly irrelevant. Dyson states that Lighthill 'lays Hardy's work in ruins, and Hardy would have enjoyed it more than anybody'.

Lindley, D. V. (1957), 'A statistical paradox', *Biometrika* **44**, 187–192.

Mentions the Soal & Bateman (1954) parapsychology experiments.

Little, J. F. & Rubin, D. B. (1987), *Statistical Analysis with Missing Data*, J. Wiley & Sons, New York.

Missing data can wreak havoc with orthodox methods because this changes the sample space, and thus changes not only the sampling distribution of the estimator, but even its analytical form; one must go back to the beginning for each such case. But however complicated the change in the sampling distribution, the change in the likelihood function is very simple. Bayesian methods accommodate missing data effortlessly; in all cases we simply include in the likelihood function all the data we have, and Bayes' theorem automatically returns the new optimal estimator for that data set.

Luce, R. D. & Raiffa, H. (1989), *Games and Decisions*, Dover Publications, NY.

Lukasiewicz, J. (1957), *Aristotle's Syllogistic from the Standpoint of Modern Formal Logic*, 2nd edn, Clarendon Press, Oxford; reprinted 1972.

Lusted, L. (1968), *Introduction to Medical Decision Making*, C. C. Thomas, Springfield, IL.

Many useful Bayesian solutions to important medical problems, with computer source codes. Lee Lusted (1923–1994) was a classmate and fellow physics major of the writer, at Cornell College many years ago. Then we followed surprisingly common paths: Lusted went into microwave radar countermeasures at the Harvard Radio Research Laboratory, the writer into radar target identification at the Naval Research Laboratory, Anacostia, DC. After World War II, Lusted enrolled in the Harvard Medical School for an MD degree, the writer in the Princeton University Graduate school for a Ph.D. Degree in Theoretical Physics; we were both interested primarily in the reasoning processes used in those fields. Then we both discovered, independently, Bayesian analysis, saw that it was the solution to our problems, and devoted the rest of our lives to it. At essentially the same time, Arnold Zellner followed a similar course, moving from physics to economics. Thus the modern Bayesian influence in three quite different fields arose from physicists, all of nearly the same age and tastes. A sociologist has complained that 'God gave the easy problems to the physicists'. While some of us would wish to qualify that, we shall add only: ' . . . and He so arranged that the solutions physicists found would help also in solving the problems of others'.

McClave, J. T. & Benson, P. G. (1988), *Statistics for Business and Economics*, 4th edn, Dellen Publ. Co., San Francisco.

Machol, R. E., Ladany, S. P. & Morrison, D. G. (eds.) (1976), *Management Science in Sports*, Vol. 4, TIMS Studies in the Management Sciences, North-Holland, Amsterdam.

Curious applications of probability theory, leading to even more curious conclusions. Similarly, Morris (1977) analyzes the game of tennis. He defines the 'importance' of a point as the probability that the server wins the game if the point is won, minus the probability that the server wins the game if the point is lost. Comparing players of about equal ability, he finds the most important point to be 30–40. Then he concludes that *by trying a little harder on the most important points, the player can greatly enhance his prospects of victory*. It is a good exercise for the reader to criticize the logic here.

Maxwell, J. C. (1860), 'Illustration of the dynamical theory of gases. Part I. On the motion and collision of perfectly elastic spheres', *Phil. Mag.* **56**.

Middleton, D. (1960), *An Introduction to Statistical Communication Theory*, McGraw-Hill Book Co., New York.

A massive work (1140 pages) with an incredible amount of mathematical material. The title is misleading, since the material really applies to statistical inference in general. Unfortunately, most of the work was done a little too early, so the outlook is that of sampling theory and Neyman–Pearson decision rules, now made obsolete by the Wald decision theory and Bayesian advances. Nevertheless, the mathematical problems – such as methods for solving singular integral equations – are independent of one's philosophy of inference, so it has much useful material applicable in our current problems. One should browse through it, and take note of what is available here.

Middleton, D. & Van Meter, D. (1955), 'Detection and extraction of signals in noise from the point of view of statistical decision theory, *J. Soc. Ind. Appl. Math.* **3**, (4), 192.

Middleton, D. & Van Meter, D. (1956), 'Detection and extraction of signals in noise from the point of view of statistical decision theory', *J. Soc. Ind. Appl. Math.* **4**, (2), 86.

Molina, E. C. (1963), *Two Papers by Bayes with Commentaries*, Hafner Publishing Co., New York.
Contains penetrating historical remarks about the relation of Laplace and Boole, noting that those who have quoted Boole in support of their attacks on Laplace may have misread Boole's intentions.

Monod, Jacques (1970), *Le Hazard et la Nécessité*, Seuil, Paris.

Morris, C. (1977), 'The most important points in tennis', in Ladany, S. P. & Machol, R. E., eds., *Studies in Management Science* and *Systems*, vol. 5 (North-Holland, Amsterdam).

Mosteller, F. (1965), *Fifty Challenging Problems in Probability with Solutions*, Addison-Wesley, Reading, MA.

Newcomb, S. (1881), 'Note on the frequency of use of the different digits in natural numbers', *Am. J. Math.* **4**, 39–40.

Neyman, J. (1950), *First Course in Probability and Statistics*, Henry Holt & Co., New York.

Neyman, J. (1952), *Lectures and Conferences on Mathematical Statistics and Probability*, Graduate School, US Dept of Agriculture.
Contains an incredible comparison of Bayesian interval estimation vs. confidence intervals. A good homework problem is to locate the error in his reasoning.

Northrop, E. P. (1944), *Riddles in Mathematics; a Book of Paradoxes*, van Nostrand, New York, pp. 181–183.

Pearson, E. S. (1967), 'Some reflections on continuity in the development of mathematical statistics 1890–94, *Biometrika* **54**, 341–355.

Pearson E. S. & Clopper C. J. (1934), 'The use of confidence in fiducial limits illustrated in the case of the binomial', *Biometrika* **26**, 404–413.

Pearson, E. S. & Kendall, M. G. (1970), *Studies in the History of Statistics and Probability*, Hafner Publishing Co., Darien, Conn.

Pearson, K. (1914–1930), *The Life, Letters and Labours of Francis Galton*, 3 vols., Cambridge University Press.
Francis Galton had inherited a modest fortune, and on his death in 1911 he endowed the Chair of Eugenics at University College, London. Karl Pearson was its first occupant; this enabled him to give up the teaching of applied mathematics to engineers and physicists, and concentrate on biology and statistics.

Pearson, K. (1920), 'Notes on the history of correlation', *Biometrika* **13**, 25–45.
Reprinted in Pearson & Kendall (1970).

Pearson, K. (1970), 'Walter Frank Raphael Weldon 1860–1906', in Pearson, E. S. & Kendall, M. G., *Studies in the History of Statistics and Probability*, London.

Penrose, O. (1979), 'Foundations of statistical mechanics', *Rep. Prog. Phys.* **42**, 1937–2006.
Published in 'Reports of progress', although it reports no progress.

Pitman, E. J. G. (1936), 'Sufficient statistics and intrinsic accuracy', *Proc. Camb. Phil. Soc.* **32**, 567–579.
Proof, almost simultaneous with Koopman (1936), of the NASC for sufficiency, now known as the Pitman–Koopman theorem.

Poincaré, H. (1899), 'L'Oeuvre mathématique de Weierstraß', *Acta. Math.* **22**, 1–18.
Contains an authoritative account of the relationship between the works of Kronecker and Weierstrasz, pointing out that the difference was more in taste than in substance; to be contrasted with that of Bell (1937), who tries to make them mortal enemies. Discussed in Appendix B.

Poincaré, H. (1904), *Science et Hypothesis*; English translation, Dover Publications, Inc., NY (1952).
Poincaré had the gift of being able to say more in a sentence than most writers can in a page. Full of quotable remarks, as true and important today as when they were written.

Poincaré, H. (1909), *Science et Méthode*; English translation, Dover Publications, Inc., New York (1952).

Like Kline (1980), a ringing indictment of the contemporary work in mathematics and logic, for which the Bourbakists have never forgiven him. However, in knowledge and judgment, Poincaré was far ahead of his modern critics, because he was better connected to the real world.

Poincaré, H. (1912), *Calcul des Probabilités*, 2nd edn, Gauthier-Villars, Paris.

Contains the first example of the assignment of a probability distribution by the principle of group invariance.

Pólya, G. (1920), 'Über den zentralen Grenzwertsatz der Wahrscheinlichkeitsrechnung und das Momentenproblem', *Math. Zeit.* **8**, 171–181.

Reprinted in Pólya (1984), Vol. IV. First appearance of the term 'central limit theorem' in print. He does not actually prove the theorem (which he attributes to Laplace), but points out a theorem on uniform convergence of a sequence of monotonic functions which can be used to shorten various proofs of it. But our proof in Chapter 7 is even simpler.

Pólya, G. (1945), *How to Solve It*, Princeton University Press. Second paperbound edition by Doubleday Anchor Books (1957).

Pólya, G. (1954), *Mathematics and Plausible Reasoning*, 2 vols, Princeton University Press.

Pólya, G. (1984), *Collected Papers*, 4 vols., ed. G-C. Rota, MIT Press, Cambridge, MA.

Volume IV contains papers on probability theory and combinatorics, several short articles on plausible reasoning, and a bibliography of 248 papers by Pólya. George Pólya always claimed that his main interest was in the mental processes for solving particular problems rather than in generalizations. Nevertheless, some of his results launched new branches of mathematics through their generalizations by others. The present work was influenced by Pólya in more ways than noted in our Preface: most of our exposition is aimed, not at expounding generalities for their own sake, but in learning how to solve specific problems – albeit by general methods.

Pólya, G. (1987), *The Pólya Picture Album: Encounters of a Mathematician*, G. L. Alexanderson, ed., Birkhäuser, Boston.

Over his lifetime, George Pólya collected a large picture album with photographs of famous mathematicians he had known, which he took delight in showing to visitors. After his death, the collection was published in this charming book, which contains about 130 photographs with commentary by Pólya, plus a biography of Pólya by the editor.

Popper, K. (1957), 'The propensity interpretation of the calculus of probability, and the quantum theory', in *Observation and Interpretation*, S. Körner, ed., Butterworth's Scientific Publications, London, pp. 65–70.

Here, Popper, who had criticized quantum theory, summarizes his views to an audience of scientists concerned with foundations of quantum theory.

Popper, K. (1959), 'The propensity interpretation of probability', *Brit. J. Phil. Sci.* **10**, pp. 25–42.

Popper, K. (1974), 'Replies to my critics', in P. A. Schilpp, ed., *The Philosophy of Karl Popper*, Open Court Publishers, La Salle.

Presumably an authoritative statement of Popper's position, since it is some years later than his best known works, and seeks to address points of criticism directly.

Popper, K. & Miller, D. W. (1983), 'A proof of the impossibility of inductive probability', *Nature*, **302**, 687–688.

They arrive at this conclusion by a process that we examined in Chapter 5; asserting an intuitive *ad hoc* principle not contained in probability theory. Written for scientists, this is like trying to prove the impossibility of heavier-than-air flight to an assembly of professional airline pilots.

Pratt, F. (1942), *Secret and Urgent; The Story of Codes and Ciphers*, 2nd edn, Blue Ribbon Books, Garden City, NY.

Pratt, J. W. (1961), 'Review of *Testing Statistical Hypotheses*' [Lehmann, 1959], *J. Am. Stat. Assoc.* **56**, pp. 163–166.

A devastating criticism of orthodox hypothesis testing theory.

Press, W. H., Teukolsky, S. A., Vetterling, W. T., and Flannery, B. P. (1986), *Numerical Recipes, The Art of Scientific Computing*, Cambridge University Press; 2nd edn, 1992.

Quetelet, L. A. (1835), *Sur L'homme et le Développement de ses Facultés, ou Essai de Physique Sociale* (Bachelier, Paris).

Republished in 1869 under the title *Physique Sociale*.

Quetelet, L. A. (1869), *Physique Sociale, ou Essai sur le Développement des Facultés, de L'homme*, C. Muquardt, Brussels.

Raiffa, H. A. & Schlaifer, R. S. (1961), *Applied Statistical Decision Theory*, Graduate School of Business Administration, Harvard University.

Raimi, R. A. (1976), 'The first digit problem', *Am. Math. Monthly* **83**, 521–538.

Review article on 'Benford's law' with many references.

Rao, M. M. (1993), *Conditional Measures and Applications*, Marcel Dekker, Inc., New York.

Noted in Appendix A as indicating how foreign the notion of conditional probability is in the Kolmogorov system.

Reid, C. (1982), *Neyman – From Life*, Springer-Verlag, New York.

Rempe, G., Walter, H. & Klein, N. (1987), *Phys. Rev. Lett.* **58**, 353–356.

Rosenkrantz, R. D. (1977), *Inference, Method, and Decision: Towards a Bayesian Philosophy of Science*, D. Reidel Publishing Co., Boston.

Reviewed by E. T. Jaynes in *J. Am. Stat. Assoc.*, Sept. 1979, pp. 740–741.

Rothman, T. (1989), *Science à la Mode*, Princeton University Press.

Accounts of what happens when scientists lose their objectivity and jump on bandwagons. We would stress that they not only make themselves ridiculous, they do a disservice to science by promoting sensational but nonproductive ideas. For example, we think it will be realized eventually that the 'chaos' bandwagon has put a stop to the orderly development of a half-dozen different fields without enabling any new predictive ability. Because, whenever chaos exists, it is surely predicted by the Hamiltonian equations of motion – just what we have been using in statistical mechanics for a century. The chaos enthusiasts cannot make any better predictions than does present statistical mechanics, because we never have the accurate knowledge of initial conditions that would require. It has always been recognized, since the time of Maxwell and Gibbs, that if we had exact knowledge of a microstate, that would enable us in principle to predict details of future 'thermal fluctuations' at present impossible; given such information, if chaos is present, its details would be predicted just as well. But in present statistical mechanics, lacking this information, we can predict only an average over all possible chaotic behaviors consistent with the information we have; and that is just the traditional thermodynamics.

Routh, E. J. (1905), *The Elementary Part of A Treatise on the Dynamics of a System of Rigid Bodies*, Macmillan, New York.

Parts 1 and 2 of a treatise on the whole subject.

Rowe, D. E. & McCleary, J., eds. (1989), *The History of Modern Mathematics*, 2 vols., Academic Press, Inc., Boston.

Royall, R. M. & Cumberland, W. G. (1981), 'The finite-population linear regression estimator and estimators of its variance – an empirical study', *J. Am. Stat. Assoc.* **76**, 924–930.

Another demonstration of the folly of randomization.

Ruelle, D. (1991), *Chance and Chaos*, Princeton University Press.

How not to use probability theory in science; see the Comments section at the end of Chapter 4.

Savage, I. R. (1961), 'Probability inequalities of the Tchebyscheff type', *J. Res. Nat. Bureau Stand.* **65B**, 211–222.

A useful collection of results, which ought to be made more accessible.

Savage, L. J. (1954), *Foundations of Statistics*, J. Wiley & Sons, NY; 2nd edn (rev.), Dover Publications, Inc., NY (1972).

This work was attacked savagely by van Dantzig (1957).

Savage, L. J. (1961), 'The foundations of statistics reconsidered', *Proceedings of the 4th Berkeley Symposium on Mathematical Statistics and Probability*, Berkeley University Press.

Savage, L. J. (1962), *The Foundations of Statistical Inference, a Discussion*, Methuen, London.

Savage, L. J. (1981), *The Writings of Leonard Jimmie Savage – A Memorial Selection*, American Association of Statistics and the Institute of Mathematical Statistics.

Jimmie Savage died suddenly and unexpectedly in 1971, and his colleagues performed an important service by putting together this collection of his writings that were scattered in many obscure places and hard to locate. Some personal reminiscences about him are in Jaynes (1984) and Jaynes (1985).

Schrödinger, E. (1948), *Statistical Thermodynamics*, Cambridge University Press.

Schuster, A. (1897), 'On lunar and solar periodicities of earthquakes', *Proc. Roy. Soc.* **61**, 455–465.

This marks the invention of the periodogram and could almost be called the origin of orthodox significance tests. Schuster undertakes to refute some claims of periodicities in earthquakes by considering only the sampling distribution for the periodogram *under the hypothesis that no periodicity exists*! He never considers what the probability is of getting the observed data if a periodicity of a certain frequency *does* exist? Orthodoxy has been following this nonsensical procedure ever since.

Schwartz, L. (1950) *Théorie des Distributions*, 2 vols., Hermann et Cie, Paris.

Seal, H. (1967), 'The historical development of the Gauss linear model', *Biometrika* **54**, 1–24.

Reprinted in Pearson & Kendall (1970).

Shannon, C. E. (1948), 'A mathematical theory of communication', *Bell Syst. Tech. J.*, **27**, 379, 623.

Reprinted in C. E. Shannon & W. Weaver, *The Mathematical Theory of Communication*, University of Illinois Press, Urbana, 1949. See also Sloane & Wyner (1993).

Shewhart, W. A. (1931), *Economic Control of Quality of Manufactured Products*, van Nostrand, New York.

Shore, J. E. & Johnson, R. W. (1980), 'Axiomatic derivation of the principle of maximum entropy and the principle of minimum cross-entropy', *IEEE Trans. Information Theory* **IT-26**, 26–37.

Many different choices of axioms all lead to the same actual algorithm for solution of problems. The authors present a different basis from the one first proposed (Jaynes, 1957a). But we stress that maximum entropy and minimum cross-entropy are not different principles; a change of variables converts one into the other.

Simon, H. A. & Rescher, N. (1966), 'Cause and Counterfactual', *Phil. Sci.* **33**, 323–340.

Simon, H. A. (1977), *Models of Discovery*, D. Reidel Publ. Co., Dordrecht, Holland.

Sims, C. A. (1988), 'Bayesian skepticism on unit root econometrics', *J. Econ. Dyn. & Control* **12**, 463–474.

Unit root hypotheses (see Chapters 7 and 17) are not well connected to economic theory, but Bayesian analysis of such models succeeds where orthodox significance tests mislead.

Sivia, D. S. & Carlile, C. J. (1992), 'Molecular spectroscopy and Bayesian spectral analysis – how many lines are there?', *J. Chem. Phys.* **96**, 170–178.

Successful resolution of noisy data into as many as nine gaussian components by a Bayesian computer program.

Sivia, D. S. (1996), *Data Analysis – A Bayesian Tutorial*, Clarendon Press, Oxford.

This small (less than 200 pages) but much-needed book contains a wealth of worked-out numerical examples of Bayesian treatments of data, expounded from a theoretical standpoint identical to ours. It should be considered an adjunct to the present work, supplying a great deal of practical advice for the beginner, at an elementary level that will be grasped readily by every science or engineering student.

Sloane, N. J. A. & Wyner, A. D. (1993), *Claude Elwood Shannon: Collected Papers*, IEEE Press, Piscataway, NJ.

Smart, W. M. (1947), *John Couch Adams and the Discovery of Neptune*, Royal Astronomical Society, Occasional Notes No. 11.

Soal, S. G. & Bateman, F. (1954), *Modern Experiments in Telepathy*, Yale University Press, New Haven.

Stigler, S. M. (1980), 'Stigler's law of eponymy', *Trans. NY Acad. Sci.* **39**, 147–159.

Stigler, S. M. (1983), 'Who discovered Bayes's Theorem?', *Am. Stat.* **37**, 290–296.

Stigler, S. M. (1986a), 'John Craig and the probability of history', *JASA* **81**, 879–887.

Stigler, S. M. (1986b), 'Translation of Laplace's 1774 memoir on 'Probability of causes', *Stat. Sci.* **1**, 359.

Stigler, S. M. (1986c), *The History of Statistics*, Harvard University Press.

A massive work of careful scholarship, required reading for all students of the subject. Gives full discussions of many topics that we touch on only briefly.

Stone, M. (1965), 'Right Harr measure for convergence in probability to quasi-posterior distributions', *Ann. Math. Stat.* **30**, 449–453.

Stone, M. (1976), 'Strong inconsistency from uniform priors', *J. Am. Stat. Assoc.* **71**, 114–116.

A random walk in Flatland.

Stone, M. (1979), 'Review and analysis of some inconsistencies related to improper priors and finite additivity', in *Proceedings of the 6th International Congress of Logic, Methodology, and Philosophy of Science*, Hanover, 1979, North Holland Press.
The tumbling tetrahedra of Chapter 15.

Stove, D. (1982), *Popper and After: Four Modern Irrationalists*, Pergamon Press, New York.

Székely, G. J. (1986), *Paradoxes in Probability Theory and Mathematical Statistics*, D. Reidel Publishing Co., Dordrecht, Holland.
Gives, on p. 64, an erroneous solution for a biased coin; no comprehension of the physics. Compare with our Chapter 10.

Takacs, L. (1958), 'On a probability problem in the theory of counters', *Ann. Math. Stat.* **29**, 1257–1263.
There are many earlier papers by Takacs on this topic.

Tax, S., ed., (1960), *Evolution After Darwin*, 3 vols., University of Chicago Press.
Volume 1: *The Evolution of Life*; Volume 2: *The Evolution of Man*; Volume 3: *Issues in Evolution*. A collection of articles and panel discussions by many workers in the field, summarizing the state of knowledge and current research directions 100 years after Darwin's original publication.

Temple, G. (1955), 'Theory of generalized functions', *Proc. Roy. Soc. Lond. Ser. A* **228**, 175–190.

Titchmarsh, E. C. (1937), *Introduction to the Theory of Fourier Integrals*, Clarendon Press, Oxford.
The 'state of the art' in Fourier analysis just before the appearance of Lighthill (1957), which made all the lengthy convergence theory nearly irrelevant. However, only a part of this classic work is thereby made obsolete; the material on Hilbert transforms, Hermite and Bessel functions, and Wiener–Hopf integral equations, remains essential for applied mathematics.

Titchmarsh, E. C. (1939), *The Theory of Functions*, 2nd edn, Oxford University Press.
In Chapter XI the reader may see – possibly for the first time – some actual examples of nondifferentiable functions. We discuss this briefly in Appendix B.

Titterington, D. M., Smith, A. F. M. & Makov, U. E. (1985), *Statistical Analysis of Finite Mixture Distributions*, Wiley, NY.

Tribus, M. (1961), *Thermostatics and Thermodynamics; An Introduction to Energy, Information and States of Matter, with Engineering Applications*, Van Nostrand, Princeton, NJ.

Tribus, M. (1969), *Rational Descriptions, Decisions and Designs*, Pergamon Press, New York.

Tribus, M. & Fitts, G. (1968), 'The widget problem revisited', *IEEE Trans.* **SSC-4**, (3), 241–248.

Tukey, J. W. (1977), *Exploratory Data Analysis*, Addison-Wesley, Reading, MA.
Introduces the word 'resistant' as a data-oriented version of 'robust'.

Tversky, A. & Kahneman, D. (1981), 'The framing of decisions and the psychology of choice', *Science* **211**, 453–458.

Ulam, S. (1957), 'Marian Smoluchowski and the theory of probabilities in physics', *Am. J. Phys.* **25**, 475–481.

Uspensky, J. V. (1937), *Introduction to Mathematical Probability*, McGraw-Hill, New York, p. 251.

Valavanis, S. (1959), *Econometrics*, McGraw-Hill, New York.
Modern students will find this useful as a documented record of what econometrics was like under orthodox statistical teaching. The demand for unbiased estimators at all costs can lead the author to throw away practically all the information in the data; he just does not think in terms of information content.

van Dantzig, D. (1957), 'Statistical priesthood (Savage on personal probabilities)', *Statistica Neerlandica* **2**, 1–16.
Younger readers who find it difficult to understand today how Bayesians could have had to fight for their viewpoint, should read this attack on the work of Jimmie Savage. But one should realize that van Dantzig was hardly alone here; his views were the ones most commonly expressed by statisticians in the 1950s and 1960s.

Venn, John (1866), *The Logic of Chance*, MacMillan & Co., London; later edns, 1876, 1888.
Picks up where Cournot and Ellis left off in the anti-Laplace cause. Some details are given in Jaynes (1986b).

Ventris, M. (1988), 'Work notes on Minoan language research and other unedited papers', Sacconi, A., ed., Edizioni dell'Ateneo, Rome.

Ventris, M. & Chadwick, J. (1956), *Documents in Mycenaean Greek*, Cambridge University Press.

Vignaux, G. A. & Robertson, B. (1996), 'Lessons from the New Evidence Scholarship', in
   G. R. Heidbreder, ed., *Maximum Entropy and Bayesian Methods*, Proceedings of the 13th
   International Workshop, Santa Barbara, California, August 1–5, 1993, pp. 391–401, Kluwer
   Academic Publishers, Dordrecht, Holland.
   A survey of the field of Bayesian jurisprudence, with many references.
von Mises, R. (1957), *Probability, Statistics and Truth*, G. Allen & Unwin, Ltd, London.
von Mises, R. (1964), in Geiringer, H., ed., *Mathematical Theory of Probability and Statistics*,
   Academic Press, New York, pp. 160–166.
von Neumann, J. & Morgenstern, O. (1953), *Theory of Games and Economic Behavior*, 2nd edn,
   Princeton University Press.
Wald, A. (1947), *Sequential Analysis*, Wiley, New York.
   Reviewed by G. A. Barnard, *J. Am. Stat. Assoc.* **42**, 668 (1947).
Wald, A. (1950), *Statistical Decision Functions*, Wiley, New York.
   Wald's final work, in which he recognized the fundamental role of Bayesian methods and called his
   optimal methods 'Bayes strategies'.
Wason, P. C. & Johnson-Laird, P. N. (1972), *Psychology of Reasoning*, Batsford, London.
Weaver, W. (1963), *Lady Luck, the Theory of Probability*, Doubleday Anchor Books, Inc., Garden
   City, NY.
Weber, B. H., Depew, D. J. & Smith, J. D., eds. (1988), *Entropy, Information, and Evolution*, MIT
   Press, Cambridge, MA.
   A collection of 16 papers given at a symposium held in 1985. An appalling display of the state into which
   evolution theory has degenerated, due to attempts to explain it in terms of the second law of
   thermodynamics – by biologists and philosophers in total disagreement and confusion over what the
   second law is. Discussed briefly in Chapter 7.
Weyl, H. (1946), *The Classical Groups*, Princeton University Press, NJ.
Weyl, H. (1949), *Philosophy of Mathematics and Natural Science*, Helmer, O., trans., Princeton
   University Press.
Weyl, H. (1961), *The Classical Groups; Their Invariants and Representations*, Princeton University
   Press.
Whittaker, E. T. & Robinson, G. (1924), *The Calculus of Observations*, Blackie & Son, London.
   Notable because the fake 'variable star' data on p. 349 were used by Bloomfield (1976), who proceeded
   to make the authors' analysis, with absurd conclusions, the centerpiece of his textbook on spectrum
   analysis.
Whittaker, E. T. & Watson, G. N. (1927), *A Course of Modern Analysis*, 4th edn, Cambridge
   University Press.
Whyte, A. J. (1980), *The Planet Pluto*, Pergamon Press, NY.
Widder, D. V. (1941), *The Laplace Transform*, Princeton University Press, Princeton, NJ.
Wiener, N. (1948), *Cybernetics*, J. Wiley & Sons, Inc., New York.
   On p. 109, Norbert Wiener reveals himself as a closet Bayesian, although we know of no work of his that
   actually uses Bayesian methods. But his conceptual understanding of the real world was in any event too
   naïve to have succeeded. On p. 46 he gets the effect of tidal forces in the Earth–Moon system backwards
   (speeding up the Earth, slowing down the Moon). The statements about the work of Gibbs on pp. 61–62 are
   pure inventions; far from introducing or assuming ergodicity, Gibbs did not mention it at all. Today it is
   clear, from the discovery of strange attractors, chaos, etc., that almost no real system is ergodic, and in any
   event ergodicity is irrelevant to statistical mechanics because it makes no functional difference in the actual
   calculations. In perceiving this, Gibbs was here a century ahead of the understanding of others.
   Unfortunately, Wiener's statements about Gibbs were quoted faithfully by other authors such as
   S. Goldman (1953) and Y. W. Lee (1960), who were in turn quoted by others, thus creating a large and still
   growing folklore. Wiener did not bother to proofread this work, and many equations are only vague hints as
   to the appearance of the correct equation.
Wiener, N. (1949), *Extrapolation, Interpolation, and Smoothing of Stationary time Series*,
   J. Wiley & Sons, Inc., New York.
   Another masterpiece of careless and obscure writing, partially deciphered by N. Levinson in the Appendix,
   and more fully in the books of S. Goldman and Y. W. Lee.

Wigner, E. P. (1931), *Gruppentheorie und ihre Anwendung auf die Quantenmechanik der Atomspektren,* Fr. Vieweg, Braunschweig.

Wigner, E. P. (1959), *Group Theory*, Academic Press, Inc., New York.

Wing, G. M. (1962), *An Introduction To Transport Theory,* John Wiley and Sons, Inc., New York.

Woodward, P. M. (1953), *Probability and Information Theory, with Applications to Radar*, McGraw-Hill, NY.

> An interesting historical document, which shows prophetic insight into what was about to happen, but unfortunately just misses the small technical details needed to make it work.

Wrinch, D. M. & Jeffreys, H. (1919), *Phil. Mag.* **38**, 715–734.

> This was Harold Jeffreys' first publication on probability theory, concerned with modifications of the Rule of Succession. He must have liked either the result or the association, because for the rest of his life he made reference back to this paper on every possible occasion. Dorothy Wrinch was a mathematician born in Argentina, who studied at Cambridge University and later taught at Smith College in the United States, and, in the words of Jeffreys, 'became a biologist'. Her photograph may be seen in Pólya (1987), p. 85. Two later papers by Wrinch and Jeffreys on the same topic are in *Phil. Mag.* **42**, 369–390 (1921); **45**, 368–374 (1923).

Zabell, S. L. (1989), 'The Rule of Succession', *Erkenntnis*, **31**, 283–321.

> A survey of the long and tangled history of the subject, with a wealth of unexpected detail and an astonishing number of references, highly recommended. His attempt to assess the past criticisms and present status of induction represents a notable advance over Popper but still fails, in our view, to recognize how induction is used in actual scientific practice. Discussed in Chapter 9.

Zellner, A. (1971), *An Introduction to Bayesian Inference in Econometrics*, J. Wiley & Sons, Inc., New York; 2nd edn, 1987, R. E. Krieger Pub. Co., Malabar, Florida.

> In spite of the word 'econometrics" in the title, this work concerns universal principles and will be highly valuable to all scientists and engineers. It may be regarded as a sequel to Jeffreys (1939, 3rd edn), carrying on multivariate problems beyond the stage reached by him. But the notation and style are the same, concentrating on the useful analytical material instead of mathematical irrelevancies. Contains a higher level of understanding of priors for linear regression than could be found in any textbook for more than 20 years thereafter.

Zellner, A. (1988), 'Optimal information processing and Bayes' theorem', *Am. Stat.* **42**, 278–284.

> With discussion. Points to the possibility of a general variational principle that includes both maximum entropy and Bayesian algorithms as solutions. Discussed in Chapter 11.

Bibliography

Akhiezer, N. I. (1965), *The Classical Moment Problem*, Hafner, New York.

Archimedes (*c* 220 BC), in Works, T. L. Heath, ed., Cambridge University Press (1897, 1912).
Paperback reprint by Dover Publications, Inc., New York, undated (*c* 1960).

Aristotle (4th century BC), Organon.
Definition of syllogisms.

Aristotle (4th century BC) *Physics*; translation with commentary by Apostle, H. G., Indiana University Press, Bloomington (1969).

Ash, B. B. (1966), *Information Theory*, John Wiley, New York.

Bacon, F. (1620), 'Novum Organum', in Spedding. J., Ellis, R. L. & Heath, D. D., eds., *The Works of Francis Bacon*, vol. 4, Longman & Co., London (1857–1858).

Barber, N. F. & Ursell, F. (1948), 'The generation and propagation of ocean waves and swell', *Phil. Trans. Roy. Soc. Lond.*, **A240**, 527–560.
Detection of chirped signals in noise.

Barlow, E. R. and Proschan, F. (1975), *Statistical Theory of Reliability and Life Testing*, Holt, Rinehart & Winston, New York.

Barndorf-Nielsen, O. (1978), *Information and Exponential Families in Statistical Theory*, J. Wiley & Sons, New York.

Barr, A. & Feigenbaum, E., eds. (1981), *The Handbook of Artificial Intelligence*, 3 vols., Wm. Kaufman, Inc., Los Altos, CA.
Contributions from over 100 authors. Volume 1 surveys search routines, one of the few aspects of AI that could be useful in scientific inference.

Barron, A. R. (1986), 'Entropy and the central limit theorem', Ann. Prob. **14**, 336–342.

Bartholomew, D. J. (1965), 'A comparison of some Bayesian and frequentist inference', *Biometrika*, **52**, 19–35.

Benford, F. (1938), 'The law of anomalous numbers', *Proc. Am. Phil. Soc.* **78**, 551–572.
Benford is probably the one referred to mysteriously by Warren Weaver (1963), p. 270. But, unknown to them, Simon Newcomb (1881) had noticed this phenomenon long before. See Raimi (1976) for many more details and references.

Berkson, J. (1977), 'My encounter with neo-Bayesianism', *Int. Stat. Rev.* **45**, 1–9.

Berkson, J. (1980), 'Minimum chi-squared, not maximum likelihood!', *Ann. Stat.* **8**, 457–487.

Bernardo, J. M. (1977) 'Inferences about the ratio of normal means: a Bayesian approach to the Fieller-Creasy problem', in Barra, J. D., *et al.*, eds., *Recent Developments in Statistics*, North Holland Press, Amsterdam, pp. 345–350.

Bernado, J. M. (1979a), 'Reference posterior distributions for Bayesian inference', *J. Roy. Stat. Soc. B* **41**, 113–147.
With discussion.

Bernado, J. M. (1979b), 'Expected information as expected utility', *Ann. Stat.* **7**, 686–690.

Bernado, J. M., de Groot, M. H., Lindley, D. V. & Smith, A. F. M., eds. (1980), *Bayesian Statistics*, Proceedings of the First Valencia International Meeting on Bayesian Statistics, Valencia, May 28–June 2, 1979, University Press, Valencia, Spain.

Bernado, J. M., de Groot, M. H. & Lindley, D. V., eds. (1985), *Bayesian Statistics 2*, Proceedings of the Second Valencia International Meeting on Bayesian Statistics, September 6–10, 1983, Elsevier Science Publishers, New York.

Billingsley, P. (1979), *Probability and Measure*, Wiley, New York.
Contains more Borel–Kolmogorov stuff that we do not go into.

Bishop, Y., Fienberg, S., & Holland, P. (1975), *Discrete Multivariate Analysis*, MIT Press, Cambridge, MA.

Blanc-Lapierre, A. & Fortet, R. (1953), *Theorie des Fonctions Aleatoires*, Masson et Cie, Paris.

Boole, G. (1916), *Collected Logical Works, Vol. 1: Studies in Logic and Probability; Vol II: An Investigation of the Laws of Thought*; Open Court, Chicago.

Borel, E. (1926), *Traité du Calcul des Probabilités*, Gauthier-Villars, Paris.
The Hausdorff paradox on congruent sets on a sphere is discussed in Tome II, Fasc. 1.

Born, M. (1964), *Natural Philosophy of Cause and Chance*, Dover, New York.

Boscovich, Roger J. (1770), *Voyage Astronomique et Geographique*, N. M. Tillard, Paris.
Adjustment of data by the criterion that the sum of the corrections is zero, the sum of their magnitudes is made a minimum.

Box, G. E. P. (1982), 'An apology for ecumenism in statistics', NRC Technical Report #2408, Mathematics Research Center, University of Wisconsin, Madison.

Box, G. E. P., Leonard, T. & Wu, C-F, eds. (1983), *Scientific Inference, Data Analysis, and Robustness*, Academic Press, Inc., Orlando, FL.
Proceedings of a conference held in Madison, Wisconsin, November 1981.

Bracewell, R. N. (1986), 'Simulating the sunspot cycle', *Nature*, **323**, 516.
Ronald Bracewell is perhaps the first author with the courage to present a definite prediction of future sunspot activity. We await the Sun's verdict with interest.

Brewster, D. (1855), *Memoirs of the Life, Writings, and Discoveries of Sir Isaac Newton*, 2 vols., Thomas Constable, Edinburgh.

Brigham, E. & Morrow, R. E. (1967), 'The fast Fourier transform', *Proc. IEEE Spectrum* **4**, 63–70.

Brillouin, L. (1956), *Science and Information Theory*, Academic Press, New York.

Bross, I. D. J. (1963), 'Linguistic analysis of a statistical controversy', *Am. Stat.* **17**, 18.
One of the most violent polemical denunciations of Bayesian methods in print – without the slightest attempt to examine the actual results they give! Should be read by all who want to understand why and by what means the progress of inference was held up for so long. Jaynes (1976) was written originally in 1963 as a reply to Bross, in circumstances explained in Jaynes (1983), p. 149.

Brown, E. E. & Duren, B. (1986), 'Information integration for decision support', *Decision Support Syst.*, **4**, (2), 321–329.

Brown, R. (1828),'A brief account of microscopical observations', *Edinburgh New Phil. J.* **5**, 358–371.
First report of the Brownian motion.

Burg, J. P. (1967), 'Maximum entropy spectral analysis', Proceedings of the 37th Meeting of the Society of Exploration Geophysicists.

Burg, J. P. (1975), 'Maximum entropy spectral analysis', Ph.D. Thesis, Stanford University.

Busnel, R. G. & Fish, J. F., eds. (1980), *Animal Sonar Systems*, NATO ASI Series, Vol. A28, Plenum Publishing Corp., New York.
A very large (1082 pp.) report of a meeting held on the island of Jersey, UK, in 1979.

Cajori, F. (1928), in *Sir Isaac Newton 1727–1927*, Waverley Press, Baltimore, pp. 127–188.

Cajori, F. (1934), *Sir Isaac Newton's Mathematical Principles of Natural Philosophy and his System of the World*, University of California Press, Berkeley.

Carnap, R. (1950), *Logical Foundations of Probability*, Routlege and Kegan Paul Ltd., London.

Chen, Wen-chen, & de Groot, M. H. (1986), 'Optimal search for new types', in Goel, P. & Zellner, A., eds. (1986), *Bayesian Inference and Decision Techniques: Essays in Honor of Bruno de Finetti*, Elsevier Science Publishers, Amsterdam, pp. 443–458.

Childers, D., ed. (1978), *Modern Spectrum Analysis*, IEEE Press, New York.
    A collection of reprints of early works on maximum entropy spectrum analysis.
Chow, Y., Robbins, H. & Siegmund, D. (1971), *Great Expectations: Theory of Optimal Stopping*, Houghton Mifflin & Co., Boston.
Cobb, L. & Watson, B. (1980), 'Statistical catastrophe theory: an overview', *Math. Modelling*, **1**, 311–317.
    We have no quarrel with this work, but wish to add two historical footnotes. (1) Their 'stochastic differential equation' is what physicists have called a 'Fokker–Planck equation' since about 1917. However, we are used to having our statistical work attributed to Kolmogorov by mathematicians. (2) Stability considerations of multiple-valued 'folded' functions of the kind associated today with the name of René Thom are equivalent to convexity properties of a single-valued entropy function, and these were given by J. Willard Gibbs in 1873.
Cohen, T. J. & Lintz, P. R. (1974), 'Long term periodicities in the sunspot cycle', *Nature*, **250**, 398.
Cooley, J. W. & Tukey, J. W. (1965), 'An algorithm for the machine calculation of complex Fourier series', *Math. Comp.*, **19**, 297–301.
Cooley, J. W., Lewis, P. A. & Welch, P. D. (1967), 'Historical notes on the fast Fourier transform', *Proc. IEEE* **55**, 1675–1677.
Cook, A. (1994), *The Observational Foundations of Physics*, Cambridge University Press.
    Notes that physical quantities are defined in terms of the experimental arrangement used to measure them. Of course, this is just the platitude that Niels Bohr emphasized in 1927.
Cox, D. R. & Hinkley, D. V. (1974), *Theoretical Statistics*, Chapman & Hall, London; reprints 1979, 1982.
    Mostly a repetition of old sampling theory methods, in a bizarre notation that can make the simplest equation unreadable. However, it has many useful historical summaries and side remarks, noting limitations or extensions of the theory, that cannot be found elsewhere. Bayesian methods are introduced only in the penultimate Chapter 10; and then the authors proceed to repeat all the old, erroneous objections to them, showing no comprehension that these were ancient misunderstandings long since corrected by Jeffreys (1939), Savage (1954), and Lindley (1956). One prominent statistician, noting this, opined that Cox and Hinkley had 'set statistics back 25 years'.
Cox, D. R. (1970), *The Analysis of Binary Data*, Methuen, London.
Cox, R. T. (1978), 'Of inference and inquiry', in Levine, R. D. & Tribus, M., eds., *The Maximum Entropy Formalism*, MIT Press, Cambridge, MA, pp. 119–167.
    Notes that, corresponding to the logic of propositions, there is a dual logic of questions. This could become very important with further development, as discussed further in Jaynes (1983), pp. 382–388.
Cozzolino, J. M. & Zahner, M. J. (1973), 'The maximum-entropy distribution of the future market price of a stock', *Operations Res.*, **21**, 1200–1211.
Creasy, M. A. (1954), 'Limits for the ratio of means', *J. Roy. Stat. Soc.* **B 16**, 175–185.
Csiszar, I. (1984), 'Sanov property, generalized I-projection and a conditional limit theorem', *Ann. Prob.*, **12**, 768–793.
Czuber, E. (1908), *Wahrscheinlichkeitsrechnung und Ihre Anwendung auf Fehlerausgleichung*, 2 vols., Teubner, Berlin.
    Some of Wolf's famous dice data may be found here. In the period roughly 1850–1890, Wolf, a Zurich astronomer, conducted and reported a mass of 'random' experiments; an account of these is given here.
Daganzo, C. (1977), *Multinomial Probit: The Theory and its Application to Demand Forecasting*, Academic Press, New York.
Dale, A. I. (1982), 'Bayes or Laplace? An examination of the origin and early applications of Bayes' theorem', *Arch. Hist. Exact Sci.* **27**, 23–47.
Daniel, C. & Wood, F. S. (1971), *Fitting Equations to Data*, John Wiley, New York.
Daniell, G. J. & Potton, J. A. (1989), 'Liquid structure factor determination by neutron scattering – some dangers of maximum entropy', in Skilling, J., ed. (1989), *Maximum Entropy and Bayesian Methods*, Proceedings of the Eighth Maximum Entropy Workshop, Cambridge, UK, August 1988, Kluwer Academic Publishers, Dordrecht, pp. 151–162.
    The 'danger' here is that a beginner's first attempt to use maximum entropy on a complex problem may be unsatisfactory because it is answering a different question than what the user had in mind. So the first effort is really a 'training exercise' which makes one aware of how to formulate the problem properly.

Davenport, W. S. & Root, W. L. (1958), *Random Signals and Noise*, McGraw-Hill, New York.

David, F. N. (1962), *Games, Gods and Gambling*, Griffin, London.
> A history of the earliest beginnings of probability theory. Notes that in archaeology, 'the farther back one goes, the more fragmentary is the evidence'. Just the kind of deep insight that we could find nowhere else.

de Finetti, B. (1958), 'Foundations of probability', in *Philosophy in the Mid-century*, La Nuova Italia Editrice, Florence, pp. 140–147.

de Groot, M. H., Bayarri, M. J. & Kadane, J. B. (1988), 'What is the likelihood function?' (with discussion), in Gupta, S. S. & Berger, J. O., eds., *Statistical Decision Theory and Related Topics IV*, Springer-Verlag, New York.

de Groot, M. H. & Cyert, R. M. (1987), *Bayesian Analysis and Uncertainty in Economic Theory*, Chapman & Hall, London.

de Groot, M. H., Fienberg, S. E. & Kadane, J. B. (1986), *Statistics and the Law*, John Wiley, New York.

Deming, W. E. (1943), *Statistical Adjustment of Data*, John Wiley, New York.

Dempster, A. P. (1963), 'On a paradox concerning inference about a covariance matrix', *Ann. Math. Stat.* **34**, 1414–1418.

Dubois, D. & Prade, H. (1988), *Possibility Theory*, Plenum Publ. Co., New York.

Dunnington, G. W. (1955), *Carl Friedrich Gauss, Titan of Science*, Hafner, New York.

Dutta, M. (1966), 'On maximum entropy estimation', *Sankhya*, ser. A, **28**, (4), 319–328.

Dyson, F. J. (1979), *Disturbing the Universe*, Harper & Row, New York.
> A collection of personal reminiscences and speculations extending over some 50 years: 90% of it is irrelevant to our present purpose; but one must persist here, because Freeman Dyson played a very important part in the development of theoretical physics in the mid-20th century. His reminiscences about this are uniquely valuable, but are unfortunately scattered in small pieces over several chapters. Unlike some of his less thoughtful colleagues, Dyson saw correctly many fundamental things about probability theory and quantum theory (but in our view missed some others equally fundamental). Reading this work is rather like reading Kepler and trying to extract the tiny nuggets of important truth.

Eddington, Sir A. (1935), *The Nature of the Physical World*, Dent, London.
> Another distinguished scientist who thinks as we do about probability.

Edwards, A. W. F. (1972), *Likelihood*, Cambridge University Press.
> Anthony Edwards was the last student of R. A. Fisher; although he understands all the technical facts pertaining to Bayesian methods as well as anybody, some mental block prevents him, as it did Fisher, from accepting their obvious consequences. So we must, sadly, part company and proceed with the constructive development of inference without him.

Edwards, A. W. F. (1992), *Nature* **352**, 386–387.
> Commentary on Bayesian methods.

Edwards, H. M. (1987), 'An appreciation of Kronecker', *Math. Intelligencer*, **9**, 28–35.

Edwards, H. M. (1988), 'Kronecker's place in history', in Aspray, W. & Kitcher, P., eds., *History and Philosophy of Modern Mathematics*, University of Minnesota Press.

Efron, B. (1975), 'Biased versus unbiased estimation', *Adv. Math.* **16**, 259–277.

Efron, B. (1978), 'Controversies in the foundations of statistics', *Am. Math. Monthly* **85**, 231–246.

Efron, B. (1979a), 'Bootstrap methods: another look at the jackknife', *Ann. Stat.* **6**, 1–26.

Efron, B. (1979b), 'Computers and the theory of statistics: thinking the unthinkable', *SIAM Rev.* **21**, 460–480.

Efron, B. & Gong, G. (1983), 'A leisurely look at the bootstrap, the jackknife, and cross-validation', *Am. Stat.* **37**, 36–48.
> Orthodox statisticians have continued trying to deal with problems of inference by inventing arbitrary *ad hoc* procedures instead of applying probability theory. Three recent examples are explained and advocated here. Of course, they all violate our desiderata of rationality and consistency; the reader will find it interesting and instructive to demonstrate this and compare their results with those of the Bayesian alternatives.

Evans, M. (1969), *Macroeconomic Forecasting*, Harper & Row, New York.

Fechner, G. J. (1860), *Elemente der Psychophysik*, 2 vols.; Vol. 1 translated as *Elements of Psychophysics*, Boring, E. G. & Howes, D. H., eds., Holt, Rinehart & Winston, New York (1966).

Fechner, G. J. (1882), *Revision der Hauptpuncte der Psychophysik*, Breitkopf u. Härtel, Leipzig.

Feinstein, A. (1958), *Foundations of Information Theory*, McGraw-Hill, New York.
> Like the work of Khinchin (1957), a mathematician's view of things, which has almost nothing in common with the physically oriented view of Goldman (1953).

Ferguson, T. S. (1982), 'An inconsistent maximum likelihood estimate', *J. Am. Stat. Assoc.* **77**, 831–834.

Fieller, E. C. (1954), 'Some problems in interval estimation', *J. Roy. Stat. Soc.* **B 16**, 175–185.
> This and the contiguous paper by Creasy (1954; this bibliography) became famous as 'the Fieller–Creasy problem' of estimating the ratio μ_1/μ_2 of means of two normal sampling distributions. It generated a vast amount of discussion and controversy because orthodox methods had no principles for dealing with it – and for decades nobody would deign to examine the Bayesian solution. It is a prime example of an estimation problem, easily stated, for which only Bayesian methods provide the technical apparatus required to solve it. It is finally considered from a Bayesian standpoint by José Bernardo (1977).

Fisher, R. A. (1912), 'On an absolute criterion for fitting frequency curves', *Messeng. Math.* **41**, 155–160.

Fisher, R. A. (1915), 'Frequency distribution of the values of the correlation coefficient in samples from an indefinitely large population', *Biometrika* **10**, 507–521.

Fisher, R. A. (1930), 'Inverse probabilities', *Proc. Camb. Phil. Soc.* **26**, 528–535.

Fisher, R. A. (1935), *The Design of Experiments*, Oliver & Boyd, Edinburgh; six later editions to 1966.

Fisher, R. A. (1938), *Statistical Tables for Biological, Agricultural and Medical Research* (with F. Yates), Oliver & Boyd, Edinburgh; five later editions to 1963.

Fisher, R. A. & Tippett, L. H. C. (1928), 'Limiting forms of the frequency distribution of the largest or smallest member of a sample', *Proc. Camb. Phil. Soc.* **24**, 180–190.

Fougeré, P. F. (1977), 'A solution to the problem of spontaneous line splitting in maximum entropy power spectrum analysis', *J. Geophys. Res.* **82**, 1051–1054.

Galton, F. (1863), *Meteorographica*, MacMillan, London.
> Here this remarkable man invents weather maps and, from studying them, discovers the 'anticyclone' circulation patterns in the northern hemisphere.

Galton, F. (1889), *Natural Inheritance*, MacMillan, London.

Gentleman, W. M. (1968), 'Matrix multiplication and fast Fourier transformations', *Bell Syst. Tech. J.*, **17**, 1099–1103.

Gillispie, C. C., ed. (1981), *Dictionary of Scientific Biography*, 16 vols., C. Scribner's Sons, New York.
> The first place to look for information on any scientist.

Glymour, C. (1980), *Theory and Evidence*, Princeton University Press.

Gnedenko, B. V. & Kolmogorov, A. N. (1954), *Limit Distributions for Sums of Independent Random Variables*, Addison-Wesley, Cambridge, MA.
> On p. 1 we find the curious statement: 'In fact, all epistomologic value of the theory of probability is based on this: that large-scale random phenomena in their collective action create strict, non-random regularity.' This was thought by some to serve a political purpose in the old USSR; in any event, the most valuable applications of probability theory today are concerned with incomplete information and have nothing to do with those so-called 'random phenomena' which are still undefined in theory and unidentified in Nature.

Goel, P. & Zellner, A. (1986), eds., *Bayesian Inference and Decision Techniques: Essays in Honor of Bruno de Finetti*, Elsevier Science Publishers, Amsterdam.

Gokhale, D. and Kullback, S. (1978), *The Information in Contingency Tables*, Marcel Dekker, New York.

Goldberg, S. (1983), *Probability in Social Science*, Birkhaeuser, Basel.

Good, I. J. (1965), *The Estimation of Probabilities*, Research Monographs #30, MIT Press, Cambridge, MA.
> Jack Good persisted in believing in the existence of 'physical probabilities' that have some kind of reality independently of human information; hence the (to us) incongruous title.

Grandy, W. T. & Schick, L. H., eds. (1991), *Maximum Entropy and Bayesian Methods*, Proceedings of the Tenth annual Maximum Entropy workshop, Kluwer Academic Publishers, Holland.

Grenander, U. & Szegö, G. (1957), *Toeplitz Forms and their Applications*, University of California Press, Berkeley.

Griffin, D. R. (1958), *Listening in the Dark*, Yale University Press, New Haven.
   See also Slaughter, R. H. & Walton, D. W., eds. (1970), *About Bats*, SMU Press, Dallas, Texas.

Gull, S. F. & Daniell, G. J. (1978), 'Image reconstruction from incomplete and noisy data', *Nature* **272**, 686.

Gull, S. F. & Daniell, G. J. (1980), 'The maximum entropy algorithm applied to image enhancement', *Proc. IEEE (E)* **5**, 170.

Gull, S. F. & Skilling, J. (1984), 'The maximum entropy method', in Roberts, J. A., ed., *Indirect Imaging*, Cambridge University Press.

Hacking, I. (1965), *Logic of Statistical Inference*, Cambridge University Press.

Hacking, I. (1973), *The Emergence of Probability*, Cambridge University Press.

Hacking, I. (1984), 'Historical models for justice', *Epistemologia, Special Issue on Probability, Statistics, and Inductive Logic*, **VII**, 191–212.

Haldane, J. B. S. (1957), 'Karl Pearson, 1857–1957', *Biometrika* **44**, 303–313.
   Haldane's writings, whatever the ostensible topic, often turned into political indoctrination for socialism. In this case it made some sense, since Karl Pearson was himself a political radical. Haldane suggests that he may have changed the spelling of his name from 'Carl' to 'Karl' in honor of Karl Marx, and from this Centenary oration we learn that V. I. Lenin quoted approvingly from Karl Pearson. Haldane was Professor of Genetics at University College, London in the 1930s, but he resigned and moved to India as a protest at the failure of the authorities to provide the financial support he felt his Department needed. It is easy to imagine that this was precisely what those authorities, exasperated at his preoccupation with left-wing politics instead of genetics, hoped to bring about. An interesting coincidence is that Haldane's sister, Naomi Haldane Mitchison, married a Labour MP and carried on the left-wing cause. James D. Watson was a guest at her home at Christmas 1951, about a year before discovering the DNA helix structure. He was so charmed by the experience that his 1968 book, *The Double Helix*, is inscribed: 'For Naomi Mitchison.'

Hampel, F. R. (1973), 'Robust estimation: a condensed partial survey', *Zeit. Wahrsch, theorie vrw. Beb.* **27**, 87–104

Hankins, T. L. (1970), *Jean d'Alembert: Science and the Enlightenment*, Oxford University Press.

Heath, D. & Sudderth, W. (1976), 'de Finetti's theorem on exchangeable variables', *Am. Stat.* **30**, 188.
   An extremely simple derivation.

Helliwell, R. A. (1965), *Whistlers and Related Ionospheric Phenomena*, Stanford University Press, Palo Alto, CA.

Hellman, M. E. (1979), 'The mathematics of public-key cryptography', *Sci. Am.* **241**, 130–139.

Hewitt, E. & Savage, L. J. (1955), 'Symmetric measures on Cartesian products', *Trans. Am. Math. Soc.* **80**, 470–501.
   A generalization of de Finetti's representation theorem to arbitrary sets.

Hirst, F. W. (1926), *Life and Letters of Thomas Jefferson*, Macmillan, New York.

Hobson, A. & Cheung, B. K. (1973), 'A comparison of the Shannon and Kullback information measures', *J. Stat. Phys.* **7**, 301–310.

Hodges, J. L. & Lehmann, E. L. (1956), 'The efficiency of some nonparametric competitors of the *t*-test', *Ann. Math. Stat.* **27**, 324–335.

Hofstadter, D. R. (1983), 'Computer tournaments of the Prisoner's dilemma suggest how cooperation evolves', *Sci. Am.* **248**, (5), 16–26.

Holbrook, J. A. R. (1981), 'Stochastic independence and space-filling curves', *Am. Math. Monthly* **88**, 426–432.

Jagers, P. (1975), *Branching Processes with Biological Applications*, John Wiley, London.

James, W. & Stein, C. (1961), 'Estimation with quadratic loss', Proc. 4th Berkeley Symp., Univ. Calif. Press, **1**, 361–380.

Jansson, P. A., ed. (1984), *Deconvolution, with Applications in Spectroscopy*, Academic Press, Orlando, FL.
   Articles by nine authors, summarizing the state of the art (mostly linear processing) as it existed just before the introduction of Bayesian and maximum entropy methods.

Jaynes, E. T. (1963), 'Review of *Noise and Fluctuations*, by D. K. C. MacDonald', *Am. J. Phys.* **31**, 946.
> Cited in Jaynes (1976) in response to a charge by Oscar Kempthorne that physicists have paid little attention to noise; notes that there is no area of physics in which the phenomenon of noise does not present itself. As a result, physicists were actively studying noise and knew the proper way to deal with it, long before there was any such thing as a statistician.

Jaynes, E. T. (1973a), 'Survey of the present status of neoclassical radiation theory', in Mandel, L. & Wolf, E., eds., *Proceedings of the 1972 Rochester Conference on Optical Coherence*, Pergamon Press, New York.

Jaynes, E. T. (1973b), 'The well-posed problem', *Found. Phys.* **3**, 477–493.
> Reprinted in Jaynes (1983).

Jaynes, E. T. (1980a), 'The minimum entropy production principle', *Ann. Rev. Phys. Chem.* **31**, 579–601.
> Reprinted in Jaynes (1983).

Jaynes, E. T. (1980b), 'What is the question?', in Bernardo, J. M., de Groot, M. H., Lindley, D. V. & Smith, A. F. M., eds., *Bayesian Statistics*, University Press, Valencia, Spain, pp. 618–629.
> Discussion of the logic of questions, as pointed out by R. T. Cox (1978), and applied to the relationship between parameter estimation and hypothesis testing. Reprinted in Jaynes (1983), pp. 382–388.

Jaynes, E. T. (1981), 'What is the problem?', in Haykin, S., ed., *Proceedings of the Second SSSP Workshop on Spectrum Analysis*, McMaster University.
> The following article is an enlarged version.

Jaynes, E. T. (1982), 'On the rationale of maximum-entropy methods', *Proc. IEEE* **70**, 939–952.

Jaynes, E. T. (1984), 'Prior information and ambiguity in inverse problems', in *SIAM-AMS Proceedings*, Vol. 14, American Mathematical Society, pp. 151–166.

Jaynes, E. T. (1985a) 'Where do we go from here?', in Smith, C. & Grandy, W. T., eds. *Maximum-Entropy and Bayesian Methods in Inverse Problems*, D. Reidel Publishing Co., Dordrecht, pp. 21–58.

Jaynes, E. T. (1985b), 'Entropy and search theory', in Smith, C. & Grandy, W. T., eds. *Maximum-Entropy and Bayesian Methods in Inverse Problems*, D. Reidel Publishing Co., Dordrecht, pp. 443–454.
> Shows that the failure of previous efforts to find a connection between information theory and search theory were due to use of the wrong entropy expression. In fact, there is a very simple and general connection, as soon as we define entropy on the deepest hypothesis space.

Jaynes, E. T. (1985c) 'Macroscopic prediction', in Haken, H., ed., *Complex Systems – Operational Approaches*, Springer-Verlag, Berlin.

Jaynes, E. T. (1985d), 'Generalized scattering', in Smith, C. & Grandy, W. T., eds., *Maximum-Entropy and Bayesian Methods in Inverse Problems*, D. Reidel Publishing Co., Dordrecht, pp. 377–398.
> Some of the remarkable physical predictions contained in the comparison of two maximum entropy distributions, before and after adding a new constraint.

Jaynes, E. T. (1986), 'Some applications and extensions of the de Finetti representation theorem', in Goel, P. & Zellner, A., eds., *Bayesian Inference and Decision Techniques: Essays in Honor of Bruno de Finetti*, Elsevier Science Publishers, Amsterdam, pp. 31–42.
> The theorem, commonly held to apply only to infinite exchangeable sequences, remains valid for finite ones if one drops the non-negativity condition on the generating function. This makes it applicable to a much wider class of problems.

Jaynes, E. T. (1988a), 'The relation of Bayesian and maximum entropy methods', in Erickson, G. J. & Smith, C. R., eds. (1988), *Maximum-Entropy and Bayesian Methods in Science and Engineering*, Vol. 1, pp. 25–29.

Jaynes, E. T. (1988b), 'Detection of extra-solar system planets', in Erickson, G. J. & Smith, C. R., eds. (1988), *Maximum-Entropy and Bayesian Methods in Science and Engineering*, Vol. 1, pp. 147–160.

Jaynes, E. T. (1991), 'Notes on present status and future prospects', in Grandy, W. T. & Schick,
    L. H., eds. (1991), *Maximum* Entropy and Bayesian Methods, Proceedings of the Tenth annual
    Maximum Entropy Workshop, Kluwer Academic Publishers, Holland.
    A general summing-up of the situation as it appeared in the summer of 1990.
Jaynes, E. T. (1993), 'A backward look to the future', in Grandy, W. T. & Milonni, P. W., eds.,
    Physics and Probability: Essays in Honor of Edwin T. Jaynes, Cambridge University Press,
    pp. 261–275.
    A response to the contributors to this *Festschrift* volume marking the writer's 70th birthday, with 22 articles
    by my former students and colleagues.
Jefferys, W. H. (1990), 'Bayesian analysis of random event generator data', *J. Sci. Expl.* **4**, 153–169.
    Shows that orthodox significance tests can grossly overestimate the significance of ESP data; Bayesian
    tests yield defensible conclusions because they do not depend on the intentions of the investigator.
Jeffreys, H. (1963), 'Review of Savage (1962)', *Technometrics* **5**, 407–410.
Jeffreys, Lady Bertha Swirles (1992), 'Harold Jeffreys from 1891 to 1940', *Notes Rec. Roy. Soc.
    Lond.* **46**, 301–308.
    A short, and puzzlingly incomplete, account of the early life of Sir Harold Jeffreys, with a photograph of
    him in his 30s. Detailed account of his interest in botany and early honors (he entered St John's College,
    Cambridge as an undergraduate, in 1910; and that same year received the Adams memorial prize for an
    essay on 'Precession and nutation'). But, astonishingly, there is no mention at all of his work in probability
    theory! In the period 1919–1939, this resulted in many published articles and two books (Jeffreys, 1931,
    1939) of very great importance to scientists today. It is, furthermore, of *fundamental* importance and will
    remain so long after all his other work recedes into history. Bertha Swirles Jeffreys was also a physicist,
    who studied with Max Born in Göttingen in the late 1920s and later became Mistress of Girton College,
    Cambridge.
Jerri, A. J. (1977), 'The Shannon sampling theorem – its various extensions and applications', *Proc.
    IEEE* **65**, 1565–1596.
    A massive tutorial collection of useful formulas, with 248 references.
Johnson, R. W. (1979), 'Axiomatic characterization of the directed divergences and their linear
    combinations', *IEEE Trans.* **IT-7**, 641–650.
Kale, B. K. (1970), 'Inadmissibility of the maximum likelihood estimation in the presence of prior
    information', *Can. Math. Bull.* **13**, 391–393.
Kalman, R. E. (1982), 'Identification from real data', in Hazewinkel, M. & Rinnooy Kan, A., eds.,
    Current Developments in the Interface: Economics, Econometrics, Mathematics, D. Reidel
    Publishing Co., Dordrecht-Holland, pp. 161–196.
Kalman, R. E. (1990), *Nine Lectures on Identification*, Lecture Notes on Economics and
    Mathematical Systems, Springer-Verlag.
Kandel, A. (1986), *Fuzzy Mathematical Techniques with Applications*, Addison-Wesley, Reading,
    MA.
Kay, S. & Marple, S. L., Jr (1979), 'Source of and remedies for spectral line splitting in
    autoregressive spectrum analysis', Proceedings of the 1979 IEEE International Conference on
    Acoustics, Speech Signal Processing, October 1978, pp. 469–471.
Kemeny, J. G. & Snell, J. L. (1960), *Finite Markov Chains*, D. van Nostrand Co., Princeton, NJ.
Kendall, M. G. (1956), 'The beginnings of a probability calculus', *Biometrika* **43**, 1–14; reprinted
    in Pearson & Kendall (1970).
    A fascinating psychological study. In the attempt to interpret the slow early development of probability
    theory as caused by the unfounded prejudices of others, he reveals inadvertently his own unfounded
    prejudices, which in our view are the major cause of retarded – even backward – progress in the 20th
    century.
Khinchin, A. I. (1949) *Mathematical Foundations of Statistical Mechanics*, Dover Publications,
    Inc., New York.
    An attempt to base the calculational techniques on the central limit theorem, not general enough for
    problems of current interest. But the treatment of the Laplace transform relation between structure
    functions and partition functions is still valuable reading today, and forms the mathematical basis for our
    own development in Part 2.

Kiefer, J. & Wolfowitz, J. (1956), 'Consistency of the maximum likelihood estimation in the presence of infinitely many incidental parameters', *Ann. Math. Stat.* **27**, 887–906.

Kindermann, R. & Snall, J. L. (1980), *Markov Random Fields*, Contemporary Mathematics Vol. 1, AMS, Providence, RI.

Kuhn, T. S. (1962), *The Structure of Scientific Revolutions*, University of Chicago Press; 2nd edn, 1970.

Kullback, S. (1959), *Information Theory and Statistics*, John Wiley, New York.
A beautiful work, never properly appreciated because it was 20 years ahead of its time.

Landau, H. J. (1983), 'The inverse problem for the vocal tract and the moment problem', *SIAM J. Math. Anal.* **14**, 1019–1035.
Modeling speech production by a reflection coefficient technique closely related to the Burg maximum entropy spectrum analysis.

Landau, H. J. (1987), 'Maximum entropy and the moment problem', *Bull. Am. Math. Soc.* **16**, 47–77.
Interprets the Burg solution in terms of more general problems in several fields. Highly recommended for a deeper understanding of the mathematics.

Legendre, A. M. (1806), 'Nouvelles méthods pour la détermination des orbits des cométes', Didot, Paris.

Leibniz, G. W. (1968), *General Investigations Concerning the Analysis of Concepts and Truths*, trans. W. H. O'Briant, University of Georgia Press.

Lessard, S., ed. (1989), *Mathematical and Statistical Developments in Evolutionary Theory*, NATO ASI Series Vol. C299, Kluwer Academic Publishers, Holland.
Proceedings of a meeting held in Montreal, Canada in 1987.

Lewis, G. N. (1930) 'The symmetry of time in physics', *Science* **71**, 569.
An early recognition of the connection between entropy and information, showing an understanding far superior to what many others were publishing 50 years later.

Lindley, D. V. (1956), 'On a measure of the information provided by an experiment', *Ann. Math.* **27**, 986–1005.

Lindley, D. V. (1958) 'Fiducial distributions and Bayes' theorem', *J. Roy. Stat. Soc.* **B20**, 102–107.

Lindley, D. V. (1971) *Bayesian Statistics: A Review*, Society for Industrial and Applied Mathemathics, Philadelphia.

Linnik, Yu. V. (1961), *Die Methode der kleinsten Quadrate in Moderner Darstellung*, Deutscher Verl. der Wiss., Berlin.

Litterman, R. B. (1985), 'Vector autoregression for macroeconomic forecasting', in Zellner, A. & Goel, P., eds., *Bayesian Inference and Decision Techniques*, North-Holland Publishers, Amsterdam.

Lukacs, E. (1960), *Characteristic Functions*, Griffin, London.

Macdonald, P. D. M. (1987), 'Analysis of length-frequency distributions', in Summerfelt, R. C. & Hall, G. E., eds., *Age and Growth of Fish*, Iowa State University Press, pp. 371–384.
A computer program for deconvolving mixtures of normal and other distributions. The program, 'MIX 3.0' is available from: Ichthus Data Systems, 59 Arkell St, Hamilton, Ontario, Canada L8S 1N6. In Chapter 7 we note that the problem is not very well-posed; Icthus acknowledges that it is 'inherently difficult' and may not work satisfactorily on the user's data. See also Titterington, Smith & Makov (1985).

Mandel, J. (1964), *The Statistical Analysis of Experimental Data*, Interscience, New York.
Straight orthodox *ad hockeries*, one of which is analyzed in Jaynes (1976).

Mandelbrot, B. (1977), *Fractals, Chance and Dimension*, W. H. Freeman & Co., San Francisco.

Marple, S. L. (1987), *Digital Spectral Analysis with Applications*, Prentice-Hall, New Jersey.

Martin, R. D. & Thompson, D. J. (1982), 'Robust-resistant spectrum estimation', *Proc. IEEE* **70**, 1097–1115.
Evidently written under the watchful eye of their mentor John Tukey, this continues his practice of inventing a succession of *ad hoc* devices based on intuition rather than probability theory. It does not even acknowledge the existence of maximum entropy or Bayesian methods. To their credit, the authors do give computer analyses of several data sets by their methods – with results that do not look very encouraging to us. It would be interesting to acquire their raw data and analyze them by methods like those of Bretthorst (1988) that do make use of probability theory; we think that the results would be quite different.

Masani, S. M. (1977), 'A paradox in admissibility', *Ann. Stat.* **5**, 544–546.

Maxwell, J. C. (1850), Letter to Lewis Campbell; reproduced in L. Campbell & W. Garrett, *The Life of James Clerk Maxwell*, Macmillan, 1881.

McColl, H. (1897) 'The calculus of equivalent statements', *Proc. Lond. Math. Soc.* **28**, 556.
    Criticism of Boole's version of probability theory.

McFadden, D. (1973), 'Conditional logit analysis of qualitative choice behavior', in Zarembka, P., ed., *Frontiers in Econometrics*, Academic Press, New York.

Mead, L. R. & Papanicolaou, N. (1984), 'Maximum entropy in the problem of moments', *J. Math. Phys.* **25**, 2404–2417.

Miller, R. G. (1974), 'The jackknife – a review', *Biometrika* **61**, 1–15.

Mitler, K. S. (1974), *Multivariate Distributions*, John Wiley, New York.

Molina, E. C. (1931), 'Bayes' theorem, an expository presentation', *Bell Syst. Tech. Publ.* Monograph B-557.
    Stands, with Keynes (1921), Jeffreys (1939) and Woodward (1953), as proof that there have always been lonely voices crying in the wilderness for a sensible approach to inference.

Moore, G. T. & Scully, M. O., eds. (1986), *Frontiers of Nonequilibrium Statistical Physics*, Plenum Press, New York.
    Here several speakers affirmed their belief, on the basis of the Bell inequality experiments, that 'atoms are not real' while maintaining the belief that probabilities *are* objectively real! We consider this a flagrant example of the mind projection fallacy, carried to absurdity.

Munk, W. H. & Snodgrass, F. E. (1957), 'Measurements of Southern Swell at Guadalupe Island', *Deep-Sea Res.* **4**, 272–286.
    This is the work which Tukey (1984) held up as the greatest example of his kind of spectral analysis, which could never have been accomplished by other methods; to which in turn Jaynes (1987) replied with chirp analysis.

Newton, Sir Isaac (1687) *Philosophia Naturalis Principia Mathematica*, trans. Andrew Motte, 1729; revised and reprinted as *Mathematical Principles of Natural Philosophy*, Florian Cajori, ed., University of California Press (1946).
    See also Cajori (1928, 1934; both this bibliography).

Neyman, J. & Pearson, E. S. (1933), 'On the problem of the most efficient test of statistical hypotheses', *Phil. Trans. Roy. Soc.* **231**, 289–337.

Neyman, J. & Pearson, E. S. (1967), *Joint Statistical Papers*, Cambridge University Press.
    Reprints of the several Neyman–Pearson papers of the 1930s, originally scattered over several different journals.

Neyman, J. (1959), 'On the two different aspects of representative method: the method of stratified sampling and the method of purposive selection', *Estadistica* **17**, 587–651.

Neyman, J. (1962) 'Two breakthroughs in the theory of statistical decision making', *Int. Stat. Rev.* **30**, 11–27.
    It is an excellent homework problem to locate and correct the errors in this.

Neyman, J. (1981), 'Egon S. Pearson (August 11, 1895–June 12, 1980)', *Ann. Stat.* **9**, 1–2.

Novák, V. (1988), *Fuzzy Sets and their Applications*, A. Hilger, Bristol.

Nyquist, H. (1924), 'Certain factors affecting telegraph speed', *Bell Syst. Tech. J.* **3**, 324.

Nyquist, H. (1928), 'Certain topics in telegraph transmission theory', *Trans. AIEE*, **47**, 617–644.

O'Hagan, A. (1977), 'On outlier rejection phenomena in Bayes inference', *J. Roy. Stat. Soc.* **B 41**, 358–367.
    Our position is that Bayesian inference has no pathological, exceptional cases and in particular no outliers. To reject any observation as an 'outlier' is a violation of the principles of rational inference, and signifies only that the problem was improperly formulated. That is, if you are able to decide that *any* observation is an outlier from the model that you specified, then that model does not properly capture your prior information about the mechanisms that are generating the data. In principle, the remedy is not to reject any observation, but to define a more realistic model (as we note in our discussion of Robustness). However, we concede that if the strictly correct procedure assigns a very low weight to the suspicious datum, its straight-out surgical removal from the data set may be a reasonable approximation, very easy to do.

Ore, O. (1953), *Cardano, the Gambling Scholar*, Princeton University Press.

Ore, O. (1960), 'Pascal and the invention of probability theory', *Am. Math. Monthly* **67**, 409–419.

Pearson, K. (1892), *The Grammar of Science*, Walter Scott, London.

Reprinted 1900, 1911 by A. & C. Black, London, and in 1937 by Everyman Press. An exposition of the principles of scientific reasoning; notably chiefly because Harold Jeffreys was much influenced by it and thought highly of it. This did not prevent him from pointing out that Karl Pearson was far from applying his own principles in his later scientific efforts. For biographical material on Karl Pearson (1857–1936) see Haldane (1957; this bibliography).

Pearson, K. (1905), 'The problem of the random walk', *Nature* **72**, 294, 342.

Pearson, K. (1921–33), *The History of Statistics in the 17'th and 18'th Centuries*, Pearson, E.S., ed., Lectures given at University College, London, Griffin, London (1978).

Penfield, W. (1958), *Proc. Natl Acad. Sci. (USA)* **44**, 59.

Accounts of observations made during brain surgery, in which electrical stimulation of a specific spots on the brain caused the conscious patient to recall various long-forgotten experiences. This undoubtedly true phenomenon is closely related to the theory of the A_p distribution in Chapter 18. But now others have moved into this field, with charges that psychiatrists are causing their patients – particularly young children – to recall things that never happened, with catastrophic legal consequences. The problem of recognizing valid and invalid recollections seems headed for a period of controversy.

Pierce, J. R. (1980) *Symbols, Signals, and Noise: An Introduction to Information Theory*, Dover Publications, Inc., New York.

An easy introduction for absolute beginners, but does not get to the currently important applications.

Poisson, S. D. (1837), *Recherches sur la Probabilité des Jugements*, Bachelier imprimeur-Libraire, Paris.

First appearance of the Poisson distribution.

Pólya, G. (1921), 'Über eine Aufgabe der Wahrscheinlichkeitsrechnung betreffend die Irrfahrt im Strassennetz', *Math. Ann.* **84**, 149–160.

It is sometimes stated that this was the first appearance of the term 'random walk'. However, we may point to Pearson (1905) and Rayleigh (1919); both in this bibliography.

Pólya, G. (1923), 'Herleitung des Gauss'schen Fehlergesetzes aus einer Funktionalgleichung', *Math. Zeit.* **18**, 96–108.

Pontryagin, L. S. (1946), *Topological Groups,* Princeton University Press, Princeton, NJ.

Popper, K. (1958), *The Logic of Scientific Discovery*, Hutchinson & Co., London.

Denies the possibility of induction, on the grounds that the prior probability of every scientific theory is zero. Karl Popper is famous mostly through making a career out of the doctrine that theories may not be proved true, only false; hence the merit of a theory lies in its falsifiability. There is an evident grain of truth here, expressed by the syllogisms of Chapter 1; and Albert Einstein also noted this in his famous remark: *'No amount of experiments can ever prove me right; a single experiment may at any time prove me wrong.'* Nevertheless, the doctrine is true only of theories which assert the existence of unobservable causes or mechanisms; any theory which asserts observable facts is a counter-example to it.

Popper, K. (1963), *Conjectures and Refutations*, Routledge & Kegan Paul, London.

Popov, V. N. (1987), *Functional Integrals and Collective Excitations*, Cambridge University Press.

Sketches applications to superfluidity, superconductivity, plasma dynamics, superradiation, and phase transitions. A useful start on understanding of these phenomena, but still lacking any coherent theoretical basis – which we think is supplied only by the principle of maximum entropy as a method of reasoning.

Prenzel, H. V. (1975), *Dynamic Trendline Charting: How to Spot the Big Stock Moves and Avoid False Signals*, Prentice-Hall, Englewood Cliffs, NJ.

Contains not a trace of probability theory or any other mathematics: merely plot the monthly ranges of stock prices, draw a few straight lines on the graph, and their intersections tell you what to do and when to do it. At least this system does enable one to see the four-year presidential election cycle, very clearly.

Press, S. J. (1989), *Bayesian Statistics: Principles, Models and Applications*, J. Wiley & Sons, Inc., New York.

Contains a list of many Bayesian computer programs now available.

Preston, C. J. (1974), *Gibbs States on Countable Sets*, Cambridge University Press.

Here we have the damnable practice of using the word *state* to denote a probability *distribution*. One cannot conceive of a more destructively false and misleading terminology.

Priestley, M. B. (1981), *Spectral Analysis and Time Series*, 2 vols., Academic Press, Inc., Orlando, FL; combined paperback edition with corrections (1983).

Puri, M. L., ed. (1975), *Stochastic Processes and Related Topics*, Academic Press, New York.

Quaster, H., ed. (1953), *Information Theory in Biology*, University Illinois Press, Urbana.

Ramsey, F. P. (1931), *The Foundations of Mathematics and Other Logical Essays*, Routledge and Kegan Paul, London.

> Frank Ramsey was First Wrangler in Mathematics at Cambridge University in 1925, then became a Fellow of Kings College, where among other activities he collaborated with John Maynard Keynes on economic theory. He would undoubtedly have become the most influential Bayesian of the 20th century, but for the fact that he died in 1930 at the age of 26. In these essays one can see the beginnings of something very much like our exposition of probability theory.

Rayleigh, Lord (1919), 'On the problem of random vibrations, and of random flights in one, two or three dimensions', *Edin. & Dublin Phil. Mag. & J. Sci.* **37**, series 6, 321–347.

Reichardt, H. (1960), *C. F. Gauss–Leben und Werk*, Haude & Spener, Berlin.

Reid, C. (1970), *Hilbert*, Springer-Verlag, New York.

Reid, C. (1959), 'On a new axiomatic theory of probability', *Acta. Math. Acad. Sci. Hung.* **6**, 285–335.

> This work has several things in common with ours, but expounded very differently.

Rihaczek, A. W. (1981), 'The maximum entropy of radar resolution', *IEEE Trans. Aerospace Electron. Syst.* **AES-17**, 144.

> Another attack on maximum entropy, still denying the possibility of so-called 'super resolution', although it had been demonstrated conclusively in both theory and practice by John Parker Burg many years before, and was by 1981 in routine use by many scientists and engineers, as illustrated by the reprint collection of Childers (1978); see this bibliography.

Rissanen, J. (1983), 'A universal prior for the integers and estimation by minimum description length', *Ann. Stat.* **11**, 416–431.

> One of the few fresh new ideas in recent decades. We think it has a bright future, but are not yet prepared to predict just what it will be.

Robbins, H. (1950), 'Asymptotically subminimax solutions of compound statistical decision problems', *Proceedings of the 2nd Berkeley Symposium of Mathematics Statistics and Probability*, University of California Press, pp. 131–148.

> An anticipation of Stein (1956); see this bibliography.

Robbins, H. (1956), 'An empirical Bayes' approach to statistics', *Proceedings of the 3rd Berkeley Symposium on Mathematics, Statistics and Probability I*, University of California Press, pp. 157–164.

Robinson, A. (1966), *Non-standard Analysis*, North-Holland, Amsterdam.

> How to do every calculation wrong.

Robinson, E. A. (1982), 'A historical perspective of spectrum estimation', *Proc. IEEE* **70**, 855–906.

Robinson, G. K. (1975), 'Some counterexamples to the theory of confidence intervals', *Biometrika* **62**, 155–162.

Rowlinson, J. S. (1970), 'Probability, information and entropy', *Nature* **225**, 1196–1198.

> An attack on the principle of maximum entropy showing a common misconception of the nature of inference. Answered in Jaynes (1978).

Sampson, A. R. & Smith, R. L. (1984), 'An information theory model for the evaluation of circumstantial evidence,' *IEEE Trans. Systems, Man, and Cybernetics* **15**, 916.

Sampson, A. R. & Smith, R. L. (1982), 'Assessing risks through the determination of rare event probabilities', *Op. Res.* **30**, 839–866.

Sanov, I. N. (1961), 'On the probability of large deviations of random variables', IMS and AMS Translations of Probability and Statistics, from *Mat. Sbornik* **42**, 1144.

Scheffé, H. (1959), *The Analysis of Variance*, John Wiley, New York.

Schendel, U. (1989) *Sparse Matrices*, J. Wiley & Sons, New York.

Schlaifer, R. (1959), *Probability and Statistics for Business Decisions: An Introduction to Managerial Economics Under Uncertainty*, McGraw-Hill Book Company, New York.

> An early recognition of the need for Bayesian methods in the real-world problems of decision; in striking contrast to the simultaneous Chernoff and Moses work (1959) on decision theory.

Schneider, T. D. (1991), 'Theory of molecular machines', *J. Theor. Biol.* **148**, 83–137.

In two parts, concerned with channel capacity and energy dissipation.

Schnell, E. E. (1960), 'Samuel Pepys, Isaac Newton and probability', *Am. Stat.* **14**, 27–30.

From this we learn that both Pascal and Newton had the experience of giving a correct solution and not being believed; the problem is not unique to modern Bayesians.

Schrödinger, E. (1945), 'Probability problems in nuclear chemistry', *Proc. Roy. Irish Acad.* **51**.

Schrödinger, E. (1947), 'The foundation of the theory of probability', *Proc. Roy. Irish Acad. (A)*, **51**, pp. 51–66, 141–146.

Valuable today because it enables us to add one more illustrious name to the list of those who think as we do. Here Schrödinger declares the 'frequentist' view of probability inadequate for the needs of science and seeks to justify the view of probability as applying to individual cases rather than 'ensembles' of cases, by efforts somewhat in the spirit of our Chapters 1 and 2. He gives some ingenious arguments but, unknown to him, these ideas had already advanced far beyond the level of his work. He was unaware of Cox's theorems and, like most scientists of that time with continental training, he had apparently never heard of Thomas Bayes or Harold Jeffreys. He gives no useful applications and obtains no theoretical results beyond what had been published by Jeffreys eight years earlier. Nevertheless, his thinking was aimed in the right direction on this and other controversial issues.

Shafer, G. (1976), *A Mathematical Theory of Evidence*, Princeton University Press, Princeton, NJ.

An attempt to develop a theory of two-valued probability, by a fanatical anti-Bayesian.

Shafer, G. (1982), 'Lindley's paradox', *J. Am. Stat. Assoc.* **77**, 325–334.

Apparently, Shafer was unaware that this was all in Jeffreys (1939, p. 194) some 20 years before Lindley. But Shafer's other work had made it clear already that he had never read and understood Jeffreys.

Shamir, A (1982), 'A polynomial time algorithm for breaking the basic Merkle–Hellman cryptosystem', in Chaum, D., Rivest, R. L. & Sherman, A. T., eds., *Advances in Cryptology: Proceedings of Crypto 82, 23–25 August 1982*, Plenum Press, New York, pp. 279–288.

Shaw, D. (1976), *Fourier Transform NMR Spectroscopy*, Elsevier, New York.

Sheynin, O. B. (1978), 'S. D. Poisson's work in probability', *Archiv. f. Hist. Exact Sci.* **18**, 245–300.

Sheynin, O. B. (1979), 'C. F. Gauss and the theory of errors', *Archiv. f. Hist. Exact Sci.* **19**, 21–72.

Shiryayev, A. N. (1978), *Optimal Stopping Rules*, Springer-Verlag, New York.

Siegmann, D. (1985) *Sequential Analysis*, Springer-Verlag, Berlin.

No mention of Bayes' theorem or optional stopping!

Simmons, G. J. (1979), 'Cryptography, the mathematics of secure communication', *The Math. Intelligencer* **1**, 233–246.

Sinai, J. G. (1982), *Rigorous Results in the Theory of Phase Transitions*, Akadémiai Kiado, Budapest.

Skilling, J., ed. (1989), *Maximum Entropy and Bayesian Methods*, Proceedings of the Eighth Maximum Entropy Workshop, Cambridge, UK, August 1988, Kluwer Academic Publishers, Dordrecht, Holland.

Smith, C. R. & Grandy, W. T., eds. (1985), *Maximum-Entropy and Bayesian Methods in Inverse Problems*, D. Reidel Publishing Co., Dordrecht, Holland.

Smith, C. R. & Erickson, G. J., eds. (1987), *Maximum-Entropy and Bayesian Spectral Analysis and Estimation Problems*, D. Reidel Publishing Co., Dordrecht, Holland.

Smith, D. E. (1959), *A Source Book in Mathematics*, McGraw-Hill Book Co., New York.

Contains the Fermat–Pascal correspondence.

Smith, W. B. (1905), 'Meaning of the Epithet Nazorean', *The Monist* **15**, 25–95.

Concludes that prior to the Council of Nicæa, 'Nazareth' was not the name of a geographical place; it had some other meaning.

Sonett, C. P. (1982), 'Sunspot time series: spectrum from square law modulation of the Hale cycle', *Geophys. Res. Lett.* **9**, 1313–1316.

Spinoza, B. (1663), 'Renati des Cartes Principiorum philosophiae pars I, & II, more geometrico demonstratae,' *Ethics*, part 2, Prop. XLIV: 'De natura Rationis no est res, ut contingentes; sed, ut necessarias, contemplari.'

Translated, this proposition reads: 'It is not in the nature of reason to regard things as contingent; instead, they should be regarded as necessary.'

Spitzer, F. (1964), *Principles of Random Walk*, van Nostrand, New York.

Background history and present status.

Stein, C. (1945), 'A two sample test for a linear hypothesis whose power is independent of the variance', *Ann. Math. Stat.* **16**, 243–258.

Stein, C. (1956), 'Inadmissibility of the usual estimator for the mean of a multivariate normal distribution', *Proceedings of the 3rd Berkeley Symposium*, vol. 1, pp. 197–206, University of California Press.

First announcement of the 'Stein shrinking' phenomenon.

Stein, C. (1959), 'An example of wide discrepancy between fiducial and confidence intervals', *Ann. Math. Stat.* **30**, 877–880.

Stein, C. (1964), 'Inadmissibility of the usual estimate for the variance of a normal distribution with unknown mean', *Ann. Inst. Stat. Math.* **16**, 155–160.

Stein's inadmissibility discoveries, while shocking to statisticians with conventional training, are not in the least surprising or disconcerting to Bayesians. They only illustrate what was already clear to us: that the criterion of admissibility, which ignores all prior information, is potentially dangerous in real problems. Here that criterion can reject as 'inadmissible' what is in fact the optimal estimator, as noted briefly in Chapter 13.

Stigler, S. M. (1974a), 'Cauchy and the Witch of Agnesi', *Biometrika* **61**, 375–380.

Stigler, S. M. (1974b), 'Gergonne's 1815 paper on the design and analysis of polynomial regression experiments', *Historia Math.* **1**, 431–477.

Stigler, S. M. (1982a), 'Poisson on the Poisson distribution', *Stat. & Prob. Lett.* **1**, 33–35.

Stigler, S. M. (1982b), 'Thomas Bayes's Bayesian inference', *J. Roy. Stat. Soc.* **A145**, 250–258.

Stone, M. & Springer-Verlag, B. G. F. (1965), 'A paradox involving quasi prior distributions', *Biometrika* **52**, 623–627.

Stromberg, K. (1979), 'The Banach-Tarski paradox', *Am. Math. Monthly* **86**, 151–160.

Congruent sets stuff.

Student (1908), 'The probable error of a mean', *Biometrika* **6**, 1–24.

Takeuchi, K., Yanai, H. & Mukherjee, B. N. (1982), *The Foundations of Multivariate Analysis*, J. Wiley & Sons, New York.

Taylor, R. L., Daffer, P. Z. & Patterson, R. F. (1985), *Limit Theorems for Sums of Exchangeable Random Variables*, Rowman & Allanheld Publishers.

Presents the known versions of limit theorems for discrete-time exchangeable sequences in Euclidean and Banach spaces.

Thomas, M. U. (1979), 'A generalized maximum entropy principle', *Operations Res.* **27**, 1188–1195.

Tikhonov, A. N. & Arsenin, V. Y. (1977), *Solutions of Ill-posed Problems*, Halsted Press, New York.

A collection of *ad hoc* mathematical recipes, in which the authors try persistently to invert operators which have no inverses. Never perceives that these are problems of *inference*, not *inversion*.

Todhunter, I. (1865), *A History of the Mathematical Theory of Probability*, Macmillan, London; reprinted 1949, 1965, by Chelsea Press, New York.

Todhunter, I. (1873), *A History of the Mathematical Theories of Attraction and the Figure of the Earth*, 2 vols., Macmillan, London; reprinted 1962 by Dover Press, New York.

Toraldo di Francia, G. (1955), 'Resolving power and information', *J. Opt. Soc. Am.* **45**, 497–501.

One of the first recognitions of the nature of a generalized inverse problem; he tries to use information theory but fails to see that (Bayesian) probability theory is the appropriate tool for his problems.

Train, K. (1986), *Qualitative Choice Analysis Theory, Econometrics, and an Application to Automobile Demand*, MIT Press, Cambridge, MA.

Truesdell, C. (1987), *Great Scientists of Old as Heretics in 'The Scientific Method'*, University Press of Virginia.

The historical record shows that some of the greatest advances in mathematical physics were made with little or no basis in experiment, in seeming defiance of the 'scientific method' as usually proclaimed. This just shows the overwhelming importance of creative hypothesis formulation as primary to inference from given hypotheses. Unfortunately, while today we have a well-developed and highly successful theory of inference, we have no formal theory at all on optimal hypothesis formulation, and very few successful recent examples of it.

Tukey, J. W. (1960), 'A survey of sampling from contaminated distributions', in Olkin, I., ed., *Contributions to Probability and Statistics: Essays in Honor of Harold Hotelling*, Stanford University Press, California, pp. 448–485.

Tukey, J. W. (1962), 'The future of data analysis', *Ann. Math. Stat.* **33**, 1–67.
  A potent object lesson for all who try to foretell the future as the realization of their own prejudices.
Tukey, J. W. (1978), 'Granger on seasonality', in Zellner, A., ed., *Seasonal Analysis of Time Series*,
  US Dept of Commerce, Washington.
  An amusing view of the nature of Bayesian inference as a sneaky way of committing indecent
  methodological sins 'while modestly concealed behind a formal apparatus'.
Tukey, J. W. (1984), 'Styles of spectrum analysis', *Scripps Inst. Oceanography Ref.* Series 84–85,
  March, pp. 100–103.
  A polemical attack on all theoretical principles, including autoregressive models, maximum entropy, and
  Bayesian methods. The 'protagonist of maximum entropy' who appears on p. 103 is none other than
  E. T. Jaynes.
Tukey, J. W., Bloomfield, P. Brillinger, D. & Cleveland, W. S. (1980), *The Practice of Spectrum
  Analysis*, University Associates, Princeton, New Jersey.
  Notes for a course given in December 1980.
Tukey, J. W. & Brillinger, D. (1982), 'Spectrum estimation and system identification relying on a
  Fourier transform', unpublished.
  This rare work was written as an invited paper for the IEEE Special Issue of September 1982 on Spectrum
  Analysis, but its length (112 pages in an incomplete version) prevented its appearing there. We hope that it
  will find publication elsewhere, because it is an important historical document. Tukey (1984; this
  bibliography) contains parts of it.
Valery-Radot, R. (1923), *The Life of Pasteur*, Doubleday, Page & Co., Garden City, New York.
Van Campenhout, J. M. & Cover, T. M. (1981), 'Maximum entropy and conditional probability',
  IEEE Trans. Info. Theor. **IT-27**, 483–489.
  A rediscovery and generalization of what physicists have, since 1928, called 'the Darwin–Fowler method
  of statistical mechanics'.
van den Bos, A. (1971), 'Alternative interpretation of maximum entropy spectral analysis', *IEEE
  Trans. Info. Theor.* **IT-17**, 493–494.
  Reprinted in Childers (1978; this bibliography). Expresses several misgivings about maximum entropy
  spectrum analysis, answered in Jaynes (1982; this bibliography).
Varian, H. (1978), *Microeconomic Analysis*, Norton & Co., New York.
Vasicek, O. (1980), 'A conditional law of large numbers', *Ann. Prob.* **8**, 142–147.
Wald, A. (1941), *Notes on the Theory of Statistical Estimation and of Testing Hypotheses*,
  Mimeographed, Columbia University.
  At this time, Wald was assuring his students that Bayesian methods were entirely erroneous and incapable
  of dealing with the problems of inference. Nine years later, his own research had led him to the opposite
  opinion.
Wald, A. (1942), *On the Principles of Statistical Inference*, Notre Dame University Press.
Wald, A. (1943), 'Sequential analysis of statistical data: theory', Restricted report dated September
  1943.
Waldmeier, M. (1961), *The Sunspot Activity in the Years 1610-1960*, Schulthes, Zürich.
  Probably the most analyzed of all data sets.
Walley, P. (1991), *Statistical Reasoning with Imprecise Probabilities*, Chapman & Hall, London.
  Worried about improper priors, he introduces the notion of a 'near-ignorance class' (NIC) of priors. Since
  then, attempts to define precisely the NIC of usable priors have occupied many authors. We propose to cut
  all this short by noting that any prior which leads to a proper posterior distribution is usable and potentially
  useful. Obviously, whether a given improper prior does or does not accomplish this is determined not by
  any property of the prior alone, but by the joint behavior of the prior and the likelihood function; that is, by
  the prior, the model, and the data. Need any more be said?
Watson, J. D. (1968), *The Double Helix*, Signet Books, New York.
  The famous account of the events leading to discovery of the DNA structure. It became a best seller
  because it inspired hysterically favorable reviews by persons without any knowledge of science, who were
  delighted by the suggestion that scientists in their ivory towers have motives just as disreputable as theirs.
  This was not the view of scientists on the scene with technical knowledge of the facts, one of whom said
  privately to the present writer: 'The person who emerges looking worst of all is Watson himself.' But that is
  ancient history; for us today, the interesting question is: would the discovery have been accelerated

appreciably if the principles of Bayesian inference, as applied to X-ray diffraction data, had been developed and reduced to computer programs in 1950? We suspect that Rosalind Franklin's first 'A-structure' photograph, which looks hopelessly confusing to the eye at first glance, if analyzed by a computer program (like those of Bretthorst (1988) but adapted to this problem), would have pointed at once to a double helix as overwhelmingly the most probable structure (at least, the open spaces which say '*helix*' were present and could be recognized by the eye after the fact). The problem is, in broad aspects, very much like that of radar target identification. For another version of the DNA story, with some different recollections of the course of events, see Crick (1988; this bibliography).

Wax, N., ed. (1954), *Noise & Stochastic Processes*, Dover Publications, Inc., New York.

Wehrl, A. (1978), 'General properties of entropy', *Rev. Mod. Phys.* **50**, 220–260.

Whittle, P. (1954), Comments on periodograms, Appendix to H. Wold (1954; this bibliography), pp. 200–227.

Whittle, P. (1957), 'Curve and periodogram smoothing', *J. Roy. Stat. Soc.* **B 19**, 38–47.

Whittle, P. (1958), 'On the smoothing of probability density functions', *J. Roy. Stat. Soc.* **B 20**, 334–343.

Wigner, E. P. (1967), *Symmetries and Reflections*, Indiana University Press, Bloomington.

From the standpoint of probability theory, the most interesting essay reprinted here is #15, 'The probability of the existence of a self-reproducing unit'. Writing the quantum-mechanical transformation from an initial state with (one living creature + environment) to a final state with (two identical ones + compatible environment), he concludes that the number of equations to be satisfied is greater than the number of unknowns, so the probability of replication is zero. Since the fact is that replication exists, the argument if correct would show only that quantum theory is invalid.

Wilbraham, H. (1854), 'On the theory of chances developed in Professor Boole's "Laws of Thought" ', *Phil. Mag. Series 4*, **7**, (48), 465–476.

Criticism of Boole's version of probability theory.

Williams, P. M. (1980), 'Bayesian conditionalisation and the principle of minimum information', *Brit. J. Phil. Sci.* **31**, 131–144.

Wilson, A. G. (1970), *Entropy and Urban Modeling*, Pion Limited, London.

Wold, H. (1954), *Stationary Time Series*, Almquist and Wiksell, Stockholm.

Yockey, H. P. (1992), *Information Theory in Molecular Biology*, Cambridge University Press.

Zabell, S. L. (1982), 'W. E. Johnson's sufficientness postulate', *Ann. Stat.* **10**, 1091–1099.

Discussed in Jaynes (1986b).

Zabell, S. L. (1988), 'Buffon, Price, and Laplace: scientific attribution in the 18'th century', *Arch. Hist. Exact Sci.* **39**, 173–181.

Zellner, A. (1984), *Basic Issues in Econometrics*, University of Chicago Press.

A collection of 17 reprints of recent articles discussing and illustrating important principles of scientific inference. Like the previous reference, this is of value to a far wider audience than one would expect from the title. The problems and examples are stated in the context of economics, but the principles themselves are of universal validity and importance. In our view they are if anything even more important for physics, biology, medicine, and environmental policy than for economics. Be sure to read Chapter 1.4, entitled 'Causality and econometrics'. The problem of deciding whether a causal influence exists is vital for physics, and one might have expected physicists to have the best analyses of it. Yet Zellner here gives a far more sophisticated treatment than anything in the literature of physics or any other 'hard' science. He makes the same points that we stress here with cogent examples showing why prior information is absolutely essential in any judgment of this.

Zubarev, D. N. (1974) *Nonequilibrium Statistical Thermodynamics*, Plenum Publishing Corp., New York.

An amazing work; develops virtually all the maximum entropy partition functional algorithm as an *ad hoc* device; but then rejects the maximum entropy principle which gives the rationale for it and explains why it works! As a result, he is willing to use the formalism only for a tiny fraction of the problems which it is capable of solving, and thus loses practically all the real value of the method. A striking demonstration of how useful applications can be paralyzed – even when all the requisite mathematics is at hand – by orthodox conceptualizing about probability.

Author index

Subject index

Printed in the United States
by Baker & Taylor Publisher Services